"This book dismantles the final, most daunting barriers to learning about moduli of higher dimensional varieties, from the point of view of the Minimal Model Program. The first chapter draws the reader in with a compelling history; a discussion of the main ideas; a visitor's trail through the subject, complete with guardrails around the most dangerous traps; and a rundown of the issues that one must overcome. The text that follows is the outcome of Kollár's monumental three-decades-long effort, with the final stones laid just in the last few years."

— Dan Abramovich, Brown University

"This is a fantastic book from János Kollár, one of the godfathers of the compact moduli theory of higher dimensional varieties. The book contains the definition of the moduli functor, the prerequisites required for the definition, and also the proof of the existence of the projective coarse moduli space. This is a stunning achievement, completing the story of 35 years of research. I expect this to become the main reference book, and also the principal place to learn about the theory for graduate students and others interested."

— Zsolt Patakfalvi, EPFL

"This excellent book provides a wealth of examples and technical details for those studying birational geometry and moduli spaces. It completely addresses several state-of-the-art topics in the field, including different stability notions, K-flatness, and subtleties in defining families of stable pairs over an arbitrary base. It will be an essential resource for both those first learning the subject and experts as it moves through history and examples before settling many of the (previously unknown) technicalities needed to define the correct moduli functor."

— Kristin DeVleming, University of Massachusetts Amherst

CAMBRIDGE TRACTS IN MATHEMATICS

General Editors

J. BERTOIN, B. BOLLOBÁS, W. FULTON, B. KRA, I. MOERDIJK,
C. PRAEGER, P. SARNAK, B. SIMON, B. TOTARO

---

231 Families of Varieties of General Type

CAMBRIDGE TRACTS IN MATHEMATICS

GENERAL EDITORS

J. BERTOIN, B. BOLLOBÁS, W. FULTON, B. KRA, I. MOERDIJK,

C. PRAEGER, P. SARNAK, B. SIMON, B. TOTARO

A complete list of books in the series can be found at www.cambridge.org/mathematics. Recent titles include the following:

196. The Theory of Hardy's Z-Function. By A. Ivić
197. Induced Representations of Locally Compact Groups. By E. Kaniuth and K. F. Taylor
198. Topics in Critical Point Theory. By K. Perera and M. Schechter
199. Combinatorics of Minuscule Representations. By R. M. Green
200. Singularities of the Minimal Model Program. By J. Kollár
201. Coherence in Three-Dimensional Category Theory. By N. Gurski
202. Canonical Ramsey Theory on Polish Spaces. By V. Kanovei, M. Sabok, and J. Zapletal
203. A Primer on the Dirichlet Space. By O. El-fallah, K. Kellay, J. Mashreghi, and T. Ransford
204. Group Cohomology and Algebraic Cycles. By B. Totaro
205. Ridge Functions. By A. Pinkus
206. Probability on Real Lie Algebras. By U. Franz and N. Privault
207. Auxiliary Polynomials in Number Theory. By D. Masser
208. Representations of Elementary Abelian p-Groups and Vector Bundles. By D. J. Benson
209. Non-homogeneous Random Walks. By M. Menshikov, S. Popov, and A. Wade
210. Fourier Integrals in Classical Analysis (Second Edition). By C. D. Sogge
211. Eigenvalues, Multiplicities and Graphs. By C. R. Johnson and C. M. Saiago
212. Applications of Diophantine Approximation to Integral Points and Transcendence. By P. Corvaja and U. Zannier
213. Variations on a Theme of Borel. By S. Weinberger
214. The Mathieu Groups. By A. A. Ivanov
215. Slenderness I: Foundations. By R. Dimitric
216. Justification Logic. By S. Artemov and M. Fitting
217. Defocusing Nonlinear Schrödinger Equations. By B. Dodson
218. The Random Matrix Theory of the Classical Compact Groups. By E. S. Meckes
219. Operator Analysis. By J. Agler, J. E. McCarthy, and N. J. Young
220. Lectures on Contact 3-Manifolds, Holomorphic Curves and Intersection Theory. By C. Wendl
221. Matrix Positivity. By C. R. Johnson, R. L. Smith, and M. J. Tsatsomeros
222. Assouad Dimension and Fractal Geometry. By J. M. Fraser
223. Coarse Geometry of Topological Groups. By C. Rosendal
224. Attractors of Hamiltonian Nonlinear Partial Differential Equations. By A. Komech and E. Kopylova
225. Noncommutative Function-Theoretic Operator Function and Applications. By J. A. Ball and V. Bolotnikov
226. The Mordell Conjecture. By A. Moriwaki, H. Ikoma, and S. Kawaguchi
227. Transcendence and Linear Relations of 1-Periods. By A. Huber and G. Wüstholz
228. Point-Counting and the Zilber–Pink Conjecture. By J. Pila
229. Large Deviations for Markov Chains. By A. D. de Acosta
230. Fractional Sobolev Spaces and Inequalities. By D. E. Edmunds and W. D. Evans

Families of Varieties of General Type

JÁNOS KOLLÁR
Princeton University, New Jersey

with the collaboration of

KLAUS ALTMANN
Freie Universität Berlin

and

SÁNDOR J. KOVÁCS
University of Washington

CAMBRIDGE
UNIVERSITY PRESS

Shaftesbury Road, Cambridge CB2 8EA, United Kingdom

One Liberty Plaza, 20th Floor, New York, NY 10006, USA

477 Williamstown Road, Port Melbourne, VIC 3207, Australia

314–321, 3rd Floor, Plot 3, Splendor Forum, Jasola District Centre,
New Delhi – 110025, India

103 Penang Road, #05–06/07, Visioncrest Commercial, Singapore 238467

Cambridge University Press is part of Cambridge University Press & Assessment,
a department of the University of Cambridge.

We share the University's mission to contribute to society through the pursuit of
education, learning and research at the highest international levels of excellence.

www.cambridge.org
Information on this title: www.cambridge.org/9781009346108
DOI: 10.1017/9781009346115

First published 2023

A catalogue record for this publication is available from the British Library.

*A Cataloging-in-Publication data record for this book is available from the Library of
Congress.*

ISBN 978-1-009-34610-8 Hardback

Contents

Preface		*page* xi
Acknowledgments		xii
Notation		xiv
Introduction		1
1	**History of Moduli Problems**	**3**
	1.1 Riemann, Cayley, Hilbert, and Mumford	4
	1.2 Moduli for Varieties of General Type	10
	1.3 From Smooth Curves to Canonical Models	18
	1.4 From Stable Curves to Stable Varieties	26
	1.5 From Nodal Curves to Stable Curves and Surfaces	33
	1.6 Examples of Bad Moduli Problems	38
	1.7 Compactifications of M_g	44
	1.8 Coarse and Fine Moduli Spaces	51
2	**One-Parameter Families**	**57**
	2.1 Locally Stable Families	58
	2.2 Locally Stable Families of Surfaces	66
	2.3 Examples of Locally Stable Families	80
	2.4 Stable Families	88
	2.5 Cohomology of the Structure Sheaf	97
	2.6 Families of Divisors I	105
	2.7 Boundary with Coefficients $> \frac{1}{2}$	110
	2.8 Local Stability in Codimension ≥ 3	114
3	**Families of Stable Varieties**	**118**
	3.1 Chow Varieties and Hilbert Schemes	120
	3.2 Representable Properties	125
	3.3 Divisorial Sheaves	129

	3.4	Local Stability	133
	3.5	Stability Is Representable I	135
4		**Stable Pairs over Reduced Base Schemes**	137
	4.1	Statement of the Main Results	138
	4.2	Examples	143
	4.3	Families of Divisors II	146
	4.4	Valuative Criteria	154
	4.5	Generically $\mathbb{Q}$-Cartier Divisors	157
	4.6	Stability Is Representable II	160
	4.7	Stable Families over Smooth Base Schemes	167
	4.8	Mumford Divisors	170
5		**Numerical Flatness and Stability Criteria**	181
	5.1	Statements of the Main Theorems	182
	5.2	Simultaneous Canonical Models and Modifications	185
	5.3	Examples	188
	5.4	Mostly Flat Families of Line Bundles	192
	5.5	Flatness Criteria in Codimension 1	196
	5.6	Deformations of SLC Pairs	203
	5.7	Simultaneous Canonical Models	206
	5.8	Simultaneous Canonical Modifications	210
	5.9	Families over Higher Dimensional Bases	213
6		**Moduli Problems with Flat Divisorial Part**	216
	6.1	Introduction to Moduli of Stable Pairs	217
	6.2	Kollár–Shepherd-Barron Stability	229
	6.3	Strict Viehweg Stability	233
	6.4	Alexeev Stability	234
	6.5	First Order Deformations	236
	6.6	Deformations of Cyclic Quotient Singularities	247
7		**Cayley Flatness**	258
	7.1	K-flatness	259
	7.2	Infinitesimal Study of Mumford Divisors	265
	7.3	Divisorial Support	273
	7.4	Variants of K-Flatness	279
	7.5	Cayley–Chow Flatness	284
	7.6	Representability Theorems	293
	7.7	Normal Varieties	294
	7.8	Hypersurface Singularities	295
	7.9	Seminormal Curves	301

8   Moduli of Stable Pairs   308
  8.1 Marked Stable Pairs   309
  8.2 Kollár–Shepherd-Barron–Alexeev Stability   312
  8.3 Stability with Floating Coefficients   317
  8.4 Polarized Varieties   326
  8.5 Canonically Embedded Pairs   330
  8.6 Moduli Spaces as Quotients by Group Actions   333
  8.7 Descent   338
  8.8 Positive Characteristic   342

9   Hulls and Husks   347
  9.1 Hulls of Coherent Sheaves   348
  9.2 Relative Hulls   350
  9.3 Universal Hulls   352
  9.4 Husks of Coherent Sheaves   355
  9.5 Moduli Space of Quotient Husks   357
  9.6 Hulls and Hilbert Polynomials   361
  9.7 Moduli Space of Universal Hulls   363
  9.8 Non-projective Versions   365

10   Ancillary Results   370
  10.1 S_2 Sheaves   370
  10.2 Flat Families of S_m Sheaves   374
  10.3 Cohomology over Non-proper Schemes   381
  10.4 Volumes and Intersection Numbers   385
  10.5 Double Points   389
  10.6 Noether Normalization   393
  10.7 Flatness Criteria   398
  10.8 Seminormality and Weak Normality   410

11   Minimal Models and Their Singularities   418
  11.1 Singularities of Pairs   418
  11.2 Canonical Models and Modifications   427
  11.3 Semi-log-canonical Pairs   430
  11.4 $\mathbb{R}$-divisors   434

  References   446
  Index   462

Preface

The aim of this book is to generalize the moduli theory of algebraic curves – developed by Riemann, Cayley, Klein, Teichmüller, Deligne, and Mumford – to higher dimensional algebraic varieties.

Starting with the theory of algebraic surfaces worked out by Castelnuovo, Enriques, Severi, Kodaira, and ending with Mori's program, it became clear that the correct higher dimensional analog of a smooth projective curve of genus ≥ 2 is a smooth projective variety with ample canonical class. We establish a moduli theory for these objects, their limits, and generalizations.

The first attempt to write a book on higher dimensional moduli theory was the 1993 Summer School in Salt Lake City, Utah. Some notes were written, but it soon became evident that, while the general aims of a theory were clear, most of the theorems were open and even many of the basic definitions unsettled.

The project was taken up again at an AIM conference in 2004, which eventually resulted in solving the moduli-theoretic problems related to singularities; these were written up in Kollár (2013b). After 30 years, we now have a complete theory, the result of the work of numerous people.

While much of the early work focused on the construction of moduli spaces, later developments in the theory of stacks emphasized families. We also follow this approach and spend most of the time understanding families. Once this is done at the right level, the existence of moduli spaces becomes a natural consequence.

Acknowledgments

Throughout the years, I learned a lot from my teachers, colleagues, and students. My interest in moduli theory was kindled by my thesis advisor T. Matsusaka, and the early influences of S. Mori and N. I. Shepherd-Barron have been crucial to my understanding of the subject.

The original 1993 group included D. Abramovich, V. Alexeev, A. Corti, A. Grassi, B. Hassett, S. Keel, S. Kovács, T. Luo, K. Matsuki, J. McKernan, G. Megyesi, and D. Morrison; many of them have been active in this area since. My students A. Corti, S. Kovács, T. Kuwata, E. Szabó, and N. Tziolas worked on various aspects of the early theory.

I gave several lecture series about moduli. The many comments and corrections of colleagues K. Ascher, G. Farkas, M. Fulger, S. Grushevsky, J. Huh, J. Li, M. Lieblich, J. Moraga, T. Murayama, A. Okounkov, R. Pandharipande, Zs. Patakfalvi, C. Raicu, J. Waldron, J. Witaszek, and students C. Araujo, G. Di Cerbo, A. Hogadi, L. Ji, D. Kim, Y. Liu, A. Sengupta, C. Stibitz, Y.-C. Tu, D. Villalobos-Paz, C. Xu, Z. Zhuang, and R. H. Zong have been very helpful.

My collaborators on these topics – K. Altmann, F. Ambro, V. Balaji, F. Bernasconi, J. Bochnak, J. Carvajal-Rojas, P. Cascini, B. Claudon, R. Cluckers, A. Corti, H. Dao, T. de Fernex, J.-P. Demailly, S. Ejiri, J. Fernandez de Bobadilla, A. Ghigi, P. Hacking, B. Hassett, A. Höring, S. Ishii, Y. Kachi, M. Kapovich, S. Kovács, W. Kucharz, K. Kurdyka, M. Larsen, R. Laza, R. Lazarsfeld, B. Lehmann, M. Lieblich, F. Mangolte, M. Mella, S. Mori, M. Mustaţă, A. Némethi, J. Nicaise, K. Nowak, M. Olsson, J. Pardon, G. Saccà, W. Sawin, K. Smith, A. Stäbler, Y. Tschinkel, C. Voisin, J. Witaszek, and L. Zhang – shared many of their ideas.

A. J. de Jong, M. Olsson, C. Skinner, and T. Y. Yu helped with several issues. M. Kim, J. Moraga, J. Peng, B. Totaro, F. Zamora, and the referees gave many comments on earlier versions of the manuscript.

xii

Moduli theory has been developed and shaped by the works of many people. Advances in minimal model theory – especially the series of papers by C. Hacon, J. McKernan, and C. Xu – made it possible to extend the theory from surfaces to all dimensions. The projectivity of moduli spaces was gradually proved by E. Viehweg, O. Fujino, S. Kovács, and Zs. Patakfalvi. After the early works of V. Alexeev and P. Hacking, many examples have been worked out by V. Alexeev and his coauthors, A. Brunyate, P. Engel, A. Knutson, R. Pardini, and A. Thompson. Recent works of K. Ascher, D. Bejleri, K. DeVleming, S. Filipazzi, G. Inchiostro, and Y. Liu give very detailed information on important examples.

The influences of V. Alexeev, A. Corti, S. Kovács, and C. Xu have been especially significant for me.

Sections 6.5–6.6 were written with K. Altmann, while S. Kovács contributed to the writing and editing of the whole book.

Partial financial support was provided to JK by the NSF grant DMS-1901855 and to SK by the NSF grant DMS-210038.

Notation

We follow the notation and conventions of Hartshorne (1977); Kollár and Mori (1998) and Kollár (2013b). Our schemes are Noetherian and separated. At the beginning of each chapter we state further assumptions. Many of the results should work over excellent base schemes, but most of the current proofs apply only in characteristic 0.

A *variety* is usually an integral scheme of finite type over a field. However, following standard usage, a *stable variety* or a *locally stable variety* is reduced, pure dimensional, but possibly reducible.

Affine n-space over a field k is denoted by $\mathbb{A}^n_k$, or by $\mathbb{A}^n(x_1, \ldots, x_n)$ or $\mathbb{A}^n_{\mathbf{x}}$ if we emphasize that the coordinates are $x_1, \ldots, x_n$. Same conventions for projective n-space $\mathbb{P}^n$.

The *canonical class* of X is denoted by K_X, and the *canonical sheaf* or *dualizing sheaf* by ω_X; see (1.23) for varieties and (11.2) for schemes. Since $\mathscr{O}_X(K_X) \simeq \omega_X$, we switch between the divisor and sheaf versions whenever it is convenient. Here K_X is more frequently used on normal varieties, and ω_X in more general settings. Functorial properties work better for ω_X.

A smooth proper variety X is of *general type* if $|mK_X|$ defines a birational map for $m \gg 1$, see (1.30). The *Kodaira dimension* of X, denoted by $\kappa(X)$, is the dimension of the image of $|mK_X|$ for m sufficiently large and divisible.

Notation Commonly Used in Birational Geometry

A *map* or *rational map* is defined on a dense set; it is denoted by $\dashrightarrow$. A *morphism* is everywhere defined; it is denoted by $\to$. A *contraction* is a proper morphism $g: X \to Y$ such that $g_* \mathscr{O}_X = \mathscr{O}_Y$.

A map $g: X \dashrightarrow Y$ between (possibly reducible) schemes is *birational* if there are nowhere dense closed subsets $Z_X \subset X$ and $Z_Y \subset Y$ such that g restricts to an isomorphism $(X \setminus Z_X) \simeq (Y \setminus Z_Y)$. The smallest such Z_X is the *exceptional set* of g, denoted by $\mathrm{Ex}(g)$. A birational map $g \colon X \dashrightarrow Y$ is *small* if $\mathrm{Ex}(g)$ has codimension ≥ 2 in X.

A *resolution* of X is a proper, birational morphism $p \colon X' \to X$, where X' is nonsingular. X has *rational singularities* if $p_*\mathscr{O}_{X'} = \mathscr{O}_X$ and $R^i p_* \mathscr{O}_{X'} = 0$ for $i > 0$; see Kollár and Mori (1998, Sec.5.1). Rational implies *Cohen–Macaulay,* abbreviated as *CM*; see (10.4).

Let $g \colon X \dashrightarrow Y$ be a birational map defined on the open set $X^\circ \subset X$. For a subscheme $W \subset X$, the closure of $g(W \cap X^\circ) \subset Y$ is the *birational transform,* provided $W \cap X^\circ$ is dense in W. It is denoted by $g_*(W)$

Following the confusion established in the literature, a *divisor on X* is either a prime divisor or a Weil divisor; the context usually makes it clear which one.

We use divisor *over X* to mean a prime divisor on some $\pi \colon X' \to X$ that is birational to X. The *center* of E on X, denoted by $\mathrm{center}_X E$, is (the closure of) $\pi(E) \subset X$.

A $\mathbb{Z}$-, $\mathbb{Q}$- *or* $\mathbb{R}$-*divisor* (more precisely, Weil $\mathbb{Z}$-, $\mathbb{Q}$- or $\mathbb{R}$-divisor) is a finite linear combinations of prime divisors $\sum a_i D_i$, where $a_i \in \mathbb{Z}$, $\mathbb{Q}$ or $\mathbb{R}$. A divisor is *reduced* if $a_i = 1$ for every i. See Section 4.3 for various versions of divisors (Weil, Cartier, etc.).

A $\mathbb{Z}$- or $\mathbb{Q}$-divisor D on a normal variety is $\mathbb{Q}$-*Cartier* if mD is Cartier for some $m > 0$. (See (11.43) for the $\mathbb{R}$ version.) The smallest $m \in \mathbb{N}$ such that mD is Cartier is called the *Cartier index* or simply *index* of D. On a nonnormal variety Y these notions make sense if Y is nonsingular at the generic points of $\mathrm{Supp}\, D$; we call these *Mumford divisors,* see (4.16.4) and Section 4.8.

The *index* of a variety Y, denoted by $\mathrm{index}(Y)$, is the Cartier index of K_Y.

Linear equivalence of $\mathbb{Z}$-divisors is denoted by $D_1 \sim D_2$. Two $\mathbb{Q}$-divisors are $\mathbb{Q}$-linearly equivalent if $mD_1 \sim mD_2$ for some $m > 0$. It is denoted by $D_1 \sim_\mathbb{Q} D_2$. (See (11.43) for the $\mathbb{R}$ version.)

Numerical equivalence of divisors D_i or curves C_i is denoted by $D_1 \equiv D_2$ and $C_1 \equiv C_2$.

The *intersection number* of $\mathbb{R}$-Cartier divisors $D_1, \dots, D_r$ on X with a proper subscheme $Z \subset X$ of dimension r is denoted by $(D_1 \cdots D_r \cdot Z)$ or $(D_1 \cdots D_r)_Z$. We omit Z if $Z = X$, and for self-intersections we use (D^r).

An $\mathbb{R}$-Cartier divisor D (resp. line bundle L) on a proper scheme X is *nef,* if $(D \cdot C) \geq 0$ (resp. $\deg(L|_C) \geq 0$) for every integral curve $C \subset X$.

Let $g \colon X \to S$ be a proper morphism. For a $\mathbb{Q}$-Cartier divisor we use g-*ample* and *relatively ample* interchangeably; see (11.51) for $\mathbb{R}$-Cartier divisors.

The *rounding down* (resp. *up*) of a real number d is denoted by $\lfloor d \rfloor$ (resp. $\lceil d \rceil$). For a divisor $D = \sum d_i D_i$ we use $\lfloor D \rfloor := \sum \lfloor d_i \rfloor D_i$, where the D_i are distinct, irreducible divisors. The *fractional part* is $\{D\} := D - \lfloor D \rfloor$.

An $\mathbb{R}$-divisor D on a proper, irreducible variety is *big* if $\lfloor mD \rfloor$ defines a birational map for $m \gg 1$.

A *pair* $(X, \Delta = \sum a_i D_i)$ consist of a scheme X and a Weil divisor Δ on it, the coefficients can be in $\mathbb{Z}$, $\mathbb{Q}$ or $\mathbb{R}$. The divisor part of a pair is frequently called the *boundary* of the pair. (Some authors call Δ a boundary only if $0 \leq a_i \leq 1$ for every i.) When we start with a scheme X and a compactification $X^* \supset X$, frequently $X^* \setminus X$ is also called a *boundary;* this usage is well entrenched for moduli spaces. (Neither agrees with the notion of "boundary" in topology.)

A *simple normal crossing* pair – usually abbreviated as *snc* pair – is a pair (X, D), where X is regular, and at each $p \in X$ there are local coordinates $x_1, \ldots, x_n$ and an open neighborhood $x \in U \subset X$ such that $U \cap \operatorname{Supp} D \subset (x_1 \cdots x_n = 0)$. We also say that D is an *snc divisor.* A scheme Y has *simple normal crossing* singularities if every point $y \in Y$ has an open neighborhood $y \in V \subset Y$ that is isomorphic to an snc divisor.

A *log resolution* of (X, Δ) is a proper, birational morphism $p \colon X' \to X$, where X' is nonsingular and $\operatorname{Supp} \pi^{-1}(\Delta) \cup \operatorname{Ex}(\pi)$ is an snc divisor.

We are mostly interested in proper pairs (X, Δ) with log canonical singularities (11.5). Such a pair is of *general type* if $K_X + \Delta$ is big. In examples, we encounter pairs with $K_X + \Delta \equiv 0$ (called (log)-Calabi–Yau pairs) or with $-(K_X + \Delta)$ ample (called (log)-Fano pairs).

In the literature, "canonical model" can refer to three different notions. We distinguish them as follows. (See Section 11.2 for pairs and for relative versions.)

Given a smooth, proper variety X, its *canonical model* is a proper variety X^c that is birational to X, has canonical singularities and ample canonical class.

Given a variety X, its *canonical modification* is a proper, birational morphism $\pi \colon X^{\mathrm{cm}} \to X$ such that X^{cm} has canonical singularities and its canonical class is π-ample.

Given a variety X with resolution $Y \to X$, the canonical model of Y is the *canonical model of resolutions* of X, denoted by X^{cr}. This is independent of Y.

Additional Conventions Used in This Book

These we follow most of the time, but define them at each occurrence.

The normalization of a scheme X is usually denoted by $\bar{X}$ or X^{n}. However, if D is a divisor on X, then usually $\bar{D}$ denotes its preimage in $\bar{X}$. Then $\bar{D}^n$ denotes the normalization of $\bar{D}$. Unfortunately, a bar is also frequently used to denote the compactification of a scheme or moduli space.

Usually, we use $S^\circ \subset S$ to denote an open, dense subset. Then sheaves or divisors on S° are usually indicated by F° or D°. If G is an algebraic group, then G° denotes the identity component.

We write moduli functors in caligraphic and moduli spaces in roman. Thus for stable varieties we have $\mathcal{SV}$ (functor) and SV (moduli space).

Let F, G be quasi-coherent sheaves on a scheme X. Then $\mathrm{Hom}_X(F, G)$ is the set of $\mathscr{O}_X$-linear sheaf homomorphisms (it is also an $H^0(X, \mathscr{O}_X)$-module), and $\mathcal{H}om_X(F, G)$ is the sheaf of $\mathscr{O}_X$-linear sheaf homomorphisms. See (9.34) for the hom-scheme $\mathbf{Hom}_S(F, G)$.

$\mathrm{Mor}_S(X, Y)$ denotes the set of S-morphisms from X to Y, and $\mathbf{Mor}_S(X, Y)$ the scheme that represents the functor $T \mapsto \mathrm{Mor}_S(X \times_S T, Y \times_S T)$ (if it exists); see (8.63). Same conventions for $\mathrm{Isom}_S(X, Y)$ and $\mathrm{Aut}_S(X)$. If X is a proper $\mathbb{C}$-scheme, then one can pretty much identify $\mathrm{Aut}_\mathbb{C}(X)$ with $\mathbf{Aut}_\mathbb{C}(X)$.

We distinguish the *Picard group* $\mathrm{Pic}(X)$ (as in Hartshorne, 1977), and the *Picard scheme* $\mathbf{Pic}(X)$ (as in Mumford, 1966).

Base change. Given morphisms $f \colon X \to S$ and $q \colon T \to S$, we write the base change diagram as

$$
\begin{array}{ccc}
X_T & \xrightarrow{\ q_X\ } & X \\
{\scriptstyle f_T}\downarrow & & \downarrow{\scriptstyle f} \\
T & \xrightarrow{\ q\ } & S.
\end{array}
$$

Objects obtained by pull-back to X_T are usually denoted either by a subscript T or by q_X^*. The fiber over a point $s \in S$ is denoted by a subscript s. However, we frequently encounter the situation that the fiber product is not the "right" pull-back and needs to be "corrected." Roughly speaking, this happens when the fiber product picks up some embedded subscheme/sheaf, and the "correct" pull-back is the quotient by it.

Thus, for divisors D on X, we let D_T denote the *divisorial pull-back* or *restriction,* which is the divisorial part of $X \times_T D$; see (4.6). We write D_T^{div} if we want to emphasize this (2.73). For coherent sheaves F on X, we frequently use the *hull bull-back,* denoted by F_T^H or $q_X^{[*]}F$; see (3.27).

Brackets are used to denote something naturally associated to an object. We use it to denote the cycle associated to a subscheme (1.3) and the point in the moduli space corresponding to a variety/pair.

The *completion* of a pointed scheme $(x \in X)$ is denoted by $\hat{X}$, or $\hat{X}_x$ if we want to emphasize the point. For $\hat{\mathbb{A}}^n$, the point is assumed the origin, unless otherwise noted. See also (10.52.6).

Numbering

We number everything consecutively. Thus, for example, (2.3) refers to item 3 in Chapter 2. References to sections are given as "Section 2.3." Tertiary numbering is consecutive within items, including lists and formulas. For example, (2.3.2) is subitem 2 in item (2.3), but within (2.3) we may use only (2) as reference.

Introduction

In the moduli theory of curves, the main objects – *stable curves* – are projective curves C that satisfy two conditions:
- (local) the singularities are nodes, and
- (global) K_C is ample.

Generalizing this, Kollár and Shepherd-Barron (1988) posited that in higher dimensions the objects of the moduli theory – *stable varieties* – are projective varieties X such that
- (local) the singularities are semi-log-canonical, and
- (global) K_X is ample.

The theory of semi-log-canonical singularities is treated in Kollár (2013b). Once the objects of a moduli theory are established, we need to describe the families that we aim to understand. For curves, the answer is clear: flat, projective morphisms whose fibers are stable curves.

By contrast, there are too many flat, projective morphisms whose fibers are stable surfaces; basic numerical invariants are not always constant in such families. The correct notion of *(locally) stable families* of surfaces was defined in Kollár and Shepherd-Barron (1988). We describe these in all dimensions, first for one-parameter families in Chapter 2, and then over an arbitrary base in Chapter 3, where seven equivalent definitions of local stability are given in Definition–Theorem 3.1.

Stable curves with weighted points also appeared in many contexts, and, correspondingly, the general objects in higher dimensions are pairs (X, Δ), where X is a variety and $\Delta = \sum a_i D_i$ is a formal linear combination of divisors with rational or real coefficients. Such a pair (X, Δ) is *stable* iff
- (local) the singularities are semi-log-canonical, and
- (global) $K_X + \Delta$ is ample.

The main aim of this book is to complete the moduli theory of stable pairs in characteristic 0.

Defining the right notion of *(locally) stable families of pairs* turned out to be very challenging. The reason is that the divisorial part Δ is not necessarily flat

1

over the base. Flatness was built into the foundations of algebraic geometry by
Grothendieck, and many new results had to be developed.

Our solution goes back to the works of Cayley (1860, 1862), who associated
a divisor in $\mathrm{Gr}(1,3)$ – the Grassmannian of lines in $\mathbb{P}^3$ – to any space curve.
More generally, given any subvariety $X^d \subset \mathbb{P}^n$ and a divisor D on X, there is
a Cayley hypersurface $\mathrm{Ca}(D) \subset \mathrm{Gr}(n-d, n)$. We declare a family of divisors
$\{D_s : s \in S\}$ *C-flat* if the corresponding Cayley hypersurfaces $\{\mathrm{Ca}(D_s) : s \in S\}$
form a flat family. This turns out to work very well over reduced base schemes,
leading to a complete moduli theory of stable families of pairs over such
bases. This is done in Chapter 4. For the rest of the book, the key result
is Theorem 4.76, which constructs the universal family of C-flat Mumford
divisors over an arbitrary base. While C-flatness is defined using a projective
embedding, it is independent of it over reduced bases, but most likely not in
general.

Chapter 5 contains numerical criteria for various fiber-wise constructions to
fit together into a flat family. For moduli theory the most important result is
Theorem 5.1: a flat, projective morphism $f : X \to S$ is stable iff the fibers are
stable and the volume of the fibers $(K_{X_s}^n)$ is locally constant on S.

Chapter 6 discusses several special cases where flatness is the right notion
for the divisor part of a family of stable pairs. This includes all the pairs
$(X, \Delta := \sum a_i D_i)$ with $a_i > \frac{1}{2}$ for every i; see Theorem 6.29.

The technical core of the book is Chapter 7. We develop the notion of
K-flatness, which is a version of C-flatness that is independent of the projective
embedding; see Definition 7.1. It has surprisingly many good properties, listed
in Theorems 7.3–7.5. We believe that this is the "correct" concept for moduli purposes. However, the proofs are rather nuts-and-bolts; a more conceptual
approach would be very desirable.

All of these methods and results are put together in Chapter 8 to arrive at
Theorem 8.1, which is the main result of the book: The notion of Kollár–
Shepherd-Barron–Alexeev stability for families of stable pairs yields a good
moduli theory, with projective coarse moduli spaces.

Section 8.8 discusses problems that complicate the moduli theory of pairs in
positive characteristic; some of these appear quite challenging.

The remaining chapters are devoted to auxiliary results. Chapter 9 discusses hulls and husks, a generalization of quot schemes, that was developed
to suit the needs of higher dimensional moduli theory. Chapter 10 collects sun-
dry results for which we could not find good references, while Chapter 11
summarizes the key concepts and theorems of Kollár (2013b), as well as the
main results of the minimal model program that we need.

1

History of Moduli Problems

The moduli spaces of smooth or stable projective curves of genus $g \geq 2$ are, quite possibly, the most studied of all algebraic varieties.

The aim of this book is to generalize the moduli theory of curves to surfaces and to higher dimensional varieties. In this chapter, we aim to outline how this is done, and, more importantly, to explain why the answer for surfaces is much more complicated than for curves. On the positive side, once we get the moduli theory of surfaces right, the higher dimensional theory works the same.

Section 1.1 is a quick review of the history of moduli problems, culminating in an outline of the basic moduli theory of curves. A'Campo et al. (2016) is a very good overview. Reading some of the early works on moduli, including Riemann, Cayley, Klein, Hilbert, Siegel, Teichmüller, Weil, Grothendieck, and Mumford gives an understanding of how the modern theory relates to the earlier works. See Kollár (2021b) for an account that emphasizes the historical connections.

In Section 1.2, we outline how the theory should unfold for higher dimensional varieties. Details of going from curves to higher dimensions are given in the next two sections. Section 1.3 introduces canonical models, which are the basic objects of moduli theory in higher dimensions. Starting from stable curves, Section 1.4 leads up to the definition of stable varieties, their higher dimensional analogs. Then we show, by a series of examples, why flat families of stable varieties are *not* the correct higher dimensional analogs of flat families of stable curves. Finding the correct replacement has been one of the main difficulties of the whole theory.

While the moduli theory of curves serves as our guideline, it also has many good properties that do not generalize. Sections 1.5–1.8 are devoted to examples that show what can go wrong with moduli theory in general, or with stable varieties in particular.

First, in Section 1.5, we show that the simple combinatorial recipe of going from a nodal curve to a stable curve has no analog for surfaces. Next we give a collection of examples showing how easy it is to end up with rather horrible moduli problems. Hypersurfaces and other interesting examples are discussed in Section 1.6, as are alternative compactifications of the moduli of curves in Section 1.7. Section 1.8 illustrates the differences between fine and coarse moduli spaces.

Two major approaches to moduli – the geometric invariant theory of Mumford and the Hodge theory of Griffiths – are mostly absent from this book. Both of these are very powerful, and give a lot of information in the cases when they apply. They each deserve a full, updated treatment of their own. However, so far neither gave a full description of the moduli of surfaces, much less of higher dimensional varieties. It would be very interesting to develop a synthesis of the three methods and gain better understanding in the future.

1.1 Riemann, Cayley, Hilbert, and Mumford

Let V be a "reasonable" class of objects in algebraic geometry, for instance, V could be all subvarieties of $\mathbb{P}^n$, all coherent sheaves on $\mathbb{P}^n$, all smooth curves or all projective varieties. The aim of the theory of moduli is to understand all "reasonable" families of objects in V, and to construct an algebraic variety (or scheme, or algebraic space) whose points are in "natural" one-to-one correspondence with the objects in V. If such a variety exists, we call it the *moduli space* of V, and denote it by M_V. The simplest, classical examples are given by the theory of linear systems and families of linear systems.

1.1 (Linear systems) Let X be a smooth, projective variety over an algebraically closed field k and L a line bundle on X. The corresponding linear system is

$$\mathcal{L}in\mathcal{S}ys(X, L) = \{\text{effective divisors } D \text{ such that } \mathcal{O}_X(D) \simeq L\}.$$

The objects in $\mathcal{L}in\mathcal{S}ys(X, L)$ are in natural one-to-one correspondence with the points of the projective space $\mathbb{P}(H^0(X, L)^\vee)$ which is traditionally denoted by $|L|$. (We follow the Grothendieck convention for $\mathbb{P}$ as in Hartshorne (1977, sec.II.7).) Thus, for every effective divisor D such that $\mathcal{O}_X(D) \simeq L$, there is a unique point $[D] \in |L|$.

Moreover, this correspondence between divisors and points is given by a universal family of divisors over $|L|$. That is, there is an effective Cartier divisor $\mathrm{Univ}_L \subset |L| \times X$ with projection $\pi\colon \mathrm{Univ}_L \to |L|$ such that $\pi^{-1}[D] = D$ for every effective divisor D linearly equivalent to L.

The classical literature never differentiates between the linear system as a set and the linear system as a projective space. There are, indeed, few reasons to distinguish them as long as we work over a fixed base field k. If, however, we pass to a field extension $K \supset k$, the advantages of viewing $|L|$ as a k-variety appear. For any $K \supset k$, the set of effective divisors D defined over K such that $\mathcal{O}_X(D) \simeq L$ corresponds to the K-points of $|L|$. Thus the scheme-theoretic version automatically gives the right answer over every field.

1.2 (Jacobians of curves) Let C be a smooth projective curve (or Riemann surface) of genus g. As discovered by Abel and Jacobi, there is a variety $\mathrm{Jac}^\circ(C)$ of dimension g whose points are in natural one-to-one correspondence with degree 0 line bundles on C. As before, the correspondence is given by a universal line bundle $L^{\mathrm{univ}} \to C \times \mathrm{Jac}^\circ(C)$, called the Poincaré bundle. That is, for any point $p \in \mathrm{Jac}^\circ(C)$, the restriction of L^{univ} to $C \times \{p\}$ is the degree 0 line bundle corresponding to p.

Unlike in (1.1), the universal line bundle L^{univ} is not unique (and need not exist if the base field is not algebraically closed). This has to do with the fact that while an automorphism of the pair $D \subset X$ that is trivial on X is also trivial on D, any line bundle $L \to C$ has automorphisms that are trivial on C: we can multiply every fiber of L by the same nonzero constant.

1.3 (Cayley forms and Chow varieties) Cayley (1860, 1862) developed a method to associate a hypersurface in the Grassmannian $\mathrm{Gr}(\mathbb{P}^1, \mathbb{P}^3)$ to a curve in $\mathbb{P}^3$. The resulting moduli spaces have been used, but did not seem to have acquired a name. Chow understood how to deal with reducible and multiple varieties, and proved that one gets a projective moduli space; see Chow and van der Waerden (1937). The name *Chow variety* seems standard, we use Cayley–Chow for the correspondence that was discovered by Cayley. See Section 3.1 for an outline and Kollár (1996, secs.I.3–4) for a modern treatment.

Let k be an algebraically closed field and X a normal, projective k-variety. Fix a natural number m. An *m-cycle* on X is a finite, formal linear combination $\sum a_i Z_i$ where the Z_i are irreducible, reduced subvarieties of dimension m and $a_i \in \mathbb{Z}$. We usually assume tacitly that all the Z_i are distinct. An m-cycle is called *effective* if $a_i \geq 0$ for every i.

The points of the *Chow variety* $\mathrm{Chow}_m(X)$ are in "natural" one-to-one correspondence with the set of effective m-cycles on X. (Since we did not fix the degree of the cycles, $\mathrm{Chow}_m(X)$ is not actually a variety, but a countable disjoint union of projective, reduced k-schemes.) The point of $\mathrm{Chow}_m(X)$ corresponding to a cycle $Z = \sum a_i Z_i$ is also usually denoted by $[Z]$.

As for linear systems, it is best to describe the "natural correspondence" by a universal family. The situation is, however, more complicated than before. There is a family (or rather an effective cycle) $\text{Univ}_m(X)$ on $\text{Chow}_m(X) \times X$ with projection $\pi\colon \text{Univ}_m(X) \to \text{Chow}_m(X)$ such that for every effective m-cycle $Z = \sum a_i Z_i$,

(1.3.1) the support of $\pi^{-1}[Z]$ is $\cup_i Z_i$, and

(1.3.2) the fundamental cycle (4.61.1) of $\pi^{-1}[Z]$ equals Z if $a_i = 1$ for every i.

If the characteristic of k is 0, then the only problem in (2) is a clash between the traditional cycle-theoretic definition of the Chow variety and the scheme-theoretic definition of the fiber, but in positive characteristic the situation is more problematic; see Kollár (1996, secs.I.3–4).

An example of a "perfect" moduli problem is the theory of *Hilbert schemes*, introduced in Grothendieck (1962, lect.IV). See Mumford (1966), (Kollár, 1996, I.1–2) or Sernesi (2006, sec.4.3) or Section 3.1 for a summary.

1.4 (Hilbert schemes) Let k be an algebraically closed field and X a projective k-scheme. Set

$$\mathcal{H}ilb(X) = \{\text{closed subschemes of } X\}.$$

Then there is a k-scheme $\text{Hilb}(X)$, called the *Hilbert scheme* of X, whose points are in a "natural" one-to-one correspondence with closed subschemes of X. The point of $\text{Hilb}(X)$ corresponding to a subscheme $Y \subset X$ is frequently denoted by $[Y]$. There is a universal family $\text{Univ}(X) \subset \text{Hilb}(X) \times X$ such that

(1.4.1) the first projection $\pi\colon \text{Univ}(X) \to \text{Hilb}(X)$ is flat, and

(1.4.2) $\pi^{-1}[Y] = Y$ for every closed subscheme $Y \subset X$.

The beauty of the Hilbert scheme is that it describes not just subschemes, but all flat families of subschemes as well. To see what this means, note that for any morphism $g\colon T \to \text{Hilb}(X)$, by pull-back we obtain a flat family of subschemes $T \times_{\text{Hilb}(X)} \text{Univ}(X) \subset T \times X$. It turns out that every family is obtained this way:

(1.4.3) For every T and closed subscheme $Z \subset T \times X$ that is flat over T, there is a unique $g_Z\colon T \to \text{Hilb}(X)$ such that $Z = T \times_{\text{Hilb}(X)} \text{Univ}(X)$.

This takes us to the functorial approach to moduli problems.

1.5 (Hilbert functor and Hilbert scheme) Let $X \to S$ be a morphism of schemes. Define the *Hilbert functor* of X/S as a functor that associates to a scheme $T \to S$ the set

$\mathcal{H}ilb_{X/S}(T) = \{$subschemes $Z \subset T \times_S X$ that are flat and proper over $T\}$.

The basic existence theorem of Hilbert schemes then says that, if $X \to S$ is quasi-projective, there is a scheme $\mathrm{Hilb}_{X/S}$ such that for any S scheme T,

$$\mathcal{H}ilb_{X/S}(T) = \mathrm{Mor}_S(T, \mathrm{Hilb}_{X/S}).$$

Moreover, there is a universal family π: $\mathrm{Univ}_{X/S} \to \mathrm{Hilb}_{X/S}$ such that the above isomorphism is given by pulling back the universal family.

We can summarize these results as follows:

Principle 1.6 π: $\mathrm{Univ}_{X/S} \to \mathrm{Hilb}_{X/S}$ *contains all the information about proper, flat families of subschemes of X/S, in the most succinct way.*

This example leads us to a general definition:

Definition 1.7 (Fine moduli spaces) Let **V** be a "reasonable" class of projective varieties (or schemes, or sheaves, or . . .). In practice "reasonable" may mean several restrictions, but for the definition we only need the following weak assumption:
(1.7.1) Let $K \supset k$ be a field extension. Then a k-variety X_k is in **V** iff $X_K := X_k \times_{\mathrm{Spec}\,k} \mathrm{Spec}\,K$ is in **V**.
Following (1.5), define the corresponding moduli functor that associates to a scheme T the set

$$\mathcal{V}arieties_{\mathbf{V}}(T) := \left\{ \begin{array}{c} \text{Flat families } X \to T \text{ such that} \\ \text{every fiber is in } \mathbf{V}, \\ \text{modulo isomorphisms over } T. \end{array} \right\} \quad (1.7.2)$$

We say that a scheme $\mathrm{Moduli}_{\mathbf{V}}$ is a *fine moduli space* for the functor $\mathcal{V}arieties_{\mathbf{V}}$, if the following holds:
(1.7.3) For every scheme T, pulling back gives an equality

$$\mathcal{V}arieties_{\mathbf{V}}(T) = \mathrm{Mor}(T, \mathrm{Moduli}_{\mathbf{V}}).$$

Applying the definition to $T = \mathrm{Moduli}_{\mathbf{V}}$ gives a universal family u: $\mathrm{Univ}_{\mathbf{V}} \to \mathrm{Moduli}_{\mathbf{V}}$. Setting $T = \mathrm{Spec}\,K$, where K is a field, we see that the K-points of $\mathrm{Moduli}_{\mathbf{V}}$ correspond to the K-isomorphism classes of objects in **V**.

We consider the existence of a fine moduli space as the ideal possibility. Unfortunately, it is rarely achieved.

Next we see what happens with the simplest case, for smooth curves.

1.8 (Moduli functor and moduli space of smooth curves) Following (1.7) we define the moduli functor of smooth curves of genus g as

$$Curves_g(T) := \left\{ \begin{array}{l} \text{Smooth, proper families } S \to T, \\ \text{every fiber is a curve of genus } g, \\ \text{modulo isomorphisms over } T. \end{array} \right\}$$

It turns out that there is no fine moduli space for curves of genus g. Every curve C with nontrivial automorphisms causes problems; there cannot be any point $[C]$ corresponding to it in a fine moduli space (see Section 1.8).

It was gradually understood that there is some kind of an object, denoted by M_g, and called the *coarse moduli space* (or simply *moduli space*) of curves of genus g, that comes close to being a fine moduli space.

For elliptic curves, we get $M_1 \simeq \mathbb{A}^1$, and the moduli map is given by the j-invariant, as was known to Dedekind and Klein; see Klein and Fricke (1892). They also knew that there is no universal family over M_1. The theory of abelian integrals due to Abel, Jacobi, and Riemann does the same for all curves, though in this case a clear moduli-theoretic interpretation seems to have been done only later; see the historical sketch at the end of Shafarevich (1974), Siegel (1969, chap.4), or Griffiths and Harris (1978, chap.2) for modern treatments. For smooth plane curves, and more generally for smooth hypersurfaces in any dimension, the invariant theory of Hilbert produces coarse moduli spaces. Still, a precise definition and proof of existence of M_g appeared only in Teichmüller (1944) in the analytic case and in Mumford (1965) in the algebraic case. See A'Campo et al. (2016) or Kollár (2021b) for historical accounts.

1.9 (Coarse moduli spaces) Mumford (1965)

As in (1.7), let **V** be a "reasonable" class. When there is no fine moduli space, we still can ask for a scheme that best approximates its properties. We look for schemes M for which there is a natural transformation

$$T_M \colon \mathcal{V}arieties_g(*) \longrightarrow \mathrm{Mor}(*, M).$$

Such schemes certainly exist: for instance, if we work over a field k, then we can take $M = \mathrm{Spec}\, k$. All schemes M for which T_M exists form an inverse system which is closed under fiber products. Thus, as long as we are not unlucky, there is a universal (or largest) scheme with this property. Though it is not usually done, it should be called the *categorical moduli space*.

This object can be rather useless in general. For instance, fix n, d and let $\mathbf{H}_{n,d}$ be the class of all hypersurfaces of degree d in $\mathbb{P}^{n+1}_k$, up to isomorphisms. We see in (1.56) that a categorical moduli space exists and it is $\mathrm{Spec}\, k$.

To get something more like a fine moduli space, we require that it give a one-to-one parametrization, at least set theoretically. Thus we say that a scheme Moduli$_V$ is a *coarse moduli space* for V if the following hold:

(1.9.1) there is a natural transformation of functors

$$\text{ModMap}\colon \mathcal{V}arieties_V(*) \longrightarrow \text{Mor}(*, \text{Moduli}_V),$$

(1.9.2) Moduli$_V$ is universal satisfying (1), and

(1.9.3) for any algebraically closed field $K \supset k$, we get a bijection

$$\text{ModMap}\colon \mathcal{V}arieties_V(\text{Spec } K) \overset{\simeq}{\longrightarrow} \text{Mor}(\text{Spec } K, \text{Moduli}_V) = \text{Moduli}_V(K).$$

1.10 (Moduli functors versus moduli spaces) While much of the early work on moduli, especially since Mumford (1965), put the emphasis on the construction of fine or coarse moduli spaces, recently the focus shifted toward the study of the families of varieties, that is, toward moduli functors and moduli stacks. The main task is to understand all "reasonable" families. Once this is done, the existence of a coarse moduli space should be nearly automatic. The coarse moduli space is not the fundamental object any longer, rather it is only a convenient way to keep track of certain information that is only latent in the moduli functor or stack.

1.11 (Compactifying M$_g$) While the basic theory of algebraic geometry is local, that is, it concerns affine varieties, most really interesting and important objects in algebraic geometry and its applications are global, that is, projective or at least proper.

The moduli spaces M$_g$ are not compact, in fact the moduli functor of smooth curves discussed so far has a definitely local flavor. Most naturally occurring smooth families of curves live over affine schemes, and it is not obvious how to write down any family of smooth curves over a projective base. For many reasons it is useful to find geometrically meaningful compactifications of M$_g$. The answer to this situation is to allow not just smooth curves, but also certain singular curves in our families.

Concentrating on one-parameter families, we have the following:

Question 1.11.1 Let B be a smooth curve, $B^\circ \subset B$ an open subset, and $\pi^\circ\colon S^\circ \to B^\circ$ a smooth family of genus g curves. Is there a "natural" extension

$$
\begin{array}{ccc}
S^\circ & \hookrightarrow & S \\
\pi^\circ \downarrow & & \downarrow \pi \\
B^\circ & \hookrightarrow & B,
\end{array}
$$

where $\pi\colon S \to B$ is a flat family of (possibly singular) curves?

There is no reason to think that there is a unique such extension. Deligne and Mumford (1969) construct one after a base change $B' \to B$, and by now it is hard to imagine a time when their choice was not the "obviously best" solution. We review their definition next. In Section 1.6 we see, by examples, why this concept has not been so obvious.

Definition 1.12 (Stable curve) A *stable curve* over an algebraically closed field k is a proper, geometrically connected k-curve C such that

(Local property) the only singularities of C are ordinary nodes, and

(Global property) the canonical class K_C is ample.

A *stable curve* over a scheme T is a flat, proper morphism $\pi \colon S \to T$ such that every geometric fiber of π is a stable curve. (The arithmetic genus of the fibers is a locally constant function on T, but we usually also tacitly assume that it is constant.) The moduli functor of stable curves of genus g is

$$\overline{Curves}_g(T) := \left\{ \begin{array}{l} \text{Stable curves of genus } g \text{ over } T, \\ \text{modulo isomorphisms over } T. \end{array} \right\}$$

Theorem 1.13 Deligne and Mumford (1969) *For every $g \geq 2$, the moduli functor of stable curves of genus g has a coarse moduli space $\overline{M}_g$. Moreover, $\overline{M}_g$ is projective, normal, has only quotient singularities, and contains M_g as an open dense subset.*

$\overline{M}_g$ has a rich and intriguing geometry, which is related to major questions in many branches of mathematics and physics; see Farkas and Morrison (2013) for a collection of surveys and Pandharipande (2018a,b) for overviews.

1.2 Moduli for Varieties of General Type

The aim of this book is to use the moduli of stable curves as a guideline, and develop a moduli theory for varieties of general type (1.30). (See (1.22) for some comments on the nongeneral type cases.)

Here we outline the main steps of the plan with some comments. Most of the rest of the book is then devoted to accomplishing these goals.

Step 1.14 (Higher dimensional analogs of smooth curves) It has been understood since the beginnings of the theory of surfaces that, for surfaces of Kodaira dimension ≥ 0 (p.xiv), the correct moduli theory should be birational, not biregular. That is, the points of the moduli space should correspond not to *isomorphism* classes of surfaces, but to *birational* equivalence classes of surfaces. There are two ways to deal with this problem.

First, one can work with smooth families, but consider two families $V \to B$ and $W \to B$ equivalent if there is a *fiber-wise birational* map between them; that is, a rational map $V \dashrightarrow W$ that induces a birational equivalence of the fibers $V_b \dashrightarrow W_b$ for every $b \in B$. This seems rather complicated technically. The second, much more useful method relies on the observation that every birational equivalence class of surfaces of Kodaira dimension ≥ 0 contains a unique *minimal model*, that is, a smooth projective surface S^m whose canonical class is nef (p.xv). Therefore, one can work with families of minimal models, modulo isomorphisms. With the works of Mumford (1965) and Artin (1974) it became clear that, for surfaces of general type, it is even better to work with the *canonical model*, which is a mildly singular projective surface S^c whose canonical class is ample. The resulting class of singularities has since been established in all dimensions; they are called *canonical singularities* (1.33).

Principle 1.14.1 In moduli theory, the main objects of study are projective varieties with ample canonical class and with canonical singularities.

Implicit in this claim is that every smooth family of varieties of general type produces a flat family of canonical models, we discuss this in (1.36).

See Section 1.3 for more details on this step.

Step 1.15 (Higher dimensional analogs of stable curves) The correct definition of the higher dimensional analogs of stable curves was much less clear. An approach through geometric invariant theory (GIT) was investigated by Mumford (1977), but never fully developed. In essence, the GIT approach starts with a particular method of construction of moduli spaces, and then tries to see for which class of varieties it works. The examples of Wang and Xu (2014) suggest that GIT is unlikely to give a good compactification for the moduli of surfaces.

A different framework was proposed in Kollár and Shepherd-Barron (1988). Instead of building on geometric invariant theory, it focuses on one-parameter families, and uses Mori's program as its basic tool.

Before we give the definition, recall a key step of the proof of (1.13) that establishes separatedness and properness of $\overline{M}_g$. (The traditional name is stable "reduction," but "extension" is more descriptive.)

1.15.1 (Stable extension for curves) Let B be a smooth curve, $B^\circ \subset B$ a dense, open subset, and $\pi^\circ : S^\circ \to B^\circ$ a flat family of smooth, projective curves of genus ≥ 2. Then there is a finite surjection $p : A \to B$ and a diagram

$$
\begin{array}{ccccc}
S^\circ \times_B A & \hookrightarrow & S_A^{ss} & \xrightarrow{\ \tau\ } & S_A^{stab} \\
\downarrow & & \downarrow{\scriptstyle \pi_A^{ss}} & & \downarrow{\scriptstyle \pi_A^{stab}} \\
B^\circ \times_B A & \hookrightarrow & A & =\!=\!= & A,
\end{array}
$$

where

(a) $\pi_A^{ss}: S_A^{ss} \to A$ is a flat family of reduced, nodal curves,

(b) $\tau: S_A^{ss} \to S_A^{stab}$ is the relative canonical model (11.26), and

(c) $\pi_A^{stab}: S_A^{stab} \to A$ is a flat family of stable curves.

A detailed proof is given in (2.51): for now we build on this to state the main theses of Kollár and Shepherd-Barron (1988) about moduli problems.

Principle 1.15.2 We should follow the proof of the Stable extension theorem (1.15.1). The resulting fibers give the right class of stable varieties.

Principle 1.15.3 As in (1.12), a connected k-scheme X is stable iff it satisfies two conditions, whose precise definitions are not important for now:
(Local property) Semi-log-canonical singularities, see (1.41).
(Global property) The canonical class K_X is ample, see (1.23).

1.15.4 (Warning about positive characteristic) The examples of Kollár (2022) suggest that, in positive characteristic, (1.15.2) gives the right families, but not quite the right objects in dimensions ≥ 3; see Section 8.8 for details.

Step 1.16 (Higher dimensional analogs of families of stable curves I) The definition (1.7) is very natural within our usual framework of algebraic geometry, but it hides a very strong supposition:

1.16.1 (Unwarranted assumption) If **V** is a "reasonable" class of varieties, then any flat family whose fibers are in **V** is a "reasonable" family.

In Grothendieck's foundations of algebraic geometry, flatness is one of the cornerstones, and there are many "reasonable" classes for which flat families are indeed the "reasonable" families. Nonetheless, even when the base of the family is a smooth curve, (1.16.1) needs arguing, but the assumption is especially surprising when applied to families over nonreduced schemes T. Consider, for instance, the case when T is the spectrum of an Artinian k-algebra. Then T has only one closed point $t \in T$. A flat family $p: X \to T$ has only one fiber X_t, and our only restriction is that X_t be in our class **V**. Thus (1.16.1) declares that we care only about X_t. Once X_t is in **V**, every flat deformation of X_t over T is automatically "reasonable."

A crucial conceptual point in the moduli theory of higher dimensional varieties is the realization that, starting with families of surfaces, flatness of the map $X \to T$ is not enough: allowing all flat families whose fibers are stable varieties leads to the wrong moduli problem.

The simple fact is that basic numerical invariants, like the self intersection of the canonical class, or even the Kodaira dimension, fail to be locally constant in flat families of stable varieties, even when the singularities are quite mild and the base is a smooth curve. We give a series of such examples in (1.42–1.47).

The difficulty of working out the correct concept has been one of the main stumbling blocks of the general theory.

Principle 1.16.2 Flat families of stable varieties $X \to T$ are **not** the correct higher dimensional analogs of flat families of stable curves (1.12).

For families over smooth curves, the Stable extension theorem (1.15.1) is again our guide to the correct definition.

1.16.3 (Stable morphisms) Let $p: Y \to B$ be a proper morphism from a normal variety to a smooth curve. Then p is *stable* iff, for every $b \in B$,

(a) Y_b has semi-log-canonical singularities,

(b) $K_{Y_b} = K_Y|_{Y_b}$ is ample, and

(c) mK_Y is Cartier for some $m > 0$, that is, K_Y is $\mathbb{Q}$-Cartier (p.xv).

This is a direct generalization of the notion of stable family of curves (1.12), except that here we have to add condition (c) for K_Y. If the K_{Y_b} are Cartier, then so is K_Y (2.6), this is why (c) was not necessary for curves. See (2.3) for other versions and (2.4) for comments on the positive characteristic cases.

Note that the K_{Y_b} are $\mathbb{Q}$-Cartier by (1.15.3), but this does not imply that K_Y is $\mathbb{Q}$-Cartier; this is a quite subtle issue with restrictions of non-Cartier divisors. We discuss this in detail in Section 2.4.

Step 1.17 (Higher dimensional analogs of families of stable curves II) Extending the definition (1.16.3) to general base schemes turned out to be very difficult. There were two main proposals in Kollár and Shepherd-Barron (1988) and Viehweg (1995). They are equivalent over reduced base schemes; we explain this in Section 3.4. However, the two versions differ for families of surfaces with quotient singularities over Spec $\mathbb{C}[\varepsilon]$ by Altmann and Kollár (2019). We treat these topics in Sections 6.2–6.3 and 6.6.

The problem becomes even harder when we treat not just stable varieties, but stable pairs. Finding the correct definition turned out to be the longest-standing open question of the theory. An answer was developed in Kollár (2019) and we devote Chapter 7 to explaining it.

Step 1.18 (Representability of moduli functors) The question is the following. Let $p: X \to S$ be an arbitrary projective morphism. Can we understand all morphisms $q: T \to S$ such that $X \times_S T \to T$ is a family in our moduli theory?

A moduli theory **M** is *representable* if, for every projective morphism $p: X \to S$, there is a morphism $j: S^{\mathbf{M}} \to S$ with the following property:

Given any $q\colon T \to S$, the pulled-back family $X \times_S T \to T$ is in **M** iff q factors uniquely as $q\colon T \to S^{\mathbf{M}} \to S$.

That is, $X \times_S S^{\mathbf{M}} \to S^{\mathbf{M}}$ is in **M** and $S^{\mathbf{M}}$ is universal with this property.

Representability is rarely mentioned for the moduli of curves, since it easily follows from general principles. The Flattening decomposition theorem (3.19) says that flatness is representable, and for proper, flat morphisms, being a family of stable curves is represented by an open subscheme.

Both of these become quite complicated in higher dimensions. Since flatness is only part of our assumptions, we need a different way of pulling back families. The theory of hulls and husks in Kollár (2008a) was developed for this reason, leading to the notion of generically Cartier pull-back, defined in Section 4.1. With these, representability is proved in Sections 3.5, 4.6, and 7.6 in increasing generality.

Representability also implies that being a stable family can be tested on 0-dimensional subschemes of T, that is, on spectra of Artinian rings. This is the reason why formal deformation theory is such a powerful tool: see Illusie (1971); Artin (1976); Sernesi (2006).

The previous steps form the basis of a good moduli theory. Once we have them, it is quite straightforward to construct the corresponding moduli space.

Step 1.19 (Two moduli spaces) Let C be a stable curve of genus $g \geq 2$. Then rK_C is very ample for $r \geq 3$, and any basis of its global sections gives an embedding $C \hookrightarrow \mathbb{P}^{r(2g-2)-g}$. Thus all stable curves of genus g appear in the Chow variety or Hilbert scheme of $\mathbb{P}^{r(2g-2)-g}$. Representability (1.18) then implies that we get a moduli space of all r-canonically embedded stable curves

$$\mathrm{EmbStab}_g \subset \mathrm{Hilb}(\mathbb{P}^{r(2g-2)-g}). \qquad (1.19.1)$$

For a fixed C, the embedding $C \hookrightarrow \mathbb{P}^{r(2g-2)-g}$ gives an orbit of $\mathbf{Aut}(\mathbb{P}^{r(2g-2)-g})$, thus we should get the moduli space as

$$\overline{M}_g = \mathrm{EmbStab}_g \, / \, \mathbf{Aut}(\mathbb{P}^{r(2g-2)-g}). \qquad (1.19.2)$$

Starting with Mumford (1965) and Matsusaka (1964), much effort was devoted to understanding quotients like (1.19.2). Already for curves the method of Mumford (1965) is quite subtle; generalizations to surfaces in Gieseker (1977) and to higher dimensions in Viehweg (1995) are quite hard. For surfaces and in higher dimensions, these approaches handle only the interior of the moduli space (where we have only canonical singularities). When GIT works, it automatically gives a quasi-projective moduli space, but Wang and Xu (2014) suggest that GIT methods do not work for the whole moduli space.

It turns out to be much easier to obtain quotients that are algebraic spaces. The general quotient theorems of Kollár (1997) and Keel and Mori (1997) take care of this question completely; see Section 8.6 for details.

The same approach works in all dimensions. We fix $r > 0$ such that rK_X is very ample, and the rest of the proof works without changes.

For curves any $r \geq 3$ works, but, starting with surfaces, a uniform choice of r is no longer possible. The strongest results say that if we fix the dimension n and the volume v (10.31), then there is an $r = r(n, v)$ such that rK_X is very ample. We discuss this in (1.21).

Once we have our moduli spaces, we start to investigate their properties. We should not expect to get moduli spaces that are as nice as those for curves. For instance, even for smooth surfaces with ample canonical class, the moduli spaces can have arbitrarily complicated singularities and scheme structures (Vakil, 2006). Nonetheless, we have two types of basic positive results.

Step 1.20 (Separatedness and properness) The valuative criteria of separatedness and properness translate to functors as follows.

We start with a smooth curve B, an open subset $B^\circ \subset B$, and a stable family $\pi^\circ : X^\circ \to B^\circ$.

1.20.1 (Separatedness) There is at most one stable extension to

$$
\begin{array}{ccc}
X^\circ & \hookrightarrow & X \\
\downarrow{\scriptstyle \pi^\circ} & & \downarrow{\scriptstyle \pi} \\
B^\circ & \hookrightarrow & B.
\end{array}
$$

We obtain a similar translation of properness, but here we have to pay attention to the difference between coarse and fine moduli spaces.

1.20.2 (Valuative-properness) There is a finite surjection $p : A \to B$ such that there is a unique stable extension

$$
\begin{array}{ccc}
X^\circ \times_B A & \hookrightarrow & X_A \\
\downarrow{\scriptstyle \pi^\circ_A} & & \downarrow{\scriptstyle \pi_A} \\
B^\circ \times_B A & \hookrightarrow & A.
\end{array}
$$

Thus the valuative criterion of properness is exactly the general version of the Stable extension theorem (1.15.1).

Step 1.21 (Discrete invariants, boundedness, and projectivity) The most important discrete invariant of a smooth projective curve C is its genus. The genus is unchanged under smooth deformations, and all smooth curves with the same genus form a single family M_g. Thus, in effect, the genus is the only discrete invariant of a smooth projective curve; it completely determines the other ones, like the Euler characteristic $\chi(C, \mathcal{O}_C) = 1 - g$, or the Hilbert polynomial $\chi(C, \mathcal{O}_C(mK_C)) = (2g - 2)m + (1 - g)$.

In a similar manner, we would like to find discrete invariants of (locally) stable varieties that are unchanged by (locally) stable deformations.

The basic such invariant is the Hilbert "polynomial" of K_X. We have to keep in mind that K_X need not be Cartier. Therefore, $m \mapsto \chi(X, \mathcal{O}_X(mK_X))$ is not a polynomial, rather a polynomial with periodic coefficients.

For stable varieties the most important invariant is $\mathrm{vol}(X) := (K_X^n)$ (where $n = \dim X$), called the *volume* (10.31) of X. This is also the leading coefficient of the Hilbert polynomial (times $n!$). The volume is positive, but it is frequently a rational number since K_X is only $\mathbb{Q}$-Cartier; it can be quite small: see Alexeev and Liu (2019a); Esser et al. (2021).

For $m = 0$ we get the Euler characteristic $\chi(X, \mathcal{O}_X)$, but it turns out that the individual groups $h^i(X, \mathcal{O}_X)$ are also deformation invariants by Kollár and Kovács (2010); see Section 2.5.

Next we would like to show that all stable varieties with fixed volume can be "parametrized" by a scheme of finite type; this is called *boundedness*. To state it, let $\mathcal{SV}^{\mathrm{set}}(n, v)$ denote the set of all stable varieties of dimension n and volume v. There are three, roughly equivalent versions.

- There is an $m = m(n, v)$ such that mK_X is very ample for $X \in \mathcal{SV}^{\mathrm{set}}(n, v)$.
- There is a $D = D(n, v)$ such that every $X \in \mathcal{SV}^{\mathrm{set}}(n, v)$ is isomorphic to a subvariety of $\mathbb{P}^D$ of degree $\leq D$.
- There is a morphism $\pi \colon U \to S$ of schemes of finite type such that every $X \in \mathcal{SV}^{\mathrm{set}}(n, v)$ is isomorphic to a fiber of π.

Proving these three turned out to be extremely difficult. For smooth varieties this was solved by Matsusaka (1972), for stable surfaces by Alexeev (1993), and the general stable case is settled in Hacon et al. (2018).

Our moduli spaces satisfy the valuative criterion of properness. Together with boundedness this implies that our moduli spaces are proper.

Once we have a proper moduli space, one would like to prove that it is projective. For surfaces this was done in Kollár (1990), and extended to higher dimensions in Fujino (2018) and Kovács and Patakfalvi (2017).

These last two topics each deserve a detailed treatment of their own; we make only a few more comments in (6.5).

1.22 (Moduli for varieties of nongeneral type) The moduli theory of varieties of nongeneral type is quite complicated.

A general problem, illustrated by abelian, elliptic, and K3 surfaces, is that a typical deformation of such an algebraic surface over $\mathbb{C}$ is a nonalgebraic complex analytic surface. Thus any algebraic theory captures only a small part of the full analytic deformation theory.

The moduli question for analytic surfaces has been studied, especially for complex tori and K3 surfaces. In both cases it seems that one needs to add some extra structure (for instance, fixing a basis in some topological homology group) in order to get a sensible moduli space. (As an example of what could happen, note that the three-dimensional space of Kummer surfaces is dense in the 20-dimensional space of all K3 surfaces.)

Even if one restricts to the algebraic case, compactifying the moduli space seems rather difficult. Detailed studies of abelian varieties and K3 surfaces show that there are many different compactifications depending on additional choices: see Kempf et al. (1973) and Ash et al. (1975).

It is only with the works of Alexeev (2002) that a geometrically meaningful compactification of the moduli of principally polarized abelian varieties became available. This relies on the observation that a pair (A, Θ) consisting of a principally polarized abelian variety A and its theta divisor Θ behaves as if it were a variety of general type.

A moduli theory for K-stable Fano varieties was developed quite recently; see Xu (2020) for an overview.

Definition 1.23 (Canonical class, bundle, and sheaf I) Let X be a smooth variety over a field k. As in Shafarevich (1974, III.6.3) or Hartshorne (1977, p.180), the *canonical line bundle* of X is $\omega_X := \wedge^{\dim X} \Omega_{X/k}$. Any divisor D such that $\mathcal{O}_X(D) \simeq \omega_X$ is called a *canonical divisor*. Their linear equivalence class is called the *canonical class*, denoted by K_X. (Both books tacitly assume that k is algebraically closed. The definition, however, works over any field k, as long as X is smooth over k.)

Let X be a normal variety over a perfect field k. Let $j: X^{\mathrm{sm}} \hookrightarrow X$ be the inclusion of the locus of smooth points. Then $X \setminus X^{\mathrm{sm}}$ has codimension ≥ 2, therefore, restriction from X to X^{sm} is a bijection on Weil divisors and on linear equivalence classes of Weil divisors. Thus there is a unique linear equivalence class K_X of Weil divisors on X such that $K_X|_{X^{\mathrm{sm}}} = K_{X^{\mathrm{sm}}}$. It is called the *canonical class* of X. The divisors in K_X need not be Cartier.

The push-forward $\omega_X := j_* \omega_{X^{\mathrm{sm}}}$ is a rank 1 coherent sheaf on X, called the *canonical sheaf* of X. The canonical sheaf ω_X agrees with the *dualizing sheaf* ω_X° as defined in Hartshorne (1977, p.241). (Note that Hartshorne (1977) defines the dualizing sheaf only if X is proper. In general, take a normal compactification $\bar{X} \supset X$ and use $\omega_{\bar{X}}^\circ|_X$ instead. For more details, see Kollár and Mori

(1998, sec.5.5), Hartshorne (1966), or Conrad (2000).) Note that ω_X satisfies Serre's condition S_2 (10.3.2), but frequently it is not locally free.

More generally, as long as X itself is normal or S_2, and ω_X is locally free outside a codimension ≥ 2 subset of X, we can work with ω_X and K_X as in the normal case. Then

$$\mathscr{O}_X(mK_X) \simeq \omega_X^{[m]} := (\omega_X^{\otimes m})^{**}. \tag{1.23.1}$$

We use this mostly when X has at worst nodes at codimension 1 points (11.35).

1.3 From Smooth Curves to Canonical Models

Here we discuss the considerations that led to Principle 1.14.1.

In the theory of curves, the basic objects are smooth projective curves. We frequently study any other curve by relating it to smooth projective curves. As a close analog, in higher dimensions, the moduli functor of smooth varieties is

$$\mathcal{S}mooth(S) := \left\{ \begin{array}{l} \text{Smooth, proper families } X \to S, \\ \text{modulo isomorphisms over } S. \end{array} \right\}$$

This, however, gives a rather badly behaved and mostly useless moduli functor already for surfaces. First of all, it is very nonseparated.

1.24 (Nonseparatedness of the moduli of smooth surfaces of general type) We construct two smooth families of projective surfaces $f_i \colon X^i \to B$ over a pointed smooth curve $b \in B$ such that

(1.24.1) all the fibers are smooth, projective surfaces of general type,

(1.24.2) $X^1 \to B$ and $X^2 \to B$ are isomorphic over $B \setminus \{b\}$,

(1.24.3) the fibers X_b^1 and X_b^2 are *not* isomorphic.

As the construction shows, this type of behavior happens every time we look at deformations of a surface that contains at least three (-1)-curves.

Let $f \colon X \to B$ be a smooth family of projective surfaces over a smooth (affine) pointed curve $b \in B$. Let $C_1, C_2, C_3 \subset X$ be three sections of f, all passing through a point $x_b \in X_b$ with independent tangent directions and disjoint elsewhere.

Set $X^1 := B_{C_1} B_{C_2} B_{C_3} X$, where we first blow up $C_3 \subset X$, then the birational transform of C_2 in $B_{C_3} X$, and finally the birational transform of C_1 in $B_{C_2} B_{C_3} X$. Similarly, set $X^2 := B_{C_1} B_{C_3} B_{C_2} X$. Since the C_i are sections, all these blow-ups give smooth families of projective surfaces over B.

Over $B \setminus \{b\}$ the curves C_i are disjoint, thus X^1 and X^2 are both isomorphic to $B_{C_1 + C_2 + C_3} X$, the blow-up of $C_1 + C_2 + C_3 \subset X$.

We claim that, by contrast, the fibers of X_b^1 and X_b^2 are not isomorphic to each other for a general choice of the C_i.

To see this, choose local analytic coordinates t at $b \in B$ and (x, y, t) at $x_b \in X$. The curves C_i are defined by equation

$$C_i = (x - a_i t - \text{(higher terms)} = y - b_i t - \text{(higher terms)} = 0).$$

The blow-up $B_{C_i}X$ is given by

$$B_{C_i}X = (u_i(x - a_i t - \text{(higher terms)}) = v_i(y - b_i t - \text{(higher terms)})) \subset X \times \mathbb{P}^1_{u_i v_i}.$$

On the fiber over b, these give the same blow-up

$$B_{x_b}(X_b) = (ux = vy) \subset X_b \times \mathbb{P}^1_{uv}.$$

Thus we see that the birational transform of C_j intersects the central fiber $(B_{C_i}X)_b = B_{x_b}(X_b)$ at the point

$$\frac{v}{u} = \frac{a_j - a_i}{b_j - b_i} \in \{x_b\} \times \mathbb{P}^1_{uv}.$$

The fibers $(B_{C_2}B_{C_3}X)_b$ and $(B_{C_3}B_{C_2}X)_b$ are isomorphic to each other since they are obtained from $B_{x_b}(X_b)$ by blowing up the same point

$$\frac{v}{u} = \frac{a_2 - a_3}{b_2 - b_3} \quad \text{resp.} \quad \frac{v}{u} = \frac{a_3 - a_2}{b_3 - b_2}.$$

When we next blow up the birational transform of C_1 on $(B_{C_2}B_{C_3}X)_b$ (resp. on $(B_{C_3}B_{C_2}X)_b$), this gives the blow-up of the point

$$\frac{a_1 - a_3}{b_1 - b_3} \quad \text{resp.} \quad \frac{a_1 - a_2}{b_1 - b_2}, \tag{1.24.4}$$

and these are different, unless $C_1 + C_2 + C_3$ is locally planar at x_b.

So far we have seen that the identity $X_b = X_b$ does not extend to an isomorphism between the fibers X_b^1 and X_b^2. If X_b is of general type, then $\operatorname{Aut}X_b$ is finite, hence, to ensure that X_b^1 and X_b^2 are not isomorphic, we need to avoid finitely many other possible coincidences in (1.24.4).

The main reason, however, why we do not study the moduli functor of smooth varieties up to isomorphism is that, in dimension two, smooth projective surfaces do not form the *smallest* basic class. Given any smooth projective surface S, one can blow up any set of points $Z \subset S$ to get another smooth projective surface $B_Z S$, which is very similar to S. Therefore, the basic object is not a single smooth, projective surface, but a whole *birational equivalence class* of smooth, projective surfaces. Thus it would be better to work with smooth, proper families $X \to S$ modulo birational equivalence over S. That is, with the moduli functor

$$\mathcal{G}en\mathcal{T}ype_{bir}(S) := \left\{ \begin{array}{c} \text{Smooth, proper families } X \to S, \\ \text{every fiber is of general type,} \\ \text{modulo birational equivalences over } S. \end{array} \right\} \qquad (1.24.5)$$

In essence this is what we end up doing – see (1.36) – but it is very cumbersome to deal with birational equivalence over a base scheme. Nonetheless, working with birational equivalence classes leads to a separated moduli functor.

Proposition 1.25 *Let $f_i \colon X^i \to B$ be two smooth families of projective varieties over a smooth curve B. Assume that the generic fibers $X^1_{k(B)}$ and $X^2_{k(B)}$ are birational, and the pluricanonical system $|mK_{X^1_{k(B)}}|$ is nonempty for some $m > 0$. Then, for every $b \in B$, the fibers X^1_b and X^2_b are birational.*

Proof Pick a birational map $\phi \colon X^1_{k(B)} \dashrightarrow X^2_{k(B)}$, and let $\Gamma \subset X^1 \times_B X^2$ be the closure of the graph of ϕ. Let $Y \to \Gamma$ be the normalization with projections $p_i \colon Y \to X^i$. Note that both of the p_i are open embeddings on $Y \setminus (\text{Ex}(p_1) \cup \text{Ex}(p_2))$. Thus if we prove that neither $p_1(\text{Ex}(p_1) \cup \text{Ex}(p_2))$ nor $p_2(\text{Ex}(p_1) \cup \text{Ex}(p_2))$ contains a fiber of f_1 or f_2, then $p_2 \circ p_1^{-1} \colon X^1 \dashrightarrow X^2$ restricts to a birational map $X^1_b \dashrightarrow X^2_b$ for every $b \in B$. (Thus the fiber Y_b contains an irreducible component that is the graph of the birational map $X^1_b \dashrightarrow X^2_b$, but it may have other components too; see (1.27.1) for such an example.)

We use the canonical class to compare $\text{Ex}(p_1)$ and $\text{Ex}(p_2)$. Since the X^i are smooth,

$$K_Y \sim p_i^* K_{X^i} + E_i, \quad \text{where } E_i \geq 0 \text{ and } \text{Supp } E_i = \text{Ex}(p_i). \qquad (1.25.1)$$

We may assume that B is affine and let $\text{Bs}\,|mK_{X^i}|$ denote the set-theoretic base locus. By assumption, $|mK_{X^i}|$ is not empty. Since B is affine, $\text{Bs}\,|mK_{X^i}|$ does not contain any of the fibers of f_i.

Every section of $\mathcal{O}_Y(mK_Y)$ pulls back from X^i, thus

$$\text{Bs}\,|mK_Y| = p_i^{-1}\left(\text{Bs}\,|mK_{X^i}|\right) + \text{Supp } E_i.$$

Comparing these for $i = 1, 2$, we conclude that

$$p_1^{-1}\left(\text{Bs}\,|mK_{X^1}|\right) + \text{Supp } E_1 = p_2^{-1}\left(\text{Bs}\,|mK_{X^2}|\right) + \text{Supp } E_2.$$

Therefore, $p_1(\text{Supp } E_2) \subset p_1(\text{Supp } E_1) + \text{Bs}\,|mK_{X^1}|$.

Since E_1 is p_1-exceptional, $p_1(E_1)$ has codimension ≥ 2 in X^1, hence it does not contain any of the fibers of f_1. We saw that $\text{Bs}\,|mK_{X^1}|$ does not contain any of the fibers either. Thus $p_1(\text{Ex}(p_1) \cup \text{Ex}(p_2))$ does not contain any of the fibers, and similarly for $p_2(\text{Ex}(p_1) \cup \text{Ex}(p_2))$. □

Remark 1.26 By Matsusaka and Mumford (1964) and Kontsevich and Tschinkel (2019), the conclusion holds even if the pluricanonical systems are empty.

The proof focuses on the role of the canonical class. It is worthwhile to go back and check that the proof works if the X^i are normal, as long as (1.25.1) holds; the latter is essentially the definition of terminal singularities. It is precisely the property (1.25.1) and its closely related variants that lead us to the correct class of singular varieties for moduli purposes.

Since it is much harder to work with a whole equivalence class, it would be desirable to find a particularly nice surface in every birational equivalence class. This is achieved by the theory of minimal models of algebraic surfaces. By a result of Enriques (Barth et al., 1984, III.4.5), every birational equivalence class of surfaces **S** contains a unique smooth projective surface whose canonical class is nef, except when **S** contains a ruled surface $C \times \mathbb{P}^1$ for some curve C. This unique surface is called the *minimal model* of **S**.

It would seem at first sight that (1.25) implies that the moduli functor of minimal models is separated. There are, however, quite subtle problems.

1.27 (Families of minimal models) Let Y be a projective 3-fold whose only singularities are ordinary nodes. Take a general pencil and blow up its base locus to get $f \colon X \to \mathbb{P}^1$. The general fiber is a smooth surface. At the nodes, in local coordinates we can write it as

$$(xy + z^2 - t^2 = (\text{higher terms})) \lhook\joinrel\longrightarrow \mathbb{A}^4_{xyzt}$$
$$f \big\downarrow \qquad\qquad\qquad\qquad \big\downarrow$$
$$\mathbb{A}^1_t =\!=\!=\!=\!=\!=\!=\!=\!=\!=\!=\!= \mathbb{A}^1_t .$$

By the Morse lemma, with a suitable analytic coordinate change we can eliminate the higher terms (10.43). Then we can blow up either of the the 2-planes $(x = z \pm t = 0)$ to get $\pi^\pm \colon X^\pm \to X$.

By explicit computation as in (10.45), we get smooth morphisms $f^\pm \colon X^\pm \to \mathbb{A}^1$, and the fiber over the origin $X_0^\pm$ is the blow-up of X_0 at the origin. However, the composite map $X^+ \to X \dashrightarrow X^-$ is not an isomorphism. Also, the exceptional set of $\pi^\pm$ is a smooth rational curve $C^\pm \subset X^\pm$.

To get a concrete example, start with a general sextic hypersurface $Y \subset \mathbb{P}^4$ that contains a 2-plane P. Let $P + Q$ be a general hyperplane section containing P. Blow up the birational transforms of P and Q in X to get $X^\pm \to X$.

1.27.1 (Nonseparatedness in the moduli of minimal models) We get two projective morphisms $f^\pm \colon X^\pm \to \mathbb{P}^1$ and a finite set $B \subset \mathbb{P}^1$ such that

(a) general fibers are smooth, canonical models,

(b) the $X^\pm$ are isomorphic over $\mathbb{P}^1 \setminus B$,

(c) the fibers X_b^+ and X_b^- are isomorphic minimal models for $b \in B$, but

(d) $X^+ \to \mathbb{P}^1$ and $X^- \to \mathbb{P}^1$ are *not* isomorphic to each other. □

Starting with a general sextic hypersurface $Y \subset \mathbb{CP}^4$ that has a single node, and using that every divisor on Y is Cartier by Cheltsov (2010), gives the next example.

1.27.2 (Nonprojective families of projective surfaces) We get two smooth, compact, complex manifolds $X^\pm$ and morphisms $f^\pm \colon X^\pm \to \mathbb{P}^1$ such that every fiber is a projective minimal model, yet the $X^\pm$ are not projective.

Proof If $X^\pm$ is projective, let $S^\pm$ be an ample divisor. We claim that $S := f^\pm(S^\pm)$ is not Cartier at the node. Indeed, since $f^\pm$ has no exceptional divisors, we must have $S^\pm = (f^\pm)^*(S)$. This is impossible since $(S^\pm \cdot C^\pm) > 0$, but $(S^\pm \cdot (f^\pm)^*(S)) = 0$. Thus, if every divisor on Y is Cartier, then the $X^\pm$ cannot be projective. □

All such problems go away when the canonical class is ample.

Proposition 1.28 *Let $f_i \colon X^i \to B$ be two smooth families of projective varieties over a smooth curve B. Assume that the canonical classes K_{X^i} are f_i-ample. Let $\phi \colon X^1_{k(B)} \simeq X^2_{k(B)}$ be an isomorphism of the generic fibers. Then ϕ extends to an isomorphism $\Phi \colon X^1 \simeq X^2$.*

Proof Let $\Gamma \subset X^1 \times_B X^2$ be the closure of the graph of ϕ. Let $Y \to \Gamma$ be the normalization, with projections $p_i \colon Y \to X^i$ and $f \colon Y \to B$. As in (1.25), we use the canonical class to compare the X^i. Since the X^i are smooth,

$$K_Y \sim p_i^* K_{X^i} + E_i \quad \text{where } E_i \text{ is effective and } p_i\text{-exceptional.} \tag{1.28.1}$$

Since $(p_i)_* \mathscr{O}_Y(mE_i) = \mathscr{O}_{X^i}$ for every $m \geq 0$, we get that

$$(f_i)_* \mathscr{O}_{X^i}(mK_{X^i}) = (f_i)_* (p_i)_* \mathscr{O}_Y(mp_i^* K_{X^i})$$
$$= (f_i)_* (p_i)_* \mathscr{O}_Y(mp_i^* K_{X^i} + mE_i) = (f_i)_* (p_i)_* \mathscr{O}_Y(mK_Y) = f_* \mathscr{O}_Y(mK_Y).$$

Since the K_{X^i} are f_i-ample, $X^i = \mathrm{Proj}_B \bigoplus_{m \geq 0} (f_i)_* \mathscr{O}_{X^i}(mK_{X^i})$. Putting these together, we get the isomorphism

$$\Phi \colon X^1 \simeq \mathrm{Proj}_B \bigoplus_{m \geq 0} (f_1)_* \mathscr{O}_{X^1}(mK_{X^1}) \simeq \mathrm{Proj}_B \bigoplus_{m \geq 0} f_* \mathscr{O}_Y(mK_Y)$$
$$\simeq \mathrm{Proj}_B \bigoplus_{m \geq 0} (f_2)_* \mathscr{O}_{X^2}(mK_{X^2}) \simeq X^2. \quad \square$$

Remark 1.29 As in (1.26), it is again worthwhile to investigate the precise assumptions behind the proof. The smoothness of the X^i is used only through the pull-back formula (1.28.1), which is weaker than (1.25.1).

If (1.28.1) holds, then, even if the K_{X^i} are not f_i-ample, we obtain an isomorphism

$$\text{Proj}_B \oplus_{m \geq 0} (f_1)_* \mathscr{O}_{X^1}(mK_{X^1}) \simeq \text{Proj}_B \oplus_{m \geq 0} (f_2)_* \mathscr{O}_{X^2}(mK_{X^2}). \qquad (1.29.1)$$

Thus it is of interest to study objects as in (1.29.1) in general.

Let us start with the absolute case, when X is a smooth projective variety over a field k. Its *canonical ring* is the graded ring

$$R(X, K_X) := \oplus_{m \geq 0} H^0(X, \mathscr{O}_X(mK_X)). \qquad (1.29.2)$$

In some cases the canonical ring tells us very little about X. For instance, if X is rational or Fano then $R(X, K_X)$ is the base field k; if X is Calabi–Yau then $R(X, K_X)$ is isomorphic to the polynomial ring $k[t]$. One should thus focus on the cases when the canonical ring is large. The following notion is due to Iitaka (1971). See (Lazarsfeld, 2004, sec.2.1.C) for a detailed treatment.

Definition 1.30 Let X be a smooth proper variety. Its *Kodaira dimension*, denoted by $\kappa(X)$, is the dimension of the image of $|mK_X|: X \dashrightarrow \mathbb{P}^{\dim |mK_X|}$ for m sufficiently large and divisible. One can also define $\kappa(X)$ by the property: the limsup of $h^0(X, \mathscr{O}_X(mK_X))/m^{\kappa(X)}$ is positive and finite. We set $\kappa(X) = -\infty$ if $|mK_X|$ is empty for all $m > 0$.

If $\kappa(X) = \dim X$, we say that X is of *general type*. In this case $|mK_X|$ defines a birational map for all $m \gg 1$, and the limit of $h^0(X, \mathscr{O}_X(mK_X))/m^{\dim(X)}$ is positive and finite. See (3.34) for more on $h^0(X, \mathscr{O}_X(mK_X))$.

Definition 1.31 (Canonical models) Let X be a smooth projective variety of general type over a field k such that its canonical ring $R(X, K_X)$ (1.29.2) is finitely generated. We define its *canonical model* as

$$X^c := \text{Proj}_k R(X, K_X).$$

If Y is a smooth projective variety birational to X, then Y^c is isomorphic to X^c. Thus X^c is also the canonical model of the whole birational equivalence class containing X. (Taking Proj of a nonfinitely generated ring may result in a quite complicated scheme. It does not seem profitable to contemplate what would happen in our case.)

The canonical ring $R(X, K_X)$ is always finitely generated in characteristic 0 (11.28), thus X^c is an irreducible, projective variety. On the other hand, X^c can be singular. Originally this was viewed as a major obstacle, but now it seems only a technical problem.

We can now give an abstract characterization of canonical models.

Theorem 1.32 *A normal proper variety Y is a canonical model iff*

(1.32.1) K_Y *is $\mathbb{Q}$-Cartier (p.xv) and ample, and*

(1.32.2) *there is a resolution $f: X \to Y$ (p.xv) and an effective, f-exceptional $\mathbb{Q}$-divisor E such that $K_X \sim_{\mathbb{Q}} f^* K_Y + E$.*

Proof For now we prove only the "if" part. For the converse, see Reid (1980) or (Kollár, 2013b, 1.15) or (11.62.2).

Choose m_0 such that $m_0 K_X$ is Cartier, then so is $m_0 E$. Note that for any $r > 0$, $f_* \mathcal{O}_X(r m_0 E) = \mathcal{O}_Y$ since E is effective and f-exceptional. Thus

$$H^0(X, \mathcal{O}_X(r m_0 K_X)) = H^0(Y, f_* \mathcal{O}_X(r m_0 K_X))$$
$$= H^0(Y, \mathcal{O}_Y(r m_0 K_Y) \otimes f_* \mathcal{O}_X(r m_0 E)) = H^0(Y, \mathcal{O}_Y(r m_0 K_Y)).$$

Therefore,

$$\mathrm{Proj} \oplus_m H^0(X, \mathcal{O}_X(m K_X)) = \mathrm{Proj} \oplus_r H^0(X, \mathcal{O}_X(r m_0 K_X))$$
$$= \mathrm{Proj} \oplus_r H^0(Y, \mathcal{O}_Y(r m_0 K_Y)) = Y. \qquad \square$$

This makes it possible to give a local definition of the singularities that occur on canonical models, using $\mathbb{Q}$-linear equivalence $\sim_{\mathbb{Q}}$ as in (p.xv).

Definition 1.33 A normal variety Y has *canonical singularities* if

(1.33.1) K_Y is $\mathbb{Q}$-Cartier, and

(1.33.2) there is a resolution $f: X \to Y$ and an effective, f-exceptional $\mathbb{Q}$-divisor E such that $K_X \sim_{\mathbb{Q}} f^* K_Y + E$.

It is easy to show that this is independent of the resolution $f: X \to Y$; see (Kollár, 2013b, sec.2.12). (One can define canonical singularities without resolutions, see (Kollár, 2013b, sec.2.1) or Luo (1987).)

Equivalently, Y has canonical singularities iff every point $y \in Y$ has an étale neighborhood which is an open subset on some canonical model.

A complete list of canonical singularities is known in dimension 2 and almost known in dimension 3; see Reid (1980). The following examples are useful to keep in mind:

(1.33.3) $(x_1 x_2 + f(x_3, \ldots, x_n) = 0)$ is canonical iff f is not identically 0.

(1.33.4) The quotient singularity $\mathbb{A}^n / \frac{1}{m}(1, m - 1, a_3, \ldots, a_n)$ (1.40.2) is canonical for $n \geq 3$ if $\gcd(m, a_3, \ldots, a_n) = 1$. Its canonical class is Cartier iff $m \mid a_3 + \cdots + a_n$.

(1.33.5) The cone $C_d(\mathbb{P}^n)$ over the d-uple Veronese embedding has a canonical singularity iff $d \le n+1$. Its canonical class is Cartier iff $d|n+1$. (See (2.35) or (Kollár, 2013b, 3.1) for the case of general cones.)

Warning 1.34 ($\mathbb{Q}$-Cartier condition) While (1.33.1) may seem like a small technical condition, in many cases it turns out to be extremely important.

First of all, one cannot pull back arbitrary divisors, so (1.33.2) does not even make sense if K_Y is not $\mathbb{Q}$-Cartier. This is a substantial problem starting with dimension 3; cf. (11.57) and (11.58).

The issue becomes more serious for families of varieties. Unexpected jumps of the Kodaira dimension happen precisely when the canonical class of the total space is not $\mathbb{Q}$-Cartier; see (1.43–1.46).

The most difficult aspects appear for nonnormal varieties. The gluing theory of (Kollár, 2013b, chap.5) is almost entirely devoted to proving that in some cases the canonical divisor is $\mathbb{Q}$-Cartier; see (11.38) for a key consequence.

Definition 1.35 The *moduli functor of canonical models* is

$$
CanMod(S) := \left\{ \begin{array}{l} \text{Flat, proper families } X \to S, \\ \text{every fiber is a canonical model,} \\ \text{modulo isomorphisms over } S. \end{array} \right\} \tag{1.35.1}
$$

This is an improved version of $GenType_{bir}$ defined in (1.24.5).

Warning. In retrospect, it seems only by luck that this definition gives the correct functor. See (1.16.2), the examples in (1.42–1.47), and (2.8).

1.36 (From $GenType$ to $CanMod$) Let $p\colon Y \to S$ be a smooth, projective morphism of varieties over a field of characteristic 0. Assume that S is reduced and the fibers Y_s are of general type. By (1.37), we get the flat family of canonical models $p^c\colon Y^c \to S$. This gives a natural transformation T_{CanMod} which, for any reduced scheme S gives a map of sets

$$
T_{CanMod}(S)\colon GenType_{bir}(S) \to CanMod(S). \tag{1.36.1}
$$

By definition, if $X_i \to S$ are two smooth, proper families of varieties of general type such that $T_{CanMod}(S)(X_1) = T_{CanMod}(S)(X_2)$, then X_1 and X_2 are birational, thus $T_{CanMod}(S)$ is injective. It is not surjective, but we have the following partial surjectivity statement.

Claim 1.36.2 Let $Y \to S$ be a flat family of canonical models. Then there is a dense open subset $S^\circ \subset S$ and a smooth, proper family of varieties of general type $Y^\circ \to S^\circ$ such that $T_{CanMod}(S^\circ)(Y^\circ) = [Y|_{S^\circ}]$. □

Theorem 1.37 *Let $p : Y \to S$ be a flat, projective morphism, whose fibers are of general type and have canonical singularities. Assume that S is reduced. Then the canonical models of the fibers form a flat, projective morphism p^{stab} : $Y^{\text{stab}} \to S$, and the natural map $Y \dashrightarrow Y^{\text{stab}}$ is fiber-wise birational.*

For surfaces, this goes back to Kodaira and Spencer (1958); the 3-fold case is proved in Kollár and Mori (1992, 12.5.1). See (2.48) for a proof using MMP. The complex analytic case is in Kollár (2021a).

The theorem implies the deformation invariance of plurigenera as in (5.1.3). Conversely, the deformation invariance of plurigenera, due to Siu (1998) and (Nakayama, 2004, chap.VI), shows that, if the Y_s have canonical models, then they form a flat family $p^{\text{stab}}: Y^{\text{stab}} \to S$.

The case when S is nonreduced is open.

1.4 From Stable Curves to Stable Varieties

Next we discuss the reasoning behind Step 1.15.

Let C be a nodal curve with normalized irreducible components C_i. We frequently view C as an object assembled from the pieces C_i. Note that the pull-back of K_C to C_i is not K_{C_i}, rather $K_{C_i} + P_i$, where $P_i \subset C_i$ are the preimages of the nodes of C.

Similarly, if X is a scheme with simple normal crossing singularities (p.xvi) and normalized irreducible components X_i, then the pull-back of K_X to X_i is not K_{X_i}, rather $K_{X_i} + D_i$, where $D_i \subset X_i$ is the divisorial part of the preimage of Sing X on X_i.

This suggests that we should develop a theory of "canonical models" where the role of the canonical class is played by a divisor of the form $K_X + D$, where D is a simple normal crossing divisor (p.xvi).

Definition 1.38 (Canonical models of pairs) Let (X, D) be a projective snc pair (p.xvi). We define the *canonical ring*[1] of the pair (X, D) as

$$R(X, K_X + D) := \oplus_{m \geq 0} H^0(X, \mathcal{O}_X(mK_X + mD)).$$

It is conjectured (but known only for $\dim X \leq 4$ in characteristic 0) that the canonical ring of a pair (X, D) is finitely generated. If this holds, then $X^c := \text{Proj}_k R(X, K_X + D)$ is a normal projective variety. We say that (X, D) is of *general type* if the natural map $\pi: X \dashrightarrow X^c$ is birational, and then $(X^c, D^c := \pi_* D)$ is called the *canonical model* of (X, D).

The proof of the "if" part of the following goes exactly as in (1.32).

---

[1] *Log* canonical ring and *log* general type is also frequently used; see (1.39.3).

Theorem 1.39 *A pair (Y, B), consisting of a proper normal variety Y and an effective, reduced, Weil divisor B, is the canonical model of a simple normal crossing pair iff*

(1.39.1) $K_Y + B$ *is $\mathbb{Q}$-Cartier, ample, and*

(1.39.2) *there is a resolution $f: X \to Y$, an effective, reduced, simple normal crossing divisor $D \subset X$ such that $f(D) = B$, and an effective, f-exceptional $\mathbb{Q}$-divisor E such that $K_X + D \sim_{\mathbb{Q}} f^*(K_Y + B) + E$.*

Warning 1.39.3 If $B = 0$, it can happen that $(X, 0)$ is the canonical model of a pair, but X is not a canonical model (1.32). To see this, choose a resolution $f: X \to Y$ and let $E_i \subset X$ be the f-exceptional divisors. Although $B = 0$, in (1.39.2) we can still take $D = \sum E_i$. Thus (1.39.2) can be rewritten as

$$K_X \sim f^* K_Y + E - \sum E_i.$$

This looks like (1.32.2), but $E - \sum E_i$ need not be effective; it can contain divisors with coefficients ≥ -1.

This is the source of some terminological problems. Originally $R(X, K_X + D)$ was called the "log canonical ring" and $\mathrm{Proj}_k R(X, K_X + D)$ the "log canonical model." Since the canonical ring is just the $D = 0$ special case of the "log canonical ring," it seems more convenient to drop the prefix "log." However, log canonical singularities are quite different from canonical singularities, so "log" cannot be omitted there. (See also p.xvi for other inconsistencies in the usage of "canonical model.")

As in (1.33), this can be reformulated as a definition. (For now we assume that every irreducible component of B appears in B with coefficient 1; later we also consider cases when the coefficients are rational or real.)

Definition 1.40 Let (Y, B) be a pair consisting of a normal variety Y and a reduced Weil divisor B. Then (Y, B) is *log canonical*, or has *log canonical singularities*, iff the conditions (1.39.1–2) are satisfied. We say that Y is log canonical if $(Y, \emptyset)$ is.

If (Y, B) is log canonical and B is $\mathbb{Q}$-Cartier then Y is also log canonical (11.5.1). However, if B is not $\mathbb{Q}$-Cartier, then K_Y is also not $\mathbb{Q}$-Cartier, so Y is not log canonical.

A complete list of log canonical singularities is known in dimension 2, see Section 2.2 or Kollár (2013b, sec.2.2). The following examples of log canonical singularities are useful to keep in mind:

1.40.1 (Simple normal crossing) $(\mathbb{A}^n, (x_1 \cdots x_r = 0))$ for any $r \leq n$.

1.40.2 (Quotient singularities) $\mathbb{A}^n / \frac{1}{m}(a_1, \ldots, a_n)$ denotes the quotient of $\mathbb{A}^n_{\mathbf{x}}$ by the action $x_i \mapsto \varepsilon^{a_i} x_i$ where ε is a primitive mth root of unity. The canonical class is Cartier iff $m \mid \sum a_i$. These are even log terminal.

1.40.3 (Cones) A cone $C(X)$ over a Calabi–Yau variety; see (2.35).

We are now ready to define the higher dimensional analogs of stable curves.

Definition 1.41 (Stable varieties) Let k be a field and Y a reduced, proper, pure dimensional scheme over k. Let $Y_i \to Y$ be the irreducible components of the normalization of Y, and $D_i \subset Y_i$ the divisorial part of the preimage of the nonnormal locus of Y. Then Y is *semi-log-canonical* – usually abbreviated as *slc* – or *locally stable* iff

(1.41.1) at codimension 1 points, Y is either smooth or has a node (11.35),

(1.41.2) each (Y_i, D_i) is log canonical, and

(1.41.3) K_Y is $\mathbb{Q}$-Cartier.

Y is a *stable variety* iff, in addition,

(1.41.4) Y is projective and K_Y is ample.

As we noted in (1.34), the $\mathbb{Q}$-Cartier condition for K_Y is quite hard to interpret in terms of the (Y_i, D_i). See (11.38) or the more detailed Kollár (2013b, chap.5). For now we only deal with examples where K_Y is obviously Cartier or $\mathbb{Q}$-Cartier.

1.41.5 (Note on terminology) This usage of "stable" has very little to do with the GIT notion of "stable" in Mumford (1965). They agree for curves, and originally there was hope of a close relationship in all dimensions. The two versions aimed to answer the same question, but from different viewpoints. They ended up quite different.

Jump of K^2 and of the Kodaira Dimension

We give examples of flat families of projective surfaces $\{S_t : t \in \mathbb{C}\}$ such that S_0 has quotient singularities and the S_t are smooth for general $t \neq 0$, but the self intersection of the canonical class $(K_{S_t}^2)$ jumps at $t = 0$. We also give examples where K_{S_t} is ample for $t = 0$, but not even big for $t \neq 0$. Among log canonical singularities, quotient singularities (1.40.2) are the mildest.

As we already noted in (1.34), such jumps happen when the canonical class of the total space is not $\mathbb{Q}$-Cartier.

Example 1.42 (Degree 4 surfaces in $\mathbb{P}^5$) There are two families of nondegenerate degree 4 smooth surfaces in $\mathbb{P}^5$. These were classified by Del Pezzo; see Eisenbud and Harris (1987) for a modern treatment.

One family consists of Veronese surfaces $\mathbb{P}^2 \subset \mathbb{P}^5$ embedded by $\mathscr{O}(2)$. The general member of the other family is $\mathbb{P}^1 \times \mathbb{P}^1 \subset \mathbb{P}^5$ embedded by $\mathscr{O}(2,1)$, special members are embeddings of the ruled surface $\mathbb{F}_2$. The two families are distinct since $(K^2_{\mathbb{P}^2}) = 9$ and $(K^2_{\mathbb{P}^1 \times \mathbb{P}^1}) = 8$. For both of these surfaces, a smooth hyperplane section is a degree 4 rational normal curve in $\mathbb{P}^4$.

Let $T_0 \subset \mathbb{P}^5$ denote the cone over the degree 4 rational normal curve in $\mathbb{P}^4$. The minimal resolution of T_0 is the ruled surface $p\colon \mathbb{F}_4 \to T_0$. Let $E, F \subset \mathbb{F}_4$ be the exceptional curve and the fiber of the ruling. Then $K_{\mathbb{F}_4} = -2E - 6F$ and $p^*(2K_{T_0}) = -3E - 12F$. Thus $2(K_{\mathbb{F}_4} + E) = p^*(2K_{T_0}) + E$ shows that T_0 has log canonical singularities. We also get that $(K^2_{T_0}) = 9$.

A key feature is that one can write T_0 as a limit of smooth surfaces in two distinct ways, corresponding to the two ways of writing the degree 4 rational normal curve in $\mathbb{P}^4$ as a hyperplane section of a surface; see (2.36).

From the first family, we get T_0 as the special fiber of a flat family whose general fiber is $\mathbb{P}^2$. This family is denoted by $\{T_t\colon t \in \mathbb{C}\}$. From the second family, we get T_0 as the special fiber of a flat family whose general fiber is $\mathbb{P}^1 \times \mathbb{P}^1$. This family is denoted by $\{T'_t\colon t \in \mathbb{C}\}$. Note that (K^2) is constant in the family $\{T_t\colon t \in \mathbb{C}\}$, but jumps at $t = 0$ in the family $\{T'_t\colon t \in \mathbb{C}\}$. (In general, one needs to worry about the possibility of getting embedded points at the vertex. However, by (2.36), in both cases the special fiber is indeed T_0.)

Alternatively, the degree 4 rational normal curve $(s\colon t) \mapsto (s^4\colon s^3 t\colon s^2 t^2\colon s t^3\colon t^4)$ can be given by determinantal equations in 2 ways, giving the families

$$T'_t = \left(\mathrm{rank} \begin{pmatrix} x_0 & x_1 & x_2 & x_3 \\ x_1 & x_2 + t x_5 & x_3 & x_4 \end{pmatrix} \le 1 \right), \quad \text{and}$$

$$T_t = \left(\mathrm{rank} \begin{pmatrix} x_0 & x_1 & x_2 \\ x_1 & x_2 + t x_5 & x_3 \\ x_2 & x_3 & x_4 \end{pmatrix} \le 1 \right).$$

These are, however, families of rational surfaces with negative canonical class, but we are interested in stable varieties.

Next we take a suitable cyclic cover (11.24) of the two families to get similar examples with ample canonical class.

Example 1.43 (Jump of Kodaira dimension I) We give two flat families of projective surfaces S_t and S'_t such that

(1.43.1) $S_0 \simeq S'_0$ has log canonical singularities and ample canonical class,

(1.43.2) S_t is a smooth surface with ample canonical class for $t \ne 0$, and

(1.43.3) S'_t is a smooth, elliptic surface with $(K^2_{S'_t}) = 0$ for $t \ne 0$.

With T_0 as in (1.42), let $\pi_0\colon S_0 \to T_0$ be a double cover, ramified along the intersection of T_0 with a general quartic hypersurface. Note that $K_{T_0} \sim_{\mathbb{Q}} -\frac{3}{2}H$, where H is the hyperplane class. Thus, by the Hurwitz formula,

$$K_{S_0} \sim_{\mathbb{Q}} \pi_0^*(K_{T_0} + 2H) \sim_{\mathbb{Q}} \tfrac{1}{2}\pi_0^* H.$$

So S_0 has ample canonical class and $(K_{S_0}^2) = 2$. Since π_0 is étale over the vertex of T_0, S_0 has two singular points, locally (in the analytic or étale topology) isomorphic to the singularity on T_0. Thus S_0 is a stable surface.

Both of the smoothings in (1.42) lift to smoothings of S_0.

From T_t we get a smoothing S_t where $\pi_t \colon S_t \to \mathbb{P}^2$ is a double cover, ramified along a smooth octic. Thus S_t is smooth, $K_{S_t} \sim_{\mathbb{Q}} \pi_t^* \mathcal{O}_{\mathbb{P}^2}(1)$ is ample and $(K_{S_t}^2) = 2$.

From T_t' we get a smoothing S_t' where $\pi_t' \colon S_t' \to \mathbb{P}^1 \times \mathbb{P}^1$ is a double cover, ramified along a smooth curve of bidegree $(8, 4)$. One of the families of lines on $\mathbb{P}^1 \times \mathbb{P}^1$ pulls back to an elliptic pencil on S_t' and $(K_{S_t'}^2) = 0$. Thus S_t' is not of general type for $t \neq 0$.

Example 1.44 (Jump of Kodaira dimension II) A similar pair of examples is obtained by working with triple covers ramified along a cubic hypersurface section of the surface families in (1.42). The family over T_t has ample canonical class and $(K^2) = 3$. As before, the family over T_t' is elliptic and has $(K^2) = 0$.

Example 1.45 (Jump of Kodaira dimension III) We construct a flat family of surfaces whose central fiber is the quotient of the square of the Fermat cubic curve by $\mathbb{Z}/3$:

$$S_F^* \simeq (u_1^3 = v_1^3 + w_1^3) \times (u_2^3 = v_2^3 + w_2^3)/\tfrac{1}{3}(1, 0, 0; 1, 0, 0), \qquad (1.45.1)$$

thus it has Kodaira dimension 0. The general fiber is $\mathbb{P}^2$ blown up at 12 points.

In $\mathbb{P}^3$, consider two lines $L_1 = (x_0 = x_1 = 0)$ and $L_2 = (x_2 = x_3 = 0)$. The linear system $|\mathcal{O}_{\mathbb{P}^2}(2)(-L_1 - L_2)|$ is spanned by the four reducible quadrics $x_i x_j$ for $i \in \{0, 1\}$ and $j \in \{2, 3\}$. They satisfy a relation $(x_0 x_2)(x_1 x_3) = (x_0 x_3)(x_1 x_2)$. Thus we get a morphism $\pi \colon B_{L_1+L_2}\mathbb{P}^3 \to \mathbb{P}^1 \times \mathbb{P}^1$, which is a $\mathbb{P}^1$-bundle whose fibers are the birational transforms of lines that intersect both of the L_i.

Let $S \subset \mathbb{P}^3$ be a cubic surface such that $\mathbf{p} := S \cap (L_1 + L_2)$ is six distinct points. Then we get $\pi_S \colon B_{\mathbf{p}}S \to \mathbb{P}^1 \times \mathbb{P}^1$.

In general, none of the lines connecting two points of $\mathbf{p}$ is contained in S; in this case π_S is a finite triple cover.

At the other extreme, we have the Fermat-type surface

$$S_F := (x_0^3 + x_1^3 = x_2^3 + x_3^3) \subset \mathbb{P}^3.$$

We can factor both sides and write its equation as $m_1 m_2 m_3 = n_1 n_2 n_3$. The nine lines $L_{ij} := (m_i = n_j = 0)$ are all contained in S_F. Let $L_{ij}' \subset B_{\mathbf{p}}S_F$ denote their birational transforms. Then the self-intersections $(L_{ij}' \cdot L_{ij}')$ equal -3 and π_{S_F} contracts these nine curves L_{ij}'. Thus the Stein factorization of π_{S_F} gives a triple cover $S_F^* \to \mathbb{P}^1 \times \mathbb{P}^1$. Here S_F^* has nine singular points of type $\mathbb{A}^2 / \frac{1}{3}(1, 1)$. We see furthermore that $-3K_{S_F} \sim \sum_{ij} L_{ij}$ and $-3K_{B_{\mathbb{P}}S_F} \sim \sum_{ij} L'_{ij}$. Thus $-3K_{S_F^*} \sim 0$.

To see that the two surfaces denoted by S_F^* are isomorphic, use the map of the surface (1.45.1) to $\mathbb{P}^1 \times \mathbb{P}^1$ given by

$$(u_1 : v_1 : w_1) \times (u_2 : v_2 : w_2) \mapsto (v_1 : w_1) \times (v_2 : w_2),$$

and the rational map to the cubic surface is given by

$$(u_1 : v_1 : w_1) \times (u_2 : v_2 : w_2) \mapsto (u_1 v_2 : u_1 w_2 : v_1 u_2 : w_1 u_2).$$

Example 1.46 (Jump of Kodaira dimension IV) The previous examples are quite typical in some sense. If S_0 is any projective rational surface with quotient singularities, then there is a flat family of surfaces $\{S_t\}$ such that S_t is a smooth rational surface for $t \neq 0$.

To see this, take a minimal resolution $S_0' \to S_0$. Since S_0' is a smooth rational surface, it can be obtained from a minimal smooth rational surface by blowing up points. We can deform S_0' by moving these points into general position (and also deforming the minimal smooth rational surface if necessary). Thus we see that if S_0 is singular, then a general deformation S_t' of S_0' is obtained by blowing up points in $\mathbb{P}^2$ or $\mathbb{P}^1 \times \mathbb{P}^1$ in general position. In the second case, if we blow up at least one point, it is also a blow-up of $\mathbb{P}^2$. There are no curves with negative self-intersection on $\mathbb{P}^1 \times \mathbb{P}^1$, and on a blow-up of $\mathbb{P}^2$ at general points, every smooth rational curve with negative self-intersection is a (-1)-curve by (de Fernex, 2005, 2.4). In particular, none of the exceptional curves of $S_0' \to S_0$ lift to S_t'.

Let H_0' be the pull-back of a very ample Cartier divisor from S_0 whose higher cohomologies vanish. Since S_0' is a smooth rational surface, $\text{Pic}(S_0') = H^2(S_0', \mathbb{Z})$, so H_0' lifts to a family of semiample Cartier divisors H_t'. As we discussed, none of the exceptional curves of $S_0' \to S_0$ lift to S_t' for general t, so H_t' is ample for general t. As before, we get a flat deformation $\{S_t\}$ such that $S_t \simeq S_t'$ for $t \neq 0$.

Many recent constructions of surfaces of general type start with a particular rational surface S_0 with quotient singularities, and show that it has a flat deformation to a smooth surface with ample canonical class; see Lee and Park (2007); Park et al. (2009a,b). Thus such an S_0 has flat deformations of general type and also flat deformations that are rational.

Even more surprisingly, a surface with ample canonical class can have nonalgebraic deformations.

Example 1.47 (Nonalgebraic deformations) (Kollár, 2021a) We construct a projective surface X_0 with a quotient singularity, ample canonical class and two deformations. An algebraic one $g^{\mathrm{alg}} : X^{\mathrm{alg}} \to D$, where g^{alg} is flat, projective, and a complex analytic one $g^{\mathrm{an}} : X^{\mathrm{an}} \to \mathbb{D}$ over the *unit disc* $\mathbb{D} \subset \mathbb{C}$, where g^{an} is flat, proper such that

(1.47.3) X_s^{alg} is a smooth, algebraic, K3 surface blown up at three points for $s \neq 0$,

(1.47.4) X_s^{an} is a smooth, nonalgebraic, K3 surface blown up at three points for very general $s \in \mathbb{D}$.

Let us start with a K3 surface $Y \subset \mathbb{P}^3$ with a hyperplane section $C \subset Y$ and three points $p_i \in C$. Blow up these points to get $\pi : Z \to Y$ with exceptional curves $E = E_1 + E_2 + E_3$. Let $C_Z \subset Z$ be the birational transform of C and $H = \pi^*C - \frac{2}{3}E$.

If the p_i are smooth points on C, then $\pi^*C = C_Z + E$, hence $H = C_Z + \frac{1}{3}E$. Since $(H \cdot C_Z) = 2$, $(H \cdot E_i) = \frac{2}{3}$ and $Z \setminus (C_Z + E) \simeq Y \setminus C$ is affine, we see that H is ample by the Nakai–Moishezon criterion.

If the p_i are double points on C, then $\pi^*C = C_Z + 2E$, hence $H = C_Z + \frac{4}{3}E$. Then $(C_Z \cdot E_i) = 2$, $(H \cdot C_Z) = 0$ and $(H \cdot E_i) = \frac{2}{3}$. So $3H$ is semiample and it contracts C_Z. Let the resulting surface be X_0 and $F_i \subset X_0$ the images of the E_i.

Note that in this case C_Z is a smooth, rational curve and $(C_Z^2) = -8$. Thus X_0 has a single quotient singularity of type $\mathbb{C}^2 / \frac{1}{8}(1, 1)$. We also get that $(F_i^2) = -\frac{1}{2}$ and $(F_i \cdot F_j) = \frac{1}{2}$ for $i \neq j$. Furthermore, $K_{X_0} \sim F_1 + F_2 + F_3$ is ample.

In order to construct the algebraic family, start with $C \subset Y$ where C is a rational curve with three nodes. The deformation is obtained by moving the points into general position. Blowing up the points we get H that is ample on the general fibers and contracts the birational transform of C in the special fiber. Thus we get $g^{\mathrm{alg}} : X^{\mathrm{alg}} \to D$.

For the analytic case, we choose a deformation $Y \to \mathbb{D}$ of Y_0 whose very general fibers are nonalgebraic K3 surfaces. Take three sections $B_i \subset Y$ that pass through the three nodes of C. Blow them up and then contract the birational transform of C. The contraction extends to the total space by Kollár and Mori (1992, 11.4). We get $g^{\mathrm{an}} : X^{\mathrm{an}} \to \mathbb{D}$ whose central fiber is X_0. The other fibers are nonalgebraic, K3 surfaces blown up at three points.

Example 1.48 (More rational surfaces with ample canonical class) (Kollár, 2008b, sec.5) Given natural numbers a_1, a_2, a_3, a_4, consider the surface

$$S = S(a_1, a_2, a_3, a_4) := (x_1^{a_1}x_2 + x_2^{a_2}x_3 + x_3^{a_3}x_4 + x_4^{a_4}x_1 = 0) \subset \mathbb{P}(w_1, w_2, w_3, w_4),$$

where $w_i' = a_{i+1}a_{i+2}a_{i+3} - a_{i+2}a_{i+3} + a_{i+3} - 1$ (with indices modulo 4) and $w_i = w_i' / \gcd(w_1', w_2', w_3', w_4')$. It is easy to see that S has only quotient singularities (at the four coordinate vertices). It is proved in (Kollár, 2008b, thm.39) that S

is rational if $\gcd(w_1', w_2', w_3', w_4') = 1$. (By Kollár, 2008b, 38, this happens with probability ≥ 0.75.)

$\mathbb{P}(w_1, w_2, w_3, w_4)$ has isolated singularities iff the $\{w_i\}$ are pairwise relatively prime. (It is easy to see that for $1 \leq a_i \leq N$, this happens for at least $c \cdot N^{4-\varepsilon}$ of the 4-tuples.) In this case $K_S = \mathcal{O}_{\mathbb{P}}(\prod a_i - 1 - \sum w_i)|_S$. From this it is easy to see that if $a_1, a_2, a_3, a_4 \geq 4$ then K_S is ample and (K_S^2) converges to 1 as $a_1, a_2, a_3, a_4 \to \infty$. See Urzúa and Yáñez (2018) for more on these surfaces.

1.5 From Nodal Curves to Stable Curves and Surfaces

We discussed stable extension for families of curves $C \to B$ over a smooth curve B in (1.15.1). Similarly, working over a higher dimensional reduced base $C \to S$ involves two main steps.

- First, we transform a given proper family of curves $C \to S$ into a proper, flat family $C_1 \to S_1$, whose fibers are reduced, nodal curves. This needs a base change $S_1 \to S$ that involves choices, and then a sequence of blow-ups that again involves choices. We can choose S_1 to be smooth.

- Once we have a proper, flat family $C_1 \to S_1$ whose fibers are reduced, nodal curves, and whose base is smooth, we take the relative canonical model (11.28) to get the stable family $C_1^{\text{stab}} \to S_1$. For MMP to work, we need S_1 to have at worst log canonical singularities.

Nonetheless, we show that one can go from flat families of nodal curves to flat families of stable curves in a functorial way over an arbitrary base.

Theorem 1.49 *For every $g \geq 2$ there is a natural transformation $C \mapsto C^{\text{stab}}$*

$$\left\{ \begin{array}{c} \textit{proper, flat families of} \\ \textit{reduced, nodal, genus } g \textit{ curves} \end{array} \right\} \longrightarrow \left\{ \begin{array}{c} \textit{stable families of} \\ \textit{genus } g \textit{ curves} \end{array} \right\},$$

such that that if C is a smooth, projective curve, then $C^{\text{stab}} = C$. (We assume that the curves are geometrically connected. By the genus of a proper nodal curve C we mean $h^1(C, \mathcal{O}_C)$.)

Proof We outline the main steps, leaving some details to the reader. We use C' to denote any irreducible component of the curve that we work with.

First, let C be a proper, reduced, nodal curve over an algebraically closed field. We start with two recipes to construct C^{stab}. With both approaches, we first obtain the largest semistable subcurve $C^{\text{ss}} \subset C$.

Step 1.a (Using MMP) Find $C' \subset C$ on which K_C has negative degree. Equivalently, $C' \simeq \mathbb{P}^1$ and it meets the rest of C in one point only. Contract (or discard) this component. Repeat if possible.

Step 1.b (Using K_C) C^{ss} is the support of the global sections of $\mathcal{O}_C(K_C)$.

Once we have C^{ss}, we continue to get C^{stab} as follows.

Step 2.a Find $C' \subset C^{ss}$ on which $K_{C^{ss}}$ has degree 0. Equivalently, $C' \simeq \mathbb{P}^1$ and it meets the rest of C^{ss} in two points only; call them p, q. Contract this component. Equivalently, discard C' and identify the points p, q. Repeat if possible.

Step 2.b (Using the canonical ring) $C^{stab} = \text{Proj} \oplus_m H^0(C^{ss}, \mathcal{O}_{C^{ss}}(mK_{C^{ss}}))$.

Once we know C^{stab}, we can also recover it in one step as follows.

Step 3 Let $\{C^i \subset C : i \in I\}$ be the irreducible components that are kept in the above process; call them *stable*. Pick nonnodal points $p^i \in C^i$ and set $L := \mathcal{O}_C(\sum p^i)$. Then, for $m \gg 1$, $H^1(C, L^m) = 0$, L^m is globally generated and maps C onto C^{stab}.

Step 4 Over an arbitrary field k with algebraic closure $\bar{k}$, we show that if C is defined over k, then $(C_{\bar{k}})^{stab}$ is also defined over k, giving us C^{stab}.

Now to the general case. Let $g : C_S \to S$ be a proper, flat family of reduced, nodal curves over an arbitrary base. We construct the stable family étale-locally; uniqueness then implies that we get a family over S.

Pick a point $s \in S$. By the arguments here, we have the stable irreducible components $C^i_s \subset C_s$. Pick nonnodal points $p^i \in C^i_s$ and let $D^i \subset C_S$ be sections that meet C_s only at p^i. (Usually this needs an étale base change.) Set $L_S := \mathcal{O}_{C_S}(\sum D^i)$. Then Step 3 shows that, for $m \gg 1$,

Step 5 $R^1 g_* L^m = 0$, $g_* L^m$ is locally free, and maps C_S onto C^{stab}_S.   □

Warning Note that Step 2.b works only for semistable curves. As an example, let $C = C_1 \cup C_2$ be a curve with a single node p with $g(C_1) \geq 2$ and $C_2 \simeq \mathbb{P}^1$. Then we have a **non**-finitely generated ring

$$\oplus_{m \geq 0} H^0(C, \mathcal{O}_C(mK_C)) = \oplus_{m \geq 0} H^0(C_1, \mathcal{O}_{C_1}(mK_{C_1} + (m-1)[p])).$$

Definition 1.50 (Stabilization functor) Trying to generalize (1.49) to higher dimensions, the best would be to have a functor from proper, flat locally stable families to stable families, that agrees with $X \to X^c$ on smooth varieties of general type. One can further restrict the singularities of the fibers and talk about stabilization functors for families of smooth varieties, simple normal crossing varieties (p.xvi), and so on.

We see here that such a stabilization functor does exist for smooth families, but not for more complicated singularities. We discuss this phenomenon in detail in Section 5.2; see especially (5.11). This is another reason why the moduli theory of higher dimensional varieties is much more complicated.

Theorem 1.51 (Stabilization functor for surfaces)

(1.51.1) *For smooth, projective surfaces of general type, $S \mapsto S^c$ is a stabilization functor.*

(1.51.2) *For projective surfaces of general type with quotient singularities, $S \mapsto S^c$ is **not** a stabilization functor.*

(1.51.3) *For projective surfaces with normal crossing singularities, $S \mapsto S^c$ is **not** a stabilization functor.*

(1.51.4) *For irreducible projective surfaces with normal crossing singularities, S^c does not even make sense in general.*

Proof For the first part, see (1.36) and (2.48). As in (1.49), more work is needed for nonreduced bases.

For (2) and (3), we run into problems even for families over smooth curves.

Consider the simplest case when we have a flat, projective morphism $p : X \to \mathbb{A}^1$ to a smooth curve such that K_X is $\mathbb{Q}$-Cartier, and the fibers are surfaces with quotient singularities only. Then we get the stable model $p^{\mathrm{stab}} : X^{\mathrm{stab}} \to \mathbb{A}^1$ as the relative canonical model (11.28).

We claim that as soon as the process involves a flip, we have an example for (2): the canonical ring of X_0 is strictly larger than the canonical ring of $(X^c)_0$.

The flip is a diagram

$$(C \subset X) - - - \overset{\phi}{- - -} \blacktriangleright (C^+ \subset X^+)$$
$$\underset{\pi}{\searrow} \quad \underset{Z}{} \quad \underset{\pi^+}{\swarrow}$$

$$(1.51.5)$$

where $-K_X$ is π-ample and K_{X^+} is π^+-ample.

Restricting it to the fiber over $0 \in \mathbb{A}^1$ we get a similar looking diagram of surfaces with quotient singularities

$$(C \subset X_0) - - - \overset{\phi_0}{- - -} \blacktriangleright (C^+ \subset X_0^+)$$
$$\underset{\pi_0}{\searrow} \quad \underset{Z_0}{} \quad \underset{\pi_0^+}{\swarrow}$$

$$(1.51.6)$$

where K_{X_0} is π_0-ample and $K_{X_0^+}$ is π_0^+-ample. The difference is that now the exceptional *curves* C, C^+ of (1.51.5) are exceptional *divisors*.

Using (1.51.8) we get the following.

Problem 1.51.7 $X_0 \mapsto X_0^+$ is **not** a step of the MMP for X_0. In fact, the canonical ring of X_0^+ is strictly smaller than the canonical ring of X_0.

Taking a suitable resolution shows that similar examples happen for families with simple normal crossing fibers.

Claim (1.51.4) is not a precise assertion, but we expect that, even over algebraically closed fields, there is no "sensible" way to associate a stable surface to every projective, normal crossing surface. For example, Kollár (2011c) constructs irreducible, projective surfaces S with normal crossing singularities for which the canonical ring $\bigoplus_{m \geq 0} H^0(S, \mathscr{O}_S(mK_S))$ is **not** finitely generated. We present a similar example in (1.53). □

Claim 1.51.8 Let $p: Y \to T$ be a proper, birational morphism of normal surfaces and $E := \mathrm{Ex}(p)$. Let D be a Cartier divisor on Y and set $D_T := p(D)$. The following are easy to see.

(a) If $-D|_E$ is ample then $p_* \mathscr{O}_Y(mD) = \mathscr{O}_T(mD_T)$ for $m \geq 1$.

(b) If $D|_E$ is ample then $p_* \mathscr{O}_Y(mD) \subsetneq \mathscr{O}_T(mD_T)$ for $m \gg 1$.

(c) If D is ample then $H^0(Y, \mathscr{O}_Y(mD)) \subsetneq H^0(T, \mathscr{O}_T(mD_T))$ for $m > 1$. □

We saw that if $p: X \to \mathbb{A}^1$ is a flat, projective family of surfaces with quotient singularities, then the relative canonical model (11.28) gives a stable family, although this is not a fiber-wise construction. The next example shows that, for families over nodal curves, there may not be any stable family.

Example 1.52 Consider any family $X \to \mathbb{A}^1_u$ as in (1.51.7), and glue it to the trivial family $q: Y := X_0 \times \mathbb{A}^1_v \to \mathbb{A}^1_v$ along the central fibers to get a locally stable family $r: X \amalg_{X_0} Y \to (uv = 0)$. Then

(1.52.1) $p^{\mathrm{stab}}: X^{\mathrm{stab}} \to \mathbb{A}^1_u$ and $q^{\mathrm{stab}}: Y^{\mathrm{stab}} \to \mathbb{A}^1_v$ both exists, yet

(1.52.2) their central fibers $(X^c)_0$ and $(Y^c)_0$ are **not** isomorphic, so

(1.52.3) $r: X \amalg_{X_0} Y \to (uv = 0)$ does **not** have a stable model.

Example 1.53 Following Kollár (2011c), we give an example of a projective, normal crossing surface whose canonical ring is not finitely generated. The key point is the following observation.

Let T be a projective surface, $C_1, C_2 \subset T$ disjoint smooth curves, and $\tau: C_1 \to C_2$ an isomorphism. Assume that $T \setminus C_1$ is smooth, T has a single node at a point $p_1 \in C_1$, and $K_T + C_1 + C_2$ is ample. Let $T/(\tau)$ be obtained from T by identifying C_1 with C_2 using τ.

Claim 1.53.1 The canonical class of $T/(\tau)$ is not $\mathbb{Q}$-Cartier. Thus its canonical ring is not finitely generated.

Proof T is smooth along C_2, hence the usual adjunction gives that

$$(K_T + C_1 + C_1)|_{C_1} = K_{C_1}.$$

T has a node along C_1. This modifies the adjunction formula to

$$(K_T + C_1 + C_2)|_{C_1} = K_{C_1} + \tfrac{1}{2}[\,p_1\,];$$

see Kollár (2013b, 4.3) for this computation. This means that we cannot match up local generating sections of the sheaf $\mathcal{O}_T(mK_T + mC_1 + mC_2)$ at the points p_1 and $\tau(\,p_1)$; see Kollár (2013b, 5.12) for the precise statement and proof. This easily implies that finite generation fails; see Kollár (2010, exc.97). □

This is almost what we want, except that T is not a normal crossing surface at the image of p_1. So next we construct a normal crossing surface and check that trying to construct its minimal model leads to a surface as needed.

We start with a smooth plane curve $C \subset \mathbb{P}^2$ of degree d and a line L intersecting C transversally. Let $c \in C \cap L$ be one of the intersection points. Fix distinct points $p, q \in \mathbb{P}^1$. In $\mathbb{P}^2_p := \{p\} \times \mathbb{P}^2$ we get C_p, L_p, and similarly for $C_q, L_q \subset \mathbb{P}^2_q$. We have the "identity" $\tau \colon C_p \simeq C_q$.

Let $\bar{S} \subset \mathbb{P}^1 \times \mathbb{P}^2$ be a surface of bidegree $(e, d+1)$ such that $\bar{S} \cap \mathbb{P}^2_p = C_p \cup L_p$ and $\bar{S} \cap \mathbb{P}^2_q = C_q \cup L_q$. We can further arrange that $\bar{S}$ is smooth, except for an ordinary node at $c_p \in C_p \cap L_p$.

Let $\bar{S}' \to \bar{S}$ be obtained by blowing up c_p and c_q. We get exceptional curves E'_p, E'_q and birational transforms C'_p and C'_q. Note that $\bar{S}'$ is smooth and $E'_p + E'_q + C'_p + C'_q$ is an snc divisor. We can now glue C'_p to C'_q using the "identity" $\tau' \colon C'_p \simeq C'_q$ to obtain the nonnormal surface $S' := \bar{S}'/(\tau')$. It has normal crossing self-intersection along a curve $C \simeq C'_p \simeq C'_q$. Note that $K_{S'} + E_p + E_q$ is a Cartier divisor.

Claim 1.53.2 The projective, normal crossing pair $(S', E_p + E_q)$ does not have a canonical model.

Proof The normalization of $(S', E_p + E_q)$ is $(\bar{S}', E'_p + E'_q + C'_p + C'_q)$, thus the only "sensible" thing to do is to construct its canonical model, and then glue the images of C'_p and C'_q together. We compute that

$$(K_{\bar{S}'} + E'_p + E'_q + C'_p + C'_q) \cdot E'_p = -1 \quad \text{and} \quad (K_{\bar{S}'} + E'_p + E'_q + C'_p + C'_q) \cdot E'_q = -1.$$

Thus we need to contract E'_p and E'_q to get $(\bar{S}, C_p + C_q)$. Note that

$$\mathcal{O}_{\bar{S}}(K_{\bar{S}}) \simeq \mathcal{O}_{\mathbb{P}^1 \times \mathbb{P}^2}(e - 2, d + 1 - 3)|_{\bar{S}},$$

which is ample for $e \geq 3, d \geq 3$. This shows that if $d \geq 4$, then $K_{\bar{S}} + C_p + C_q$ is ample. Therefore, the only possible choice for the canonical model of $(S', E'_p + E'_q)$ is $\bar{S}/(\tau)$. Now (1.53.1) shows that the canonical ring is not finitely generated. □

1.6 Examples of Bad Moduli Problems

Now we turn to a more general overview of moduli problems. The aim of this section is to present examples of moduli problems that seem quite reasonable at first sight, but turn out to have rather bad properties. We start with the moduli of hypersurfaces.

The Chow and Hilbert varieties describe families of hypersurfaces in a fixed projective space $\mathbb{P}^n$. For many purposes, it is more natural to consider the moduli functor of hypersurfaces modulo isomorphisms. We consider what kind of "moduli spaces" one can obtain in various cases.

Definition 1.54 (Hypersurfaces modulo linear isomorphisms) Over an algebraically closed field k, we consider hypersurfaces $X \subset \mathbb{P}^n_k$ where $X_1, X_2 \subset \mathbb{P}^n_k$ are considered isomorphic if there is an automorphism $\phi \in \mathrm{Aut}(\mathbb{P}^n_k)$ such that $\phi(X_1) = X_2$.

Over an arbitrary base scheme S, we consider pairs $(X \subset P)$ where P/S is a $\mathbb{P}^n$-bundle for some n and $X \subset P$ is a closed subscheme, flat over S such that every fiber is a hypersurface. There are two natural invariants: the relative dimension of P and the degree of X. Thus for any given n, d we get a functor

$$\mathcal{H}yp\mathcal{S}ur_{n,d}(S) := \left\{ \begin{array}{c} \text{Flat families } X \subset P \\ \text{such that } \dim_S P = n, \deg X = d, \\ \text{modulo isomorphisms over } S. \end{array} \right\}$$

One can also consider subfunctors, where we allow only reduced, normal, canonical, log canonical, or smooth hypersurfaces; these are indicated by the superscripts red, norm, c, lc, or sm.

Our aim is to investigate what the "coarse moduli spaces" of these functors look like. Our conclusion is that in many cases there cannot be any scheme or algebraic space that is a coarse moduli space: any "coarse moduli space" would have to have very strange topology.

Let $\mathcal{H}yp\mathcal{S}ur^*_{n,d}$ be any subfunctor of $\mathcal{H}yp\mathcal{S}ur_{n,d}$, and assume that it has a coarse moduli space $\mathrm{HypSur}^*_{n,d}$. By definition, the set of k-points of $\mathrm{HypSur}^*_{n,d}$ is $\mathcal{H}yp\mathcal{S}ur^*_{n,d}(\mathrm{Spec}\,k)$. We can also get some idea about the Zariski topology of $\mathrm{HypSur}^*_{n,d}$ using various families of hypersurfaces.

For instance, we can study the closure $\bar{U}$ of a subset $U \subset \mathrm{HypSur}^*_{n,d}(\mathrm{Spec}\,k)$ using the following observation:

- Assume that there is a flat family of hypersurfaces $\pi \colon X \to S$ and a dense open subset $S^\circ \subset S$ such that $[X_s] \in U$ for every $s \in S^\circ(k)$. Then $[X_s] \in \bar{U}$ for every $s \in S(k)$.

Next we write down flat families of hypersurfaces $\pi\colon X \to \mathbb{A}^1$ in $\mathcal{H}\!yp\mathcal{S}\!ur^*_{n,d}$ such that for $t \neq 0$ the fibers X_t are isomorphic to each other, but X_0 is not isomorphic to them. Such a family corresponds to a morphism $\tau\colon \mathbb{A}^1 \to$ $\mathrm{HypSur}^*_{n,d}$ such that $\tau(\mathbb{A}^1 \setminus \{0\}) = [X_1]$, but $\tau(\{0\}) = [X_0]$. This implies that the point $[X_1]$ is not closed, and its closure contains $[X_0]$.

This is not very surprising in a scheme, but note that X_1 itself is defined over our base field k, so $[X_1]$ is supposed to be a k-point. On a k-scheme, k-points are closed. Thus we conclude that if there is any family as listed, the moduli space $\mathrm{HypSur}^*_{n,d}$ cannot be a k-scheme, not even a quasi-separated algebraic space (Stacks, 2022, tag 08AL).

The simplest way to get such families is by the following construction.

Example 1.55 (Deformation to cones I) Let $f(x_0, \ldots, x_n)$ be a homogeneous polynomial of degree d and $X := (f = 0)$ the corresponding hypersurface. For some $0 \leq i < n$ consider the family of hypersurfaces

$$\mathbf{X} := (f(x_0, \ldots, x_i, tx_{i+1}, \ldots tx_n) = 0) \subset \mathbb{P}^n \times \mathbb{A}^1_t \qquad (1.55.1)$$

with projection $\pi\colon \mathbf{X} \to \mathbb{A}^1_t$. If $t \neq 0$ then the substitution

$$x_j \mapsto x_j \quad \text{for } j \leq i, \text{ and} \quad x_j \mapsto t^{-1}x_j \quad \text{for } j > i$$

shows that the fiber X_t is isomorphic to X. If $t = 0$ then we get the cone over $X \cap (x_{i+1} = \cdots = x_n = 0)$:

$$X_0 = (f(x_0, \ldots, x_i, 0, \ldots, 0) = 0) \subset \mathbb{P}^n.$$

This is a hypersurface iff $f(x_0, \ldots, x_i, 0, \ldots, 0)$ is not identically 0.

More generally, any algebraic variety has a similar deformation to a cone over its hyperplane section, see (2.36).

Already these simple deformations show that various moduli spaces of hypersurfaces have very few closed points.

Corollary 1.56 *The sole closed point of* $\mathrm{HypSur}_{d,n}$ *is* $[(x_0^d = 0)]$.

Proof Take any $X = (f = 0) \subset \mathbb{P}^n$. After a general change of coordinates, we can assume that x_0^d appears in f with nonzero coefficient. For $i = 0$ consider the family (1.55.1).

Then $X_0 = (x_0^d = 0)$, hence $[X]$ cannot be a closed point unless $X \simeq X_0$. It is quite easy to see that if $X \to S$ is a flat family of hypersurfaces whose generic fiber is a d-fold plane, then every fiber is a d-fold plane. This shows that $[(x_0^d = 0)]$ is a closed point. $\qquad \square$

Corollary 1.57 *The only closed points of* $\text{HypSur}_{d,n}^{\text{red}}$ *are* $[(f(x_0, x_1) = 0)]$ *where f has no multiple roots.*

Proof If X is a reduced hypersurface of degree d, there is a line that intersects it in d distinct points. We can assume that this is the line $(x_2 = \cdots = x_n = 0)$. For $i = 1$, consider the family (1.55.1).

Then $X_0 = (f(x_0, x_1, 0, \ldots, 0) = 0)$ where $f(x_0, x_1)$ has d distinct roots. Since X_0 is reduced, we see that none of the other hypersurfaces correspond to closed points.

It is not obvious that the points corresponding to $(f(x_0, x_1, 0, \ldots, 0) = 0)$ are closed, but this can be established by studying the moduli of d points in $\mathbb{P}^1$; see (Mumford, 1965, chap.3) or (Dolgachev, 2003, sec.10.2). □

A similar argument establishes the normal case:

Corollary 1.58 *The only closed points of* $\text{HypSur}_{d,n}^{\text{norm}}$ *are* $[(f(x_0, x_1, x_2) = 0)]$ *where $(f(x_0, x_1, x_2) = 0) \subset \mathbb{P}^2$ is a nonsingular curve.* □

In these examples the trouble comes from cones. Cones can be normal, but they are very singular by other measures; they have a singular point whose multiplicity equals the degree of the variety. So one could hope that high multiplicity points cause the problems. This is true to some extent as the next theorems and examples show. For proofs, see Mumford (1965, sec.4.2) and Dolgachev (2003, sec.10.1).

Theorem 1.59 *Each of the following functors has a coarse moduli space which is a quasi-projective variety.*

(1.59.1) *The functor of smooth hypersurfaces* $\mathcal{H}yp\mathcal{S}ur_{n,d}^{\text{sm}}$.

(1.59.2) *For $d \geq n + 1$, the functor* $\mathcal{H}yp\mathcal{S}ur_{n,d}^{c}$ *of hypersurfaces with canonical singularities.*

(1.59.3) *For $d > n + 1$, the functor* $\mathcal{H}yp\mathcal{S}ur_{n,d}^{\text{lc}}$ *of hypersurfaces with log canonical singularities.*

(1.59.4) *For $d > n + 1$, the functor* $\mathcal{H}yp\mathcal{S}ur_{n,d}^{\text{low-mult}}$ *of those hypersurfaces that have only points of multiplicity* $< \frac{d}{n+1}$.

Example 1.60 Consider the family of even degree d hypersurfaces

$$((x_0^{d/2} + t^d x_1^{d/2}) x_1^{d/2} + x_2^d + \cdots + x_n^d = 0) \subset \mathbb{P}^n \times \mathbb{A}_t^1.$$

For $t \neq 0$ the substitution $(x_0 : x_1 : x_2 : \cdots : x_n) \mapsto (t x_0 : t^{-1} x_1 : x_2 : \cdots : x_n)$ transforms the equation of X_t to

$$X := ((x_0^{d/2} + x_1^{d/2})x_1^{d/2} + x_2^d + \cdots + x_n^d = 0) \subset \mathbb{P}^n.$$

X has a single singular point which is at $(1{:}0{:}\cdots{:}0)$ and has multiplicity $d/2$.
For $t = 0$ we obtain the hypersurface

$$X_0 := (x_0^{d/2} x_1^{d/2} + x_2^d + \cdots + x_n^d = 0).$$

X_0 has two singular points of multiplicity $d/2$, hence it is not isomorphic to X.

Thus we conclude that $[X]$ is not a closed point of the "moduli space" of those hypersurfaces of degree d that have only points of multiplicity $\le d/2$.

This is especially interesting when $d \le n$ since in this case X_0 has canonical singularities (1.33).

Thus we see that for $d \le n$, the functor $\mathcal{H}yp\mathcal{S}ur^c_{n,d}$ parametrizing hypersurfaces with canonical singularities does not have a coarse moduli space. By contrast, for $d > n$ the coarse moduli scheme $\mathrm{HypSur}^c_{n,d}$ exists and is quasi-projective by (1.59).

Example 1.61 One could also consider hypersurfaces modulo isomorphisms which do not necessarily extend to an isomorphism of the ambient projective space. It is easy to see that smooth hypersurfaces can have such nonlinear isomorphisms only for $(d, n) \in \{(3, 2), (4, 3)\}$. A smooth cubic curve in $\mathbb{P}^2$ has an infinite automorphism group, but only finitely many extend to an automorphism of $\mathbb{P}^2$. Similarly, a smooth quartic surface in $\mathbb{P}^3$ can have an infinite automorphism group as in (1.66), but only finitely many extend to an automorphism of $\mathbb{P}^3$. See (1.66) or Shimada and Shioda (2017); Oguiso (2017) for examples of isomorphisms of smooth quartic surfaces in $\mathbb{P}^3$.

The nonseparated examples produced so far all involved ruled or uniruled varieties. Next we consider some examples where the varieties are not uniruled. The bad behavior is due to the singularities, not to the global structure.

Example 1.62 (Double covers of $\mathbb{P}^1$) Let $f(x, y)$ and $g(x, y)$ be two cubic forms without multiple roots, neither divisible by x or y. Set

$$S_1 := (f(x_1, y_1)g(t^2 x_1, y_1) = z_1^2) \subset \mathbb{P}(1, 1, 3) \times \mathbb{A}^1, \quad \text{and}$$
$$S_2 := (f(x_2, t^2 y_2)g(x_2, y_2) = z_2^2) \subset \mathbb{P}(1, 1, 3) \times \mathbb{A}^1.$$

Note that $K_{S_i/\mathbb{A}^1}$ is relatively ample and the general fiber of $\pi_1 : S_i \to \mathbb{A}^1$ is a smooth curve of genus 2.

The central fibers are $(f(x_1, y_1)g(0, y_1) = z_1^2)$ resp. $(f(x_2, 0)g(x_2, y_2) = z_2^2)$. By assumption, $g(0, y_1) = a_1 y_1^3$ and $f(x_2, 0) = a_2 x_2^3$ where the $a_i \ne 0$. Setting $z_1 = a_1^{1/2} w_1 y_1$ and $z_2 = a_2^{1/2} w_2 x_2$ gives the normalizations. Hence the central fibers are elliptic curves with a cusp. Their normalization is isomorphic to $(f(x_1, y_1)y_1 = w_1^2)$ resp. $(x_2 g(x_2, y_2) = w_2^2)$. These are, in general, not isomorphic to each other.

This also shows that along the central fibers, the only singularities are at $(1{:}0{:}0; 0)$ and at $(0{:}1{:}0; 0)$, with local equations $g(t^2, y_1) = z_1^2$ and $f(x_2, t^2) = z_2^2$. (These are simple elliptic. The minimal resolution contains a single smooth elliptic curve of self intersection -1.) Hence the S_i are normal surfaces, each having one simple elliptic singular point.

Finally, the substitution $(x_1 : y_1 : z_1; t) = (x_2 : t^2 y_2 : t^3 z_2; t)$ transforms $f(x_1, y_1)g(t^2 x_1, y_1) - z_1^2$ into

$$f(x_2, t^2 y_2)g(t^2 x_2, t^2 y_2) - t^6 z_2^2 = t^6(f(x_2, t^2 y_2)g(x_2, y_2) - z_2^2),$$

thus the two families are isomorphic over $\mathbb{A}^1 \setminus \{0\}$.

Let us end our study of hypersurfaces with a different type of example. This shows that the moduli problem for hypersurfaces usually includes smooth limits that are not hypersurfaces. These pose no problem for the general theory, but they show that it is not always easy to see what schemes one needs to include in a moduli space.

Example 1.63 (Smooth limits of hypersurfaces) (Mori, 1975) Fix integers $a, b > 1$ and $n \geq 2$. We construct a family of smooth n-folds X_t such that X_t is a smooth hypersurface of degree ab for $t \neq 0$ and X_0 is not isomorphic to a smooth hypersurface.

It is not known if similar examples exist for $n \geq 3$ and $\deg X$ a prime number; see Ottem and Schreieder (2020) for the cases $\deg X \leq 7$.

Start with the weighted projective space $\mathbb{P}(1^{n+1}, a)_{\mathbf{x}, \mathbf{z}}$. Let f_a, g_{ab} be general homogeneous forms of degree a (resp. ab) in $x_0, \ldots, x_n$. Consider the family of complete intersections

$$X_t := (tz - f_a(x_0, \ldots, x_n) = z^b - g_{ab}(x_0, \ldots, x_n) = 0) \subset \mathbb{P}(1^{n+1}, a).$$

For $t \neq 0$, we can eliminate z to obtain a degree ab smooth hypersurface

$$X_t \simeq (f_a^b(x_0, \ldots, x_n) = t^b g_{ab}(x_0, \ldots, x_n)) \subset \mathbb{P}^{n+1}.$$

For $t = 0$, we see that $\mathscr{O}_{X_0}(1)$ is not very ample, but realizes X_0 as a b-fold cyclic cover (11.24) of the degree a smooth hypersurface $(f_a(x_0, \ldots, x_n) = 0)$. In particular, X_0 is not isomorphic to a smooth hypersurface.

The next example shows that seemingly equivalent moduli problems may lead to different moduli spaces.

Example 1.64 We start with the moduli space P_{n+1} of $n + 1$ points in $\mathbb{C}$ up to translations. We can view such a point set as the zeros of a unique polynomial of degree $n + 1$ whose leading term is x^{n+1}. We can use a translation to kill the coefficient of x^n, and the universal polynomial is then given by

$$x^{n+1} + a_2 x^{n-1} + \cdots + a_{n+1}.$$

Thus $P_{n+1} \simeq \mathbb{C}^n$ with coordinates $a_2, \ldots, a_{n+1}$.

Let us now look at those point sets where n of the points coincide. There are two ways to formulate this as a moduli problem:

(1.64.1) unordered point sets $p_0, \ldots, p_n \in \mathbb{C}$ where at least n of the points coincide, up to translations, or

(1.64.2) unordered point sets $p_0, \ldots, p_n \in \mathbb{C}$ plus a point $q \in \mathbb{C}$ such that $p_i = q$ at least n-times, up to translations.

If $n \geq 2$ then q is uniquely determined by the points $p_0, \ldots, p_n$, so it would seem that the two formulations are equivalent. We claim, however, that the two versions have nonisomorphic moduli spaces.

If the n-fold point is at t then the corresponding polynomial is $(x-t)^n(x+nt)$. By expanding it we get that

$$a_i = t^i \left[(-1)^i \binom{n}{i} + (-1)^{i-1} n \binom{n}{i-1} \right] \quad \text{for } i = 2, \ldots, n + 1.$$

This shows that the space $R_{n+1} \subset P_{n+1}$ of polynomials with an n-fold root is a cuspidal rational curve given as the image of the map

$$t \mapsto \left(a_i = t^i \left[(-1)^i \binom{n}{i} + (-1)^{i-1} n \binom{n}{i-1} \right] : i = 2, \ldots, n + 1 \right).$$

So the moduli space R_{n+1} of the first variant (1) is a cuspidal rational curve.

By contrast, the space $\bar{R}_{n+1}$ of the second variant (2) is a smooth rational curve, the isomorphism given by

$$(p_0, \ldots, p_n; q) \mapsto \Sigma_i (p_i - q) \in \mathbb{C}.$$

Not surprisingly, the map that forgets the n-fold root gives $\pi: \bar{R}_{n+1} \to R_{n+1}$ which is the normalization map.

Next we have two examples of moduli functors that are not representable (1.18). They suggest that varieties whose canonical class is not ample present special challenges.

Example 1.65 Let $S \subset \mathbb{P}^3$ be a smooth surface of degree 4 over $\mathbb{C}$, with an infinite discrete automorphism group, for example as in (1.66).

Let $\mathbf{S} \to W$ be the universal family of smooth degree 4 surfaces in $\mathbb{P}^3$. The isomorphisms classes of the pairs $(S, \mathscr{O}_S(1))$ correspond to the $\text{Aut}(\mathbb{P}^3)$-orbits in W. We see that the fibers isomorphic to S form countably many $\mathrm{Aut}(\mathbb{P}^3)$-orbits.

For any $g \in \mathrm{Aut}(S)$, $g^* \mathcal{O}_S(1)$ gives another embedding of S into $\mathbb{P}^3$. Two such embeddings are projectively equivalent iff $g^* \mathcal{O}_S(1) \simeq \mathcal{O}_S(1)$, that is, when $g \in \mathrm{Aut}(S, \mathcal{O}_S(1))$. The latter can be viewed as the group of automorphisms of $\mathbb{P}^3$ that map S to itself. Thus $\mathrm{Aut}(S, \mathcal{O}_S(1))$ is a closed subvariety of $\mathrm{Aut}(\mathbb{P}^3) \simeq \mathrm{PGL}_4$. Since $\mathrm{Aut}(S)$ is discrete, this implies that $\mathrm{Aut}(S, \mathcal{O}_S(1))$ is finite. Hence the fibers of $\mathbf{S} \to W$ that are isomorphic to S lie over countably many $\mathrm{Aut}(\mathbb{P}^3)$-orbits, corresponding to $\mathrm{Aut}(S)/\mathrm{Aut}(S, \mathcal{O}_S(1))$.

Example 1.66 (Surfaces with infinite discrete automorphism group) Let us start with a smooth genus 1 curve E defined over a field K. Any point $q \in E(K)$ defines an involution τ_q where $\tau_q(p)$ is the unique point such that $p + \tau_q(p) \sim 2q$. (Equivalently, we can set q as the origin, then $\tau_q(p) = -p$.) The first formulation shows that if L/K is a quadratic extension, then any $Q \in E(L)$ also defines an involution τ_Q where $\tau_Q(p)$ is the unique point such that $p + \tau_Q(p) \sim Q$.

Given points $q_1, q_2 \in E(K)$, we see that $p \mapsto \tau_{q_2} \circ \tau_{q_1}(p)$ is translation by $2q_1 - 2q_2$. Similarly, given $Q_i \in E(L_i)$, $p \mapsto \tau_{Q_2} \circ \tau_{Q_1}(p)$ is translation by $Q_1 - Q_2$. Usually these translations have infinite order.

Let $g: S \to C$ be a smooth, minimal, elliptic surface. Then, any section or double section of g gives an involution of S, and two involutions usually generate an infinite group of automorphisms of S.

As a concrete example, let $S \subset \mathbb{P}^3$ be a smooth quartic that contains three lines L_i. The pencil of planes through L_1 gives an elliptic fibration with L_2, L_3 as sections. Thus these K3 surfaces usually have an infinite automorphism group.

1.7 Compactifications of $\mathbf{M}_g$

Here we consider what happens if we try to define other compactifications of $\mathbf{M}_g$. First, we give a complete study of a compactified moduli functor of genus 2 curves that uses only irreducible curves.

Definition 1.67 Working over $\mathbb{C}$, let $\mathcal{M}_2^{\mathrm{irr}}$ be the moduli functor of flat families of irreducible curves of arithmetic genus 2 that are either

(1.67.1) smooth,

(1.67.2) nodal,

(1.67.3) rational with two cusps, or

(1.67.4) rational with a triple point whose complete local ring is isomorphic to $\mathbb{C}[[x, y, z]]/(xy, yz, zx)$.

The aim of this subsection is to prove the following; see Mumford (1965, chap.3) or Dolgachev (2003, sec.10.2) for the relevant background on GIT quotients.

Proposition 1.68 *Let M_2^{irr} be the moduli functor defined at (1.67). Then*

(1.68.1) *the coarse moduli space M_2^{irr} exists and equals the geometric invariant theory quotient (8.59) of the symmetric power $\mathrm{Sym}^6 \mathbb{P}^1 // \mathrm{Aut}(\mathbb{P}^1)$, but*

(1.68.2) M_2^{irr} *is a very bad moduli functor.*

Proof A smooth curve of genus 2 can be uniquely written as a double cover $\tau \colon C \to \mathbb{P}^1$, ramified at six distinct points $p_1, \ldots, p_6 \in \mathbb{P}^1$, up to automorphisms of $\mathbb{P}^1$. Thus, M_2 is isomorphic to the space of six distinct points in $\mathbb{P}^1$, modulo the action of $\mathrm{Aut}(\mathbb{P}^1)$. If some of the six points coincide, we get singular curves as double covers.

It is easy to see the following; see Mumford (1965, chap.3), Dolgachev (2003, sec.10.2).

(1.68.3) A point set is semistable iff it does not contain any point with multiplicity ≥ 4. Equivalently, if the genus 2 cover has only nodes and cusps.

(1.68.4) The properly semistable point sets are of the form $3p_1 + p_2 + p_3 + p_4$ where the p_2, p_3, p_4 are different from p_1, but may coincide with each other. Equivalently, the corresponding genus 2 cover has at least one cusp.

(1.68.5) Point sets $2p_1 + 2p_2 + 2p_3$, where the p_1, p_2, p_3 are different from each other. The double cover is reducible, with two smooth rational components meeting each other at three points.

In the properly semistable case, generically the double cover is an elliptic curve with a cusp over p_1. As a special case, we can have $3p_1 + 3p_2$, giving as double cover a rational curve with two cusps. Note that the curves of this type have a one-dimensional moduli (the cross ratio of the points p_1, p_2, p_3, p_4 or the j-invariant of the elliptic curve), but they all correspond to the same point in $\mathrm{Sym}^6 \mathbb{P}^1 // \mathrm{Aut}(\mathbb{P}^1)$. (See (1.62) for an explicit construction.) Our definition (1.67) aims to remedy this nonuniqueness by always taking the most degenerate case; a rational curve with two cusps (1.67.3).

In case (5), write the reducible double cover as $C = C_1 + C_2$. The only obvious candidate to get an irreducible curve is to contract one of the two components C_i. We get an irreducible rational curve; denote it by C_j' where $j = 3 - i$. Note that C_j' has one singular point which is analytically isomorphic to the three coordinate axes in $\mathbb{A}^3$. The resulting singular rational curves C_j' are isomorphic to each other. These are listed in (1.67.4).

Let $p\colon X \to S$ be any flat family of irreducible, reduced curves of arithmetic genus 2. The trace map (Barth et al., 1984, III.12.2) shows that $R^1 p_* \omega_{X/S} \simeq \mathscr{O}_S$. Thus, by cohomology and base change, $p_* \omega_{X/S}$ is locally free of rank 2. Set $P := \mathbb{P}_S(p_* \omega_{X/S})$. Then P is a $\mathbb{P}^1$-bundle over S, and we have a rational map $\pi\colon X \to P$. If X_s has only nodes and cusps, then ω_{X_s} is locally free and generated by global sections, thus π is a morphism along X_s.

If X_s is as in (1.67.4), then ω_{X_s} is not locally free, and π is not defined at the singular point. $\pi|_{X_s}$ is birational and the three local branches of X_s at the singular point correspond to three points on $\mathbb{P}(H^0(X_s, \omega_{X_s}))$.

The branch divisor of π is a degree 6 multisection of $P \to S$, all of whose fibers are stable point sets. Thus we have a natural transformation

$$\mathcal{M}_2^{\mathrm{irr}}(*) \to \mathrm{Mor}(*, \mathrm{Sym}^6 \mathbb{P}^1 /\!/ \mathrm{Aut}(\mathbb{P}^1)).$$

We have already seen that we get a bijection

$$\mathcal{M}_2^{\mathrm{irr}}(\mathbb{C}) \simeq (\mathrm{Sym}^6 \mathbb{P}^1 /\!/ \mathrm{Aut}(\mathbb{P}^1))(\mathbb{C}).$$

Since $\mathrm{Sym}^6 \mathbb{P}^1 /\!/ \mathrm{Aut}(\mathbb{P}^1)$ is normal, we conclude that it is the coarse moduli space. This completes the proof of (1.68.1).

The assertion (1.68.2) is more a personal opinion. There are three main things "wrong" with the functor $\mathcal{M}_2^{\mathrm{irr}}(*)$. Let us consider them one at a time.

1.68.6 (Stable extension questions)

At the set-theoretic level, we have $\mathrm{M}_2^{\mathrm{irr}} = \mathrm{Sym}^6 \mathbb{P}^1 /\!/ \mathrm{Aut}(\mathbb{P}^1)$, but what about at the level of families?

The first indications are good. Let $\pi_B\colon S_B \to B$ be a stable family of genus 2 curves. Assume that no fiber is of type (1.68.5). Let $b_i \in B$ be the points corresponding to fibers with two components of arithmetic genus 1. Let $p\colon A \to B$ be a double cover ramified at the points b_i. Consider the pull-back family $\pi_A\colon S_A \to A$. Set $a_i = p^{-1}(b_i)$ and let $s_i \in \pi_A^{-1}(a_i)$ be the point where the two components meet. Since we took a ramified double cover, each $s_i \in S_A$ is a double point. Thus if we blow up every s_i, the exceptional curves appear in the fiber with multiplicity 1. We can now contract the birational transforms of the elliptic curves to get a family where all these reducible fibers are replaced by a rational curve with two cusps. We have proved the following analog of (1.15.1):

Claim 1.68.6.a Let $\pi\colon S \to B$ be a stable family of genus 2 curves such that no fiber has two smooth rational components. Then, after a suitable double cover $A \to B$, the pull-back $S \times_B A$ is birational to another family where each reducible fiber is replaced by a rational curve with two cusps. □

This solved our problem for one-parameter families, but, as it turns out, not over higher dimensional bases. In particular, there is no universal family over any base scheme Y that finitely dominates $\mathrm{Sym}^6 \, \mathbb{P}^1 /\!/ \mathrm{Aut}(\mathbb{P}^1)$, not even locally in any neighborhood of the properly semistable point. Indeed, this would give a proper, flat family of curves of arithmetic genus 2 over a three-dimensional base $\pi\colon X \to Y$ where only finitely many of the fibers (the ones over the unique properly semistable point) have cusps. However, there is no such family.

To see this we use that, by (2.27), every flat deformation of a cusp is induced by pull-back from the two-parameter family

$$
\begin{array}{ccc}
(y^2 = x^3 + ux + v) & \lhook\joinrel\longrightarrow & \mathbb{A}^2_{xy} \times \mathbb{A}^2_{uv} \\
p\downarrow & & \downarrow \\
\mathbb{A}^2_{uv} & =\!=\!=\!=\!=\!=\!= & \mathbb{A}^2_{uv}.
\end{array}
\qquad (1.68.6.\mathrm{b})
$$

Thus our family π gives an analytic morphism $\tau\colon Y \to \mathbb{A}^2_{uv}$ (defined in some neighborhood of $0 \in Y$), and $C = \tau^{-1}(0,0) \subset Y$ is a curve along which the fiber has a cusp.

1.68.7 (Failure of representability)
Following (1.68.6.b), consider the universal deformation of the rational curve with two cusps. This is given as

$$
\begin{array}{ccc}
(z^2 = (x^3 + uxy^2 + vy^3)(y^3 + syx^2 + tx^3)) & \lhook\joinrel\longrightarrow & \mathbb{P}^2(1,1,3) \times \mathbb{A}^4_{uvst} \\
p\downarrow & & \downarrow \\
\mathbb{A}^4_{uvst} & =\!=\!=\!=\!=\!=\!= & \mathbb{A}^4_{uvst}.
\end{array}
$$

Let us work in a neighborhood of $(0,0,0,0) \in \mathbb{A}^4$, where the two factors $x^3 + uxy^2 + vy^3$ and $y^3 + syx^2 + tx^3$ have no common roots. There are three types of fibers: $p^{-1}(0,0,0,0)$ is a rational curve with two cusps, $p^{-1}(a,b,0,0)$ and $p^{-1}(0,0,a,b)$ are irreducible with exactly one cusp if $(a,b) \neq (0,0)$, and $p^{-1}(a,b,c,d)$ is irreducible with at worst nodes otherwise.

Thus the curves that we allow in our moduli functor $\mathcal{M}_2^{\mathrm{irr}}$ do not form a representable family. Even worse, the subfamily

$$
(z^2 = (x^3 + uxy^2 + vy^3)y^3) \to \mathrm{Spec}\, k[[u,v]]
$$

is not allowed in our moduli functor $\mathcal{M}_2^{\mathrm{irr}}$, but the family

$$
(z^2 = (x^3 + uxy^2 + vy^3)(y^3 + u^n yx^2 + v^n x^3)) \to \mathrm{Spec}\, k[[u,v]]
$$

is allowed. Over $\mathrm{Spec}\, k[u,v]/(u^n, v^n)$ the two families are isomorphic. Since deformation theory is essentially a study of families over Artinian rings, this means that the usual methods cannot be applied to understand the functor $\mathcal{M}_2^{\mathrm{irr}}$.

1.68.8 (Unusual nonseparatedness) A quite different type of problem arises at the curve corresponding to $2p_1 + 2p_2 + 2p_3$.

Write the double cover as $C = C_1 + C_2$. As before, if we contract one of the two components C_i, we get an irreducible rational curve C'_j, where $j = 3 - i$ as in (1.67.4).

Since the curves C'_1 and C'_2 are isomorphic, from the set-theoretic point of view this is a good solution. However, as in (1.27), something strange happens with families. Let $p: S \to \mathbb{A}^1$ be a family of stable curves whose central fiber $S_0 := p^{-1}(0)$ is isomorphic to $C = C_1 + C_2$. We have two ways to construct a family with an irreducible central fiber: contract either of the two irreducible components C_i. Thus we get two families

$$S \xrightarrow{\pi_i} S_i \xrightarrow{p_i} \mathbb{A}^1 \quad \text{with } p_i^{-1}(0) \simeq C'_{3-i}.$$

Over $\mathbb{A}^1 \setminus \{0\}$ the two families are naturally isomorphic to $S \to \mathbb{A}^1$, hence to each other, yet this isomorphism does not extend to an isomorphism of S_1 and S_2. Indeed, the closure of the graph of the resulting birational map is given by the image $(\pi_1, \pi_2): S \to S_1 \times_{\mathbb{A}^1} S_2$. Thus the corresponding moduli functor is not separated.

We claimed in (1.68.1) that, by contrast, the coarse moduli space is $\overline{M}_2$, hence separated. A closer study reveals the source of this discrepancy: we have been thinking of schemes instead of algebraic spaces. The occurrence of such problems in moduli theory was first observed by Artin (1974). The aim of the next paragraph is to show how such examples arise.

1.68.9 (Bug-eyed covers) (Artin, 1974); (Kollár, 1992a) A non-separated scheme always has "extra" points. The typical example is when we take two copies of a scheme $X \times \{i\}$ for $i = 0, 1$, an open dense subscheme $U \subsetneq X$, and glue $U \times \{0\}$ to $U \times \{1\}$ to get $X \amalg_U X$. The non-separatedness arises from having two points in $X \amalg_U X$ for each point in $X \setminus U$.

By contrast, an algebraic space can be nonseparated by having no extra points, only extra tangent directions. The simplest example is the following.

On $\mathbb{A}^1_t$ consider two equivalence relations. The first is $R_1 \rightrightarrows \mathbb{A}^1$ given by

$$(t_1 = t_2) \cup (t_1 = -t_2) \subset \mathbb{A}^1_{t_1} \times \mathbb{A}^1_{t_2}.$$

Then $\mathbb{A}^1_t / R_1 \simeq \mathbb{A}^1_u$ where $u = t^2$.

The second is the étale equivalence relation $R_2 \rightrightarrows \mathbb{A}^1$ given by

$$\mathbb{A}^1 \xrightarrow{(1,1)} \mathbb{A}^1 \times \mathbb{A}^1 \quad \text{and} \quad \mathbb{A}^1 \setminus \{0\} \xrightarrow{(1,-1)} \mathbb{A}^1 \times \mathbb{A}^1.$$

(Note that we take the disconnected union of the two components, instead of their union as two lines in $\mathbb{A}^1 \times \mathbb{A}^1$ intersecting at the origin.)

One can also obtain $\mathbb{A}^1_t/R_2$ by taking the quotient of the nonseparated scheme $\mathbb{A}^1 \amalg_{\mathbb{A}^1\setminus\{0\}} \mathbb{A}^1$ by the (fixed point free) involution that interchanges $(t, 0)$ and $(-t, 1)$.

The morphism $\mathbb{A}^1_t \to \mathbb{A}^1_t/R_2$ is étale, thus $\mathbb{A}^1_t/R_2 \neq \mathbb{A}^1_t/R_1$. Nonetheless, there is a natural morphism $\mathbb{A}^1_t/R_2 \to \mathbb{A}^1_t/R_1$ which is one-to-one and onto on closed points. The difference between the two spaces is seen by the tangent spaces. The tangent space of $\mathbb{A}^1_t/R_2$ at the origin is spanned by $\partial/\partial t$ while the tangent space of $\mathbb{A}^1_t/R_1$ at the origin is spanned by $\partial/\partial u = (2t)^{-1}\partial/\partial t$.

1.69 Our attempt to replace the moduli functor of stable curves of genus 2 with another one that parametrizes only irreducible curves was not successful, but some of the problems seemed to have arisen from the symmetry that forced us to make artificial choices. We can avoid such choices for other values of the genus using the following observation.

Let $\pi\colon S \to B$ be a flat family of curves with smooth general fiber and reduced special fibers. If $C_b := \pi^{-1}(b)$ is a singular fiber and C_{bi} are the irreducible components of its normalization, then

$$\sum_i h^1(C_{bi}, \mathcal{O}_{C_{bi}}) \leq h^1(C_b, \mathcal{O}_{C_b}) = h^1(C_{gen}, \mathcal{O}_{C_{gen}}),$$

where C_{gen} is the general smooth fiber. In particular, there can be at most one irreducible component with geometric genus $> \frac{1}{2}g(C_{gen})$.

From this it is easy to prove the following:

Claim 1.69.1 Let B be a smooth curve and $S^\circ \to B^\circ$ a smooth family of genus g curves over an open subset of B. Then there is at most one normal surface $S \to B$ extending S° such that every fiber of $S \to B$ is irreducible and of geometric genus $> \frac{1}{2}g(C_{gen})$.

Moreover, if $S_{stab} \to B$ is a stable family extending S° and every fiber of $S_{stab} \to B$ contains an irreducible curve of geometric genus $> \frac{1}{2}g(C_{gen})$, then we obtain S from S_{stab} by contracting all connected components of curves of geometric genus $< \frac{1}{2}g(C_{gen})$ that are contained in the fibers. (It is not hard to show that $S \to B$ exists, at least as an algebraic space.)

In fact, this way we obtain a partial compactification $M_g \subset M'_g$ such that

- M'_g parametrizes smoothable irreducible curves of arithmetic genus g and geometric genus $> \frac{1}{2}g$.
- Let $M_g \subset M''_g \subset \overline{M}_g$ be the largest open subset parametrizing curves that contain an irreducible component of geometric genus $> \frac{1}{2}g$. Then there is a natural morphism $M''_g \to M'_g$.

So far so good, but, as we see next, we cannot extend M'_g to a compactification in a geometrically meaningful way. This happens for every $g \geq 3$; the following example with $g = 13$ is given by simple equations.

This illustrates a general pattern: one can easily propose partial compactifications that work well for some families, but lead to contradictions for some
others. (See Schubert, 1991; Hassett and Hyeon, 2013; Smyth, 2013 for a
search for geometrically meaningful compactifications of M_g.)

Example 1.70 Consider the surface $F := (x^8 + y^8 + z^8 = t^2) \subset \mathbb{P}^3(1,1,1,4)$
and on it the curve $C := F \cap (xyz = 0)$. C has three irreducible components
$C_x = (x = 0), C_y = (y = 0), C_z = (z = 0)$, which are smooth curves of genus 3.
C itself has arithmetic genus 13.

We work with a three-parameter family of deformations

$$T := (xyz - ux^3 - vy^3 - wz^3 = 0) \subset F \times \mathbb{A}^3_{uvw}. \qquad (1.70.1)$$

For general $uvw \neq 0$ the fiber of the projection $\pi: T \to \mathbb{A}^3$ is a smooth curve
of genus 13. If one of the u, v, w is zero, then generically we get a curve with
two nodes, hence with geometric genus 11.

If two of the coordinates are zero, say $v = w = 0$, then we have a family

$$T_x := (x(yz - ux^2) = 0) \subset F \times \mathbb{A}^1_u.$$

For $u \neq 0$, the fiber $C_{(u,0,0)}$ has two irreducible components. One is $C_x = (x = 0)$, the other is $(yz - ux^2 = 0)$ which is a smooth genus 7 curve.

Thus the proposed rule says that we should contract $C_x \subset C_{(u,0,0)}$.

Similarly, by working over the v and the w-axes, the rule tells us to contract
$C_y \subset C_{(0,v,0)}$ for $v \neq 0$ and $C_z \subset C_{(0,0,w)}$ for $w \neq 0$.

It is easy to see that over $\mathbb{A}^3 \setminus \{(0,0,0)\}$ these contractions can be performed
(at least among algebraic spaces). Thus we obtain

$$
\begin{array}{ccc}
T \setminus \{\pi^{-1}(0,0,0)\} & \xrightarrow{\ \ p^\circ\ \ } & S^\circ \\
\Big\downarrow & & \Big\downarrow{\scriptstyle \tau^\circ} \\
\mathbb{A}^3 \setminus \{(0,0,0)\} & =\!=\!=\!= & \mathbb{A}^3 \setminus \{(0,0,0)\}
\end{array}
\qquad (1.70.2)
$$

where τ° is flat with irreducible fibers.

Claim 1.70.3 There is no proper family of curves $\tau: S \to \mathbb{A}^3$ that extends τ°.
(We do not require τ to be flat.)

Proof Assume to the contrary that $\tau: S \to \mathbb{A}^3$ exists, and let $\Gamma \subset T \times_{\mathbb{A}^3} S$ be
the closure of the graph of p°. Since p° is a morphism on $T \setminus \{\pi^{-1}(0,0,0)\}$,
we see that the first projection $\pi_1: \Gamma \to T$ is an isomorphism away from
$\pi^{-1}(0,0,0)$. Since $T \times_{\mathbb{A}^3} S \to \mathbb{A}^3$ has two-dimensional fibers, we conclude that

$\dim \pi_1^{-1}(\pi^{-1}(0,0,0)) \le 2$. T is, however, a smooth 4-fold, hence the exceptional set of any birational map to T has pure dimension 3. Thus $\Gamma \simeq T$ and so p° extends to a morphism $p \colon T \to S$.

Now the rule lands us in a contradiction over the origin $(0,0,0)$. Here all three components $C_x, C_y, C_z \subset C_{(0,0,0)} = C$ should be contracted. This is impossible to do since this would give that the central fiber of $S \to \mathbb{A}^3$ is a point.   □

1.8 Coarse and Fine Moduli Spaces

As in (1.7), let $\mathbf{V}$ be a "reasonable" class of projective varieties and *Varieties*$_\mathbf{V}$ the corresponding functor. The aim of this section is to study the difference between coarse and fine moduli spaces, mostly through a few examples. We are guided by the following:

Principle 1.71 *Let* $\mathbf{V}$ *be a "reasonable" class as above, and assume that it has a coarse moduli space* Moduli$_\mathbf{V}$. *Then* Moduli$_\mathbf{V}$ *is a fine moduli space iff* **Aut**(V), *the group of automorphisms of V (8.63), is trivial for every $V \in \mathbf{V}$.*

From the point of view of algebraic stacks, a precise version is given in Laumon and Moret-Bailly (2000, 8.1.1). In positive characteristic, one should pay attention to the scheme structure of **Aut**(V). Our construction of the moduli spaces shows that this principle is true for polarized varieties, see Section 8.7, but a precise version needs careful attention to the difference between schemes and algebraic spaces.

Let L be a field and $X_L \in \mathbf{V}$ an L-variety. Let $[X] \in$ Moduli$_\mathbf{V}$ be the corresponding point with residue field $K := k([X])$. If Moduli$_\mathbf{V}$ is fine, then the resulting map Spec $K \to$ Moduli$_\mathbf{V}$ corresponds to a K-variety X_K such that $X_L \simeq X_K \times \operatorname{Spec} L$. Moreover, X_K is the unique K-variety with this property.

If Moduli$_\mathbf{V}$ is not a fine moduli space, then it is not clear how to define this field K. X_K may not be unique and may not exist. We study these questions, mostly through examples.

1.72 (Field of moduli) Let $X \subset \mathbb{P}^n$ be a projective variety defined over an algebraically closed field K. Any set of defining equations involves only finitely many elements of K, thus X can be defined over a finitely generated subfield of K. It is a natural question to ask: Is there a smallest subfield $F \subset K$ such that X can be defined by equations over F? There are two variants of this question.

1.72.1 (Embedded version) Fix coordinates on $\mathbb{P}^n_K$ and view X as a specific subvariety. In this case a smallest subfield F exists; see Weil (1946, sec.I.7) or Kollár et al. (2004, sec.3.4). This is a special case of the existence of Hilbert schemes (1.5). More generally, the same holds if $\mathbb{P}^n$ is replaced by any $\mathbb{Z}$-scheme. We can also think of this as a Galois invariance property. If $\sigma \in \text{Aut}(K)$ then $\sigma(X) = X$ iff σ is the identity on F. If char $K = 0$, this property characterizes F, but otherwise only its purely inseparable closure F^{ins}.

1.72.2 (Absolute version) No embedding of X is fixed. Thus we are looking for a field $F \subset K$ and an F-variety X_F such that $X \simeq (X_F)_K$. It turns out that there is no smallest field in general. As a first approximation, we call the intersection of all such fields F the *field of moduli* of X. As the examples (1.76) show, this naive version can be unexpectedly small.

The situation is better if K_X is ample, but in (1.75) we construct a hyperelliptic curve whose field of moduli is $\mathbb{Q}$, yet it cannot be defined over $\mathbb{R}$. The first such examples are in Earle (1971); Shimura (1972).

To get the right notion, we instead look for *isotrivial* families with fiber X, defined over some subfield $F \subset K$. That is, flat, projective morphisms $u : U_Z \to Z$ (defined over F), whose every geometric fiber is isomorphic to X.

We say that $u: U_Z \to Z$ is *universal* if every isotrivial family $v: U_S \to S$ with fiber X is locally the pull-back of $u: U_Z \to Z$. That is, there is an open cover $S = \cup_i S_i$ and morphisms $\sigma_i: S_i \to Z$ such that the restriction $v_i: U_{S_i} \to S_i$ is isomorphic to the pull-back $U \times_{u,\sigma_i} S_i \to S_i$.

We see in (1.73) that universal isotrivial families exist and they are defined over the same subfield $F_X \subset K$, giving the right notion of field of moduli. How is this connected with moduli theory?

Let $\mathbf{V}$ be a class of varieties with a coarse moduli space $\text{Moduli}_{\mathbf{V}}$. Let $u: U_Z \to Z$ be an isotrivial family with fiber X defined over $F \subset K$. By the definition of a coarse moduli space, there is a morphism $Z \to \text{Moduli}_{\mathbf{V}}$, whose image must be the point $[X] \in \text{Moduli}_{\mathbf{V}}$ corresponding to X. In particular, we get an injection of the residue field $k([X])$ into F.

If $\text{Moduli}_{\mathbf{V}}$ is a fine moduli space, then X can be defined over $k([X])$, and (1.73.2) shows that $k([X]) = F_X$.

The construction of the moduli spaces of stable varieties shows that the extension $F_X/k([X])$ is purely inseparable, hence trivial in characteristic 0.

Proposition 1.73 *Let K be an algebraically closed field of characteristic 0 and X a projective K-variety with ample canonical class. Then there is a unique smallest field $F_X \subset K$ – called the field of moduli of X – such that there is a geometrically irreducible, universal, isotrivial family $u: U \to Z$ with fiber X, defined over F_X. Moreover, $X \simeq X^\sigma$ for every $\sigma \in \text{Gal}(K/F_X)$.*

Proof Fix m such that $|mK_X|$ is very ample, giving an embedding $X \hookrightarrow \mathbb{P}^N$. The image depends on a choice of a basis in $H^0(X, \mathscr{O}_X(mK_X))$, so instead of getting a point in $\text{Chow}(\mathbb{P}^N)$ or $\text{Hilb}(\mathbb{P}^N)$, we get a whole $\mathbf{Aut}(\mathbb{P}^N)$-orbit. Denote it by Z (it depends on X and m). Over it we have a universal family $u \colon U_Z \to Z$, which is isotrivial with fiber X.

The closure of Z is now a closed subvariety of the $\mathbb{Z}$-schemes $\text{Chow}(\mathbb{P}^N)$ or $\text{Hilb}(\mathbb{P}^N)$, thus it has a smallest field of definition by (1.72.1). This is our F_X.

To see that $u \colon U_Z \to Z$ is universal, let $v \colon V_S \to S$ be an isotrivial family with fiber X. Then $v_* \mathscr{O}_{V_S}(mK_{V_S/S})$ is locally free. Choose an open trivializing cover $S = \cup_i S_i$. These define embeddings $V_{S_i} \hookrightarrow \mathbb{P}^N \times S_i$, hence morphisms $\sigma_i \colon S_i \to Z$.

For a subvariety $X \subset \mathbb{P}^n_K$, let $[X] \in \text{Hilb}(\mathbb{P}^N)$ denote the corresponding point. Then $[X^\sigma] = \sigma[X]$, hence the last claim is a reformulation of the Galois invariance property noted in (1.72.1). $\square$

1.74 (Field of moduli for hyperelliptic curves) Let A be a smooth hyperelliptic curve of genus ≥ 2. Over an algebraically closed field, A has a unique degree 2 map to $\mathbb{P}^1$. Let $B \subset \mathbb{P}^1$ be the branch locus, that is, a collection of $2g + 2$ points in $\mathbb{P}^1$. If the base field k is not closed, then A has a unique degree 2 map to a smooth genus 0 curve Q. (One can always think of Q as a conic in $\mathbb{P}^2$.) Thus A is defined over a field k iff the pair $(B \subset \mathbb{P}^1)$ can be defined over k.

The latter problem is especially transparent if A is defined over $\mathbb{C}$, and we want to know if it is defined over $\mathbb{R}$ or if its field of moduli is contained in $\mathbb{R}$.

Up to isomorphism, there are two real forms of $\mathbb{P}^1$. One is $\mathbb{P}^1$, corresponding to the antiholomorphic involution $(x{:}y) \mapsto (\bar{x}{:}\bar{y})$, which, after a coordinate change, can also be written as $\sigma_1 \colon (x{:}y) \mapsto (\bar{y}{:}\bar{x})$. (In the latter, the real points are the unit circle.) The other is the "empty" conic, corresponding to the antiholomorphic involution $\sigma_2 \colon (x{:}y) \mapsto (-\bar{y}{:}\bar{x})$.

Thus let $A \to \mathbb{CP}^1$ be a smooth hyperelliptic curve of genus ≥ 2 over $\mathbb{C}$ with branch locus $B \subset \mathbb{CP}^1$. Then (1.72.5) gives that

(1.74.1) A can be defined over $\mathbb{R}$ iff there is a $g \in \text{Aut}(\mathbb{CP}^1)$ such that gB is invariant under σ_1 or σ_2, and

(1.74.2) the field of moduli of A is contained in $\mathbb{R}$ iff there is an $h \in \text{Aut}(\mathbb{CP}^1)$ such that hB equals B^{σ_1} or B^{σ_2}.

Note that if $(gB)^\sigma = gB$ then $B^\sigma = (g^\sigma)^{-1}gB$ shows that (1) $\Rightarrow$ (2). Conversely, if $B^\sigma = hB$ and we can write $h = (g^\sigma)^{-1}g$ then $(gB)^\sigma = gB$.

Example 1.75 Here is an example of a hyperelliptic curve C whose field of moduli is $\mathbb{Q}$, but C cannot be defined over $\mathbb{R}$.

Pick $\alpha = a + ib$ where a, b are rational. Consider the hyperelliptic curve

$$C(\alpha) := \left(z^2 - (x^8 - y^8)(x^2 - \alpha y^2)(\bar{\alpha}x^2 + y^2) = 0 \right) \subset \mathbb{P}^2(1, 1, 6).$$

Its complex conjugate is

$$C(\bar{\alpha}) := \left(z^2 - (x^8 - y^8)(x^2 - \bar{\alpha}y^2)(\alpha x^2 + y^2) = 0 \right) \subset \mathbb{P}^2(1, 1, 6).$$

$C(\alpha)$ and $C(\bar{\alpha})$ are isomorphic, as shown by the substitution $(x, y, z) \mapsto (iy, x, z)$. So, over $\mathrm{Spec}_{\mathbb{Q}} \, \mathbb{Q}[t]/(t^2 + 1)$ we have a curve

$$C(a, b) := \left(z^2 - (x^8 - y^8)(x^2 - (a + tb)y^2)((a - tb)x^2 + y^2) = 0 \right) \subset \mathbb{P}^2(1, 1, 6)$$

whose geometric fibers are isomorphic to $C(\alpha)$. Thus the field of moduli of $C(\alpha)$ is $\mathbb{Q}$ by (1.72.5).

We claim that, for sufficiently general a, b, the curve $C(\alpha)$ cannot be defined over $\mathbb{Q}$, not even over $\mathbb{R}$. By (1.74) we need to show that there is no anti-holomorphic involution that maps the branch locus to itself. In the affine chart $y \neq 0$, the ramification points of $C(\alpha) \to \mathbb{P}^1$ are:

(1.75.1) the 8th roots of unity corresponding to $x^8 - y^8$, and

(1.75.2) the four points $\pm\beta, \pm i/\bar{\beta}$ where $\beta^2 = \alpha$.

The anti-holomorphic automorphisms of the Riemann sphere map circles to circles. Out of our 12 points, the 8th roots of unity lie on the circle $|z| = 1$, but no other 8 can lie on a circle. Thus any antiholomorphic automorphism that maps our configuration to itself, must fix the unit circle $|z| = 1$ and map the 8th roots of unity to each other.

The only such antiholomorphic involutions are

(1.75.3) reflection on the line $\mathbb{R} \cdot \varepsilon$, where ε is a 16th root of unity, and

(1.75.4) $z \mapsto 1/\bar{z}$ or $z \mapsto -1/\bar{z}$.

A short analysis shows that $C(\alpha)$ is not isomorphic (over $\mathbb{C}$) to a real curve, as long as β^{16} is not a positive real number.

Example 1.76 We give an example of a smooth projective surface S such that if S is defined over a field extension $K/\mathbb{C}$ then $\mathrm{trdeg}_{\mathbb{C}} K = 2$, but the intersection of all such fields of definition is $\mathbb{C}$.

Let X be a projective surface such that $\mathrm{Aut}(X)$ is discrete and contains finite subgroups G_1, G_2 such that $\langle G_1, G_2 \rangle$ has a Zariski dense orbit on X.

One such example is $B_0(E \times E)$, the blow-up of the square of an elliptic curve at a point, as shown by the subgroups generated by the matrices

$$\begin{pmatrix} 0 & -1 \\ 1 & 1 \end{pmatrix} \quad \text{and} \quad \begin{pmatrix} 0 & -1 \\ 1 & 0 \end{pmatrix}.$$

There are also K3 surfaces with infinite automorphism group generated by two involutions (1.66).

Let $\Delta \subset X \times X$ be the diagonal and, using one of the projections, consider the family of smooth varieties $f : Y := B_\Delta(X \times X) \to X$. Our example is $K = \mathbb{C}(X)$ and Y_K the generic fiber of $Y \to X$.

Note that $Y \to X$ is the universal family of the varieties of the form $B_x X$ for $x \in X$. This shows that Y_K cannot be obtained by base change from a variety over a field of smaller transcendence degree over $\mathbb{C}$.

Let $G \subset \mathrm{Aut}(X)$ be a finite subgroup. There is an open subset $U_G \subset X$ such that G operates on U_G without fixed points. Thus $f/G : Y/G \to X/G$ is a family of smooth varieties over U_G/G and $Y|_{U_G} \simeq Y/G \times_{X/G} U_G$. Thus Y_K can be defined over $\mathbb{C}(X/G) = K^G$ for every finite subgroup $G \subset \mathrm{Aut}(X)$.

On the other hand, the intersection $K^{G_1} \cap K^{G_2}$ is $\mathbb{C}$. Indeed, any function in $K^{G_1} \cap K^{G_2}$ is constant on every G_1-orbit and also on every G_2-orbit, hence on a dense set by our assumptions.

This phenomenon is also connected with the behavior of ample line bundles on $\pi_i : X \to X/G_i$. Although both of the X/G_i are projective, there are no ample line bundles L_i on X/G_i such that $\pi_1^* L_1 \simeq \pi_2^* L_2$.

1.77 (Openness of the fine locus) Let **V** be a "reasonable" class of varieties with a coarse moduli space $\mathrm{Moduli_V}$.

If $\mathbf{Aut}(X) = \{1\}$ is an open condition in flat families with fibers in **V**, then there is an open subscheme $\mathrm{Moduli_V^{rigid}} \subset \mathrm{Moduli_V}$ that is a coarse moduli space for varieties in **V** without automorphisms. By (1.71), $\mathrm{Moduli_V^{rigid}}$ should be a fine moduli space. In many cases, $\mathrm{Moduli_V^{rigid}}$ is dense in $\mathrm{Moduli_V}$, thus one can understand much about the whole $\mathrm{Moduli_V}$ by studying the fine moduli space $\mathrm{Moduli_V^{rigid}}$.

Let $X \to S$ be a flat family with fibers in **V** and $\pi : \mathbf{Aut}_S(X) \to S$ the scheme representing automorphisms of the fibers (8.63). If **V** satisfies the valuative criterion of separatedness (1.20), and all automorphism groups are finite, then π is proper. More careful attention to the scheme structure of the automorphism groups shows that in fact $\mathbf{Aut}(X) = \{1\}$ is an open condition.

However, automorphism groups of smooth, projective surfaces can jump unexpectedly. For example, the automorphism group of a general Enriques surface is infinite, but there are special Enriques surfaces with finite automorphism group. A more elementary example is the following:

Example 1.77.1 Let ζ be a primitive mth root of unity. Then $\tau(x{:}y{:}z) = (\zeta x{:}y{:}z)$ defines a $\mathbb{Z}/m$-action on $\mathbb{P}^2$. For $t \neq 0$, let S_t be the surface obtained by blowing up the m points $(\zeta^i t{:}t{:}1)$.

What should $\lim_{t\to 0} S_t$ be? A natural candidate is to blow up first $(0{:}0{:}1)$ and then the m intersection points p_i of the exceptional curve E with the birational transforms of the lines $L_i := (x = \zeta^i y)$. The resulting S_0 has a $\mathbb{Z}/m$-action, but we blew up $m + 1$-times, so there is no family of smooth surfaces with fibers $\{S_t : t \in \mathbb{C}\}$.

As in (1.24), for any $j \in \mathbb{Z}/m$ we can get a smooth family of surfaces with central fiber S_0^j, obtained by blowing up first $(0{:}0{:}1)$ and then all the p_i for $i \neq j$. These give m distinct families, and we do not have a $\mathbb{Z}/m$-action on any of these S_0^j.

2

One-Parameter Families

In Kollár (2013b) we studied in detail canonical and semi-log-canonical varieties, especially their singularities; a summary of the main results is given in Section 11.1. These are the objects that correspond to the points in a moduli functor/stack of canonical and semi-log-canonical varieties. We start the study of the general moduli problem with one-parameter families.

In traditional moduli theory – for instance, for curves, smooth varieties or sheaves – the description of all families over one-dimensional regular schemes pretty much completes the story: the definitions and theorems have obvious generalizations to families over an arbitrary base. The best examples are the valuative criteria of separatedness and properness; we discussed these in (1.20). In our case, however, much remains to be done in order to work over arbitrary base schemes.

Two notions of locally stable or semi-log-canonical families are introduced in Section 2.1; their equivalence is proved in characteristic 0. For surfaces, one can give a rather complete étale-local description of all locally stable families; this is worked out in Section 2.2.

A series of higher dimensional examples is presented in Section 2.3. These show that stable degenerations of smooth projective varieties can get rather complicated.

Next we turn to global questions and define our main objects, stable families, in Section 2.4. The main result says that stable families satisfy the valuative criteria of separatedness and properness.

Cohomological properties of stable families are studied in Section 2.5. In particular, we show that in a proper, locally stable family $f : X \to C$, the basic numerical invariants $h^i(X_c, \mathscr{O}_{X_c})$ and $h^i(X_c, \omega_{X_c})$ are independent of $c \in C$. We also show that X_c being Cohen–Macaulay (10.4) is also independent of $c \in C$.

In the next two sections, we turn to a key problem of the theory: understanding the difference between the divisor-theoretic and the scheme-theoretic restriction of divisors, equivalently, the role of embedded points. The general theory is outlined in Section 2.6. Then in Section 2.7 we show that if all the coefficients of the boundary divisor are $> \frac{1}{2}$, then we need not worry about embedded points in moduli questions.

Checking local stability is easier in codimension ≥ 3, we discuss this and its relation to Grothendieck–Lefschetz-type theorems in Section 2.8.

From now on we use many definitions and results about log canonical and semi-log canonical pairs as in Kollár (2013b). The most important ones are summarized in Section 11.1.

Assumptions The basic definitions in Section 2.1 are formulated for schemes. In the rest of Sections 2.1–2.5 and 2.7, we work in characteristic 0, unless a more general set-up is specified.

In Section 2.6 we work with arbitrary Noetherian schemes.

2.1 Locally Stable Families

Following the pattern established in Section 1.4, we expect that the definition of a stable family $f : (X, \Delta) \to S$ consists of some local conditions describing the singularities of f, and a global condition, that $K_{X/S} + \Delta$ be f-ample. We are now ready to formulate the correct local condition, at least for one-parameter families.

Note on $\mathbb{R}$-divisors From now on, we state definitions and results for $\mathbb{R}$-divisors, which seems the natural level of generality; see Section 11.4 for a detailed treatment. However, there will be no major differences in the proofs between $\mathbb{Q}$- and $\mathbb{R}$-divisors until Chapter 6.

We already defined stable varieties in (1.41). The basic objects of our moduli theory are their generalizations.

Definition 2.1 (Stable and locally stable pairs) A *locally stable pair* (X, Δ) over a field k consists of a pure dimensional, geometrically reduced k-scheme X and an effective $\mathbb{R}$-divisor Δ such that (X, Δ) has semi-log-canonical (abbreviated as slc) singularities (11.37).

(X, Δ) is a *stable pair* if, in addition, X is proper and $K_X + \Delta$ is an ample $\mathbb{R}$-Cartier divisor (11.51). Thus a locally stable pair is the same as an slc pair; we usually use the former terminology for fibers of families.

If $\Delta = 0$, we have a *stable variety* as in (1.41).

Definition 2.2 Let C be a regular one-dimensional scheme. A *family of varieties* over C is a flat morphism of finite type $f : X \to C$, whose fibers are pure dimensional and geometrically reduced. We also call this a *one-parameter family*. For $c \in C$, let $X_c := f^{-1}(c)$ denote the fiber of f over c.

A *family of pairs* over C is a family of varieties $f\colon X \to C$ plus an effective Mumford $\mathbb{R}$-divisor Δ (p.xv) on X, That is, for every $c \in C$, the support of Δ does not contain any irreducible component of X_c and none of the irreducible components of $X_c \cap \operatorname{Supp}\Delta$ is contained in $\operatorname{Sing}X_c$; see (4.16.4) and Section 4.8 for details. This condition holds if the fibers are slc pairs. It turns out to be technically crucial, so it is much easier to assume it from the beginning.

The assumptions imply that X is regular at the generic points of $X_c \cap \operatorname{Supp}\Delta$. We can thus define Δ_c as the closure of the restriction of Δ to $X_c \setminus \operatorname{Sing}X_c$.

Warning For non-Cartier divisors, the divisor-theoretic restriction is a divisor, but the scheme-theoretic restriction $\Delta \cap X_c$ may have extra embedded points. This becomes quite important starting from Section 2.6.

Our main interest is in families with demi-normal (11.36) fibers, but we also want to understand to what extent this follows from other assumptions. However, we do not wish to get bogged down in technicalities, so we almost always assume the following conditions, both of which hold if the fibers are demi-normal.

(2.2.1) X satisfies Serre's condition S_2. Since the fibers are assumed reduced, X is S_2 iff the generic fiber X_g is S_2.

(2.2.2) The canonical sheaf ω_{X_c} of the fiber X_c is locally free at codimension 1 points for every $c \in C$. Equivalently, the relative canonical sheaf $\omega_{X/C}$ (2.5) is locally free at codimension 1 points of X_c. Thus the relative canonical class exists; we denote it by $K_{X/C}$ (2.5).

We can now define local stability for one-parameter families in characteristic 0. (We define stable families in (2.46).)

Definition–Theorem 2.3 Let C be a one-dimensional, regular scheme over a field of characteristic 0 and $f\colon (X, \Delta) \to C$ a family of pairs satisfying (2.2.1–2). We say that f is *locally stable* or *semi-log-canonical* at a point $p \in X_c$, if the following equivalent conditions hold:

(2.3.1) $K_{X/C} + \Delta$ is $\mathbb{R}$-Cartier at p and (X_c, Δ_c) is slc at p.

(2.3.2) $K_{X/C} + \Delta$ is $\mathbb{R}$-Cartier at p and $(\bar{X}_c, \operatorname{Diff}_{\bar{X}_c}(\Delta))$ (11.14) is log canonical at $\pi^{-1}(p)$, where $\pi\colon \bar{X}_c \to X_c$ denotes the normalization.

(2.3.3) $(X, X_c + \Delta)$ is slc at p.

(2.3.4) There is an open neighborhood $p \in X^\circ \subset X$ such that $(X, X_{f(q)} + \Delta)$ is slc at q for every $q \in X^\circ$.

Proof If (2) holds, then inversion of adjunction (11.17) shows that $(X, X_c + \Delta)$ is slc at p. The converse also holds since (11.17) works both ways. Thus (2) $\Leftrightarrow$ (3) and Kollár (2013b, 4.10) shows that (3) $\Leftrightarrow$ (4).

Since X_c is a Cartier divisor in X, the restriction Δ_c equals the different $\mathrm{Diff}_{X_c}(\Delta)$ by (11.15). Furthermore, by (11.14.5)

$$K_{\bar{X}_c} + \mathrm{Diff}_{\bar{X}_c}(\Delta) = \pi^*(K_{X_c} + \mathrm{Diff}_{X_c}(\Delta)).$$

Thus (11.37) shows that (1) $\Rightarrow$ (2). Note that (11.37) is an equivalence, but in order to apply it we need to know that X_c is demi-normal.

By assumption, X_c is geometrically reduced. A local computation shows that X_c is either smooth or has nodes at codimension 1 points; see Kollár (2013b, 2.33). Thus it remains to prove that X_c is S_2.

This is actually quite subtle. We outline three different approaches, all of which provide valuable insight.

First, if the generic fiber is klt, then, by (2.15), (X, Δ) is klt. Thus X is CM (10.4) by (11.18), so is every fiber X_c. In general, however, (X, Δ) is not klt and X is not CM. However, CM is much more than we need.

We should look carefully at weaker versions of CM that still imply that the fibers are S_2. Since the X_c are Cartier divisors in X, it would be enough to prove that X is S_3. However, as noted in Kollár (2013b, 3.6), X is not S_3 in general. Fortunately this is not a problem for us. If $g \in C$ is the generic point, then a local ring of X_g is also a local ring of X, hence X_g is S_2 if X is S_2. Therefore, (X_g, Δ_g) is slc. If $c \in C$ is a closed point and $p \in X_c$ has codimension ≥ 2, then $p \in X$ has codimension ≥ 3, thus $\mathrm{depth}_p \mathscr{O}_X \geq 3$ by (11.21), hence $\mathrm{depth}_p \mathscr{O}_{X_c} \geq 2$. Thus again X_c is S_2.

Third, we know that X_c is a Cartier divisor on a demi-normal scheme. A local version of the Enriques–Severi–Zariski lemma (2.93) implies that if $p \in X_c$ is a point of codimension ≥ 2, then $\hat{X}_{c,p} \setminus \{p\}$ is connected, where $\hat{X}_{c,p}$ denotes the completion of X_c at p.

Furthermore, X_c is the union of log canonical centers of $(X, X_c + \Delta)$. Therefore, X_c is seminormal by (11.12.2). These two observations together imply that X_c is S_2, hence demi-normal. $\square$

Comment 2.3.5 For proofs, the versions (2.3.3–4) are the most useful, but it is not clear how they could be generalized to families over higher dimensional bases. By contrast, the variants (2.3.1–2) are harder to use directly, but they make sense in general. This observation leads to the general definition of our moduli functor in Chapters 6–8.

2.4 (Positive characteristic) For arbitrary regular, one-dimensional schemes C, the conditions (2.3.1–4) are equivalent if the relative dimension of X/C is

1, and are expected to be equivalent if the relative dimension of X/C is 2. However, the examples of Kollár (2022) show that they are **not** equivalent if the relative dimension of X/C is ≥ 3. We discuss this in Section 8.8.

Here we adopt (2.3.4) as the definition of local stability in positive and mixed characteristics. This is dictated by the proof of (2.51), but few of the arguments work in full generality; see (2.15), (2.50), and (2.55).

2.5 (The relative canonical or dualizing sheaf I) Let C be a regular scheme of dimension 1 and $f: X \to C$ a flat morphism of finite type. Then the relative canonical or dualizing sheaf $\omega_{X/C}$ exists; see (2.68) or (11.2) for discussions.

If C is a smooth curve over a field, then $\omega_{X/C} = \omega_X \otimes f^* \omega_C^{-1}$.

If each ω_{X_c} is locally free in codimension 1 (for example, the fibers are normal or demi-normal) then $\omega_{X/C}$ is also locally free in codimension 1 and determines the relative canonical class $K_{X/C}$.

By (11.13), for $c \in C$ there is a Poincaré residue (or adjunction) map

$$\mathcal{R} : \omega_{X/C}|_{X_c} \to \omega_{X_c}. \tag{2.5.1}$$

The map exists for any flat morphism $f: X \to C$. General duality theory implies that it is an isomorphism if the fibers are CM, see (2.68.2). It is, however, not an isomorphism in general, but we prove in (2.67) that, for locally stable morphisms, the adjunction map is an isomorphism. Thus $\omega_{X/C}$ can be thought of as a flat family of the canonical sheaves of the fibers.

The isomorphism in (2.5.1) is easy to prove if the fibers are dlt, or if $K_{X/C}$ is $\mathbb{Q}$-Cartier (2.79.2). For the general case, see Section 2.5.

It is also worth noting that the reflexive powers (3.25) of the residue map

$$\mathcal{R}^m : \omega_{X/C}^{[m]}|_{X_c} \to \omega_{X_c}^{[m]} \tag{2.5.2}$$

are isomorphisms for locally stable maps if $\Delta = 0$, but not in general; see (2.79.2) and (2.44).

In (2.3.1) we make a fiber-wise assumption, that (X_c, Δ_c) be slc, and a total space assumption, that $K_{X/C} + \Delta$ be $\mathbb{R}$-Cartier. As in Section 1.4, usually (2.3.1) cannot be reformulated as a condition about the fibers of f only.

However, if ω_{X_c} is locally free then (2.5.1) implies that $\omega_{X/C}$ is also locally free along X_c. Thus (2.67) and (2.3) imply the following.

Lemma 2.6 *Let C be a smooth curve over a field of characteristic 0 and $f: X \to C$ a flat morphism of finite type such that X_c is slc and ω_{X_c} is locally free for some $c \in C$. Then $\omega_{X/C}$ is locally free along X_c and f is locally stable near X_c.* □

Note that (2.6) is a special property of slc varieties. Analogous claims fail both for normal varieties (2.45) and for pairs (X, D). To see the latter, consider a flat family X_c of smooth quadrics in $\mathbb{P}^3$ becoming a quadric cone for $c = 0$. Let $D_c \subset X_c$ be two disjoint lines that degenerate to a pair of distinct lines on X_0. Then K_{X_c} and D_c are both Cartier divisors for every c, but on the total space X they give a divisor $K_X + D$ that is not even $\mathbb{Q}$-Cartier.

In Section 1.4, we saw families of surfaces with quotient singularities where $K_{X/C}$ is not $\mathbb{R}$-Cartier, but the situation gets better in dimension ≥ 3.

Theorem 2.7 (Kollár, 2013a, Thm.18) *Let C be a smooth curve over a field of characteristic 0 and $f : (X, \Delta) \to C$ a family of pairs over C satisfying (2.2.1–2). Let $c \in C$ be a closed point and $Z_c \subset X_c$ a closed subset of codimension ≥ 3. Assume that*

(2.7.1) f is locally stable along $X_c \setminus Z_c$, and

(2.7.2) $(\bar{X}_c, \mathrm{Diff}_{\bar{X}_c}(\Delta))$ (11.14) is log canonical.

Then f is locally stable along X_c.

Note that $\mathrm{Diff}_{\bar{X}_c}(\Delta)$ is the closure of $\mathrm{Diff}_{\bar{X}_c \setminus \bar{Z}_c}(\Delta)$, which is defined by (2.7.1). We prove this in Section 2.8; see (5.6) for higher dimensonal base spaces.

If X_c is canonical, then K_{X_c} is Cartier in codimension 2. We can thus use (2.6) in codimension 2 and then (2.7) to obtain the next result.

Corollary 2.8 (Families with canonical fibers) *Let C be a smooth curve over a field of characteristic 0 and $f : X \to C$ a flat morphism of finite type such that X_c has canonical singularities for some $c \in C$. Then K_X is $\mathbb{Q}$-Cartier along X_c and f is locally stable near X_c.* □

Next we study permanence properties of local stability. We start with the invariance of local stability for morphisms that are quasi-étale, that is, étale outside a subset of codimension ≥ 2.

Lemma 2.9 *Let C be a smooth curve over a field of characteristic 0 and $f : (X, \Delta) \to C$ a family of pairs over C satisfying (2.2.1). Let $\pi : Y \to X$ be quasi-étale, where Y is S_2. If f is locally stable then so is $f \circ \pi$. The converse also holds if π is surjective.*

Proof This follows directly from (2.3) and (11.23.3). □

Note that $\pi_c : Y_c \to X_c$ need not be quasi-étale, but codimension 1 ramification can occur only along the singular locus of X_c. A typical example is given by $\mathbb{A}^2_{xy} \xrightarrow{\pi} \mathbb{A}^2/\frac{1}{n}(1,-1) \xrightarrow{\tau} \mathbb{A}^1$, where $\pi \circ \tau(x,y) = xy$.

Next we consider base changes $C' \to C$.

Proposition 2.10 *Let C be a smooth curve over a field of characteristic 0 and $g: C' \to C$ a quasi-finite morphism. If $f: (X,\Delta) \to C$ is locally stable, then so is the pull-back*

$$g^*f: (X',\Delta') := (X \times_C C', \Delta \times_C C') \to C'.$$

Proof We may assume that $g: (c',C') \to (c,C)$ is a finite, local morphism, étale away from c'. Set $D := X_c$ and $D' := X'_{c'}$. By (11.23.5), $(X, D + \Delta)$ is lc iff $(X', D' + \Delta')$ is. The rest follows from (2.3).   □

The following is useful for dimension induction.

Lemma 2.11 *Let C be a smooth curve over a field of char 0 and $f: (X, D + \Delta) \to C$ a locally stable morphism, where D is a $\mathbb{Z}$-divisor with normalization $n: \bar{D} \to D$. Then $f \circ n: (\bar{D}, \mathrm{Diff}_{\bar{D}} \Delta) \to C$ is also locally stable.*

Proof For any $c \in C$, the fiber X_c is a Cartier divisor, thus

$$\mathrm{Diff}_{\bar{D}}(\Delta + X_c) = (\mathrm{Diff}_{\bar{D}} \Delta) + X_c|_{\bar{D}} = (\mathrm{Diff}_{\bar{D}} \Delta) + \bar{D}_c.$$

Together with adjunction (11.17), this shows that $f_D: (\bar{D}, \mathrm{Diff}_{\bar{D}} \Delta) \to C$ is locally stable.   □

Complement 2.11.1 Since $K_{\bar{D}} + \mathrm{Diff}_{\bar{D}} \Delta \sim_{\mathbb{Q}} n^*(K_X + D + \Delta)$ and $\bar{D} \to D$ is finite, if $K_X + D + \Delta$ is f-ample, then $K_{\bar{D}} + \mathrm{Diff}_{\bar{D}}$ is $f \circ n$-ample. Thus if f is stable (2.46), then so is $f \circ n$.

The following result shows that one can usually reduce questions about locally stable families to the special case when X is normal; see also (2.54).

Proposition 2.12 *Let C be a smooth curve over a field of characteristic 0 and $f: (X, \Delta) \to C$ a family of pairs over C. Assume that X is demi-normal and let $\pi: \bar{X} \to X$ denote the normalization with conductor $\bar{D} \subset \bar{X}$ (11.36).*

(2.12.1) *If $f: (X, \Delta) \to C$ is locally stable, then so is $f \circ \pi: (\bar{X}, \bar{D} + \bar{\Delta}) \to C$.*

(2.12.2) *If $K_X + \Delta$ is $\mathbb{R}$-Cartier and $f \circ \pi: (\bar{X}, \bar{D} + \bar{\Delta}) \to C$ is locally stable, then so is $f: (X, \Delta) \to C$.*

Proof Fix a closed point $c \in C$. By (11.38) or Kollár (2013b, 5.38), if $K_X + \Delta$ is $\mathbb{R}$-Cartier, then $(X, X_c + \Delta)$ is slc iff $(\bar{X}, \bar{X}_c + \bar{D} + \bar{\Delta})$ is lc.   □

The next result allows us to pass to hyperplane sections. This is quite useful in proofs that use induction on the dimension. (As with many Bertini-type theorems, the characteristic 0 assumption is essential.)

Proposition 2.13 (Bertini theorem for local stability) *Let C be a smooth curve over a field of* char *0 and* $f: (X, \Delta) \to C$ *a locally stable morphism. Fix a point* $c \in C$ *and let H be a general divisor in a basepoint-free linear system on X. Then there is an open* $c \in C^\circ \subset C$ *such that the following morphisms are also locally stable over* C°:

(2.13.1) $f: (X, H + \Delta) \to C$,

(2.13.2) $f|_H: (H, \Delta|_H) \to C$, *and*

(2.13.3) *the composite* $f \circ \pi: (Y, \pi^{-1}(\Delta)) \to C$ *where* $\pi: Y \to X$ *is a* μ_m*-cover ramified along H; see (11.24).*

Proof As we noted in (2.12), we can assume that X is normal. Let $p: Y \to X$ be a log resolution (p.xvi) of (X, Δ) such that $p^{-1}(\operatorname{Supp}\Delta) + \operatorname{Ex}(p) + Y_c$ is an snc divisor. Pick H such that $p^{-1}(H) = p_*^{-1}(H)$ and

$$p^{-1}(H) + p^{-1}(\operatorname{Supp}\Delta) + \operatorname{Ex}(p) + Y_c$$

is an snc divisor. Then every exceptional divisor of p has the same discrepancy with respect to $(X, X_c + \Delta)$ and $(X, X_c + H + \Delta)$. Therefore, $(X, X_c + H + \Delta)$ is slc near X_c. Thus $f: (X, H + \Delta) \to C$ is locally stable over some $C^\circ \subset C$, proving (1). By adjunction, this implies that $(H, H_{c'} + \Delta|_H)$ is slc for every $c' \in C^\circ$, proving (2). By (11.23),

$$(Y, Y_{c'} + \pi^{-1}(\Delta)) \quad \text{is slc} \quad \Leftrightarrow \quad (X, X_{c'} + (1 - \tfrac{1}{m})H + \Delta) \quad \text{is slc}.$$

The latter holds since even $(X, X_{c'} + H + \Delta)$ is slc for every $c' \in C^\circ$. $\square$

2.14 (Inverse Bertini theorem, weak form) Let $H \subset X$ be any Cartier divisor. If $f|_H: (H, \Delta|_H) \to C$ is locally stable, then $f: (X, H + \Delta) \to S$, and hence also $f: (X, \Delta) \to S$, are locally stable in a neighborhood of H by (11.17). Stronger results are in (2.7) and (5.7).

The following simple result shows that if $f: (X, \Delta) \to C$ is locally stable, then (X, Δ) behaves as if it were *canonical*, as far as divisors over closed fibers are concerned. In some situations, for instance in (2.50), this is a very useful observation, but at other times the technical problems caused by log canonical centers in the generic fiber are hard to overcome.

Proposition 2.15 *Let* $f: (X, \Delta) \to C$ *be a locally stable morphism. Let E be a divisor over X (p.xv) such that* center$_X E \subset X_c$ *for some closed point* $c \in C$.

Then $a(E, X, \Delta) \geq 0$. Therefore, every log center (11.11) of (X, Δ) dominates C. In particular, if the generic fiber is klt (resp. canonical) then (X, Δ) is also klt (resp. canonical).

Proof Since $(X, X_c + \Delta)$ is slc, $a(E, X, X_c + \Delta) \geq -1$. Let $\pi\colon Y \to X$ be a proper birational morphism such that E is a divisor on Y and let b_E denote the coefficient of E in $\pi^*(X_c)$. Then b_E is an integer and it is positive since $\mathrm{center}_X E \subset X_c$. Thus,

$$a(E, X, \Delta) = a(E, X, X_c + \Delta) + b_E \geq -1 + b_E \geq 0.$$

In particular, none of the log centers of (X, Δ) are contained in X_c. □

2.16 (Some results in positive characteristic) As we already noted, very few of the previous theorems are known in positive characteristic, but the following partial results are sometimes helpful.

(2.16.1) Let (X, Δ) be a pair and $g\colon Y \to X$ a smooth morphism. By Kollár (2013b, 2.14.2), if (X, Δ) is slc, lc, klt, ... then so is $(Y, g^*\Delta)$.

(2.16.2) As a special case of Kollár (2013b, 2.14.4) we see that if (X, Δ) is slc then, for every smooth curve C, the trivial family $(X, \Delta) \times C \to C$ is locally stable.

(2.16.3) The proof of (2.15) works in any characteristic. Applying this to a trivial family will have useful consequences in (8.64).

(2.16.4) Let (X_i, Δ_i) be two pairs that are slc, lc, klt, ... Then their product $(X_1 \times X_2, X_1 \times \Delta_2 + \Delta_1 \times X_2)$ is also slc, lc, klt, ... This is a generalization of (2.16.2) and can be proved by the same method as in Kollár (2013b, 2.14.2), using Kollár (2013b, 2.22).

(2.16.5) Assume that $f\colon (X, \Delta) \to C$ is locally stable and let $g\colon C' \to C$ be a tamely ramified morphism. Then $g^*f\colon (X \times_C C', \Delta \times_C C') \to C'$ is also locally stable. This follows from (11.23.3) as in (2.10); see Kollár (2013b, 2.42) for details.

(2.16.6) Neither the wildly ramified nor the inseparable case of (2.16.5) is known. By Hu and Zong (2020), the inseparable case would imply the wildly ramified one. The case when all fibers are snc divisors is treated in (2.55).

The dualizing sheaf plays a very special role in algebraic geometry, thus it is natural to focus on understanding the powers of the relative dualizing sheaf. The next result, closely related to Lee and Nakayama (2018, 7.18), says that the relative dualizing sheaf is the "best" deformation of the dualizing sheaf.

Proposition 2.17 *Let C be a smooth curve over a field of characteristic 0 and $f\colon X \to C$ a flat morphism of finite type such that X_c is slc for some $c \in C$.*

Let L be a rank 1, reflexive sheaf on X such that a reflexive power $L^{[n]}$ (3.25) is locally free for some $n > 0$ and $L|_{X_c \setminus Z} \simeq \omega_{X_c \setminus Z}$ for some closed subset $Z \subset X_c$ of codimension ≥ 2.

Then there is a line bundle M such that $L \simeq \omega_{X/C} \otimes M$, near X_c.

Proof We may assume that X is local, hence $L^{[n]}$ is free. By (11.24) we can take a cyclic cover $\pi\colon Y \to X$, giving direct sum decompositions into μ_n-eigensheaves $\pi_* \mathcal{O}_Y = \bigoplus_{i=0}^{n-1} L^{[-i]}$ and

$$\pi_* \omega_{Y/C} \simeq \mathcal{H}om_X(\pi_* \mathcal{O}_Y, \omega_{X/C}) = \bigoplus_{i=0}^{n-1} L^{[i]} [\otimes] \omega_{X/C},$$

where $[\otimes]$ is the reflexive tensor product (3.25.1).

The resulting $g\colon Y \to C$ is locally stable by (2.9) and ω_{Y_c} is locally free. Therefore, $\omega_{Y/C}$ is locally free by (2.6), hence free since Y is semilocal. Thus $\pi_* \omega_{Y/C} \simeq \pi_* \mathcal{O}_Y$, so one of the summands $L^{[i]} [\otimes] \omega_{X/C}$ is free. Restriction to X_c tells us that in fact $i = n - 1$. Next note that

$$\omega_{X/C} \simeq \omega_{X/C} [\otimes] L^{[n-1]} [\otimes] L [\otimes] L^{[-n]} \simeq \mathcal{O}_X \otimes L \otimes \mathcal{O}_X \simeq L,$$

where at the end we changed to the usual tensor product, since the tensor product of a reflexive sheaf and of a line bundle is reflexive. □

2.2 Locally Stable Families of Surfaces

In this section, we develop a rather complete local picture of slc families of surfaces. That is, we start with a pointed, local slc pair $(x \in X_0, \Delta_0)$ and aim to describe all locally stable deformations over local schemes $0 \in S$

$$
\begin{array}{ccc}
(X_0, \Delta_0) & \hookrightarrow & (X_S, \Delta_S) \\
\downarrow & & \downarrow \\
0 & \in & S.
\end{array}
$$

In the study of singularities it is natural to work étale-locally.

Definition 2.18 Following Stacks (2022, tag 02LD), an étale morphism $\pi\colon (s', S') \to (s, S)$ is called *elementary étale* if the induced map on the residue fields $\pi^*\colon k(s) \to k(s')$ is an isomorphism. (This notion is also called strictly étale or strongly étale in the literature.) The inverse limit of all elementary étale morphisms is the *Henselisation* of (s, S), denoted by (s^h, S^h).

The inverse limit of all étale morphisms is the *strict Henselisation* of (s, S), denoted by $(s^{\mathrm{sh}}, S^{\mathrm{sh}})$. See Stacks (2022, tag 0BSK) for details.

For deformation purposes, two pointed schemes $(x_1 \in X_1)$ and $(x_2 \in X_2)$ are considered the "same" if they have isomorphic Henselisations. Equivalently, there is a third pointed scheme $(x_3 \in X_3)$ and elementary étale morphisms

$$(x_1 \in X_1) \xleftarrow{\pi_1} (x_3 \in X_3) \xrightarrow{\pi_2} (x_2 \in X_2).$$

Since we have not yet defined the notion of a locally stable family in general, we concentrate on the case when S is the spectrum of a DVR.

We start by recalling the classification of lc surface singularities. This has a long history, starting with Du Val (1934). For simplicity, we work over an algebraically closed field. It turns out that lc surface singularities have a very clear description using their dual graphs and this is independent of the characteristic. (By contrast, the equations of the singularities depend on the characteristic.)

Definition 2.19 (Dual graph) Let $(0 \in S)$ be a normal surface singularity over an algebraically closed field and $f : S' \to S$ the minimal resolution with irreducible exceptional curves $\{C_i\}$. We associate to this a *dual graph* $\Gamma = \Gamma(0 \in S)$ whose vertices correspond to the C_i. We use the *negative* of the self-intersection number $(C_i \cdot C_i)$ to represent a vertex and connect two vertices C_i, C_j by r edges iff $(C_i \cdot C_j) = r$. In the lc cases, the C_i are almost always smooth rational curves and $(C_i \cdot C_j) \leq 1$, so we get a very transparent picture.

The *intersection matrix* of the resolution is $(-(C_i \cdot C_j))$. This matrix is positive definite (essentially by the Hodge index theorem). Its determinant is denoted by $\det(\Gamma) := \det(-(C_i \cdot C_j))$.

Let B be a curve on S and B_i the local analytic branches of B that pass through $0 \in S$. The *extended dual graph* (Γ, B) has an additional vertex for each B_i, represented by $\bullet$, and it is connected to C_j by r edges if $(f_*^{-1} B_i \cdot C_j) = r$.

Definition 2.20 A connected graph is a *twig* if all vertices have ≤ 2 edges. Thus such a graph is of the form

$$c_1 \text{———} c_2 \text{———} \cdots \text{———} c_n$$

Here $\det(\Gamma)$ is also the numerator of the continued fraction (6.70.4).

A connected graph is a *tree with one fork* if there is a vertex (the root) with three edges and all other vertices have ≤ 2 edges. Such a dual graph is of the form

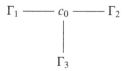

where each Γ_i is a twig joined to c_0 at an end vertex.

Next we list the dual graphs of all lc pairs $(0 \in S, B)$, starting with the terminal and canonical ones. For proofs see Alexeev (1993) or Kollár (2013b, Sec.3.3).

2.21 (List of log canonical surface singularities) Here $(0 \in S)$ is a normal surface singularity over an algebraically closed field and $B \subset S$ a curve (with coefficient 1).

CASE 2.21.1 (Terminal). $(0 \in S, B)$ is terminal iff $B = \emptyset$ and S is smooth.

CASE 2.21.2 (Canonical). $(0 \in S, B)$ is canonical iff either B and S are both smooth at 0, or $B = \emptyset$ and Γ is one of the following. The corresponding singularities are called *Du Val* singularities or *rational double points* or *simple surface singularities*. See Durfee (1979) for more information. The following equations are correct only in characteristic 0; see Artin (1977), in general.

A_n: $x^2 + y^2 + z^{n+1} = 0$, with $n \geq 1$ curves in the dual graph:

$$2 \text{———} 2 \text{———} \cdots \text{———} 2 \text{———} 2$$

D_n: $x^2 + y^2 z + z^{n-1} = 0$, with $n \geq 4$ curves in the dual graph:

E_n: with n curves in the dual graph:

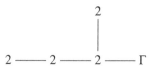

There are 3 possibilities:

E_6: $x^2 + y^3 + z^4 = 0$ and $\Gamma = 2 - 2$,

E_7: $x^2 + y^3 + yz^3 = 0$ and $\Gamma = 2 - 2 - 2$,

E_8: $x^2 + y^3 + z^5 = 0$ and $\Gamma = 2 - 2 - 2 - 2$.

CASE 2.21.3 (Purely log terminal) The names here reflect that, at least in characteristic 0, these singularities are obtained as the quotient of $\mathbb{C}^2$ by the indicated type of group. See Brieskorn (1967/1968) and (6.65).
Subcase 2.21.3.1 (Cyclic quotient) B is smooth at 0 (or empty) and (Γ, B) is

$$\bullet \overline{\quad} c_1 \overline{\quad} \cdots \overline{\quad} c_n \quad \text{or} \quad c_1 \overline{\quad} \cdots \overline{\quad} c_n$$

We discuss these in detail in (6.65–6.70).
Subcase 2.21.3.2 (Dihedral quotient)

Subcase 2.21.3.3 (Other quotient) The dual graph is a tree with one fork (2.20) with three possibilities for $(\det(\Gamma_1), \det(\Gamma_2), \det(\Gamma_3))$:

(Tetrahedral) (2,3,3)
(Octahedral) (2,3,4)
(Icosahedral) (2,3,5).

CASE 2.21.4 (Log canonical with $B = 0$)
Subcase 2.21.4.1 (Simple elliptic) There is a unique exceptional curve E; it is smooth and of genus 1. If the self-intersection $r := -(E^2)$ is ≥ 3 then the singularity is isomorphic to the cone over the elliptic normal curve $E \subset \mathbb{P}^{r-1}$ of degree r.
Subcase 2.21.4.2 (Cusp) The dual graph is a circle of smooth rational curves

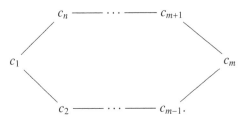

The cases $n = 1, 2$ are exceptional. For $n = 2$, we have two smooth rational curves meeting at two points, and for $n = 1$, the unique exceptional curve is a rational curve with a single node. We can draw the dual graphs as

$$c_1 \doubleequals c_2 \quad \text{and} \quad \bigcirc c_1.$$

For example the dual graphs of the three singularities $(z(xy - z^2) = x^4 + y^4)$, $(z^2 = x^2(x + y^2) + y^7)$, and $(z^2 = x^2(x^2 + y^2) + y^5)$ are

$$3 \doubleequals 4, \quad \bigcirc 1 \quad \text{and} \quad \bigcirc 2.$$

Subcase 2.21.4.3 ($\mathbb{Z}/2$-quotient of a cusp)

(For $n = 1$, it is a $\mathbb{Z}/2$-quotient of a simple elliptic singularity.)

Subcase 2.21.4.4 (Simple elliptic quotient) The dual graph is a tree with one fork (2.20) with three possibilities for $(\det(\Gamma_1), \det(\Gamma_2), \det(\Gamma_3))$:

($\mathbb{Z}/3$-quotient) (3,3,3)

($\mathbb{Z}/4$-quotient) (2,4,4)

($\mathbb{Z}/6$-quotient) (2,3,6).

CASE 2.21.5 (Log canonical with $B \neq 0$)

Subcase 2.21.5.1 (Cyclic) B has two smooth branches meeting transversally at 0 and (Γ, B) is

Subcase 2.21.5.2 (Dihedral)

2.22 (List of slc surface singularities) The dual graphs are very similar to the previous ones, but there are two possible changes due to the double curve of the surface S passing through the chosen point $0 \in S$.

In the normal case, the local picture represented by an edge is

$$(xy = 0) \subset \mathbb{A}^2, \quad \text{denoted by} \quad \circ - \circ \quad \text{or} \quad \bullet - \circ,$$

where $(y = 0)$ is an exceptional curve and $(x = 0)$ is either an exceptional curve or a component of B. We can now have a nonnormal variant

$$(xy = z = 0) \subset (xy = 0) \subset \mathbb{A}^3, \quad \text{denoted by} \quad \circ \overset{d}{-} \circ \quad \text{or} \quad \bullet \overset{d}{-} \circ,$$

where the d over the edge signifies that the two curves denoted by $\circ$ or $\bullet$ (here $(x = z = 0)$ and $(y = z = 0)$) meet at a point that is also on a double curve of the surface (here $(x = y = 0)$).

The local picture represented by • — ∘ also has another nonnormal variant where (as long as char ≠ 2) we create a pinch point by identifying the points $(0, y) \leftrightarrow (0, -y)$. The local equation is

$$(xy = z = 0) \subset (z^2 = xy^2) \subset \mathbb{A}^3, \quad \text{denoted by} \quad \mathbf{p} \text{ — } \circ,$$

where $(y = z = 0)$ is the double curve of the surface and $(x = z = 0)$ an exceptional curve.

CASE 2.22.1 (Semi-plt)
Subcase 2.22.1.1 (Higher pinch points) These are obtained from the cyclic dual graph of (2.21.3.1) by replacing • — ∘ by $\mathbf{p}$ — ∘.
   The simplest one is the pinch point, whose dual graph is $\mathbf{p}$ — 1. The equation of the pinch point is $(x^2 = zy^2)$; it is its own semi-resolution Kollár (2013b, sec.10.4).
   As another example, start with the A_n singularity $(xy = z^{n+1})$ and pinch it along the line $(x = z = 0)$. The dual graph is

$$\mathbf{p} \text{ — } 2 \text{ — } \cdots \text{ — } 2$$

with two occurring n-times. As a subring of $k[x, y, z]/(xy - z^{n+1})$, the coordinate ring is generated by (x, z, y^2, xy, yz), but $xy = z^{n+1}$. Thus $u_1 = x, u_2 = z, u_3 = y^2, u_4 = yz$ gives an embedding into $\mathbb{A}^4$. The image is a triple point whose equations can be written as

$$\text{rank} \begin{pmatrix} u_2^n & u_4 & u_3 \\ u_1 & u_2^2 & u_4 \end{pmatrix} \le 1.$$

Subcase 2.22.1.2 The dual graph is $\Gamma_1 \overset{\mathbf{d}}{\text{—}} \Gamma_2$, where the Γ_i are twigs such that $\det(\Gamma_1) = \det(\Gamma_2)$. Note that here we allow $\Gamma_i = \{1\}$ and $1 \overset{\mathbf{d}}{\text{—}} 1$ corresponds to $(xy = 0) \subset \mathbb{A}^3$. Similarly $2 \overset{\mathbf{d}}{\text{—}} 2$ corresponds to

$$(x_1y - z_1^2 = x_2 = z_2 = 0) \cup (x_2y - z_2^2 = x_1 = z_1 = 0) \subset \mathbb{A}^5.$$

Aside It is a good exercise to check that if $\det(\Gamma_1) \ne \det(\Gamma_2)$ then the canonical class of the resulting surface is not $\mathbb{Q}$-Cartier. The case $2 \overset{\mathbf{d}}{\text{—}} 1$ is easy to compute by hand. The key in general is to compute the different (11.14) on the double curve; see Kollár (2013b, 5.18) for details. This is one of the special cases of (11.38).

CASE 2.22.2 (Slc and $K_S + B$ Cartier)
Subcase 2.22.2.1 (Degenerate cusp) Here $B = 0$ and these are obtained from the dual graph of a cusp (2.21.4.2) by replacing some of the edges $\circ$ — $\circ$ with $\circ \overset{d}{\text{—}} \circ$.

The cases $n = 1, 2$ are again exceptional. For $n = 2$ we can replace either of the edges $\circ$ — $\circ$ with $\circ \overset{d}{\text{—}} \circ$. For example, $(z^2 = x^2 y^2)$ and $(z^2 = x^2 y^2 + y^5)$ correspond to the dual graphs

$$1 \overset{d}{\underset{d}{=\!\!=\!\!=}} 1 \quad \text{and} \quad 2 \overset{d}{=\!\!=\!\!=} 2.$$

For $n = 1$ the unique exceptional curve is a rational curve with a single node. We can think of the dual graph as

$$d \bigg(c_1.$$

For example the singularities $(z^2 = x^2(x + y^2))$ and $(z^2 = x^2(x^2 + y^2))$ give the dual graphs

$$d \bigg(1 \quad \text{and} \quad d \bigg(2.$$

Subcase 2.22.2.2 These are obtained from the cyclic dual graph of (2.21.5.1) by replacing some of the edges $\circ$ — $\circ$ with $\circ \overset{d}{\text{—}} \circ$.

CASE 2.22.3 (Slc and $2(K_S + B)$ Cartier)
Subcase 2.22.3.1 Here $B = 0$ and these are obtained from the dual graph of a $\mathbb{Z}/2$-quotient of a cusp (2.21.4.3) by replacing some of the horizontal edges $\circ$ — $\circ$ with $\circ \overset{d}{\text{—}} \circ$.
Subcase 2.22.3.2 These are obtained from the cyclic dual graph of (2.21.5.1) by replacing at least one of $\bullet$ — $\circ$ by $\mathbf{p}$ — $\circ$ and replacing some of the edges $\circ$ — $\circ$ with $\circ \overset{d}{\text{—}} \circ$.
Subcase 2.22.3.3 These are obtained from the dual graph of (2.21.5.2) by replacing $\bullet$ — $\circ$ by $\mathbf{p}$ — $\circ$ and replacing some of the horizontal edges $\circ$ — $\circ$ with $\circ \overset{d}{\text{—}} \circ$.

This completes the list of all slc surface singularities and now we turn to describing their locally stable deformations. An slc surface can be singular along a curve and the transversal hyperplane sections are nodes. Deformations of nodes are described in (11.35).

The situation is much more complicated for surfaces, so we start with the case $\Delta_0 = 0$. It would be natural to first try to understand all flat deformations of $(x \in X_0)$ and then decide which of these are locally stable. However, in many interesting cases, flat deformations are rather complicated, but a good description of all locally stable deformations can be obtained by relating them to locally stable deformations of certain cyclic covers of X (11.24).

Proposition 2.23 *Let k be a field and (X, D) a local, slc scheme over k with D reduced. Assume that $\omega_X^{[m]}(mD) \simeq \mathcal{O}_X$ for some $m \geq 1$ that is not divisible by* char k. *Let $\pi \colon (\tilde{X}, \tilde{D}) \to (X, D)$ be a corresponding μ_m-cover (11.24). Let R be a complete DVR with residue field k and set $S = \operatorname{Spec} R$.*

Taking μ_m-invariants establishes a bijection between the sets:

(2.23.1) *Flat, local, slc morphisms $\tilde{f} \colon (\tilde{X}_S, \tilde{D}_S) \to S$ such that $(\tilde{X}_0, \tilde{D}_0) \simeq (\tilde{X}, \tilde{D})$, plus a μ_m-action on $(\tilde{X}_S, \tilde{D}_S)$ extending the μ_m-action on $(\tilde{X}, \tilde{D})$.*

(2.23.2) *Flat, local, slc morphisms $f \colon (X_S, D_S) \to S$ such that $(X_0, D_0) \simeq (X, D)$.*

Note that $\omega_{\tilde{X}}(\tilde{D})$ is locally free, and, in many cases, this makes $(\tilde{X}, \tilde{D})$ much simpler than (X, D). This reduction step is especially useful when $D = 0$, in which case $\omega_{\tilde{X}}$ is locally free. As we saw in (2.6), then all flat deformations of $\tilde{X}$ are slc. For surfaces, this leads to an almost complete description of all slc deformations.

Aside 2.24 (Deformations of quotients) Let $\tilde{X}$ be a scheme and G a finite group acting on it. The proof of (2.23) shows that G-equivariant deformations of $\tilde{X}$ always induce flat deformations of $X := \tilde{X}/G$ provided the characteristic does not divide $|G|$.

The converse is, however, quite subtle, and usually deformations of X are not related to any deformation of $\tilde{X}$. As an example, consider the family $(xy - z^n - tz^m = 0)$ for $m < n$. For $t = 0$, the fiber is isomorphic to $\mathbb{C}^2/\mathbb{Z}_n$ and, for $t \neq 0$, the fiber has a singularity (analytically) isomorphic to $\mathbb{C}^2/\mathbb{Z}_m$. There is no relation between the corresponding degree n cover of the central fiber and the (local analytic) degree m cover of a general fiber. However, if G acts freely outside a subset of codimension ≥ 3 and $\tilde{X}$ is S_3, then every deformation of X arises from a deformation of $\tilde{X}$ (Kollár, 1995a, 12.7).

The following two examples show that the codimension ≥ 3 condition is not enough, not even for μ_m-covers.

(2.24.1) Let E be an elliptic curve and S a K3 surface with a fixed point free involution τ. Set $Y = E \times S$ and $X = Y/\sigma$ where σ is the involution $(-1, \tau)$. Note that $p \colon Y \to X$ is an étale double cover, $h^1(Y, \mathcal{O}_Y) = 1$ and $h^1(X, \mathcal{O}_X) = 0$. Let H_X be a smooth ample divisor on X and H_Y its pull-back to Y. Consider the cones (2.35) and general projections

$$C_a(Y, H_Y) \xrightarrow{\ pc\ } C_a(X, H_X)$$

$$\pi_Y \downarrow \qquad\qquad\qquad \downarrow \pi_X$$

$$\mathbb{A}^1 =\!=\!=\!=\!=\!=\!= \mathbb{A}^1.$$

Since $h^1(X, \mathscr{O}_X) = 0$, the central fiber of π_X is the cone over H_X by (2.36). By contrast, the central fiber F_0 of π_Y is not S_2 since $h^1(Y, \mathscr{O}_Y) \neq 0$, again by (2.36). Thus, although the normalization of F_0 is the cone over H_Y, it is not isomorphic to it.

(2.24.2) Let $g \colon X \to B$ be a smooth projective morphism to a smooth curve, H an ample line bundle on X and choose m large enough. Then the direct images $g_* \mathscr{O}_X(rmH)$ commute with base change for every $r \in \mathbb{N}$, hence the cones $C_a(X_b, \mathscr{O}_{X_b}(mH|_{X_b}))$ form a flat family.

The cones $C_a(X_b, \mathscr{O}_{X_b}(H|_{X_b}))$ are μ_m-covers of the cones $C_a(X_b, \mathscr{O}_{X_b}(mH|_{X_b}))$, but they form a flat family only if $g_* \mathscr{O}_X(rH)$ commutes with base change for every r. That is, we get the required examples whenever $H^0(X_b, \mathscr{O}_{X_b}(H|_{X_b}))$ jumps for special values of b. The latter is easy to arrange, even on a family of smooth curves, as long as $\deg H|_{X_b} < 2g - 2$.

Proof of 2.23 Let us start with $f \colon (X_S, D_S) \to S$. Since $\omega_{X_S}^{[m]}(mD_S)$ is locally free, the restriction map

$$\omega_{X_S}^{[m]}(mD_S) \twoheadrightarrow \omega_{X_0}^{[m]}(mD_0) \simeq \mathscr{O}_{X_0}$$

is surjective. Since X_S is affine, the constant 1 section lifts to a nowhere zero section $s \colon \mathscr{O}_{X_S} \simeq \omega_{X_S}^{[m]}(mD_S)$. Let $\tilde{f} \colon (\tilde{X}_S, \tilde{D}_S) \to S$ be the corresponding μ_m-cover (11.24).

The map $\tilde{f}$ is also locally stable by (2.9). By (2.3), this implies that $\tilde{X}_0$ is S_2, hence it agrees with the μ_m-cover of (X_0, D_0).

To see the converse, let $g \colon Y \to S$ be any flat, affine morphism and G a reductive group (or group scheme) acting on Y with quotient $g/G \colon Y/G \to S$. Then $(g/G)_* \mathscr{O}_{Y/G} = (g_* \mathscr{O}_Y)^G$ is a direct summand of $g_* \mathscr{O}_Y$, hence g/G is also flat. Taking invariants commutes with base change since G is reductive. This shows that (1) $\Rightarrow$ (2).   □

Assumptions For the rest of this section, we work in characteristic 0, though almost everything works in general as long as the characteristic does not divide m in (2.25), but very little has been proved otherwise.

2.25 (Classification plan) We establish an étale-local description of all slc deformations of surface singularities in four steps.

(2.25.1) Classify all slc surface singularities $(0, \tilde{S})$ with $\omega_{\tilde{S}}$ locally free.

(2.25.2) Classify all flat deformations of these $(0, \tilde{S})$.

(2.25.3) Classify all μ_m-actions on these surfaces and decide which ones correspond to our μ_m-covers.

(2.25.4) Describe the μ_m-actions on the deformation spaces of the $(0, \tilde{S})$.

The first task was already accomplished in (2.21–2.22); we have Du Val singularities (2.21.2), simple elliptic singularities and cusps (2.21.4.1–2), and degenerate cusps (2.22.1). We can thus proceed to the next step (2.25.2).

2.26 (Deformations of slc surface singularities with K_S Cartier)
2.26.1 (Du Val singularities) It is easy to work out the miniversal deformation space from the equations and (2.27). For each of the A_n, D_n, E_n cases the dimension of the miniversal deformation space is exactly n. For instance, for A_n we get (in char 0)

$$
\begin{array}{ccc}
(xy + z^{n+1} = 0) \hookrightarrow (xy + z^{n+1} + \sum_{i=0}^{n-1} t_i z^i = 0) \hookrightarrow \mathbb{A}^3_{xyz} \times \mathbb{A}^n_t \\
\downarrow \qquad\qquad\qquad\qquad \downarrow \qquad\qquad\qquad\qquad \downarrow \\
0 \qquad \in \qquad\qquad \mathbb{A}^n_t =\!=\!=\!=\!= \mathbb{A}^n_t.
\end{array}
$$

2.26.2 (*Elliptic/cusp/degenerate cusp*) Let $(0 \in S)$ be one of these singularities and C_i the exceptional curves of the minimal (semi)resolution. Set $m = -(\sum C_i)^2$ and write $(0 \in S_m)$ to indicate such a singularity.

If $m = 1, 2, 3$ then $(0 \in S_m)$ is (isomorphic to) a singular point on a surface in $\mathbb{A}^3$ by Saito (1974); Laufer (1977). Their deformations are completely described by (2.27).

If $m = 4$, then $(0 \in S_4)$ is (isomorphic to) a singular point on a surface in $\mathbb{A}^4$ that is a complete intersection of two hypersurfaces. The miniversal deformation space of a complete intersection can be described in a manner similar to (2.27); see Artin (1976); Looijenga (1984); or Hartshorne (2010).

If $m = 5$ then the deformations are fully described by the method of Buchsbaum and Eisenbud (1977); see Hartshorne (2010, sec.9).

If $m \geq 3$ and $(0 \in S_m)$ is simple elliptic, then it is (isomorphic to) the singular point of a projective cone $\bar{S}_m \subset \mathbb{P}^m$ over an elliptic normal curve $E_m \subset \mathbb{P}^{m-1}$. By Pinkham (1974, sec.9), every deformation of $(0 \in S_m)$ is the restriction of a deformation of $\bar{S}_m \subset \mathbb{P}^m$. In particular, any smoothing corresponds to a smooth surface of degree m in $\mathbb{P}^m$. The latter have been fully understood classically: these are the del Pezzo surfaces embedded by $|-K|$. In particular, a simple elliptic singularity $(0 \in S_m)$ is smoothable only for $m \leq 9$ Pinkham (1974, sec.9).

The $m = 9$ case is especially interesting. Given an elliptic curve E, a degree 9 embedding $E_9 \hookrightarrow \mathbb{P}^8$ is given by global sections of a line bundle L_9 of degree 9 on E. Embeddings of E into $\mathbb{P}^2$ are given by line bundles L_3 of degree 3. If we take $(E \hookrightarrow \mathbb{P}^2)$ given by L_3, and then embed $\mathbb{P}^2$ into $\mathbb{P}^9$ by $\mathcal{O}_{\mathbb{P}^2}(3)$, then E is mapped to E_9 iff $L_3^{\otimes 3} \simeq L_9$. For a fixed L_9 this gives nine choices of L_3. Thus a given $E_9 \hookrightarrow \mathbb{P}^8$ is a hyperplane section of a $\mathbb{P}^2 \hookrightarrow \mathbb{P}^9$ in nine different ways. Correspondingly, the deformation space $(0 \in S_9)$ has nine smoothing components. (This was overlooked in Pinkham (1974, sec.9).) The automorphism group of $(0 \in S_9)$ permutes these nine components. See Looijenga and Wahl (1986, sec.6) for another description.

For $m \geq 6$, the deformation theory of cusps is much harder, see Gross et al. (2015). Degenerate cusps are all smoothable; Stevens (1998).

2.27 (Deformations of hypersurface singularities) For general references, see Artin (1976); Looijenga (1984); Arnol'd et al. (1985); Hartshorne (2010).

Let $0 \in X \subset \mathbb{A}_{\mathbf{x}}^n$ be a hypersurface singularity defined by an equation $(f(\mathbf{x}) = 0)$. Choose polynomials p_i that give a basis of

$$k[[x_1, \ldots, x_n]]/\left(f, \frac{\partial f}{\partial x_1}, \ldots, \frac{\partial f}{\partial x_n}\right). \tag{2.27.1}$$

If $(0 \in X)$ is an isolated singularity, then the quotient has finite length, say N. In this case, the miniversal deformation of $(0 \in X)$ is given by

$$
\begin{array}{ccccc}
X & \hookrightarrow & (f(\mathbf{x}) + \sum_i t_i p_i(\mathbf{x}) = 0) & \hookrightarrow & \mathbb{A}_{\mathbf{x}}^n \times \mathbb{A}_{\mathbf{t}}^N \\
\downarrow & & \downarrow & & \downarrow \\
0 & \in & \mathbb{A}_{\mathbf{t}}^N & = = & \mathbb{A}_{\mathbf{t}}^N.
\end{array}
$$

In particular, the miniversal deformation space $\mathrm{Def}(X)$ is smooth.

If the quotient in (2.27.1) has infinite length, then it is best to think of the resulting infinite dimensional deformation space as an inverse system of deformations over Artinian rings whose embedding dimension goes to infinity.

The next step (2.25.3) in the classification is to describe all μ_m-actions, but it is more transparent to consider reductive commutative groups. These are of the form $G \times \mathbb{G}_m^r$ where G is a finite, commutative group and $\mathbb{G}_m = \mathrm{GL}(1)$ the multiplicative group of scalars, see Humphreys (1975, sec.16).

2.28 (Commutative groups acting on Du Val singularities) The action of a reductive commutative group on $\mathbb{A}^n$ can be diagonalized. Thus let $S \subset \mathbb{A}^3$ be a Du Val singularity that is invariant under a diagonal group action on $\mathbb{A}^3$. It is easy to work through any one of the standard classification methods (for instance, the one in Kollár and Mori (1998, 4.24)) to obtain the following normal forms. In each case we describe first the maximal connected group actions and then the maximal nonconnected group actions.

Main series: $\mathbb{G}_m$-actions.

A_n $(xy + z^{n+1} = 0)$ and $\mathbb{G}_m^2$ acts with character $(1, -1, 0), (0, n + 1, 1)$.
D_n $(x^2 + y^2 z + z^{n-1} = 0)$ and $\mathbb{G}_m$ acts with character $(n - 1, n - 2, 2)$.
E_6 $(x^2 + y^3 + z^4 = 0)$ and $\mathbb{G}_m$ acts with character $(6, 4, 3)$.
E_7 $(x^2 + y^3 + yz^3 = 0)$ and $\mathbb{G}_m$ acts with character $(9, 6, 4)$.
E_8 $(x^2 + y^3 + z^5 = 0)$ and $\mathbb{G}_m$ acts with character $(15, 10, 6)$.

Twisted versions: $\mu_r \times \mathbb{G}_m$-actions.

A_n $(x^2 + y^2 + z^{n+1} = 0)$. If $n+1$ is odd, then $\mathbb{G}_m$ acts with character $(n+1, n+1, 2)$ and μ_2 acts with character $(0, 1, 0)$. If $n + 1$ is even, then $\mathbb{G}_m$ acts with character $(\frac{n+1}{2}, \frac{n+1}{2}, 1)$ and μ_2 acts with character $(0, 1, 0)$.
D_n $(x^2 + y^2 z + z^{n-1} = 0)$, $\mathbb{G}_m$ acts with character $(n - 1, n - 2, 2)$ and μ_2 acts with character $(1, 1, 0)$.
D_4 $(x^2 + y^3 + z^3 = 0)$, $\mathbb{G}_m$ acts with character $(3, 2, 2)$ and μ_3 acts with character $(0, 1, 0)$.
E_6 $(x^2 + y^3 + z^4 = 0)$ and $\mathbb{G}_m$ acts with character $(6, 4, 3)$ and μ_2 acts with character $(1, 0, 0)$.

Example 2.29 (Locally stable deformations of surface quotient singularities)
Let $(0 \in S)$ be a surface quotient singularity with Du Val cover $(0 \in \tilde{S}) \to (0 \in S)$. By (2.23), the classification of locally stable deformations of all such $(0 \in S)$ is equivalent to classifying all cyclic group actions on Du Val singularities $(0 \in \tilde{S})$ that are free outside the origin and whose action on $\omega_{\tilde{S}} \otimes k(0)$ is faithful. This is straightforward, though somewhat tedious, using (2.28). Alternatively, one can use the classification of finite subgroups of $GL(2)$ as in Brieskorn (1967/1968).

Thus the miniversal locally stable deformation space, which we denote by $\mathrm{Def}_{KSB}(S)$ (6.64), is the fixed point set of the corresponding cyclic group action on $\mathrm{Def}(\tilde{S})$, hence it is also smooth.

A_n-**series** $(xy + z^{n+1} = 0)/\frac{1}{m}(1, (n+1)c - 1, c)$ for any m where $((n+1)c - 1, m) = 1$. These are equivariantly smoothable only if $m|(n + 1)c$.
D_n-**series** $(x^2 + y^2 z + z^{n-1} = 0)/\frac{1}{2k+1}(n - 1, n - 2, 2)$ where $(2k + 1, n - 2) = 1$. These are not equivariantly smoothable, but, for instance, if $2k + 1|n - 1$, they deform to the quotient singularity $\mathbb{A}^2/\frac{1}{2k+1}(-1, 2)$.

E_6-series $(x^2 + y^3 + z^4 = 0)/\frac{1}{m}(6, 4, 3)$ for $(m, 6) = 1$. For $m > 1$ all equivariant deformations are trivial, save for $m = 5$, when there is a one-parameter family $(x^2 + y^3 + z^4 + \lambda yz = 0)/\frac{1}{5}(1, 4, 3)$.

E_7-series $(x^2 + y^3 + yz^3 = 0)/\frac{1}{m}(9, 6, 4)$ for $(m, 6) = 1$. For $m > 1$ all equivariant deformations are trivial, save for $m = 5$ and $m = 7$, when there are one-parameter families $(x^2 + y^3 + yz^3 + \lambda xz = 0)/\frac{1}{5}(4, 1, 4)$ and $(x^2 + y^3 + yz^3 + \lambda z = 0)/\frac{1}{7}(2, 6, 4)$.

E_8-series $(x^2 + y^3 + z^5 = 0)/\frac{1}{m}(15, 10, 6)$ for $(m, 30) = 1$. For $m > 1$ all equivariant deformations are trivial, save for $m = 7$, when there is a one-parameter family $(x^2 + y^3 + z^5 + \lambda yz = 0)/\frac{1}{7}(1, 3, 6)$.

A_n-twisted $(x^2 + y^2 + z^{n+1} = 0)/\frac{1}{4m}(n+1, n+1+2m, 2)$ for any $(2m, n+1) = 1$. These are never equivariantly smoothable.

D_4-twisted $(x^2 + y^3 + z^3 = 0)/\frac{1}{18k+9}(9k + 6, 1, 6k + 4)$. All equivariant deformations are trivial.

Example 2.30 (Quotients of simple elliptic and cusp singularities) Let $(0 \in S)$ be a simple elliptic, cusp or degenerate cusp singularity with minimal resolution (or semi-resolution) $f \colon T \to S$ and exceptional curves $C = \sum C_i$. Then $\omega_T(C) \simeq f^*\omega_S$, which gives a canonical isomorphism $\omega_S \otimes k(0) \simeq H^0(C, \omega_C)$. Since C is either a smooth elliptic curve or a cycle of rational curves, $\mathrm{Aut}(C)$ is infinite, but a finite index subgroup acts trivially on $H^0(C, \omega_C)$.

For cusps and for most simple elliptic singularities this leaves only μ_2-actions. The corresponding quotients are listed in (2.21.4.3), see Simonetti (2022) for their deformations. When the elliptic curves have extra automorphisms, one can have μ_3, μ_4 and μ_6-actions as in (2.21.4.4).

The following is one of the simplest degenerate cusp quotients.

Example 2.31 (Deformations of the double pinch point) Let $(0 \in S)$ be the *double pinch point* singularity, defined by $(\bar{S} = \mathbb{A}^2, \bar{D} = (xy = 0), \tau = (-1, -1))$. Here ω_S is not locally free, but $\omega_S^{[2]}$ is,

$$S \simeq \tilde{S}/\tfrac{1}{2}(1, 1, 1), \quad \text{where} \quad \tilde{S} = (z^2 - x^2 y^2 = 0) \subset \mathbb{A}^3.$$

A local generator of $\omega_{\tilde{S}}$ is given by $z^{-1} dx \wedge dy$, which is anti-invariant. Thus ω_S has index 2 and $\tilde{S} \to S$ is the index 1 cover. Thus every locally stable deformation of S is obtained as the μ_2-quotient of an equivariant deformation of $\tilde{S}$. By (2.27) the miniversal deformation space is given by

$$(z^2 - x^2 y^2 + u_0 + u_1 xy + \textstyle\sum_{i \geq 1} v_i x^{2i} + \sum_{j \geq 1} w_j y^{2j} = 0)/\tfrac{1}{2}(1, 1, 1).$$

When $u_0 = u_1 = v_1 = w_1 = 0$, we get equimultiple deformations to μ_2-quotients of cusps.

The slc deformations of pairs (X, Δ) are more complicated, even if Δ is a $\mathbb{Z}$-divisor. One difficulty is that $\omega_S(D)$ is locally free for every pair

$$(S, D) := (\mathbb{A}^2, (xy = 0))/\tfrac{1}{n}(1, q)$$

since $\frac{dx}{x} \wedge \frac{dy}{y}$ is invariant. Thus we would need to describe the deformations of every such pair (S, D) by hand. The following is one of the simplest examples, and it already shows that the answer is likely to be subtle.

Example 2.32 (Deformations of $(\mathbb{A}^2, (xy = 0))/\tfrac{1}{n}(1, 1)$)   Flat deformations of the quotient singularity $H_n := \mathbb{A}^2/\tfrac{1}{n}(1, 1)$ are quite well understood; see Pinkham (1974). H_n can be realized as the affine cone over the rational normal curve $C_n \subset \mathbb{P}^n$ and all local deformations are induced by deformations of the projective cone $C_p(C_n) \subset \mathbb{P}^{n+1}$. If $n \neq 4$ then the deformation space is irreducible and the smooth surfaces in it are minimal ruled surfaces of degree n in $\mathbb{P}^{n+1}$. (For $n = 4$, there is another component, corresponding to the Veronese embedding $\mathbb{P}^2 \hookrightarrow \mathbb{P}^5$.)

Since $(xy)^{-1} dx \wedge dy$ is invariant under the group action, it descends to a 2-form on H_n with poles along the curve $D_n := (xy = 0)/\tfrac{1}{n}(1, 1)$. Thus $K_{H_n} + D_n \sim 0$ and the pair (H_n, D_n) is lc. Our aim is to understand which deformations of H_n extend to a deformation of the pair (H_n, D_n).

Claim 2.32.1 Fix $n \geq 7$ and let $\pi: X \to \mathbb{A}^1$ be a general smoothing of H_n. Then the divisor D_n cannot be extended to a divisor D_X such that $\pi: (X, D_X) \to \mathbb{A}^1$ is locally stable. However, there are special smoothings $\pi: X' \to \mathbb{A}^1$ for which such a divisor D'_X exists.

Proof For $m \in \mathbb{N}$, let $\mathbb{F}_m$ denote the ruled surface $\mathrm{Proj}_{\mathbb{P}^1}(\mathscr{O}_{\mathbb{P}^1} + \mathscr{O}_{\mathbb{P}^1}(-m))$. Let $E_m \subset \mathbb{F}_m$ denote the section with self intersection $-m$ and $F \subset \mathbb{F}_m$ denote a fiber. Note that $K_{\mathbb{F}_m} \sim -(2E_m + (m + 2)F)$.

For $a \geq 1$ set $A_{ma} := E + (m + a)F$. Then A_{ma} is very ample with self intersection $n := m + 2a$ and it embeds $\mathbb{F}_m$ into $\mathbb{P}^{n+1}$ as a surface of degree n. Denote the image by S_{ma}. A general hyperplane section of S_{ma} is a rational normal curve $C_n \subset \mathbb{P}^n$. Consider the affine cones $X_{ma} := C_a(S_{ma})$ and $H_n := C_a(C_n)$. We can choose coordinates such that

$$X_{ma} \subset \mathbb{A}^{n+2}_{x_1, \ldots, x_{n+2}} \quad \text{and} \quad H_n = (x_{n+2} = 0).$$

The last coordinate projection gives $\pi: X_{ma} \to \mathbb{A}^1$ which is a flat deformation (in fact a smoothing) of H_n. By Kollár (2013b, 3.14.5),

$$H^0(X_{ma}, \mathscr{O}_{X_{ma}}(-K_{X_{ma}})) = \sum_{i \in \mathbb{Z}} x_0^i \cdot H^0(S_{ma}, \mathscr{O}_{S_{ma}}(-K_{S_{ma}} + iA_{ma}))$$

$$= \sum_{i \in \mathbb{Z}} x_0^i \cdot H^0(S_{ma}, \mathscr{O}_{S_{ma}}((2 + i)E_m + (m + 2 + im + ia)F)).$$

The lowest degree terms in the sum depend on m and a. For $i < -2$, we get 0. For $i = -2$, we have

$$H^0(S_{ma}, \mathcal{O}_{S_{ma}}((2 - m - 2a)F)) = H^0(S_{ma}, \mathcal{O}_{S_{ma}}((2 - n)F)).$$

This is 0, unless $n = 2$, that is, when X is the quadric cone in $\mathbb{A}^3$. Then D_2 is a Cartier divisor H_2 and so every deformation of H_2 extends to a deformation of the pair (H_2, D_2). Thus assume next that $n \geq 3$.

For $i = -1$ we have the summand $H^0(S_{ma}, \mathcal{O}_{S_{ma}}(E_m + (2 - a)F))$. This is again 0 if $a \geq 3$, but for $a = 1$ we get a pencil $|E_m + F|$ (whose members are pairs of intersecting lines) and for $a = 2$ we get a unique member E_m (which is a smooth conic in $\mathbb{P}^{n+1}$). This shows the following.

Claim 2.32.2 For $a = 1, 2$ and any $m \geq 0$, the anticanonical class of the 3-fold X_{ma} contains a (possibly reducible) quadric cone $D \subset X_{ma}$ and $\pi\colon (X_{ma}, D) \to \mathbb{A}^1$ is locally stable.   □

For $a \geq 3$, we have to look at the next term $H^0(S_{ma}, \mathcal{O}_{S_{ma}}(2E_m + (m + 2)F))$ for a nonzero section. The corresponding linear system consists of reducible curves of the form $E_m + G_m$ where $G_m \in |E_m + (m + 2)F|$. These curves have 2 nodes and arithmetic genus 1. Let $B \subset X_{ma}$ denote the cone over any such curve. Then (X_{ma}, B) is log canonical, but $\pi\colon (X_{ma}, B) \to \mathbb{A}^1$ is not locally stable since the restriction of B to H_n consists of $n + 2$ lines through the vertex. Thus we have proved:

Claim 2.32.3 For $a \geq 3$ and $m \geq 0$, the anticanonical class of X_{ma} does not contain any divisor D for which $\pi\colon (X_{ma}, D) \to \mathbb{A}^1$ is locally stable.   □

Note finally that the surfaces S_{ma} with $n = m + 2a$ form an irreducible family. General points correspond to the largest possible value $a = \lfloor (n - 1)/2 \rfloor$. The surfaces with $a \leq 2$ correspond to a closed subset, which is a two-dimensional subspace of the versal deformation space of H_n.

2.3 Examples of Locally Stable Families

The aim of this section is to investigate, mostly through examples, fibers of locally stable morphisms. If (S, Δ) is slc then, for any smooth curve C, the projection $\pi\colon (S \times C, \Delta \times C) \to C$ is locally stable with fiber (S, Δ). Thus, in general we can only say that fibers of locally stable morphisms are exactly the slc pairs.

The question becomes, however, quite interesting, if we look at special fibers of locally stable morphisms whose general fibers are "nice," for instance smooth or canonical. The main point is thus to probe the difference between arbitrary slc pairs and those slc pairs that occur on locally stable degenerations of smooth varieties. We focus on two main questions.

Question 2.33 Let $f\colon X \to T$ be a locally stable morphism over a pointed curve $(0 \in T)$ such that X_t is smooth for $t \neq 0$.

(2.33.1) Is X_0 CM (10.4)?

(2.33.2) Are the irreducible components of X_0 CM?

(2.33.3) Is the normalization of X_0 CM?

Question 2.34 Let $f\colon (X, \Delta) \to T$ be a locally stable morphism over a pointed curve $(0 \in T)$ such that X_t is smooth and Δ_t is snc for $t \neq 0$.

(2.34.1) Do the supports of $\{\Delta_t : t \in T\}$ form a flat family of divisors?

(2.34.2) Are the sheaves $\mathscr{O}_{X_0}(mK_{X_0} + \lfloor m\Delta_0 \rfloor)$ CM?

(2.34.3) Do the sheaves $\{\mathscr{O}_{X_t}(mK_{X_t} + \lfloor m\Delta_t \rfloor) : t \in T\}$ form a flat family?

A normal surface is always CM, and the (local analytic) irreducible components of an slc surface are CM. The latter follows from the classification of slc surfaces given in Kollár (2013b, sec.2.2). Starting with dimension 3, there are lc singularities that are not CM. The simplest examples are cones over abelian varieties; see (2.35). On the other hand, canonical and log terminal singularities are CM and rational (p.xv) in characteristic 0 by (11.18).

Let us note next that the answer to (2.33.1) is positive, that is, X_0 is CM. Indeed, X is canonical by (2.15) and hence CM by (11.18). Therefore, X_0 is also CM. A more complete answer to (2.33.1), without assuming that X_t is smooth or canonical for $t \neq 0$, is given in (2.66).

For locally stable families of pairs, the boundary provides additional sheaves whose CM properties are important to understand; this motivates (2.34). Unlike for (2.33), the answers to all of these are negative already for surfaces. The first convincing examples were discovered by Hassett (2.41). As a consequence, we see that we cannot think of the deformations of (S, Δ) as a flat deformation of S and a flat deformation of Δ that are compatible in certain ways. In general it is imperative to view (S, Δ) as a single object. See, however, Section 2.7 for many cases where viewing (S, Δ) as a pair does work well.

Our examples will be either locally or globally cones and we need some basic information about them.

2.35 (Cones) Let X be a projective scheme with an ample line bundle L. The *affine cone* over X with conormal bundle L is

$$C_a(X, L) := \operatorname{Spec}_k \oplus_{m \geq 0} H^0(X, L^m).$$

Away from the vertex $v \in C_a(X, L)$, the cone is locally isomorphic to $X \times \mathbb{A}^1$, but the vertex is usually more complicated. If X is normal then so is $C_a(X, L)$ and its canonical class is Cartier (resp. $\mathbb{Q}$-Cartier) iff $\mathcal{O}_X(K_X) \sim L^m$ for some $m \in \mathbb{Z}$ (resp. $\mathcal{O}_X(rK_X) \sim L^m$ for some $r, m \in \mathbb{Z}$ with $r \neq 0$).

The following results are straightforward; see Kollár (2013b, sec.3.1).

(2.35.1) $H_v^{i+1}(C_a(X, L), \mathcal{O}_{C_a(X,L)}) \simeq \oplus_{m \in \mathbb{Z}} H^i(X, \mathcal{O}_X(L^m))$ for every i.

Over a field of char 0, assume that X has rational singularities.

(2.35.2) If $-K_X$ is ample then $C_a(X, L)$ is CM and has rational singularities. If $-K_X$ is nef (for instance, $K_X \equiv 0$), then $C_a(X, L)$ is CM $\Leftrightarrow H^i(X, \mathcal{O}_X) = 0$ for $0 < i < \dim X$, and $C_a(X, L)$ has rational singularities $\Leftrightarrow H^i(X, \mathcal{O}_X) = 0$ for $0 < i \leq \dim X$.

Next let (X, Δ) be a projective, slc pair and L an ample line bundle on X. Let $\Delta_{C_a(X,L)}$ denote the $\mathbb{R}$-divisor corresponding to Δ on $C_a(X, L)$. Assume that $K_X + \Delta \sim_{\mathbb{Q}} r \cdot L$ for some $r \in \mathbb{R}$. Then $(C_a(X, L), \Delta_{C_a(X,L)})$ is

(2.35.3) terminal iff $r < -1$ and (X, Δ) is terminal,

(2.35.4) canonical iff $r \leq -1$ and (X, Δ) is canonical,

(2.35.5) klt iff $r < 0$ (that is, $-(K_X + \Delta)$ is ample) and (X, Δ) is klt,

(2.35.6) dlt iff either $r < 0$ and (X, Δ) is dlt or $(X, \Delta) \simeq (\mathbb{P}^n, (\prod x_i = 0))$ and the cone is $(\mathbb{A}^{n+1}, (\prod x_i = 0))$.

(2.35.7) lc iff $r \leq 0$ (that is, $-(K_X + \Delta)$ is nef) and (X, Δ) is lc,

(2.35.8) slc iff $r \leq 0$ and X is slc.

Aside The failure of (2.35.2) in positive characteristic has significant consequences for the moduli problem; see Section 8.8.

2.36 (Deformation to cones II) Let $X \subset \mathbb{P}^n$ be a closed subscheme and $H \subset \mathbb{P}^n$ a hyperplane. Thinking of $\mathbb{P}^n \subset \mathbb{P}^{n+1}$ as the hyperplane at infinity, let $C_p(X) \subset \mathbb{P}^{n+1}$ be the projective cone over X with vertex v.

If $H^0(\mathbb{P}^n, \mathcal{O}_{\mathbb{P}^n}(r)) \to H^0(X, \mathcal{O}_X(r))$ is surjective for every r, then $C_p(X) \setminus X$ is the affine cone $C_a(X)$.

Let $|L|$ be the pencil of hyperplanes in $\mathbb{P}^{n+1}$ that contain $H \subset \mathbb{P}^n$. If $v \notin L_t$ then projection from v shows that $C_p(X) \cap L_t \simeq X$.

There is a unique $L_0 \in |L|$ such that $v \in L_0$. Then $C_p(X) \cap L_0$ is isomorphic to $C_p(X \cap H)$ away from v. If X is pure and $\dim(X \cap H) = \dim X - 1$, then the two are isomorphic iff $H^1(X, \mathcal{O}_X(r)) = 0$ for every r; see (2.35.1) or Kollár (2013b, 3.10).

If all these hold then blowing up H we get a flat morphism $\pi: B_H C_p(X) \to \mathbb{P}^1$. There is a unique fiber of π that is isomorphic to $C_p(X \cap H)$, all other fibers are isomorphic to X.

Example 2.37 (Counterexample to 2.33.2) Let $Q_0 \subset \mathbb{P}^4$ be the quadric cone $(xy - uv = 0)$. Let $|A|$ and $|B|$ be the two families of planes on Q_0 and $H \sim A + B$ the hyperplane class. Let $S_1 \in |2A + H|$ be a general member. Note that S_1 is smooth away from the vertex of Q_0 and at the vertex it has two local analytic components intersecting at a single point. In particular, S_1 is nonnormal and non-CM. (The easiest way to see these is to blow up a plane $B_1 \in |B|$. Then $B_{B_1} Q_0 \to Q_0$ is a small resolution whose exceptional set E is a smooth rational curve. The birational transform of $|2A + H|$ is a very ample linear system whose general member is a smooth surface that intersects E in two points. This is the normalization of the surface S_1.)

Let B_1, B_2 be planes in the other family. Then $X_0 := S_1 + B_1 + B_2 \sim 3H$, thus X_0 is a $(2) \cap (3)$ complete intersection in $\mathbb{P}^4$. We can thus write X_0 as the limit of a smooth family of $(2) \cap (3)$ complete intersections X_t. The general X_t is a smooth K3 surface.

On the other hand, X_0 can also be viewed as a general member of a flat family whose special fiber is $A_1 + A_2 + B_1 + B_2 + H$. The latter is slc by (2.35), thus X_0 is also slc. Hence $\{X_t : t \in T\}$ is a locally stable family such that X_t is a smooth K3 surface for $t \neq 0$. Moreover, the irreducible component $S_1 \subset X_0$ is not CM.

In this case, the source of the problem is easy to explain. At its singular point, S_1 is analytically reducible. The local analytic branches of S_1 and the normalization of S_1 are both smooth.

One can, however, modify this example to get analytically irreducible non-CM examples, albeit in dimension 3. To see this, let

$$Y_0 := C(X_0) = C(S_1) + C(B_1) + C(B_2) \subset \mathbb{P}^5$$

be the cone over X_0. It is still a $(2) \cap (3)$ complete intersection, thus we can write Y_0 as the limit of a smooth family of $(2) \cap (3)$ complete intersections Y_t. The general Y_t is a smooth Fano 3-fold.

By (2.35), Y_0 is slc, thus $\{Y_t : t \in T\}$ is a stable family such that Y_t is a smooth 3-fold for $t \neq 0$. Since S_1 is irreducible, the cone $C(S_1)$ is analytically irreducible at its vertex. It is nonnormal along a line and non-CM.

One can check that the normalization of $C(S_1)$ is CM.

Example 2.38 (Counterexample to 2.33.3) As in (2.37), let $Q_0 \subset \mathbb{P}^4$ be the singular quadric $(xy - uv = 0)$. On it, take a divisor

$$D_0 := A_1 + A_2 + \tfrac{1}{2}(B_1 + \cdots + B_4) + \tfrac{1}{2}H_4$$

where the A_i are planes in one family, the B_i are planes in the other family and H_4 is a general quartic section.

Note that (Q_0, D_0) is lc (2.35) and $2D_0$ is the intersection of Q_0 with an octic hypersurface. We can thus write (Q_0, D_0) as the limit of a family (Q_t, D_t) where Q_t is a smooth quadric and $2D_t$ a smooth octic hypersurface section of Q_t.

Let us now take the double covers of Q_t ramified along $2D_t$ (11.24). We get a family of $(2) \cap (8)$ complete intersections $X_t \subset \mathbb{P}(1^5, 4)$. The general X_t is smooth with ample canonical class. The special fiber is irreducible, slc, but not normal along $A_1 + A_2$, which is the union of two planes meeting at a point.

Let $\pi \colon \bar{X}_0 \to Q_0$ denote the projection of the normalization of X_0. Then

$$\pi_* \mathscr{O}_{\bar{X}_0} = \mathscr{O}_{Q_0} + \mathscr{O}_{Q_0}(4H - A_1 - A_2).$$

It is easy to compute that $\mathscr{O}_{Q_0}(4H - A_1 - A_2)$ is not CM (see, for instance, Kollár (2013b, 3.15)), so we conclude that $\bar{X}_0$ is not CM.

It is also interesting to note that the preimage of $A_1 + A_2$ in $\bar{X}_0$ is the union of two elliptic cones meeting at their common vertex. These are quite complicated lc centers.

Example 2.39 (Counterexample to 2.33.2–3) Here is an example of a locally stable family of smooth projective varieties $\{Y_t : t \in T\}$ such that

(2.39.1) the canonical class K_{Y_t} is ample and Cartier for every t,

(2.39.2) Y_0 is slc and CM,

(2.39.3) the irreducible components of Y_0 are normal, but

(2.39.4) one of the irreducible components of Y_0 is not CM.

Let Z be a smooth Fano variety of dimension $n \geq 2$ such that $-K_Z$ is very ample, for instance $Z = \mathbb{P}^2$. Set $X := \mathbb{P}^1 \times Z$ and view it as embedded by $|-K_X|$ into $\mathbb{P}^N$ for suitable N. Let $C(X) \subset \mathbb{P}^{N+1}$ be the cone over X.

Let $M \in |-K_Z|$ be a smooth member and consider the following divisors

$$D_0 := \{(0 : 1)\} \times Z, \; D_1 := \{(1 : 0)\} \times Z, \quad \text{and} \quad D_2 := \mathbb{P}^1 \times M.$$

Note that $D_0 + D_1 + D_2 \sim -K_X$. Let $E_i \subset C(X)$ denote the cone over D_i. Then $E_0 + E_1 + E_2$ is a hyperplane section of $C(X)$ and $(C(X), E_0 + E_1 + E_2)$ is lc by (2.35). For some $m > 0$, let $H_m \subset C(X)$ be a general intersection with a degree m hypersurface. Then

$$(C(X), E_0 + E_1 + E_2 + H_m)$$

is snc outside the vertex and is lc at the vertex. Set $Y_0 := E_0 + E_1 + E_2 + H_m$. Since $\mathscr{O}_{C(X)}(Y_0) \sim \mathscr{O}_{C(X)}(m + 1)$, as in (2.36), we can view Y_0 as an slc limit of a family of smooth hypersurface sections $Y_t \subset C(X)$.

The cone over X is CM by (2.35), hence its hyperplane section $E_0 + E_1 + E_2 + H_m$ is also CM. However, E_2 is not CM. To see this, note that E_2 is the cone over $\mathbb{P}^1 \times M$ and, by the Künneth formula,

$$H^i(\mathbb{P}^1 \times M, \mathcal{O}_{\mathbb{P}^1 \times M}) = H^i(M, \mathcal{O}_M) = \begin{cases} k & \text{if } i = 0, n-1, \\ 0 & \text{otherwise.} \end{cases}$$

Thus E_2 is not CM by (2.35).

Example 2.40 (Easy counterexamples to 2.34) There are some obvious problems with all of the questions in (2.34) if the D_t contain divisors with different coefficients. For instance, let C be a smooth curve and $D', D'' \subset \mathbb{A}^1 \times C =: S$ two sections of the 1st projection π_1. Set $D := \frac{1}{2}(D' + D'')$. Then $\pi_1 : (S, D) \to \mathbb{A}^1$ is a stable family of one-dimensional pairs. For general t, the sections D', D'' intersect C_t at two different points and then $\mathcal{O}_{C_t}(K_{C_t} + \lfloor D_t \rfloor) \simeq \mathcal{O}_C(K_C)$. If, however, D', D'' intersect C_t at the same point $p_t \in C_t$, then $\mathcal{O}_{C_t}(K_{C_t} + \lfloor D_t \rfloor) \simeq \mathcal{O}_C(K_C)(p_t)$.

Similarly, the support of D_t is 2 points for general t, but only one point for special values of t.

One can correct for these problems in relative dimension 1 by a more careful bookkeeping of the different parts of the divisor D_t. However, starting with relative dimension 2, no correction seems possible, except when all the coefficients are $> \frac{1}{2}$ (2.81).

The following example is due to Hassett (unpublished).

Example 2.41 (Counterexample to 2.34.1–3) We start with the already studied example of deformations of the cone $S \subset \mathbb{P}^5$ over the degree 4 rational normal curve (1.42), but here we add a boundary to it. Fix $r \geq 1$ and let D_S be the sum of $2r$ lines. Then $(S, \frac{1}{r}D_S)$ is lc and $(K_S + \frac{1}{r}D_S)^2 = 4$.

As in (1.42), there are two different deformations of the pair (S, D_S). (2.41.1) First, set $P := \mathbb{P}^2$ and let D_P be the sum of r general lines. Then $(P, \frac{1}{r}D_P)$ is lc (even canonical if $r \geq 2$) and $(K_P + \frac{1}{r}D_P)^2 = 4$. The usual smoothing of $S \subset \mathbb{P}^5$ to the Veronese surface gives a family $f : (X, D_X) \to \mathbb{P}^1$ with general fiber (P, D_P) and special fiber (S, D_S). We can concretely realize this as deforming $(P, D_P) \subset \mathbb{P}^5$ to the cone over a general hyperplane section. Note that for any general D_S there is a choice of lines D_P such that the limit is exactly D_S.

The total space (X, D_X) is the cone over (P, D_P) (blown up along a curve) and X is $\mathbb{Q}$-factorial. Thus by (11.18) the structure sheaf of an effective divisor on X is CM. In particular, D_S is a flat limit of D_P. Since the D_P is a plane curve of degree r, we conclude that

$$\chi(\mathcal{O}_{D_S}) = \chi(\mathcal{O}_{D_P}) = -\frac{r(r-3)}{2}.$$

(2.41.2) Second, set $Q := \mathbb{P}^1 \times \mathbb{P}^1$ and let A, B denote the classes of the two rulings. Let D_Q be the sum of r lines from the A-family. Then $(Q, \frac{1}{r}D_Q)$ is canonical and $(K_Q + \frac{1}{r}D_Q)^2 = 4$. The usual smoothing of $S \subset \mathbb{P}^5$ to $\mathbb{P}^1 \times \mathbb{P}^1$ embedded by $H := A + 2B$ gives a family $g: (Y, D_Y) \to \mathbb{P}^1$ with general fiber (Q, D_Q) and special fiber (S, D_S). We can concretely realize this as deforming $(Q, D_Q) \subset \mathbb{P}^5$ to the cone over a general hyperplane section.

The total space (Y, D_Y) is the cone over (Q, D_Q) (blown up along a curve) and Y is not $\mathbb{Q}$-factorial. However, $K_Q + \frac{1}{r}D_Q \sim_{\mathbb{Q}} -H$, thus $K_Y + \frac{1}{r}D_Y$ is $\mathbb{Q}$-Cartier and $(Y, S + \frac{1}{r}D_Y)$ is lc by inversion of adjunction (11.17) and so is $(Y, \frac{1}{r}D_Y)$.

In this case, however, D_S is not a flat limit of D_Q for $r > 1$. This follows, for instance, from comparing their Euler characteristic:

$$\chi(\mathcal{O}_{D_S}) = -\frac{r(r-3)}{2} \quad \text{and} \quad \chi(\mathcal{O}_{D_Q}) = r.$$

(2.41.3) Because of their role in the canonical ring, we are also interested in the sheaves $\mathcal{O}(mK + \lfloor \frac{m}{r}D \rfloor)$.

Let H_P be the hyperplane class of $P \subset \mathbb{P}^5$ (that is, $\mathcal{O}_{\mathbb{P}^2}(2)$) and write $m = br + a$ where $0 \le a < r$. Then $mK_P + \lfloor \frac{m}{r}D_P \rfloor + nH_P \sim (2n - 2m - a)L$, so

$$\chi(P, \mathcal{O}_P(mK_P + \lfloor \tfrac{m}{r}D_P \rfloor + nH_P)) =$$
$$\binom{2n-2m-a+2}{2} = \binom{2n-2m+2}{2} - a(2n - 2m + 1) + \binom{a}{2}.$$

Again by (11.18), $\mathcal{O}_X(mK_X + \lfloor \frac{m}{r}D_X \rfloor)$ is CM, hence its restriction to the central fiber S is $\mathcal{O}_S(mK_S + \lfloor \frac{m}{r}D_S \rfloor)$ as in (2.75). In particular,

$$\chi(S, \mathcal{O}_S(mK_S + \lfloor \tfrac{m}{r}D_S \rfloor + nH_S)) = \binom{2n-2m+2}{2} - a(2n - 2m + 1) + \binom{a}{2}.$$

The other deformation again behaves differently. Write $m = br + a$ where $0 \le a < r$. Then, for $H_Q \sim A + 2B$, we see that

$$mK_Q + \lfloor \tfrac{m}{r}D_Q \rfloor + nH_Q \sim (n - m - a)A + (2n - 2m)B,$$

and therefore

$$\chi(Q, \mathcal{O}(mK_Q + \lfloor \tfrac{m}{r}D_Q \rfloor + nH_Q)) = \binom{2n-2m+2}{2} - a(2n - 2m + 1).$$

From this we conclude that the restriction of $\mathcal{O}_Y(mK_Y + \lfloor mD_Y \rfloor)$ to the central fiber S agrees with $\mathcal{O}_S(mK_S + \lfloor mD_S \rfloor)$ only if $a \in \{0, 1\}$, that is when $m \equiv 0, 1 \mod r$. The if part was clear from the beginning. Indeed, if $a = 0$ then $\mathcal{O}_Y(mK_Y + \lfloor mD_Y \rfloor) = \mathcal{O}_Y(mK_Y + mD_Y)$ is locally free and if $a = 1$ then

$$\mathcal{O}_Y(mK_Y + \lfloor mD_Y \rfloor) = \mathcal{O}_Y(K_Y) \otimes \mathcal{O}_Y((m - 1)K_Y + (m - 1)D_Y)$$

is $\mathscr{O}_Y(K_Y)$ tensored with a locally free sheaf. Both of these commute with restrictions. In the other cases we only get an injection

$$\mathscr{O}_Y(mK_Y + \lfloor mD_Y \rfloor)|_S \hookrightarrow \mathscr{O}_S(mK_S + \lfloor mD_S \rfloor)$$

whose quotient is a torsion sheaf of length $\binom{a}{2}$ supported at the vertex.

In the next example, nonflatness appears in codimension 3.

Example 2.42 On $\mathbb{P}^5$ denote coordinates by $x_1, x_2, x_3, x_1', x_2', x_3'$. Set

$$S := (x_1 x_1' = x_2 x_2' = x_3 x_3' = 0) \subset \mathbb{P}^5.$$

It is a reducible K3 surface, a union of eight planes.

Pick constants a_1, a_2, a_3 and a_1', a_2', a_3' such that $a_i a_j' \neq a_j a_i'$ for $i \neq j$. Set

$$X := (\textstyle\sum a_i x_i x_i' = \sum a_i' x_i x_i' = 0) \subset \mathbb{P}^5.$$

By direct computation, X is singular only at the six coordinate vertices, and it has ordinary double points there. Furthermore, $S \sim -K_X$.

Set $Y := X \cap (\sum(x_i + x_i') = 0) \subset \mathbb{P}^4$ and $C := Y \cap S$. Then Y is a smooth, degree 4 Del Pezzo surface and $C \sim -2K_Y$. Thus $(Y, \frac{1}{2}C)$ is a log CY pair. Let $(X_0, \frac{1}{2}S_0) \subset \mathbb{P}^5$ denote the cone over $(Y, \frac{1}{2}C)$. Deformation to the cone (2.36) gives $\pi : (\mathbf{X}, \frac{1}{2}\mathbf{S}) \to \mathbb{A}^1$, whose central fiber is $(X_0, \frac{1}{2}S_0)$. The other fibers are isomorphic to $(X, \frac{1}{2}S)$.

Note that S contains the pair of disjoint planes $P := (x_1 = x_2 = x_3 = 0)$ and $P' := (x_1' = x_2' = x_3' = 0)$. Their specializations P_0, P_0' meet only at the vertex. This is a nonflat deformation of $P \cup P'$.

Example 2.43 (Counterexample to 2.34.1) As in (2.39), let Z be a smooth Fano variety of dimension $n \geq 2$ such that $-K_Z$ is very ample. Set $X := \mathbb{P}^1 \times Z$, but now view it as embedded by global sections of $\mathscr{O}_{\mathbb{P}^1}(1) \otimes \mathscr{O}_Z(-K_Z)$ into $\mathbb{P}^N$ for suitable N. Let $C(X) \subset \mathbb{P}^{N+1}$ be the cone over X.

Fix $r \geq 1$ and let D_r be the sum of r distinct divisors of the form $\{\text{point}\} \times Z \subset X$. Let $H \subset X$ be a general hyperplane section. Then $H \sim_{\mathbb{Q}} -(K_X + \frac{1}{r}D_r)$, that is, $(X, \frac{1}{r}D_r)$ is (numerically) anticanonically embedded. Thus, by (2.35), $(C(H), \frac{1}{r}C(H \cap D_r))$ is lc and there is a locally stable family with general fiber $(X, \frac{1}{r}D_r)$ and special fiber $(C(H), \frac{1}{r}C(H \cap D_r))$.

However, $C(H \cap D_r)$ is not a flat deformation of D_r. Indeed, if $D_{ri}(\simeq Z)$ is any irreducible component of D_r, then $C(H \cap D_{ri})$ is a flat deformation of D_{ri}. Thus $\amalg_i C(H \cap D_{ri})$ is a flat deformation of $D_r = \amalg_i D_{ri}$. Note further that $\amalg_i C(H \cap D_{ri})$ is the normalization of $C(H \cap D_r)$, and the normalization map is $r : 1$ over the vertex of the cone. Thus

$$\chi(D_r, \mathscr{O}_{D_r}) = \textstyle\sum_i \chi(D_{ri}, \mathscr{O}_{D_{ri}}) = \sum_i \chi(C(H \cap D_{ri}), \mathscr{O}_{C(H \cap D_{ri})})$$
$$\geq \chi(C(H \cap D_r), \mathscr{O}_{C(H \cap D_r)}) + (r - 1).$$

Therefore, $C(H \cap D_r)$ cannot be a flat deformation of D_r for $r > 1$. We pick up at least $r - 1$ embedded points.

Example 2.44 (Counterexample to 2.34.3) Set $X := C_a(\mathbb{P}^1 \times \mathbb{P}^n, \mathscr{O}_{\mathbb{P}^1 \times \mathbb{P}^n}(1, a))$ for some $0 < a < n + 1$. Let $D \subset X$ be the cone over a smooth divisor in $|\mathscr{O}_{\mathbb{P}^1 \times \mathbb{P}^n}(1, n + 1 - a)|$. Then (X, D) is canonical and $K_X + D$ is Cartier.

Let $\pi\colon (X, D) \to \mathbb{A}^1$ be a general projection. Then π is locally stable and its central fiber is the cone $X_0 = C_a(H, \mathscr{O}_{\mathbb{P}^1 \times \mathbb{P}^n}(1, a)|_H)$ where $H \in |\mathscr{O}_{\mathbb{P}^1 \times \mathbb{P}^n}(1, a)|$ is a smooth divisor.

We claim that if $2a > n + 1$ then $\mathcal{R}_m\colon \omega_{X/\mathbb{A}^1}^{[m]}\big|_{X_0} \to \omega_{X_0}^{[m]}$ is not surjective for $m \gg 1$. Indeed, $\mathcal{R}_m$ is a sum, for $r \geq 0$ of the restriction maps

$$H^0(\mathbb{P}^1 \times \mathbb{P}^n, \mathscr{O}(r - 2m, ra - (n + 1)m)) \to H^0(H, \mathscr{O}(r - 2m, ra - (n + 1)m)|_H),$$

and $\mathcal{R}_m$ is surjective iff $H^1(\mathbb{P}^1 \times \mathbb{P}^n, \mathscr{O}_{\mathbb{P}^1 \times \mathbb{P}^n}(r - 2m, ra - (n + 1)m)) = 0$ for every $r \geq -1$. Choose $r = 2m - 2$. By the Künneth formula, this group is

$$H^1(\mathbb{P}^1, \mathscr{O}_{\mathbb{P}^1}(-2)) \otimes H^0(\mathbb{P}^n, \mathscr{O}_{\mathbb{P}^n}(2a(m - 1) - m(n - 1))).$$

This is nonzero iff $2a \geq \frac{m}{m-1}(n - 1)$.

The following example, related to Patakfalvi (2013), shows that the relative dualizing sheaf does not commute with base change in general.

Example 2.45 Let S be a smooth, projective surface with K_S ample and $p_g = q = 0$. Let C be a smooth, projective curve with K_C ample. For $[L] \in \mathrm{Pic}^\circ(C)$ set $L_X := \omega_{S \times C} \otimes \pi_C^* L$, where π_C is the projection to C. Note that $H^0(S \times C, L_X) = 0$ and, for $m \geq 2$, $h^0(S \times C, L_X^m) = \chi(S \times C, L_X^m)$ is independent of L. Thus the cones $X_L := \mathrm{Spec} \oplus_{m \geq 0} H^0(S \times C, L_X^m)$ form a flat family over $\mathbf{Pic}^\circ(C)$.

By (2.35), K_{X_L} is Cartier iff $L \simeq \mathscr{O}_C$ and $\mathbb{Q}$-Cartier iff $[L] \in \mathrm{Pic}^\circ(C)$ is a torsion point.

2.4 Stable Families

Next we define the notion of stable families over regular, one-dimensional base schemes and establish the valuative criteria of separatedness and properness.

Definition 2.46 Let $f : (X, \Delta) \to C$ be a family of pairs (2.2) over a regular one-dimensional scheme C of characteristic 0. We say that $f : (X, \Delta) \to C$ is *stable* if f is locally stable (2.3), proper and $K_{X/C} + \Delta$ is f-ample.

Note that if f is locally stable then $K_X + \Delta$ is $\mathbb{R}$-Cartier, so f-ampleness makes sense. By (2.10), being stable is preserved by base change $C' \to C$.

More generally, whenever the notion of local stability is defined later over a scheme S, then $f : (X, \Delta) \to S$ is called *stable* if the conditions in the definition are satisfied. (Thus we have to make sure that local stability implies that $K_{X/S} + \Delta$ makes sense and is $\mathbb{R}$-Cartier.)

The relationship between locally stable morphisms and stable morphisms parallels the connection between smooth varieties and their canonical models.

Proposition 2.47 *Let $f : (Y, \Delta) \to B$ be a locally stable, proper morphism over a one-dimensional regular scheme B of characteristic 0. Assume that the generic fibers are normal, of general type and f has a relative canonical model $f^c : (Y^c, \Delta^c) \to B$. Then $f^c : (Y^c, \Delta^c) \to B$ is stable.*

Furthermore, taking the relative canonical model commutes with flat base changes $\pi : B' \to B$.

Proof First, $K_{Y^c} + \Delta^c$ is f^c-ample by definition (1.38) and (Y^c, Δ^c) is lc.

Let $b \in B$ be any closed point and Y_b (resp. Y_b^c) the fibers over b. Since f is locally stable, $(Y, Y_b + \Delta_Y)$ is lc. Since any fiber is f-linearly trivial, we conclude using Kollár (2013b, 1.28) that $(Y^c, Y_b^c + \Delta^c)$ is also lc. Thus f^c is locally stable, hence stable.

In characteristic 0, being locally stable is preserved by base change (2.10), thus the last assertion follows from (11.40). □

Remark 2.48 In most cases, the fibers of f^c are *not* the canonical models of the fibers of f; see Section 1.5 and (5.10).

A significant exception is when $\Delta = 0$ and Y_b has canonical singularities. Then (Y, Y_b) is canonical by (11.17) and so is (Y^c, Y_b^c) by Kollár and Mori (1998, 3.51). Thus Y_b^c also has canonical singularities by (11.17), it is thus the canonical model of Y_b.

2.49 (Separatedness and Properness) Let C be a regular scheme of dimension 1, and $C^\circ \subset C$ an open and dense subscheme. Let $f^\circ : (X^\circ, \Delta^\circ) \to C^\circ$ be a stable morphism. We aim to prove the following two properties.

Separatedness $f^\circ : (X^\circ, \Delta^\circ) \to C^\circ$ has at most one extension to a stable morphism $f : (X, \Delta) \to C$.

Properness There is a finite surjection $\pi \colon B \to C$ such that the pull back

$$\pi^* f^\circ \colon (X^\circ \times_C B, \Delta^\circ \times_C B) \to B^\circ := \pi^{-1}(C^\circ)$$

extends to a stable morphism $f_B \colon (X_B, \Delta_B) \to B$.

Next we show that separatedness holds in general and properness holds in characteristic 0. In both cases, the proof relies on theorems that we state in general forms in Section 11.3.

Proposition 2.50 (Separatedness for stable maps) *Let $f_i \colon (X^i, \Delta^i) \to B$ be two stable morphisms over a one-dimensional, regular scheme B. Let*

$$\phi \colon (X^1_{k(B)}, \Delta^1_{k(B)}) \simeq (X^2_{k(B)}, \Delta^2_{k(B)})$$

be an isomorphism of the generic fibers. Then ϕ extends to an isomorphism

$$\Phi \colon (X^1, \Delta^1) \simeq (X^2, \Delta^2).$$

Proof Note that ϕ always extends to an isomorphism over an open, dense subset $B^\circ \subset B$. We can now apply (11.40), whose assumptions are satisfied by (2.15). □

Example 2.50.1 Regularity of B is needed here. As a simple example, let $\bar{B}$ be a smooth curve and $\bar{f} \colon \bar{S} \to \bar{B}$ a smooth, projective family of curves of genus ≥ 2. Assume that we have points $b_1, b_2 \in \bar{B}$ such that the fibers $C_i := \bar{f}^{-1}(b_i)$ are isomorphic. Let B be the nodal curve obtained by identifying b_1 and b_2. We can then descend the family to $f \colon S \to B$ using an isomorphism $C_1 \simeq C_2$. The number of different choices is $|\operatorname{Aut}(C_1)|$. Thus the family over $\bar{B} \setminus \{b_1, b_2\}$ may have several stable extensions over the nodal curve B.

Remark 2.50.2 As a consequence of (2.50) we obtain that $\mathbf{Aut}(X, \Delta)$ is finite for a stable pair (X, Δ) in arbitrary characteristic, using (2.16.2). We prove a more general form of it in (8.64).

Theorem 2.51 (Valuative-properness for stable maps) *Let C be a smooth curve over a field of characteristic 0 and $C^\circ \subset C$ an open and dense subset. Let $f^\circ \colon (X^\circ, \Delta^\circ) \to C^\circ$ be a stable morphism.*

Then there is a finite surjection $\pi \colon B \to C$ such that the pull back

$$f_B^\circ := \pi^* f^\circ \colon (X^\circ \times_C B, \Delta^\circ \times_C B) \to \pi^{-1}(C^\circ)$$

extends to a stable morphism $f_B \colon (X_B, \Delta_B) \to B$.

Proof We begin with the case when X° is normal. Start with $f^\circ : (X^\circ, \Delta^\circ) \to C^\circ$ and extend it to a proper flat morphism $f_1 : (X_1, \Delta_1) \to C$ where X_1 is normal. In general (X_1, Δ_1) is no longer lc.

By Kollár (2013b, 10.46), there is a log resolution (p.xvi) $g_1 : Y_1 \to X_1$ such that $(g_1^{-1})_* \Delta_1 + \mathrm{Ex}(g_1) + Y_{1c}$ is an snc divisor for every $c \in C$. In general, the fibers of $f_1 \circ g_1 : Y_1 \to C$ are not reduced, hence $g_1 : (Y_1, (g_1^{-1})_* \Delta_1 + \mathrm{Ex}(g_1)) \to C$ is not locally stable.

Let B be a smooth curve and $\pi : B \to C$ a finite surjection. Let $X_2 \to X_1 \times_C B$ and $Y_2 \to Y_1 \times_C B$ denote the normalizations and $g_2 : Y_2 \to X_2$ the induced morphism. Let Δ_2 be the pull back of $\Delta_1 \times_C B$ to X_2. Note that

$$f_2 \circ g_2 : (Y_2, (g_2^{-1})_* \Delta_2 + \mathrm{Ex}(g_2)) \to B$$

is a log resolution over the points where π is étale, but Y_2 need not be smooth. However, by (2.52), $(Y_2, (g_2^{-1})_* \Delta_2 + \mathrm{Ex}(g_2) + \mathrm{red}\, Y_{2b})$ is lc for every $b \in B$.

By (2.53), one can choose $\pi : B \to C$ such that every fiber of $f_2 \circ g_2$ is reduced. With such a choice, $f_2 \circ g_2$ is locally stable.

If the generic fiber $(X_g^\circ, \Delta_g^\circ)$ is klt, then, using (2.15) and after shrinking C°, we may assume that (X°, Δ°) is klt. Pick $0 < \varepsilon \ll 1$. Then $(Y_2, \Delta_2 + (1 - \varepsilon)\mathrm{Ex}(g_2))$ is also klt and so it has a canonical model $f_B : (X_B, \Delta_B) \to B$ by (11.28.1), which is stable by (2.47).

We are almost done, except that, by construction, $f_B : (X_B, \Delta_B) \to B$ is isomorphic to the pull-back of $f^\circ : (X^\circ, \Delta^\circ) \to C^\circ$ only over a possibly smaller dense open subset. However, by (2.50), this implies that this isomorphism holds over the entire C°.

The argument is the same if (X°, Δ°) is lc, but we need to take the canonical model of $(Y_2, \Delta_2 + \mathrm{Ex}(g_2))$. Here we rely on (11.28.2).

Next we show how the slc case can be reduced to the lc case.

Let $\bar{X}^\circ \to X^\circ$ be the normalization with conductor $\bar{D}^\circ \subset \bar{X}^\circ$. As we noted in (2.12), we get a stable morphism

$$\bar{f}^\circ : (\bar{X}^\circ, \bar{\Delta}^\circ + \bar{D}^\circ) \to C^\circ. \tag{2.51.4}$$

By the already completed normal case, we get $B \to C$ such that the pull-back of (2.51.4) extends to a stable morphism

$$\bar{f}_B : (\bar{X}_B, \bar{\Delta}_B + \bar{D}_B) \to B. \tag{2.51.5}$$

Finally, (2.54) shows that (2.51.5) is the normalization of a stable morphism $f_B : (X_B, \Delta_B) \to B$, which is the required extension of f_B°. $\square$

We have used the following three lemmas during the proof.

Lemma 2.52 *Let C be a smooth curve over a field of characteristic 0, $f \colon X \to C$ a flat morphism and Δ an $\mathbb{R}$-divisor on X. Assume that $(X, \operatorname{red} X_c + \Delta)$ is lc for every $c \in C$. Let B be a smooth curve, $g \colon B \to C$ a quasi-finite morphism, $g_Y \colon Y \to X \times_C B$ the normalization and $\Delta_Y := g_Y^* \Delta$.*
Then $(Y, \operatorname{red} Y_b + \Delta_Y)$ is lc for every $b \in B$.

Proof Pick $c \in C$ and let $b_i \in B$ be its preimages. By the Hurwitz formula

$$K_Y + \Delta_Y + \textstyle\sum_i \operatorname{red} Y_{b_i} = g_X^*(K_X + \Delta + \operatorname{red} X_c).$$

By assumption, $(X, \Delta + \operatorname{red} X_c)$ is lc for every $c \in C$. Hence, by (11.23.3), $(Y, \Delta_Y + \sum_i \operatorname{red} Y_{b_i})$ is also lc. □

Lemma 2.53 *Let $f \colon X \to T$ be a flat morphism from a normal scheme to a one-dimensional regular scheme T. Let S be another one-dimensional regular scheme and $\pi \colon S \to T$ a quasi-finite morphism. Let $Y \to X \times_T S$ be the normalization and $f_Y \colon Y \to S$ the projection. Assume that π is tamely ramified and, for every $s \in S$, the multiplicity of every irreducible component of $X_{\pi(s)}$ divides the ramification index of π at s.*
Then every fiber of $f_Y \colon Y \to S$ is reduced.

Proof The claim is local, so pick points $0_S \in S$ and $0_T := \pi(0_S) \in T$ with local parameters $t \in \mathscr{O}_T$ and $s \in \mathscr{O}_S$.

We want to study how the multiplicities of the irreducible components of the fiber over 0_T change under base extension. We can focus on one such irreducible component and pass to any open subset of X that is not disjoint from the chosen component. By Noether normalization (10.51), we can think of X as a hypersurface $X \subset \mathbb{A}_T^n$ defined by an equation $f \in \mathscr{O}_T[x_1, \ldots, x_n]$. The central fiber X_0 is defined by $\bar{f} = 0$ where $\bar{f}$ is the mod t reduction of f. By focusing at a generic point of X_0, after an étale coordinate change we may assume that $\bar{f} = x_1^m$ where m is the multiplicity of X_0. We can thus write $f = x_1^m - t \cdot u(\mathbf{x}, t)$. Since X is normal (hence regular) at the generic point of X_0, we see that u is not identically zero along X_0.

We can write $\pi^* t = s^e v(s)$ where e is the ramification index of π at 0_S and v is a unit at 0_S. Consider now the fiber product $X_S := X \times_T S \to S$. It is defined by the equation $x_1^m = s^e \cdot u(\mathbf{x}, s^e v(s)) \cdot v(s)$. Note that X_S is not normal along $(s = x_1 = 0)$ if $m, e > 1$.

Constructing the normalization is especially simple if e is a multiple of m. Write $e = md$ and set $x_1' := x s^{-d}$. Then we get $Y \subset \mathbb{A}_S^n$ (with coordinates $x_1', x_2, \ldots, x_n$) defined by $x_1'^m = u(x_1' s^d, x_2, \ldots, x_n, s^e v(s)) \cdot v(s)$, and the central fiber Y_0 is defined by the equation $x_1'^m = u(0, x_2, \ldots, x_n, 0) \cdot v(0)$, where the right-hand side is not identically zero.

If the characteristic of $k(0_S)$ does not divide m, then the projection $Y_0 \to \mathbb{A}^{n-1}_{x_2,\dots,x_n}$ is generically étale and Y_0 is smooth at its generic points. In this case, Y is the normalization of X_S (at least generically along Y_0) and the central fiber of $Y \to S$ has multiplicity 1. □

Aside 2.53.1 If $p := \text{char}\, k(0_S)$ divides m, then $Y_0 \to \mathbb{A}^{n-1}_{x_2,\dots,x_n}$ is inseparable. If $u(0, x_2, \dots, x_n, 0)$ is not a pth power over the algebraic closure of $k(0_S)$, then Y_0 is geometrically integral, hence generically nonsingular. In this case, Y is the normalization of X_S and the central fiber of $Y \to S$ has multiplicity 1.

If $u(0, x_2, \dots, x_n, 0)$ is a pth power, then Y_0 is not generically reduced. In this case, Y need not be normal and further blow-ups may be needed to reach the normalization. The situation is rather complicated, even for families of curves. A weaker result is in (2.60).

At the end of the proof of (2.51), we needed to construct an slc pair from its normalization. The following is a special case of (11.41), whose assumptions hold by (2.15).

Lemma 2.54 *Let B be a smooth curve over a field of characteristic 0 and $B^\circ \subset B$ a dense open subset. Let $f^\circ : (X^\circ, \Delta^\circ) \to B^\circ$ be a stable morphism. Let $\bar{f}^\circ : (\bar{X}^\circ, \bar{\Delta}^\circ + \bar{D}^\circ) \to B^\circ$ be the normalization with conductor $\bar{D}^\circ \subset \bar{X}^\circ$. Assume that $\bar{f}^\circ$ extends to a stable morphism $\bar{f} : (\bar{X}, \bar{\Delta} + \bar{D}) \to B$. Then f° also extends to a stable morphism $f : (X, \Delta) \to B$.* □

As we noted in (2.16), it is not known whether being locally stable is preserved by base change in positive characteristic. However, the next result shows that this holds for all families obtained as in (2.51).

Theorem 2.55 *Let $h : C' \to C$ be a quasi-finite morphisms of regular schemes of dimension 1 and $f : X \to C$ a proper morphism from a regular scheme X to C whose fibers are geometrically reduced, simple normal crossing divisors. Then $X' := X \times_C C'$ has canonical singularities and*

$$\oplus_{m \geq 0} f'_* \omega^{\otimes m}_{X'/C'} \simeq h^* \oplus_{m \geq 0} f_* \omega^{\otimes m}_{X/C}. \tag{2.55.1}$$

Proof Note that (2.55.1) is just the claim that push-forward commutes with flat base change $h : C' \to C$. The substantial part is the assertion that X' has canonical singularities, hence the proj of $\oplus_{m \geq 0} f'_* \omega^{\otimes m}_{X'/C'}$ is also the relative canonical model of any resolution of X'.

Pick a point $x \in X$ and set $c = f(x)$. We may assume that C and C' are the spectra of DVRs with local parameters t and s. Thus the Henselisation of (x, X)

can be given as a hypersurface $(x_1 \cdots x_m = t) \subset (0, \mathbf{A}_C^n)$, where $\mathbf{A}_C^n$ denotes the Henselisation of $\mathbb{A}_C^n$ at $(0,0)$.

If $h^*t = \phi(s)$ then (x', X') can be given as a hypersurface

$$(x_1 \cdots x_m = \phi(s)) \subset (0, \mathbf{A}_{C'}^n). \tag{2.55.2}$$

Thus the main claim is that the singularity defined by (2.55.2) is canonical.

If we are over a field, then (2.55.2) defines a toric singularity. We check that although there is no torus action on the base C, we can compute the simplest blow-ups suggested by toric geometry and everything works out as expected.

(Note that, although the pair $(\mathbb{A}_k^n, (x_1 \cdots x_n = 0))$ is lc, this is not a completely toric question. We need to understand all exceptional divisors over $\mathbb{A}_k^n$, not just the toric ones; see Kollár (2013b, 2.11).) $\Box$

Lemma 2.56 *Let T be a DVR with local parameter t, residue field k and $\mathbf{A}_T^n$ the Henselisation of $\mathbb{A}_T^n$ at $(0,0)$. Let $m \leq n$ and e be natural numbers and ϕ a regular function on $\mathbf{A}_T^n$. Set*

$$X := X(m, n, e, \phi) = (x_1 \cdots x_m = t^e + t^{e+1}\phi(x_1, \ldots, x_n)) \subset (0, \mathbf{A}_T^n), \tag{2.56.1}$$

and let D be the divisor $(t = 0) \subset X$. Then the pair (X, D) is log canonical and X is canonical.

Proof If $\operatorname{char} k = 0$, this immediately follows from (2.10), so the main point is that it also holds for any DVR.

If $m = 0$ or $e = 0$, then X is empty and we are done. Otherwise we can set $x_m' := x_m(1 + t\phi)^{-1}$ to get the simpler equation $x_1 \cdots x_m = t^e$. For inductive purposes we introduce a new variable s and work with

$$\begin{aligned} X &:= \left(x_1 \cdots x_m - s^e = x_{m+1} \cdots x_{m+r}s - t = 0\right) \subset (0, \mathbf{A}_T^{n+1}), \\ D &:= (t = 0), \quad \text{where } 0 \leq r \leq n - m. \end{aligned} \tag{2.56.2}$$

The case $r = 0$ corresponds to (2.56.1). We use induction on m and e.

Let E be an exceptional divisor over X and v the corresponding valuation. Assume first that $v(x_1) \geq v(s)$. We blow up $(x_1 = s = 0)$. In the affine chart where $x_1' := x_1/s$, we get the new equations

$$x_1' x_2 \cdots x_m - s^{e-1} = x_{m+1} \cdots x_{m+r}s - t = 0$$

defining (X', D'). A local generator of $\omega_{X/T}(D)$ is

$$\frac{1}{t} \cdot \frac{dx_2 \wedge \cdots \wedge dx_n}{x_2 \cdots x_{m+r}}, \tag{2.56.3}$$

which is unchanged by pull-back.

Such operations reduce e, until we reach a situation where $v(x_i) < v(s)$ for every i. If $v(x_i) = 0$ for some i and $i \neq m$ then x_i is nonzero at the generic point of center$_X E$. Thus we can set $x'_m := x_i x_m$ and reduce the value of m. Thus we may assume that $v(x_i) > 0$ for $i = 1, \ldots, m$. Since $\sum v(x_i) = e \cdot v(s)$, we conclude that $e < m$. If $e \geq 2$ then we may assume that $v(x_e)$ is the smallest. Set $x'_i = x_i/x_e$ for $i = 1, \ldots, e - 1$ and $s' := s/x_e$. We get new equations

$$x'_1 \cdots x'_{e-1} x_{e+1} \cdots x_m - (s')^e = x_e x_{m+1} \cdots x_{m+r} s' - t = 0 \tag{2.56.4}$$

defining (X', D') and the value of m dropped. The pull-back of (2.56.3) is

$$
\begin{aligned}
\frac{1}{t} & \cdot \frac{d(x_e x'_2) \wedge \cdots \wedge d(x_e x'_{e-1}) \wedge dx_e \wedge \cdots \wedge dx_n}{(x_e x'_2) \cdots (x_e x'_{e-1}) x_e \cdots x_{m+r}} \\
&= \frac{1}{t} \cdot \frac{dx'_2 \wedge \cdots dx'_{e-1} \wedge dx_e \wedge \cdots \wedge dx_n}{x'_2 \cdots x'_{e-1} x_e \cdots x_{m+r}},
\end{aligned}
\tag{2.56.5}
$$

which is again a local generator of $\omega_{X'/T}(D')$.

Eventually we reach the situation where $e = 1$. We can now eliminate s and, after setting $r + m \mapsto m$, rewrite the system as

$$
\begin{aligned}
X &:= (x_1 \cdots x_m = t) \subset (0, \mathbf{A}^n_T), \\
D &:= (t = 0).
\end{aligned}
\tag{2.56.6}
$$

Now X is regular: this case was treated in Kollár (2013b, 2.11). $\qquad \square$

We discuss a collection of other results about extending one-parameter families of varieties or pairs. These can be useful in many situations.

2.57 (Extending a stable family without base change) Let C be a smooth curve over a field of characteristic 0, $C^\circ \subset C$ an open and dense subscheme, and $f^\circ : (X^\circ, \Delta^\circ) \to C^\circ$ a stable morphism. Here we consider the question of how to extend f° to a proper morphism $f : X \to C$ in a "nice" way without base change. For simplicity, assume that X° is normal.

We can take any extension of f° to a proper morphism $f_1 : X_1 \to C$, then take a log resolution of $(X_2, \Delta_2) \to (X_1, \Delta_1)$, and finally the canonical model of (X_2, Δ_2) using (11.28). We have proved:

Claim 2.57.1 There is a unique extension $f : (X, \Delta) \to C$ such that (X, Δ) is lc and $K_X + \Delta$ is f-ample. $\qquad \square$

This model has the problem that its fibers over $C \setminus C^\circ =: \{c_1, \ldots, c_r\}$ can be pretty complicated. A slight twist improves the fibers considerably. Instead of starting with the (X_1, Δ_1), we take a log resolution $(X_2, \Delta_2 + \sum \operatorname{red} X_{2,c_i})$ of $(X_1, \Delta_1 + \sum \operatorname{red} X_{1,c_i})$ and its canonical model over C. We need to apply

(11.28) to $(X_2, \Delta_2 + \sum \mathrm{red}\, X_{2,c_i} - \varepsilon \sum X_{2,c_i})$ and use (11.28.2) to obtain the following.

Claim 2.57.2 There is a unique extension $f : (X, \Delta) \to C$ such that $(X, \Delta + \sum \mathrm{red}\, X_{c_i})$ is lc and $K_X + \Delta + \sum \mathrm{red}\, X_{c_i}$ is f-ample. By adjunction, in this case $(\mathrm{red}\, X_{c_i}, \mathrm{Diff}\, \Delta)$ is slc.   □

A variant of this starts with any extension (X_1, Δ_1) and then takes a dlt modification of $(X_1, \Delta_1 + \sum \mathrm{red}\, X_{1,c_i})$ as in Kollár (2013b, 1.36).

Claim 2.57.3 There is a dlt modification $(Y^\circ, \Delta_Y^\circ) \to (X^\circ, \Delta^\circ)$ and an extension of it to $g : (Y, \Delta_Y) \to C$ such that $(Y, \Delta + \sum \mathrm{red}\, Y_{c_i})$ is dlt.   □

Taking a minimal model of $g : (Y, \Delta_Y + \sum \mathrm{red}\, Y_{c_i}) \to C$ yields another useful version.

Claim 2.57.4 There is a dlt modification $(Y^\circ, \Delta_Y^\circ) \to (X^\circ, \Delta^\circ)$ and an extension of it to $g : (X, \Delta_X) \to C$ such that $(X, \Delta_X + \sum \mathrm{red}\, X_{c_i})$ is dlt and $K_X + \Delta_X + \sum \mathrm{red}\, X_{c_i}$ is f-nef.   □

Finally, if we are willing to change X° drastically, Kollár (2013b, 10.46) gives the following.

Claim 2.57.5 There is a log resolution $(Y^\circ, \Delta_Y^\circ) \to (X^\circ, \Delta^\circ)$ and an extension of it to $g : (Y, \Delta_Y) \to C$ such that $(Y, \Delta_Y + \mathrm{red}\, Y_c)$ is snc for every $c \in C$.   □

Let us also mention the following very strong variant of (2.57.5), traditionally called the "semistable reduction theorem." We do not use it, and one of the points of our proof of (2.51) was to show that the much easier (2.52) and (2.53) are enough for our purposes.

Theorem 2.58 (Kempf et al., 1973) *Let C be a smooth curve over a field of characteristic 0, $f : X \to C$ a flat morphism of finite type and D a divisor on X. Then there is a smooth curve B, a finite surjection $\pi : B \to C$ and a log resolution $g : Y \to X \times_C B$ such that for every $b \in B$,*

(2.58.1) $g_*^{-1}(D \times_C B) + \mathrm{Ex}(g) + Y_b$ *is an snc divisor and*

(2.58.2) Y_b *is reduced.*   □

The positive or mixed characteristic analogs of (2.58) are not known, but the following result on "semi-stable alterations" holds in general.

Theorem 2.59 (de Jong, 1996) *Let T be a one-dimensional regular scheme and $f : X \to T$ a flat morphism of finite type whose generic fiber is geometrically reduced. Then there is a one-dimensional regular scheme S, a finite* *surjection* $\pi \colon S \to T$ *and a generically finite, separable, proper morphism* $g \colon Y \to X \times_T S$ *such that* Y_s *is a reduced snc divisor for every* $s \in S$. □

The following variant of (2.53) is an easy consequence of (2.59).

Corollary 2.60 *Let* $f \colon X \to T$ *be a flat morphism of finite type from a pure dimensional scheme to a one-dimensional regular scheme* T. *Then there is a one-dimensional regular scheme* S *and a finite morphism* $\pi \colon S \to T$ *such that every fiber of the projection of the normalization* $\overline{X \times_T S} \to S$ *is reduced.* □

2.5 Cohomology of the Structure Sheaf

In studying moduli questions, it is very useful to know that certain numerical invariants are locally constant. In this section, we study the deformation invariance of (the dimension of) certain cohomology groups. The key to this is the Du Bois property of slc pairs. The definition of Du Bois singularities is rather complicated, but fortunately for our applications we need to know only the following two facts.

2.61 (Properties of Du Bois singularities) Let M be a complex analytic variety. Since constant functions are analytic, there is an injection of sheaves $\mathbb{C}_M \hookrightarrow \mathscr{O}_M^{\mathrm{an}}$. Taking cohomologies we get

$$H^i(M, \mathbb{C}) \to H^i(M, \mathscr{O}_M^{\mathrm{an}}).$$

If X is projective over $\mathbb{C}$ and X^{an} is the corresponding analytic variety, then, by the GAGA theorems of Serre (1955–1956), $H^i(X^{\mathrm{an}}, \mathscr{O}_X^{\mathrm{an}}) \simeq H^i(X, \mathscr{O}_X)$.

If X is also smooth, Hodge theory tells us that

$$H^i(X^{\mathrm{an}}, \mathbb{C}) \to H^i(X^{\mathrm{an}}, \mathscr{O}_X^{\mathrm{an}}) \simeq H^i(X, \mathscr{O}_X)$$

is surjective. Du Bois singularities were essentially defined to preserve this surjectivity; see Du Bois (1981) and Steenbrink (1983). (There does not seem to be a good definition of Du Bois singularities in positive characteristic; see however Kollár and Kovács (2020).) Thus we have the following.

Property 2.61.1 (Du Bois (1981)) Let X be a proper variety over $\mathbb{C}$ with Du Bois singularities. Then the natural maps

$$H^i(X^{\mathrm{an}}, \mathbb{C}) \to H^i(X^{\mathrm{an}}, \mathscr{O}_X^{\mathrm{an}}) \simeq H^i(X, \mathscr{O}_X) \quad \text{are surjective.} \qquad \square$$

Next we need to know which singularities are Du Bois. Over a field of characteristic 0, rational singularities are Du Bois; see Kollár (1995b, 12.9) and Kovács (1999), but for our applications the key result is the following.

Property 2.61.2 (Kollár and Kovács (2010, 2020)) Let (X, Δ) be an slc pair over $\mathbb{C}$. Then X has Du Bois singularities. □

These are the only facts we need to know about Du Bois singularities. The main use of (2.61.1) is through a base-change theorem, due to Du Bois and Jarraud (1974); Du Bois (1981).

Theorem 2.62 *Let S be a Noetherian scheme over a field of characteristic 0 and $f: X \to S$ a flat, proper morphism. Assume that the fiber X_s is Du Bois for some $s \in S$. Then there is an open $s \in S° \subset S$ such that, for all i,*

(2.62.1) $R^i f_* \mathcal{O}_X$ *is locally free and commutes with base change over $S°$, and*

(2.62.2) $s \mapsto h^i(X_s, \mathcal{O}_{X_s})$ *is a locally constant function on $S°$.*

Proof By Cohomology and Base Change, the theorem is equivalent to proving that the restriction maps

$$\phi_s^i : R^i f_* \mathcal{O}_X \to H^i(X_s, \mathcal{O}_{X_s}) \tag{2.62.3}$$

are surjective for every i. By the Theorem on Formal Functions, it is enough to prove this when S is replaced by any Artinian local scheme S_n, whose closed point is s.

Thus assume from now on that we have a flat, proper morphism $f_n: X_n \to S_n$, $s \in S_n$ is the only closed point, and X_s is Du Bois. Then $H^0(S_n, R^i f_* \mathcal{O}_X) = H^i(X_n, \mathcal{O}_{X_n})$, hence we can identify the ϕ_s^i with the maps

$$\psi^i : H^i(X_n, \mathcal{O}_{X_n}) \to H^i(X_s, \mathcal{O}_{X_s}). \tag{2.62.4}$$

By the Lefschetz principle, we may assume that $k(s) \simeq \mathbb{C}$. Then both sides of (2.62.4) are unchanged if we replace X_n by the corresponding analytic space X_n^{an}. Let $\mathbb{C}_{X_n}$ (resp. $\mathbb{C}_{X_s}$) denote the sheaf of locally constant functions on X_n (resp. X_s) and $j_n: \mathbb{C}_{X_n} \to \mathcal{O}_{X_n}$ (resp. $j_s: \mathbb{C}_{X_s} \to \mathcal{O}_{X_s}$) the natural inclusions. We have a commutative diagram

$$
\begin{CD}
H^i(X_n, \mathbb{C}_{X_n}) @>{\alpha^i}>> H^i(X_s, \mathbb{C}_{X_s}) \\
@V{j_n^i}VV @VV{j_s^i}V \\
H^i(X_n, \mathcal{O}_{X_n}) @>{\psi^i}>> H^i(X_s, \mathcal{O}_{X_s}).
\end{CD}
$$

Note that α^i is an isomorphism since the inclusion $X_s \hookrightarrow X_n$ is a homeomorphism, and j_s^i is surjective since X_s is Du Bois. Thus ψ^i is also surjective. □

Complement 2.62.5 This proof also works if $f\colon X \to S$ is a flat, proper morphism of complex analytic spaces and X_s is an algebraic space with Du Bois singularities.

Definition 2.63 A scheme Y is said to be *potentially slc* or *slc-type* if, for every $y \in Y$, there is an effective $\mathbb{R}$-divisor Δ_y on Y such that (Y, Δ_y) is slc at y.

Let $f\colon X \to S$ be a flat morphism. We say that f has *potentially slc fibers over closed points* if the fiber X_s is potentially slc for every closed point $s \in S$.

One can similarly define the notion *potentially klt,* and so on.

In our final applications, the Δ_s usually come as the restriction of a global divisor Δ to X_s, but we do not assume this.

If (X_s, Δ_s) is slc then X_s is Du Bois by (2.61.2), hence (2.62) implies the following.

Corollary 2.64 *Let S be a Noetherian scheme over a field of characteristic 0, and $f\colon X \to S$ a proper and flat morphism with potentially slc fibers over closed points. Then, for all i,*

(2.64.1) *$R^i f_* \mathcal{O}_X$ is locally free and compatible with base change, and*

(2.64.2) *if S is connected, then $h^i(X_s, \mathcal{O}_{X_s})$ is independent of $s \in S$.* □

We can derive similar results for other line bundles from (2.64). A line bundle L on X is called f-*semi-ample* if there is an $m > 0$ such that L^m is f-generated by global sections. That is, the natural map $f^*(f_*(L^m)) \to L^m$ is surjective. Equivalently, L^m is the pull-back of a relatively ample line bundle by a suitable morphism $X \to Y$.

Corollary 2.65 *Let S be a Noetherian, connected scheme over a field of characteristic 0 and $f\colon X \to S$ a proper and flat morphism with potentially slc fibers over closed points. Let L be an f-semi-ample line bundle on X. Then*

(2.65.1) *$R^i f_*(L^{-1})$ is locally free and compatible with base change, and*

(2.65.2) *$h^i(X_s, L_{X_s}^{-1})$ is independent of $s \in S$ for all i.*

Proof The question is local on S, thus we may assume that S is local with closed point s. Chose $m > 0$ such that L^m is f-generated by global sections. Since S is local, L^m is generated by global sections. By (2.13), there is a finite morphism $\pi\colon Y \to X$ such that $\pi_* \mathcal{O}_Y = \bigoplus_{r=0}^{m-1} L^{-r}$ and $f \circ \pi\colon (Y, \pi^{-1}\Delta) \to S$ also has potentially slc fiber over s. Thus, by (2.64),

$$R^i(f \circ \pi)_* \mathcal{O}_Y = \bigoplus_{r=0}^{m-1} R^i f_*(L^{-r})$$

is locally free and compatible with arbitrary base change. Thus the same holds for every summand. □

Warning 2.65.4 Note that we assume that L is f-semi-ample, not only that L is semi-ample on fibers over closed points. The Poincaré bundle on $E \times \hat{E}_0 \to \hat{E}_0$ shows that the latter is not enough, where E is an elliptic curve and $\hat{E}_0$ is the localization of its dual at $0 = [\mathscr{O}_E]$.

Corollary 2.66 (Kollár and Kovács, 2010, 2020) *Let S be a Noetherian, connected scheme over a field of characteristic 0 and $f \colon X \to S$ a proper, flat morphism of finite type. Assume that all fibers are potentially slc and X_s is CM for some $s \in S$. Then all fibers of f are CM.*

For an arbitrary flat morphisms $\pi \colon X \to S$, the set of points $x \in X$ such that the fiber $X_{\pi(x)}$ is CM at x is open (10.11), but usually not closed. (Many such examples can be constructed using Kollár (2013b, 3.9–11).) If π is proper, then the set $\{s \in S : X_s \text{ is CM}\}$ is open in S (10.12). Thus the key point of (2.66) is to show that, in our case, this set is also closed.

More generally, under the assumptions of (2.66), if one fiber of f is S_k (10.3.2) for some k, then all fibers of f are S_k; see Kollár and Kovács (2020).

Proof We prove the projective case; see Kollár and Kovács (2020) for the proper one.

Let L be an f-ample line bundle on X. If X_s is CM for some $s \in S$, then, by Kollár and Mori (1998, 5.72), $H^i(X_s, L_{X_s}^{-r}) = 0$ for $r \gg 1$ and $i < \dim X_s$. Thus by (2.65), the same vanishing holds for every $s \in S$. Hence, using Kollár and Mori (1998, 5.72) again, we conclude that X_s is CM for every $s \in S$. $\qquad\square$

Theorem 2.67 (*Kollár and Kovács, 2010, 2020*) *Let S be a Noetherian scheme over a field of characteristic 0 and $f \colon X \to S$ a flat morphism of finite type with potentially slc fibers over closed points. Then $\omega_{X/S}$ is*

(2.67.1) *flat over S with S_2 fibers, and*

(2.67.2) *compatible with base change. That is, for any $g \colon T \to S$, the natural map $g_X^* \omega_{X/S} \to \omega_{X_T/T}$ is an isomorphism, where $g_X \colon X_T := X \times_S T \to X$ is the first projection.*

We give a detailed proof of the projective case; this is sufficient for almost all applications in this book. For the general case, we refer to Kollár and Kovács (2020).

The existence of $\omega_{X/S}$ is easy and, as we see in (2.68.1–3), it holds under rather weak restrictions. Compatibility with base change is not automatic; see Patakfalvi (2013) and (2.45) for some examples.

As we explain in (2.68.4–5), once the definition of $\omega_{X/S}$ is set up right, (2.67.2) and the flatness claim in (2.67.1) become easy consequences of (2.65).

Once these hold, the fiber of $\omega_{X/S}$ over $s \in S$ is ω_{X_s}; and the latter is S_2 as in Kollár and Mori (1998, 5.69).

2.68 (The relative dualizing sheaf II) The best way to define the relative dualizing sheaf is via general duality theory as in Hartshorne (1966); Conrad (2000) or (Stacks, 2022, tag 0DWE); see also (11.2). It is, however, worthwhile to observe that a slight modification of the treatment in Hartshorne (1977) gives the relative dualizing sheaf in the following cases.

Assumptions S is an arbitrary Noetherian scheme and $f \colon X \to S$ a projective morphism of pure relative dimension n (2.71). (We do not assume flatness.)

2.68.1 (Weak duality for $\mathbb{P}^n_S$) Let $P = \mathbb{P}^n_S$ with projection $g \colon P \to S$ and set $\omega_{P/S} := \wedge^n \Omega_{P/S}$.

The proof of Hartshorne (1977, III.7.1 or III.Exc.8.4) shows that there is a natural isomorphism, called the *trace map*, $t \colon R^n g_* \omega_{P/S} \simeq \mathcal{O}_S$ and, for any coherent sheaf F on X, there is a natural isomorphism

$$g_* \mathcal{H}om_P(F, \omega_{P/S}) \simeq \mathcal{H}om_S(R^n g_* F, \mathcal{O}_S).$$

Note that if S is a point then $g_* \mathcal{H}om_P = \operatorname{Hom}_P$, thus we recover the usual formulation of Hartshorne (1977, III.7.1).

2.68.2 (Construction of $\omega_{X/S}$) Let $f \colon X \to S$ be a projective morphism of pure relative dimension n. We construct $\omega_{X/S}$ first locally over S. Once we establish weak duality, the proof of Hartshorne (1977, III.7.2) shows that a relative dualizing sheaf is unique up to unique isomorphism, hence the local pieces glue together to produce $\omega_{X/S}$. Working locally over S, we can assume that there is a finite morphism $\pi \colon X \to P = \mathbb{P}^n_S$. Set

$$\omega_{X/S} := \mathcal{H}om_P(\pi_* \mathcal{O}_X, \omega_{P/S}).$$

If f is flat with CM fibers over S then $\pi_* \mathcal{O}_X$ is locally free and so is $\pi_* \omega_{X/S}$. Thus $\omega_{X/S}$ is also flat over S with CM fibers and it commutes with base change. We discuss a local version of this in (2.68.7).

2.68.3 (Weak duality for X/S) Let $f \colon X \to S$ be a projective morphism of pure relative dimension n (2.71). Use Hartshorne (1977, Exc.III.6.10) to show that there is a trace map $t \colon R^n f_* \omega_{X/S} \to \mathcal{O}_S$, and for any coherent sheaf F on X there is a natural isomorphism

$$f_* \mathcal{H}om_X(F, \omega_{X/S}) \simeq \mathcal{H}om_S(R^n f_* F, \mathcal{O}_S).$$

If F is locally free, this is equivalent to the isomorphism

$$f_*(\omega_{X/S} \otimes F^{-1}) \simeq \mathcal{H}om_S(R^n f_* F, \mathcal{O}_S).$$

(Note that $M \mapsto \mathcal{H}om_S(M, \mathcal{O}_S)$ is a duality for locally free, coherent $\mathcal{O}_S$-sheaves, but not for all coherent sheaves. In particular, the torsion in $R^n f_* F$ is invisible on the left-hand side $f_*(\omega_{X/S} \otimes F^{-1})$.)

2.68.4 (Flatness of $\omega_{X/S}$) Let L be relatively ample on X/S. By (3.20), $\omega_{X/S}$ is flat over S iff $f_*(\omega_{X/S} \otimes L^m)$ is locally free for $m \gg 1$. If this holds then $\omega_{X/S}$ is the coherent $\mathcal{O}_X$-sheaf associated to $\oplus_{m \geq m_0} f_*(\omega_{X/S} \otimes L^m)$, as a module over the $\mathcal{O}_S$-algebra $\sum_{m \geq 0} f_*(L^m)$.

Applying weak duality with $F = L^{-m}$, we see that these hold if $R^n f_*(L^{-m})$ is locally free for $m \gg 1$. The latter is satisfied in two important cases.

(a) $f \colon X \to S$ is flat with CM fibers. Then $R^i f_*(L^{-m}) = 0$ for $i < n$ and $m \gg 1$, hence $R^n f_*(L^{-m})$ is locally free of rank $(-1)^n \chi(X_s, L_s^{-m})$ for $m \gg 1$.

(b) $f \colon X \to S$ is flat with potentially slc fibers. Then $R^n f_*(L^{-m})$ is locally free for $m \geq 0$ by (2.65).

2.68.5 (Base change properties of $\omega_{X/S}$) Let $f \colon X \to S$ be a projective morphism of pure relative dimension n and L relatively ample. We claim that the following are equivalent:

(a) $\omega_{X/S}$ commutes with base change as in (2.67.2).

(b) $R^n f_*(L^{-m})$ is locally free for $m \gg 0$.

By (2.68.3–4), $\omega_{X/S}$ commutes with base change iff $\mathcal{H}om_S(R^n f_*(L^{-m}), \mathcal{O}_S)$ is locally free and commutes with base change for $m \gg 0$. Finally, show that a coherent sheaf M is locally free iff $\mathcal{H}om_S(M, \mathcal{O}_S)$ is locally free and commutes with base change.

2.68.6 (Warning on general duality) If F is locally free, then we get

$$R^i f_*(\omega_{X/S} \otimes F^{-1}) \times R^{n-i} f_*(F) \to R^n f_* \omega_{X/S} \to \mathcal{O}_S,$$

but this is not a perfect pairing, unless both sheaves on the left are locally free and commute with base change.

2.68.7 (More on the CM case) Let $f \colon X \to S$ be a projective morphism of pure relative dimension n. We already noted in (2.68.2) that if f is flat with CM fibers over S, then the same holds for $\omega_{X/S}$. We consider what happens if f is not everywhere CM. By (10.11), there is a largest open subset $X^{\mathrm{cm}} \subset X$ such that $f|_{X^{\mathrm{cm}}}$ is flat with CM fibers. Assume for simplicity that $X_s \cap X^{\mathrm{cm}}$ is dense in X_s and $s \in S$ is local. Then, for every $x \in X_s \cap X^{\mathrm{cm}}$ one can choose a finite morphism $\pi \colon X \to P = \mathbb{P}_S^n$ such that $\pi^{-1}(\pi(x)) \subset X^{\mathrm{cm}}$. Thus $\pi_* \mathcal{O}_X$ is locally free at $\pi(x)$ and so is $\pi_* \omega_{X/S}$. Thus we have proved that the restriction of $\omega_{X/S}$ to X^{cm} is flat over S with CM fibers and commutes with base change.

This is actually true for all finite type morphisms, one just needs to find a local analog of the projection π (see Section 10.6) and show that (2.68.2.a) holds if π is finite; see Conrad (2000) for details.

Corollary 2.69 *Let S be a connected, Noetherian scheme over a field of characteristic 0 and $f : X \to S$ a proper and flat morphism with potentially slc fibers over closed points. Let L be an f-semi-ample line bundle on X. Then*

(2.69.1) $R^i f_*(\omega_{X/S} \otimes L)$ *is locally free and compatible with base change, and*

(2.69.2) $h^i(X_s, \omega_{X_s} \otimes L_s)$ *is independent of $s \in S$ for all i.*

In particular, for $L = \mathscr{O}_X$ we get that

(2.69.3) $R^i f_* \omega_{X/S}$ *is locally free and compatible with base change, and*

(2.69.4) $h^i(X_s, \omega_{X_s})$ *is independent of $s \in S$ for all i.*

If the fibers X_s are CM, then $H^i(X_s, \omega_{X_s} \otimes L_s)$ is dual to $H^{n-i}(X_s, L_s^{-1})$, so (2.69) follows from (2.65). If the fibers X_s are not CM, the relationship between (2.69) and (2.65) is not so clear. See (8.16) for a more general version.

Proof Let us start with the case $i = 0$. By weak duality (2.68.3),

$$f_*(\omega_{X/S} \otimes L) \simeq \mathcal{H}om_S(R^n f_*(L^{-1}), \mathscr{O}_S),$$

where $n = \dim(X/S)$. By (2.65), $R^n f_*(L^{-1})$ is locally free and compatible with base change, hence so is $f_*(\omega_{X/S} \otimes L)$. Thus (2.69.1) holds for $i = 0$. Next we use this and induction on n to get the $i > 0$ cases.

Choose M very ample on X such that $R^i f_*(\omega_{X/S} \otimes L \otimes M) = 0$ for $i > 0$, and this also holds after any base change. Working locally on S, as in the proof of (2.65), let $H \subset X$ be a general member of $|M|$ such that $H \to S$ is also flat with potentially slc fibers (2.13). The push-forward of the sequence

$$0 \to \omega_{X/S} \otimes L \to \omega_{X/S} \otimes L \otimes M \to \omega_{H/S} \otimes L \to 0$$

gives isomorphisms

$$R^i f_*(\omega_{X/S} \otimes L) \simeq R^{i-1} f_*(\omega_{H/S} \otimes L) \quad \text{for } i \geq 2.$$

Using induction, these imply that (2.69.1) holds for $i \geq 2$.

The beginning of the push-forward is an exact sequence

$$0 \to f_*(\omega_{X/S} \otimes L) \to f_*(\omega_{X/S} \otimes L \otimes M) \to f_*(\omega_{H/S} \otimes L) \to R^1 f_*(\omega_{X/S} \otimes L) \to 0.$$

We already proved that the first three terms are locally free. In general, this does not imply that the last term is locally free, but this implication holds if S is the spectrum of an Artinian ring (2.70).

In general, pick any point $s \in S$ with maximal ideal sheaf m_s. Set $A_n :=$ $\mathscr{O}_{s,S}/m_s^n$ and $X_n := \mathrm{Spec}(\mathscr{O}_X/f^*m_s^n)$. Then $H^1(X_n, (\omega_{X/S} \otimes L)|_{X_n})$ is a free A_n-module by the above considerations, and the restriction maps

$$H^1(X_n, (\omega_{X/S} \otimes L)|_{X_n}) \otimes_{A_n} k(s) \to H^1(X_s, \omega_{X_s} \otimes L_s)$$

are isomorphisms. By the Theorem on Formal Functions, this implies that $R^1 f_*(\omega_{X/S} \otimes L)$ is locally free and commutes with base change. □

2.70 Let (A, m) be a local Artinian ring. Let F be a free A-module and $j: A \hookrightarrow F$ an injection. We claim that $j(A)$ is a direct summand of F. Indeed, let $r \geq 1$ be the smallest natural number such that $m^r A = 0$. Note that $m^{r-1}m = 0$. If $j(A) \subset mF$ then $m^{r-1}A = 0$, a contradiction. Thus $j(A)$ is a direct summand of F. By induction this shows that any injection between free A-modules is split. This also implies that if $0 \to M_1 \to \cdots \to M_n \to 0$ is an exact sequence of A-modules and all but one of them are free, then they are all free.

2.71 (Pure dimensional morphisms) A finite type morphism $f: X \to S$ is said to have *pure relative dimension* n if, for every integral scheme T and every $h: T \to S$, every irreducible component of $X \times_S T$ has dimension $\dim T + n$. We also say that f is *pure dimensional* if it is pure of relative dimension n for some n. It is enough to check this property for all cases when T is the spectrum of a DVR.

Applying the definition when T is a point shows that if f has pure relative dimension n, then every fiber of f has pure dimension n, but the converse does not always hold. For instance, let C be a curve and $\pi: \bar{C} \to C$ the normalization. If C is nodal then π does not have pure relative dimension 0 since $\bar{C} \times_C \bar{C}$ contains two isolated points. However, the converse does hold in several important cases.

Claim 2.71.1 Let $f: X \to S$ be a finite type morphism whose fibers have pure dimension n. Then f has pure relative dimension n iff it is universally open. Thus both properties hold if f is flat.

Proof Both properties can be checked after base change to spectra of DVRs. In the latter case the equivalence is clear and flatness implies both. □

2.71.2 (Chevalley's criterion) (Grothendieck, 1960, IV.14.4.1) Let $f: X \to S$ be a finite type morphism whose fibers have pure dimension n. Assume that S is normal (or geometrically unibranch) and X is irreducible. Then f is universally open.

Proof By an easy limit argument, it is enough to check openness after base change for finite type, affine morphisms $S' \to S$; see Grothendieck (1960,

IV.8.10.1). We may thus assume that $S' \subset \mathbb{A}_S^n$ for some n. The restriction of an open morphism to the preimage of a closed subset is also open, thus it is enough to show that the natural morphism $f^{(n)}: \mathbb{A}_Y^n \to \mathbb{A}_S^n$ is open for every n. If S is normal then so is $\mathbb{A}_S^n$, thus it is enough to show that all maps as in (2.71.2) are open.

To see openness, let $U \subset X$ be an open set and $x \in U$ a closed point. We need to show that $f(U)$ contains an open neighborhood of $s := f(x)$. Let $x \in W \subset X$ be an irreducible component of a complete intersection of n Cartier divisors such that x is an isolated point of $W \cap X_s$. It is enough to prove that $f(U \cap W)$ contains an open neighborhood of s. After extending $W \to S$ to a proper morphism and Stein factorization, we are reduced to showing that (2.71.2) holds for finite morphisms.

Since $f(U)$ is constructible, it is open iff it is closed under generalization. The latter holds by the going-down theorem. $\square$

2.6 Families of Divisors I

We saw in (2.67) that for locally stable morphisms $g: (X, \Delta) \to C$, the relative dualizing sheaf $\omega_{X/C}$ commutes with base change. We also saw in (2.44) that its powers $\omega_{X/C}^{[m]}$ usually do not commute with base change. Here we consider this question for a general divisor D: What does it mean to restrict a divisor D on X to a fiber X_c? How are the two sheaves $\mathscr{O}_X(D)|_{X_c}$ and $\mathscr{O}_{X_c}(D|_{X_c})$ related?

2.72 (Comments on Serre's conditions) For the definition of S_m, see (10.3) or (Stacks, 2022, tag 033P). The following variant will be useful for us.

Let X be a scheme, $Z \subset X$ a closed subset, and F a coherent sheaf on X. We say that F is S_m *along* Z if (10.3.2) holds whenever $x \in Z$.

The following is the key example for us. Let T be a regular one-dimensional scheme, $f: X \to T$ a proper morphism, and F a coherent sheaf on X, flat over T. Assume that every fiber F_t is S_m. If $x \in X$ is contained in a closed fiber, then $\mathrm{depth}_x F \geq \min\{m+1, \mathrm{codim}(x, \mathrm{Supp}\, F)\}$, but for points in the generic fiber we can only guarantee that $\mathrm{depth}_x F \geq \min\{m, \mathrm{codim}(x, \mathrm{Supp}\, F)\}$. Thus F is not S_{m+1}, but it is S_{m+1} *along closed fibers.*

2.73 (One-parameter families of divisors) Let T be a regular, one-dimensional scheme and $f: X \to T$ a flat, proper morphism. For simplicity, assume for now that X is normal. Let D be an effective Weil divisor on X. Under what conditions can we view D as giving a "reasonable" family of Weil divisors on the fibers of f?

We can view D as a subscheme of X and, if $\operatorname{Supp} D$ does not contain any irreducible component of any fiber X_t, then $f|_D \colon D \to T$ is flat, hence the fibers $D|_{X_t}$ form a flat family of subschemes of the fibers X_t. The $D|_{X_t}$ may have embedded points; ignoring them gives a well-defined effective Weil divisor on the fiber X_t. We will eventually denote it by D_t, but use D_t^{div} or the more precise $D|_{X_t}^{\mathrm{div}}$ if we want to emphasize its construction; see also (2.77).

Understanding the difference between the subscheme $D|_{X_t}$ and the divisor D_t^{div} is the key to dealing with many issues. As a rule of thumb, D defines a "nice" family of divisors iff $D_t^{\mathrm{div}} = D|_{X_t}$ for every $t \in T$.

It can happen that $D \cap X_t$ is contained in $\operatorname{Sing} X_t$ for some t. These are the cases when the correspondence between Weil divisors and rank 1 reflexive sheaves breaks down. Fortunately, this does not happen for locally stable families. Thus we can focus on the cases when D is a relative Mumford divisor (p.xiv).

It is now time to drop the normality assumption and work with divisorial subschemes (4.16.2) in one of the following general settings. (Further generalizations will be considered in Sections 5.4 and 9.3.) We start with the absolute version.

(1.a) X is a pure dimensional, reduced scheme and $H \subset X$ a Cartier divisor. Assume that H is S_2; equivalently, X is S_3 along H (2.72).

(1.b) There is a closed subscheme $Z \subset X$ such that $D|_{X \setminus Z}$ is a Cartier divisor and $\operatorname{codim}_H(H \cap Z) \geq 2$.

(1.c) D is a Mumford divisor along H, that is, $\operatorname{Supp} D$ does not contain any irreducible component of H, and H is regular at generic points of $H \cap D$; see (4.16.4).

In the relative version, we assume the following.

(2.a) T is a regular, one-dimensional, irreducible scheme and $f \colon X \to T$ is a flat, pure dimensional morphism whose fibers are reduced and S_2.

(2.b) There is a closed subscheme $Z \subset X$ such that $D|_{X \setminus Z}$ is a Cartier divisor and $\operatorname{codim}_{X_t}(X_t \cap Z) \geq 2$ for every $t \in T$.

(2.c) D is a relative Mumford divisor (4.68).

Under these conditions, the *divisorial restriction* D_H^{div} (resp. D_t^{div}) is defined as the unique divisorial subscheme (4.16.2) on H (resp. X_t) that agrees with the restriction of the Cartier divisor $D|_{X \setminus Z}$ to $H \setminus Z$ (resp. $X_t \setminus Z$).

Proposition 2.74 *Notation and assumptions as in (2.73.1.a–c). The following conditions are equivalent:*

(2.74.1) $\mathscr{O}_X(-D)$ *is* S_3 *along* $H \cap Z$.

(2.74.2) $\mathscr{O}_X(-D)$ *is* S_3 *along* H.

(2.74.3) *The restriction map r_H: $\mathscr{O}_X(-D)|_H \to \mathscr{O}_H(-D_H^{div})$ is an isomorphism.*
(2.74.4) *The following sequence is exact:*

$$0 \to \mathscr{O}_X(-D-H) \to \mathscr{O}_X(-D) \to \mathscr{O}_H(-D_H^{div}) \to 0.$$

If D is effective, these are further equivalent to:
(2.74.5) *$\mathscr{O}_D$ has depth ≥ 2 at every point of $H \cap Z$.*
(2.74.6) *$\mathscr{O}_D$ is S_2 along H.*
(2.74.7) *$D \cap H = D_H^{div}$ (as schemes).*

Proof Since we assume that X is S_3 along H, (2) and (6) hold outside Z. Thus (1) $\Leftrightarrow$ (2) and (5) $\Leftrightarrow$ (6).

Since $\mathscr{O}_X(-D)$ is S_2, r_H is an injection and an isomorphism outside Z. Since $\mathscr{O}_H(-D_H^{div})$ is S_2 by definition, it is the S_2-hull of $\mathscr{O}_X(-D)|_H$; see (9.3.4). Thus r_H is surjective $\Leftrightarrow$ r_H is an isomorphism $\Leftrightarrow$ $\mathscr{O}_X(-D)|_H$ is S_2. This proves (2) $\Leftrightarrow$ (3) while (3) $\Leftrightarrow$ (4) is clear.

Since $\mathscr{O}_X$ has depth ≥ 3 at codimension ≥ 2 points of H, the sequence

$$0 \to \mathscr{O}_X(-D) \to \mathscr{O}_X \to \mathscr{O}_D \to 0,$$

and an easy lemma (10.28) show that (5) $\Leftrightarrow$ (1).

Let s be a local equation of H. Then s is not a zero divisor on $\mathscr{O}_D$ and $\mathscr{O}_{D \cap H} = \mathscr{O}_D/(s)$. Thus (6) $\Leftrightarrow$ (7). $\qquad\square$

Proposition 2.75 (Relative version) *Using the notation and assumptions of (2.73.2.a–c), let $0 \in T$ be a closed point and $g \in T$ the generic point.*
(2.75.1) *The conditions (2.74.1–7) are equivalent for $H = X_0$.*
If f is projective and L is f-ample, then these are also equivalent to:
(2.75.2) *$\chi(X_0, L_0^m(-D_0^{div})) = \chi(X_g, L_g^m(-D_g))$ for all $m \in \mathbb{Z}$.*
If $\dim(X_0 \cap Z) = 0$, then these are further equivalent to:
(2.75.3) *$\chi(X_0, \mathscr{O}_{X_0}(-D_0^{div})) = \chi(X_g, \mathscr{O}_{X_g}(-D_g))$.*

Proof The first claim follows from (2.74). If f is projective and $\mathscr{O}_X(-D)$ is flat over T, then

$$\chi(X_g, L_g^m(-D_g)) = \chi(X_0, L^m(-D)|_{X_0}).$$

Hence the difference of the two sides in (2) is $\chi(X_0, L_0^m \otimes Q)$, where Q is the cokernel of r_0: $\mathscr{O}_X(-D)|_{X_0} \to \mathscr{O}_{X_0}(-D_0^{div})$. Thus $Q = 0$ iff equality holds in (2). If $\dim(X_0 \cap Z) = 0$ then Q has 0-dimensional support, thus

$$\chi(X_0, L_0^m \otimes Q) = \chi(X_0, Q) = H^0(X_0, Q),$$

so, in this case, (2) is equivalent to (3). $\qquad\square$

Note that (2.75) shows that one can go rather freely between effective divisors and their ideal sheaves when studying restrictions. Much of the results here on ideal sheaves generalize to arbitrary sheaves; these are worked out in Sections 5.4 and 9.3.

The conditions (2.75) are all preserved by linear equivalence. However, they are not preserved by sums of divisors.

Example 2.76 Consider a family of smooth quadrics $Q \subset \mathbb{P}^3 \times \mathbb{A}^1$ degenerating to the quadric cone Q_0. Take four families of lines L^i, M^i such that $L_0^1, L_0^2, M_0^1, M_0^2$ are four distinct lines in Q_0, $L_c^1 \neq L_c^2$ are in one family of lines on Q_c and $M_c^1 \neq M_c^2$ are in the other family for $c \neq 0$. Note that

$$(Q, \tfrac{1}{2}(L^1 + L^2 + M^1 + M^2)) \to \mathbb{A}^1$$

is a locally stable family.

Each of the four families of lines L^i, M^i is a flat family of Weil divisors.

For pairs of lines, flatness is more complicated. $L^1 + L^2$ is *not* a flat family (the flat limit has an embedded point at the vertex), but $L^i + M^j$ is a flat family for every i, j. The union of any three of them, for instance $L^1 + L^2 + M^1$ is again a flat family, and so is $L^1 + L^2 + M^1 + M^2$.

Notation 2.77 Let C be a regular, one-dimensional scheme and $f : X \to C$ a flat, pure dimensional morphism with reduced, S_2 fibers. Let Δ be a relative Mumford $\mathbb{R}$-divisor (4.68). From now on we use Δ_c to denote the *divisorial fiber* (instead of Δ_c^{div} or $\Delta|_{X_c}^{\mathrm{div}}$ as in (2.73)).

Thus the fiber of a pair (X, Δ) over $c \in C$ is denoted by (X_c, Δ_c).

This notation is harmless for $\mathbb{R}$-divisors, but there is a potential for confusion when used for prime divisors. Then we use the longer $D|_{X_c}$ for the scheme-theoretic fiber and D_c^{div} or $D|_{X_c}^{\mathrm{div}}$ for the divisor-theoretic fiber.

2.78 The main source of divisors D and divisorial sheaves $\mathcal{O}_X(D)$ that satisfy the equivalent conditions of (2.75) is (11.20).

Let (X, Δ) be an slc pair. The conditions of (2.75) are local on X, we can thus assume that $K_X + \Delta \sim_{\mathbb{R}} 0$. Then

$$mK_X + \lfloor m\Delta \rfloor + \{m\Delta\} \sim_{\mathbb{R}} 0 \qquad (2.78.1)$$

for any $m \in \mathbb{Z}$. If $\Delta = \sum a_i D_i$ and $\{ma_i\} \leq a_i$ for every i, then $\{m\Delta\} \leq \Delta$, hence $-mK_X - \lfloor m\Delta \rfloor$ satisfies the assumptions of (11.20).

Furthermore, if $B \leq \lfloor \Delta \rfloor$ is an effective $\mathbb{Z}$-divisor, then we can also work with

$$(mK_X + \lfloor m\Delta \rfloor - B) + (\{m\Delta\} + B) \sim_{\mathbb{R}} 0. \qquad (2.78.2)$$

If $a_1 + a_2 = 1$ and the a_i are irrational, then $\{ma_1\} \leq a_1$ and $\{ma_2\} \leq a_2$ hold only for $m = 0$, but (2.78.2) can be useful, relying on (11.50).

However, the numerical conditions $\{ma_i\} \leq a_i$ hold in many other cases; we list some of them in (2.79). These results are generalized to reduced base schemes in (4.33). They influence the definition of various moduli theories in Chapters 6 and 8.

Proposition 2.79 *Let $f \colon (X, \Delta = \sum a_i D_i) \to C$ be a locally stable morphism to a smooth curve over a field of characteristic 0 and $c \in C$ a closed point. Let D be a relative Mumford $\mathbb{Z}$-divisor (4.68). Then*

$$\mathscr{O}_X(D)|_{X_c} \simeq \mathscr{O}_{X_c}(D_c) := \mathscr{O}_{X_c}(D_c^{div}) \qquad (*)$$

in any of the following cases:

(2.79.1) *D is $\mathbb{Q}$-Cartier.*

(2.79.2) *$\Delta = 0$ and $D \sim_{\mathbb{Q}} mK_{X/C}$ for any $m \in \mathbb{Z}$.*

(2.79.3) *$m\Delta$ is a $\mathbb{Z}$-divisor and $D \sim_{\mathbb{Q}} mK_{X/C} + m\Delta$.*

(2.79.4) *$m\Delta$ is a $\mathbb{Z}$-divisor and $D \sim_{\mathbb{Q}} (m + 1)K_{X/C} + m\Delta$.*

(2.79.5) *$\Delta = \sum(1 - \frac{1}{r_i})D_i$ for some $r_i \in \mathbb{N}$, and $D \sim_{\mathbb{Q}} mK_{X/C} + \lfloor m\Delta \rfloor$ for any $m \in \mathbb{Z}$.*

(2.79.6) *$\Delta = \sum c_i D_i$, $D \sim_{\mathbb{Q}} mK_{X/C} + \lfloor m\Delta \rfloor$ and $1 - \frac{1}{m} \leq c_i \leq 1$ for every i.*

(2.79.7) *The set $\{m \in \mathbb{N} \colon (*)$ holds for $D \sim_{\mathbb{Q}} mK_{X/C} + \sum \lfloor ma_i \rfloor D_i\}$ has positive density.*

(2.79.8) *In (1–6) we may replaced D by $D - B$ for any effective relative Mumford $\mathbb{Z}$-divisor $B \leq \lfloor \Delta \rfloor$.*

Proof Let D be a Weil divisor on X as in (2.73.2–4). Assume that there is an effective $\mathbb{R}$-divisor $\Delta' \leq \Delta$ and an $\mathbb{R}$-Cartier $\mathbb{R}$-divisor L such that $D \sim_{\mathbb{R}} \Delta' + L$. Then $\mathscr{O}_X(-D)$ satisfies the equivalent conditions of (2.75) by (11.20).

For (1) we can use $\Delta' = L = 0$, in cases (1–3) we can take $\Delta' = 0$ and $L := -m(K_{X/C} + \Delta)$, and in case (4) we use $\Delta' = \Delta$ and $L := -(m + 1)(K_{X/C} + \Delta)$.

For (5–6) we employ $\Delta' = m\Delta - \lfloor m\Delta \rfloor$ and $L := -m(K_{X/C} + \Delta)$. The assumptions on the coefficients of Δ ensure that $\Delta' \leq \Delta$. (Note that if $m\Delta - \lfloor m\Delta \rfloor \leq \Delta$ for every m then in fact every coefficient of Δ is of the form $1 - \frac{1}{r}$ for some $r \in \mathbb{N}$.) Claim (7) follows from (11.50).

Finally, if $B \leq \lfloor \Delta \rfloor$ then $\{m\Delta\} + B \leq \Delta$, giving (8). $\qquad \square$

These results are close to being optimal. For instance, under the assumptions of (2.79.3), if n is different from m and $m + 1$ then the two sheaves

$$\left(\omega_{X/C}^{[n]}(mD)\right)\big|_{X_c} \quad \text{and} \quad \omega_{X_c}^{[n]}(mD|_{X_c})$$

are frequently different, see (2.41.3). In general, as shown by (2.44), even the two sheaves $(\omega_{X/C}^{[m]})|_{X_c}$ and $\omega_{X_c}^{[m]}$ can be different if $\Delta \neq 0$. However, a considerable generalization of the cases (2.79.5–6) is proved in Section 2.7.

2.7 Boundary with Coefficients $> \frac{1}{2}$

Consider a locally stable morphism $f\colon (X, \Delta = \sum a_i D^i) \to C$ to a smooth curve C. It is very tempting to think of each fiber (X_c, Δ_c) as a compound object $(X_c, D_c^i\colon i \in I, a_i\colon i \in I)$ consisting of the scheme X_c, the divisors D_c^i, and their coefficients a_i. Two issues muddy up this simple picture.

- Different D_c^i may have an irreducible component E_c in common. The definition of a pair treats E_c as a divisor with coefficient $\sum_{i \in I} \operatorname{coeff}_{E_c} D_c^i$. The individual D_c^i do not seem to be part of the data any more.
- Should we just ignore the embedded points of $X_c \cap D^i$?

One could hope that the first is just a matter of bookkeeping, but this does not seem to be the case, as shown by the examples (2.76) and (2.41). In both cases the coefficients in Δ were $\leq \frac{1}{2}$.

The aim of this section is to show that these examples were optimal; these complications do not occur if the coefficients in Δ are all $> \frac{1}{2}$. We start with the case when the coefficients are 1.

Given a locally stable map $f\colon (X, \Delta) \to C$, usually the lc centers of the fibers (X_c, Δ_c) do not form a flat family. Indeed, there are many cases when the generic fiber is smooth, but a special fiber is not klt. However, as we show next, the specialization of an lc center on the generic fiber becomes a union of lc centers on a special fiber. Set theoretically, this follows from adjunction (11.17) and (11.12.4), but now we prove this even scheme theoretically.

Theorem 2.80 *Let C be a smooth curve over a field of characteristic 0, $f\colon (X, \Delta) \to C$ a locally stable morphism and $Z \subset X$ any union of lc centers of (X, Δ). Then $f|_Z\colon Z \to C$ is flat with reduced fibers and the fiber Z_c is a union of lc centers of (X_c, Δ_c) for every $c \in C$.*

Proof Z is reduced and every irreducible component of Z dominates C by (2.15). Thus $f|_Z\colon Z \to C$ is flat. We can write its fibers as $Z_c = X_c \cap Z$. Since $X_c + Z$ is a union of lc centers of $(X, X_c + \Delta)$, it is seminormal (11.12.2), so $X_c \cap Z$ is reduced by (11.12.3). The last claim follows from (11.10.3). $\square$

When the coefficients are in $(\frac{1}{2}, 1]$, we start with a simple result.

2.81 (Restriction and rounding down) Let $f\colon (X, \Delta = \sum_{i \in I} a_i D^i) \to C$ be a locally stable family over a one-dimensional, regular scheme.

By (2.3), (X_c, Δ_c) is slc, hence every component of Δ_c appears with coefficient ≤ 1. For a divisor $A \subset X_c$,

$$1 \ge \operatorname{coeff}_A \Delta_c = \sum_{i \in I} a_i \cdot \operatorname{coeff}_A D_c^i.$$

Since the $\operatorname{coeff}_A D_c^i$ are natural numbers, we get the following properties.

(2.81.1) If $a_i > \frac{1}{2}$, then every irreducible component of D_c^i has multiplicity 1.

(2.81.2) If $a_i + a_j > 1$ and $i \ne j$, then the divisors D_c^i and D_c^j have no irreducible components in common.

Next let $\Theta = \sum_j b_j B^j$ be an $\mathbb{R}$-divisor on X. If every irreducible component of B_c^j has multiplicity 1, and the different restrictions have no irreducible components in common, then combining (1–2) we get:

Claim 2.81.3 Assume that $\operatorname{Supp} \Theta \subset \operatorname{Supp}(\Delta^{>1/2})$ (11.1). Then $\operatorname{coeff}(\Theta|_H) \subset$ coeff Θ and $\lfloor \Theta|_H \rfloor = \lfloor \Theta \rfloor|_H$.   □

Applying this to $\Theta = m\Delta$ gives the following.

Corollary 2.81.4. If coeff $\Delta \subset (\frac{1}{2}, 1]$ then $\lfloor m\Delta_c \rfloor = \lfloor m\Delta \rfloor_c$ for every m.   □

The next result of Kollár (2014) solves the embedded point question when all the occurring coefficients are $> \frac{1}{2}$. Examples (2.41 and 2.42) show that the strict inequality is necessary.

Theorem 2.82 *Let $f\colon (X, \Delta = \sum_{i \in I} a_i D_i) \to C$ be a locally stable morphism to a smooth curve over a field of characteristic 0. Let $J \subset I$ be any subset such that $a_j > \frac{1}{2}$ for every $j \in J$. Set $D_J := \cup_{j \in J} D_j$. Then*

(2.82.1) $f|_{D_J}\colon D_J \to C$ *is flat with reduced fibers,*

(2.82.2) D_J *is S_2 along every closed fiber, and*

(2.82.3) $\mathscr{O}_X(-D_J)$ *is S_3 along every closed fiber.*

Proof Note that each D_i is a log center of (X, Δ) (11.11) and $\operatorname{mld}(D_i, X, \Delta) = 1 - a_i$ by (11.8). Thus $\operatorname{mld}(D_J, X, \Delta) < \frac{1}{2}$.

Let X_c be any fiber of f. Then $(X, X_c + \Delta)$ is slc and

$$\operatorname{mld}(D_i, X, X_c + \Delta) = \operatorname{mld}(D_i, X, \Delta) < \tfrac{1}{2},$$

since none of the D_i is contained in X_c. Each irreducible component of X_c is a log canonical center of $(X, X_c + \Delta)$ (11.10), thus $\operatorname{mld}(X_c, X, X_c + \Delta) = 0$. Therefore, $\operatorname{mld}(D_J, X, X_c + \Delta) + \operatorname{mld}(X_c, X, X_c + \Delta) < \frac{1}{2}$.

We can apply (11.12.3) to $(X, X_c + \Delta)$ with $W = D_J$ and $Z = X_c$ to conclude that $X_c \cap D_J$ is reduced. This proves (1) which implies (2–3) by (2.75).   □

For the plurigenera, we have the following generalization of (2.79.5–6).

Theorem 2.83 (Kollár, 2018a) *Let C be a smooth curve over a field of characteristic 0 and $f : (X, \Delta) \to C$ a locally stable morphism with normal generic fiber. Assume that* $\operatorname{coeff} \Delta \subset (\frac{1}{2}, 1]$. *Then, for every $c \in C$ and $m \in \mathbb{Z}$,*

$$\omega_{X/C}^{[m]}(\lfloor m\Delta \rfloor)|_{X_c} \simeq \omega_{X_c}^{[m]}(\lfloor m\Delta \rfloor_c). \tag{2.83.1}$$

Complement 2.83.2 If $\operatorname{coeff} \Delta \subset [\frac{1}{2}, 1]$ then (2.83.1) still holds, but needs a more careful case analysis, see Kollár (2018a). Note also that $\lfloor m\Delta \rfloor_c = \lfloor m\Delta_c \rfloor$ if $\operatorname{coeff} \Delta \subset (\frac{1}{2}, 1]$ by (2.81.3), but they may be different if some coefficients equal $\frac{1}{2}$ and m is odd.

Method of proof If $mK_X + \lfloor m\Delta \rfloor$ is $\mathbb{Q}$-Cartier, then this follows from (2.79.1). Thus we aim to construct a birational modification $X' \to X$ such that $mK_{X'} + \lfloor m\Delta' \rfloor$ is $\mathbb{Q}$-Cartier, and then descend from X' to X.

More generally, let $g \colon Y \to X$ be a proper morphism of normal varieties, F a coherent sheaf on Y, $H \subset X$ a Cartier divisor, and $H_Y := g^* H$. Assuming that F is S_m along H_Y, we would like to understand when $g_* F$ is S_m along H. If (the local equation of) H_Y is not a zero divisor on F, then the sequence

$$0 \to F(-H_Y) \to F \to F|_{H_Y} \to 0 \tag{2.83.3}$$

is exact. By push-forward we get the exact sequence

$$0 \to g_* F(-H_Y) \to g_* F \to g_*(F|_{H_Y}) \to R^1 g_* F(-H_Y), \tag{2.83.4}$$

and $R^1 g_* F(-H_Y) \simeq \mathscr{O}_X(-H) \otimes R^1 g_* F$. Thus, by (10.28), $g_* F$ is S_m along H if $R^1 g_* F = 0$, and $g_*(F|_{H_Y})$ is S_{m-1} along H. (In many cases, for instance if g is an isomorphism outside H_Y, these conditions are also necessary.)

We choose $F = \mathscr{O}_{X'}(mK_{X'} + \lfloor m\Delta' \rfloor)$. Then we need that

(5.a) $R^1 g_* \mathscr{O}_{X'}(mK_{X'} + \lfloor m\Delta' \rfloor) = 0$,

(5.b) $g_*(\mathscr{O}_{X'}(mK_{X'} + \lfloor m\Delta' \rfloor)|_{H_Y})$ is S_2 along H, and

(5.c) $g_* \mathscr{O}_{X'}(mK_{X'} + \lfloor m\Delta' \rfloor) \simeq \mathscr{O}_X(mK_X + \lfloor m\Delta \rfloor)$.

For us, (5.c) will be easy to satisfy. Using a Kodaira-type vanishing theorem, (5.a) needs some semipositivity condition on $(m-1)K_{X'} + \lfloor m\Delta' \rfloor$. By contrast, (11.61) suggests that (5.b) needs some negativity condition on $mK_{X'} + \lfloor m\Delta' \rfloor$.

The next result grew out of trying to satisfy the assumptions of both the relative Kodaira-type vanishing theorem and (11.61). The proof of (2.83) is then given in (2.85).

Proposition 2.84 *Let $(X, S + \Delta)$ be an lc pair where S is $\mathbb{Q}$-Cartier. Let B be a Weil $\mathbb{Z}$-divisor that is Mumford along S (4.68) and Θ an effective $\mathbb{R}$-divisor such that*

(2.84.1) $B \sim_{\mathbb{R}} -\Theta$,

(2.84.2) $\text{Supp} \, \Theta \leq \text{Supp}(\Delta^{(>1/2)})$, *and*

(2.84.3) $\lfloor \Theta \rfloor \leq \lfloor \Delta \rfloor$.

Then $\mathcal{O}_X(B)$ is S_3 along S.

Proof Assume first that $\lfloor \Theta \rfloor = 0$. A suitable cyclic cover, as in (11.25), reduces us to the case when S is Cartier. We assume this from now on.

(X, Δ) is also an lc pair and none of its lc centers are contained in S by (11.10.7). If B is $\mathbb{Q}$-Cartier then $\mathcal{O}_X(B)$ is S_3 along S by (11.20), applied with $\Delta' = 0$.

If B is not $\mathbb{Q}$-Cartier, we use (11.32) to obtain $\pi : (X', S' + \Delta') \to (X, S + \Delta)$. Note that B' is $\mathbb{Q}$-Cartier by (11.32.1), $(X', S' + \Delta')$ is lc and none of the lc centers of $(X', S' + \Delta' - \varepsilon \Theta')$ are contained in $\text{Ex}(\pi)$. In particular, S' is smooth at the generic points of all exceptional divisors of $\pi_S := \pi|_{S'} : S' \to S$. Thus B' is also Mumford along S', hence, as we proved at the beginning, $\mathcal{O}_{X'}(B')$ is S_3 along S'. Thus the sequence

$$0 \to \mathcal{O}_{X'}(B' - S') \to \mathcal{O}_{X'}(B') \to \mathcal{O}_{S'}(B'|_{S'}) \to 0 \qquad (2.84.4)$$

is exact by (2.74). Since $R^1 \pi_* \mathcal{O}_{X'}(B') = 0$ by (11.32.5), pushing (2.84.4) forward and using (11.32.4) gives an exact sequence

$$0 \to \mathcal{O}_X(B - S) \to \mathcal{O}_X(B) \to (\pi_S)_* \mathcal{O}_{S'}(B'|_{S'}) \to 0. \qquad (2.84.5)$$

Again by (2.74), $\mathcal{O}_X(B)$ is S_3 along S iff $(\pi_S)_* \mathcal{O}_{S'}(B'|_{S'})$ is S_2. The latter is equivalent to

$$(\pi_S)_* \mathcal{O}_{S'}(B'|_{S'}) \overset{?}{=} \mathcal{O}_S(B|_S). \qquad (2.84.6)$$

Now we apply (11.61) with $-N := B'|_{S'} + \Theta'|_{S'}$, which is numerically π_S-trivial. This gives that

$$(\pi_S)_* \mathcal{O}_{S'}(B'|_{S'} + \lfloor \Theta'|_{S'} \rfloor) = \mathcal{O}_S(B|_S). \qquad (2.84.7)$$

We are done if $\lfloor \Theta'|_{S'} \rfloor = 0$. This is where assumption (2) enters, in a seemingly innocent way. Indeed, (2.81.3) guarantees that $\lfloor \Theta'|_{S'} \rfloor = \lfloor \Theta' \rfloor|_{S'} = 0$ and $\lfloor \Theta' \rfloor = 0$ by our assumption (3).

The proof is similar if $\lfloor \Theta \rfloor \neq 0$, see Kollár (2018a, prop.28). □

2.85 (Proof of 2.83) We may assume that X is affine and $K_X + \Delta \sim_{\mathbb{R}} 0$. Pick a fiber X_c and let $x \in X_c$ be a point of codimension 1. Then either X_c and X are both smooth at x or X_c has a node and $x \notin \operatorname{Supp} \Delta$. Thus $mK_X + \lfloor m\Delta \rfloor$ is Cartier at x, hence a general divisor $B \sim mK_X + \lfloor m\Delta \rfloor$ is Mumford along X_c. We apply (2.84) to B with $\Theta := m\Delta - \lfloor m\Delta \rfloor$. Thus

$$B \sim mK_X + \lfloor m\Delta \rfloor = m(K_X + \Delta) - \Theta \sim_{\mathbb{R}} -\Theta.$$

By assumption $\Theta \leq \lceil \Delta^{(>1/2)} \rceil = \operatorname{Supp} \Delta$. So the assumptions of (2.84) are satisfied and $\mathscr{O}_X(mK_X + \lfloor m\Delta \rfloor) \simeq \mathscr{O}_X(B)$ is S_3 along X_c. By (2.75) this implies (2.83). $\square$

2.8 Local Stability in Codimension ≥ 3

In this section, we prove (2.7). If $K_X + D + \Delta$ is $\mathbb{R}$-Cartier, then (11.17) implies that f is locally stable. The $\mathbb{R}$-divisor case is reduced to the $\mathbb{Q}$-divisor case using (11.47). So from now on we may assume that Δ is a $\mathbb{Q}$-divisor. We need to show that $K_X + D + \Delta$ is $\mathbb{Q}$-Cartier.

We discuss three, increasingly general cases. The last one, treated in (2.88.5), implies (2.7).

2.86 Using the notation of (2.7), assume also that $(X_c, \operatorname{Diff}_{X_c} \Delta)$ is slc. (This holds for flat families of stable pairs.)

After localizing at a generic point of Z_c, we may assume that $Z_c = \{x\}$ is a point. Thus there is an $m > 0$ such that $m(K_X + D + \Delta)$ is a Cartier divisor on $X \setminus \{x\}$, whose restriction to $X_c \setminus \{x\}$ extends to a Cartier divisor on X_c. Since $\operatorname{codim}_{X_c} \{x\} \geq 3$, (2.91) implies that $m(K_X + D + \Delta)$ is a Cartier divisor.

2.87 Here we assume (2.7.2) and apply (11.42) to $(\bar{X}_c, \operatorname{Diff}_{\bar{X}_c} \Delta) \to X_c$. The conclusion is that there is an slc pair (X_c', Δ_c') and a finite morphism $\tau \colon X_c' \to X_c$, that is an isomorphism over $X_c \setminus Z_c$.

If X_c is S_2, then $X_c' \simeq X_c$, so $(X_c, \operatorname{Diff}_{X_c} \Delta)$ is slc, as in (2.86).

If X_c is not S_2, then, after localizing, we may assume that τ is an isomorphism, except at a point $x \in X_c$. Since $\tau^{-1}(x) \subset X_c'$ is finite, $m(K_{X_c'} + \operatorname{Diff}_{X_c'} \Delta)$ is trivial in a neighborhood of $\tau^{-1}(x)$ for some $m > 0$. Thus $m(K_{X_c} + \operatorname{Diff}_{X_c} \Delta)$ is trivial in a punctured neighborhood of x.

As before, $m(K_X + D + \Delta)$ is in the kernel of $\operatorname{Pic}^{\mathrm{loc}}(x, X) \to \operatorname{Pic}^{\mathrm{loc}}(x, X_c)$, but (2.91) guarantees its triviality only if $\operatorname{depth}_x X_c \geq 2$.

If $\operatorname{depth}_x X_c = 1$, then typically the kernel of $\operatorname{Pic}^{\mathrm{loc}}(x, X) \to \operatorname{Pic}^{\mathrm{loc}}(x, X_c)$ is a positive dimensional vector space; see Bhatt and de Jong (2014, 1.14) and Kollár (2016a, thm.7) for precise statements. Thus the kernel is p-torsion in char $p > 0$, but torsion free in char 0.

It is better to discuss this case in the more general setting of the following conjecture, where X_c is replaced by D.

Conjecture 2.88 *Let* $(X, D + \Delta)$ *be a demi-normal pair, where D is a reduced, $\mathbb{Q}$-Cartier divisor that is demi-normal in codimension 1, whose normalization* $(\bar{D}, \mathrm{Diff}_{\bar{D}} \Delta)$ *is lc. Let* $W \subset X$ *be a closed subset such that* $\mathrm{codim}_D(W \cap D) \geq 3$ *and* $(X \setminus W, (D + \Delta)_{X \setminus W})$ *is slc. Then the following are equivalent:*
(2.88.1) $(X, D + \Delta)$ *is slc in a neighborhood of D.*
(2.88.2) $(D, \mathrm{Diff}_D \Delta)$ *is slc.*
(2.88.3) $(\bar{D}, \mathrm{Diff}_{\bar{D}} \Delta)$ *is lc.*

The main difference between (2.7) and (2.88) is that in the latter we do not assume that $K_X + D + \Delta$ is $\mathbb{R}$-Cartier on $X \setminus D$.

Known implications Note that (1) $\Rightarrow$ (2) $\Rightarrow$ (3) follow from (11.17). If $K_X + D + \Delta$ is $\mathbb{R}$-Cartier, then (3) $\Rightarrow$ (1) also follows from (11.17). The arguments of (2.87) show that (3) $\Rightarrow$ (2) if D is S_2.

Thus it remains to show that if (3) holds, then $K_X + D + \Delta$ is $\mathbb{R}$-Cartier. As here, the $\mathbb{R}$-divisor case is reduced to the $\mathbb{Q}$-divisor case using (11.47), so from now on we assume that Δ is a $\mathbb{Q}$-divisor.

Special case 2.88.5 Assume that $K_X + D + \Delta$ is $\mathbb{R}$-Cartier on $X \setminus D$. Applying (5.41) gives a small, birational morphism $f \colon Y \to X$ such that $D_Y := f^{-1}(D) \to D$ is birational, $f(\mathrm{Ex}(f)) \subset D$, and it has codimension ≥ 3. There are two ways to get a contradiction from this.

First, note that the relative canonical divisor of $D_Y / \bar{D}$ is ample and is supported on the exceptional divisor of $D_Y \to D$. This cannot happen by (2.90).

Second, we use reduction to char p as in Bhatt and de Jong (2014) or Kollár and Mori (1998, p.14). In char $p > 0$, Bhatt and de Jong (2014, 1.14) shows that $K_{X_p} + D_p + \Delta_p$ is $\mathbb{Q}$-Cartier. By itself, this does not imply that $K_X + D + \Delta$ is $\mathbb{Q}$-Cartier.

However, if $K_X + D + \Delta$ is not $\mathbb{Q}$-Cartier, then $f \colon Y \to X$ is not an isomorphism. So $f_p \colon Y_p \to X_p$ is also not an isomorphism. By (2.89) this implies that $K_{X_p} + D_p + \Delta_p$ is not $\mathbb{Q}$-Cartier, a contradiction.   □

Special case 2.88.6 Assume that we are in a situation where the conclusion of (5.41) holds and X is a variety over a field of char 0.

As before, (5.41) gives a small, birational morphism $f \colon Y \to X$ such that $f(\mathrm{Ex}(f)) \cap D$ has codimension ≥ 3. The relative canonical divisor of $D_Y / \bar{D}$ is ample and is supported on the exceptional divisor of $D_Y \to D$. This gives a contradiction using (2.90).   □

Lemma 2.89 *Let* $\pi\colon Y \to X$ *be a proper birational morphism of normal schemes. Assume that* $Z := \mathrm{Ex}(\pi) \subset Y$ *has codimension* ≥ 2. *Let* M_Y *be a* π-*ample line bundle on* Y *and* M_X *a line bundle on* X *such that* $M_Y|_{Y\setminus Z} \simeq \pi^* M_X|_{Y\setminus Z}$. *Then* π *is an isomorphism.*

Proof Since Z has codimension ≥ 2, the assumed isomorphism extends to $M_Y \simeq \pi^* M_X$. If π contracts any curve C, then $0 < (C \cdot M_Y) = (C \cdot \pi^* M_X) = 0$ gives a contradiction. $\qquad\square$

We have used the following two theorems. The methods of the proofs would take us in other directions, so we give only some comments and references.

Proposition 2.90 (Kollár, 2013a, Prop.22) *Let* $f\colon Y \to X$ *be a projective, birational morphism of varieties over a field of* char 0. *Let* $D \subset X$ *be a Cartier divisor. Assume that* $f^{-1}(D) \to D$ *is birational and there is a nonzero (but not necessarily effective)* $\mathbb{Q}$-*Cartier divisor* E *on* $f^{-1}(D)$ *such that* $\dim f(\mathrm{Supp}\, E) \leq \dim D - 3$. *Then* $\mathrm{Ex}(f)$ *has codimension 1 in* Y.

Outline of proof The argument is topological over $\mathbb{C}$. Since the claim is algebraic, it would be very good to find a proof that works for arbitrary schemes.

We may assume that $x := f(\mathrm{Supp}\, E)$ is a point. Let V denote an open neighborhood of $f^{-1}(x) \subset f^{-1}(D)$ that retracts to $f^{-1}(x)$. The assumptions imply that, for $n := \dim D$, the cup product pairing

$$H^2(V, \partial V, \mathbb{Q}) \times H^{2n-2}(V, \mathbb{Q}) \to H^{2n}(V, \partial V, \mathbb{Q}) \qquad (2.90.1)$$

is nonzero. If $\mathrm{Ex}(f)$ has codimension ≥ 2, then f is small over a small deformation of D. This can be used to compute that (2.90.1) is zero, giving a contradiction. $\qquad\square$

We have the following Grothendieck–Lefschetz-type theorem, where, for a pointed scheme (x, X), we set $\mathrm{Pic}^{\mathrm{loc}}(x, X) := \mathrm{Pic}(\mathrm{Spec}_X \mathscr{O}_{x,X} \setminus \{x\})$.

Theorem 2.91 *Let* $(x \in X)$ *be an excellent, local scheme of pure dimension* ≥ 4 *such that* $\mathrm{depth}_x \mathscr{O}_X \geq 3$. *Let* $x \in D \subset X$ *be a Cartier divisor. Then we have an injective restriction map*

$$r_D^X\colon \mathrm{Pic}^{\mathrm{loc}}(x, X) \hookrightarrow \mathrm{Pic}^{\mathrm{loc}}(x, D). \qquad (2.91.1)$$

The original version (Grothendieck, 1968, XI.3.16), applies if $\mathrm{depth}_x \mathscr{O}_X \geq 4$. The current form was conjectured in Kollár (2013a) and proved there in the lc case. After Bhatt and de Jong (2014) and Kollár (2016a), the most general version is (Stacks, 2022, tag 0F2B).

The next results are very useful when dealing with Cartier divisors.

2.92 (Flat maps and Cartier divisors) Let $p\colon X \to Y$ be a morphism and D an effective Cartier divisor on Y. Under mild conditions p^*D is an effective Cartier divisor on Y. The converse also holds for flat morphisms.

Claim 2.92.1 Let $(R, m_R) \to (S, m_S)$ be a flat extension of local rings and $I_R \subset R$ an ideal. Then I_R is principal iff $I_R S$ is principal.

Proof One direction is clear. Conversely, assume that $I_R S$ is principal, thus $I_R S / m_S I_R S \simeq S / m_S$. Let $r_1, \ldots, r_n$ be generators of I_R. They also generate $I_R S$, hence at least one of them, say r_1, is not contained in $m_S I_R S$. Thus $(r_1) \subset I_R$ is a sub-ideal such that $r_1 S = I_R S$. Since $(R, m_R) \to (S, m_S)$ is faithfully flat, this implies that $(r_1) = I_R$. □

Pushing forward Cartier divisors is more problematic. For example, consider the natural map $\mathbb{P}^1_{\mathbb{Q}(i)} \to \mathbb{P}^1_{\mathbb{Q}}$. The points $(1{:}1)$ and $(i{:}1)$ are linearly equivalent, but their scheme-theoretic images have different degrees.

It is better to work with line bundles. Let $\pi\colon X \to Y$ be a finite, flat morphism of degree d. Let L be a line bundle on X. There are two natural ways of getting a line bundle on Y: the determinant of π_*L and the *norm,* denoted by $\mathrm{norm}_{X/Y}(L)$ as in Stacks (2022, tag 0BCX). The two are related by

$$\det(\pi_*L) \simeq (\mathrm{norm}_{X/Y} L) \otimes_Y \det(\pi_*\mathscr{O}_X).$$

The norm gives a group homomorphism $\mathrm{norm}_{X/Y}\colon \mathrm{Pic}(X) \to \mathrm{Pic}(Y)$ and there is a natural isomorphism $\mathrm{norm}_{X/Y}(\pi^*M) \simeq M^d$ for any line bundle M on Y.

Lemma 2.93 (Grothendieck, 1968, XIII.2.1) *Let* $(x \in X)$ *be a Noetherian, local scheme and* $x \in D \subset X$ *the support of a Cartier divisor. Assume that* $X \setminus Z$ *is connected for every closed subset* Z *of dimension* $\leq i + 1$. *Then* $D \setminus Z$ *is connected for every closed subset* Z *of dimension* $\leq i$. □

3

Families of Stable Varieties

We have defined stable and locally stable families over one-dimensional regular schemes in Sections 2.1 and 2.4. The first task in this chapter is to define these notions for families over more general base schemes. It turns out that this is much easier if there is no boundary divisor Δ. Since this case is of considerable interest, we treat it here before delving into the general setting in the next chapter. While restricting to the special case saves quite a lot of foundational work, the key parts of the proofs of the main theorems stay the same. To avoid repetition, we outline the proofs here, but leave detailed discussions to Chapter 4.

In Section 3.1 we review the theory of Chow varieties and Hilbert schemes. In general these suggest different answers to what a "family of varieties" or a "family of divisors" should be. The main conclusions, (3.11) and (3.13), can be summarized in the following principles:

- A *family of* S_2 *varieties* should be a flat morphism whose geometric fibers are reduced, connected, and satisfy Serre's condition S_2.
- Flatness is not the right condition for divisors on the fibers.

As in (2.46), a morphism $f\colon (X, \Delta) \to S$ is stable iff it is locally stable, proper and $K_{X/S} + \Delta$ is f-ample. Thus the key question is the right concept of local stability. There are many equivalent ways to define it when $\Delta = 0$.

Definition–Theorem 3.1 (Local stability over reduced schemes) Let S be a reduced scheme over a field of characteristic 0 and $f\colon X \to S$ a flat morphism of finite type whose fibers are semi-log-canonical (slc). Then f is *locally stable* iff the following equivalent conditions are satisfied:

(3.1.1) $K_{X/S}$ is $\mathbb{Q}$-Cartier.

(3.1.2) $\omega_{X/S}^{[m]}$ is an invertible sheaf for some $m > 0$.

(3.1.3) $\omega_{X/S}^{[m]}$ is flat with S_2 fibers for every $m \in \mathbb{Z}$.

(3.1.4) The restriction $\omega_{X/S}^{[m]} \to \omega_{X_s}^{[m]}$ is surjective for every $s \in S$ and $m \in \mathbb{Z}$.

(3.1.5) For every reduced W and morphism $q : W \to S$, the natural map

$$q_X^*(\omega_{X/S}^{[m]}) \to \omega_{X_W/W}^{[m]} \quad \text{is an isomorphism for every } m \in \mathbb{Z}.$$

(3.1.6) For every spectrum of a DVR T and morphism $q : T \to S$, the pull-back $f_T : X_T \to T$ satisfies the above (1–5).

(3.1.7) There is a closed subset $Z \subset X$ such that codim$(Z \cap X_s, X_s) \geq 3$ for every $s \in S$, and $f|_{X \setminus Z} : (X \setminus Z) \to S$ satisfies the above (1–6).

We prove the equivalence in (3.37). Over nonreduced bases, local stability is defined by (3.1.3); see (3.40). It implies all the other properties in (3.1), but is not equivalent to them; see Section 6.6 for such examples. The situation turns out to be much more complicated when $\Delta \neq 0$. Chapters 4 and 7 are entirely devoted to finding the right answers.

Let now $f : X \to S$ be a flat, projective family of S_2 varieties. It turns out that, starting in relative dimension 3, the set of points

$$\{s \in S : X_s \text{ is semi-log-canonical}\}$$

is not even locally closed; see (3.41) for an example. In order to describe the situation, in Section 3.2 we study functors that are representable by a locally closed decomposition (10.83).

We start the study of families of non-Cartier divisors in Section 3.3. As we have noted, this is one of the key new technical issues of the theory.

In Section 3.4 we use a representability theorem (3.36) to clarify the definition of stable and locally stable families, leading to the proof of (3.1). In Section 3.5 we bring these results together in (3.42) to prove the next main theorem of the chapter.

Theorem 3.2 (Local stability is representable) *Let S be a scheme over a field of characteristic 0 and $f : X \to S$ a projective morphism. Then there is a locally closed partial decomposition (10.83) $j : S^{ls} \to S$ such that the following holds.*

Let W be a scheme and $q : W \to S$ a morphism. Then the family obtained by base change $f_W : X_W \to W$ is locally stable iff q factors as $q : W \to S^{ls} \to S$.

Since ampleness is an open condition for a $\mathbb{Q}$-Cartier divisor, (3.2) implies the following.

Corollary 3.3 (Stability is representable) *Let S be a scheme over a field of characteristic 0 and $f : X \to S$ a projective morphism. Then there is a locally closed partial decomposition $j : S^{stab} \to S$ such that the following holds.*

Let W be a reduced scheme and $q: W \to S$ a morphism. Then the family obtained by base change $f_W: X_W \to W$ is stable iff q factors as $q: W \to S^{stab} \to S$.   □

Aside from some generalities, we have all the ingredients in place to construct the coarse moduli space of stable varieties. To formulate it, let $\mathcal{SV}$ (for stable varieties) denote the functor that associates to a scheme S the set of all stable families $f: X \to S$, up to isomorphism.

In order to get a moduli space of finite type, we fix the relative dimension n and the volume $v = \mathrm{vol}(K_{X_s}) := (K_{X_s}^n)$ of the fibers. This gives the subfunctor $\mathcal{SV}(n, v)$. The proof of the following is given in (6.18).

Theorem 3.4 (Moduli space of stable varieties) *Let S be a base scheme of characteristic 0 and fix n, v. Then the functor $\mathcal{SV}(n, v)$ has a coarse moduli space $\mathrm{SV}(n, v) \to S$, which is projective over S.*

Assumptions We work over arbitrary schemes in Sections 3.1–3.3, but over a field of characteristic 0 starting with Section 3.4.

3.1 Chow Varieties and Hilbert Schemes

What is a good family of algebraic varieties? Historically, two answers emerged to this question. The first one originates with Cayley (1860, 1862).[1] The corresponding moduli space is usually called the Chow variety. The second one is due to Grothendieck (1962); it is the theory of Hilbert schemes. For both of them, see Kollár (1996, chap.I), Sernesi (2006), or the original sources for details.

For the purposes of the following general discussion, a variety is a proper, geometrically reduced, and pure dimensional k-scheme.

The theory of Chow varieties suggests the following.

Definition 3.5 A *Cayley–Chow family* of varieties over a reduced base scheme S is a proper, pure dimensional (2.71) morphism $f : X \to S$, whose fibers X_s are generically geometrically reduced for every $s \in S$.

This is called an *algebraic family of varieties* in Hartshorne (1977, p.263). More general Cayley–Chow families are defined in Kollár (1996, sec.I.3).

It seems hard to make a precise statement, but one can think of Cayley–Chow families as being "topologically flat." That is, any topological

---

[1] The two papers have identical titles

consequence of flatness also holds for Cayley–Chow families. This holds for the Zariski topology, but also for the Euclidean topology if we are over $\mathbb{C}$.

There are two disadvantages of Cayley–Chow families. First, basic numerical invariants, for example, the arithmetic genus of curves, can jump in a Cayley–Chow family. Second, the topological nature of the definition implies that we completely ignore the nilpotent structure of S. In fact, it really does not seem possible to define what a Cayley–Chow family should be over an Artinian base scheme S.

The theory of Hilbert schemes was introduced to solve these problems. It suggests the following definition.

Definition 3.6 A *Hilbert–Grothendieck family* of varieties is a proper, flat morphism $f : X \to S$ whose fibers X_s are geometrically reduced and pure dimensional. (Here S is allowed to be nonreduced.)

Every Hilbert–Grothendieck family is also a Cayley–Chow family, and technically it is much better to have a Hilbert–Grothendieck family than a Cayley–Chow family. However, there are many Cayley–Chow families that are not flat.

3.7 (Universal families) Both Cayley–Chow and Hilbert–Grothendieck families are preserved by pull-backs, thus they form a functor. In both cases, this functor has a fine moduli space if we work with families that are subvarieties of a given scheme Y/S.

Let us thus fix a scheme Y that is projective over a base scheme S. For general existence questions, the key case is $Y = \mathbb{P}_S^N$. For any closed subscheme $Y \subset \mathbb{P}_S^N$, the Chow variety (resp. the Hilbert scheme) of Y is naturally a subvariety (resp. subscheme) of the Chow variety (resp. the Hilbert scheme) of $\mathbb{P}_S^N$. The corresponding universal family is obtained by restriction. (See (3.15) or (Kollár, 1996, sec.I.5) for some cases when Y/S is not projective.)

3.7.1 (*Chow variety*) (See Section 4.8 or Kollár (1996, sec.I.3) for details, and (3.14) for comments on seminormality.) There is a seminormal S-scheme $\mathrm{Chow}^\circ(Y/S)$ and a universal family $\mathrm{Univ}^\circ(Y/S) \to \mathrm{Chow}^\circ(Y/S)$ that represents the functor

$$\mathit{Chow}^\circ(Y/S)(T) := \left\{ \begin{array}{c} \text{closed subsets } X \subset Y \times_S T \text{ such that} \\ X \to T \text{ is a Cayley–Chow family of varieties} \end{array} \right\}$$

on seminormal S-schemes $q : T \to S$. ($\mathrm{Chow}^\circ(Y/S)$ is the "open" part of the full $\mathrm{Chow}(Y/S)$, as defined in Kollár (1996, Sec.I.3).) If we also fix a relatively very ample line bundle $\mathcal{O}_Y(1)$, then we can write

$$\text{Chow}^\circ(Y/S) = \amalg_n \text{Chow}_n^\circ(Y/S) = \amalg_{n,d} \text{Chow}_{n,d}^\circ(Y/S).$$

Here Chow_n° parametrizes varieties of dimension n and $\text{Chow}_{n,d}^\circ$ varieties of dimension n and of degree d. Each $\text{Chow}_{n,d}^\circ(Y/S)$ is of finite type, but usually still reducible.

3.7.2 (*Hilbert scheme*) (See Kollár (1996, Sec.I.1) or Sernesi (2006).) There is a universal family $\text{Univ}^\circ(Y/S) \to \text{Hilb}^\circ(Y/S)$ that represents the functor of Hilbert–Grothendieck families

$$\mathcal{H}ilb^\circ(Y/S)(T) := \left\{ \begin{array}{c} \text{closed subschemes } X \subset Y \times_S T \text{ such that} \\ X \to T \text{ is a flat family of varieties} \end{array} \right\}.$$

More generally, there is a universal family $\text{Univ}(Y/S) \to \text{Hilb}(Y/S)$ that represents the functor

$$\mathcal{H}ilb(Y/S)(T) := \left\{ \begin{array}{c} \text{closed subschemes } X \subset Y \times_S T \\ \text{such that } X \to T \text{ is flat} \end{array} \right\}.$$

We can write $\text{Hilb}(Y/S) = \amalg_n \text{Hilb}_n(Y/S) = \amalg_H \text{Hilb}_H(Y/S)$. Here Hilb_n parametrizes subschemes of (not necessarily pure) dimension n, and Hilb_H subschemes with Hilbert polynomial $H(t)$. Each $\text{Hilb}_H(Y/S)$ is projective, but usually still reducible.

3.8 (Comparing Chow and Hilb) Given a subscheme $X \subset Y$ of dimension $\le n$, we get an n dimensional cycle $[X] = \sum_i m_i[X_i]$, where X_i are the n-dimensional irreducible components and m_i is the length of $\mathcal{O}_X$ at the generic point of X_i. (Thus we completely ignore the lower dimensional irreducible components.)

If $m_i = 1$ for every i, then $[X] = \sum_i [X_i]$ can be identified with a point in $\text{Chow}^\circ(Y/S)$. In order to make this map everywhere defined, we need to extend the notion of Cayley–Chow families to allow fibers that are formal linear combinations of varieties; see Kollár (1996, sec.I.3) for details. The end result is an everywhere defined, set-theoretic map $\text{Hilb}_n(Y/S) \dashrightarrow \text{Chow}_n(Y/S)$. Since $\text{Hilb}_n(Y/S)$ is a scheme, but $\text{Chow}_n(Y/S)$ is a seminormal variety, it is better to think of it as a morphism defined on the seminormalization

$$\mathcal{R}_C^H : \text{Hilb}_n(Y/S)^{\text{sn}} \to \text{Chow}_n(Y/S). \tag{3.8.1}$$

This is a very complicated morphism. As written, its fibers have infinitely many irreducible components for $n \ge 1$, since we can just add disjoint zero-dimensional subschemes to any variety $X \subset Y$ to get new subschemes with the same underlying variety. Even if we restrict to pure dimensional subschemes, we get fibers with infinitely many irreducible components. This happens, for instance, for the fiber over $m[L] \in \mathrm{Chow}_{1,m}(\mathbb{P}^3)$, where $L \subset \mathbb{P}^3$ is a line and $m \geq 2$.

It is much more interesting to understand what happens on

$$\overline{\mathrm{Hilb}}_n^{\circ}(Y/S) := \text{closure of } \mathrm{Hilb}_n^{\circ}(Y/S) \text{ in } \mathrm{Hilb}_n(Y/S). \tag{3.8.2}$$

That is, $\overline{\mathrm{Hilb}}_n^{\circ}(Y/S)$ parametrizes n-dimensional subschemes that occur as limits of varieties. It turns out that the restriction of the Hilbert-to-Chow map

$$\mathcal{R}_C^H : \overline{\mathrm{Hilb}}_n^{\circ}(Y/S)^{\mathrm{sn}} \to \mathrm{Chow}_n(Y/S) \tag{3.8.3}$$

is a local isomorphism at many points. For smooth varieties this is quite clear from the definition of Chow-forms. Classical writers seem to have been fully aware of various equivalent versions, but I did not find an explicit formulation. The normal case, due to Hironaka (1958), is quite surprising; see Hartshorne (1977, III.9.11) for its usual form and (10.72) for a stronger version. These imply the following comparison of Hilbert schemes and Chow varieties.

Theorem 3.9 *Using the notation of (3.8), let $s \in S$ be a point and $X_s \subset Y_s$ a geometrically normal, projective subvariety of dimension n. Then the Hilbert-to-Chow morphism*

$$\mathcal{R}_C^H : \overline{\mathrm{Hilb}}_n^{\circ}(Y/S)^{\mathit{sn}} \to \mathrm{Chow}_n(Y/S)$$

is a local isomorphism over $[X_s] \in \mathrm{Chow}_n(Y/S)$. □

Informally speaking, for normal varieties, the Cayley–Chow theory is equivalent to the Hilbert–Grothendieck theory, at least over seminormal base schemes.

By contrast, $\mathrm{Hilb}(Y/S)$ and $\mathrm{Chow}(Y/S)$ are different near the class of a singular curve. For example, let $B \subset \mathbb{P}^3$ be a planar, nodal cubic. Then $[B]$ is contained in one irreducible component of $\mathrm{Hilb}_1(\mathbb{P}^3)$, but in two different irreducible components of $\mathrm{Chow}_1(\mathbb{P}^3)$. A general member of one component is a planar, smooth cubic. This component parametrizes flat deformations. A general member of the other component is a smooth, rational, nonplanar cubic. The arithmetic genus jumps, so these deformations are not flat. $\mathcal{R}_C^H$ is not a local isomorphism over $[B] \in \mathrm{Chow}_1(\mathbb{P}^3)$, but this is explained by the change of the genus. Once we correct for the genus change, (3.9) becomes stronger.

Definition 3.10 Let $X \subset \mathbb{P}^N$ be a closed subscheme of pure dimension n. The *sectional genus* of X is $1 - \chi(X \cap L, \mathcal{O}_{X \cap L})$, where $X \cap L$ is the intersection of X with $n - 1$ general hyperplanes. Knowing the degree of X and its sectional genus is equivalent to knowing the two highest coefficients of its Hilbert polynomial.

It is easy to see that the sectional genus is a constructible and upper semi-continuous function on $\mathrm{Chow}_n^\circ(Y/S)$; see (5.36). Thus there are locally closed subschemes $\mathrm{Chow}_{n,*,g}^\circ(Y/S) \subset \mathrm{Chow}_n^\circ(Y/S)$ that parametrize geometrically reduced cycles with sectional genus g; see (10.83). (The $*$ stands for the degree which we ignore in these formulas.) We can now define the Chow variety parametrizing families with locally constant sectional genus as

$$\mathrm{Chow}_n^{sg}(Y/S) := \amalg_{n,g} \mathrm{Chow}_{n,*,g}^\circ(Y/S)^{sn},$$

the disjoint union of the seminormalizations of the $\mathrm{Chow}_{n,*,g}^\circ(Y/S)$.

The sectional genus is constant in a flat family, thus we get the following strengthening of (3.9); see (5.36) and (10.71).

Theorem 3.11 *Using the notation of (3.8), let $s \in S$ be a point and $X_s \subset Y_s$ a geometrically reduced, projective, S_2 subvariety of pure dimension n. Then the Hilbert-to-Chow map*

$$\mathcal{R}_C^H : \overline{\mathrm{Hilb}_n^\circ(Y/S)}^{sn} \to \mathrm{Chow}_n^{sg}(Y/S)$$

is a local isomorphism over $[X_s] \in \mathrm{Chow}_n^{sg}(Y/S)$.

We can informally summarize these considerations as follows.

Principle 3.12 *For geometrically reduced, pure dimensional, projective, S_2 varieties, the Cayley–Chow theory is equivalent to the Hilbert–Grothendieck theory over seminormal base schemes, once we correct for the sectional genus.*

We are studying not just varieties, but slc pairs (X, Δ). The underlying variety is demi-normal, hence geometrically reduced and S_2. Thus (3.12) says that even if we start with the more general Cayley–Chow families, we end up with flat morphisms $f : X \to S$ with S_2 fibers. The latter is a class that is well-behaved over arbitrary base schemes.

However, the divisorial part is harder to understand. Although we have seen only a few examples supporting it, the following counterpart of (3.12) turns out to give the right picture.

Principle 3.13 *For stable families of slc pairs (X, Δ), the Hilbert–Grothendieck theory is optimal for the underlying variety X, but the Cayley–Chow theory is the "right" one for the divisorial part Δ.*

3.14 (Comment on seminormality) Hilbert schemes work well over any base scheme, but in Kollár (1996) the theory of Cayley–Chow families is developed only over seminormal bases. Following the methods of Section 4.8, it is possible to work out the Cayley–Chow theory of geometrically reduced cycles over reduced bases. In characteristic 0 this works for all cycles by Barlet (1975); Barlet and Magnússon (2020), but examples of Nagata (1955) suggest that, in positive characteristic, the restriction to seminormal bases may be necessary.

3.15 (Nonprojective cases) Let Y be an algebraic space over S. We define $\mathcal{H}ilb(Y/S)(T)$ as the set of all subspaces $X \subset Y \times_S T$ that are proper and flat over T. Artin (1969) proves that if $Y \to S$ is locally of finite presentation then the Hilbert functor is represented by an algebraic space $\mathrm{Hilb}(Y/S) \to S$ that is also locally of finite presentation.

Most likely similar results hold for $\mathrm{Chow}(Y/S)$; see Kollár (1996, sec.I.5). The complex analytic case is worked out in Barlet and Magnússon (2020).

3.2 Representable Properties

Let $\mathcal{P}$ be a property of schemes. For a morphism $f : X \to S$ consider the set $S(\mathcal{P}) := \{s \in S : X_s \text{ satisfies } \mathcal{P}\}$. Note that $S(\mathcal{P})$ depends on $f : X \to S$, so we use the notation $S(\mathcal{P}, X/S)$ if the choice of $f : X \to S$ is not clear.

In nice situations, $S(\mathcal{P})$ is an open or closed subset of S. For example satisfying Serre's condition S_m is an open condition for proper, flat morphisms by (10.12), and being singular is a closed condition.

Similarly, if $f : X \to S$ is a proper morphism of relative dimension 1, then $\{s \in S : X_s \text{ is a stable curve}\}$ is an open subset of S. However, we see in (3.41), that if $f : X \to S$ is a proper, flat morphism of relative dimension ≥ 3 then $\{s \in S : X_s \text{ is a stable variety}\}$ is not even a locally closed subset of S.

We already noted in Section 1.4 that flat morphisms with stable fibers do not give the right moduli problem in higher dimensions. One should look at stable families instead. Thus our main interest is in the class of morphisms $q : T \to S$ for which the pulled-back family $f_T : X_T \to T$ is stable. We then hope to prove that this happens in a predictable way. The following definition formalizes this.

Definition 3.16 Let $\mathcal{P}$ be a property of morphisms that is preserved by pull-back. That is, if $X \to S$ satisfies $\mathcal{P}$ and $q: T \to S$ is a morphism, then $f_T: X_T \to T$ also satisfies $\mathcal{P}$. Depending on the situation, pull–back can mean the usual fiber product $X_T := X \times_S T$, the hull pull-back to be defined in (3.27), the divisorial pull-back to be defined in (4.6), or the Cayley–Chow pull-back of Kollár (1996, I.3.18).

The functor of $\mathcal{P}$-pull-backs is defined for morphisms $T \to S$ by setting

$$Property(\mathcal{P})(T) := \begin{cases} \{\emptyset\} & \text{if } X_T \to T \text{ satisfies } \mathcal{P}, \text{ and} \\ \emptyset & \text{otherwise.} \end{cases} \qquad (3.16.1)$$

(That is, $Property(\mathcal{P})(T)$ is either empty or consists of a single element.) Thus a morphism $i_P : S^P \to S$ *represents* $\mathcal{P}$-pull-backs iff the following hold:

(3.16.2) $f^P : X^P := X_{S^P} \to S^P$ satisfies $\mathcal{P}$, and

(3.16.3) if $f_T : X_T \to T$ satisfies $\mathcal{P}$, then q factors as $q : T \to S^P \to S$, and the factorization is unique.

It is also of interest to understand what happens if we focus on special classes of bases. Let $\mathcal{R}$ be a property of schemes. We say that $i_P : S^P \to S$ *represents* $\mathcal{P}$-pull-backs for $\mathcal{R}$-schemes if S^P satisfies $\mathcal{R}$ and (3.16.3) holds whenever T satisfies $\mathcal{R}$. In this section we are mostly interested in the properties $\mathcal{R} =$ (reduced), $\mathcal{R} =$ (seminormal), and $\mathcal{R} =$ (normal).

If (3.16.3) holds for all $T =$ (spectrum of a field), then $i_P : S^P \to S$ is geometrically injective (10.82). If (3.16.3) holds for all Artinian schemes, then i_P is a monomorphism (10.82).

In many cases of interest, $\mathcal{P}$ is invariant under base field extensions. Then $i_P : S^P \to S$ also preserves residue fields (10.82).

If $X \to S$ is projective, then we are frequently able to prove that $i_P : S^P \to S$ is a locally closed partial decomposition (10.83).

If $i_P : S^P \to S$ represents $\mathcal{P}$-pull-backs and i_P is of finite type (this will always be the case for us), then $S(\mathcal{P}) = \{s : X_s \text{ satisfies } \mathcal{P}\} = i_P(S^P)$ is a constructible subset of S. Constructibility is much weaker than representability, but we frequently need it in our proofs of representability.

Example 3.17 (Simultaneous normalization) Sometimes it is best to focus not on a property of a morphism, but on a property of its "improvement." We say that $f : X \to S$ has *simultaneous normalization* if there is a finite morphism $\pi : \bar{X} \to X$ such that $\pi_s : \bar{X}_s \to X_s$ is the normalization for every $s \in S$ and $f \circ \pi : \bar{X} \to S$ is flat. For example, consider the family of quadrics

$$X := (x_0^2 - x_1^2 + u_2 x_2^2 + u_3 x_3^2 = 0) \subset \mathbb{P}_{\mathbf{x}}^3 \times \mathbb{A}_{\mathbf{u}}^2.$$

Then $\{(0,0)\} \amalg (\mathbb{A}_{\mathbf{u}}^2 \setminus \{(0,0)\}) \to \mathbb{A}_{\mathbf{u}}^2$ represents the functor of simultaneous normalizations. In general, we have the following result, due to Chiang-Hsieh and Lipman (2006) and Kollár (2011b).

Claim 3.17.1 Let $f : X \to S$ be a proper morphism whose fibers X_s are generically geometrically reduced. Then there is a morphism $\pi : S^{sn} \to S$ such that, for any $g : T \to S$, the fiber product $X \times_S T \to T$ has a simultaneous normalization iff g factors through $\pi : S^{sn} \to S$. □

Definition 3.18 Let $f : X \to S$ be a morphism and F a coherent sheaf on X. Given any $q : W \to S$, we get

$$
\begin{array}{ccc}
X \times_S W =: X_W & \xrightarrow{\;q_X\;} & X \\
{\scriptstyle f_W}\downarrow & & \downarrow{\scriptstyle f} \\
W & \xrightarrow{\;q\;} & S.
\end{array}
\qquad (3.18.1)
$$

As in (3.16.1), we have the *functor of flat pull-backs* $Flat(F)(*)$.

One of the most useful representation theorems is the following; see Mumford (1966, Lect.8) and Artin (1969).

Theorem 3.19 (Flattening decomposition) *Let $f : X \to S$ be a proper morphism and F a coherent sheaf on X. Then the functor of flat pull-backs $Flat(F)(*)$ is represented by a monomorphism $i^{flat} : S^{flat} \to S$ that is locally of finite type. If f is projective then i^{flat} is a locally closed decomposition.* □

Example 3.19.1 As a trivial special case, assume that $X = S$ is affine. Write F as the cokernel of a map of free sheaves $g : \mathcal{O}_S^n \to \mathcal{O}_S^m$. Then F is free of rank $m - r$ precisely on the subscheme (rank $g \leq r$) \ (rank $g \leq r - 1$).

One can frequently check flatness using the following numerical criterion which is proved, but not fully stated, in Hartshorne (1977, III.9.9).

Theorem 3.20 *Let $f : X \to S$ be a projective morphism with relatively ample $\mathcal{O}_X(1)$ and F a coherent sheaf on X. The following are equivalent:*

(3.20.1) *F is flat over S.*

(3.20.2) *$f_*(F(m))$ is locally free for $m \gg 1$.*

If S is reduced then these are also equivalent to the following:

(3.20.3) *$s \mapsto \chi(X_s, F_s(m))$ is a locally constant function on S.* □

In Chapter 8 we need the following results.

Proposition 3.21 *Let $f : X \to S$ be a proper morphism and G a coherent sheaf on X, flat over S. The following properties of morphisms $q : T \to S$ are representable by locally closed subschemes:*

(3.21.1) *$(f_T)_* q_X^* G$ is locally free of rank r and commutes with base change.*

(3.21.2) $(f_T)_*q_X^*G$ *is locally free of rank r, commutes with base change, and* q_X^*G *is relatively globally generated.*

Proof Using the notation of (3.24.1), locally we can write d_1 in as a matrix with entries in $\mathscr{O}_S$. Then $(\operatorname{rank} d_1 \leq r) \subset S$ is the subscheme defined by the vanishing of the determinants of all $(r+1) \times (r+1)$-minors. With this definition we see that $(\operatorname{rank} d_1 \leq \operatorname{rank} K^0 - r) \setminus (\operatorname{rank} d_1 \leq \operatorname{rank} K^0 - r - 1)$ represents the functor (3.21.1).

For (3.21.2) we may assume that f_*G is locally free of rank r. Then (2) is represented by the open subscheme $S \setminus f(\operatorname{Supp} \operatorname{coker}(f^*f_*G \to G))$.   □

Corollary 3.22 *Let* $f : X \to S$ *be a proper morphism and* G *a coherent sheaf on* X, *flat over* S. *Assume that* $H^0(X_s, \mathscr{O}_{X_s}) \simeq k(s)$ *for* $s \in S$.

Then there is a locally closed subscheme $S' \hookrightarrow S$ *such that, a morphism* $q : T \to S$ *factors through* S' *iff* q_X^*G *is isomorphic to the pull-back of a line bundle from* T.

Proof If q_X^*G is isomorphic to the the pull-back of a line bundle from T, then $\mathscr{O}_{X_T}$ is locally isomorphic to q_X^*G, hence X_T is flat over T. Thus S' factors through the flattening decomposition of f (3.19). We may thus assume that f is flat and S is affine. Since $H^0(S, \mathscr{O}_S) \to H^0(X_s, \mathscr{O}_{X_s})$ is surjectve, so is $H^0(X, \mathscr{O}_X) \to H^0(X_s, \mathscr{O}_{X_s})$, hence $f_*\mathscr{O}_X \simeq \mathscr{O}_S$ by cohomology and base change. So we are in the $r = 1$ case of (3.21.2).   □

Remark 3.23 Being pure dimensional is an open property for flat, proper morphisms. Thus, using (3.19) we obtain that for any projective morphism $f : X \to S$ we have a locally closed partial decomposition $S^{fp} \to S$ that represents flat and pure dimensional pull-backs of f. Next let $\mathcal{P}$ be a property that implies flat and pure dimensional. Assume that $q : T \to S$ is a morphism such that $f_T : X_T \to T$ satisfies $\mathcal{P}$. Then $f_T : X_T \to T$ is also flat and pure dimensional, hence $q : T \to S$ factors through f^{fp}. Thus $S^{P} = (S^{fp})^{P}$.

In particular, if we want to prove that $S^{P} \to S$ exists for all projective morphisms, then it is enough to show that it exists for all flat, pure dimensional and projective morphisms. More generally, if $\mathcal{P}_1 \Rightarrow \mathcal{P}_2$ and S^{P_2} exists, then

$$S^{P_1} = (S^{P_2})^{P_1}. \tag{3.23.1}$$

3.24 (Semicontinuity) Let $f : X \to S$ be a proper morphism and G a coherent sheaf on X, flat over S. By a version of the semicontinuity theorem, there is a finite complex of locally free sheaves on S

$$K^{\bullet} := 0 \to K^0 \xrightarrow{d_1} K^1 \xrightarrow{d_2} \cdots \xrightarrow{d_{n-1}} K^n \to 0, \tag{3.24.1}$$

such that, for every morphism $h : T \to S$,

$$R^i(f_T)_* h_X^* G \simeq H^i(h^* K^\bullet). \tag{3.24.2}$$

(This form is stated and proved in Mumford (1970, §5); Hartshorne (1977, III.12.2) has a weaker statement, but the proof works to give this.)
    This can be used to define

$$\det R^\bullet f_* G := (\textstyle\prod_{\text{even}} \det K^i) \otimes (\textstyle\prod_{\text{odd}} \det K^i)^{-1}. \tag{3.24.3}$$

This is independent of the choices made. If $R^i f_* G = 0$ for $i > 0$, then $\det R^\bullet f_* G = \det f_* G$. This is the main case that we use.

3.3 Divisorial Sheaves

We frequently have to deal with divisors $D \subset X$ that are not Cartier, hence the corresponding sheaves $\mathscr{O}_X(D)$ are not always locally free. Understanding families of such sheaves is a key aspect of the moduli problem. Many of the results proved here are developed for arbitrary coherent sheaves in Chapter 9.

Definition 3.25 (Divisorial sheaves) A coherent sheaf L on a scheme X is called a *divisorial sheaf* if L is S_2 and there is a closed subset $Z \subset X$ of codimension ≥ 2 such that $L|_{X \setminus Z}$ is locally free of rank 1.
    We are mostly interested in the cases when X itself is demi-normal, but the definition makes sense in general, although with unexpected properties. For example, $\mathscr{O}_X$ is a divisorial sheaf iff X is S_2.
    Set $U := X \setminus Z$ and let $j : U \hookrightarrow X$ denote the natural injection. Then $L = j_*(L|_U)$ by (10.6), thus L is uniquely determined by $L|_U$.
    If $\dim X = 1$, then $Z = \emptyset$, so a divisorial sheaf is invertible. If D is a Mumford divisor, then $\mathscr{O}_X(D)$ is a divisorial sheaf. If X is demi-normal, then the $\omega_X^{[m]}$ are divisorial sheaves. Divisorial sheaves form a group, with

$$L_{[\otimes]} M := j_*(L|_U \otimes M|_U). \tag{3.25.1}$$

For powers, we use the notation $L^{[m]} := (L^{\otimes m})^{[**]}$.
    Let $H \subset X$ be a general member (depending on L, M) of a base point free linear system. Then $L|_H, M|_H$ are divisorial sheaves and $(L_{[\otimes]} M)|_H = L|_H {}_{[\otimes]} M|_H$; see (10.18).
    Let $f : X \to S$ be a morphism. A coherent sheaf L on X is a *flat family of divisorial sheaves,* if L is flat over S and its fibers are divisorial sheaves. (L need not be a divisorial sheaf on X.)

Given any $q : T \to S$ with induced $q_X : X \times_S T \to X$, the pull-back $q_X^* L$ is again a flat family of divisorial sheaves.

Let $f : X \to S$ be a morphism. We frequently need to deal with properties that hold not everywhere, but only on open subsets of each fiber.

Definition 3.26 Let $f : X \to S$ be a morphism and F a coherent sheaf on X. We say that F is *generically flat* (resp. *mostly flat*) over S, if there is a dense, open subset $j : U \hookrightarrow X$ such that

(3.26.1) $F|_U$ is flat over S, and

(3.26.2) Supp $F_s \setminus U$ has codimension ≥ 1 (resp. ≥ 2) in Supp F_s for $s \in S$. We usually set $Z := X \setminus U$.

A subscheme $Y \subset X$ is generically (resp. mostly) flat iff $\mathcal{O}_Y$ is.

Definition 3.27 (Hull and hull pull-back) With $j : U \hookrightarrow X$ as in (3.26), let F be a mostly flat family of coherent sheaves. Assume that $F|_U$ has S_2 fibers. We imagine that F is the "correct" object over U, but a mistake may have been made over $Z = X \setminus U$. We correct F by replacing it with its *hull*

$$F^H := j_*(F|_U).\tag{3.27.1}$$

Under mild conditions (for example, when X is excellent), F^H is a coherent sheaf on X; see Chapter 9 for a detailed treatment of hulls.

Let $q : W \to S$ be a morphism. We get a fiber product diagram as in (3.18.1). Then $F_W := q_X^* F$ has S_2 fibers over $q_X^{-1}(U)$. Its hull F_W^H is called the *hull pull-back* of F. If confusion is likely, we use $(F_W)^H$ to denote the hull of the pull-back and $(F^H)_W$ to denote pull-back of the hull F^H.

We are especially interested in the maps

$$r_W^S : (F^H)_W \to (F_W)^H.\tag{3.27.2}$$

We have already encountered these in (2.75) when $W = \{s\}$ is a point. For applications the key is to understand when F^H is flat. The following basic observations guide us:

(3.27.3) F^H is flat with S_2 fibers over a dense, open $S^\circ \subset S$ by (10.11).

(3.27.4) We see in (9.36) that F^H is flat with S_2 fibers $\Leftrightarrow r_W^S$ is an isomorphisms for every $q : W \to S \Leftrightarrow r_s^S$ is surjective for every $s \in S$.

Definition 3.28 Using the notation of (3.26), F is a *mostly flat family of S_2 sheaves* if $F|_U$ is flat with S_2 fibers and $F = F^H$.

L is a *mostly flat family of divisorial sheaves* if L is invertible on U.

For now, we study these problems for divisorial sheaves. The first main result is the following special case of (9.40), the second is (4.32).

Theorem 3.29 *Let* $f : X \to S$ *be a projective morphism and* L *a mostly flat family of divisorial sheaves on* X *(3.28). Then there is a locally closed decomposition* $j : S^{H\text{-}flat} \to S$ *such that, for every morphism* $q : W \to S$, *the hull pull-back* L_W^H *is a flat family of divisorial sheaves (3.25) on* X_W, *iff* q *factors as* $q : W \to S^{H\text{-}flat} \to S$. □

Corollary 3.30 *Let* $f : X \to S$ *be a flat, projective morphism with* S_2 *fibers and* L *a mostly flat family of divisorial sheaves on* X. *Then there is a locally closed partial decomposition* $j : S^{inv} \to S$ *such that, for every morphism* $q : W \to S$, *the hull pull-back* L_W^H *is invertible, iff* q *factors as* $q : W \to S^{inv} \to S$.

Proof For flat morphisms with S_2 fibers, an invertible sheaf is also a flat family of divisorial sheaves. Thus if L_W^H is invertible, then q factors through $S^{H\text{-}flat} \to S$. So, by (3.23.1), $S^{inv} = (S^{H\text{-}flat})^{inv}$. For a flat family of sheaves, being invertible is an open condition, thus S^{inv} is open in $S^{H\text{-}flat}$. □

The following variant turns out to be very useful in (3.42) and (6.24).

Proposition 3.31 *Let* $f : X \to S$ *be a flat, projective morphism with* S_2 *fibers and* $N_1, \ldots, N_s, L_1, \ldots, L_r$ *mostly flat families of divisorial sheaves. Then there is a locally closed partial decomposition* $S^{NL} \to S$ *such that, a morphism* $q : T \to S$ *factors through* S^{NL} *iff the following hold:*
(3.31.1) *The hull pull-backs* $(L_j)_T^H$ *are invertible, and*
(3.31.2) *the* $(N_i \otimes L_1^{[m_1]} \otimes \cdots \otimes L_r^{[m_r]})_T^H$ *are flat families of divisorial sheaves for every* $m_i \in \mathbb{Z}$.

Proof We apply (3.29) to each N_i and (3.30) to each L_j to get locally closed partial decompositions $S^{N_i} \to S$ and $S^{L_j} \to S$ that represent the functors of flat hull pull-backs with S_2 fibers for N_i and L_j, plus invertibility for the L_j. Let $S^* \to S$ denote the fiber product of all of them.

It is clear that S^{NL} factors through S^*. Tensoring with an invertible sheaf preserves flat families of divisorial sheaves, thus $S^{NL} = S^*$. □

The following analog of (3.20) is a special case of (9.36), where for polynomials we use the ordering $f(*) \le g(*) \Leftrightarrow f(t) \le g(t) \, \forall t \gg 1$ as in (5.14).

Theorem 3.32 *Let* S *be a reduced scheme,* $f : X \to S$ *a projective morphism with ample* $\mathcal{O}_X(1)$ *and* L *a mostly flat family of divisorial sheaves on* X. *Then*

(3.32.1) $s \mapsto h^0(X_s, L_s^H)$ is constructible and upper semi-continuous,

(3.32.2) $s \mapsto \chi(X_s, L_s^H(*))$ is constructible, upper semi-continuous, and

(3.32.3) L is a flat family of divisorial sheaves (3.25) iff $s \mapsto \chi(X_s, L_s^H(*))$ is locally constant on S.   □

Remark 3.32.4. Recall that by (3.20) a coherent sheaf G is flat over S iff $s \mapsto \chi(X_s, G_s(*))$ is locally constant on S. However, the assumptions of (3.32) are quite different since L_s^H is not assumed to be the fiber of L over s. In fact, usually there is no coherent sheaf on X whose fiber over s is isomorphic to L_s^H for every $s \in S$. The map $r_s^S : L_s \to L_s^H$ is an isomorphism over U_s, but both its kernel and the cokernel can be nontrivial. They have opposite contributions to the Euler characteristic.

3.33 (Hilbert function of divisorial sheaves) Let X be a proper scheme of dimension n and L, M line bundles on X. The Hirzebruch–Riemann–Roch theorem computes $\chi(X, L \otimes M^r)$ as a polynomial of r. Its leading terms are

$$\chi(X, L \otimes M^r) = \frac{r^n}{n!}(M^n) + \frac{r^{n-1}}{2(n-1)!}\left((\tau_1(X) + 2L) \cdot M^{n-1}\right) + O(r^{n-2}), \quad (3.33.1)$$

where τ_1 is the first Todd class.

Assume next that L is invertible only outside a subset $Z \subset X$ of codimension ≥ 2. By blowing up L, we get a proper birational morphism $\pi : X' \to X$ and a line bundle L' such that $\pi_*L' = L$. Thus we can compute $\chi(X, L \otimes M^r)$ as $\chi(X', L' \otimes \pi^*M^r)$, modulo an error term which involves the sheaves $R^i\pi_*L'$. These are supported on Z, hence the $\chi(X, R^i\pi_*L' \otimes M^r)$ all have degree $\leq n - 2$. Thus we again obtain (3.33.1), and, if X is demi-normal, then $\tau_1(X) = -K_X$.

If, in addition, $L^{[m]}$ is locally free for some $m > 0$, then applying (3.33.1) to $L \mapsto L^{[a]}$ for all $0 \leq a < m$ and $M = L^{[m]}$ we end up with the expected formula

$$\chi(X, L^{[r]}) = \frac{r^n}{n!}(L^n) - \frac{r^{n-1}}{2(n-1)!}(K_X \cdot L^{n-1}) + O(r^{n-2}). \quad (3.33.2)$$

Further note that $\chi(X, L^{[r]})$ is a polynomial on any translate of $m\mathbb{Z}$, so one can write the $O(r^{n-2})$ summand as $\sum_{i=0}^{n-2} a_i(r)r^i$, where the $a_i(r)$ are periodic functions that depend on X and L.

3.34 (Hilbert function of slc varieties) Let X be a proper, slc variety of dimension n. We are especially interested in $r \mapsto \chi(X, \omega_X^{[r]})$, which we call the *Hilbert function* of ω_X. By (3.33), we can write it as

$$\chi(X, \omega_X^{[r]}) = \frac{r^n}{n!}(K_X^n) - \frac{r^{n-1}}{2(n-1)!}(K_X^n) + \sum_{i=0}^{n-2} a_i(r)r^i, \quad (3.34.1)$$

where the $a_i(r)$ are periodic functions with period $= \text{index}(X)$.

By (11.34), if ω_X is ample and the characteristic is 0, then, for $i, r \geq 2$,

$$h^0(X, \omega_X^{[r]}) = \chi(X, \omega_X^{[r]}), \quad \text{and} \quad h^i(X, \omega_X^{[r]}) = 0. \tag{3.34.2}$$

Comment on the terminology It might seem natural to call $r \mapsto h^0(X, \omega_X^{[r]})$ the Hilbert function. However, (3.34.1) is not a polynomial in general. For stable varieties the two variants differ only for $r = 1$ by (3.34.2).

3.4 Local Stability

Definition 3.35 (Relative canonical class) Let $f : X \to S$ be a flat, projective morphism with demi-normal fibers. The relative canonical sheaf $\omega_{X/S}$ was constructed in (2.68).

Let $Z \subset X$ be the subset where the fibers are neither smooth nor nodal. Set $j : U := X \setminus Z \hookrightarrow X$. Then $X_s \cap Z$ has codimension ≥ 2 for every fiber X_s and $\omega_{U/S}$ is locally free. Thus $\omega_{X/S}$ is a mostly flat family of divisorial sheaves. The corresponding divisor class is denoted by $K_{X/S}$. As in (3.25), we define its reflexive powers by the formula

$$\omega_{X/S}^{[m]} := j_*(\omega_{U/S}^m) \simeq \mathscr{O}_X(mK_{X/S}). \tag{3.35.1}$$

All these also hold for flat, finite type morphisms (that are not necessarily projective) by (2.68.7).

If the fibers of $f : X \to S$ are slc, then $\omega_{X/S}$ is a flat family of divisorial sheaves by (2.67). However, its reflexive powers are usually only mostly flat over S. Applying (3.30) to $\omega_{X/S}^{[m]}$ gives the following, which turns out to be the key to our treatment of local stability over reduced schemes.

Corollary 3.36 *Let $f : X \to S$ be a flat, projective family of demi-normal varieties and fix $m \in \mathbb{Z}$. Then there is a locally closed decomposition $j : S^{[m]} \to S$ such that the following holds.*

Let $q : W \to S$ be a morphism. Then $\omega_{X_W/W}^{[m]}$ is a flat family of divisorial sheaves iff q factors as $q : W \to S^{[m]} \to S$. □

In applications of (3.36), a frequent problem is that $S^{[m]}$ depends on m, even if we choose m to be large and divisible; see (2.45) for such an example.

3.37 (Proof of 3.1) Assertions (3.1.1) and (3.1.2) say the same using different terminology. The equivalences of (3.1.3–5) follow from (9.17).

Assume (3.1.3) and pick $s \in S$. Since X_s is slc, $\omega_{X_s}^{[m_s]}$ is locally free for some $m_s > 0$. In a flat family of sheaves being invertible is an open condition, thus $\omega_{X/S}^{[m_s]}$ is invertible in an open neighborhood $X_s \subset U_s \subset X$. Finitely many of these U_{s_i} cover X. Then $m = \mathrm{lcm}\{m_{s_i}\}$ works for (3.1.2).

It is clear that (3.1.1) implies (3.1.6) and for (3.1.6) $\Rightarrow$ (3.1.3) we argue as follows. We need to prove that $\omega_{X/S}^{[m]}$ is a flat family of divisorial sheaves. This is a local question on S, hence we may assume that $(0 \in S)$ is local.

Let us discuss first the case when f is projective. By (3.36), the property

$$\mathcal{P}^{[m]}(W) := (\omega_{X_W/W}^{[m]} \text{ is a flat family of divisorial sheaves})$$

is representable by a locally closed decomposition $i_m : S^{[m]} \to S$. We aim to prove that i_m is an isomorphism.

For each generic point $g_i \in S$, choose a local morphism $q_i : (0_i \in T_i) \to (0 \in S)$ that maps the generic point $t_i \in T_i$ to g_i. By assumption, $X_{T_i} \to T_i$ is locally stable, hence $\omega_{X_{T_i}/T_i}^{[m]}$ is a flat family of divisorial sheaves by (2.79.2). Thus q_i factors through $i_m : S^{[m]} \to S$. Therefore, $i_m : S^{[m]} \to S$ is an isomorphism by (10.83.2), completing the proof for projective morphisms.

This argument also works in the nonprojective case, provided $i_m : S^{[m]} \to S$ exists. As we discuss in Section 9.8, the latter is unlikely. However, if S is local, complete, and we aim to represent flat hull pull-backs for local morphisms, then $i_m : S^{[m]} \to S$ exists; see (9.44) for details. The rest of the argument now works as before; see also (3.38).

Finally, if any of the properties (3.1.1–6) holds for X, then it also holds for $X \setminus Z$. The surprising part is the converse. By using (3.1.6) both for X and for $X \setminus Z$, it is enough to see that (3.1.7) $\Rightarrow$ (3.1.1) holds when S is the spectrum of a DVR. The latter is proved in (2.7). □

Corollary 3.38 *Let S be a reduced scheme over a field of characteristic 0 and $f : X \to S$ a flat family of demi-normal varieties. Let $T \to S$ be faithfully flat. Then $X \to S$ is locally stable iff $X_T \to T$ is.* □

Corollary 3.39 *Let $f : X \to S$ be a flat, proper morphism of finite type with demi-normal fibers such that $K_{X/S}$ is $\mathbb{Q}$-Cartier. Then*

$$S^{\mathrm{slc}} := \{s : X_s \text{ is slc }\} \subset S \quad \text{is open.} \tag{3.39.1}$$

Proof By (10.14), a set $U \subset S$ is open iff it is closed under generalization and U contains a dense open subset of $\bar{s}$ for every $s \in U$.

For S^{slc}, the first of these follows from (2.3). In order to see the second, assume first that X_s is lc. Then mK_{X_s} is Cartier for some $m > 0$, hence $mK_{X/S}$ is Cartier over an open neighborhood of $s \in U_s \subset \bar{s}$. Next consider a log resolution $p_s : Y_s \to X_s$. It extends to a simultaneous log resolution $p° : Y° \to X°$ over a suitable $U_s° \subset \bar{s}$. Thus, if $E° \subset Y°$ is any exceptional divisor, then $a(E_t, X_t) = a(E°, X°) = a(E_s, X_s)$ for every $t \in U_s°$. This shows that all fibers over $U_s°$ are lc.

If X_s is not normal, one can use either a simultaneous semi-log-resolution (Kollár, 2013b, sec.10.4) or normalize first, apply the above argument, and descend to X, essentially by definition (11.37). □

3.5 Stability Is Representable I

Focusing on the property (3.1.3), over nonreduced bases we get the definition of local stability, due to Kollár and Shepherd-Barron (1988).

Definition 3.40 (Local stability and stability II) Let S be a scheme over a field of characteristic 0 and $f : X \to S$ a flat morphism of finite type with demi-normal fibers. Then f is *locally stable* iff the fibers X_s are slc and $\omega_{X/S}^{[m]}$ is a flat family of divisorial sheaves (3.25) for every $m \in \mathbb{Z}$.

Furthermore, f is *stable* iff, in addition, f is proper and $\omega_{X/S}$ is f-ample.

The next example shows that being stable is not a locally closed condition.

Example 3.41 In $\mathbb{P}_{\mathbf{x}}^5 \times \mathbb{A}_{st}^2$, consider the family of varieties

$$X := \left(\operatorname{rank} \begin{pmatrix} x_0 & x_1 & x_2 \\ x_1 + sx_4 & x_2 + tx_5 & x_3 \end{pmatrix} \le 1 \right).$$

We claim that the fibers X_{st} are normal, projective with rational singularities and for every s, t the following equivalences hold:

(3.41.1) X_{st} is lc $\Leftrightarrow$ X_{st} is klt $\Leftrightarrow$ $K_{X_{st}}$ is $\mathbb{Q}$-Cartier $\Leftrightarrow$ $3K_{X_{st}}$ is Cartier $\Leftrightarrow$ either $(s, t) = (0, 0)$ or $st \ne 0$.

All these become clear once we show that there are three types of fibers.

(3.41.2) If $st \ne 0$ then, after a linear coordinate change, we get that

$$X_{st} \simeq X_{11} \simeq \left(\operatorname{rank} \begin{pmatrix} x_0 & x_1 & x_2 \\ x_4 & x_5 & x_3 \end{pmatrix} \le 1 \right).$$

This is the Segre embedding of $\mathbb{P}^1 \times \mathbb{P}^2$, hence smooth. The self-intersection of its canonical class is -54.

(3.41.3) If $s = t = 0$ then we get the fiber

$$X_{00} := \left(\operatorname{rank} \begin{pmatrix} x_0 & x_1 & x_2 \\ x_1 & x_2 & x_3 \end{pmatrix} \le 1 \right).$$

This is the cone (with $\mathbb{P}^1$ as vertex-line) over the rational normal curve $C_3 \subset \mathbb{P}^3$. The singularity along the vertex-line is isomorphic to $\mathbb{A}^2/\frac{1}{3}(1,1) \times \mathbb{A}^1$, hence log terminal. The canonical class of X_{00} is $-\frac{8}{3}H$, where H is the hyperplane class and its self-intersection is $-512/9 < -54$.

(3.41.4) Otherwise either s or t (but not both) are zero. After possibly permuting s, t, and a linear coordinate change, we get the fiber

$$X_{0t} \simeq X_{01} \simeq \left(\mathrm{rank} \begin{pmatrix} x_0 & x_1 & x_2 \\ x_1 & x_4 & x_3 \end{pmatrix} \leq 1 \right).$$

This is the cone over the degree 3 surface $S_3 \simeq \mathbb{F}_1 \hookrightarrow \mathbb{P}^4$. Its canonical class is not $\mathbb{Q}$-Cartier at the vertex, so this is not lc.

This is a locally stable example. Taking a general double cover ramified along a general, sufficiently ample hypersurface gives a stable example.

Thus the best one can hope for is that local stability is representable. From now on the base scheme is assumed to be over a field of characteristic 0.

3.42 (Proof of 3.2) Being flat is representable by (3.19) and being demi-normal is an open condition for flat morphisms by (10.42). So, using (3.23.1), we may assume that $f : X \to S$ is flat, of pure relative dimension n and its fibers are demi-normal.

Now we come to a surprisingly subtle part of the argument. If X_s is slc then K_{X_s} is $\mathbb{Q}$-Cartier, thus the next natural step would be the following.

Question 3.42.1 Is $\{s \in S : K_{X_s}$ is $\mathbb{Q}$-Cartier$\}$ a constructible subset of S?

We saw in (2.45) that this is not the case, not even for families of normal varieties. The key turns out to be the following immediate consequence of (4.44); the latter is the hardest part of the proof.

Claim 3.42.2 Let $f : X \to S$ be a flat, proper family of demi-normal varieties. Then $\{\mathrm{index}(X_s) : X_s$ is slc$\}$ is a finite set. $\qquad\square$

We can now complete (3.2). Let M be a common multiple of the indices of the slc fibers. We apply (3.31) with $N_i := \omega_{X/S}^{[i]}$ for $1 \leq i < M$ and $L_1 := \omega_{X/S}^{[M]}$. We get $S^{NL} \to S$ such that the $\omega_{X^{NL}/S^{NL}}^{[m]}$ are flat families of divisorial sheaves for every m, and $\omega_{X^{NL}/S^{NL}}^{[M]}$ is invertible. Finally (3.39) gives that S^{ls} is an open subscheme of S^{NL}. $\qquad\square$

4

Stable Pairs over Reduced Base Schemes

So far we have identified stable pairs (X, Δ) as the basic objects of our moduli problem, defined stable and locally stable families of pairs over one-dimensional regular schemes in Chapter 2, and in Chapter 3 we treated families of varieties over reduced base schemes. Here we unite the two by discussing stable and locally stable families over reduced base schemes.

After stating the main results in Section 4.1, we give a series of examples in Section 4.2. The technical core of the chapter is the treatment of various notions of families of divisors given in Section 4.3. Valuative criteria are proved in Section 4.4 and the behavior of generically $\mathbb{R}$-Cartier divisors is studied in Section 4.5.

In Section 4.6, we finally define stable and locally stable families over reduced base schemes (4.7) and prove that local stability is a representable property. Families over a smooth base scheme are especially well behaved; their properties are discussed in the short Section 4.7.

The universal family of Mumford divisors is constructed in Section 4.8; this is probably the main technical result of the chapter. The correspondence between (not necessarily flat) families of Mumford divisors and flat families of Cayley–Chow hypersurfaces – established over reduced bases in Theorem 4.69 – leads to the fundamental notion of Cayley flatness in Chapter 7.

At the end, we have all the ingredients needed to treat the moduli functor $\mathcal{SP}^{\mathrm{red}}$, which associates to a reduced scheme S the set of all stable families $f \colon (X, \Delta) \to S$, up to isomorphism. (Here the superscript ^{red} indicates that we work with reduced base schemes only.)

To be precise, we fix the dimension n of the fibers, a finite set of allowed coefficients $\mathbf{c} \subset [0, 1]$ and the volume v. Our families are $f \colon (X, \Delta) \to S$, where $X \to S$ is flat and projective, Δ is a Weil $\mathbb{R}$-divisor on X whose coefficients are in $\mathbf{c}$, $K_{X/S} + \Delta$ is $\mathbb{R}$-Cartier, and the fibers (X_s, Δ_s) are stable pairs of dimension n with $\mathrm{vol}(K_{X_s} + \Delta_s) := ((K_{X_s} + \Delta_s)^n) = v$. This gives the functor

$\mathcal{SP}^{\mathrm{red}}(\mathbf{c}, n, v)$: {reduced S-schemes} $\to$ {sets}.

We can now state one of the main consequence of the results of this chapter.

Theorem 4.1 (Moduli theory of stable pairs I) *Let S be an excellent base scheme of characteristic 0 and fix $n, \mathbf{c}, v$. Then $\mathcal{SP}^{red}(\mathbf{c}, n, v)$ is a good moduli theory (6.10), which has a projective, coarse moduli space* $\mathrm{SP}^{red}(\mathbf{c}, n, v) \to S$.

Moreover, $\mathrm{SP}^{\mathrm{red}}(\mathbf{c}, n, v)$ is the reduced subscheme of the "true" moduli space $\mathrm{SP}(\mathbf{c}, n, v)$ of marked, stable pairs, to be constructed in Chapter 8.

Assumptions In the foundational Sections 4.1–4.5 we work with arbitrary schemes, but for Sections 4.6 and 4.7 we need to assume that the base scheme is over a field of characteristic 0.

4.1 Statement of the Main Results

In the study of locally stable families of pairs over reduced base schemes, the key step is to give the "correct" definition for the divisorial component

Temporary Definition 4.2 A *family of pairs* (with $\mathbb{Z}$-coefficients) of dimension n over a reduced scheme is an object

$$f \colon (X, D) \to S, \tag{4.2.1}$$

consisting of a morphism of schemes $f \colon X \to S$ and an effective Weil divisor D satisfying the following properties.

4.2.2 (Flatness for X) *The morphism $f \colon X \to S$ is flat, of finite type, of pure relative dimension n, with geometrically reduced fibers.* This is the expected condition from the point of view of moduli theory, following the Principles (3.12) and (3.13).

4.2.3 (Equidimensionality for Supp D) *Every irreducible component $D_i \subset$ Supp D dominates an irreducible component of S and all nonempty fibers of* Supp $D \to S$ *have pure dimension $n - 1$.* In particular, Supp D does not contain any irreducible component of any fiber of f. If S is normal then Supp $D \to S$ has pure relative dimension $n - 1$ by (2.71.2), but in general our assumption is weaker. We noted in (2.41) that $D \to S$ need not be flat for locally stable families. So we start with this weak assumption and strengthen it later.

4.2.4 (Mumford condition) *The morphism f is smooth at generic points of $X_s \cap$ Supp D for every $s \in S$.* Equivalently, for each $s \in S$, none of the irreducible components of $X_s \cap \text{Supp } D$ is contained in $\text{Sing}(X_s)$.

This condition was first codified in Mumford's observation that, in order to get a good moduli theory of pointed curves (C, P), the marked points $P = \{p_1, \ldots, p_n\}$ should be smooth points of C; see Section 4.8 for details.

If (X, Δ) is an slc pair, then X is smooth at all generic points of $\text{Supp } \Delta$. So if D is an effective divisor supported on $\text{Supp } \Delta$, then this conditions is satisfied.

It turns out that such generic smoothness is a crucial condition technically. So we make it part of the definition for families of pairs.

A big advantage is that, if S is reduced, then X is regular at the generic points of $\text{Supp } D$. Thus, as for normal varieties, we can harmlessly identify Mumford divisors with divisorial subschemes; see (4.16.6–7) for details.

Next we come to the heart of the matter: we would like the notion of families of pairs to give a functor. So, for any morphism $g : W \to S$, we need to define the pulled-back family. We have a fiber product diagram

$$X \times_S W \xrightarrow{q_X} X$$
$$f_W \downarrow \qquad \downarrow f \qquad\qquad (4.2.5)$$
$$W \xrightarrow{q} S.$$

It is clear that we should take $X_W := X \times_S W$, with morphism $f_W : X_W \to W$. The definition of the divisor part D_W is less clear, since pull-backs of Cartier and of Weil divisors are not compatible in general.

4.2.6 (Weil-divisor pull-back) For any subscheme $Z \subset X$ and morphism $h : Y \to X$, define the *Weil-divisor pull-back* as the Weil divisor $\text{Weil}(h^{-1}(Z))$ associated to the subscheme $h^{-1}(Z) \subset Y$; see (4.16.6) for formal definitions.

Let D, X be as in (4.2.1) and $g : W \to S$ a morphism. Using the Mumford condition we can view D as a subscheme of X. Then set

$$g^*_{\text{Wdiv}}(D) := \text{Weil}(g_X^{-1}(D)).$$

In particular, if $\tau : \{s\} \to S$ is a point, we get the *Weil-divisor fiber*, denoted by $\tau^*_{\text{Wdiv}}(D)$.

If $H \subset X$ is a relative Cartier divisor and $g^*_X H$ does not contain any codimension ≤ 1 associated points of $g_X^{-1}(D)$, then

$$g^*_{\text{Wdiv}}(D \cap H) = g^*_{\text{Wdiv}}(D) \cap g^*_X H.$$

Warning The Weil-divisor fiber is always defined, but frequently not functorial, not even additive. If D', D'' are two divisors on X then $\tau^*_{\text{Wdiv}}(D' + D'')$

and $\tau^*_{\text{Wdiv}}(D') + \tau^*_{\text{Wdiv}}(D'')$ have the same support, but the multiplicities can be different, even in étale locally trivial families as in (4.14). If D', D'' satisfy (4.2.4), then $\tau^*_{\text{Wdiv}}(D' + D'') \le \tau^*_{\text{Wdiv}}(D') + \tau^*_{\text{Wdiv}}(D'')$, but otherwise the inequality can go the other way; see (4.12) and (4.13).

4.2.7 (Generically Cartier divisor and pull-back) Assume that D is a relative Cartier divisor (4.20) on an open subset $U \subset X$ such that $g_X^{-1}(U \cap D)$ is dense in $g_X^{-1}(D)$. We can then define the *generically Cartier pull-back* of D as

$$g^{[*]}(D) := \text{ the closure of } g_X^{-1}(D|_U) \subset X_W.$$

If f has S_2 fibers then $\mathscr{O}_{X_W}(-g^{[*]}(D))$ is the hull pull-back of $\mathscr{O}_X(-D)$ (3.27). The generically Cartier pull-back is clearly functorial, but not always defined. If it is defined, then $g^*_{\text{Wdiv}}(D)$ is the Weil divisor corresponding to $g^{[*]}(D)$, so the two notions are equivalent; see (4.6).

4.2.8 (Well-defined pull-backs) We say that $f : (X, D) \to S$ has *well-defined Weil-divisor pull-backs* if it satisfies the assumptions (4.2.2–4) and the Weil-divisor pull-back (4.2.6) is a functor for reduced schemes. That is,

$$h^*_{\text{Wdiv}}(g^*_{\text{Wdiv}}(D)) = (g \circ h)^*_{\text{Wdiv}}(D)$$

for all morphisms of reduced schemes $h: V \to W$ and $g: W \to S$.

In any concrete situation, the conditions (4.2.2–4) should be easy to check, but (4.2.8) requires computing $g^*_{\text{Wdiv}}(D)$ for all morphisms $W \to S$. The following variant is much easier to verify.

4.2.9 (Well-defined specializations) We say that $f: (X, D) \to S$ has *well-defined specializations* if (4.2.8) holds whenever W is the spectrum of a DVR.

The good news is that, over reduced schemes, the three versions (4.2.6–9) are equivalent to each other and also to other natural conditions. The common theme is that we need to understand only the codimension 1 behavior of $f: (X, D) \to S$.

Theorem-Definition 4.3 (Well-defined families of pairs I) *Let S be a reduced scheme. A family of pairs $f : (X, D) \to S$ satisfying (4.2.2–4) is well defined if the following equivalent conditions hold.*

(4.3.1) *The family has well-defined Weil-divisor pull-backs (4.2.8).*

(4.3.2) *The family has well-defined specializations (4.2.9).*

(4.3.3) *D is a relative, generically Cartier divisor (4.2.7).*

(4.3.4) *$D \to S$ is flat at the generic points of $X_s \cap \text{Supp}\, D$ for every $s \in S$.*

If f is projective then these are also equivalent to

(4.3.5) *$s \mapsto \deg(X_s \cap D)$ is a locally constant function on S.*

The theorem is proved in (4.25). The next result says that, if S is normal, then the conditions (4.2.2–4) imply that $f: (X, D) \to S$ is well defined. It follows from (4.21) by setting $W := \mathrm{Sing}\, S$.

Theorem 4.4 (Ramanujam (1963); Samuel (1962)) *Let S be a normal scheme, $f: X \to S$ a smooth morphism and D a Weil divisor on X. Assume that D does not contain any irreducible component of a fiber. Then D is a Cartier divisor, hence a relative Cartier divisor.*

Over nonnormal base schemes it is usually easy to check well-definedness using the normalization.

Corollary 4.5 *Let S be a reduced scheme with normalization $\bar{S} \to S$. Let $f: (X, D) \to S$ be a projective family of pairs satisfying the assumptions (4.2.2–4) and*

$$\bar{f} : (\bar{X}, \bar{D}) := (X, D) \times_S \bar{S} \to \bar{S}$$

the corresponding family over $\bar{S}$. Then D is a relative, generically Cartier divisor in either of the following cases.

(4.5.1) $\tau^*_{W div}(D) = \bar{\tau}^*_{W div}(\bar{D}) = \bar{\tau}^{[*]}(\bar{D})$ *for every geometric point $\tau: \{s\} \to S$ and for every lifting $\bar{\tau} : \{s\} \to \bar{S}$.*

(4.5.2) S *is weakly normal and* $\bar{\tau}^*_{W div}(\bar{D}) = \bar{\tau}^{[*]}(\bar{D})$ *is independent of the lifting* $\bar{\tau} : s \to \bar{S}$ *for every geometric point* $\tau : \{s\} \to S$.

Proof Note first that $\bar{D}$ is a relative, generically Cartier divisor by (4.4), so $\bar{\tau}^*_{W div}(\bar{D}) = \bar{\tau}^{[*]}(\bar{D})$.

Let $g \in S$ be a generic point. Then $(\bar{D})_g = D_g$ and $\deg \bar{\tau}^*_{W div}(\bar{D}) = \deg(\bar{D})_g$ by (4.3) applied to $\bar{f} : (\bar{X}, \bar{D}) \to \bar{S}$. Together with (1) this shows that (4.3.5) holds for $f : (X, D) \to S$.

For (2), we explain in (4.25) how to reduce everything to the special case when f has relative dimension 1. Then (10.64) shows that D is flat over S. $\quad\square$

Next we turn to the case that we are really interested in, when the boundary Δ is a $\mathbb{Q}$ or $\mathbb{R}$-divisor. The right choice is to work with the relative, generically Cartier condition.

Definition 4.6 (Divisorial pull-back) Let S be a scheme, $f: X \to S$ a morphism and Δ a $\mathbb{Z}, \mathbb{Q}$ or $\mathbb{R}$-divisor on X. For $q: W \to S$, consider the fiber product as in (4.2.5). We define *relatively, generically Cartier* divisors and their *divisorial pull-backs*, denoted by Δ_W, in three steps as follows.

(4.6.1) Δ is a relatively, generically Cartier $\mathbb{Z}$-divisor if it is Cartier at the generic points of $X_s \cap \operatorname{Supp} D$ for every $s \in S$. Δ_W is then defined as in (4.2.7).

(4.6.2) Δ is a relatively, generically $\mathbb{Q}$-Cartier $\mathbb{Q}$-divisor iff $m\Delta$ is a relatively, generically Cartier $\mathbb{Z}$-divisor for some $m > 0$. Then we set $\Delta_W := \frac{1}{m}((m\Delta)_W)$.

This is independent of m, but there is a subtle point. We prove in (4.39) that, if the characteristic is 0, then a $\mathbb{Z}$-divisor is relatively, generically $\mathbb{Q}$-Cartier iff it is relatively, generically Cartier. So we can choose m to be the common denominator of the coefficients in Δ. However, this is not true in positive characteristic; see (8.75–8.76).

(4.6.3) Δ is a relatively, generically $\mathbb{R}$-Cartier $\mathbb{R}$-divisor iff one can write $\Delta = \sum c_i \Delta_i$ where the Δ_i are relatively, generically $\mathbb{Q}$-Cartier $\mathbb{Q}$-divisors. Then we set $\Delta_W := \sum c_i (\Delta_i)_W$.

This is independent of the choice of c_i and Δ_i. We may assume that the c_i are $\mathbb{Q}$-linearly independent. Then Δ is relatively, generically $\mathbb{R}$-Cartier iff the Δ_i are relatively, generically $\mathbb{Q}$-Cartier by (11.43.2).

Let $f \colon (X, \Delta) \to S$ be a well-defined family of pairs as in (4.3). In (3.1) we gave seven equivalent definitions of locally stable families of varieties. Some of these extend to families of pairs. See (2.41) for some negative examples and Section 8.2 for some solutions.

Definition–Theorem 4.7 Let S be a reduced scheme, $f \colon X \to S$ a flat morphism of finite type and $f \colon (X, \Delta) \to S$ a well-defined family of pairs. Assume that (X_s, Δ_s) is slc for every $s \in S$. Then $f \colon (X, \Delta) \to S$ is *locally stable* or *slc* if the following equivalent conditions hold.

(4.7.1) $K_{X/S} + \Delta$ is $\mathbb{R}$-Cartier.

(4.7.2) For every spectrum of a DVR T and morphism $q \colon T \to S$, the pullback $f_T \colon (X_T, \Delta_T) \to T$ is locally stable, as in (2.3).

(4.7.3) There is a closed subset $Z \subset X$ such that $\operatorname{codim}(Z \cap X_s, X_s) \geq 3$ for every $s \in S$ and $f|_{X \setminus Z} \colon (X \setminus Z) \to S$ satisfies the above (1–2).

Such a family is called *stable* if, in addition, f is proper and $K_{X/S} + \Delta$ is f-ample.

Proof The arguments are essentially the same as in (3.37). It is clear that (4.7.1) $\Rightarrow$ (4.7.2). If (4.7.2) holds then $K_{X_T} + \Delta_T$ is $\mathbb{R}$-Cartier for every $q : T \to S$. Thus $K_{X/S} + \Delta$ is $\mathbb{R}$-Cartier by (4.35).

Finally, if any of the properties (4.7.1–2) holds for X, then it also holds for $X \setminus Z$. Using (4.7.2) both for X and for $X \setminus Z$, reduces us to checking (4.7.3) $\Rightarrow$ (4.7.2) when S is the spectrum of a DVR; which is (2.7). $\qquad\square$

Let $f\colon (X, \Delta) \to S$ be a family of pairs. It turns out that, starting in relative dimension 3, the set of points $\{s \in S \; : \; (X_s, \Delta_s) \text{ is slc}\}$ is neither open nor closed; see (3.41) for an example. Thus the strongest result one can hope for is the following.

Theorem 4.8 (Local stability is representable) *Let S be a reduced, excellent scheme over a field of characteristic 0 and $f\colon (X, \Delta) \to S$ a well-defined, projective family of pairs. Assume that Δ is an effective, relative, generically $\mathbb{R}$-Cartier divisor. Then there is a locally closed partial decomposition $j : S^{ls} \to S$ such that the following holds.*

Let W be any reduced scheme and $q\colon W \to S$ a morphism. Then the family obtained by base change $f_W\colon (X_W, \Delta_W) \to W$ is locally stable iff q factors as $q\colon W \to S^{ls} \to S$.

A stable morphism is locally stable and stability is an open condition for a locally stable morphism. Thus (4.8) implies the following.

Corollary 4.9 (Stability is representable) *Using the notation and assumptions as in (4.8), there is a locally closed partial decomposition $j : S^{stab} \to S$ such that the following holds.*

Let W be any reduced scheme and $q : W \to S$ a morphism. Then the family obtained by base change $f_W : (X_W, \Delta_W) \to W$ is stable iff q factors as $q : W \to S^{stab} \to S$. □

4.2 Examples

We start with a series of examples related to (4.3).

Example 4.10 Let $S = (xy = 0) \subset \mathbb{A}^2$ and $X = (xy = 0) \subset \mathbb{A}^3$. Consider the divisors $D_x := (y = z - 1 = 0)$ and $D_y := (x = z + 1 = 0)$. We get a family $f : (X, D_x + D_y) \to S$ that satisfies the assumptions (4.2.2–4).

We compute the "fiber" of the family over the origin in three different ways and get three different results.

First, restrict the family to the x-axis. The pull-back of X becomes the plane $\mathbb{A}^2_{xz}$. The divisor D_x pulls back to $(z - 1 = 0)$, but the pull-back of the ideal sheaf of D_y is the maximal ideal $(x, z + 1)$. It has no divisorial part, so restriction to the x-axis gives the pair $(\mathbb{A}^2_{xz}, (z - 1 = 0)) \to \mathbb{A}^1_x$. Similarly, restriction to the y-axis gives the pair $(\mathbb{A}^2_{yz}, (z + 1 = 0)) \to \mathbb{A}^1_y$. If we restrict these to the origin, we get $(\mathbb{A}^1_z, (z - 1 = 0))$ and $(\mathbb{A}^1_z, (z + 1 = 0))$.

Finally, if we restrict to the origin of S in one step then we get the pair $(\mathbb{A}^1_z, (z-1=0) + (z+1=0))$. Thus we have three different pairs that can claim to be the fiber of $f : (X, D_x + D_y) \to S$ over the origin.

In this example the problem is visibly set-theoretic, but there can be problems even when the set theory works out.

Example 4.11 Set $C := (xy(x-y) = 0) \subset \mathbb{A}^2_{xy}$ and $X := (xy(x-y) = 0) \subset \mathbb{A}^3_{xyz}$. For any $c \in k$ consider the divisor

$$D_c := (x = z = 0) + (y = z = 0) + (x - y = z - cx = 0).$$

The pull-back of D_c to any of the irreducible components of X is Cartier, it intersects the central fiber at the origin of the z-axis and with multiplicity 1. Nonetheless, we claim that D_c is Cartier only for $c = 0$.

Indeed, assume that $h(x, y, z) = 0$ is a local equation of D_c. Then $h(x, 0, z) = 0$ is a local equation of the x-axis and $h(0, y, z) = 0$ is a local equation of the y-axis. Thus $h = az + $ (higher terms). Restricting to the $(x - y = 0)$ plane we get that $c = 0$.

Note also that if $\operatorname{char} k = 0$ and $c \neq 0$ then no multiple of D_c is a Cartier divisor. To see this note that if $f(x, y, z) = 0$ is a local defining equation of mD_c on X then $\partial^{m-1} f / \partial z^{m-1}$ vanishes on D_c. Its restriction to the z-axis vanishes at the origin with multiplicity 1. We proved above that this is not possible.

However, if $\operatorname{char} k = p > 0$, then $z^p - c^p x y^{p-1} = 0$ shows that pD_c is a Cartier divisor.

Example 4.12 Consider the cusp $C := (x^2 = y^3) \subset \mathbb{A}^2_{xy}$ and the trivial curve family $Y := C \times \mathbb{A}^1_z \to C$. Let $D \subset Y$ be the Cartier divisor given by the equation $y = z^2$. Then $D \to C$ is flat of degree 2. Furthermore, D is reducible with irreducible components $D^{\pm} := $ image of $t \mapsto (t^3, t^2, \pm t)$.

Note that $D^{\pm} \simeq \mathbb{A}^1_t$ and the projections $D^{\pm} \to C$ corresponds to the ring extension $k[t^3, t^2] \hookrightarrow k[t]$. Thus the projections $D^{\pm} \to C$ are not flat and the Weil-divisorial fiber of $D^{\pm} \to C$ over the origin has length 2.

However, the Weil-divisorial fiber of $D = D^+ \cup D^- \to C$ over the origin is again the point $(0, 0, 0)$ with multiplicity 2.

Arguing as in (4.11) shows that the $D^{\pm}$ are not $\mathbb{Q}$-Cartier in characteristic 0, but $pD^+ = (xy^{(p-3)/2} = z^p)$ shows that it is $\mathbb{Q}$-Cartier in characteristic $p > 0$.

The next example shows the importance of the Mumford condition.

Example 4.13 Set $X = (x^2 - y^2 = u^2 - v^2) \subset \mathbb{A}^4$, $D = (x - u = y - v = 0) \cup (x + u = y + v = 0)$ and $f : (X, D) \to \mathbb{A}^2_{uv}$, the coordinate projection. The fiber X_{uv} is a pair of intersecting lines if $u^2 = v^2$ and a smooth conic otherwise.

The irreducible components of D intersect only at the origin and D is not Cartier there. The divisorial fiber D_{uv} consists of 2 distinct smooth points if $(u, v) \neq (0, 0)$, but D_{00} is the origin with multiplicity 3.

Let L_c be the line $(v = cu)$ for some $c \neq \pm 1$. Restricting the family to L_c we get $X_c = (x^2 - y^2 = (1 - c^2)u^2) \subset \mathbb{A}^3$ and the divisor becomes $D_c = (x - u = y - cu = 0) \cup (x + u = y + cu = 0)$. Observe that D_c is a Cartier divisor with defining equation $cx = y$. (Note that base change does not commute with union, so $D \times_{\mathbb{A}^2} L_c$ has an embedded point at the origin.)

Thus although D is not Cartier at the origin, after base change to a general line we get a Cartier divisor. For all of these base changes, D_c has multiplicity 2 at the origin. (These also hold after base change to any smooth curve.)

However, the origin is a singular point of the fiber. If we restrict D_c to the fiber over the origin, the resulting scheme structure varies with c.

This would be a very difficult problem to deal with, but for a stable pair (X, Δ) we are in a better situation since the irreducible components of Δ are not contained in Sing X.

Example 4.14 Let B be a smooth projective curve of genus ≥ 1 with an involution σ and $b_1, b_2 \in B$ a pair of points interchanged by σ. Let C' be another smooth curve with two points $c'_1, c'_2 \in C'$. Start with the trivial family $(B \times C', \{b_1\} \times C' + \{b_2\} \times C') \to C'$ and then identify $c'_1 \sim c'_2$ and $(b, c'_1) \sim (\sigma(b), c'_2)$ for every $b \in B$. We get an étale locally trivial stable morphism $(S, D_1 + D_2) \to C$. Here C is a nodal curve with node $\tau : \{c\} \to C$. The fiber over the node is $(B, [b_1] + [b_2])$. However, the fiber of each D_i over c is $[b_1] + [b_2]$, hence

$$\tau^*_{\text{Wdiv}}(D_1) + \tau^*_{\text{Wdiv}}(D_1) = (B, 2[b_1] + 2[b_2]) \neq (B, [b_1] + [b_2]) = \tau^*_{\text{Wdiv}}(D_1 + D_2).$$

The next examples discuss the variation of the $\mathbb{Q}$-Cartier property in families of divisors. Related positive results are in Section 4.6.

Example 4.15 Let $C \subset \mathbb{P}^2$ be a smooth cubic curve and $S_C \subset \mathbb{P}^3$ the cone over it. For $p \in C$ let $L_p \subset S_C$ denote the ruling over p. Note that L_p is $\mathbb{Q}$-Cartier iff p is a torsion point, that is, $3m[p] \sim \mathscr{O}_C(m)$ for some $m > 0$. The latter is a countable dense subset of the moduli space of the lines $\text{Chow}_{1,1}(S_C) \simeq C$.

In the above example the surface is not $\mathbb{Q}$-factorial and the curve L_p is sometimes $\mathbb{Q}$-Cartier, sometimes not. Next we give a similar example of a flat family of lc surfaces $S \to B$ such that $\{b : S_b$ is $\mathbb{Q}$-factorial$\} \subset B$ is a countable set of points. Thus being $\mathbb{Q}$-factorial is not a constructible condition.

Let $C \subset \mathbb{P}^2$ be a smooth cubic curve. Pick 11 points $P_1, \ldots, P_{11} \in C$ and set $P_{12} = -(P_1 + \cdots + P_{11})$. Then there is a quartic curve D such that $C \cap D = P_1 + \cdots + P_{12}$. Thus the linear system $|\mathscr{O}_{\mathbb{P}^2}(4)(-P_1 - \cdots - P_{12})|$ blows up the points P_i and contracts C. Its image is a degree 4 surface $S = S(P_1, \ldots, P_{11})$ in $\mathbb{P}^3$ with a single simple elliptic singularity. If $C = (f_3(x, y, z) = 0)$ and $D = (f_4(x, y, z) = 0)$ then

$$S \simeq (f_3(x, y, z)w + f_4(x, y, z) = 0) \subset \mathbb{P}^3.$$

At the point $(x = y = z = 0)$, the singularity of S is analytically isomorphic to the cone S_C and S is smooth elsewhere iff the points $P_1, \ldots, P_{12}$ are distinct. If this holds, then the class group of S is generated by the image L of a line in $\mathbb{P}^2$ and the images $E_1, \ldots, E_{12}$ of the 12 exceptional curves. They satisfy a single relation $3L = E_1 + \cdots + E_{12}$. Note that E_i is $\mathbb{Q}$-Cartier iff P_i is a torsion point.

If we vary $P_1, \ldots, P_{11} \in C$, we get a flat family of lc surfaces parametrized by $\pi : \mathbf{S} \to C^{11} \setminus$ (diagonals), with universal divisors $\mathbf{E}_i \subset \mathbf{S}$. We see that

(4.15.1) $E_i(P_1, \ldots, P_{11})$ is $\mathbb{Q}$-Cartier iff P_i is a torsion point and

(4.15.2) $S(P_1, \ldots, P_{11})$ is $\mathbb{Q}$-factorial iff P_i is a torsion point for every i.

4.3 Families of Divisors II

At least three different notions of effective divisors are commonly used in algebraic geometry, but our discussions show that other variants are also necessary.

4.16 (Five notions of effective divisors) Let X be an arbitrary scheme.

(4.16.1) An effective *Cartier divisor* is a subscheme $D \subset X$ such that, for every $x \in D$, the ideal sheaf $\mathscr{O}_X(-D)$ is locally generated by a non-zero divisor $s_x \in \mathscr{O}_{x,X}$, called a *local equation* of D.

(4.16.2) A *divisorial subscheme* is a subscheme $D \subset X$ such that $\mathscr{O}_D$ has no embedded points and Supp D has pure codimension 1 in X.

(4.16.3) A divisorial subscheme D is called an effective, *generically Cartier divisor* if it is Cartier at its generic points. These are called almost Cartier divisors in Hartshorne (1986) and Hartshorne and Polini (2015).

(4.16.4) A divisorial subscheme D is called an effective *Mumford divisor* if X is regular at generic points of D. More generally, D is *Mumford along Z*, if X and Z are both regular at every generic point of $Z \cap$ Supp D.

(4.16.5) A *Weil divisor* is a formal, finite linear combination $D = \sum_i m_i D_i$ where $m_i \in \mathbb{Z}$ and the D_i are integral subschemes of codimension 1 in X. We say that D is effective if $m_i \geq 0$ for every i.

If A is an abelian group then a *Weil A-divisor* is a formal, finite linear combination $D = \sum_i a_i D_i$ where $a_i \in A$. We will only use the cases $A = \mathbb{Z}, \mathbb{Q}, \mathbb{R}$. Thus Weil $\mathbb{Z}$-divisor = traditional Weil divisor; we use the terminology "Weil $\mathbb{Z}$-divisor" if the coefficient group is not clear. (A Weil $\mathbb{Z}$-divisor is sometimes called an integral Weil divisor, but the latter could also mean the Weil divisor corresponding to an integral subscheme of codimension 1.)

Note that usually divisorial subschemes and Weil divisors are used only when X is irreducible or at least pure dimensional, but the definition makes sense in general.

If X is smooth then the five variants are equivalent to each other, but in general they are different.

Usually we think of Cartier divisor as the most restrictive notion. If X is S_2 then every effective Cartier divisor is a divisorial subscheme. However, if X is not S_2, then there are Cartier divisors $D \subset X$ such that D is not a divisorial subscheme, and the natural map from Cartier divisors to divisorial subschemes is not injective; see (4.16.9). These are good to keep in mind, but they will not cause problems for us.

Let $W \subset X$ be a closed subscheme. We can associate to it both a divisorial subscheme and a Weil divisor by the rules

$$
\begin{aligned}
\mathrm{Div}(W) &:= \text{pure } W := \mathrm{Spec}(\mathcal{O}_W/(\text{torsion})), \quad \text{and} \\
\mathrm{Weil}(W) &:= \textstyle\sum_i \mathrm{length}_{g_i}(\mathcal{O}_{g_i,W}) \cdot [D_i],
\end{aligned}
\qquad (4.16.6)
$$

where, in the first case, we take the quotient by the subsheaf of those sections whose support has codimension ≥ 2 in X (see also (10.1)). In the second case, $D_i \subset \mathrm{Supp}\, W$ are the irreducible components of codimension 1 in X with generic points $g_i \in D_i$. In particular, this associates an effective Weil divisor to any effective Cartier divisor or divisorial subscheme.

Thus, if X is S_2, then we have the basic relations among effective divisors

$$
\begin{pmatrix} \text{Cartier} \\ \text{divisors} \end{pmatrix} \subset \begin{pmatrix} \text{Mumford} \\ \text{divisors} \end{pmatrix} \subset \begin{pmatrix} \text{generically} \\ \text{Cartier divisors} \end{pmatrix} \subset \begin{pmatrix} \text{divisorial} \\ \text{subschemes} \end{pmatrix}.
$$

Assume next that X is regular at a codimension 1 point $g \in X$. Then $\mathcal{O}_{g,X}$ is a DVR, hence its ideals are uniquely determined by their colength. Thus we have the following.

Claim 4.16.7 If X is a normal scheme then four of the notions agree for effective divisors

$$\begin{pmatrix} \text{Mumford} \\ \text{divisors} \end{pmatrix} = \begin{pmatrix} \text{generically} \\ \text{Cartier divisors} \end{pmatrix} = \begin{pmatrix} \text{divisorial} \\ \text{subschemes} \end{pmatrix} = \begin{pmatrix} \text{Weil} \\ \text{divisors} \end{pmatrix}.$$

We are mainly interested in slc pairs (X, Δ), thus the underlying schemes X are demi-normal. Fortunately, X is smooth at the generic points of Δ. Thus, for our purposes, we can always imagine that the identifications (4.16.7) hold.

Convention 4.16.8 Let X be a scheme and $W \subset X$ a subscheme. Assume that X is regular at all generic points of W. Then we will frequently identify $\mathrm{Div}(W)$, the divisorial subscheme associated to W and $\mathrm{Weil}(W)$, the Weil divisor associated to W. We denote this common object by $[W]$.

We can thus usually harmlessly identify divisorial subschemes and Weil divisors. However – and this is one of the basic difficulties of the theory – it is quite problematic to keep the identification between *families* of divisorial subschemes and *families* of Weil divisors.

Example 4.16.9 Let $S \subset \mathbb{A}^4$ be the union of the planes $(x_1 = x_2 = 0)$ and $(x_3 = x_4 = 0)$. For $c \neq 0$, consider the Cartier divisors $D_c := (x_1 + cx_3 = 0)$. For any c, the corresponding divisorial subscheme is the union of the lines $(x_1 = x_2 = x_3 = 0) \cup (x_1 = x_3 = x_4 = 0)$, hence independent of c. However the D_c are different Cartier divisors for different $c \in k$. Indeed, $(x_1 + c'x_3)/(x_1 + cx_3)$ is a nonregular rational function that is constant c'/c on the first plane and 1 on the second. Note that S is seminormal and the D_c are Mumford.

Corresponding to the five notions of divisors, there are five notions of families. We discuss four of these next, leaving Mumford divisors to Section 4.8.

Relative Weil divisors

Definition 4.17 Let $f : X \to S$ be a morphism whose fibers have pure dimension n. A Weil divisor $W = \sum m_i W_i$ is called a *relative Weil divisor* if the fibers of $f|_{W_i} : W_i \to f(W_i)$ have pure dimension $n-1$ for every i.

We are interested in defining the divisorial fibers of $W \to S$. A typical example is (4.13), where the multiplicity of the scheme-theoretic fiber jumps over the origin. It is, however, quite natural to say that the "correct" fiber is the origin with multiplicity 2; the only problem we have is that scheme theory miscounts the multiplicity. The following theorem, proved in Kollár (1996, 3.17), says that this is indeed frequently the case.

Theorem 4.18 *Let S be a normal scheme, $f : X \to S$ a projective morphism, and $Z \subset X$ a closed subscheme such that $f|_Z : Z \to S$ has pure relative dimension m. Then there is a section $\sigma_Z : S \to \mathrm{Chow}_m(X/S)$ with the following properties.*

(4.18.1) *Let $g \in S$ be the generic point. Then $\sigma_Z(g) = [Z_g]$, the cycle associated to the generic fiber of $f|_Z : Z \to S$ as in (3.8).*

(4.18.2) $\mathrm{Supp}(\sigma_Z(s)) = \mathrm{Supp}(Z_s)$ *for every $s \in S$.*

(4.18.3) $\sigma_Z(s) = [Z_s]$ *if $f|_Z$ is flat at all generic points of Z_s.*

(4.18.4) $s \mapsto (\sigma_Z(s) \cdot L^m)$ *is a locally constant function of $s \in S$, for any line bundle L on X.* □

Example (4.10) shows that (4.18) does not hold if S is only seminormal. The notion of well-defined families of algebraic cycles is designed to avoid similar problems, leading to the definition of the Cayley–Chow functor; see Kollár (1996, sec.I.3–4) for details.

Flat Families of Divisorial Subschemes

Let $X \to S$ be a morphism and $D \subset X$ a subscheme. If $\mathrm{Supp}\,D$ does not contain any irreducible component of a fiber X_s, then $\mathscr{O}_{D \cap X_s}/(\text{torsion})$ is (the structure sheaf of) a divisorial subscheme of X_s. This notion, however, frequently does not have good continuity properties, as illustrated by (4.13).

We would like to have a notion of flat families of divisorial subschemes, where both the structure sheaf $\mathscr{O}_D$ and the ideal sheaf $\mathscr{O}_X(-D)$ are "well behaved." This seems possible only if $X \to S$ is "well behaved," but then the two aspects turn out to be equivalent.

Definition–Lemma 4.19 Let $f : X \to S$ be a flat morphism of pure relative dimension n with S_2 fibers and $D \subset X$ a closed subscheme of relative dimension $n-1$ over S. We say that $f|_D : D \to S$ is a *flat family of divisorial subschemes* if the following equivalent conditions hold.

(4.19.1) $f|_D : D \to S$ is flat with pure fibers of dimension $n-1$ (10.1).

(4.19.2) $\mathscr{O}_X(-D)$ is flat over S with S_2 fibers.

If f is projective and pure D_s denotes the largest pure subscheme as in (10.1), these are further equivalent to:

(4.19.3) $s \mapsto \chi(X_s, \mathscr{O}_{\text{pure}\,D_s}(*))$ is locally constant on S.

(4.19.4) $s \mapsto \chi(X_s, \mathscr{O}_{X_s}(-\text{pure}\,D_s)(*))$ is locally constant on S.

Proof We have a surjection $\mathscr{O}_X \to \mathscr{O}_D$ and if both of these sheaves are flat then so is the kernel $\mathscr{O}_X(-D)$. If the kernel is flat then $\mathscr{O}_{X_s}(-D_s) \simeq \mathscr{O}_X(-D)|_{X_s}$

is also the kernel of $\mathcal{O}_{X_s} \to \mathcal{O}_{D_s}$. Since $\mathcal{O}_{X_s}$ is S_2, we see that $\mathcal{O}_{X_s}(-D_s)$ is S_2 iff $\mathcal{O}_{D_s}$ is pure of dimension $n-1$.

Conversely, assume (2). For any $T \to S$ the pull-back map $q_T^* \mathcal{O}_X(-D) \to q_T^* \mathcal{O}_X$ is an isomorphism over $X_T \setminus D_T$. Since $\mathcal{O}_X(-D)$ is flat with S_2 fibers, $q_T^* \mathcal{O}_X(-D)$ does not have any sections supported on D_T. Thus the pulled-back sequence

$$0 \to q_T^* \mathcal{O}_X(-D) \to q_T^* \mathcal{O}_X \to q_T^* \mathcal{O}_D \to 0$$

is exact. Therefore, $\mathrm{Tor}_1^S(\mathcal{O}_T, \mathcal{O}_D) = 0$ hence $\mathcal{O}_D$ is flat over S and we already noted that then it has pure fibers of dimension $n-1$.

The last two claims are proved as in (2.75).   □

Relative Cartier Divisors

Definition–Lemma 4.20 Let $f: X \to S$ be a flat morphism with S_2 fibers, $x \in X$ a point, and $s := f(x)$. A subscheme $D \subset X$ is a *relative Cartier divisor* at $x \in X$ if the following equivalent conditions hold.

(4.20.1) D is flat over S at x and $D_s := D|_{X_s}$ is a Cartier divisor on X_s at x.

(4.20.2) D is a Cartier divisor on X at x and a local equation $g_x \in \mathcal{O}_{x,X}$ of D restricts to a non-zerodivisor on the fiber X_s.

(4.20.3) D is a Cartier divisor on X at x and it does not contain any irreducible component of X_s that passes through x.

If these hold for all $x \in D$ then D is a *relative Cartier divisor.* If $f: X \to S$ is also proper then the functor of relative Cartier divisors is represented by an open subscheme of the Hilbert scheme $\mathrm{CDiv}(X/S) \subset \mathrm{Hilb}(X/S)$; see Kollár (1996, I.1.13) for the easy details.

If (2) holds then D is flat by (4.19). The other nontrivial claim is that (1) implies that D is a Cartier divisor on X at x. We may assume that $(x \in X)$ is local. A defining equation g_s of D_s lifts to an equation g of D. We have the exact sequence

$$0 \to I_D/(g) \to \mathcal{O}_X/(g) \to \mathcal{O}_D \to 0.$$

Here $\mathcal{O}_X/(g)$ and $\mathcal{O}_D$ are both flat, hence so is $I_D/(g)$. Restricting to X_s we get

$$0 \to (I_D/(g))_s \to \mathcal{O}_{X_s}/(g_s) \xrightarrow{\sim} \mathcal{O}_{D_s} \to 0.$$

Thus $I_D/(g) = 0$ by Nakayama's lemma and g is a defining equation of D.   □

Relative Cartier divisors form a very well behaved class, but in applications we frequently have to handle two problems. It is not always easy to see which divisors are Cartier, and we also need to deal with divisors that are not Cartier.

On a smooth variety every divisor is Cartier, thus if X itself is smooth then a divisor D is relatively Cartier iff its support does not contain any of the fibers. In the relative setting, we usually focus on properties of the morphism f. Thus we would like to have similar results for smooth morphisms. (See (4.36) and (4.41) for closely related results.)

Theorem 4.21 *Let* $f : X \to S$ *be a smooth morphism and* $W \subset S$ *a closed subset such that* $\text{depth}_W S \geq 2$. *Let* D° *be a Cartier divisor on* $X \setminus f^{-1}(W)$ *and* $D \subset X$ *its closure. Assume that* $\text{Supp}\, D$ *does not contain any irreducible component of any fiber. Then* D *is Cartier, hence a relative Cartier divisor.*

Proof Assume first that f has relative dimension 1. Then $f|_D : D \to S$ is quasi-finite, so $f|_D$ is flat by (10.63), so D is a Cartier divisor by (4.20.1).

For the general case, pick a closed point $x \in D$. Since f is smooth, locally it factors through an étale morphism $\tau : (x, X) \to ((0, s), \mathbb{A}^n_S)$. Composing with any linear projection we locally factor f as

$$f : (x, X) \xrightarrow{g} ((0, s), \mathbb{A}^{n-1}_S) \to S,$$

where g is smooth of relative dimension 1. If D does not contain the fiber of g passing through x, then D is a Cartier divisor by the already discussed one-dimensional case.

To find such a g, assume first that $k(s)$ is infinite. Let $L \subset \mathbb{A}^n_s$ be a general line through the origin. Then $\pi_s^{-1}(L) \not\subset D_s$. Thus if we choose the projection $\mathbb{A}^n_S \to \mathbb{A}^{n-1}_S$ to have kernel L over s, then the argument proves that D is a Cartier divisor at x.

If $k(s)$ is finite then consider the trivial lifting $f^{(1)} : X \times \mathbb{A}^1 \to S \times \mathbb{A}^1$. By the previous argument $D \times \mathbb{A}^1$ is a Cartier divisor at the generic point of $\{x\} \times \mathbb{A}^1$, hence D is a Cartier divisor at x by (2.92.1). $\qquad\square$

Examples 4.22 We give two examples showing that in (4.21) we do need depth assumptions on S.

Set $S_n := \text{Spec}\, k[x, y]/(xy)$ and $X_n = \text{Spec}\, k[x, y, z]/(xy)$. Then (x, z) defines a Weil divisor which is not Cartier.

Set $S_c := \text{Spec}\, k[x^2, x^3]$ and $X_c = \text{Spec}\, k[x^2, x^3, y]$. Then $(y^2 - x^2, y^3 - x^3)$ defines a Weil divisor which is not Cartier.

Lemma 4.23 *Let* X *be a pure dimensional,* S_2 *scheme,* $D \subset X$ *a Cartier divisor and* $W \subset D$ *a subscheme such that* $\text{codim}_D W \geq 2$. *Let* L *be a rank 1, torsion-free sheaf on* X *that is locally free along* $D \setminus W$. *Let* s *be a section of*

L such that $s|_{D \setminus W}$ is nowhere zero. Then L is trivial and s is nowhere zero in a neighborhood of D.

Proof The section s gives an exact sequence

$$0 \to \mathscr{O}_X \xrightarrow{s} L \to Q \to 0.$$

By (10.7) every associated prime of Q has codimension 1 in X. Thus $D \cap$ Supp Q has codimension 1 in D. Therefore, D is disjoint from Supp Q and L is trivial on $X \setminus$ Supp Q.   □

Relative Generically Cartier Divisors

This is the most important class for moduli purposes.

Definition 4.24 Let $f: X \to S$ be a morphism. A subscheme $D \subset X$ is a *relative, generically Cartier, effective divisor* or a *family of generically Cartier, effective divisors* over S if there is an open subset $U \subset X$ such that

(4.24.1) f is flat over U with S_2 fibers,

(4.24.2) $\mathrm{codim}_{X_s}(X_s \setminus U) \geq 2$ for every $s \in S$,

(4.24.3) $D|_U$ is a relative Cartier divisor (4.20), and

(4.24.4) D is the closure of $D|_U$.

If $U \subset X$ denotes the largest open set with these properties then $Z := X \setminus U$ is the *non-Cartier locus* of D.

Thus $\mathscr{O}_X(mD)$ is a mostly flat family of divisorial sheaves on X (3.28) for any $m \in \mathbb{Z}$. Conversely, if L is a mostly flat family of divisorial sheaves on X and h a global section of it that does not vanish on any irreducible component of any fiber, then $(h = 0)$ is a family of generically Cartier, effective divisors over S.

4.25 (Proof of 4.3) All five conditions are local on S; the first four are local on X. All of them can be checked on a general relative hyperplane section of X; see (4.2.6), (4.26) and (10.56).

Thus we may assume that $X \to S$ has relative dimension 1, hence f is smooth along Supp D. We view D as a divisorial subscheme of X. After an étale base change we may assume that $D \to S$ is finite.

Applying (3.20) to $F := f_* \mathscr{O}_D$ (with $X = S$) we see that (4.3.5) holds iff $\mathscr{O}_D$ is flat over S. By (4.20) the latter holds iff D a relative Cartier divisor. Thus (4.3.3) $\Leftrightarrow$ (4.3.4) $\Leftrightarrow$ (4.3.5).

As we noted in (4.24), these imply (4.3.1), and (4.3.1) $\Rightarrow$(4.3.2) is clear. It remains to show that (4.3.2) $\Rightarrow$(4.3.4).

To see this, fix a point $\tau\colon \{s\} \to S$ and let T be the spectrum of a DVR and $h\colon T \to S$ a morphism that maps the closed point to $\tau(s)$ and the generic point of T to a generic point $g \in S$. Then $h^*_{\mathrm{Wdiv}}D$ is flat over T of degree $\deg_{k(g)} \mathcal{O}_{D_g}$. Thus if $\bar{\tau}\colon \bar{s} \to T$ is a lifting of τ and (4.3.2) holds, then

$$\deg \tau^*_{\mathrm{Wdiv}}D = \deg \bar{\tau}^*_{\mathrm{Wdiv}} h^*_{\mathrm{Wdiv}}D = \deg_{k(g)} \mathcal{O}_{D_g}.$$

Thus $D \to S$ is flat by (3.20). □

The following Bertini-type results are frequently useful. The first claim is an immediate consequence of (10.56) and the second follows from (10.20).

Proposition 4.26 *Let* $(0 \in S)$ *be a local scheme,* $X \subset \mathbb{P}^N_S$ *a quasi-projective* S-scheme with fibers of pure dimension ≥ 2, and $D \subset X$ a relative divisorial subscheme. Then, for general $H \in |\mathcal{O}_X(1)|$,

(4.26.1) D *is a generically Cartier family of divisors on* X *iff* $D|_H$ *is a generically Cartier family of divisors on* H, *and*

(4.26.2) $\mathcal{O}_X(D)|_H \simeq \mathcal{O}_H(D|_H)$. □

Representability Theorems

4.27 (Representability of the generically Cartier condition) There are two versions of this question. Let $f\colon X \to S$ be a flat, projective morphism and $D \subset X$ a divisorial subscheme.

The traditional problem is to study those morphisms $q\colon W \to S$ for which q^*D is a generically Cartier divisor on X_W. This gives a representable functor. This will be used during the construction of the moduli of Mumford divisors, so we treat it there (4.77).

From the point of view of Section 4.1, it may seem more natural to study those morphisms $q\colon W \to S$ for which the Weil-divisor pull-back $q^*_{\mathrm{Wdiv}}D$ is a generically Cartier divisor on X_W. This, however, does not give a representable functor; see (4.13). This variant is actually not well posed, since the Weil-divisor pull-back is not functorial in general.

Fortunately, it turns out to be relatively easy to ensure the generically Cartier condition. So we focus on studying additional properties of such families.

As a first problem, we start with a family of generically Cartier divisors, and study those morphisms $q\colon W \to S$ for which the generically Cartier pull-back D_W is flat or relatively Cartier.

The first result is a reformulation of (3.29) and (3.30).

Theorem 4.28 *Let S be a scheme, $f : X \to S$ a flat, projective morphism with S_2 fibers, and $D \subset X$ a family of generically Cartier divisors. Then there is a locally closed decomposition $j^{H\text{-}flat} : S^{H\text{-}flat} \to S$ (resp. a locally closed partial decomposition $j^{car} : S^{car} \to S$) such that, for every morphism $q \colon W \to S$, the divisorial pull-back $D_W = q^{[*]}D$ is flat (resp. Cartier) iff q factors through $S^{H\text{-}flat}$ (resp. S^{car}).* □

This leads to a valuative criterion for Cartier divisors in (4.34).

As we saw in (4.15), the set of fibers where a divisor is $\mathbb{Q}$-Cartier need not be constructible. So the straightforward $\mathbb{Q}$-Cartier version of (4.28) fails. However, this failure of constructibility is the only obstruction.

Proposition 4.29 *Let S be a reduced scheme, $f : X \to S$ a flat, projective morphism with S_2 fibers, and D a family of generically $\mathbb{Q}$-Cartier (resp. $\mathbb{R}$-Cartier) divisors on X. Let $S^* \subset S$ be a constructible subset. Assume that D_s is $\mathbb{Q}$-Cartier (resp. $\mathbb{R}$-Cartier) for every point $s \in S^*$.*

Then there is a locally closed partial decomposition $j^{qcar} \colon S^{qcar} \to S$ (resp. $j^{rcar} \colon S^{rcar} \to S$) such that the following holds.

(4.29.1) *Let $q : W \to S$ be a reduced S-scheme such that $q(W) \subset S^*$. Then the divisorial pull-back $D_W \subset X_W$ is $\mathbb{Q}$-Cartier (resp. $\mathbb{R}$-Cartier) iff q factors though S^{qcar} (resp. S^{rcar}).*

Proof We may assume that S^* is dense in S and start with the $\mathbb{Q}$-Cartier case. By (4.28) there are maximal open subsets $S_1^{car} \subset S_2^{car} \subset \cdots$ such that $r! \cdot D$ is Cartier over S_r^{car}. By assumption, S_r^{car} is dense for $r \gg 1$ and the union of all of them is the open stratum of $S^{qcar} \to S$. Noetherian induction then gives the other strata.

In the $\mathbb{R}$-Cartier case, we write $D = \sum d_i D^i$ where the D^i are $\mathbb{Q}$-divisors and the $d_i \in \mathbb{R}$ are linearly independent over $\mathbb{Q}$. We already have locally closed partial decompositions $j_i^{qcar} \colon S_i^{qcar} \to S$ using D^i, and $j^{rcar} : S^{rcar} \to S$ is their fiber product over S using (11.43.2). □

4.4 Valuative Criteria

We aim to show that various properties of morphisms can be checked after base change to one-dimensional, regular schemes, equivalently, to spectra of DVRs. We aim to use as few DVRs as possible.

Definition 4.30 A morphism $q\colon (x, X) \to (y, Y)$ of local schemes is *local* if $q(x) = y$. A morphism of schemes $q\colon X \to Y$ is *component-wise dominant* if every generic point of X is mapped to a generic point of Y. If X, Y are irreducible, then component-wise dominant is the same as dominant.

We are especially interested in local, component-wise dominant morphisms $q\colon (t, T) \to (s, S)$ from the spectrum of a DVR to S. To construct these, let $S_1 \subset S$ be an irreducible component and $\pi\colon B_s S_1 \to S_1$ the blow-up of s. The exceptional divisor has pure codimension 1. Let $\eta \in \mathrm{Ex}(\pi)$ be a generic point and $\mathscr{O}_\eta$ its local ring. If S is excellent, we can take T to be the normalization of $\mathrm{Spec}\, \mathscr{O}_\eta$. Then $(\eta, T) \to (s, S_1)$ is essentially of finite type. In general, we need to take T to be one of the irreducible components of the normalization of the completion of $\mathscr{O}_\eta$. Then T is excellent, but q is not essentially of finite type.

Lemma 4.31 *Let (s, S) be a local scheme and $g\colon S' \to S$ a locally closed partial decomposition (10.83). Then g is an isomorphism iff every local, component-wise dominant morphism $q\colon (t, T) \to (s, S)$ from the spectrum of an excellent DVR to S factors through g.*

Proof We see that g is proper and dominant by (10.78.1), hence an isomorphism by (10.83.2). □

Theorem 4.32 (Valuative criterion for divisorial sheaves) *Let (s, S) be a reduced, local scheme and $f\colon X \to S$ a flat morphism of finite type with S_2 fibers. Let L be a mostly flat family of divisorial sheaves on X (3.28). Assume that either f is projective or S is excellent. The following are equivalent.*
(4.32.1) *L is flat over S with S_2 fibers.*
(4.32.2) *For every local, component-wise dominant morphism $q\colon (t, T) \to (s, S)$ from the spectrum of an excellent DVR to S, the hull pull-back (3.27) L_T^H is flat over T with S_2 fibers.*

Proof It is clear that (1) implies (2). For the converse, we use the theory of hulls and husks from Chapter 9.

Assume first that f is projective. Consider the locally closed decomposition $j\colon \mathrm{Hull}(L) \to S$ given by (9.40). By assumption, every $q\colon (t, T) \to (s, S)$ factors through j, so j is an isomorphism by (4.31). Thus L is its own hull, hence it is flat over S with S_2 fibers.

This is the main case that we use. The argument in the nonprojective case is similar, but relies on (9.44).

Pick any point $x \in X$ and its image $s := f(x)$. Let $\hat{S}$ denote the completion of S at s; it is reduced since S is excellent. Then L is flat over S with S_2 fibers at x iff this holds after base change to $\hat{S}$. Thus it is enough to show that (2) $\Rightarrow$ (1) whenever $s \in S$ is complete, in which case the hull of L is represented by a subscheme $i : S^u \hookrightarrow S$ for local, Artinian S-algebras by (9.44).

Let (R, m) be a complete DVR and $q \colon \operatorname{Spec} R \to (s, S)$ a local morphism. By assumption (2), the hull pull-back L_R^H is flat over R with S_2 fibers. Thus the same holds for $\operatorname{Spec}(R/m^n)$ for every n, hence the restriction of q to $\operatorname{Spec}(R/m^n)$ factors through $i : S^u \hookrightarrow S$. Since this holds for every $n \in \mathbb{N}$, q factors through $i : S^u \hookrightarrow S$. We conclude that $S^u = S$. So, as before, L is its own hull, hence it is flat over S with S_2 fibers. □

Putting together (2.79), (2.82) and (4.32) gives the following higher dimensional version.

Corollary 4.33 *Let $f \colon (X, \Delta) \to S$ be a locally stable morphism to a reduced scheme over a field of characteristic 0. Let D be a relative Mumford $\mathbb{Z}$-divisor (4.68). Assume that either f is projective or S is excellent. Then, in any of the cases (2.79.1–8) and (2.82),*

(4.33.1) *$\mathscr{O}_X(D)$ is flat over S with S_2 fibers, and*

(4.33.2) *$\mathscr{O}_X(D)|_{X_s} \simeq \mathscr{O}_{X_s}(D_s)$ for $s \in S$.*   □

We can restate (4.32) for Cartier divisors as follows.

Corollary 4.34 (Valuative criterion for Cartier divisors) *Let (s, S) be a reduced, local scheme, $f \colon X \to S$ a flat morphism of finite type with S_2 fibers, and D a relative, generically Cartier divisor on X. Assume that either f is projective or S is excellent. Then the following are equivalent.*

(4.34.1) *D is a relative Cartier divisor.*

(4.34.2) *For every local, component-wise dominant morphism $q \colon (t, T) \to (s, S)$ from the spectrum of an excellent DVR to S, the divisorial pull-back $D_T \subset X_T$ is a Cartier divisor.*   □

Reduction to the Cartier case as in (4.29) gives the following.

Corollary 4.35 (Valuative criterion for $\mathbb{Q}$- and $\mathbb{R}$-Cartier divisors) *Let (s, S) be a reduced, local scheme, $f \colon X \to S$ a flat morphism of finite type with S_2 fibers, and D a family of generically $\mathbb{Q}$-Cartier (resp. $\mathbb{R}$-Cartier) divisors on X. Assume that either f is projective or S is excellent. Then the following are equivalent.*

(4.35.1) *D is a $\mathbb{Q}$-Cartier (resp. $\mathbb{R}$-Cartier) divisor.*

(4.35.2) *For every local, component-wise dominant morphism* $q: (t, T) \to$ (s, S) *from the spectrum of an excellent DVR to* S, *the divisorial pull-back* D_T *is* $\mathbb{Q}$-*Cartier (resp.* $\mathbb{R}$-*Cartier).* □

The following two consequences of (4.34) are important; see (4.41.1) for a more direct proof of the first one.

Corollary 4.36 *Let* S *be a reduced scheme,* $f : X \to S$ *a smooth morphism, and* D *a relative, generically Cartier divisor on* X. *Assume that either* f *is projective or* S *is excellent. Then* D *is a relative Cartier divisor.*

Proof Let $q : T \to S$ be a morphism from the spectrum of a DVR to S. Then X_T is regular, hence D_T is Cartier. So D is Cartier by (4.34). □

Theorem 4.37 *Let* (s, S) *be a reduced, local, excellent scheme,* $f: X \to S$ *a flat morphism of finite type with* S_2 *fibers, and* D *a relative, generically Cartier divisor on* X. *Then* D *is Cartier* $\Leftrightarrow$ D *is* $\mathbb{Q}$-*Cartier,* D_s *is Cartier, and* D_g *is Cartier for all generic points* $g \in S$.

Proof The necessity is clear. By (4.34) it is enough to prove the converse after base change to T whenever $q: (t, T) \to (s, S)$ is a local, component-wise dominant morphism from the spectrum of an excellent DVR to S. The assumptions are preserved.

Let $Z \subset X_t$ be the locus where D_T is not known to be Cartier. After localizing at the generic point of Z, we are in the situation of (2.91). Thus D_T is Cartier and so is D. □

Another valuative criterion is the following local version of (3.20).

Theorem 4.38 (Grothendieck, 1960, IV.11.6, IV.11.8) *Let* (s, S) *be a reduced, local scheme,* $f : X \to S$ *a morphism of finite type, and* F *a coherent sheaf on* X. *Let* T *be a disjoint union of spectra of DVRs and* $q: T \to S$ *a dominant, local morphism. Then* F *is flat over* S *at* $x \in X_s$ *iff* $q_X^* F$ *is flat over* T *along* $q_X^{-1}(x)$. □

4.5 Generically $\mathbb{Q}$-Cartier Divisors

In the study of lc and slc pairs, $\mathbb{Q}$-Cartier divisors are more important than Cartier divisors. We have seen many examples of Weil $\mathbb{Z}$-divisors that are $\mathbb{Q}$-Cartier, but not Cartier. By contrast, we show that if a relative Weil $\mathbb{Z}$-divisor is generically $\mathbb{Q}$-Cartier, then it is generically Cartier in characteristic 0.

Let $f : (X, D) \to S$ be a family of pairs and D a relative Weil $\mathbb{Z}$-divisor.

Since we are interested in generic properties, we can focus on a generic point x of $D \cap X_s$. If the assumption (4.2.4) holds then f is smooth at x. Thus we may as well assume that f is smooth (but not proper).

If S is normal then D is a Cartier divisor by (4.4), thus here our main interest is in those cases where S is reduced, but not normal. As (4.10) shows, D need not be Cartier in general. However, the next result shows that if some multiple of D is Cartier, then so is D, at least in characteristic 0.

Positive characteristic counter examples are given in (4.11) and (4.12).

Theorem 4.39 *Let S be a reduced scheme, $f : X \to S$ a smooth morphism of relative dimension ≥ 1, and D a relative Weil $\mathbb{Z}$-divisor on X. Assume that mD is Cartier at a point $x \in X$ and $\operatorname{char} k(x) \nmid m$. Then D is Cartier at x.*

Proof By Noetherian induction and shrinking X, we may assume that D is Cartier on $X \setminus \{x\}$ and $mD \sim 0$.

By (11.24), $mD \sim 0$ determines a cyclic cover $\tilde{X} \to X$ that is étale over $X \setminus \{x\}$ whenever $\operatorname{char} k(x) \nmid m$. This gives a correspondence between torsion in $\operatorname{Pic}^{\mathrm{loc}}(x, X)$ and torsion in the abelian quotient of the fundamental group $\hat{\pi}_1(X \setminus \{x\})$. There are now two ways to finish the proof.

In characteristic 0, we may work over $\mathbb{C}$. After replacing X with a suitable Euclidean neighborhood $x \in U \subset X$, it is enough to prove that $\pi_1(U \setminus \{x\})$ is trivial. We do this in (4.40).

In general, let $X^{\mathrm{wn}} \to X$ be the weak normalization (10.74). We prove in (4.41) that $\operatorname{Pic}^{\mathrm{loc}}(x^{\mathrm{wn}}, X^{\mathrm{wn}})$ is free of finite rank. It remains to show that $K^{\mathrm{wn}} := \ker[\operatorname{Pic}^{\mathrm{loc}}(x, X) \to \operatorname{Pic}^{\mathrm{loc}}(x^{\mathrm{wn}}, X^{\mathrm{wn}})]$ does not contain prime-to-p torsion in characteristic $p > 0$.

Since $X^{\mathrm{wn}} \to X$ is finite and purely inseparable, it is a factor of a power F_q of the Frobenius (10.78.2). This gives pull-back maps

$$\operatorname{Pic}^{\mathrm{loc}}(x, X) \to \operatorname{Pic}^{\mathrm{loc}}(x^{\mathrm{wn}}, X^{\mathrm{wn}}) \to \operatorname{Pic}^{\mathrm{loc}}(x_q, X_q),$$

where the composite is $L \mapsto L^q$. So K^{wn} is q-torsion.

Alternatively, one can use Grothendieck (1971, I.11), which implies that $X^{\mathrm{wn}} \setminus \{x^{\mathrm{wn}}\} \to X \setminus \{x\}$ induces an isomorphism of the fundamental groups. □

4.40 (Links and smooth morphisms) Let $f : X \to S$ be a smooth morphism of complex spaces of relative dimension $n \geq 1$. We describe the topology of the link of a point $x \in X$ in terms of the topology of the link of $s := f(x) \in S$.

We can write $S \subset \mathbb{C}_z^N$ such that s is the origin and $X \subset S \times \mathbb{C}_t^n$ where x is the origin. Intersecting S with a sphere of radius ε centered at s, we get L_S, the link of $s \in S$. The intersection of S with the corresponding ball of radius ε is homeomorphic to the cone C_S over L_S.

The link L_X of $x \in X$ can be obtained as the intersection of X with the level set $\max\{\sum|z_i|^2, \sum|t_j|^2\} = \varepsilon^2$. Thus L_X is homeomorphic to the amalgamation of

$$
\begin{aligned}
L_S \times \mathbb{D}^{2n} &= \{(\mathbf{z},\mathbf{t}) : \textstyle\sum|z_i|^2 = \varepsilon^2,\ \sum|t_j|^2 \leq \varepsilon^2\} && \text{and of} \\
C_S \times \mathbb{S}^{2n-1} &= \{(\mathbf{z},\mathbf{t}) : \textstyle\sum|z_i|^2 \leq \varepsilon^2,\ \sum|t_j|^2 = \varepsilon^2\}, && \text{glued along} \\
L_S \times \mathbb{S}^{2n-1} &= \{(\mathbf{z},\mathbf{t}) : \textstyle\sum|z_i|^2 = \varepsilon^2,\ \sum|t_j|^2 = \varepsilon^2\}.
\end{aligned}
$$

Let L_S^i be the connected components of L_S. Note that $\pi_1(L_S^i \times \mathbb{S}^{2n-1}) \simeq \pi_1(L_S^i) \times \pi_1(\mathbb{S}^{2n-1})$. The first factor gets killed in $\pi_1(C_S \times \mathbb{S}^{2n-1})$; the second is trivial if $n \geq 2$ and gets killed in $\pi_1(L_S^i \times \mathbb{D}^{2n})$ if $n = 1$. Thus L_X is simply connected for $n \geq 1$.

The cohomology of L_X can be computed from the Mayer–Vietoris sequence. Using that $H^i(L_S \times \mathbb{D}^{2n}, \mathbb{Z}) = H^i(L_S, \mathbb{Z})$ and $H^i(C_S \times \mathbb{S}^{2n-1}, \mathbb{Z}) = H^i(\mathbb{S}^{2n-1}, \mathbb{Z})$, for H^2 the key pieces are

$$
\begin{aligned}
&\longrightarrow H^1(L_S,\mathbb{Z}) \oplus H^1(\mathbb{S}^{2n-1}, \mathbb{Z}) \xrightarrow{\ \sigma^1\ } H^1(L_S \times \mathbb{S}^{2n-1}, \mathbb{Z}) \\
&\longrightarrow H^2(L_X, \mathbb{Z}) \longrightarrow H^2(L_S, \mathbb{Z}) \oplus H^2(\mathbb{S}^{2n-1}, \mathbb{Z}) \xrightarrow{\ \sigma^2\ } H^2(L_S \times \mathbb{S}^{2n-1}, \mathbb{Z}).
\end{aligned}
$$

The Künneth formula gives that the σ^i are injections and σ^1 is an isomorphism if $n \geq 2$. In this case $H^2(L_X, \mathbb{Z}) = 0$. If $n = 1$, then

$$
\begin{aligned}
H^2(L_X, \mathbb{Z}) &\simeq \operatorname{coker}[H^1(\mathbb{S}^1, \mathbb{Z}) \to H^0(L_S, \mathbb{Z}) \otimes H^1(\mathbb{S}^1, \mathbb{Z})] \\
&\simeq H^0(L_S, \mathbb{Z})/\mathbb{Z}.
\end{aligned} \tag{4.40.1}
$$

We have thus proved the following.

Claim 4.40.2 Let $f : X \to S$ be a smooth morphism of complex spaces, L_X the link of a point $x \in X$, and $s := f(x)$. Assume that $\dim_x X > \dim_s S \geq 1$.

Then L_X is simply connected. Furthermore, $H^2(L_X, \mathbb{Z}) = 0$ iff either $n \geq 2$ or the link of $s \in S$ is connected. □

Next we compute the local Picard groups in more detail in the weakly normal case.

Theorem 4.41 *Let $(s \in S)$ be a local, weakly normal pair (10.74) and $f : X \to S$ a smooth morphism. Let $x \in X_s$ be a point. Then,*

(4.41.1) *if $\operatorname{codim}(x \in X_s) \geq 2$ then $\operatorname{Pic}^{\mathrm{loc}}(x, X) = 0$, and*

(4.41.2) *if $\operatorname{codim}(x \in X_s) = 1$ then $\operatorname{Pic}^{\mathrm{loc}}(x, X)$ is free of finite rank.*

Proof Set $d = \dim X_s$ and let $\pi\colon X \to \mathbb{A}_S^{d-1}$ be a general projection. Then π is generically quasi-finite along the closure of x. Let (w, W) be the strict

Henselization of $\pi(x) \in \mathbb{A}_S^{d-1}$ (2.18). By base change, we have a smooth morphism $\pi' : (x', X') \to (w, W)$ of relative dimension 1, where $x' \in X', w \in W$ are closed points.

By (2.92.1), $\mathrm{Pic}^{\mathrm{loc}}(x, X) \hookrightarrow \mathrm{Pic}^{\mathrm{loc}}(x', X')$, thus it is enough to prove (1–2) for $\mathrm{Pic}^{\mathrm{loc}}(x', X')$.

Every class in $\mathrm{Pic}^{\mathrm{loc}}(x', X')$ can be represented by an effective divisor D that does not contain X'_w. Then $\pi'|_D : D \to W$ is finite and flat over $W \setminus \{w\}$.

Let $\{W_i : i \in I\}$ be the connected components of $W \setminus \{w\}$. Then $[D] \mapsto (\mathrm{rank}_{W_i} \pi'_* \mathscr{O}_D : i \in I)$ gives a map

$$\mathrm{Pic}^{\mathrm{loc}}(x', X') \to \mathbb{Z}^{|I|} \to \mathbb{Z}^{|I|}/\mathbb{Z}(1, \dots, 1).$$

We claim that it is an injection. Indeed, if $\pi'_* \mathscr{O}_D$ has constant rank d then $\pi'|_D$ is flat by (10.64), hence D is Cartier by (4.20). This proves (2).

If $\mathrm{codim}(x \in X_s) \geq 2$ then $g(x)$ is not the generic point $\eta_s \in \mathbb{A}_S^{d-1}$. Thus every irreducible component of $\mathbb{A}_S^{d-1}$ contains η_s, and this continues to hold after strict Henselization. Thus $W \setminus \{w\}$ is connected and we get (1). □

Complement 4.41.3 The proof shows that in case (2) the rank is bounded by $r - 1$, where r is the maximum number of connected components of $S' \setminus \{s'\}$ for all étale $(s', S') \to (s, S)$. It is $\leq$ the number of geometric points over s on the normalization of S.

4.6 Stability Is Representable II

Assumption. In this section we work over a field of characteristic 0.

The main result of this section is the following. Eventually we remove the reduced assumption by introducing the notion of K-flatness in Chapter 7.

Theorem 4.42 *Let* $f : (X, \Delta) \to S$ *be a projective, well-defined family of pairs. Then the functor of locally stable pull-backs is represented, for reduced schemes, by a locally closed partial decomposition* $i^{lst} : S^{lst} \to S$.

Since ampleness is an open condition for an $\mathbb{R}$-Cartier divisor (11.54.2), (4.42) implies the analogous result for stable morphisms.

Corollary 4.43 *Let* $f : (X, \Delta) \to S$ *be a projective, well-defined family of pairs. Then the functor of stable pull-backs is represented, for reduced schemes, by a locally closed partial decomposition* $i^{stab} : S^{stab} \to S$. □

We start the proof of (4.42), which will be completed in (4.46), with a weaker version.

Lemma 4.44 *Let $f: (X, \Delta) \to S$ be a proper, well-defined family of pairs. Then there is a finite collection of locally closed subschemes $S_i \subset S$ such that*

(4.44.1) $f_i : (X_{S_i}, \Delta_{S_i}) \to S_i$ *is locally stable for every i, and*

(4.44.2) *a fiber (X_s, Δ_s) is slc iff $s \in \cup_i S_i$.*

In particular, $\{s : (X_s, \Delta_s)$ is slc$\} \subset S$ is constructible.

Proof Being demi-normal is an open condition by (10.42) and slc implies demi-normal by definition. So we may assume that all fibers are demi-normal and S is irreducible with generic point g. Throughout the proof we use $S^\circ \subset S$ to denote a dense open subset which we shrink whenever necessary.

First, we treat morphisms whose generic fiber X_g is normal.

Case 1: (X_g, Δ_g) is lc. Then $K_{X_g} + \Delta_g$ is $\mathbb{R}$-Cartier, hence $K_{X/S} + \Delta$ is $\mathbb{R}$-Cartier over an open neighborhood of g. Next consider a log resolution $p_g : Y_g \to X_g$. It extends to a simultaneous log resolution $p^\circ : Y^\circ \to X^\circ$ over a suitable $S^\circ \subset S$. Thus, if $E^\circ \subset Y^\circ$ is any exceptional divisor, then $a(E_s, X_s, \Delta_s) = a(E^\circ, X^\circ, \Delta^\circ) = a(E_g, X_g, \Delta_g)$. This shows that all fibers over S° are lc.

Case 2: (X_g, Δ_g) is not lc. Note that the previous argument works if $K_{X_g} + \Delta_g$ is $\mathbb{R}$-Cartier. Indeed, then there is a divisor E with $a(E_g, X_g, \Delta_g) < -1$ and this shows that $a(E_s, X_s, \Delta_s) < -1$ for $s \in S^\circ$. However, if $K_{X_g} + \Delta_g$ is not $\mathbb{R}$-Cartier, then the discrepancy $a(E_g, X_g, \Delta_g)$ is not defined. We could try to prove that $K_{X_s} + \Delta_s$ is not $\mathbb{R}$-Cartier for $s \in S^\circ$, but this is not true in general; see (4.15).

Thus we use the notion of numerically Cartier divisors (4.48) instead. If $K_{X_g} + \Delta_g$ is not numerically Cartier, then, by (4.51), $K_{X_s} + \Delta_s$ is also not numerically Cartier over an open subset $S^\circ \ni g$. Thus (X_s, Δ_s) is not lc for $s \in S^\circ$.

If $K_{X_g} + \Delta_g$ is numerically Cartier, then the notion of discrepancy makes sense (4.48) and, again using (4.51), the arguments show that if (X_g, Δ_g) is numerically lc (resp. not numerically lc) then the same holds for (X_s, Δ_s) for s in a suitable open subset $S^\circ \ni g$. We complete Case 2 by noting that being numerically lc is equivalent to being lc by (4.50).

An alternative approach to the previous case is the following. By (11.30), the log canonical modification (5.15) $\pi_g : (Y_g, \Theta_g) \to (X_g, \Delta_g)$ exists and it extends to a simultaneous log canonical modification $\pi : (Y, \Theta) \to (X, \Delta)$ over an open subset $S^\circ \subset S$. By the arguments of Case 1, (Y_s, Θ_s) is lc for $s \in S^\circ$ and the relative ampleness of the log canonical class is also an open condition. Thus $\pi_s : (Y_s, \Theta_s) \to (X_s, \Delta_s)$ is the log canonical modification for $s \in S^\circ$. By assumption, π_g is not an isomorphism, so none of the π_s are isomorphisms. Therefore, none of the fibers over S° are lc.

If X_g is not normal, the proofs mostly work the same using a simultaneous semi-log-resolution (Kollár, 2013b, Sec.10.4). However, for Case 2 it is more convenient to use the following argument.

Let $\pi_g \colon \bar{X}_g \to X_g$ denote the normalization. Over an open subset $S^\circ \ni g$ it extends to a simultaneous normalization $(\bar{X}, \bar{D} + \bar{\Delta}) \to S$. If $(\bar{X}_g, \bar{D}_g + \bar{\Delta}_g)$ is not lc then $(\bar{X}_s, \bar{D}_s + \bar{\Delta}_s)$ is not lc for $s \in S^\circ$, hence (X_s, Δ_s) is not slc, essentially by definition; see Kollár (2013b, 5.10).

Using the already settled normal case, it remains to deal with the situation when $(\bar{X}_s, \bar{D}_s + \bar{\Delta}_s)$ is lc for every $s \in S^\circ$. By Kollár (2013b, 5.38), (X_s, Δ_s) is slc iff $\mathrm{Diff}_{\bar{D}_s^n} \bar{\Delta}_s$ is τ_s-invariant. The different can be computed on any log resolution as the intersection of the birational transform of $\bar{D}_s$ with the discrepancy divisor. Thus $\mathrm{Diff}_{\bar{D}_s^n} \bar{\Delta}_s$ is also locally constant over an open set S°. Therefore, if $\mathrm{Diff}_{\bar{D}_g^n} \bar{\Delta}_g$ is not τ_g-invariant then $\mathrm{Diff}_{\bar{D}_s^n} \bar{\Delta}_s$ is also not τ_s-invariant for $s \in S^\circ$. Hence (X_s, Δ_s) is not slc for every $s \in S^\circ$.

In both cases we complete the proof by Noetherian induction. □

The following consequence of (4.44) is quite useful, though it could have been proved before it as in (3.39).

Corollary 4.45 *Let $f \colon (X, \Delta) \to S$ be a proper, well-defined family of pairs such that $K_{X/S} + \Delta$ is $\mathbb{R}$-Cartier. Then $\{s : (X_s, \Delta_s)$ is slc $\} \subset S$ is open.*

Proof By (4.44), this set is constructible. A constructible set $U \subset S$ is open iff it is closed under generalization, that is, $x \in U$ and $x \in \bar{y}$ implies that $y \in U$. This follows from (2.3). □

4.46 (Proof of 4.42) Let $S_i \subset S$ be as in (4.44). We apply (4.29) to the family $f \colon (X, K_{X/S} + \Delta) \to S$ to obtain $S^{\mathrm{rcar}} \to S$ such that, for every reduced S-scheme $q \colon T \to S$ satisfying $q(T) \subset \cup_i S_i$, the pulled-back divisor $K_{X_T/T} + \Delta_T$ is $\mathbb{R}$-Cartier iff q factors as $q \colon T \to S^{\mathrm{rcar}} \to S$.

Assume now that $f_T \colon (X_T, \Delta_T) \to T$ is slc. Then $K_{X_T/T} + \Delta_T$ is $\mathbb{R}$-Cartier, hence q factors through $S^{\mathrm{rcar}} \to S$. As we observed in (3.23), this implies that $S^{\mathrm{slc}} = (S^{\mathrm{rcar}})^{\mathrm{slc}}$. By definition $K_{X^{\mathrm{rcar}}/S^{\mathrm{rcar}}} + \Delta$ is $\mathbb{R}$-Cartier, thus (4.45) implies that $S^{\mathrm{slc}} = (S^{\mathrm{rcar}})^{\mathrm{slc}}$ is an open subscheme of S^{rcar}. □

We showed in (4.15) that being $\mathbb{Q}$-Cartier or $\mathbb{R}$-Cartier is not a constructible condition. The next result shows that the situation is better for boundary divisors of lc pairs.

Corollary 4.47 *Let* $f: (X, \Delta) \to S$ *be a proper, flat family of pairs with slc fibers. Let D be an effective divisor on X. Assume that*

(4.47.1) *either* $\operatorname{Supp} D \subset \operatorname{Supp} \Delta$,

(4.47.2) *or* $\operatorname{Supp} D$ *does not contain any of the log canonical centers of any of the fibers* (X_s, Δ_s).

Then $\{s : D_s$ *is* $\mathbb{R}$-*Cartier*$\} \subset S$ *is constructible.*

Proof Over an open subset, we have a simultaneous log resolution of $(X, D + \Delta)$. Choose $0 < \varepsilon \ll 1$. In the first case, $(X_s, \Delta_s - \varepsilon D_s)$ is slc iff D_s is $\mathbb{R}$-Cartier. In the second case, $(X_s, \Delta_s + \varepsilon D_s)$ is slc iff D_s is $\mathbb{R}$-Cartier. Thus, in both cases, (4.44) implies our claim. $\square$

Numerically Cartier Divisors

Definition 4.48 Let $g: Y \to S$ be a proper morphism. An $\mathbb{R}$-Cartier divisor D is called *numerically g-trivial* if $(C \cdot D) = 0$ for every curve $C \subset Y$ that is contracted by g.

Let X be a demi-normal scheme. A Mumford $\mathbb{R}$-divisor D on X is called *numerically $\mathbb{R}$-Cartier* if there is a proper, birational contraction $p: Y \to X$ and a numerically p-trivial $\mathbb{R}$-Cartier Mumford divisor D_Y on Y such that $p_*(D_Y) = D$.

It follows from (11.60) that such a D_Y is unique. If D is a $\mathbb{Q}$-divisor then D_Y is also a $\mathbb{Q}$-divisor since its coefficients are solutions of a linear system of equations. Such a D is called *numerically $\mathbb{Q}$-Cartier*.

If $p' : Y' \to X$ is a proper, birational contraction and Y' is $\mathbb{Q}$-factorial, then being numerically $\mathbb{R}$-Cartier can be checked on Y'.

Being numerically $\mathbb{R}$-Cartier is preserved by $\mathbb{R}$-linear equivalence, but the exceptional part $D_Y - p_*^{-1}D$ depends on $D \in |D|$.

For $K_X + \Delta$, we can make a canonical choice. Thus we see that $K_X + \Delta$ is numerically $\mathbb{R}$-Cartier iff there is a p-exceptional $\mathbb{R}$-divisor $E_{K+\Delta}$ such that $E_{K+\Delta} + K_Y + p_*^{-1}\Delta$ is numerically p-trivial.

If $K_X + \Delta$ is numerically $\mathbb{R}$-Cartier, then one can define the *discrepancy* of any divisor E over X by

$$a(E, X, \Delta) := a(E, Y, E_{K+\Delta} + p_*^{-1}\Delta).$$

We can thus define when a demi-normal pair (X, Δ) is *numerically lc* or *slc*.

If $g: X \to S$ is proper, then a numerically $\mathbb{R}$-Cartier divisor D is called *numerically g-trivial* if D_Y is numerically $(g \circ p)$-trivial on Y.

Examples 4.49 On a normal surface, every divisor is numerically $\mathbb{R}$-Cartier.

The divisor $(x = z = 0)$ is not numerically $\mathbb{R}$-Cartier on the demi-normal surface $(xy = 0) \subset \mathbb{A}^3$.

If X has rational singularities, then a numerically $\mathbb{R}$-Cartier divisor is also $\mathbb{R}$-Cartier by Kollár and Mori (1992, 12.1.4).

Assume that dim $X \geq 3$ and D is Cartier except at a point $x \in X$. There is a local Picard scheme $\mathbf{Pic}^{\mathrm{loc}}(x, X)$, which is an extension of a finitely generated local Néron–Severi group with a connected algebraic group $\mathbf{Pic}^{\mathrm{loc}-\circ}(x, X)$; see Boutot (1978) or Kollár (2016a) for details. Then D is numerically $\mathbb{R}$-Cartier iff $[D] \in \mathbf{Pic}^{\mathrm{loc}-\tau}(x, X)$ where $\mathbf{Pic}^{\mathrm{loc}-\tau}(x, X) / \mathbf{Pic}^{\mathrm{loc}-\circ}(x, X)$ is the torsion subgroup of the local Néron–Severi group.

There are many divisors that are numerically $\mathbb{R}$-Cartier, but not $\mathbb{R}$-Cartier, however, the next result says that the notion of numerically slc pairs does not give anything new.

Theorem 4.50 (Hacon and Xu, 2016, 1.4) *A numerically slc pair is slc.*

Outline of the proof　This is surprisingly complicated, using many different ingredients. We start with the normal, numerically $\mathbb{Q}$-Cartier case.

For clarity, let us concentrate on the very special case when (X, Δ) is dlt, except at a single point $x \in X$. All the key ideas appear in this case, but we avoid a technical inductive argument.

Starting with a thrifty log resolution (Kollár, 2013b, 2.79), the method of (Kollár, 2013b, 1.34) gives a $\mathbb{Q}$-factorial, dlt modification $f : (Y, E + \Delta_Y) \to (X, \Delta)$ such that $K_Y + E + \Delta_Y$ is numerically f-trivial, where E is the exceptional divisor dominating x and Δ_Y is the birational transform of Δ. Let $\Delta_E := \mathrm{Diff}_E \Delta_Y$. Then (E, Δ_E) is a semi-dlt pair such that $K_E + \Delta_E$ is numerically trivial. Next we need a global version of the theorem.

Claim 4.50.1 Let (E, Δ_E) be a projective semi-slc pair such that $K_E + \Delta_E$ is $\mathbb{Q}$-Cartier and numerically trivial. Then $K_E + \Delta_E \sim_{\mathbb{Q}} 0$.

The first general proof is in Gongyo (2013), but special cases go back to Kawamata (1985) and Fujino (2000). We discuss a very special case: E is smooth and $\Delta = 0$. The following argument is from Campana et al. (2012) and Kawamata (2013).

We assume that $\mathscr{O}_E(K_E) \in \mathrm{Pic}^\tau(E)$, but after passing to an étale cover of E we have that $\mathscr{O}_E(K_E) \in \mathrm{Pic}^\circ(E)$. Serre duality shows that if $[L] \in \mathrm{Pic}^\tau(E)$ and $h^n(E, L) = 1$, then $L \simeq \mathscr{O}_E(K_E)$.

Next we use a theorem of Simpson (1993) which says that the cohomology groups of line bundles in Pic° jump precisely along torsion translates of abelian subvarieties. Thus $[K_E]$ is a torsion translate of a trivial abelian subvariety, hence a torsion element of Pic°(E). □

It remains to lift information from the exceptional divisor E to the dlt model Y. To this end consider the exact sequence

$$0 \to \mathscr{O}_Y(m(K_Y + E + \Delta_Y) - E) \to \mathscr{O}_Y(m(K_Y + E + \Delta_Y)) \to \mathscr{O}_E(m(K_E + \Delta_E)) \to 0.$$

Note that $D := m(K_Y + E + \Delta_Y) - E - (K_Y + \Delta_Y) \equiv_f 0$. We apply (Kollár, 2013b, 10.38.1) or the even stronger (Fujino, 2014, 1.10) to conclude that

$$R^1 f_*(\mathscr{O}_Y(m(K_Y + E + \Delta_Y) - E)) = R^1 f_*(\mathscr{O}_Y(D + K_Y + \Delta_Y)) = 0.$$

Hence a nowhere zero global section of $\mathscr{O}_E(m(K_E + \Delta_E))$ lifts back to a global section of $\mathscr{O}_Y(m(K_Y + E + \Delta_Y))$ that is nowhere zero near E. Thus $\mathscr{O}_X(m(K_X + \Delta)) \simeq f_*\mathscr{O}_Y(m(K_Y + E + \Delta_Y))$ is free in a neighborhood of x. Thus completes the numerically $\mathbb{Q}$-Cartier case.

The $\mathbb{R}$-Cartier case is reduced to the numerically $\mathbb{Q}$-Cartier setting using (11.47) as follows.

Let $f\colon (Y, \Delta_Y) \to (X, \Delta)$ be a log resolution. Pick curves C_m that span $N_1(Y/X)$ and apply (11.47) to (Y, Δ_Y). Thus for $n \gg 1$ we get $K_Y + \Delta_Y = \sum_j \lambda_j(K_Y + \Delta_Y^j)$ where the Δ_Y^j are $\mathbb{Q}$-divisors and (Y, Δ_Y^j) is lc. Also, since $(C_m \cdot (K_Y + \Delta_Y)) = 0$, (11.47.6.a) shows that $(C_m \cdot (K_Y + \Delta_Y^j)) = 0$. Thus each $(X, f(\Delta_Y^j))$ is a numerically $\mathbb{Q}$-Cartier lc pair. They are thus lc and so is (X, Δ) by (11.4.4). The demi-normal case now follows using (11.38). □

The advantage of the concept of numerically $\mathbb{R}$-Cartier divisors is that we have better behavior in families.

Proposition 4.51 *Let $f\colon X \to S$ be a proper morphism with normal fibers over a field of characteristic 0 and D a generically Cartier family of divisors on X. Then there is a finite collection of locally closed subschemes $S_i \subset S$ such that*

(4.51.1) *D_s is numerically $\mathbb{R}$-Cartier iff $s \in \cup_i S_i$, and*

(4.51.2) *the pull-back of D to $X \times_S S_i$ is numerically $\mathbb{R}$-Cartier for every i.*

In particular, $\{s \in S : D_s$ is numerically $\mathbb{R}$-Cartier$\} \subset S$ is constructible.

Proof Let $g \in S$ be a generic point. We show that if D_g is numerically $\mathbb{R}$-Cartier (resp. not numerically $\mathbb{R}$-Cartier) then the same holds for all D_s in an open neighborhood $g \in S° \subset S$. Then we finish by Noetherian induction.

To see our claim, consider a log resolution $p_g \colon Y_g \to X_g$. It extends to a simultaneous log resolution $p^\circ \colon Y^\circ \to X^\circ$ over a suitable open neighborhood $g \in S^\circ \subset S$.

If D_g is numerically $\mathbb{R}$-Cartier then there is a p_g-exceptional $\mathbb{R}$-divisor E_g such that $E_g + (p_g)_*^{-1} D_g$ is numerically p_g-trivial. This E_g extends to a p-exceptional $\mathbb{R}$-divisor E and $E + p_*^{-1}D$ is numerically p-trivial over an open neighborhood $g \in S^\circ \subset S$ by (4.52). Thus D_s is numerically $\mathbb{R}$-Cartier for $s \in S^\circ$.

Assume next that D_g is not numerically $\mathbb{R}$-Cartier. Let E_g^i be the p-exceptional divisors. Then there are proper curves $C_g^j \subset Y_g$ that are contracted by p_g and such that $(p_g)_*^{-1} D_g$, viewed as a linear function on $\oplus_j \mathbb{R}[C_g^j]$, is linearly independent of the E_g^i. Both the divisors E_g^i and the curves C_g^j extend to give divisors E_s^i and curves C_s^j over an open neighborhood $g \in S^\circ \subset S$. Thus $(p_s)_*^{-1} D_s$, viewed as a linear function on $\oplus_j \mathbb{R}[C_s^j]$, is linearly independent of the E_s^i, hence D_s is not numerically $\mathbb{R}$-Cartier for $s \in S^\circ$. $\square$

Lemma 4.52 *Let $p \colon Y \to X$ be a morphism of proper S-schemes and D an $\mathbb{R}$-Cartier divisor on Y. Then*

$$S^{nt} := \{ s \in S : D_s \text{ is numerically } p_s\text{-trivial} \}$$

is an open subset of S.

Proof We check Nagata's openness criterion (10.14).

Let us start with the special case when $X = S$. Pick points $s_1 \in \overline{s_2} \subset S$. A curve $C_2 \subset Y_{s_2}$ specializes to $C_1 \subset Y_{s_1}$ and if $(D_{s_1} \cdot C_1) = 0$ then $(D_{s_2} \cdot C_2) = 0$.

Next assume that D_{s_2} is numerically p_{s_2}-trivial. By (11.43.2), $D_{s_2} = \sum a_i A_{s_2}^i$ where the $A_{s_2}^i$ are numerically p_{s_2}-trivial $\mathbb{Q}$-divisors. Thus each $m A_{s_2}^i$ is algebraically equivalent to 0 for some $m > 0$; see Lazarsfeld (2004, I.4.38). We can spread out this algebraic equivalence to obtain that there is an open subset $U \subset \overline{s_2}$ such that $m D_s$ is algebraically (and hence numerically) equivalent to 0 on all fibers $s \in U$.

Applying this to $Y \to X$ shows that

$$X^{nt} := \{ x \in X : D_x \text{ is numerically trivial on } Y_x \}$$

is an open subset of X. Thus $S^{nt} = S \setminus \pi_X(X \setminus X^{nt})$ is an open subset of S, where $\pi_X \colon X \to S$ is the structure map. $\square$

4.53 (Warning on intersection numbers) In general, one cannot define intersection numbers of numerically $\mathbb{R}$-Cartier divisors with curves. This would need the stronger property: $(Z \cdot D_Y) = 0$ for every (not necessarily effective) 1-cycle Z on Y such that $p_*[Z] = 0$.

To see that this is indeed a stronger requirement, let $E \subset \mathbb{P}^2$ be a smooth cubic and $S \subset \mathbb{P}^3$ the cone over it. For $x \in E$ let $L_x \subset S$ denote the line over x. Set $X := S \times E$ and consider the divisors D_1, swept out by the lines $L_{x_0} \times \{x\}$ for some fixed $x_0 \in E$, and D_2, swept out by the lines $L_x \times \{x\}$ for $x \in E$. Let $p \colon Y \to X$ be the resolution obtained by blowing up the singular set, with exceptional divisor $F \simeq E \times E$. Then $p_*^{-1}(D_1 - D_2)$ shows that $D_1 - D_2$ is numerically Cartier.

Set $C := F \cap p_*^{-1}(D_1 - D_2)$. It is a section minus the diagonal on $E \times E$. Thus $p_*[C] = 0$, but $(C \cdot p_*^{-1}(D_1 - D_2)) = -2$.

4.7 Stable Families over Smooth Base Schemes

All the results of the previous sections apply to families $p \colon (X, \Delta) \to S$ over a smooth base scheme, but the smooth case has other interesting features as well. The following can be viewed as a direct generalization of (2.3).

Theorem 4.54 *Let $(0 \in S)$ be a smooth, local scheme and $D_1 + \cdots + D_r \subset S$ an snc divisor such that $D_1 \cap \cdots \cap D_r = \{0\}$. Let $p : (X, \Delta) \to (0 \in S)$ be a pure dimensional morphism and Δ an $\mathbb{R}$-divisor on X such that $\operatorname{Supp} \Delta \cap \operatorname{Sing} X_0$ has codimension ≥ 2 in X_0. The following are equivalent:*

(4.54.1) $p : (X, \Delta) \to S$ *is slc.*

(4.54.2) $K_{X/S} + \Delta$ *is $\mathbb{R}$-Cartier, p is flat and (X_0, Δ_0) is slc.*

(4.54.3) $K_{X/S} + \Delta$ *is $\mathbb{R}$-Cartier, X is S_2 and $(\operatorname{pure}(X_0), \Delta_0)$ (10.1) is slc.*

(4.54.4) $(X, \Delta + p^*D_1 + \cdots + p^*D_r)$ *is slc.*

Proof Note that (1) $\Rightarrow$ (2) holds by definition and (2) $\Rightarrow$ (3) since both S and X_0 are S_2 (10.10). If (3) holds, then (10.72) shows that p is flat and X_0 is pure, hence (3) $\Rightarrow$ (2). Next we show that (2) $\Leftrightarrow$ (4) using induction on r. Both implications are trivial if $r = 0$.

Assume (4) and pick a point $x \in X_0$. Then $K_X + \Delta + p^*D_1 + \cdots + p^*D_r$ is $\mathbb{R}$-Cartier at x hence so is $K_X + \Delta$. Set $D_Y := p^*D_r$. By (11.17),

$$(D_Y, \Delta|_{D_Y} + p^*D_1|_{D_Y} + \cdots + p^*D_{r-1}|_{D_Y})$$

is slc at x, hence (X_0, Δ_0) is slc at x by induction. The local equations of the p^*D_i form a regular sequence at x by (4.58), hence p is flat at x.

Conversely, assume that (2) holds. By induction,

$$(D_Y, \Delta|_{D_Y} + p^*D_1|_{D_Y} + \cdots + p^*D_{r-1}|_{D_Y})$$

is slc at x, hence inversion of adjunction (11.17) shows that $(X, \Delta + p^*D_1 + \cdots + p^*D_r)$ is slc at x. □

Corollary 4.55 *Let S be a smooth scheme and p: $(X, \Delta) \to S$ a morphism. Then $p : (X, \Delta) \to S$ is locally stable iff the pair $(X, \Delta + p^*D)$ is slc for every snc divisor $D \subset S$.* □

Corollary 4.56 *Let S be a smooth, irreducible scheme and $p : (X, \Delta) \to S$ a locally stable morphism. Then every log center of (X, Δ) dominates S.*

Proof Let E be a divisor over X such that $a(E, X, \Delta) < 0$ and let $Z \subset S$ denote the image of E in S. If $Z \neq S$ then, possibly after replacing S by an open subset, we may assume that Z is contained in a smooth divisor $D \subset S$. Thus $(X, \Delta + p^*D)$ is slc by (4.55). However, $a(E, X, \Delta + p^*D) \leq a(E, X, \Delta) - 1 < -1$, a contradiction. □

Corollary 4.57 *Let S be a smooth scheme and p: $(X, \Delta) \to S$ a projective, locally stable morphism with normal generic fiber. Let p^c: $(X^c, \Delta^c) \to S$ denote the canonical model of $p : (X, \Delta) \to S$ and $p^w : (X^w, \Delta^w) \to S$ a weak canonical model as in Kollár and Mori (1998, 3.50). Then*

(4.57.1) $p^w : (X^w, \Delta^w) \to S$ *is locally stable and*

(4.57.2) $p^c : (X^c, \Delta^c) \to S$ *is stable.*

Warning 4.57.3 As in (2.47.1), the fibers of p^c are *not* necessarily the canonical models of the fibers of p.

Proof Let $D \subset S$ be an snc divisor. By (4.55), $(X, \Delta + p^*D)$ is lc and p^w : $(X^w, \Delta^w + (p^*D)^w) \to S$ is also a weak canonical model over S by Kollár (2013b, 1.28). Thus $(X^w, \Delta^w + (p^*D)^w)$ is also slc, where $(p^*D)^w$ is the pushforward of p^*D. Next we claim that $(p^*D)^w = (p^w)^*D$. This is clear away from the exceptional set of $(p^w)^{-1}$ which has codimension ≥ 2 in X^w. Thus $(p^*D)^w$ and $(p^w)^*D$ are two divisors that agree outside a codimension ≥ 2 subset, hence they agree. Now we can use (4.55) again to conclude that $p^w : (X^w, \Delta^w) \to S$ is locally stable.

A weak canonical model is a canonical model iff $K_{X^w/S} + \Delta^w$ is p^w-ample and the latter is also what makes a locally stable morphism stable. □

Lemma 4.58 *Let $(y \in Y, \Delta + D_1 + \cdots + D_r)$ be slc. Assume that the D_i are Cartier divisors with local equations $(s_i = 0)$. Then the s_i form a regular sequence.*

Proof We use induction on r. Since Y is S_2, s_r is a non-zerodivisor at y. By adjunction $(y \in D_r, \Delta|_{D_r} + D_1|_{D_r} + \cdots + D_{r-1}|_{D_r})$ is slc, hence the restrictions $s_1|_{D_r}, \ldots, s_{r-1}|_{D_r}$ form a regular sequence at x. Thus $s_1, \ldots, s_r$ is a regular sequence at y. □

The following result of Karu (2000) is a generalization of (2.51) from one-dimensional to higher-dimensional bases.

Theorem 4.59 *Let U be a k-variety and $f_U \colon (X_U, \Delta_U) \to U$ a stable morphism. Then there is projective, generically finite, dominant morphism $\pi \colon V \to U$ and a compactification $V \hookrightarrow \bar{V}$ such that the pull-back $(X_U, \Delta_U) \times_U V$ extends to a stable morphism $f_{\bar{V}} \colon (X_{\bar{V}}, \Delta_{\bar{V}}) \to \bar{V}$.*

Proof We may assume that U is irreducible with generic point g.

Assume first that the generic fiber of f_U is normal and geometrically irreducible. Let $(Y_g, \Delta_g^Y) \to (X_g, \Delta_g)$ be a log resolution. It extends to a simultaneous log resolution $(Y_{U_0}, \Delta_{U_0}^Y) \to (X_{U_0}, \Delta_{U_0})$ over an open subset $U_0 \subset U$. By Abramovich and Karu (2000) (see also Adiprasito et al. (2019)), there is a projective, generically finite, dominant morphism $\pi \colon V_0 \to U_0$ and a compactification $V_0 \hookrightarrow \bar{V}$ such that the pull-back $(Y_{U_0}, \Delta_{U_0}^Y) \times_{U_0} V_0$ extends to a locally stable morphism $g_{\bar{V}} \colon (Y_{\bar{V}}, \Delta_{\bar{V}}^Y) \to \bar{V}$.

We can harmlessly replace $\bar{V}$ by a resolution of it. Thus we may assume that $\bar{V}$ is smooth and there is an open subset $V \subset \bar{V}$ such that the rational map $\bar{\pi}|_V \colon V \dashrightarrow U$ is a proper morphism.

Since $g_{\bar{V}}$ is a projective, locally stable morphism, the relative canonical model $f_{\bar{V}} \colon (X_{\bar{V}}, \Delta_{\bar{V}}) \to \bar{V}$ of $g_{\bar{V}} \colon (Y_{\bar{V}}, \Delta_{\bar{V}}^Y) \to \bar{V}$ exists by Hacon and Xu (2013) and it is stable by (4.57.2).

By construction, $(X_{\bar{V}}, \Delta_{\bar{V}})$ and $(X_U, \Delta_U) \times_U V$ are isomorphic over $V_0 \subset V$, but (11.40) implies that in fact they are isomorphic over V. This completes the case when the generic fiber of f_U is normal.

In general, we can first pull back everything to the Stein factorization of $X^n \to U$ where X^n is the normalization of X. The previous step now gives $f_{\bar{V}} \colon (X_{\bar{V}}^n, \Delta_{\bar{V}}^n) \to \bar{V}$. Finally, (4.56) shows that (11.41) applies and we get $f_{\bar{V}} \colon (X_{\bar{V}}, \Delta_{\bar{V}}) \to \bar{V}$. □

Corollary 4.60 *Let k be a field of characteristic 0 and assume that the coarse moduli space of stable pairs SP exists, is separated, and locally of finite type. Then every irreducible component of SP is proper over k.*

Proof Let M be an irreducible component of SP with generic point g_M. By assumption, there is a field extension $K \supset k(g_M)$ and a stable K-variety (X_K, Δ_K) corresponding to g_M.

Since it takes only finitely many equations to define a stable pair, we may also assume that $K/k(g_M)$ is finitely generated, hence so is K/k.

By (4.59), there is a smooth, projective k-variety $\bar{V}$ and a stable family $\bar{f}$: $(\bar{Y}, \bar{\Delta}_Y) \to \bar{V}$ such that $k(\bar{V})$ is a finite field extension of K and the generic fiber of $\bar{f}$ is isomorphic to $(X_K, \Delta_K)_{k(\bar{V})}$.

The image of the corresponding moduli morphism $\phi : \bar{Y} \to$ SP contains g_M and it is proper. It is thus the closure of g_M, which is M. So M is proper.   □

4.8 Mumford Divisors

On a normal variety, our basic objects are Weil divisors. On a nonnormal variety, we work with Weil divisors whose irreducible components are not contained in the singular locus. It has been long understood that these give the correct theory of generalized Jacobians of curves; see Serre (1959). Their first appearance in the moduli theory of curves may be Mumford's definition of pointed stable curves given in Knudsen (1983, Def.1.1).

Here we consider the relative version that is compatible with Cayley–Chow forms in a very strong way (4.69). This enables us to construct a universal family of Mumford divisors (4.76), which is a key step in the construction of the moduli space of stable pairs.

We start by recalling the foundational properties of Chow varieties, as treated in Kollár (1996, secs.I.3–4), and then discuss the ideal of Chow equations. We focus on the classical theory, which is over fields. A closer inspection reveals that the theory works for Mumford divisors over arbitrary bases. The end result is that the functor of Mumford divisors (4.69) is representable over reduced bases (4.76).

Definition 4.61 A *d-cycle* on a scheme X is a finite linear combination $Z :=$ $\sum_i m_i[V_i]$, where $m_i \in \mathbb{Z}$ and the V_i are d-dimensional irreducible, reduced subschemes. We usually tacitly assume that the V_i are distinct and $m_i \neq 0$. Then the V_i are called the *irreducible components* of Z and the m_i the *multiplicities*. A d-cycle is called *effective* if $m_i \geq 0$ for every i and *reduced* if all its multiplicities equal 1.

To a subscheme $W \subset X$ of dimension $\leq d$, we associate a d-cycle, called the *fundamental cycle*

$$[W] := \sum_i (\mathrm{length}_{w_i} \mathscr{O}_W) \cdot [W_i], \qquad (4.61.1)$$

where $W_i \subset W$ are the d-dimensional irreducible components with generic points $w_i \in W_i$. If W is reduced and pure dimensional then $[W]$ determines W; we will not always distinguish them clearly. However, if W is nonreduced, then it carries much more information than $[W]$. The only exception is when W is a Mumford divisor.

If X is projective and L is an ample line bundle on X, then the *degree* of a d-cycle $Z = \sum_i m_i[V_i]$ is defined as $\deg_L Z := \sum_i m_i \deg_L V_i = \sum_i m_i(L^d \cdot V_i)$.

Assume that X is a scheme of finite type over a field k and K/k a field extension. If $V \subset X$ is a d-dimensional irreducible, reduced subvariety then $V_K \subset X_K$ is a d-dimensional subscheme which may be reducible and, if char $k > 0$, may be nonreduced. If $Z = \sum m_i V_i$ is a d-cycle, we set

$$Z_K := \sum m_i[(V_i)_K]. \tag{4.61.2}$$

Z is called *geometrically reduced* if $Z_{\bar{k}}$ is reduced. If char $k = 0$ then reduced is the same as geometrically reduced.

Given an embedding $X \hookrightarrow \mathbb{P}^n$, every d-cycle on X is also a d-cycle on $\mathbb{P}^n$. Thus Cayley–Chow theory focuses primarily on cycles in $\mathbb{P}^n$

4.62 (Cayley–Chow correspondence over fields I) Fix a projective space $\mathbb{P}^n$ over a field k with dual projective space $\check{\mathbb{P}}^n$. Points in $\check{\mathbb{P}}^n$ are hyperplanes in $\mathbb{P}^n$. For $d \leq n - 1$ we have an incidence correspondence

$$\mathbf{I}^{(n,d)} := \{(p, H_0, \ldots, H_d) : p \in H_0 \cap \cdots \cap H_d\} \subset \mathbb{P}^n \times (\check{\mathbb{P}}^n)^{d+1}, \tag{4.62.1}$$

which comes with the coordinate projections

$$\mathbb{P}^n \xleftarrow{\pi_1} \mathbf{I}^{(n,d)} \xrightarrow{\pi_2} (\check{\mathbb{P}}^n)^{d+1} \xrightarrow{\sigma_i} (\check{\mathbb{P}}^n)^d, \tag{4.62.2}$$

where π_1 is a $(\check{\mathbb{P}}^{n-1})^{d+1}$-bundle and σ_i deletes the ith component. The projection π_2 is a $\mathbb{P}^{n-d-1}$-bundle over a dense open subset. For a closed subscheme $Y \subset \mathbb{P}^n$ set $\mathbf{I}_Y^{(n,d)} := \pi_1^{-1}(Y)$.

Let $Z \subset \mathbb{P}^n$ be an irreducible, geometrically reduced, closed subvariety of dimension d. Its *Cayley–Chow hypersurface* is defined as

$$\begin{aligned} \mathrm{Ch}(Z) &:= \pi_2(\mathbf{I}_Z^{(n,d)}) \\ &= \{(H_0, \ldots, H_d) \in (\check{\mathbb{P}}^n)^{d+1} : Z \cap H_0 \cap \cdots \cap H_d \neq \emptyset\}. \end{aligned} \tag{4.62.3}$$

An equation of $\mathrm{Ch}(Z)$ is called a *Cayley–Chow form*. Next note that

$$\mathbf{I}_Z^{(n,d)} \cap \pi_2^{-1}(H_0, \ldots, H_d) = Z \cap H_0 \cap \cdots \cap H_d. \tag{4.62.4}$$

In particular, a general $H_0 \cap \cdots \cap H_d$ is disjoint from Z and a general $H_0 \cap \cdots \cap H_d$ containing a smooth point $p \in Z$ meets Z only at p (scheme theoretically). Thus we see the following.

Claim 4.62.5 Let Z be a geometrically reduced d-cycle. Then $\pi_2 : \mathbf{I}_Z^{(n,d)} \to$ $\mathrm{Ch}(Z)$ is birational and $\mathrm{Ch}(Z)$ is a hypersurface in $(\check{\mathbb{P}}^n)^{d+1}$.   □

For any $H_0, \ldots, H_{d-1}$ the fiber of the coordinate projection $\sigma_d : \mathrm{Ch}(Z) \to$ $(\check{\mathbb{P}}^n)^d$ is $\check{\mathbb{P}}^n$ if $\dim(Z \cap H_0 \cap \cdots \cap H_{d-1}) \geq 1$; otherwise it is the set of hyperplanes that contain one of the points of $Z \cap H_0 \cap \cdots \cap H_{d-1}$. Similarly for all the other σ_i. Thus we proved the following.

Claim 4.62.6 Let Z be a geometrically reduced d-cycle of degree r. Then a general geometric fiber of any of the projections $\sigma_i : \mathrm{Ch}(Z) \to (\check{\mathbb{P}}^n)^d$ is the union of r distinct hyperplanes in $\check{\mathbb{P}}^n$. In particular, the projections are geometrically reduced and $\mathrm{Ch}(Z)$ has multidegree $(r, \ldots, r)$.   □

For $p \in \mathbb{P}^n$, let $\check{p}$ denote the set of hyperplanes passing through p. Then $p \in Z$ iff $\check{p} \times \cdots \times \check{p} \subset \mathrm{Ch}(Z)$. This leads us to the definition of the inverse of the map $Z \mapsto \mathrm{Ch}(Z)$. Let $D \subset (\check{\mathbb{P}}^n)^{d+1}$ be a geometrically reduced subscheme. (In practice, D will always be a hypersurface.) Define

$$\mathrm{Ch}_{\mathrm{set}}^{-1}(D) := \{p : \check{p} \times \cdots \times \check{p} \subset D\} \subset \mathbb{P}^n. \tag{4.62.7}$$

For now we will view $\mathrm{Ch}_{\mathrm{set}}^{-1}(D)$ as a reduced subscheme; scheme-theoretic versions will be discussed in (4.71).

It is easy to see that $\dim \mathrm{Ch}_{\mathrm{set}}^{-1}(D) \leq d$ and an irreducible hypersurface D is of *Cayley–Chow type* if $\dim \mathrm{Ch}_{\mathrm{set}}^{-1}(D) = d$. An arbitrary hypersurface D is of *Cayley–Chow type* if all of its irreducible components are. The basic correspondence of Cayley–Chow theory is the following; see Kollár (1996, I.3.24.5).

Claim 4.62.8 Fix n, d, r, and a base field k. Then the maps Ch and $\mathrm{Ch}_{\mathrm{set}}^{-1}$ provide a one-to-one correspondence between

$$\left\{ \begin{array}{c} \text{geometrically reduced} \\ d\text{-cycles of degree } r \text{ in } \mathbb{P}^n \end{array} \right\} \Leftrightarrow \left\{ \begin{array}{c} \text{geometrically reduced} \\ \text{Cayley–Chow type hypersurfaces of} \\ \text{degree } (r, \ldots, r) \text{ in } (\check{\mathbb{P}}^n)^{d+1} \end{array} \right\}.$$

Proof We already saw the $\Rightarrow$ part. To see the converse, observe the inclusion $\mathrm{Ch}(\mathrm{Ch}_{\mathrm{set}}^{-1}(D)) \subset D$. Thus if $Z \subset \mathrm{Ch}_{\mathrm{set}}^{-1}(D)$ is any subvariety of dimension d, then $\mathrm{Ch}(Z) \subset D$, hence $\mathrm{Ch}(Z)$ is an irreducible component of D. Thus $D = \mathrm{Ch}(\mathrm{Ch}_{\mathrm{set}}^{-1}(D))$.   □

Let $Z \subset \mathbb{P}^n$ be a pure dimensional subscheme or a cycle. The Chow equations are the "most obvious" equations of Z. They generate a homogeneous ideal (or an ideal sheaf), which was studied in various forms in Catanese (1992), Dalbec and Sturmfels (1995), and Kollár (1999). Its relationship with the scheme-theoretic $\mathrm{Ch}^{-1}_{\mathrm{sch}}$ will be given in (4.73).

4.63 (Element-wise power) Let R be a ring, $I \subset R$ an ideal, and $m \in \mathbb{N}$. Set

$$I^{[m]} := (r^m : r \in I).$$

These ideals have been studied mostly when $\mathrm{char}\, k = p > 0$ and q is a power of p; one of the early occurrences is in Kunz (1976). In these cases, $I^{[q]}$ is called a *Frobenius power* of I. Other values of the exponent are also interesting. Of the following properties, (1) is clear and (4.63.2–3) are implied by (4.63.4–5). We assume for simplicity that R is a k-algebra.

(4.63.1) If I is principal then $I^{[m]} = I^m$.
(4.63.2) If $\mathrm{char}\, k = 0$ then $I^{[m]} = I^m$.
(4.63.3) If $m < \mathrm{char}\, k$ then $I^{[m]} = I^m$.
(4.63.4) If k is infinite then $(r_1, \dots, r_n)^{[m]} = ((\sum c_i r_i)^m : c_i \in k)$.
Note that (3) is close to being optimal. For example, if $I = (x, y) \subset k[x, y]$ and $\mathrm{char}\, k = p \geq 3$ then $(x, y)^{[p+1]} = (x^{p+1}, x^p y, x y^p, y^{p+1}) \subsetneq (x, y)^{p+1}$.

Claim 4.63.5 Let k be an infinite field. Then

$$\langle (c_1 x_1 + \cdots + c_n x_n)^m : c_i \in k \rangle = \langle x_1^{i_1} \cdots x_n^{i_n} : \binom{m}{i_1 \dots i_n} \neq 0 \rangle.$$

Here $\binom{m}{i_1 \dots i_n}$ denotes the coefficient of $x_1^{i_1} \cdots x_n^{i_n}$ in $(x_1 + \cdots + x_n)^m$.

Proof The containment $\subset$ is clear. If the two sides are not equal then the left-hand side is contained in some hyperplane of the form $\sum \lambda_I x^I = 0$, but this would give a nontrivial polynomial identity $\sum \binom{m}{i_1 \dots i_n} \lambda_I c^I = 0$ for the c_i. $\qquad\square$

4.64 (Ideal of Chow equations) Let Z be a d-cycle of degree r in $\mathbb{P}^n$. Let $\varrho : \mathbb{P}^n \dashrightarrow \mathbb{P}^{d+1}$ be a projection that is defined along Z. Then $\varrho_*(Z)$ is a d-cycle in $\mathbb{P}^{d+1}$, thus it can be identified with a hypersurface; hence with a homogeneous polynomial $\phi(Z, \varrho)$ of degree r. Its pull-back to $\mathbb{P}^n$ is a homogeneous polynomial $\Phi(Z, \varrho)$ of degree r. Together they generate the *ideal sheaf of Chow equations* $I^{\mathrm{ch}}(Z) \subset \mathcal{O}_{\mathbb{P}^n}$.

Over a finite field k there may not be any projections defined along Z. The definition gives $I^{\mathrm{ch}}(Z)$ over $\bar{k}$ and it is clearly defined over k.

$^0{}_*$ This is not related to the symbolic power, frequently denoted by $I^{(m)}$.

The embedded primes of $I^{ch}(Z)$ are quite hard to understand, so frequently we focus on the *Chow hull* of the cycle Z:

$$\text{CHull}(Z) := \text{pure}(\text{Spec } \mathcal{O}_{\mathbb{P}^n} / I^{ch}(Z)).$$

Any Zariski dense set of projections generate $I^{ch}(Z)$. That is, if $P \subset \text{Gr}(n - d, n + 1)$ is Zariski dense then $I^{ch}(Z) = (\Phi(Z, \varrho) : \varrho \in P)$. It is enough to show that this holds in every Artinian quotient $\sigma : \mathcal{O}_{\mathbb{P}^n} \twoheadrightarrow A$. Let $B \subset A$ be the ideal generated by $\sigma(\Phi(Z, \varrho) : \varrho \in P)$. All the $\sigma(\Phi(Z, \varrho))$ are points of an irreducible subvariety $G \subset A$ obtained as an image of $\text{Gr}(n - d, n + 1)$. By assumption, $G \cap B$ contains the points $\sigma(\Phi(Z, \varrho) : \varrho \in P)$, hence it is dense in G. So $G \subset B$, since B is Zariski closed, if we think of A as a k-vectorspace.

Claim 4.64.1 Let Z be a geometrically reduced cycle. Then $I^{ch}(Z) \subset I_Z$ and the two agree along the smooth locus of Z.

Proof Let $p \in Z$ be a smooth point and $v \in T_p \mathbb{P}^n \setminus T_p Z$. A general projection $\varrho : \mathbb{P}^n \dashrightarrow \mathbb{P}^{d+1}$ maps $\langle T_p Z, v \rangle$ isomorphically onto $T_{\varrho(p)} \mathbb{P}^{d+1}$. Then $d\Phi(Z, \varrho)$ is nonzero on v. Thus the $\Phi(Z, \varrho)$ generate I_Z in a neighborhood of p.   □

For the nonreduced case, we need a definition.

Definition–Lemma 4.65 Let $Z \subset \mathbb{P}^n$ be an irreducible, d-dimensional subscheme such that red Z is geometrically reduced. Its *width* is defined in the following equivalent ways.

(4.65.1) The *projection width* of Z is the generic multiplicity of $\pi(Z)$ for a general projection $\pi : \mathbb{P}^n \dashrightarrow \mathbb{P}^{d+1}$.

(4.65.2) The *power width* of Z is the smallest m such that $I_{\text{red}\,Z}^{[m]} \cdot \mathcal{O}_Z$ is generically 0 along Z.

In general, we first take a purely inseparable field extension K/k such that $\text{red}(Z_K)$ is geometrically reduced and define the width of Z as the width of Z_K.

For example, it is easy to see that the width of $\text{Spec } k[x, y]/(x, y)^m$ is m and the width of $\text{Spec } k[x, y]/(x^m, y^m)$ is $2m - 1$.

Proof For a general projection $\pi : \mathbb{P}^n \dashrightarrow \mathbb{P}^{d+1}$ let ϕ_π be an equation of $\pi(\text{red } Z)$ and Φ_π its pull-back to $\mathbb{P}^n$. Then Z has projection width m iff $\Phi_\pi^m \cdot \mathcal{O}_Z$ is generically 0 for every π, and m is the smallest such. Since the Φ_π generically generate $I_{\text{red}\,Z}$, this holds iff $I_{\text{red}\,Z}^{[m]} \cdot \mathcal{O}_Z$ is generically 0 and m is the smallest. Thus the projection width equals the power width.   □

Proposition 4.66 *Let $Z_i \subset \mathbb{P}^n$ be distinct, geometrically irreducible cycles of the same dimension. Then* $\text{CHull}(\sum m_i Z_i) = \text{pure}(\text{Spec } \mathcal{O}_{\mathbb{P}^n} / \cap_i I(Z_i)^{[m_i]})$.

Proof The equations of the projections $\phi(\sum Z_i, \varrho)$ (as in (4.64)) generate $I_{\sum Z}$ at its smooth points. So if $p \in Z_i$ is a smooth point of $\sum Z$, then $I(Z_i)^{[m_i]}$ agrees with $I^{ch}(\sum m_i Z_i)$ at p by (4.63.4). □

The following consequence of (4.66) is key to our study of Mumford divisors.

Corollary 4.67 *Let k be an infinite field, $X \subset \mathbb{P}^n_k$ a reduced subscheme of pure dimension $d + 1$ and $D \subset X$ a Mumford divisor, viewed as a divisorial subscheme. Then* $\mathrm{pure}(X \cap \mathrm{CHull}(D)) = D$.

Proof The containment $\supset$ is clear, hence equality can be checked after a field extension. Write $D = \sum m_i D_i$ where the D_i are geometrically irreducible and reduced. Then $\mathrm{CHull}(D) = \mathrm{pure}(\mathrm{Spec}\, \mathscr{O}_{\mathbb{P}^n_k} / \cap_i I(D_i)^{[m_i]})$ by (4.66). Let $g_i \in D_i$ be the generic point and R_i its local ring in $\mathbb{P}^n_k$. Let $J_i \subset R_i$ be the ideal defining X and (J_i, h_i) the ideal defining D_i. The ideal defining the left-hand side of (4.67.1) is then $(J_i + (J_i, h_i)^{[m_i]})/J_i$. This is the same as $(h_i)^{[m_i]}$, as an ideal in R_i/J_i, which equals $(h_i^{m_i})$ by (4.63.1). □

Relative Mumford Divisors

Definition 4.68 Let S be a scheme and $f: X \to S$ a morphism of pure relative dimension n that is mostly flat (3.26). A *relative Mumford divisor* on X is a relative, generically Cartier divisor D (4.24) such that, for every $s \in S$, the fiber X_s is smooth at all generic points of D_s.

Let S' be another scheme and $h: S' \to S$ a morphism. Then the pull-back $h^{[*]}D$ is again a relative Mumford divisor on $X \times_S S' \to S'$. This gives the *functor of Mumford divisors,* denoted by

$$M\mathcal{D}iv(X/S)(*): \{S\text{-schemes}\} \to \{\text{sets}\}. \tag{4.68.1}$$

We prove in (4.76) that if f is projective, then the functor of effective Mumford divisors is represented by an S-scheme

$$\mathrm{Univ}^{md}(X/S) \to \mathrm{MDiv}(X/S), \tag{4.68.2}$$

whose connected components are quasi-projective over S.

We will see that relative, effective Mumford divisors form the right class for moduli purposes over a reduced base, but not in general. Fixing this problem leads to the notion of K-flatness in Chapter 7.

The following result – whose proof will be given after (4.76.5) – turns a relative, effective Mumford divisor into a flat family of Cartier divisors on another morphism, leading to the existence of $\mathrm{MDiv}(X/S)$ in (4.76).

Theorem 4.69 *Let S be a reduced scheme, $f : X \to S$ a projective morphism that is mostly flat (3.26), and $j : X \hookrightarrow \mathbf{P}_S$ an embedding into a $\mathbb{P}^N$-bundle. Then the maps* Ch *and* Ch_X^{-1} *– to be defined in (4.70) and (4.75.2) – provide a one-to-one correspondence*

$$\left\{ \begin{array}{c} \textit{relative Mumford} \\ \textit{divisors on } X \end{array} \right\} \leftrightarrow \left\{ \begin{array}{c} \textit{flat Cayley–Chow forms of} \\ \textit{Mumford divisors on } X \end{array} \right\}. \qquad (4.69.1)$$

Comments 4.69.2 There are two remarkable aspects of this equivalence. First, the left-hand side depends only on $X \to S$, while the right-hand side is defined in terms of an embedding $j : X \hookrightarrow \mathbf{P}_S$.

Second, on the left we have families that are usually not flat, on the right families of hypersurfaces in a product of projective spaces; these are the simplest possible flat families.

The correspondence (4.69.1) fails very badly over nonreduced bases. We see in (7.14) that, in an analogous local setting, the left-hand side is locally infinite dimensional for $S = \mathrm{Spec}\,\mathbb{C}[\varepsilon]$, but the right-hand side is locally finite dimensional. Nonetheless, we will be guided by (4.69.1). The rough plan is that we declare the right-hand side to give the correct answer and then work backwards to see what additional conditions this imposes on relative Mumford divisors. This leads us to the notion of C-flatness (7.37). Independence of the embedding $j : X \hookrightarrow \mathbf{P}_S$ then becomes a major issue in Chapter 7.

4.70 (Definition of Ch) In order to construct $\mathrm{Chow}_{d,r}(\mathbb{P}_S^n)$, the Chow variety of degree r cycles of dimension d in $\mathbb{P}_S^n$, we start with the incidence correspondence as in (4.62)

$$(4.70.1)$$

Note that here $\sigma = \sigma_{n,d,r}$ is a $(\check{\mathbb{P}}^{n-1})^{d+1}$-bundle. The fibers of $\tau = \tau_{n,d,r}$ are linear spaces of dimension $\geq n - d - 1$ and τ is a $\mathbb{P}^{n-d-1}$-bundle over a dense open subset.

Let now $D \subset \mathbb{P}_S^n$ be a generically flat family of d-dimensional subschemes (3.26). Assume also that the generic embedding dimension of D_s is $\leq d + 1$ for every $s \in S$. (This is satisfied iff each D_s is a Mumford divisor on some $X \subset \mathbb{P}_S^n$; a more general definition is in (7.46).) Set $\mathrm{Ch}(D) := \tau_*(\sigma^{-1}(D))$.

Claim 4.70.2 The map, $\tau : \sigma^{-1}(D) \to \mathrm{Ch}(D)$ is a local isomorphism on the preimage of a dense open subset $U \subset D$ such that $U \cap D_s$ is dense in D_s for every $s \in S$.

Proof Pick $p \in D_s$ such that T_{D_s} has dimension $d+1$ at p. If $L_s \supset p$ is a general linear subspace of dimension $n-d-1$, then $L_s \cap D_s = \{p\}$, scheme theoretically. This is exactly the fiber of $\tau : \sigma^{-1}(D) \to \mathrm{Ch}(D)$ over any $(H_0, \ldots, H_d)$ for which $L_s = H_0 \cap \cdots \cap H_n$. □

Corollary 4.70.3 $\mathrm{Ch}(D)$ is a generically flat family of Cartier divisors. If S is reduced, then $\mathrm{Ch}(D)$ is flat over S.

Proof By assumption, D is a generically flat family, hence so is $\sigma^{-1}(D)$ since σ is smooth. The first part is now immediate from (4.70.2). The second claim then follows from (4.36). □

4.71 (Definition of $\mathrm{Ch}_{\mathrm{sch}}^{-1}$) Although $\mathrm{Ch}(D)$ is not a flat family of Cartier divisor in general, we decide that from now on we are only interested in the cases when it is flat. Thus let $H^{\mathrm{cc}} \subset (\check{\mathbb{P}}_S^n)^{d+1}$ be a relative hypersurface of multidegree $(r, \ldots, r)$. We first define its *scheme-theoretic Cayley–Chow inverse*, denoted by $\mathrm{Ch}_{\mathrm{sch}}^{-1}(H^{\mathrm{cc}})$. It is a first approximation of the "correct" Cayley–Chow inverse.

Working with (4.70.1) consider the restriction of the left-hand projection

$$\sigma^{\mathrm{cc}} : (\mathrm{Inc}_S \cap \tau^{-1}(H^{\mathrm{cc}})) \to \mathbb{P}_S^n. \qquad (4.71.1)$$

Fix $s \in S$ and a point $p_s \in \mathbb{P}_s^n$. Note that the preimage of p_s consists of all $(d+1)$-tuples $(H_0, \ldots, H_d)$ such that $p_s \in H_i$ for every i and $(H_0, \ldots, H_d) \in H_s^{\mathrm{cc}}$. In particular, if Z is a d-cycle of degree r on $\mathbb{P}_s^n$ and $H^{\mathrm{cc}} = \mathrm{Ch}(Z)$ is its Cayley–Chow hypersurface, then σ^{cc} is a $(\check{\mathbb{P}}_S^{n-1})^{d+1}$-bundle over $\mathrm{Supp}\, Z$.

The key observation is that this property alone is enough to define $\mathrm{Ch}_{\mathrm{sch}}^{-1}$ and to construct the Chow variety. So we define $\mathrm{Ch}_{\mathrm{sch}}^{-1}(H^{\mathrm{cc}}) \subset \mathbb{P}_S^n$ as the unique, largest, closed subscheme over which σ^{cc} is a $(\check{\mathbb{P}}^{n-1})^{d+1}$-bundle. (Its existence is a special case of (3.19), but we derive its equations in (4.72.2).)

The set-theoretic behavior of the projection $\varrho : \mathrm{Ch}_{\mathrm{sch}}^{-1}(H^{\mathrm{cc}}) \to S$ is described in (4.62). The fibers have dimension $\leq d$ and $Z_s \subset \mathbb{P}_s^n$ is a d-dimensional irreducible component iff $\mathrm{Ch}(Z_s)$ is an irreducible component of H_s^{cc}. It is not hard to see that there is a maximal closed subset $S(H^{\mathrm{cc}}) \subset S$ over which H^{cc} is the Cayley–Chow hypersurface of a family of d-cycles; see Kollár (1996, I.3.25.1).

However, we do not yet have the "correct" scheme structure on $S(H^{\mathrm{cc}})$, since the scheme structure of the fibers of $\varrho : \mathrm{Ch}_{\mathrm{sch}}^{-1}(H^{\mathrm{cc}}) \to S$ is not the "correct" one. Before we move ahead, we need to understand this scheme structure.

4.72 (Scheme structure of $\mathrm{Ch}_{\mathrm{sch}}^{-1}(H^{cc})$) Let S be a scheme and $H^{cc} := (F = 0) \subset (\check{\mathbb{P}}^n)_S^{d+1}$ a hypersurface of multidegree $(r, \ldots, r)$. We aim to write down equations for $\mathrm{Ch}_{\mathrm{sch}}^{-1}(F = 0)$.

Choose coordinates $(x_0 : \cdots : x_n)$ on $\mathbb{P}_S^n$ and dual coordinates $(\check{x}_{0j} : \cdots : \check{x}_{nj})$ on the jth copy of $\check{\mathbb{P}}_S^n$ for $j = 0, \ldots, d$. So $F = F(\check{x}_{ij})$ is a homogeneous polynomial of multidegree $(r, \ldots, r)$. For notational simplicity we compute in the affine chart $\mathbb{A}_S^n = \mathbb{P}_S^n \setminus (x_0 = 0)$.

For $(x_1, \ldots, x_n) \in \mathbb{A}_S^n$, the hyperplanes H in the jth copy of $\check{\mathbb{P}}_S^n$ that pass through $(x_1, \ldots, x_n)$ are all written as $(-\sum_{i=1}^n x_i \check{x}_{ij} : \check{x}_{1j} : \cdots : \check{x}_{nj})$.

Let $M(\check{x}_{ij})$ be all the monomials in the $\check{x}_{ij}$ and write

$$F(-\sum_{i=1}^n x_i \check{x}_{i0} : \check{x}_{10} : \cdots : \check{x}_{n0}; \cdots ; -\sum_{i=1}^n x_i \check{x}_{id} : \check{x}_{1d} : \cdots : \check{x}_{nd})$$
$$=: \sum_M F_M(x_1, \ldots, x_n) M(\check{x}_{ij}). \qquad (4.72.1)$$

Since the monomials $M(\check{x}_{ij})$ are linearly independent, this vanishes for all $\check{x}_{ij}$ iff $F_M = 0$ for every M. Equivalently:

Claim 4.72.2 The subscheme $\mathrm{Ch}_{\mathrm{sch}}^{-1}(F = 0) \cap \mathbb{A}_S^n$ is given by the equations $F_M(x_1, \ldots, x_n) = 0$ for all monomials M, with F_M as in (4.72.1). □

Assume that $(F = 0) = \mathrm{Ch}(Y)$. If we fix $\check{x}_{ij} = c_{ij}$, then these give the matrix of a linear projection $\pi_c : \mathbb{A}_S^n \to \mathbb{A}_S^{d+1}$. The corresponding Chow equation of Y is $\sum_M F_M(x_1, \ldots, x_n) M(c_{ij}) = 0$. Thus we proved the following.

Theorem 4.73 *Let $Z \subset \mathbb{P}_k^n$ be a d-cycle of degree r. Then $\mathrm{Ch}_{\mathrm{sch}}^{-1}(\mathrm{Ch}(Z)) \subset \mathbb{P}_k^n$ is the subscheme defined by the ideal of Chow equations $I^{ch}(Z)$.* □

Note that we proved a little more. If the residue field of S is infinite, then $I^{\mathrm{ch}}(Y)|_{\mathbb{A}_S^n}$ is generated by the Chow equations of the linear projections $\pi_c : \mathbb{A}_S^n \to \mathbb{A}_S^{n+1}$. A priori we would need to use the more general projections (7.34.4), but this is just a matter of choosing the hyperplane at infinity.

Combining (4.73) and (4.67) gives the following.

Corollary 4.74 *Let k be a field, $X \subset \mathbb{P}_k^n$ a subscheme of pure dimension $d + 1$, and $D \subset X$ a Mumford divisor. Then $\mathrm{pure}(X \cap \mathrm{Ch}_{\mathrm{sch}}^{-1}(\mathrm{Ch}(D))) = D$.* □

4.75 (Construction of $\mathrm{MDiv}(X/S)$) As we noted in (4.69.2), we construct $\mathrm{MDiv}(X/S)$ by starting on the right-hand side of (4.69.1)

Let S be a scheme, $f : X \to S$ a mostly flat, projective morphism of pure dimension d, and $j : X \hookrightarrow \mathbb{P}_S^n$ an embedding.

We fix the intended degree to be r and let $\mathbf{P}_{n,d,r} = |\mathscr{O}_{(\check{\mathbb{P}}^n)^{d+1}}(r,\ldots,r)|$ be the linear system of hypersurfaces of multidegree $(r,\ldots,r)$ in $(\check{\mathbb{P}}^n)^{d+1}$, with universal hypersurface $\mathbf{H}^{cc}_{n,d,r} \subset (\check{\mathbb{P}}^n)^{d+1} \times \mathbf{P}_{n,d,r}$. Thus (4.70.1) extends to

$$
\begin{array}{ccc}
& \mathrm{Inc}_S\big(\mathrm{point}, (\check{\mathbb{P}}^n)^{d+1}\big) \times_S \mathbf{P}_{n,d,r} & \\
& \swarrow_{\sigma_{n,d,r}} \qquad \searrow^{\tau_{n,d,r}} & \\
\mathbb{P}^n_S \times_S \mathbf{P}_{n,d,r} & & (\check{\mathbb{P}}^n)^{d+1}_S \times_S \mathbf{P}_{n,d,r}.
\end{array}
\tag{4.75.1}
$$

As in (4.71), we get $\mathrm{Ch}^{-1}_{\mathrm{sch}}(\mathbf{H}^{cc}_{n,d,r}) \subset \mathbb{P}^n_S \times_S \mathbf{P}_{n,d,r}$. We are interested in d-cycles that lie on X, so we should take

$$
\mathrm{Ch}^{-1}_X(\mathbf{H}^{cc}_{n,d,r}) := \mathrm{Ch}^{-1}_{\mathrm{sch}}(\mathbf{H}^{cc}_{n,d,r}) \cap (X \times_S \mathbf{P}_{n,d,r}) \subset \mathbb{P}^n_S \times_S \mathbf{P}_{n,d,r}.
\tag{4.75.2}
$$

By (4.74), if $D_s \subset X_s$ is a Mumford divisor of degree r then the fiber of the coordinate projection $\varrho_{n,d,r} : \mathrm{Ch}^{-1}_X(\mathbf{H}^{cc}_{n,d,r}) \to \mathbf{P}_{n,d,r}$ over $[\mathrm{Ch}(D_s)]$ is D_s (aside from possible embedded points).

This leads us to the following. Recall the difference between mostly flat (in codimension ≤ 1) and generically flat (in codimension 0) as in (3.26).

Theorem 4.76 *Let S be a scheme, $f : X \to S$ a mostly flat, projective morphism of pure relative dimension $d + 1$, and $j : X \hookrightarrow \mathbb{P}^n_S$ an embedding. Then the functor of generically flat families of degree r Mumford divisors on X is represented by a locally closed subscheme $\mathrm{MDiv}_r(X/S)$ of $\mathbf{P}_{n,d,r}$ (4.75). Over $\mathrm{MDiv}_r(X)$ we have*

(4.76.1) $\mathrm{Univ}^{md}_r(X/S) \subset X \times_S \mathrm{MDiv}_r(X/S)$, *a universal, generically flat family of degree r Mumford divisors on X, and*

(4.76.2) $\mathbf{H}^{cc}_r \subset (\check{\mathbb{P}}^n)^{d+1} \times_S \mathrm{MDiv}_r(X/S)$, *a flat family of multidegree $(r,\ldots,r)$ hypersurfaces,*

that correspond to each other under Ch *and* Ch^{-1}_X.

Proof As we noted in (4.62), every fiber of $\varrho_{n,d,r}$ has dimension $\leq d$. So

$$
\{H^{cc}_s : \dim(\mathrm{Sing}\, X_s \cap \mathrm{Supp}\, \mathrm{Ch}^{-1}_X(H^{cc}_s)) \leq d - 1\}
$$

defines a closed subset of $\mathbf{P}_{n,d,r}$; let $\mathbf{P}^\circ_{n,d,r}$ denote its complement. Thus $[H^{cc}_s] \in \mathbf{P}^\circ_{n,d,r}$ iff the divisorial part of $\mathrm{Ch}^{-1}_X(H^{cc}_s)$ satisfies the Mumford condition.

Now apply (4.77) to $\mathrm{Ch}^{-1}_X(H^{cc})$ over $\mathbf{P}^\circ_{n,d,r}$ to get a locally closed decomposition $j^{\mathrm{flat}} : \mathbf{P}^{\mathrm{flat}}_{n,d,r} \to \mathbf{P}^\circ_{n,d,r}$, representing the functor of generically flat pull-backs of $\mathrm{Ch}^{-1}_X(H^{cc})$ as in (4.77). Over each connected component of $\mathbf{P}^{\mathrm{flat}}_{n,d,r}$, the degree of the d-dimensional part is locally constant. The union of those connected components where this degree equals r is $\mathrm{MDiv}_r(X/S)$. $\square$

Warning 4.76.3 In the nonreduced case the resulting Mdiv(X) a priori depends on the projective embedding $j: X \hookrightarrow \mathbb{P}^n_S$. We write Mdiv($X \subset \mathbb{P}^n_S$) if we want to emphasize this. In Chapter 7 we construct a subscheme Kdiv(X) $\subset$ Mdiv($X \subset \mathbb{P}^n_S$), that does not depend on the embedding. The two have the same underlying reduced structure and a positive answer to Question 7.42 would imply that in fact Mdiv($X \subset \mathbb{P}^n_S$) = Kdiv(X).

We have used the following variant of (3.19).

Proposition 4.77 *Let $f: X \to S$ be a projective morphisms and F a coherent sheaf on X such that $\operatorname{Supp} F \to S$ has fiber dimension $\leq d$. Then there is a locally closed decomposition $j_F^{flat}: S_F^{flat} \to S$ such that F_W is flat at d-dimensional points of the fibers iff $W \to S$ factors through j_F^{flat}.*

Proof We may replace X by the scheme-theoretic support SSupp F. The question is local on S. By (10.46.1), we may assume that there is a finite morphism $\pi: X \to \mathbb{P}^d_S$. Note that F_W is flat at d-dimensional points iff the same holds for $(\pi_W)_*F_W$. We may thus assume that $X = \mathbb{P}^d_S$; the important property is that now $f: X \to S$ is flat with integral geometric fibers. By (3.19.1) we get a decomposition $\amalg_i X_i \to X$, where $F|_{X_i}$ is locally free of rank i.

Let $Z \subset X$ be a closed subscheme. Applying (3.19) to the projection $\mathcal{O}_Z$, we see that there is a unique largest subscheme $S(Z) \subset S$ such that $f^{-1}(S(Z)) \subset Z$, scheme theoretically. For a locally closed subscheme $Z \subset X$ set $S(Z) = S(\bar{Z}) \setminus S(\bar{Z} \setminus Z)$, where $\bar{Z}$ denotes the closure of Z. Note that $S(Z)$ is the largest subscheme $T \subset S$ with the following property:

(4.77.1) There is an open subscheme $X_T^\circ \subset X_T$ that contains the generic point of X_t for every $t \in T$ and $X_T^\circ \subset Z$, scheme theoretically.

We claim that $S_F^{flat} = \amalg_i S(X_i)$. One direction is clear. $F|_{X_i}$ is locally free of rank i, so the restriction of F to $S(X_i) \times_S X$ is locally free, hence flat, at the generic point of every fiber.

Conversely, let W be a connected scheme and $q: W \to S$ a morphism such that F_W is generically flat over W the fiber dimension of $\operatorname{Supp} F_W \to S$ is n. Since X_w is integral, F_w is generically free for every $w \in W$, so F_W is locally free at the generic point of every fiber. Let $X_W^\circ \subset X_W$ be the open set where F_W is locally free.

By assumption, X_W° contains the generic point of every fiber X_w, so X_W° is connected. Thus F_W has constant rank, say i, on X_W°. Therefore, the restriction of $q_X: X_W \to X$ to X_W° lifts to $q_X^\circ: X_W^\circ \to X_i$. By (4.77.1), this means that q factors as $q: W \to S(X_i) \to S$. $\square$

5

Numerical Flatness and Stability Criteria

The aim of this chapter is to prove several characterizations of stable and locally stable families $f\colon (X, \Delta) \to S$. An earlier result, established in (3.1), has two assumptions:

- every fiber (X_s, Δ_s) is semi-log-canonical, and
- $K_{X/S} + \Delta$ is $\mathbb{Q}$-Cartier.

In many applications, the first of these is given, but the second one can be quite subtle.

Note that such difficulties arise already for surfaces, even if $\Delta = 0$. Indeed, we saw in Section 1.4 that there are flat, projective families $g\colon X \to C$ of surfaces with quotient singularities that are not locally stable. In these cases every fiber is log terminal, but $K_{X/C}$ is not $\mathbb{Q}$-Cartier, although its restriction to every fiber $K_{X/C}|_{X_c} = K_{X_c}$ is $\mathbb{Q}$-Cartier.

In all the examples in Section 1.4, this unexpected behavior coincides with a jump in the self-intersection number of the canonical class of the fiber. Our aim is to prove that this is always the case, as shown by the following simplified version of (5.4). The main part of its proof is in Section 5.4.

Theorem 5.1 (Numerical criteria of stability) *Let S be a connected, reduced scheme over a field of characteristic 0, and $f\colon X \to S$ a flat, proper morphism of pure relative dimension n. Assume that all fibers are semi-log-canonical with ample canonical class K_{X_s}. The following are equivalent:*

(5.1.1) *f is stable.*

(5.1.2) *$K_{X/S}$ is $\mathbb{Q}$-Cartier and f-ample.*

(5.1.3) *$h^0(X_s, \omega_{X_s}^{[m]})$ is independent of $s \in S$ for every $m > 0$.*

(5.1.4) *$f_*(\omega_{X/S}^{[m]})$ is locally free for every $m > 0$.*

(5.1.5) *$(K_{X_s}^n)$ is independent of $s \in S$.*

Proof Note that once $K_{X/S}$ is $\mathbb{Q}$-Cartier, it is f-ample since the K_{X_s} are ample. Then (5.1.1) $\Rightarrow$ (5.1.2) follows from (3.1.1).

The $m = 1$ case of (5.1.3) is proved in (2.69); the $m \geq 2$ cases follow from (3.1.3) and the vanishing of higher cohomologies (11.34). Next (5.1.3) $\Rightarrow$ (5.1.4) by Grauert's theorem. By Riemann–Roch (3.33), $(K_{X_s}^n)$ is the leading term of $h^0(X_s, \omega_{X_s}^{[m]})$, thus (5.1.3) $\Rightarrow$ (5.1.5). Finally (5.1.5) $\Rightarrow$ (5.1.1) is a special case of (5.4). □

If $f \colon X \to S$ is stable, then $K_{X/S}$ is $\mathbb{Q}$-Cartier, hence $(K_{X_s}^n)$ is clearly independent of $s \in S$, but the converse is surprising. General theory says that stability holds iff the Hilbert function $\chi(X_s, \mathcal{O}_{X_s}(mK_{X_s}))$ is independent of $s \in S$. Thus (5.1.2) asserts that if the leading coefficient of the Hilbert function is independent of s, then the same holds for the whole Hilbert function. We collect many similar results in this chapter; see Kollár (2015) for other such examples.

The main theorems are stated in Section 5.1. Related results on simultaneous canonical models and modifications are discussed in Section 5.2. The key claim is that, for families of slc pairs, local stability can fail only in relative codimension 2, and it can be characterized by the constancy of just one intersection number. A similar numerical condition characterizes Cartier divisors on flat families.

A series of examples in Section 5.3 shows that the assumptions of the theorems are likely to be optimal in characteristic 0.

Numerical criteria for stability in codimension ≤ 1 are discussed in Section 5.5. For all of the main theorems the key step is to establish them for families over smooth curves. This is done in Section 5.6. The numerical criterion of global stability, and a weaker version of local stability are derived in Section 5.6. The existence of simultaneous canonical models is studied in Section 5.7, and we treat simultaneous canonical modifications in Section 5.8.

Going from families over smooth curves to families over higher dimensional singular bases turns out to be quite quick, but several of the arguments, presented in Section 5.9, rely heavily on the techniques and results of Chapter 9.

Assumptions For all the main theorems of this chapter we work with varieties over a field of characteristic 0, but the background results worked out in Section 5.4 are established for excellent schemes.

5.1 Statements of the Main Theorems

We develop a series of criteria to characterize stable and locally stable (4.7) morphisms using a few, simple, numerical invariants of the fibers.

We follow the general set-up of (5.1), but we strengthen it in three ways:
• We add a boundary divisor Δ.
• We assume only that f is flat in codimension 1 on each fiber. The reason for this is that many natural constructions (for instance flips, taking cones or ramified covers) do not preserve flatness. Thus we frequently end up with morphisms that are not known to be flat everywhere.
• We deal with local stability as well. A weak variant, involving several intersection numbers, is quite similar to the global case, but the sharper form requires different considerations.

For the main results of this chapter we work with the following set-up, which is a slight generalization of (3.28) and (4.2).

Notation 5.2 Let $f: X \to S$ be a proper morphism of pure relative dimension n (2.71), and $Z \subset X$ a closed subset with complement $U := X \setminus Z$ such that
(5.2.1) $\mathrm{codim}_{X_s}(Z \cap X_s) \geq 2$ for every $s \in S$,
(5.2.2) $f|_U: U \to S$ is flat, and
(5.2.3) $\mathrm{depth}_Z X \geq 2$.
We also consider effective $\mathbb{R}$-divisors $\Delta = \sum b_i B_i$ on X, where the B_i are generically Cartier divisors (4.24). (Sheaf versions are studied in Section 5.4.)

We are mainly interested in cases where each fiber (X_s, Δ_s) is slc, but it turns out to be very useful to work with the following more general set-up.

Assumption 5.3 Given $f: (X, \Delta) \to S$ as in (5.2), we assume the following:
(5.3.1) $f|_U: (U, \Delta|_U) \to S$ is locally stable,
(5.3.2) (X_g, Δ_g) is slc for all generic points $g \in S$, and
(5.3.3) every fiber has lc normalization $\pi_s: (\bar{X}_s, \bar{D}_s + \bar{\Delta}_s) \to (X_s, \Delta_s)$.
Note that $(\bar{X}_s, \bar{D}_s + \bar{\Delta}_s)$ is defined over U_s by (1), and this determines $\bar{D}_s + \bar{\Delta}_s$ since $X_s \setminus U_s$ has codimension ≥ 2. Thus it makes sense to ask whether $(\bar{X}_s, \bar{D}_s + \bar{\Delta}_s)$ is lc or not.

The next result generalizes (5.1) to pairs. Its proof is given in (5.42).

Theorem 5.4 (Numerical criterion of stability) *We use the notation of (5.2). In addition to (5.3.1–3) assume that S is a reduced scheme over a field of char 0, and $K_{\bar{X}_s} + \bar{D}_s + \bar{\Delta}_s$ is ample for every $s \in S$. Then*
(5.4.1) $v(s) := ((K_{\bar{X}_s} + \bar{D}_s + \bar{\Delta}_s)^n)$ *is an upper semi-continuous function, and*
(5.4.2) $f: (X, \Delta) \to S$ *is stable iff $v(s)$ is locally constant on S.*

The local version is the following, to be proved in (5.27) and (5.54).

Theorem 5.5 (Numerical criterion of local stability) *We use the notation of (5.2). In addition to (5.3.1–3) assume that S is a reduced scheme over a field of* char 0, *and H is a relatively ample Cartier divisor class on X. Then*

(5.5.1) $v_2(s) := (\pi_s^* H^{n-2} \cdot (K_{\bar{X}_s} + \bar{D}_s + \bar{\Delta}_s)^2)$ *is upper semi-continuous, and*

(5.5.2) $f : (X, \Delta) \to S$ *is locally stable iff* $v_2(s)$ *is locally constant on S.*

Note that the functions $(\pi_s^* H^n)$ and $(\pi_s^* H^{n-1} \cdot (K_{\bar{X}_s} + \bar{D}_s + \bar{\Delta}_s))$ are always locally constant, but $(\pi_s^* H^{n-i} \cdot (K_{\bar{X}_s} + \bar{D}_s + \bar{\Delta}_s)^i)$ are neither upper nor lower semi-continuous for $i \geq 3$.

A key part of the proof of (5.5) is to show that local stability is essentially a two-dimensional question. The following, proved in (5.54), generalizes (2.7).

Theorem 5.6 (Local stability in codimension ≥ 3) (Kollár, 2013a, Thm.18) *Using the notation and assumptions of (5.2–5.3), let S be a reduced scheme of* char 0. *Assume also that* $\text{codim}_{X_s}(Z \cap X_s) \geq 3$ *for every* $s \in S$.
Then $f : (X, \Delta) \to S$ *is locally stable.*

One can also restate this as a converse of the Bertini-type result (2.13).

Corollary 5.7 *Notation and assumptions as in (5.2–5.3), let S be a reduced scheme of* char 0. *Assume in addition that the relative dimension is* $n \geq 3$ *and* $f|_H : (H, \Delta|_H) \to S$ *is locally stable, where* $H \subset X$ *is a relatively ample Cartier divisor. Then* $f : (X, \Delta) \to S$ *is also locally stable.* □

Comment As we noted in (2.14), (11.17) implies that $f : (X, H + \Delta) \to S$, and hence also $f : (X, \Delta) \to S$, are locally stable in a neighborhood of H. The unexpected new claim is that local stability holds everywhere.

A variant of (5.4) holds for arbitrary divisors and for non-slc fibers, but we have to assume that f is flat with S_2 fibers. On the other hand, this holds over any base scheme.

Theorem 5.8 (Numerical criterion for relative line bundles) Kollár (2016a) *Let S be a reduced scheme,* $f : X \to S$ *a flat, proper morphism of pure relative dimension* n *with* S_2 *fibers, and* $Z \subset X$ *a closed subset such that* codim_{X_s} $(Z \cap X_s) \geq 2$ *for every* $s \in S$. *Let A be an* f-*ample line bundle on X.*
Let L_U *be a line bundle on* $U := X \setminus Z$ *and assume that, for every* $s \in S$, *the restriction* $L_U|_{U_s}$ *extends to a line bundle* L_s *on* X_s. *Then*

(5.8.1) $d_2(s) := (A_s^{n-2} \cdot L_s^2)$ *is an upper semi-continuous function on S, and*

(5.8.2) L_U *extends to a line bundle on X iff* $d_2(s)$ *is locally constant on S.*
Furthermore, if L_s *is ample for every s, then*

(5.8.3) $d(s) := (L_s^n)$ is an upper semi-continuous function on S, and
(5.8.4) L_U extends to a line bundle on X iff $d(s)$ is locally constant on S.

5.2 Simultaneous Canonical Models and Modifications

Given a morphism $f \colon X \to S$, we would like to know when the canonical models (or the canonical modifications) of the fibers form a flat family; see (5.9) and (5.15) for the precise definitions.

As we discussed in Section 1.5, the canonical models of the fibers need not form a flat family, not even for locally stable morphisms. We develop numerical criteria, after some definitions.

Definition 5.9 (Simultaneous canonical model) Let $f \colon (X, \Delta) \to S$ be a morphism as in (5.2). Assume that (5.3.1) holds, and every fiber has log canonical normalization $\pi_s \colon (\bar{X}_s, \bar{\Delta}_s) \to (X_s, \Delta_s)$. Its *simultaneous canonical model* is a diagram

$$
\begin{array}{ccc}
X & \dashrightarrow^{\phi} & X^{\mathrm{sc}} \\
 & {}_{f}\searrow \quad \swarrow_{f^{\mathrm{sc}}} & \\
 & S &
\end{array}
\tag{5.9.1}
$$

where $f^{\mathrm{sc}} \colon (X^{\mathrm{sc}}, \Delta^{\mathrm{sc}}) \to S$ is stable, and $\phi_s \circ \pi_s \colon (\bar{X}_s, \bar{\Delta}_s) \dashrightarrow (X_s^{\mathrm{sc}}, \Delta_s^{\mathrm{sc}})$ is the canonical model (over s), as in (11.26), for every $s \in S$.

Warning A "simultaneous" canonical model is not the same as a "relative" canonical model (11.26). For both notions $K + \Delta$ is relatively ample, but the former requires the singularities of the fibers to be lc, the latter the singularities of the total space to be lc. Neither implies the other.

For a pure dimensional, proper morphism $f \colon X \to S$, the *simultaneous canonical model of resolutions* $f^{\mathrm{scr}} \colon X^{\mathrm{scr}} \to S$ is defined analogously. Here we require that each $\phi_s \colon X_s \dashrightarrow X_s^{\mathrm{scr}}$ be obtained by first taking a resolution $X_s^{\mathrm{r}} \to \mathrm{red}\, X_s$, and then the disjoint union of the canonical models of those irreducible components that are of general type.

5.9.2 (Some known cases) Let $f \colon X \to S$ be flat, projective with S reduced. Assume that X_s is of general type and has canonical singularities for some $s \in S$. Then a simultaneous canonical model exists over an open neighborhood $s \in S^\circ \subset S$; see (1.37). The $\Delta \neq 0$ case is more subtle, see (5.20) and (5.48).

[0]* See Comment 5.9.3.

5.9.3 (Comment on the conductor) Note that we do not add the conductor of π_s to $\bar{\Delta}_s$. If the fibers are normal in codimension 1 then D_s (the divisorial part of the conductor) is 0, hence our notion is the only sensible one. In general, however, one has a choice, and the simultaneous slc model, to be defined in (5.51), seems a better concept when $D_s \neq 0$.

We give criteria for the existence of simultaneous canonical models in terms of the volume (10.31) of the canonical class of the fibers. Note that if Y is a proper scheme of dimension n then $\mathrm{vol}(K_{Y^r})$ is independent of the choice of the resolution $Y^r \to Y$, and it equals the self-intersection number $((K_{Y^{cr}})^n)$. Similarly, if (Y, Δ) is log canonical then $\mathrm{vol}(K_Y + \Delta) = ((K_{Y^c} + \Delta^c)^n)$.

Theorem 5.10 (Numerical criterion for simultaneous canonical models I) *Let S be a seminormal scheme of* char 0, *and $f : X \to S$ a proper morphism of pure relative dimension n. Then*

(5.10.1) $v(s) := \mathrm{vol}(K_{(X_s)^r})$ *is a lower semicontinuous function on S, and*

(5.10.2) $f : X \to S$ *has a simultaneous canonical model of resolutions iff $v(s)$ is locally constant (and positive).*

The key case, when S is a smooth curve, is settled in (5.44); the general case is in (5.55). This is a surprising result on two accounts. First, cohomology groups almost always vary *upper* semicontinuously; the *lower* semicontinuity in this setting was first observed and proved by Nakayama (1986, 1987). Second, usually it is easy to generalize similar proofs from smooth varieties to klt or lc pairs, but here adding any boundary can ruin the argument and the conclusion. Example 5.19 shows that S needs to be seminormal.

The following is a similar result for normal lc pairs, but the lower semicontinuity of (5.10) changes to upper semicontinuity.

Theorem 5.11 (Numerical criterion for simultaneous canonical models II) *Let S be a seminormal scheme of* char 0, *and $f : (X, \Delta) \to S$ as in (5.2). Assume furthermore that*

(5.11.1) $f|_U : U \to S$ *is smooth with irreducible fibers,*

(5.11.2) *every fiber has lc normalization $\pi_s : (\bar{X}_s, \bar{\Delta}_s) \to (X_s, \Delta_s)$, and*

(5.11.3) *the canonical models $\phi_s : (\bar{X}_s, \bar{\Delta}_s) \dashrightarrow (\bar{X}_s^c, \bar{\Delta}_s^c)$ exist.*

Then

(5.11.4) $v(s) := \mathrm{vol}(K_{\bar{X}_s} + \bar{\Delta}_s)$ *is an upper semicontinuous function on S, and*

(5.11.5) $f : (X, \Delta) \to S$ *has a simultaneous canonical model iff $v(s)$ is locally constant.*

The proof is given in (5.46), and (5.55).

One should think of (5.11) as a generalization of (5.4), but there are differences. In (5.11) we allow only fibers that are smooth in codimension 1, and S is assumed seminormal. (The extra assumption (3) is expected to hold always.) However, the key difference is in the proofs given in Section 5.9. While the proof of (5.4) uses only the basic theory of hulls and husks, we rely on the existence of moduli spaces of pairs in order to establish (5.11).

Both (5.10) and (5.11) apply to $f\colon X \to S$ iff the normalizations of the fibers have canonical singularities. In this case, f is locally stable (2.8), and the plurigenera – and hence the volume – are locally constant (1.37).

A key ingredient of the proofs of (5.10–5.11) is the following characterization of canonical models. We prove a more general version of it in (10.36).

Proposition 5.12 *Let X be a smooth proper variety of dimension n. Let Y be a normal, proper variety birational to X, and D an effective $\mathbb{R}$-divisor on Y such that $K_Y + D$ is $\mathbb{R}$-Cartier, nef and big. Then*

(5.12.1) $\operatorname{vol}(K_X) \le \operatorname{vol}(K_Y + D) = ((K_Y + D)^n)$, *and*

(5.12.2) *equality holds iff $D = 0$ and Y has canonical singularities.*

For surfaces, the existence criterion of simultaneous canonical modifications is proved in Kollár and Shepherd-Barron (1988, Sec.2). In higher dimensions, we need to work with a sequence of intersection numbers and with their lexicographic ordering.

Definition 5.13 Let X be a proper scheme of dimension n, and A, B $\mathbb{R}$-Cartier divisors on X. Their *sequence of intersection numbers* is

$$I(A, B) := ((A^n), \ldots, (A^{n-i} \cdot B^i), \ldots, (B^n)) \in \mathbb{R}^{n+1}.$$

Definition 5.14 The *lexicographic* ordering of length $n + 1$ real sequences is

$$(a_0, \ldots, a_n) \le (b_0, \ldots, b_n).$$

This holds if either $a_i = b_i$ for every i, or there is an $r \le n$ such that $a_i = b_i$ for $i < r$, but $a_r < b_r$. For polynomials we define an ordering

$$f(t) \le g(t) \iff f(t) \le g(t) \; \forall t \gg 0.$$

We use $\equiv$ to denote identity of sequences or polynomials. Note that

$$\textstyle\sum_i a_i t^{n-i} \le \sum_i b_i t^{n-i} \iff (a_0, \ldots, a_n) \le (b_0, \ldots, b_n).$$

If we have proper schemes X, X' of dimension n, $\mathbb{R}$-Cartier divisors A, B on X and A', B' on X', then

$$I(A, B) \le I(A', B') \iff (tA + B)^n \le (tA' + B')^n.$$

We will consider functions that associate a sequence or a polynomial to all points of a scheme X. Using the above definitions, it makes sense to ask if such a function is *upper/lower semicontinuous* for $\le$ or not.

Definition 5.15 (Simultaneous canonical modification) Let $f \colon X \to S$ be a morphism of pure relative dimension n, and $\Delta = \sum a_i D_i$ a generically $\mathbb{Q}$-Cartier effective divisor on X. A *simultaneous canonical modification* is a proper morphism $p^{\mathrm{scm}} \colon (X^{\mathrm{scm}}, \Delta^{\mathrm{scm}}) \to (X, \Delta)$ such that $f \circ p^{\mathrm{scm}} \colon (X^{\mathrm{scm}}, \Delta^{\mathrm{scm}}) \to S$ is locally stable, and

$$p_s^{\mathrm{scm}} \colon ((X^{\mathrm{scm}})_s, (\Delta^{\mathrm{scm}})_s) \to (X_s, \Delta_s)$$

is the canonical modification (11.29) for every $s \in S$.

A *simultaneous log canonical modification* $p^{\mathrm{slcm}} \colon (X^{\mathrm{slcm}}, \Delta^{\mathrm{slcm}}) \to (X, \Delta)$ is defined analogously.

In the following result we need to assume that the base scheme is seminormal; see (5.21) for some examples.

Theorem 5.16 (Numerical criterion for simultaneous canonical modification) *We use the notation of (5.2). In addition to (5.3.1), assume that S is a seminormal scheme of* char 0*, and H is a relatively ample Cartier divisor class on X. For $s \in S$ let $p_s^{cm} \colon (X_s^{cm}, \Delta_s^{cm}) \to (X_s, \Delta_s)$ denote the canonical modification of the fiber (X_s, Δ_s). Then*

(5.16.1) $I(s) := I(\pi_s^* H_s, K_{X_s^{cm}} + \Delta_s^{cm})$ *is lower semi-continuous for* $\le$*, and*

(5.16.2) $f \colon (X, \Delta) \to S$ *has a simultaneous canonical modification iff $I(s)$ is locally constant.*

There is also a similar condition for simultaneous log canonical and semi-log-canonical modifications (5.52), but these only apply when $K_{X/S} + \Delta$ is $\mathbb{Q}$-Cartier.

5.3 Examples

Here we present a series of examples that show that the assumptions of the Theorems in Sections 5.1–5.2 are close to being optimal, except that the characteristic 0 assumption is probably superfluous.

The following is the simplest example illustrating the difference between being Cartier and fiber-wise Cartier.

Example 5.17 Consider the family of quadrics

$$X = (x^2 - y^2 + z^2 - t^2 w^2 = 0) \subset \mathbb{P}^3_{xyzw} \times \mathbb{A}_t \quad \text{and} \quad D = (x - y = z - tw = 0).$$

Here X_0 is a quadric cone, and X_t is a smooth quadric for $t \neq 0$. The divisor D is Cartier, except at the origin, where it is not even $\mathbb{Q}$-Cartier. However D_0 is a line on a quadric cone, hence $2D_0 = (x - y = t = 0)$ is Cartier. It is easy to compute that

$$L = \mathscr{O}_X(-2D) = (x - y, z - tw)^2 \cdot \mathscr{O}_X$$

is locally free outside the origin and not locally free at the origin, but the reflexive hull of its restriction

$$L_0^H := \mathscr{O}_{X_0}(-2D_0) = (x - y) \cdot \mathscr{O}_{X_0}$$

is locally free. The natural restriction map gives an identification

$$\mathscr{O}_X(-2D)|_{X_0} = (x, y, z) \cdot \mathscr{O}_{X_0}(-2D_0) \subset \mathscr{O}_{X_0}(-2D_0).$$

Note that the self-intersection number of the fibers of D also jumps. For $t \neq 0$ we have $(D_t^2) = 0$, but $(D_0^2) = 1/2$.

It is harder to get examples where the self-intersections in (5.8) are locally constant, yet the divisor is not Cartier, but, as we see next, this can happen even for the canonical class. Thus in (5.8) one needs to assume that the fibers of f are S_2, and in (5.4) that the fibers are slc.

Example 5.18 (See (2.35) or Kollár (2013b, 3.8–14) for the notation and basic results on cones.) Let $X \subset \mathbb{P}^N$ be a smooth, projective variety of dimension n and $L_X = \mathscr{O}_X(1)$. Let $C(X) := C_p(X, L_X)$ denote the projective cone over X with vertex v and natural ample line bundle $L_{C(X)}$. Let $H \subset X$ be a smooth hyperplane section, and $C(H) := C_p(H, L_H)$ the projective cone over H. Note that

$$(L_X^n) = (L_{C(X)}^{n+1}) = (L_H^{n-1}) = (L_{C(H)}^n).$$

The canonical class of $C(X)$ is Cartier iff $K_X \sim m c_1(L_X)$ for some $m \in \mathbb{Z}$. In this case $K_{C(X)} \sim (m - 1)c_1(L_{C(X)})$.

We can think of H as sitting in $X \subset C(X)$. The pencil of hyperplanes containing $H \subset C(X)$ gives a morphism of the blow-up $p\colon Y := B_H C(X) \to \mathbb{P}^1$ such that $Y_t \simeq X$ for $t \neq 0$, and the normalization $\bar{Y}_0$ of Y_0 is isomorphic to $C(H)$. However, if $H^1(X, \mathscr{O}_X) \neq 0$ then Y_0 is not normal. For instance, this happens if X is the product of nonhyperelliptic curves of genus ≥ 2 with its canonical embedding. Thus, if these hold, then

(5.18.1) Y_t is smooth and K_{Y_t} is ample for $t \neq 0$,

(5.18.2) Y_0 is not normal, the normalization $\bar{Y}_0 \rightarrow Y_0$ is an isomorphism except at v, $K_{\bar{Y}_0}$ is locally free and ample and

(5.18.3) $(K_{Y_t}^n) = (K_{\bar{Y}_0}^n)$ (where $n = \dim X$).

The next example shows that (5.10) fails if S is not seminormal.

Example 5.19 Let S be a local, reduced, nonseminormal scheme with semi-normalization $S' \rightarrow S$. Choose an embedding of S' into the moduli space of automorphism-free curves of genus g for some g. Let $p'\colon X' \rightarrow S'$ be the resulting smooth family. This induces a family $p\colon X' \rightarrow S' \rightarrow S$ that satisfies the assumptions of (5.10). However, there is no simultaneous canonical model since $p'\colon X' \rightarrow S'$ does not descend to $p\colon X \rightarrow S$.

The next examples show that there does not seem to be a log version of (5.10) for families with reducible fibers, not even for families of curves.

Example 5.20 Let $g\colon S \rightarrow C$ be a smooth family of curves, and $D_i \subset S$ a set of n disjoint sections. Set $\Delta := \sum d_i D_i$. Pick a point $0 \in C$, the fiber over it is $(S_0, \sum d_i[p_i])$ where $p_i = S_0 \cap D_i$. The "log volume" is $2g(S_0) - 2 + \sum d_i$.

Let $\pi\colon S^1 \rightarrow S$ be the blow up of all the points p_i with exceptional curves E_i and set $\Delta^1 := \pi_*^{-1}\Delta$. The central fiber of $g^1\colon (S^1, \Delta^1) \rightarrow C$ is $(S_0^1, 0) + \sum_i(E_i, d_i[p_i'])$. Its normalization consists of S_0 (with no boundary points) and $E_i \simeq \mathbb{P}^1$, each with one marked point of multiplicity d_i. Thus the "log volume" of the central fiber is now $2g(S_0) - 2$; the effect of the boundary vanished.

One can try to compensate for this by adding the double point divisor $\bar{D}_0$. This variant of the "log volume" is now $2g(S_0) - 2 + n$. This formula remembers only the number of the sections, not their coefficients. Even worse, we can blow up m other points on S_0, then the "log volume" formula gives $2g(S_0) - 2 + n + m$.

In general, there does not seem to be a sensible and birationally invariant way do define the "log volume" of degenerations.

In (5.16), the base scheme is assumed to be seminormal. The reason for this is that canonical modifications do have unexpected infinitesimal deformations.

Example 5.21 (Deformation of canonical modifications) We give an example of a normal, projective variety with isolated singularities and canonical modification $X^{\mathrm{cm}} \rightarrow X$ such that the trivial deformation of X can be lifted to a nontrivial deformation of X^{cm}.

Consider the isolated hypersurface singularity

$$X := X_{n,r} := (x_1^r + \cdots + x_n^r + x_{n+1}^{r+1} = 0) \subset \mathbb{A}_k^{n+1}.$$

Let $p \colon Y := B_0 X \to X$ denote the blow-up of the origin. Then Y is smooth, the exceptional divisor is the cone $E \simeq (x_1^r + \cdots + x_n^r = 0) \subset \mathbb{P}^n$ and $N_{E|Y} \simeq \mathcal{O}_E(-1)$. We compute that $a(E, X_{n,r}, 0) = n - r$. Thus $X_{n,r}$ is canonical iff $r \le n$, and Y is the canonical modification for $r > n$.

We claim that $p \colon Y \to X$ has a nontrivial deformation over $X \times_k \operatorname{Spec} k[\varepsilon]$. The trivial deformation is obtained by blowing up

$$(x_1 = \cdots = x_{n+1} = 0) \subset X \times_k \operatorname{Spec} k[\varepsilon].$$

The nontrivial deformation is obtained by blowing up

$$Z := (x_1 = \cdots = x_n = x_{n+1} - \varepsilon = 0) \subset X \times_k \operatorname{Spec} k[\varepsilon].$$

We need to check that X is equimultiple along the blow-up center. Introducing a new coordinate $y := x_{n+1} - \varepsilon$, the equations become

$$Z := (x_1 = \cdots = x_n = y = 0) \subset (x_1^r + \cdots + x_n^r + y^{r+1} + (r + 1)\varepsilon y^r = 0),$$

thus $X \times_k \operatorname{Spec} k[\varepsilon]$ is clearly equimultiple along Z.

Note that $E \subset Y$ has a unique extension E_ε to a deformation Y_ε of Y since $H^1(E, N_{E|Y}) = 0$. The blow-up ideal is then the push-forward of the ideal sheaf of E_ε. Thus different blow-up ideals give different deformations of Y.

The following examples show that the existence of simultaneous canonical modifications is more complicated for pairs.

Example 5.22 In $\mathbb{P}^2$ consider a line $L \subset \mathbb{P}^2$ and a family of degree 8 curves C_t such that C_0 has four nodes on L plus an ordinary 6-fold point outside L, and C_t is smooth and tangent to L at four points for $t \ne 0$.

Let $\pi_t \colon S_t \to \mathbb{P}^2$ denote the double cover of $\mathbb{P}^2$ ramified along C_t. Note that $K_{S_t} = \pi_t^* \mathcal{O}(1)$, thus $(K_{S_t}^2) = 2$. For each t, the preimage $\pi_t^{-1}(L)$ is a union of two curves $D_t + D_t'$. Our example is the family of pairs (S_t, D_t). We claim that,

(5.22.1) there is a log canonical modification $(S_t^{\mathrm{lcm}}, D_t^{\mathrm{lcm}}) \to (S_t, D_t)$, and

(5.22.2) $((K_{S_t^{\mathrm{lcm}}} + D_t^{\mathrm{lcm}})^2) = 1$, yet

(5.22.3) there is no simultaneous log canonical modification.

If $t \ne 0$ then S_t and D_t are smooth. Furthermore D_t, D_t' meet transversally at four points, thus $(D_t \cdot D_t') = 4$. Using $((D_t + D_t')^2) = 2$, we obtain that $(D_t^2) = -3$. Thus $((K_{S_t} + D_t)^2) = 1$.

If $t = 0$ then S_0 is singular at 5 points. D_0, D_0' meet transversally at four singular points of type A_1, thus $(D_0 \cdot D_0') = 2$. This gives that $(D_0^2) = -1$. Thus $((K_{S_0} + D_0)^2) = 3$. The pair (S_0, D_0) is lc away from the preimage of the 6-fold point. Let $q \colon T_0 \to S_0$ denote the minimal resolution of this point. The exceptional curve E is smooth, has genus 2 and

$(E^2) = -2$. Thus $K_{T_0} = q^* K_{S_0} - 2E$ hence $(T_0, E + D_0)$ is the log canonical modification of (S_0, D_0), and

$$((K_{T_0} + E + D_0)^2) = ((q^* K_{S_0} - E + D_0)^2) = ((K_{S_0} + D_0)^2) + (E^2) = 1.$$

Thus $((K_{S_t^{\mathrm{lcm}}} + D_t^{\mathrm{lcm}})^2) = 1$ for every t.

Nonetheless, the log canonical modifications do not form a flat family. Indeed, such a family would be a family of surfaces with ordinary nodes, so the relative canonical class would be a Cartier divisor. However, $(K_{S_t}^2) = 2$ for $t \neq 0$, but $(K_{T_0}^2) = ((q^* K_{S_0} - 2E)^2) = -6$.

Example 5.23 We start with a family of quadric surfaces $Q_t \subset \mathbb{P}^3$ where Q_0 is a cone, and Q_t is smooth for $t \neq 0$. We take six families of lines L_t^i such that for $t = 0$ we have six distinct lines, and for $t \neq 0$ two of them – L_t^1, L_t^2 – are from one ruling of the quadric, the other four from the other ruling. S_t denotes the double cover of Q_t ramified along the six lines $L_t^1 + \cdots + L_t^6$.

For $t \neq 0$ the surface S_t has ordinary nodes and $(K_{S_t}^2) = 0$. For $t = 0$ the surface S_0 has a unique singular point. Its minimal resolution $q \colon T_0 \to S_0$ is a double cover of $\mathbb{F}_2$ ramified along six fibers. Thus $(K_{T_0}^2) = -4$. Thus the canonical modifications do not form a flat family. The log canonical modification of S_0 is (T_0, E_0) where E_0 is the q-exceptional curve. Thus $((K_{T_0} + E_0)^2) = 0$.

The numerical condition is satisfied, but the log canonical modifications do not form a flat family since $T_0 = S_0^{\mathrm{lcm}}$ is smooth, but $S_t^{\mathrm{lcm}} = S_t$ is singular for $t \neq 0$. However, there is a flat family that is a weaker variant of a simultaneous log canonical modification.

This is obtained by replacing the singular quadric Q_0 with its resolution $Q_0' \simeq \mathbb{F}_2$. Let $E \subset \mathbb{F}_2$ denote the -2-section, and $|F|$ the ruling. One can arrange that L_t^1, L_t^2 degenerate to $F^i + E$ for $F^i \in |F|$, and the others degenerate to fibers F^j. This way a flat limit of the double cover S_t is obtained as the double cover of $\mathbb{F}_2$ ramified along $F^1 + \cdots + F^6 + 2E$. This is a semi-log-canonical surface whose normalization is the log canonical modification of S_0.

5.4 Mostly Flat Families of Line Bundles

We investigate sheaves that are known to be invertible in codimension 1; a topic we already encountered in Section 2.6. This leads to the proofs of (5.5) and (5.8). Many of the results proved here are developed for arbitrary coherent sheaves in Chapter 9.

Definition 5.24 (Mostly flat family of line bundles) Let $f \colon X \to S$ be a morphism and L a mostly flat divisorial sheaf (3.28). We say that L is a *mostly flat family of line bundles* if the hull L_s^H of L_s (3.27.1) is locally free over the hull $\mathscr{O}_{X_s}^H$ of $\mathscr{O}_{X_s}$. (In most cases of interest f has S_2 fibers, and then $\mathscr{O}_{X_s}^H = \mathscr{O}_{X_s}$.)

A mostly flat family of line bundles L on X is called *fiber-wise ample* if L_s^H is ample for every $s \in S$. See (5.17) for typical examples.

Our aim is to find conditions to ensure that a mostly flat family of line bundles is a flat family of line bundles.

Lemma 5.25 *Let* $f \colon X \to S$ *be a proper morphism of pure relative dimension* n, A *a relatively ample line bundle on* X, *and* L *a mostly flat family of fiber-wise ample line bundles. Then*

(5.25.1) $s \mapsto (A_s^i \cdot (L_s^H)^{n-i})$ *is upper semi-continuous for every* i, *and*

(5.25.2) *if* $s \mapsto ((L_s^H)^n)$ *is constant, then so is every* $(A_s^i \cdot (L_s^H)^{n-i})$.

Proof As we noted in (5.24), there is a dense open subset $S^\circ \subset \operatorname{red} S$ such that $L|_{X^\circ}$ is a line bundle. Thus the functions $s \mapsto (A_s^i \cdot (L_s^H)^{n-i})$ are locally constant on S°, hence constructible on S by Noetherian induction.

It remains to check upper semicontinuity when $(0 \in S)$ is the spectrum of a DVR. We may assume that X is S_2.

L_0 is also S_1, hence $L_0 \to L_0^H$ is an injection. By semicontinuity we have $h^0(X_0, L_0^H) \geq h^0(X_0, L_0) \geq h^0(X_g, L_g)$. Applying this to powers of L and taking the limit, we obtain that $\operatorname{vol}(L_0^H) \geq \operatorname{vol}(L_g)$ by (10.31). If L is fiber-wise ample, then volume equals the self-intersection number, so $((L_0^H)^n) \geq ((L_g^H)^n)$. This shows upper semicontinuity for $i = 0$.

For $i > 0$, we prove (1) by induction on n. We may assume that S is local, and A is relatively very ample. Let $Y \subset X$ be a hypersurface cut out by a general section of A. By (4.26), the restriction $L|_Y$ is a mostly flat family of fiber-wise ample line bundles on $Y \to S$. Furthermore

$$(A_s^i \cdot (L_s^H)^{n-i}) = (Y_s \cdot A_s^{i-1} \cdot (L_s^H)^{n-i}) = ((A|_Y)_s^{i-1} \cdot ((L|_Y)_s^H)^{n-i}); \qquad (5.25.3)$$

the latter is constructible and upper semi continuous by induction.

In order to see (2), note that $L^m \otimes A^{-1}$ is also a mostly flat family of fiber-wise ample line bundles for $m \gg 1$, and

$$m^n((L_s^H)^n) = \textstyle\sum_i \binom{n}{i}(A_s^i \cdot ((L^m \otimes A^{-1})_s^H)^{n-i}). \qquad (5.25.4)$$

By (1), all summands on the right are constructible and upper semi continuous. Therefore, if the sum is constant as a function of s, then so is every summand.

Finally note that

$$(((L^m \otimes A^{-1})_s^H)^n) = \sum_i (-1)^i m^{n-i} \binom{n}{i} (A_s^i \cdot (L_s^H)^{n-i}). \tag{5.25.5}$$

If the left side is constant for $m \gg 1$, as a function of s, then every summand on the right is constant. □

Remark 5.25.6 Let $f : X \to S$ be a proper morphism of pure relative dimension n, and L a line bundle on X. It is not well understood when the function $s \mapsto \mathrm{vol}(L_s)$ is constructible; see Lesieutre (2014); Pan and Shen (2013).

5.26 (Proof of 5.8) The assertions (5.8.1) and (5.8.3) are proved in (5.25.1). Furthermore, (5.25.2) shows that (5.8.2) implies (5.8.4).

Thus it remains to prove (5.8.2). We start with the case when S is the spectrum of a DVR; this implies the general case by (4.34).

Our argument has three parts. The first step, when the relative dimension is 2, is done in (5.28).

The next step is induction on the dimension. We may assume that S is local and A is relatively very ample. Let $Y \subset X$ be a general hypersurface cut out by a general section of A. Then (4.26) ensures that $L^H|_Y = (L|_Y)^H$. The restriction $L|_Y$ is a mostly flat family of fiber-wise ample line bundles on $Y \to S$ and, as we noted in (5.25.3),

$$(A_s^{n-2} \cdot (L_s^H)^2) = ((A|_Y)_s^{n-3} \cdot ((L|_Y)_s^H)^2).$$

Thus, by induction, $L^H|_Y$ is a line bundle. This implies that L^H is a line bundle along Y. So L^H is a line bundle, except possibly at finitely many points $Z \subset X$.

Finally we need to exclude this finite set Z when the fiber dimension is at least 3. This follows from (2.91). □

5.27 (Start of the proof of 5.5) Note that (5.5.1) follows from (5.25.1). For (5.5.2), the general setting is postponed to (5.54). Here we consider the case when $S = C$ is one-dimensional and regular.

As a first step, we replace (X, Δ) by its normalization. This leaves the assumptions and the numerical conclusion unchanged. By (2.54), a demi-normal pair $(X, \Delta) \to C$ with slc generic fibers is slc iff its normalization is lc. Thus the conclusion is also unchanged.

It would seem that we should use (5.8). However, a key assumption of (5.8) is that every fiber is S_2; this is true, but not obvious in our case. Thus we consider two separate cases.

If $n = 2$, then the weak numerical criterion (5.43) implies (5.5). For $n \geq 3$, the weak numerical criterion involves the terms $(\pi_c^* H^{n-i} \cdot (K_{\bar{X}_c} + \bar{D}_c + \bar{\Delta}_c)^i)$ for $i \geq 3$; these are unknown to us.

Instead, using the already established $n = 2$ case and (4.26) as in (5.26), we may assume that $f: (X, \Delta) \rightarrow C$ is locally stable outside a subset of codimension ≥ 3. We can now apply (2.7) to complete the argument. □

Proposition 5.28 *Let T be an irreducible, regular, one-dimensional scheme, and $f: X \rightarrow T$ a flat, proper morphism of relative dimension 2 with S_2 fibers. Let L be a mostly flat family of line bundles on X. Then*

(5.28.1) $d(t) := (L_t^H \cdot L_t^H)$ *is upper semicontinuous, and*

(5.28.2) *L is locally free on X iff $d(t)$ is constant on T.*

Proof If L is locally free then $(L_t^H \cdot L_t^H) = (L \cdot L \cdot [X_t])$ is independent of $t \in T$. To see the converse we may assume that T is local with closed point $0 \in T$ and generic point $g \in T$. Note that L is locally free, except possibly at a finite set $Z_0 \subset X_0$, and $L_g^H \simeq L_g$.

For each $t \in T$, the Euler characteristic is a quadratic polynomial

$$\chi(X_t, (L_t^H)^{\otimes m}) = a_t m^2 + b_t m + c_t,$$

and we know from Riemann–Roch that $a_t = \frac{1}{2}(L_t^H \cdot L_t^H)$ and $c_t = \chi(X_t, \mathcal{O}_{X_t})$. Furthermore, (9.36.4) implies that

$$a_0 m^2 + b_0 m + c_0 \geq a_g m^2 + b_g m + c_g \quad \text{for every } m \in \mathbb{Z}. \tag{5.28.3}$$

For $m \gg 1$, the quadratic terms dominate, which gives that

$$(L_0^H \cdot L_0^H) = 2a_0 \geq 2a_g = (L_g \cdot L_g). \tag{5.28.4}$$

Assume now that $(L_0^H \cdot L_0^H) = (L_g \cdot L_g)$. Then $a_0 = a_g$ thus (5.28.3) implies that

$$b_0 m + c_0 \geq b_g m + c_g \quad \text{for every } m \in \mathbb{Z}. \tag{5.28.5}$$

For $m \gg 1$, this implies that $b_0 \geq b_g$, and for $m \ll -1$ that $-b_0 \geq -b_g$. Thus $b_0 = b_g$ and $c_0 = c_g$ also holds since f is flat. Therefore we have equality in (5.28.3). Thus L is a flat family of locally free sheaves by (3.32). □

The following turns out to be quite elementary; see Stacks (2022, tag 0F29) for a subtle local version.

Proposition 5.29 *Let T be the spectrum of a DVR with closed point $0 \in T$ and generic point $g \in T$. Let $f: X \rightarrow T$ be a projective morphism with S_2 fibers, and L a mostly flat family of line bundles such that $L^{[m]}$ is locally free for some $m > 0$. Then L is locally free.*

Proof We claim an equality of the Hilbert polynomials

$$\chi(X_0, (L_0^H)^{\otimes r}) = \chi(X_g, L_g^{\otimes r}). \tag{5.29.1}$$

Since both sides are polynomials in r, it is sufficient to prove that they are equal for all multiples of m.

Note that $L^{[m]}|_{X_0}$ and $(L_0^H)^{\otimes m}$ are both locally free sheaves that agree outside a codimension 2 subset, hence they are isomorphic. Thus

$$\begin{aligned}\chi(X_0, (L_0^H)^{\otimes rm}) &= \chi(X_0, (L^{[m]}|_{X_0})^{\otimes r}) \\ &= \chi(X_g, (L^{[m]}|_{X_g})^{\otimes r}) = \chi(X_g, L_g^{\otimes rm}),\end{aligned} \tag{5.29.2}$$

where the last equality holds since L_g is a line bundle (5.24). In particular we conclude that $\chi(X_0, L_0^H) = \chi(X_g, L_g)$. Let $\mathcal{O}_{X/S}(1)$ be an f-ample invertible sheaf. We can apply the same argument to any $L(t)$ to obtain that $\chi(X_0, L_0^H(t)) = \chi(X_g, L_g(t))$ for every t. By (3.32) this implies that L is locally free.   □

5.5 Flatness Criteria in Codimension 1

Let $f: X \to S$ be a projective morphism with f-ample $\mathcal{O}_X(1)$, and F a coherent sheaf on X. Assume that S is reduced. By (3.20), the polynomial valued function $s \mapsto \chi(X_s, F_s(*))$ is

- upper semicontinuous on S, and
- it is locally constant iff F is flat over S.

In Sections 5.1–5.2, we discussed numerous situations where we first associate some other object to each (X_s, F_s), and then compute a numerical invariant. Usually these objects cannot be realized as fibers of some morphism. However, we still would like to show that the numerical invariant is an upper or lower semicontinuous function on S. Furthermore, if the numerical invariant is locally constant on S, then we would like to prove that the objects fit together into a flat family over S.

As a typical example – generalizing (5.8) – consider the Hilbert polynomial of the reflexive hulls $\chi(X_s, (F_s)^{[**]}(*))$. Assume that X, S are normal, and so are the fibers of f. Note that (3.20) does not apply, since usually there is no coherent sheaf on X whose fibers are $(F_s)^{[**]}$. There is a natural map

$$r_s: (F^{[**]})_s \to (F_s)^{[**]},$$

but frequently it is neither injective nor surjective. So we do not get any comparison between the Hilbert polynomials of $(F^{[**]})_s$ and $(F_s)^{[**]}$. It is also not clear what should happen if $s \mapsto \chi(X_s, (F_s)^{[**]}(*))$ is locally constant.

Next we outline a method to study such problems in three steps.

• Show that the numerical function is upper or lower semicontinuous.
• If the function is locally constant, construct a candidate $f': (X', F') \to S'$ for the flat model.
• Prove that, under suitable assumptions, $S' \simeq S$ and (X', F') as expected.

The details are simple, but at the end they lead to interesting consequences.

5.30 (How to prove semicontinuity?) Let $\phi(\ )$ be a function that associates to certain pairs (X, F) (consisting of a proper scheme over a field k, and a coherent sheaf on it, plus possibly some other data) an element in a partially ordered set. Typical examples for us are $\phi_1(X) := (K_X^n)$, $\phi_2(X, F) := \chi(X, F)$, or $\phi_3(X, F) := \chi(X, \text{pure } F(*))$ (if we also have an ample line bundle $\mathcal{O}_X(1)$). We always assume that $\phi(\ )$ is invariant under base field extensions.

Let $f: X \to S$ be a morphism, and F a coherent sheaf on X. We would like to prove that $s \mapsto \phi(X_s, F_s)$ is upper or lower semicontinuous. In many cases, this can be done in two stages.

(5.30.1) Prove that $s \mapsto \phi(X_s, F_s)$ is constant on a nonempty open subset $S° \subset S$. If this works inductively for closed subsets of S, then Noetherian induction shows that $s \mapsto \phi(X_s, F_s)$ is constructible.

A constructible function is upper (resp. lower) semicontinuous iff it is upper (resp. lower) semicontinuous after base change to any DVR $T \to S$. Thus it remains to do:

(5.30.2) Let T be the spectrum of a DVR with closed point 0_T, generic point g_T, and $\pi: T \to S$ a morphism. Prove that

$$\phi(X_{0_T}, F_{0_T}) \geq (\text{resp. } \leq) \phi(X_{g_T}, F_{g_T}).$$

(Frequently $k(0_T) \neq k(\pi(0_T))$, this is why $\phi(\ )$ should be invariant under base field extensions.)

If we want to prove that $s \mapsto \phi(X_s, F_s)$ is locally constant, then we need only the following.

(5.30.3) Let T be the spectrum of a DVR with closed point 0_T, generic point g_T, and $\pi: T \to S$ a morphism that maps g_T to a generic point of S. Prove that

$$\phi(X_{0_T}, F_{0_T}) = \phi(X_{g_T}, F_{g_T}).$$

The generic point property of π is helpful if we have extra information about the generic fibers of $X \to S$.

5.31 (How to construct a candidate?) This is usually the hard part. If our objects $\phi(\ )$ are subvarieties, then $\phi(X_s, F_s)$ is a point in the Hilbert scheme or the Chow variety. Thus we have a set-theoretic map

$$\sigma^{\text{set}} \colon (\text{points of } S) \to \begin{cases} (\text{points of Hilb}(X/S)), & \text{or} \\ (\text{points of Chow}(X/S)). \end{cases}$$

If the objects $\phi(\ )$ are non-embedded varieties, we get a point in some moduli space. For the case of reflexive hulls we have considered, we need the moduli space of husks, which we discuss in Section 9.5.

Usually there are several choices for the moduli theory, and the proofs need the "correct" one to work. At the end we have a set-theoretic map

$$\sigma^{\text{set}} \colon (\text{points of } S) \to (\text{points of some moduli space } \mathbf{M}).$$

Then the only sensible thing is to let S' be the closure of the image if σ^{set}; it comes with a natural map $\pi \colon S' \to S$.

If $\mathbf{M}$ is a coarse moduli space, then we have to make sure that there is a universal family over S', which usually means that we have to eliminate possible automorphisms (1.71); see (5.55–5.56) for such examples.

If this works out, then we have our candidate family $f' \colon (X', F') \to S'$, and a natural morphism $\pi \colon S' \to S$.

Then we need to show that $S' \simeq S$, and X', F' are as expected. The key is usually the isomorphism $S' \simeq S$. We typically know that π is proper, and an isomorphism over the generic points of S.

5.32 (How to check isomorphism?) Let $\pi \colon S' \to S$ be a proper morphism, $W \subset S$ a nowhere dense, closed subset, and $W' := \pi^{-1}(W) \subset S'$. Assume that $\pi \colon (S' \setminus W') \to (S \setminus W)$ is an isomorphism, and S' (resp. S) has no associated points in W' (resp. W). Then π is an isomorphism in the following cases:

(5.32.1) $\pi^{-1}(w) \simeq w$ for $w \in W$ (by Nakayama's lemma),

(5.32.2) $k(\operatorname{red} \pi^{-1}(w))/k(w)$ is purely inseparable for $w \in W$, and (W, S) is weakly normal (by definition (10.74)),

(5.32.3) $k(\operatorname{red} \pi^{-1}(w)) = k(w)$ for $w \in W$, and (W, S) is seminormal (by definition (10.74)),

(5.32.4) $\operatorname{depth}_W S \geq 2$ (by (10.6)),

(5.32.5) S is normal.

We illustrate the method in the simplest case, when we look at the reduced structure of the fibers of a morphism. Being reduced is invariant under separable ground field extensions. Thus working with $X_s \mapsto \operatorname{red}(X_s)$ is sensible in characteristic 0, but in general it is better to work with the reduced structure of the geometric fibers.

Theorem 5.33 *Let* $f : X \to S$ *be a projective morphism of pure relative dimension n with f-ample* $\mathscr{O}_X(1)$. *Assume that X is reduced and S is weakly normal. For* $s \in S$, *let* $X_{\bar{s}}$ *denote the corresponding geometric fiber. Then*

(5.33.1) $\chi(s) := \chi(\mathrm{red}(X_{\bar{s}}), \mathscr{O}(*))$ *is lower semicontinuous, and*

(5.33.2) f *is flat with geometrically reduced fibers iff the generic fibers are geometrically reduced, and* $\chi(s)$ *is locally constant on S.*

We prove this in (5.37), but first some consequences and variants. If we understand only the leading coefficient of $\chi(\mathrm{red}(X_{\bar{s}}), \mathscr{O}(*))$, we still get very useful information about f as in Kollár (1996, I.6.5).

Corollary 5.34 (Smoothness criterion in codimension 0) *Let* $f : X \to S$ *be a projective morphism of pure relative dimension n, and H an f-ample divisor class. Assume that X is reduced and S is weakly normal. For* $s \in S$ *let* $X_{\bar{s}}$ *denote the geometric fiber. Then*

(5.34.1) $s \mapsto \deg_H(\mathrm{red}(X_{\bar{s}}))$ *is lower semicontinuous, and*

(5.34.2) f *is smooth on a dense subset of each fiber iff* $s \mapsto \deg_H(\mathrm{red}(X_{\bar{s}}))$ *is locally constant and f is generically smooth. (The latter is automatic in characteristic 0.)*

Proof Repeated application of (10.56) reduces the proof to $n = 0$, which is a special case of (5.33).   □

It turns out that codimension 0 is the hardest part of (5.33), and we have stronger results in higher codimensions. The following is proved in (5.37).

Theorem 5.35 *Let* $f : X \to S$ *be a projective morphism of pure relative dimension n with f-ample* $\mathscr{O}_X(1)$. *Assume that X, S are reduced, and f is smooth at the generic points of each fiber. Then f is flat with geometrically reduced fibers iff* $s \mapsto \chi(\mathrm{red}(X_s), \mathscr{O}(*))$ *is locally constant.*

As a consequence, we get one part of (3.11) about the Hilbert-to-Chow map.

Corollary 5.36 (Flatness criterion in codimension 1) *Let* $f : X \to S$ *be a projective morphism of pure relative dimension n, and H an f-ample divisor class. Assume that X, S are reduced, and f is smooth at the generic points of each fiber. Then*

(5.36.1) *the sectional genus (3.10) of the fibers is a lower semi-continuous function on S, and*

(5.36.2) f *is flat with reduced fibers at codimension 1 points of each fiber iff the sectional genus is locally constant.*

Proof Repeated application of (10.56) reduces the proof to the case $n = 1$, which is a special case of (5.35). □

We can thus expect that, for families that are locally stable in codimension 1, there are results connecting the intersection numbers $((\pi_0^*H)^{n-i} \cdot (K_{\bar{X}_0} + \bar{D}_0)^i)$ with the higher codimension behavior of f. There are two surprising twists.

- The *lower* semicontinuity in (5.34) and (5.36) switches to *upper* semicontinuity for $i = 2$.

- In most cases we need only *one* more intersection number to take care of all codimensions.

5.37 (Proof of 5.33 and 5.35) For (5.33.1), we follow (5.30). After a purely inseparable base change $S' \to S$, the generic fiber of $\mathrm{red}(X \times_S S') \to S$ is geometrically reduced, hence $s \mapsto \chi(\mathrm{red}\, X_{\bar{s}}, \mathscr{O}(*))$ is locally constant on a dense open set by generic flatness. This gives constructibility as in (5.30.1).

Continuing with (5.30.2), let T be the spectrum of a DVR with closed point t, generic point g, and $\pi\colon T \to S$ a morphism mapping t to $s \in S$. Set $Y := \mathrm{red}(X \times_S T)$, and assume that Y_g is geometrically reduced. Since f has pure relative dimension, $Y \to T$ is flat, hence

$$\chi(\mathrm{red}(Y_{\bar{t}}), \mathscr{O}(*)) \le \chi(\mathrm{red}(Y_t), \mathscr{O}(*)) \le \chi(Y_t, \mathscr{O}(*)) = \chi(Y_g, \mathscr{O}(*)). \quad (5.37.1)$$

By (5.30), this proves (5.33.1) since $\chi(\mathrm{red}(Y_{\bar{t}}), \mathscr{O}(*)) = \chi(\mathrm{red}(X_{\bar{s}}), \mathscr{O}(*))$. We also see that the two sides of (5.37.1) are equal iff Y_t is also geometrically reduced.

If f is flat with geometrically reduced fibers then (5.33.2) is clear. For the converse we may assume that S is connected, so $p(*) := \chi(\mathrm{red}\, X_{\bar{s}}, \mathscr{O}(*))$ is independent of s.

For both (5.33) and (5.35), the relative Hilbert scheme now gives a clear choice for the candidate as in (5.31). Indeed, $\pi\colon \mathrm{Hilb}_p(X/S) \to S$ parametrizes subschemes of the fibers with Hilbert polynomial $p(*)$.

We claim that $\pi\colon \mathrm{Hilb}_p(X/S) \to S$ is an isomorphism. In both theorems we assume that the generic fibers are generically smooth, hence geometrically generically reduced. They are also S_1, hence geometrically S_1. A generically reduced S_1 scheme is reduced, so the generic fibers are geometrically reduced. The latter is an open property by (10.12). Thus π is an isomorphism over the dense open subset $S^\circ \subset S$ where f is flat with geometrically reduced fibers. The question is, what happens over other points.

The easy case is (5.35). By (5.38.4) $\pi^{-1}(s) = \mathrm{Hilb}_p(X_s) \simeq s$, thus π is an isomorphism by (5.32.1).

For (5.33) the argument is more circuituous. Let $\text{Hilb}'_p(X/S) \subset \text{Hilb}_p(X/S)$ denote the closure of $\pi^{-1}(S^\circ)$ with projection $\pi'\colon \text{Hilb}'_p(X/S) \to S$. First, we claim that π' is geometrically injective.

To see this pick any $s \in S$, and let $\tau\colon T \to S$ be a DVR that maps the closed point $t \in T$ to s and the generic point $g \in T$ to S°. We have a lifting $\tau'\colon T \to \text{Hilb}'_p(X/S)$, and we check in (5.38.5) that $\tau'(t) = [\text{red}(X_t)] \in \text{Hilb}'_p(X/S)$.

Since S is assumed weakly normal, (5.31.2) implies that $\pi'\colon \text{Hilb}'_p(X/S) \to S$ is an isomorphism.

We have $\text{Univ}'_p(X/S) \to \text{Hilb}'_p(X/S)$, and $u'\colon \text{Univ}'_p(X/S) \to X$, which is a closed embedding on each fiber. Thus u' is a closed embedding by (10.54), hence an isomorphism since X is reduced.

Therefore $X \simeq \text{Univ}'_d(X/S)$ is flat over S. In particular, $\text{Hilb}_p(X/S) = \text{Hilb}'_p(X/S) \simeq S$. ◻

5.38 (Uniqueness of red X) A scheme X uniquely determines red X, but what about in families? What if we know only the Hilbert polynomial of red X?

We start with two negative examples, followed by two positive results.

Example 5.38.1 Let X be the scheme $\text{Spec}\, k[x, y]/(x^2, xy, y^2(y - 1))$. Then red $X = \text{Spec}\, k[x, y]/(x, y(y - 1))$ has length 2, but so are the subschemes $\text{Spec}\, k[x, y]/(x^2, xy, y^2, x - cy)$. Thus $\text{Hilb}_2\, X \simeq \mathbb{P}^1_k \amalg \text{Spec}\, k$.

Example 5.38.2 $\text{Spec}\, k$ is the only subscheme of length 1 of $\text{Spec}\, k[x]/(x^2)$. However, consider the trivial family $\pi\colon \text{Spec}\, k[x, t]/(x^2, t^2) \to \text{Spec}\, k[t]/(t^2)$. Then for every $c \in k$, the subscheme $\text{Spec}\, k[x, t]/(x^2, t^2, x + ct)$ is flat over $\text{Spec}\, k[t]/(t^2)$. Thus $\text{Hilb}_1\, \text{Spec}\, k[x]/(x^2) \simeq \text{Spec}\, k[t]/(t^2)$.

Claim 5.38.3 Let $(0 \in A)$ be the spectrum of a local Artinian k-algebra, and $f\colon Y \to A$ a projective morphism with f-ample $\mathcal{O}_X(1)$. Let F be a coherent sheaf on Y, and set $p(*) := \chi(Y_0, \text{pure}(F_0)(*))$. Then F has at most one quotient $q\colon F \twoheadrightarrow Q$ that is flat over A with Hilbert polynomial $p(*)$. If Q exists then $Q = \text{pure}\, F$.

Proof Set $n = \dim F_0$. If $q'\colon F_0 \to Q'$ is any map that is surjective at n-dimensional points, then $\chi(Y_0, F_0(*))$ and $\chi(Y_0, Q'(*))$ have the same leading coefficient iff $\dim(\ker q') < n$. Also, if $\chi(Y_0, Q'(*)) = \chi(Y_0, \text{pure}(F_0)(*))$, then $Q' = \text{pure}(F_0)$.

Thus, if Q is flat over A with Hilbert polynomial $p(*)$, then $\ker q \subset F$ is the largest subsheaf whose support has dimension $< n$. This shows that $Q = \text{pure}\, F$ is the only possibility. ◻

Corollary 5.38.4 Let X be a proper scheme of pure dimension n over a field k. Assume that X is geometrically reduced at its generic points. Set $p(*) := \chi(\text{red } X, \mathscr{O}_{\text{red } X}(*))$. Then $\text{Hilb}_p(X) \simeq \text{Spec } k$.

Proof In this case $\text{red}(X_K) = \text{Spec}_{X_K}(\text{pure } \mathscr{O}_{X_K})$ for any field extension $K \supset k$. The rest follows from (5.38.3). □

Claim 5.38.5 Let T be the spectrum of a DVR, and $f : Y \to T$ a projective morphism of pure relative dimension n with f-ample $\mathscr{O}_X(1)$. Assume that Y_g is reduced and set $p(*) := \chi(Y_g, \mathscr{O}(*))$.

If $\chi(\text{red}(Y_0), \mathscr{O}(*)) = p(*)$, then $\text{red } Y \subset Y$ is the unique subscheme that is flat over T with Hilbert polynomial $p(*)$.

Proof Let $Z \subset Y$ be such a subscheme. Then $Z_g = Y_g$. Since f has pure relative dimension, the closure of Y_g contains Y_0, thus $\text{red } Y_0 \subset Z_0$. These have the same Hilbert polynomial $p(*)$, hence they are equal. □

Examples 5.39 The following series of examples show that the assumptions in (5.33–5.35) are necessary.

(5.39.1) Let C be a cuspidal curve with normalization $p : \bar{C} \to C$. Then p is not flat, but $\text{red } p^{-1}(c) \simeq c$ for every $c \in C$. Here C is not seminormal. Over imperfect fields, (10.75.3) gives similar examples where C is seminormal, but not weakly normal.

(5.39.2) Set $S := (uv = 0)$ and let $X \subset S \times \mathbb{A}_w^1$ be the union of two curves $(u = v - (w - 1)^2 = 0) \cup (v = u - (w + 1)^2 = 0)$, with projection $\pi : X \to S$. Then $\deg(k[\text{red } X_s]/k(s)) = 2$ for every $s \in S$, but π is not flat, and it does not have pure relative dimension 0.

(5.39.3) A more complicated example of relative dimension 1 is the following. Set $S := (uv = 0)$ and let $X \subset S \times \mathbb{P}_x^3$ be a reduced subscheme with three irreducible components as follows.

Over the u-axis we take a planar smooth cubic E_u degenerating to a cuspidal cubic E_0, for example $X_1 := (x_1 = ux_0^3 + x_2^3 - x_0x_3^2 = 0)$. We also add the line $X_3 := (u; 0{:}1{:}0{:}0)$.

Over the v-axis we take a smooth twisted cubic C_v degenerating to E_0. For example, X_2 can be the image of $(v; s{:}t) \mapsto (s^3{:}vs^2t{:}st^2{:}t^3)$. (The flat limit C_0 has an embedded point at the cusp.)

If $v \neq 0$ then $X_{0,v}$ is a smooth rational cubic, so $\chi(X_{0,v}, \mathscr{O}(m)) = 3m + 1$. If $u \neq 0$ then $X_{u,0}$ is a smooth elliptic cubic plus a disjoint point, so again $\chi(X_{u,v}, \mathscr{O}(m)) = 3m + 1$. Finally, $X_{0,0}$ is nonreduced, but $\text{red } X_{0,0}$ is a singular planar cubic plus a disjoint point, so $\chi(\text{red } X_{u,v}, \mathscr{O}(m)) = 3m + 1$.

However, the projection $\pi\colon X \to S$ is not flat. Here π is not pure dimensional, and $X_{0,0}$ has two subschemes with Hilbert polynomial $3m + 1$. One is red $X_{0,0}$ the other is the one-dimensional irreducible component of $X_{0,0}$.

It is straightforward to generalize (5.35) from $\mathscr{O}_X$ to an arbitrary coherent sheaf F. The only change is that, instead of the Hilbert scheme $\mathrm{Hilb}(X/S)$, we use the quot-scheme $\mathrm{Quot}(F)$ (9.33). Thus we get the following.

Theorem 5.40 *Let $f\colon X \to S$ be a projective morphism with f-ample $\mathscr{O}_X(1)$ with S is reduced. Let F be a coherent sheaf on X that is generically flat (3.26) over S. Assume that $\mathrm{Supp}\, F \to S$ has pure relative dimension n, and F does not have embedded points. Then F is flat over S with pure fibers iff $s \mapsto \chi(X_s, \mathrm{pure}(F_s)(*))$ is locally constant.* □

5.6 Deformations of SLC Pairs

So far we have focused on locally stable deformations of slc pairs. The next result, due to Kollár and Shepherd-Barron (1988), connects arbitrary flat deformations (X_t, Δ_t) of an slc pair (X_0, Δ_0) to locally stable deformations of a suitable birational modification $f_0\colon (Y_0, \Delta_0^Y) \to (X_0, \Delta_0)$. We then compare various numerical invariants of (X_0, Δ_0) and of (X_t, Δ_t) by going through (Y_0, Δ_0^Y). This implies a weaker version of (5.5).

Theorem 5.41 *Kollár and Shepherd-Barron (1988) Let $(X, D + \Delta)$ be a normal pair, where D is a reduced, $\mathbb{Q}$-Cartier divisor that is demi-normal in codimension 1, and whose normalization $(\bar{D}, \mathrm{Diff}_{\bar{D}} \Delta)$ is lc. Assume also[1] that*
(5.41.1) *either $(\bar{D}, \mathrm{Diff}_{\bar{D}} \Delta)$ is klt,*
(5.41.2) *or $K_X + \Delta$ is $\mathbb{R}$-Cartier on $X \setminus D$.*
Then, in a neighborhood of D, the following hold.
(5.41.3) *The log canonical modification $f\colon (Y, D_Y + \Delta_Y + E) \to (X, D + \Delta)$ exists, and it is small, that is, $E = 0$.*
(5.41.4) *$(Y, D_Y + \Delta_Y)$ is lc.*
(5.41.5) *D_Y is normal at the generic point of every f_0-exceptional divisor $F \subset D_Y$, and $a(F, \bar{D}, \mathrm{Diff}_{\bar{D}} \Delta) < 0$.*
(5.41.6) *$f(\mathrm{Ex}(f))$ is precisely the locus where $K_X + \Delta$ is not $\mathbb{R}$-Cartier.*

---

[1] Conjecturally, these are not needed; see (11.29).

Proof Let $h\colon X' \to X$ be a log resolution with exceptional divisor E'. Set $\mathrm{discrep}(\bar{D}, \mathrm{Diff}_{\bar{D}}\,\Delta) = -1 + \varepsilon$. Let $f\colon (Y, D_Y + \Delta_Y + (1 - \varepsilon)E) \to (X, D + \Delta)$ be the relative canonical model of $(X', h_*^{-1}(D + \Delta) + (1 - \varepsilon)E')$. This is the same as the relative canonical model of $(X', h_*^{-1}(D + \Delta) + (1 - \varepsilon)E' - \eta h^* D)$ since $h^* D$ is numerically h-trivial.

If $(\bar{D}, \mathrm{Diff}_{\bar{D}}\,\Delta)$ is klt then $\varepsilon > 0$, hence $(X', h_*^{-1}(D + \Delta) + (1 - \varepsilon)E' - \eta h^* D)$ is klt, and the relative canonical model exists by (11.28.2).

If $\varepsilon = 0$ then we note that $(X', h_*^{-1}(D + \Delta) + (1 - \varepsilon)E' - \eta h^* D)$ has no lc centers over D for $\eta > 0$, hence the relative canonical model exists by (11.30) and (11.28.2).

Let $\pi_X\colon \bar{D} \to D$ and $\pi_Y\colon \bar{D}_Y \to D_Y$ be the normalizations. Then f_0 lifts to $\bar{f}_0\colon \bar{D}_Y \to \bar{D}$. Write $K_{\bar{D}_Y} + \Delta_{\bar{D}_Y} \sim_\mathbb{R} \bar{f}_0^*(K_{\bar{D}} + \mathrm{Diff}_{\bar{D}}\,\Delta)$. By adjunction,

$$\pi_Y^*(K_Y + D_Y + \Delta_Y + (1 - \varepsilon)E) \sim_\mathbb{R} K_{\bar{D}_Y} + \mathrm{Diff}_{\bar{D}_Y}(\Delta_Y + (1 - \varepsilon)E)$$

$$\sim_\mathbb{R} \bar{f}_0^*(K_{\bar{D}} + \mathrm{Diff}_{\bar{D}}\,\Delta) + (\mathrm{Diff}_{\bar{D}_Y}(\Delta_Y + (1 - \varepsilon)E) - \Delta_{\bar{D}_Y}).$$

Since D has only nodes at codimension 1 points, X is canonical at codimension 1 points of D (11.35), and f is an isomorphism near these points. Thus $\mathrm{Diff}_{\bar{D}_Y}(\Delta_Y + (1 - \varepsilon)E) - \Delta_{\bar{D}_Y}$ is $\bar{f}_0$-exceptional, and $\bar{f}_0$-ample. By (11.60) this implies that every $\bar{f}_0$-exceptional divisor appears in $\mathrm{Diff}_{\bar{D}_Y}(\Delta_Y + (1 - \varepsilon)E) - \Delta_{\bar{D}_Y}$ with strictly negative coefficient.

Every divisor in $D_Y \cap E$ appears in $\mathrm{Diff}_{\bar{D}_Y}(\Delta_Y + (1 - \varepsilon)E)$ with coefficient $\geq 1 - \varepsilon$ by (11.16). On the other hand, every exceptional divisor appears in $\Delta_{\bar{D}_Y}$ with coefficient $\leq 1 - \varepsilon$ by our choice of ε. Thus the divisors in $D_Y \cap E$ appear in $\mathrm{Diff}_{\bar{D}_Y}(\Delta_Y + (1 - \varepsilon)E) - \Delta_{\bar{D}_Y}$ with coefficient $\geq (1 - \varepsilon) - (1 - \varepsilon) = 0$. We noted above that these coefficients are strictly negative, so $D_Y \cap E = \emptyset$.

Hence, after shrinking X, there are no exceptional divisors in $f\colon Y \to X$, so f is small, $D_Y = f^* D$, and $(Y, \Delta_Y + D_Y)$ is lc.

Let $\bar{F} \subset \bar{D}_Y$ be any $\bar{f}_0$-exceptional divisor. Since it appears in $\mathrm{Diff}_{\bar{D}_Y}(\Delta_Y) - \Delta_{\bar{D}_Y}$ with negative coefficient, it must appear in $\Delta_{\bar{D}_Y}$ with positive coefficient, and in $\mathrm{Diff}_{\bar{D}_Y}(\Delta_Y)$ with coefficient < 1. By (11.16), the latter implies that D_Y is smooth at the generic point of $\pi_Y(\bar{F})$, proving (5).

Finally let $x \in X \setminus D$ be a point where $K_X + \Delta$ is $\mathbb{R}$-Cartier. Since f is small, $K_Y + \Delta_Y \sim_\mathbb{R} f^*(K_X + \Delta)$ over a neighborhood of x. Since $K_Y + \Delta_Y$ is f-ample, f is an isomorphism over a neighborhood of x. $\qquad\square$

Complement 5.41.7 If $(\bar{D}, \mathrm{Diff}_{\bar{D}}\,\Delta)$ is klt then D is normal. This was used in Kollár and Shepherd-Barron (1988) to get a description of the deformation space of log terminal surface singularities. The cone over an elliptic scroll gives examples where D is not normal, but its normalization has a simple elliptic singularity, see Mumford (1978).

See also Sato and Takagi (2022) for closely related results.

5.42 (Proof of 5.4) We prove (5.4) when the base S is the spectrum of a DVR. By (4.7), this implies the case when S is higher dimensional, provided f is assumed to be flat with S_2 fibers.

As a preliminary step, we replace (X, Δ) by its normalization. This leaves the assumptions and the numerical conclusion (5.4.1) unchanged. Then (2.54), shows that the conclusion in (5.4.2) is also unchanged.

Thus assume that X is normal. The conclusions are local on C, so pick a point $0 \in C$, and let $f : (Y, \Delta^Y + Y_0) \to (X, X_0 + \Delta)$ be the log canonical modification as in (5.41). Let $\pi_Y : \bar{Y}_0 \to Y_0$ be the normalization and $\bar{f}_0 : \bar{Y}_0 \to \bar{X}_0$ the induced birational morphism. We apply (10.32.3–4) to

$$D_Y := K_{\bar{Y}_0} + \mathrm{Diff}_{\bar{Y}_0} \Delta^Y \quad \text{and} \quad D_X := K_{\bar{X}_0} + \mathrm{Diff}_{\bar{X}_0} \Delta = K_{\bar{X}_0} + \bar{D}_0 + \bar{\Delta}_0.$$

The assumptions are satisfied since $(\bar{f}_0)_*(K_{\bar{Y}_0} + \mathrm{Diff}_{\bar{Y}_0} \Delta^Y) = K_{\bar{X}_0} + \mathrm{Diff}_{\bar{X}_0} \Delta$, and $K_{\bar{Y}_0} + \mathrm{Diff}_{\bar{Y}_0} \Delta^Y$ is $\bar{f}_0$-ample. Using the volume (10.31), this implies that

$$(K_{\bar{X}_0} + \mathrm{Diff}_{\bar{X}_0} \Delta)^n = \mathrm{vol}(K_{\bar{X}_0} + \mathrm{Diff}_{\bar{X}_0} \Delta) \geq \mathrm{vol}(K_{\bar{Y}_0} + \mathrm{Diff}_{\bar{Y}_0} \Delta^Y),$$

and equality holds iff $\bar{f}_0$ is an isomorphism. Since $K_Y + \Delta^Y$ is $\mathbb{Q}$-Cartier,

$$\mathrm{vol}(K_{\bar{Y}_0} + \mathrm{Diff}_{\bar{Y}_0} \Delta^Y) \geq \mathrm{vol}(K_{\bar{Y}_c} + \Delta^Y|_{\bar{Y}_c}) = ((K_{\bar{Y}_c} + \bar{\Delta}_c)^n)$$

for general $c \neq 0$, and $(\bar{Y}_c, \bar{\Delta}_c) = (X_c, \Delta_c)$ by (5.41.4). Combining the inequalities shows that $((K_{\bar{X}_0} + \bar{D}_0 + \bar{\Delta}_0)^n) \geq ((K_{\bar{X}_c} + \Delta_c)^n)$ for general $c \neq 0$, and equality holds iff $\bar{f}_0$, and hence f, are isomorphisms over $0 \in C$. □

The same method can be used to prove a weaker version of the numerical criterion of local stability over smooth curves. This establishes (5.5) for families of surfaces over a smooth curve. It is not clear how to use these methods to complete the proof of (5.5) for higher dimensional families. We will derive (5.5) from (5.8) instead; see (5.27) for the key step.

Proposition 5.43 (Weak numerical criterion of local stability) *Let C be a smooth curve of* char 0, *and* $f : (X, \Delta) \to C$ *a morphism satisfying the assumptions (5.5.1–3). Then*

(5.43.1) $I(c) := I(\pi_c^* H, K_{\bar{X}_c} + \bar{D}_c + \bar{\Delta}_c)$ *is upper semi-continuous for $\leq$, and*

(5.43.2) $f : (X, \Delta) \to C$ *is locally stable iff $I(c)$ is locally constant on C.*

Note that the first two numbers in the sequence $I(\pi_c^* H, K_{\bar{X}_c} + \bar{D}_c + \bar{\Delta}_c)$ equal $(H^n \cdot X_c)$ and $(H^{n-1} \cdot (K_X + \Delta) \cdot X_c)$, hence they are always locally constant. The first interesting number is $(\pi_c^* H^{n-2} \cdot (K_{\bar{X}_c} + \bar{D}_c + \bar{\Delta}_c)^2)$ which is thus an upper semicontinuous function on C by (1).

Proof As in (5.42) we may assume that X is normal. Let $f : (Y, \Delta^Y + Y_0) \to (X, X_0 + \Delta)$ be the log canonical modification, and $\bar{f}_0 : \bar{X}_0 \to \bar{Y}_0$ the induced birational morphism between the normalizations. Here we apply (10.32.1–2) to $K_{\bar{Y}_0} + \mathrm{Diff}_{\bar{Y}_0} \Delta^Y$ and $K_{\bar{X}_0} + \mathrm{Diff}_{\bar{X}_0} \Delta$ to obtain that

$$I(\pi_0^* H, K_{\bar{X}_0} + \mathrm{Diff}_{\bar{X}_0} \Delta) \geq I(\bar{f}_0^* \pi_0^* H, K_{\bar{Y}_0} + \mathrm{Diff}_{\bar{Y}_0} \Delta^Y),$$

and equality holds iff $\bar{f}_0$ is an isomorphism. Since $K_Y + \Delta^Y$ is a $\mathbb{Q}$-Cartier divisor,

$$I(\bar{f}_0^* \pi_0^* H, K_{\bar{Y}_0} + \mathrm{Diff}_{\bar{Y}_0} \Delta^Y) = I(\pi_c^* H, K_{\bar{Y}_c} + \Delta^Y|_{\bar{Y}_c}) = I(\pi_c^* H, K_{\bar{X}_c} + \bar{\Delta}_c)$$

for general $c \neq 0$. Thus $I(\pi_0^* H, K_{\bar{X}_0} + \bar{D}_0 + \bar{\Delta}_0) \geq I(\pi_c^* H, K_{\bar{X}_c} + \bar{\Delta}_c)$ for general $c \neq 0$, and equality holds iff $\bar{f}_0$, and hence f, are isomorphisms.   □

5.7 Simultaneous Canonical Models

In this section, we consider the existence of simultaneous canonical models.

5.44 (Proof of (5.10) over curves) Let B be a smooth curve of char 0, and $f : X \to B$ a morphism of pure relative dimension n.

First, we prove that $b \mapsto \mathrm{vol}(K_{X_b^\tau})$ is a lower semicontinuous function on B.

If we replace X by a resolution $X^\tau \to X$ then $\mathrm{vol}(K_{X_b^\tau})$ is unchanged for general fibers, and it can only increase for special fibers. There are two sources for an increase. First, the resolution may introduce new divisors of general type. Second, if X is not normal, an irreducible component of a fiber may be replaced by a finite cover of it. The latter increases the volume by (10.38).

Thus it is enough to check lower semicontinuity when X is smooth, and all fibers are snc. If the volume of the general fiber is 0, then the volume of every fiber is 0 by (5.45), so assume that general fibers are of general type.

Fix a fiber $F = X_b$. By shrinking B we may assume that all other fibers are smooth. Let $f^c : X^c \to B$ be the relative canonical model of $(X, \mathrm{red}\, F) \to B$ as in (2.57.2). An irreducible component $E \subset F$ may get contracted. However, when this happens, then $K_E + (\mathrm{red}\, F - E)|_E = (K_X + \mathrm{red}\, F)|_E$ is negative on the fibers of the contraction, and so is K_E. Such divisors contribute 0 to the volume. Thus we can check lower semicontinuity on $f^c : X^c \to B$.

Write $F^c = \sum e_i E_i$, and let $\pi_i: \bar{E}_i \to E_i$ be the normalizations. As in (11.14), write $\pi_i^*(K_{X^c} + \text{red}\, F^c) = K_{\bar{E}_i} + \bar{D}_i$, where $\bar{D}_i = \text{Diff}_{\bar{E}_i}(\sum_{j \neq i} E_j)$. Let $g \in B$ be a general point. Then F^c is disjoint from X_g^c, and we have

$$
\begin{aligned}
(K_{X_g^c})^n &= ((K_{X^c} + \text{red}\, F^c)^n \cdot X_g^c) = ((K_{X^c} + \text{red}\, F^c)^n \cdot F^c) \\
&= \sum_i e_i((K_{\bar{E}_i} + \bar{D}_i)^n) \geq \sum_i ((K_{\bar{E}_i} + \bar{D}_i)^n).
\end{aligned}
\tag{5.44.1}
$$

Next we use (5.12) to obtain that $((K_{\bar{E}_i} + \bar{D}_i)^n) \geq \text{vol}(K_{E_i^\tau})$, hence

$$
\text{vol}(K_{X_g}) = ((K_{X_g^c})^n) \geq \sum_i \text{vol}(K_{E_i^\tau}) = \text{vol}(\text{red}\, F^\tau),
$$

proving the lower semicontinuity assertion. Furthermore, by (5.12), equality holds iff $D_i = 0$, the E_i have canonical singularities and $e_i = 1$ for every i. If $D_i = 0$, then E_i is the only irreducible component of its fiber by (11.16). Thus F^c is reduced and irreducible and has canonical singularities. So $f^c \colon X^c \to B$ is the simultaneous canonical model of $f \colon X \to B$. □

Lemma 5.45 *Let $f \colon X \to B$ be a projective morphism to a smooth curve B such that $\text{vol}(K_{X_g^\tau})$ is zero for the generic fiber X_g. Then $\text{vol}(K_{F^\tau})$ is zero for every fiber F of f.*

Proof The proof in (5.44) gives this if a resolution of X has a minimal model over B. This is not fully known, so we have to find a way to go around it.

As in (5.44), we can reduce to the case when X is smooth and F is an snc divisor. Let H be a general, smooth relatively ample divisor such that $K_X + H$ is f-ample. Using the continuity of the volume (Lazarsfeld, 2004, 2.2.44), there is a largest $0 \leq c < 1$ such that $\text{vol}(K_{X_g} + cH_g) = 0$. Fix some $c' > c$ and run the MMP for $(X, \text{red}\, F + c'H) \to B$. Then $K_{X_g} + c'H_g$ is big, so (11.28) applies, and we get a relative canonical model $(X^c, \text{red}\, F^c + c'H^c) \to B$. Let $\pi: \bar{F}^c \to \text{red}\, F^c$ denote the normalization, and set $\bar{H}^c = \pi^* H^c$. As in (5.44.1), we get that

$$
\text{vol}(K_{X_g} + c'H_g^c) \geq \text{vol}(K_{\bar{F}^c} + c'\bar{H}^c) \geq \text{vol}(K_{\bar{F}^c}).
$$

Letting $c' \to c$ gives that $0 = \text{vol}(K_{X_g^c} + cH_g^c) \geq \text{vol}(K_{F^\tau})$, as required. □

5.46 (Proof of (5.11) over curves) Let B be a smooth curve over a field of char 0, and $f \colon (X, \Delta) \to B$ a flat morphism whose fibers are irreducible and smooth outside a codimension ≥ 2 subset. We may replace X by its normalization. Thus we may assume that X is normal, and the generic fiber is lc.

Assume first that f is locally stable. We prove that $b \mapsto \mathrm{vol}(K_{X_b} + \Delta_b)$ is an upper semicontinuous function on S, and $f \colon (X, \Delta) \to B$ has a simultaneous canonical model iff this function is locally constant.

To see these, let $f^c \colon (X^c, \Delta^c) \to B$ denote the canonical model of $f \colon (X, \Delta) \to B$ (11.28). For every $b \in B$ we need to understand the difference between

- $((X^c)_b, (\Delta^c)_b)$, the fiber of f^c over b, and
- $((X_b)^c, (\Delta_b)^c)$, the canonical model of the fiber (X_b, Δ_b) of f.

These two are the same for general $g \in B$, but they can be different for some special points in B.

Let $\phi \colon X \dashrightarrow X^c$ denote the natural birational map. Since the fibers of f are irreducible, they cannot be contracted, thus ϕ induces birational maps $\phi_b \colon X_b \dashrightarrow (X^c)_b$. Let Z_b denote the normalization of the closure of the graph of ϕ_b with projections $X_b \xleftarrow{g} Z_b \xrightarrow{h} (X^c)_b$. The key computation in (5.47), shows that $g^*(K_{X_b} + \Delta_b) \sim_{\mathbb{R}} h^*(K_{(X^c)_b} + (\Delta^c)_b) + F_b$, where F_b is effective. Thus

$$\mathrm{vol}(K_{X_b} + \Delta_b) = \mathrm{vol}(g^*(K_{X_b} + \Delta_b)) \geq \mathrm{vol}(h^*(K_{(X^c)_b} + (\Delta^c)_b)) = \mathrm{vol}(K_{(X^c)_b} + (\Delta^c)_b).$$

Note further that since $f^c \colon (X^c, \Delta^c) \to B$ is flat, and $K_{X^c} + \Delta^c$ is f^c-ample, its restrictions to the various fibers have the same volume. Therefore

$$\mathrm{vol}(K_{(X^c)_b} + (\Delta^c)_b) = \mathrm{vol}(K_{(X^c)_g} + (\Delta^c)_g) = \mathrm{vol}(K_{X_g} + \Delta_g)$$

for generic $g \in B$. Thus $\mathrm{vol}(K_{X_b} + \Delta_b) \geq \mathrm{vol}(K_{X_g} + \Delta_g)$, and, by (10.39), equality holds iff F_b is h-exceptional. Then $((X^c)_b, (\Delta^c)_b)$ is the canonical model of (X_b, Δ_b). This proves both claims.

In the general case, when $f \colon (X, \Delta) \to B$ is not locally stable, we first use (5.41) to construct $h \colon (\bar{X}, \bar{\Delta}) \to (X, \Delta)$ such that the composite $f \circ h \colon (\bar{X}, \bar{\Delta}) \to B$ is locally stable. Thus $\mathrm{vol}(K_{\bar{X}_b} + \bar{\Delta}_b) \geq \mathrm{vol}(K_{X_g} + \Delta_g)$.

Note that $h_b \colon (\bar{X}_b, \bar{\Delta}_b) \to (X_b, \Delta_b)$ is birational by (5.41), and $K_{\bar{X}_b} + \bar{\Delta}_b$ is h_b-ample. Thus $\mathrm{vol}(X_b, \Delta_b) \geq \mathrm{vol}(\bar{X}_b, \bar{\Delta}_b)$ by (10.32.1). Putting these together shows the upper semicontinuity of the volume.

It remains to show that if equality holds then there is a simultaneous canonical model. We already proved that if $\mathrm{vol}(K_{\bar{X}_b} + \bar{\Delta}_b) = \mathrm{vol}(K_{X_g} + \Delta_g)$ then $f \circ h \colon (\bar{X}, \bar{\Delta}) \to B$ has a simultaneous canonical model, which is also the simultaneous canonical model of $f \colon (X, \Delta) \to B$ if $\mathrm{vol}(X_b, \Delta_b) = \mathrm{vol}(\bar{X}_b, \bar{\Delta}_b)$. Then $(\bar{X}_b, \bar{\Delta}_b)$ and (X_b, Δ_b) have isomorphic canonical models. The latter follows from (10.39), but it can also be obtained by applying the simpler (10.32) to the (normalization of the closure of the) graph of $(\bar{X}_b, \bar{\Delta}_b) \dashrightarrow (X^c_b, \Delta^c_b)$.   □

Lemma 5.47 *Let* $(X, D + \Delta)$ *be lc where D is a reduced Weil divisor, and* $\Delta = \sum a_i D_i$ *is an $\mathbb{R}$-divisor. Let $f: X \to S$ be a proper morphism, and* $\phi: (X, D + \Delta) \dashrightarrow (X^c, D^c + \Delta^c)$ *the relative canonical model. If none of the irreducible components of D are contracted by ϕ, we get a birational map*

$$\phi_{\bar{D}}: (\bar{D}, \mathrm{Diff}_{\bar{D}} \Delta) \dashrightarrow (\bar{D}^c, \mathrm{Diff}_{\bar{D}^c} \Delta^c).$$

Moreover, $a(E, \bar{D}, \mathrm{Diff}_{\bar{D}} \Delta) \le a(E, \bar{D}^c, \mathrm{Diff}_{\bar{D}^c} \Delta^c)$ *for every divisor E over $\bar{D}$ and* $(\phi_{\bar{D}})_* \mathrm{Diff}_{\bar{D}} \Delta \ge \mathrm{Diff}_{\bar{D}^c} \Delta^c$.

Proof Let Y be the normalization of the main component of the fiber product $X \times_S X^c$ with projections $X \xleftarrow{g} Y \xrightarrow{h} X^c$. By definition,

$$g^*(K_X + D + \Delta) \sim_{\mathbb{R}} h^*(K_{X^c} + D^c + \Delta^c) + F$$

where F is effective. Let D_Y denote the birational transform of D on Y. Restricting to D_Y we get that

$$(g|_{D_Y})^*(K_D + \mathrm{Diff}_D \Delta) \sim_{\mathbb{R}} (h|_{D_Y})^*(K_{D^c} + \mathrm{Diff}_{D^c} \Delta^c) + F|_{D_Y}$$

and $F|_{D_Y}$ is also effective. This proves (1) and (2) is a special case. □

The existence of simultaneous canonical models is part of the following.

Question 5.48 Let $(X, D + \Delta)$ be an lc pair, and $(X^c, D^c + \Delta^c)$ its canonical model. What is the relationship between the canonical model of $(D, \mathrm{Diff}_D \Delta)$ and $(D^c, \mathrm{Diff}_{D^c} \Delta^c)$?

The following smooth example shows that these two are usually different. Start with a smooth variety X', a smooth divisor $D' \subset X'$, and another smooth divisor $C' \subset D'$. Assume that $K_{X'} + D'$ is ample. Set $X := B_{C'} X'$ with exceptional divisor E, and let $D \subset X$ denote the birational transform of D'. Then $(X, D + E)$ is an lc pair whose canonical model is (X', D'), and $(D', 0)$ is its own canonical model. However, $(D, \mathrm{Diff}_D E) \simeq (D', C') \ne (D', 0)$.

The following is proved in Ambro and Kollár (2019, Thm.7).

Theorem 5.49 *Let* $(X, D + \Delta)$ *be an lc pair that is projective over a base scheme S with relatively ample divisor H, where all divisors in D appear with coefficient 1. Set $(X^0, D^0 + \Delta^0) := (X, D + \Delta)$, and, for $i = 1, \ldots, m$, let*

$$\phi^i: (X^{i-1}, D^{i-1} + \Delta^{i-1}) \dashrightarrow (X^i, D^i + \Delta^i)$$

be the steps of the $(X, D+\Delta)$-MMP with scaling of H. Assume that the intersection of D with the exceptional locus of $\phi^m \circ \cdots \circ \phi^1: X \dashrightarrow X^m$ does not contain any log center (11.11) of $(X, D + \Delta)$. Let $\varrho: \bar{D} \to D$ be the normalization.

Then the induced maps

$$\phi_{\bar{D}}^i \colon (\bar{D}^{i-1}, \mathrm{Diff}_{\bar{D}}\,\Delta^{i-1}) \dashrightarrow (\bar{D}^i, \mathrm{Diff}_{\bar{D}}\,\Delta^i)$$

form the steps of the MMP starting with $(\bar{D}^0, \mathrm{Diff}_{\bar{D}}\,\Delta^0) := (\bar{D}, \mathrm{Diff}_{\bar{D}}\,\Delta)$ *and with scaling of* $\varrho^* H$. $\qquad\qquad\qquad\qquad\qquad\qquad\qquad$ □

5.8 Simultaneous Canonical Modifications

If S is smooth, then the simultaneous canonical modification of $f \colon (X, \Delta) \to S$ is also the canonical modification of (X, Δ) by (4.56). Thus, over a smooth curve, we consider the canonical modification of (X, Δ), and aim to prove that it is a simultaneous canonical modification.

5.50 (Proof of (5.16) over curves) Let C be a smooth curve, and $f \colon (X, \Delta) \to C$ a flat, projective morphism of pure relative dimension n that satisfies the assumptions of (5.16).

Each $c \mapsto \left(\pi_c^* H_c^{n-i} \cdot (K_{X_c^{\mathrm{cm}}} + \Delta_c^{\mathrm{cm}})^i\right)$ is a constructible function on C. Thus, in order to prove (5.16.1) we may assume that C is the spectrum of a DVR with closed point $0 \in C$ and generic point $g \in C$. We may also assume that X is reduced, thus f is flat.

By (5.34), $(\pi_0^* H_0^n) \le (\pi_g^* H_g^n)$, and equality holds iff X_0 is generically reduced. It is thus enough to deal with the latter case. Then X is generically normal along X_0, and we can replace X by its normalization without changing any of the assumptions or conclusions. We may now also assume that X is irreducible. Let $\pi \colon (Y, \Delta^Y = \pi_*^{-1}\Delta) \to (X, \Delta)$ denote the canonical modification.

Write $Y_0 = \sum_i e_i E_i$ where $e_0 = 1$, and E_0 is the birational transform of X_0. (For now E_0 is allowed to be reducible.) Set $E := \mathrm{red}\, Y_0 = \sum E_i$. Let $\tau \colon \bar{E}_0 \to E_0$ denote the normalization, and write $\tau^*(K_Y + E + \Delta^Y) = K_{\bar{E}_0} + D_0$ where $D_0 = \mathrm{Diff}_{\bar{E}_0}(E - E_0 + \Delta^Y)$ as in (11.14). Choose $m \ge 0$ such that $K_Y + E + \Delta^Y + m\pi^* H$ is ample over C. We claim that

$$\left((K_{X_g^{\mathrm{cm}}} + \Delta_g^{\mathrm{cm}} + m\pi_g^* H)^n\right)$$
$$= \left((K_{Y_g} + \Delta_g^Y + m\pi_g^* H)^n\right) = \left((K_Y + \Delta^Y + m\pi^* H)^n \cdot [Y_g]\right)$$
$$= \left((K_Y + E + \Delta^Y + m\pi^* H)^n \cdot [Y_g]\right) = \left((K_Y + E + \Delta^Y + m\pi^* H)^n \cdot [Y_0]\right)$$
$$= \sum_i e_i\left(((K_Y + E + \Delta^Y + m\pi^* H)|_{E_i})^n\right) \ge \left((K_{\bar{E}_0} + D_0 + m\pi_0^* H)^n\right)$$
$$\ge \mathrm{vol}(K_{X_0^{\mathrm{cm}}} + \Delta_0^{\mathrm{cm}} + m\pi_0^* H) = \left((K_{X_0^{\mathrm{cm}}} + \Delta_0^{\mathrm{cm}} + m\pi_0^* H)^n\right).$$

The first equality holds since (Y_g, Δ_g^Y) is the canonical modification of (X_g, Δ_g), hence $\Delta_g^{cm} = \Delta_g^Y$. The second equality is clear. We are allowed to add E in the fourth row since it is disjoint from Y_g. We can then replace Y_g by Y_0 since they are algebraically equivalent, and compute the latter one component at a time. $K_Y + E + \Delta^Y + m\pi^*H$ is ample, thus if we keep only the summands corresponding to E_0, we get the first inequality, which is an equality iff $Y_0 = E_0$.

The second inequality follows from (10.36), once we check that $\sigma_*^{-1}\Delta_0 \leq D_0$ where $\sigma := \pi_0 \circ \tau \colon \bar{E}_0 \to \bar{X}_0$ is the natural map. Since D_0 is effective, this is clear for σ-exceptional divisors. Otherwise, either π is an isomorphism over the generic point of a divisor D_0^i (hence D_0^i has the same coefficients in $\sigma_*^{-1}\Delta_0$ and D_0) or $\sigma_*^{-1}D_0^i$ is contained in another irreducible component of red Y_0. In this case $\sigma_*^{-1}D_0^i$ appears in D_0 with coefficient 1, and in $\sigma_*^{-1}\Delta_0$ with coefficient ≤ 1 by assumption. This proves the second inequality and, by (10.36), if equality holds then $D_0 = \sigma_*^{-1}\Delta_0$. The last equality is a general property of ample divisors.

As we noted in (5.14), the inequality proved in (5.50.1) is equivalent to $I(\pi_g^*H_g, K_{X_g^{cm}} + \Delta_g^{cm}) \geq I(\pi_0^*H_0, K_{X_0^{cm}} + \Delta_0^{cm})$, which proves (5.16.1).

If equality holds everywhere in (5.50.1) then $Y_0 = E_0$, $D_0 = \sigma_*^{-1}\Delta_0$, and $(\bar{E}_0, D_0)$ is canonical. On the other hand, D_0 is the sum of $\sigma_*^{-1}\Delta_0$ and of the conductor of $\bar{E}_0 \to E_0 = Y_0$. So the conductor is 0, Y_0 is normal in codimension 1, $D_0 = (\pi_0)_*^{-1}\Delta_0$, and $(Y_0, (\pi_0)_*^{-1}\Delta_0)$ is canonical in codimension 1. Thus Y_0 is normal and $(Y_0, (\pi_0)_*^{-1}\Delta_0)$ is canonical by (2.3). Since $K_{Y_0} + D_0$ is ample over X_0, these show that $(Y_0, (\pi_0)_*^{-1}\Delta_0)$ is the canonical modification of (X_0, Δ_0). Thus the canonical modification of (X, Δ) is also the simultaneous canonical modification, proving (5.16.2) over curves. □

In analogy with (5.15), we can define simultaneous slc modifications.

Definition 5.51 Let (X, Δ) be a pair over a field k that is slc in codimension 1. Its *semi-log-canonical modification* is a proper, birational morphism $\pi \colon (X^{slcm}, \Delta^{slcm}) \to (X, \Delta)$ such that π is an isomorphism over codimension 1 points of X, $\Delta^{slcm} = \pi_*^{-1}\Delta + E$ where E contains every π-exceptional divisor with coefficient 1, $K_{X^{slcm}} + \Delta^{slcm}$ is π-ample, and $(X^{slcm}, \Delta^{slcm})$ is slc.

If X is normal, then the SLC modification is automatically normal, and it agrees with the log canonical modification.

In general, lc modifications are conjectured to exist, but there are slc pairs without slc modification: see Kollár (2013b, 1.40). In both cases, existence is known when $K_X + \Delta$ is $\mathbb{R}$-Cartier; see Odaka and Xu (2012).

Let $f\colon (X,\Delta) \to S$ be a morphism as in (5.2) that satisfies the condition (5.3.1). A *simultaneous slc modification* is a proper morphism $\pi\colon (Y,\Delta^Y) \to (X,\Delta)$ such that $f\circ\pi\colon (Y,\Delta^Y) \to S$ is locally stable, and $\pi_s\colon (Y_s,\Delta_s^Y) \to (X_s,\Delta_s)$ is the slc modification for every $s \in S$.

We get the following variant of (5.16).

Theorem 5.52 *Let C be a smooth curve, $f\colon (X,\Delta) \to C$ a projective morphism as in (5.2) that satisfies the condition (5.3.1). Assume that $K_X + \Delta$ is $\mathbb{R}$-Cartier, and the slc modification $\pi_c\colon (X_c^{slcm}, \Delta_c^{slcm}) \to (X_c, \Delta_c)$ exists for every $c \in C$. Then*

(5.52.1) $c \mapsto I(\pi_c^* H_c^{n-2}, K_{X_c^{slcm}} + \Delta_c^{slcm})$ *is lower semi-continuous for $\le$, and*

(5.52.2) $f\colon (X,\Delta) \to C$ *has a simultaneous slc modification iff this function is locally constant.*

Proof Using (2.54), we may assume that X is normal. Next we closely follow the proof of (5.50).

Let $\pi\colon (Y,\Delta^Y) \to (X,\Delta)$ denote the log canonical modification; this exists by (11.29). Note that here $\Delta^Y = \pi_*^{-1}\Delta + F$ where F is the sum of all π-exceptional divisors that dominate C.

Write $Y_0 = \sum_i e_i E_i$ where $e_0 = 1$ and E_0 is the birational transform of X_0. Let $\tau\colon \bar{E}_0 \to E_0$ denote the normalization, and write $\tau^*(K_Y + Y_0 + \Delta^Y) = K_{\bar{E}_0} + D_0$. Choose $m \ge 0$ such that $K_Y + Y_0 + \Delta^Y + m\pi^* H$ is ample over C. As in the proof of (5.50), we get that

$$((K_{X_g^{lcm}} + \Delta_g^{lcm} + m\pi_g^* H)^n) \ge ((K_{\bar{E}_0} + D_0 + m\pi_0^* H)^n) \quad \text{and}$$
$$\mathrm{vol}(K_{X_0^{lcm}} + \Delta_0^{lcm} + m\pi_0^* H) = (K_{X_0^{lcm}} + \Delta_0^{lcm} + m\pi_0^* H)^n.$$

It remains to prove that $(K_{\bar{E}_0} + D_0 + m\pi_0^* H)^n \ge \mathrm{vol}(K_{X_0^{lcm}} + \Delta_0^{lcm} + m\pi_0^* H)$.

We have $\sigma\colon \bar{E}_0 \to X_0$, and we can apply (10.37) if every σ-exceptional divisor $\bar{F}_0 \subset \bar{E}_0$ appears in D_0 with coefficient 1.

By the definition of lc modifications, every divisor F_i that is exceptional for $Y \to X$ appears in Δ^Y with coefficient 1. If $K_X + \Delta$ is $\mathbb{R}$-Cartier then the exceptional set of $Y \to X$ has pure codimension 1. In this case, $\tau(\bar{F}_0)$ is contained in a divisor that is exceptional for $Y \to X$. Thus, by adjunction, $\bar{F}_0$ appears in D_0 with coefficient 1.

If (X_0, Δ_0) is slc at a point x_0 then (X,Δ) is also slc at x_0 by inversion of adjunction (11.17), hence π is a local isomorphism over x_0. Thus $\pi_0\colon (Y_0, \Delta_0^Y) \to (X_0, \Delta_0)$ is an isomorphism over codimension 1 points of X_0.

The rest of the proof works as before. $\square$

If $K_X + \Delta$ is not $\mathbb{R}$-Cartier then it can happen that an exceptional divisor $\bar{F}_0 \subset \bar{E}_0$ is not contained in any exceptional divisor of $X^{\mathrm{lcm}} \to X$. In such cases we lose control of the coefficient of $\bar{F}$ in D_0. This occurs in (5.22) over the 4 singular points that lie on D_0.

5.9 Families over Higher Dimensional Bases

Here we complete the proofs of Theorems 5.4–5.16. In all cases, the first part asserts that a certain constructible function on the base scheme S is upper or lower semicontinuous. As in (5.30), for constructible functions, semicontinuity can be checked along spectra of DVRs, and this was already done in all cases.

The remaining part is to show that if our functions are locally constant on S, then certain constructions produce a flat family of varieties or sheaves. In all cases, we have already checked that this holds when the base is a smooth curve.

Going to arbitrary reduced bases is quickest in the following example.

5.53 (Proof of 5.1) We already proved the case when S is the spectrum of a DVR in (5.42). As we noted in (5.30), this implies (5.1.1) in general. Thus it remains to prove that if $s \mapsto (K_{X_s}^n)$ is constant then $f: X \to S$ is stable.

In view of (5.42), we know that $f_T: X_T \to T$ is stable for every $T \to S$ where T is the spectrum of a DVR. Thus $f: X \to S$ is stable by (4.7). □

We aim to argue similarly for Theorems 5.4, 5.5 and 5.6. Note that in these cases we cannot apply (5.8) since f is not assumed to be flat, and its fibers are not assumed to be S_2. We follow (5.31). For (5.4–5.6) this needs the theory of hulls and husks, to be explained in Chapter 9.

5.54 (Proof of 5.4–5.6) Let $\pi\colon \mathrm{Hull}(\mathscr{O}_X/S) \to S$ denote the hull (9.39) of $\mathscr{O}_X$. We aim to show that π is an isomorphism.

By (9.40), π is a locally closed decomposition (10.83).

Let T be the spectrum of a DVR, and $g\colon T \to S$ a morphism that maps the generic point of T to a generic point of S. We apply (5.42) or (5.27) to the divisorial pull-back $f_T\colon (X_T, \Delta_T) \to T$ to conclude that it is stable (resp. locally stable). For (5.6) we use (2.88.5).

Thus $g\colon T \to S$ factors uniquely through $\pi\colon \mathrm{Hull}(\mathscr{O}_X/S) \to S$, hence π is proper. $\pi\colon H \to S$ is an isomorphism by (10.83.2). In particular, $f\colon X \to S$ is flat with S_2 fibers. Thus the fibers are slc by assumption and (11.37).

Now we can apply (4.35) to conclude that $K_{X/S} + \Delta$ is $\mathbb{R}$-Cartier, hence $f\colon (X, \Delta) \to S$ is stable (resp. locally stable). □

For the remaining cases, (5.31) needs the moduli space of pairs with an artificial, but efficient, rigidification.

5.55 (Proof of 5.10–5.11) Both claims were already established over the spectrum of a DVR, see (5.44) and (5.46). This implies the semicontinuity assertions in both cases.

It remains to show that if the volume is constant then $f \colon X \to S$ (resp. $f \colon (X, \Delta) \to S$) has a simultaneous canonical model.

Consider the moduli space of marked stable pairs $\pi \colon \mathrm{SP}^{\mathrm{red}} \to S$; since S is reduced, the version in (4.1) is sufficient for our purposes. Set

$$S' := \{ (X_s^c, \Delta_s^c) : s \in S \} \subset \mathrm{SP}^{\mathrm{red}} .$$

In order to prove that S' is a closed subset, first we claim that it is constructible. This is clear since the canonical model over a generic point of S extends to a canonical model over an open subset of S, and we can finish by Noetherian induction. Thus closedness needs to be checked over spectra of DVRs, and the latter follows from (5.44) and (5.46).

Thus S' is a scheme, and the projection π induces a geometric bijection $S' \to S$ which is finite by (5.44) and (5.46). Thus $S' \to S$ is an isomorphism since we assumed that S is seminormal.

If each (X_s^c, Δ_s^c) is rigid, then $S' \subset \mathrm{SP}^{\mathrm{rigid}}$, and there is a universal family $\mathrm{Univ}^{\mathrm{rigid}} \to \mathrm{SP}^{\mathrm{rigid}}$ by (8.71). Therefore the pull-back of the universal family $\mathrm{Univ}^{\mathrm{rigid}}$ to S' gives the simultaneous canonical model over $S \simeq S'$.

We have no reason to assume that the (X_s^c, Δ_s^c) are rigid, but we can make the proof work by rigidifying $f \colon (X, \Delta) \to S$.

The simultaneous canonical model is unique, hence it is enough to construct it étale locally. After replacing S by an étale neighborhood of a point $0 \in S$, we may assume that there are r sections $\sigma_i \colon S \to X$ such that $(X_0, \Delta_0, \sigma_1(0), \dots, \sigma_r(0))$ is rigid, and the $\sigma_i(0)$ are smooth points of $X_0 \setminus \mathrm{Supp}\, \Delta_0$ such that $(X_0, \Delta_0) \dashrightarrow (X_0^c, \Delta_0^c)$ is a local isomorphism at these points.

By (8.65), after further shrinking S we may assume that the same holds at every point $s \in S$. Using the moduli of marked, pointed stable pairs MpSP (8.44) and (8.71.1), we can run the previous argument for

$$S' := \{ (X_s^c, \Delta_s^c, \sigma_1(s), \dots, \sigma_r(s)) : s \in S \} \subset \mathrm{MpSP}^{\mathrm{rigid}}$$

to prove that the simultaneous canonical model exists over S. $\square$

5.56 (Proof of 5.16) The proof follows very closely the arguments in (5.55). Both claims were already established over the spectrum of a DVR, see (5.50). This implies the semicontinuity assertion in general.

It remains to show that if $I(s) = I(\pi_s^* H_s, K_{X_s^{\mathrm{cm}}})$ is constant, then $f : (X, \Delta) \to S$ has a simultaneous canonical modification. Since the simultaneous canonical modification is unique, it is sufficient to construct it étale locally over S. So pick a point $s_0 \in S$, in the sequel we are free to replace S by smaller neighborhoods of s_0.

Choose $m > 0$ such that $K_{X_s^{\mathrm{cm}}} + m\pi_s^* H_s$ is ample for every $s \in S$. Next choose a general $D \in |mH|$ such that $(X_{s_0}^{\mathrm{cm}}, \Delta_{s_0}^{\mathrm{cm}} + \pi_{s_0}^* D_{s_0})$ is log canonical. We claim that, possibly after shrinking S, $(X_s^{\mathrm{cm}}, \Delta_s^{\mathrm{cm}} + \pi_s^* D_s)$ is log canonical for every $s \in S$. By (4.44) this condition defines a constructible subset of S and, by (5.50), it contains every generalization of s_0. Thus it contains an open neighborhood of s_0. Thus $(X_s^{\mathrm{cm}}, \Delta_s^{\mathrm{cm}} + \pi_s^* D_s)$ is a stable pair for every $s \in S$.

Consider the moduli space of marked stable pairs $\pi : \mathrm{SP} \to S$, and set

$$S' := \{(X_s^{\mathrm{cm}}, \Delta_s^{\mathrm{cm}} + \pi_s^* D_s) : s \in S\} \subset \mathrm{SP}.$$

In order to prove that S' is a closed subset, first we claim that it is constructible. This is clear since the canonical modification over a generic point of S extends to a canonical modification over an open subset of S, and we can finish by Noetherian induction. Thus closedness needs to be checked over spectra of DVRs, and the latter follows from (5.50).

Thus S' is a scheme, and the projection π induces a geometric bijection $S' \to S$ which is finite by (5.50). Thus $S' \to S$ is an isomorphism since S is assumed seminormal.

For general D, the pairs $(X_s^{\mathrm{cm}}, \Delta_s^{\mathrm{cm}} + \pi_s^* D_s)$ should be rigid, and then the pull-back of the universal family to S' gives the simultaneous canonical modification over $S \simeq S'$. Technically it may be easier to rigidify using étale-local sections as in (5.55). □

6

Moduli Problems with Flat Divisorial Part

So far we have identified stable pairs (X, Δ) as the basic objects of our moduli problem, the one-parameter families that we want to allow, and worked out the reduced part of the moduli spaces. Now we come to the next step of identifying the stable families over an arbitrary base scheme.

In this chapter we consider several special cases that are easier to handle, since we are able to treat the underlying variety X and the boundary divisor Δ as separate objects that are both flat over the base. This is achieved by imposing one of four different types of restrictions on the coefficients occurring in Δ.

- (No boundary) Stable varieties X with $\Delta = 0$.
- (Standard coefficients) The coefficients in Δ are in the "diminished standard coefficient" set $\{1 - \frac{1}{3}, 1 - \frac{1}{4}, 1 - \frac{1}{5}, \dots, 1\}$.
- (Major coefficients) The coefficients in Δ are all $> \frac{1}{2}$.
- (Generic coefficients) The coefficients in Δ are $\mathbb{Q}$-linearly independent.

These examples cover many cases; the most jarring omission is that none of these allow $\frac{1}{2}$ as a coefficient.

After a general discussion of moduli problems in Section 6.1, we treat two notions of stability for stable varieties in Sections 6.2–6.3. The first of these – introduced in Kollár and Shepherd-Barron (1988) – starts with the proposal that all plurigenera should be deformation invariant. The second – introduced in Viehweg (1995) – posits that all sufficiently divisible plurigenera should be deformation invariant. The two versions agree over reduced base schemes.

Both of these versions can be extended to pairs (X, Δ), as long as Δ is a standard or major boundary as above.

In Section 6.4 we discuss another variant – due to Alexeev (2006, 2015) – that works if the coefficients in Δ are sufficiently general. This is especially natural when the boundary arises as a small perturbation of a basic situation.

The infinitesimal deformation theory of stable varieties is not yet well understood, but a large part of the first order theory for surfaces is treated in Altmann and Kollár (2019). After a general discussion of first order deformations of singular varieties in Section 6.5, we work out in detail the theory for cyclic quotient surface singularities in Section 6.6. These are the simplest noncanonical singularities and they show that the two versions outlined in Sections 6.2–6.3 differ from each other already over Spec $k[\varepsilon]$.

Assumptions In this chapter we work over a $\mathbb{Q}$-scheme, but the definitions are set up in full generality. See Section 8.8 for a discussion of some problems in positive characteristic.

6.1 Introduction to Moduli of Stable Pairs

Based on the outline in Section 1.2, we discuss the plan that we use to treat many moduli problems in algebraic geometry. The following version is designed to work best for the moduli of stable pairs (X, Δ).

The method first deals with stable pairs with an embedding into a fixed projective space and then removes the effect of the embedding.

Step 6.1 (Objects of the moduli problem) At the beginning we have to decide which objects and families our moduli problem should cover. This is usually done in three stages.

6.1.1 (Interior objects over algebraically closed fields) As the very first step, we have to decide what kind of objects we want to parametrize. Probably the first nonlinear moduli problem considered was elliptic curves, followed by smooth projective curves of higher genus and their close relatives, abelian varieties. The study of the moduli of higher dimensional smooth projective varieties was systematically undertaken first by Matsusaka. His approach focuses on polarized pairs (X, L), where X is a variety and L an ample divisor or divisor class. Here our main aim is to study canonical models of varieties and pairs of general type.

It is expected that, once we understand the moduli of varieties, it should be relatively easy to work out the moduli theory of related compound objects. For example varieties with a group action, pointed varieties, maps between varieties, or various combinations of these.

6.1.2 (Boundary objects over algebraically closed fields) By now the answers are mostly well established, but historically this was a difficult and very nontrivial step. The compactification of the moduli of smooth curves by stable curves was discovered by Deligne and Mumford (1969).

For surfaces, the need to work with canonical models (instead of minimal models) seems to have become clear early, but the choice of stable surfaces for boundary points was proposed only in Kollár and Shepherd-Barron (1988).

It should be noted that the distinction between interior and boundary points is not always clear cut. While everyone agrees that smooth curves give the interior points and nodal curves the boundary points of $\overline{M}_g$, for surfaces one may view either canonical models or only smooth canonical models as interior points.

Although historically the development went in the other direction, for a logical treatment of a moduli problem it is better to settle on the right class of interior and boundary objects at the beginning. Then gradually prove that they have the required properties.

6.1.3 (Objects over arbitrary fields) For stable pairs, the definitions of (6.1.1–2) carry over to arbitrary fields, but in a few examples new questions emerge.

For pointed schemes $(X, p_1, \ldots, p_r)$ it may be better to replace the set of closed points $\{p_1, \ldots, p_r\}$ by a 0-dimensional subscheme $Z \subset X$ of length r. A more subtle problem appears for polarizations, due to the difference between $\mathrm{Pic}(X_k)$ and $\mathbf{Pic}(X_k)(k)$, where $\mathbf{Pic}(X_k)(k)$ is the set of k-points of the Picard scheme of X_k; see Bosch et al. (1990, sec.8.1) for a discussion. This will not be a major issue for us. There are also problems caused by inseparable extensions in positive characteristic.

Conclusion 6.1.4 We are working with stable varieties (1.41) and, more generally, with stable pairs (X, Δ) as defined in (2.1). There seems to be full agreement about these being the right objects in characteristic 0.

Step 6.2 (Families of the moduli problem) In many moduli problems, it is considered obvious that the families are determined by the objects: one should work with flat families whose fibers are among our objects. Then the traditional approach is to determine families over $\mathrm{Spec}\, k[\varepsilon]$, and, more generally, over Artinian base schemes. This is usually called *obstruction theory*; see Artin (1976), Sernesi (2006), and Hartshorne (2010) for introductions to various cases.

However, for stable varieties and pairs, flat families with stable fibers do *not* give a sensible moduli theory. We need to proceed differently.

6.2.1 (Families over DVRs) In Chapter 2, we defined and described stable families over smooth curves and one-dimensional regular schemes. The advantage of this setting is that the total space of a family is also a locally stable pair, so minimal model theory can be applied both to the fibers and to the total space.

6.2.2 (*Families over reduced bases*) For stable varieties, we proved in (3.1) that stable families over DVRs determine stable families over reduced base schemes. We needed to work quite a bit harder to extend the theory to stable families of pairs over reduced base schemes in Chapter 4, but the end result is the same, at least in characteristic 0: the families over DVRs determine the families over reduced base schemes.

6.2.3 (Families over arbitrary bases) This is where the picture becomes rather complicated. For stable varieties, there have been different proposals for about 30 years; we discuss these in Sections 6.2–6.3. These were proved to be non-equivalent in Altmann and Kollár (2019); see Section 6.6.

We believe that the notion of KSB stability – to be treated in Section 6.2 – gives the optimal answer for stable varieties.

For pairs, the problem is that, while KSB stability has a natural generalization to pairs, not all stable families over smooth curves satisfy it; see (2.41). Thus insisting on it frequently gives nonproper moduli spaces. Still, the strongest version of KSB stability is expected to work well for pairs (X, Δ) if all the coefficients in Δ are $> \frac{1}{2}$; we discuss these in (6.24) and (6.29).

Another approach, outlined in Alexeev (2006, 2015), gives a good theory if the coefficients in Δ are sufficiently general real numbers; see (6.40).

However, there was not even a plausible proposal for the general theory before Kollár (2019). We work out the details of it in Chapter 7.

Conclusion 6.2.4 We are not aware of any other proposed definition that might work in general, but it is too soon to tell whether the theory of Chapter 7 is the final word on the subject. We comment on some of the issues next.

Once we have settled on the right objects and families, we need to start working on producing all families and constructing the moduli spaces.

We would like to have a "sensible" way to obtain all stable varieties, pairs, and their stable families. It is not a priori clear what this means.

For example, every variety of dimension n is obtained as the normalization of a hypersurface in $\mathbb{P}^{n+1}$. We can thus start working through all hypersurfaces and describe their normalizations.

For curves, this is not a bad approach. Classical authors developed much of the theory by thinking of smooth curves as normalizations of plane curves with nodes. However, this becomes harder as the genus increases. The problem is that even if a curve is general, the nodal sets of its plane representatives are always in special position.

There are some cases of surfaces where such a description is useful. For example, Enriques obtained his namesake surfaces in 1896 as sextics in $\mathbb{P}^3$

that are double along the edges of a tetrahedron. However, for most surfaces, projection to $\mathbb{P}^3$ introduces very complicated singular sets that hide the geometry of the surface. There is no "optimal" representation and it is quite hard to decide when the normalizations of two hypersurfaces are isomorphic to each other. This approach does not seem very helpful in general; see, however, the proof of Noether's formula in Griffiths and Harris (1978, sec.4.6).

Thus we aim to find projective embeddings of varieties that do not depend on too many auxiliary choices.

Step 6.3 (Rigidification by embedding) A global coordinate system on a space V is a way of associating a string of numbers (called coordinates) to any point of V. Equivalently, a choice of a map from V to $\mathbb{R}^n$ or $\mathbb{C}^n$. We prefer to work with projective objects, so for us the natural choice is to use homogeneous coordinates. Equivalently, we fix an algebraic morphism $X \to \mathbb{P}^N$. (There is a slight notational issue here. Although we almost always construct $\mathbb{P}^N$ as $\operatorname{Proj} k[x_0, \ldots, x_N]$, we usually emphasize that there are no natural coordinates on it. By contrast, with rigidification we do think of the target $\mathbb{P}^N$ as having fixed coordinates.)

For varieties, the most frequently used approach is to use an embedding $(X \hookrightarrow \mathbb{P}^N)$, though sometimes finite maps $X \to \mathbb{P}^N$ or maps to other targets – weighted projective spaces or $\mathbb{P}^N$-bundles over curves – give better insight.

Thus we choose a very ample line bundle L on X, a subspace $V^{N+1} \subset H^0(X, L)$ and a basis of V^{N+1} (up to a multiplicative constant). In practice, it is much better to eliminate the second of these choices by taking $V = H^0(X, L)$. That is, we work with embeddings $(X \hookrightarrow \mathbb{P}^N)$ whose image is *linearly normal*. The rigidification involves two types of choices.

6.3.1 (Discrete choice) A very ample line bundle L. (We use this terminology although **Pic**(X) is not always discrete).

If C is a stable curve, then ω_C^r is very ample for $r \geq 3$. If S is a canonical model of a surface of general type, then ω_S is an ample line bundle and ω_S^r is very ample for $r \geq 5$ by Bombieri (1973), and Ekedahl (1988). Thus again we get an embedding of S into a projective space whose dimension depends only on the coefficients of the Hilbert polynomial $\chi(\omega_S^r)$, namely (K_S^2) and $\chi(\mathcal{O}_S)$.

The situation is more complicated for stable surfaces. These can have singularities where ω_S is not locally free. Even worse, for any $m \in \mathbb{N}$ there are stable surfaces S_m and canonical 3-folds X_m such that $\omega_{S_m}^{[m]}$ (resp. $\omega_{X_m}^{[m]}$) is not locally free at some point $x_m \in S_m$. Thus every section of $\omega_{S_m}^{[m]}$ vanishes at x_m and $H^0(X, \omega_{S_m}^{[m]})$ gives a rational map that is not defined at x_m.

We skirt this problem by fixing $m > 0$ and aiming to construct a moduli space for those stable varieties for which $\omega_S^{[m]}$ is locally free, very ample and has no higher cohomologies. Similarly, if (X, Δ) is a stable pair and Δ is a $\mathbb{Q}$-divisor, we can take $L = \omega_X^{[m]}(m\Delta)$ for some $m > 0$. Thus L is indeed a discrete choice for us.

Then we show in Step 6.8 that, if m is sufficiently divisible (depending on other numerical invariants), then the theory we get is independent of m.

There does not seem to be a similarly natural choice of L if Δ is an $\mathbb{R}$-divisor. We have to work around this in Section 8.2.

6.3.2 (Continuous choice) Different bases in $H^0(X, L)$ are equivalent to each other under the natural group action by $\mathrm{GL}(H^0(X, L))$. We eliminate the effect of this choice in Step 6.5.

Aside For smooth varieties over $\mathbb{C}$, the use of topological rigidifiers can be very powerful; leading to the Teichmüller space for curves and to Griffiths's theory of period domains. These work well for smooth varieties, but have many problems for their degenerations. For flat families of stable varieties $f \colon X \to S$, the topological type, or even dimension of $H^*(X_s(\mathbb{C}), \mathbb{C})$ need not be a locally constant function on S. It does not seem to be possible to make sense of a topological rigidification in general.

6.3.3 (Moduli of embedded varieties) Once we have a rigidification, we construct moduli spaces of more general embedded objects. Instead of embedded stable varieties $(X \hookrightarrow \mathbb{P}^N)$ of dimension n, one can work either with n-cycles (Cayley–Chow approach) or, which works better for us, with all subschemes $(X \subset \mathbb{P}^N)$ (Hilbert–Grothendieck approach). Thus we start with the universal family over the Hilbert scheme of n-dimensional subschemes

$$\pi \colon \mathrm{Univ}_n(\mathbb{P}^N) \to \mathrm{Hilb}_n(\mathbb{P}^N).$$

We encounter a severe difficulty when we try to work with pairs (X, Δ).

6.3.4 (Moduli of embedded pairs) We need to construct the universal family of relative Mumford divisors (6.13)

$$\mathrm{MDiv}(\mathrm{Univ}_n(\mathbb{P}^N)/\mathrm{Hilb}_n(\mathbb{P}^N)) \to \mathrm{Hilb}_n(\mathbb{P}^N).$$

The traditional approaches try to obtain this as a subscheme of either

- $\mathrm{Hilb}_{n-1}(\mathrm{Univ}_n(\mathbb{P}^N)/\mathrm{Hilb}_n(\mathbb{P}^N))$, or of
- $\mathrm{Chow}_{n-1}(\mathrm{Univ}_n(\mathbb{P}^N)/\mathrm{Hilb}_n(\mathbb{P}^N))$.

By (4.76), the Chow version works over reduced schemes, but neither works in general.

Conclusion 6.3.5 We have the universal family of embedded varieties, but we hit a problem with pairs. This was a long-standing conundrum in the theory; K-flatness – to be worked out in Chapter 7 – was introduced to solve it. Here we take an easier path, and in Sections 6.2–6.4 we consider several cases when the Hilbert scheme variant works in (6.3.4); see (6.13) for details.

Assume now that the above steps have been completed. Then, instead of our original moduli problem, we have solved a related one that also includes a rigidification and has many more objects. In order to get back to our original problem, we need to remove the nonstable objects and then see how to undo the effects of rigidification.

Step 6.4 (Representability) Assume first that $\Delta = 0$ and let us go back to $\mathrm{Hilb}_n(\mathbb{P}^N)$ as in (6.3.3). As we saw in Section 3.5, the set of stable fibers of π is not even a locally closed subset of $\mathrm{Hilb}_n(\mathbb{P}^N)$. Nonetheless, as we proved in Section 3.5, stable families are parametrized by a locally closed partial decomposition of $\mathrm{Hilb}(\mathbb{P}^N)$. By the choice we made in Step 6.3.1, we aim to work only with those stable subvarieties $X \subset \mathbb{P}^N$ for which $\mathscr{O}_X(1) \simeq \omega_X^{[m]}$. This is again a representable condition by (9.42). Thus we get the moduli space of m-canonically embedded stable subvarieties of dimension n in $\mathbb{P}^N$

$$\mathrm{C}^m\mathrm{ESV}(n, *, \mathbb{P}^N) \to \mathrm{Hilb}_n(\mathbb{P}^N). \tag{6.4.1}$$

(Here $*$ stands for the not-yet-specified volume.)

For pairs, we start with the case when Δ is a $\mathbb{Q}$-divisor, which we write as $\Delta = \sum a_i D_i$ for some fixed $\mathbf{a} := (a_1, \ldots, a_r)$, where the D_i are effective $\mathbb{Z}$-divisors. (This will be called a *marking* in Section 8.1; see (8.21) for real coefficients.) Once we solve the questions raised in Step 6.3.4, the results of Section 4.6 give the moduli space of m-canonically embedded stable pairs

$$\mathrm{C}^m\mathrm{ESP}(\mathbf{a}, n, *, \mathbb{P}^N) \to \mathrm{Hilb}_n(\mathbb{P}^N). \tag{6.4.2}$$

Conclusion 6.4.3 For each $m > 0$, we have obtained universal families of m-canonically embedded stable varieties and pairs. However, m and the embedding are artificial choices; we still need to undo their effect.

(In practice we need to be more precise here and control various properties of the embedding – like linear normality, vanishing of certain cohomology groups – but these turn out to be technical issues; see Section 8.4.)

Step 6.5 (Quotients by group actions) Let us deal next with the continuous choice in the rigidification, which is a basis in $H^0(X, L)$. As we noted in (6.3.2), the different continuous choices are equivalent to each other under a GL-action.

This gives a group action on the moduli of rigidified objects, and the moduli space of the nonrigidified objects is the space of orbits of this action

$$C^m ESP(\mathbf{a}, n, *, \mathbb{P}^N)/PGL_{N+1}. \tag{6.5.1}$$

We discuss in Section 8.6 that such quotients have a natural algebraic space structure. So, aside from the slight difference between schemes and algebraic spaces, we consider the quotient problem solved.

6.6 (Conclusion of Steps 6.1–6.5) As in Step 6.4, fix a rational coefficient vector $\mathbf{a}$. Let $\mathcal{SP}(\mathbf{a}, n, v)$ denote the functor of stable pairs $(X, \Delta = \sum a_i D_i)$ of dimension n and volume v.

We check in Step 6.8 that there is an $m = m(\mathbf{a}, n, v)$ such that $\omega_X^{[m]}(m\Delta)$ is locally free, very ample, and has no higher cohomologies. Thus $P_m(X, \Delta) := H^0(X, \omega_X^{[m]}(m\Delta))$ is a locally constant function on stable families in $\mathcal{SP}(\mathbf{a}, n, v)$. Using (8.62), we obtain the coarse moduli space of $\mathcal{SP}(\mathbf{a}, n, v)$ as the union of geometric quotients

$$SP(\mathbf{a}, n, v) = \amalg_i C^m ESP(\mathbf{a}, n, v, \mathbb{P}^{N_i})//PGL_{N_i+1},$$

where $N_i + 1$ runs through the possible values of $P_m(X, \Delta)$. (See (8.21) for real coefficients.)

Now that we have constructed our moduli spaces $\mathcal{SP}(\mathbf{a}, n, v)$, we should study their properties.

Step 6.7 (Separatedness and valuative-properness) Since these notions depend only on families over DVRs, these will always hold for us. The discussion in (1.20) needs no amplification.

The next two topics merit a treatment of their own; here we give only a few comments and the main references to the literature.

Step 6.8 (Boundedness) We aim to prove that $SP(\mathbf{a}, n, v)$ is actually of finite type, hence proper. Equivalently, that $SP(\mathbf{a}, n, v) = SP(\mathbf{a}, n, v, m)$ for some m (depending on $\mathbf{a}, n, v$).

We discussed stable varieties in (1.21), but there are some changes for pairs. The Hilbert function $\chi(X, \omega_X^{[r]}(\lfloor r\Delta \rfloor))$ is no longer deformation invariant, but its (rescaled) leading coefficient $\text{vol}(X, \Delta) = (K_X + \Delta)^{\dim X}$, and the constant coefficient $\chi(X, \mathcal{O}_X)$ are. This is why we use only the volume in the definition of $SP(\mathbf{a}, n, v)$ in (6.5.1).

An infinite union is of finite type only if it eventually stabilizes, so one can formulate our question independent of moduli theory as follows. It was proved by Alexeev (1993) for surfaces and by Hacon et al. (2018) in general.

6.8.1 (Boundedness theorem, rational coefficients) Assume that the a_i are rational. Then there is an $m = m(\mathbf{a}, n, v)$ such that $mK_X + m\Delta$ is a very ample Cartier divisor for every $(X, \Delta) \in \mathcal{SP}(\mathbf{a}, n, v)$.

If some of the a_i are irrational, then usually $mK_X + m\Delta$ is never a $\mathbb{Z}$-divisor. The natural correction would be to use $mK_X + \lfloor m\Delta \rfloor$, but there are examples when it is never Cartier (11.50.3). Thus we need a different form.

6.8.2 (Boundedness theorem, real coefficients) Assume that the a_i are real. Fix an algebraically closed field k of characteristic 0. Then there is a k-scheme of finite type S and a stable morphism $p\colon (X^S, \Delta^S) \to S$ such that every $(X, \Delta) \in \mathcal{SP}(\mathbf{a}, n, v)(k)$ appears among the fibers of p.

The two versions are equivalent for rational coefficients by (6.14).

The following variant is much easier to prove (4.60) and is sufficient for most applications.

6.8.3 (Weak boundedness theorem) Every irreducible component of SP$(\mathbf{a}, n, v)$ is of finite type. $\qquad\qquad\square$

6.8.4 (Hints to the proof for real coefficients) (Based on suggestions of C. Xu.) Hacon et al. (2014) proves that there is a smooth k-scheme of finite type S and a projective, log smooth morphism $p\colon (Y, E + D) \to S$ such that, for every $(X, \Delta) \in \mathcal{SP}(\mathbf{a}, n, v)(k)$, there is a log resolution $(X', E' + \Delta') \to (X, \Delta)$ and an $s \in S$ such that $(Y_s, E_s + D_s) \simeq (X', E' + \Delta')$. Therefore, if $p\colon (Y, E + D) \to S$ has a simultaneous canonical model $p^c\colon (Y^c, D^c) \to S$, then every $(X, \Delta) \in \mathcal{SP}(\mathbf{a}, n, v)(k)$ appears among the fibers of p^c, proving boundedness. If $\mathbf{a} \subset \mathbb{Q}$, the latter is proved in Hacon et al. (2018).

In the irrational case, we argue as follows. Pick any (X, Δ) and choose convex rational approximations (X, Δ_j) for $j = 1, \ldots, r$ as in (11.47), so that they have the same dlt modifications (11.47.9).

Choose $s \in S$ such that $(Y_s, E_s + D_s) \simeq (X', E' + \Delta')$. Working in an étale neighborhood of s, there is a bijection between the irreducible components of D and the irreducible components of D_s, hence the irreducible components of Δ. Thus the Δ_j determine $\mathbb{Q}$-divisors D_j.

The aim is to show that applying Hacon et al. (2018) to any one of the $p\colon (Y, E + D_j) \to S$, we get $p^c\colon (Y^c, D^c) \to S$.

To see this, note that the fiber of $p\colon (Y, E + D_1) \to S$ over s is a log resolution of (X, Δ_1). Thus Hacon et al. (2018) gives a simultaneous, minimal, $\mathbb{Q}$-factorial model $p^m\colon (Y^m, E^m + D_1^m) \to S$.

By our choice, the fiber over s is also a $\mathbb{Q}$-factorial, dlt model for the other (X, Δ_j). Since Y^m is $\mathbb{Q}$-factorial, the other $p^m \colon (Y^m, E^m + D_j^m) \to S$ are also locally stable, possibly after shrinking S.

The contraction $Y_s^m \to X$ now extends to a neighborhood of Y_s^m, giving a morphism $p^c \colon Y^c \to S$ such that, $p^c \colon (Y^c, D_j^c) \to S$ is stable for every j, again possibly after shrinking S. Thus all the $K_{Y^c/S} + D_j^c$ are $\mathbb{Q}$-Cartier.

Since Δ is a convex linear combination of the Δ_j, $K_{Y^c/S} + \Delta^c$ is $\mathbb{R}$-Cartier, hence $p^c \colon (Y^c, \Delta^c) \to S$ is stable by (11.4.4), as needed.

This takes care of an open neighborhood of $s \in S$; we finish by Noetherian induction. $\qquad\square$

Step 6.9 (Projectivity) Once we know that the connected (or irreducible) components are proper, we would like to show that they are projective. In cases when GIT works, it gives (quasi)projectivity right away, but the general quotient theorems of Kollár (1997) and Keel and Mori (1997) do not give projectivity; in fact there are many quotients that are not quasi-projective Kollár (2006).

So we need to find some ample line bundles on our moduli spaces. Let $f \colon X \to S$ be a stable morphism. The only divisorial sheaves that we can always write down on X are $\omega_{X/S}^{[m]}$; these give the sheaves $\det f_* \omega_{X/S}^{[m]}$ on S. It is not hard to work out that these are actually line bundles, so let us hope that some of these are ample.

It was Iitaka who realized that the sheaves $f_* \omega_{X/S}^{[m]}$ should always have semi-positivity properties, at least in characteristic 0, Iitaka (1972). These properties were established and applied to prove Iitaka's conjectures by many authors; see Mori (1987) for a survey. These methods were used to prove projectivity statements for the moduli of stable surfaces in Kollár (1990). Extending these results to higher dimensions turned out to be quite difficult. It was done by Fujino (2018) for stable varieties and by Kovács and Patakfalvi (2017) for stable pairs. The situation is more complicated in positive characteristic, but the surface case was settled by Patakfalvi (2014, 2017).

Conclusion 6.9.1 In all cases, the outcome is that every proper subset of the moduli space is projective. Thus we consider the projectivity question solved.

Let us now summarize the properties that we would like to see.

6.10 (Good moduli theories) A moduli theory **M** is given by specifying the objects over fields and the families. We are mainly studying those cases whose objects are various subsets of all stable pairs.

For example, the most classical example is **M** = Curves, whose objects are stable curves and whose families are all flat, proper morphisms with stable curves as fibers.

In Chapter 4, we established the optimal definitions for families of stable pairs over reduced base spaces and proved many properties. However, unlike for curves, there seem to be several natural, but nonequivalent, moduli theories of stable pairs over nonreduced base schemes.

We say that **M** is a *good moduli theory* if the following hold:

(6.10.1) **M** is separated (1.20). Since this depends only on families over DVRs, this always holds for us by (2.50).

(6.10.2) **M** is valuative-proper (1.20). The positive answer is given by (2.51), but we need to check that the central fiber also satisfies the additional assumptions that we have in **M**.

(6.10.3) Embedded moduli spaces exist (6.3.3–4). Having a flat divisorial part makes this much simpler; see (6.12–6.13) for details.

(6.10.4) Representability as in (6.4).

(6.10.5) Boundedness in the weaker form (6.8.3). Together with valuative-properness, this means that the irreducible components of the corresponding moduli spaces are proper.

For the main results of this chapter we work with the following set-up, which is a slight generalization of (3.28) and (4.2).

6.11 (Basic set-up for Chapter 6) We consider flat families of demi-normal schemes with flat families of Mumford divisors. That is, our objects are proper morphisms $f\colon X \to S$ of pure relative dimension n and subschemes $\{D_i \subset X\colon i \in I\}$ satisfying the following conditions:

(6.11.1) f is flat with demi-normal (11.36) fibers,

(6.11.2) the D_i are relative Mumford divisors (4.68), and

(6.11.3) the $D_i \to S$ are flat with divisorial subschemes (4.16) as fibers.

Next, fix distinct, positive real numbers $\{a_i\colon i \in I\}$. Then $f\colon (X, \sum a_i D_i) \to S$ is family of pairs as in (5.2).

We already treated stable families over reduced bases in Chapter 4, so assume that $f\colon (X, \sum a_i D_i) \to S$ is stable or locally stable over red S. The main question we aim to address is the following.

Question 6.11.4 If S is nonreduced, what additional restrictions should be imposed in order to get a stable (resp. locally stable) family over S?

Comments 6.11.5 There may be several different good answers to this question. These in turn give different moduli spaces, though all of them have the same underlying reduced subspace.

Also, as we noted in (2.41–2.44), requiring the D_i to be flat over S means that we do not even get all stable families over smooth curves when $a_i < \frac{1}{2}$. So, while our answers cover many important special cases, substantially new ideas will be needed to get the full theory.

6.12 (Advantages of flat divisorial parts) The cases considered in this chapter have four major technical advantages. The first three come from using the flatness option for the divisorial part in (6.3.4).

(6.12.1) One can define the families using only flatness; thus we avoid the notion of K-flatness, which is defined and studied in Chapter 7.

(6.12.2) Hilbert schemes give a quick way to write down the universal family of Mumford divisors.

(6.12.3) The pluricanonical sheaves commute with base change, as in (2.79) and (4.33). This is not crucial, but it helps us avoid some artificial choices. The last one may be an accidental consequence of our choices.

(6.12.4) There is a natural way of writing the boundary as a linear combination of $\mathbb{Z}$-divisors, thus we avoid the notion of marking, to be introduced in Section 8.5.

The key advantage turns out to be (6.12.2), which takes care of Step 6.3.4. So let us discuss it in detail.

6.13 (Universal family of flat Mumford divisors) Let $g\colon X \to S$ be a flat, projective morphism. Consider the relative Hilbert scheme $\mathrm{Hilb}(X/S)$. It parametrizes flat families of closed subschemes of $X \to S$. Thus it has a largest open subscheme that parametrizes subschemes $B_s \subset X_s$ of pure codimension 1, without embedded points, such that X_s is regular at the generic points of B_s. This is the universal family of flat, Mumford divisors on X/S, denoted by

$$\mathrm{MDiv}(X/S) \to S.$$

When we wish to parametrize r such divisors, the universal family is given by the r-fold fiber product

$$\mathrm{MDiv}(X/S) \times_S \cdots \times_S \mathrm{MDiv}(X/S),$$

which we abbreviate as $\times_S^r \mathrm{MDiv}(X/S)$.

We want to apply this to the Hilbert scheme of n-dimensional subschemes of $\mathbb{P}^N$, with its universal family

$$u\colon \operatorname{Univ}_n(\mathbb{P}^N_S) \to \operatorname{Hilb}_n(\mathbb{P}^N_S).$$

Although not strictly necessary, it is convenient to pass to the largest open subscheme $\operatorname{Hilb}^\circ(\mathbb{P}^N_S) \subset \operatorname{Hilb}(\mathbb{P}^N_S)$ over which the fibers of u are demi-normal and of pure dimension n. Thus we have

$$u^\circ\colon \operatorname{Univ}_n^\circ(\mathbb{P}^N_S) \to \operatorname{Hilb}_n^\circ(\mathbb{P}^N_S). \tag{6.13.1}$$

The universal family of flat, Mumford divisors is

$$\operatorname{MDiv}(\operatorname{Univ}_n^\circ(\mathbb{P}^N_S)/\operatorname{Hilb}_n^\circ(\mathbb{P}^N_S)) \to \operatorname{Hilb}_n^\circ(\mathbb{P}^N_S). \tag{6.13.2}$$

If we need r such divisors, the universal family we want is given by the r-fold fiber product

$$\times^r_{\operatorname{Hilb}_n^\circ(\mathbb{P}^N_S)} \operatorname{MDiv}(\operatorname{Univ}_n^\circ(\mathbb{P}^N_S)/\operatorname{Hilb}_n^\circ(\mathbb{P}^N_S)) \to \operatorname{Hilb}_n^\circ(\mathbb{P}^N_S). \tag{6.13.3}$$

As in (6.3.3), we can now use (4.43) to show that the functor of stable pairs is representable by a monomorphism. (9.42) takes care of the condition of being embedded by a given multiple of $K_X + \Delta$.

The schemes in (6.13.3) have infinitely many irreducible components, but once we bound the degrees of the underlying varieties and of the divisors, we get a quasi-projective parameter space.

The following was used in (6.8).

Lemma 6.14 *Fix n, v, a rational vector $\mathbf{a}$ and the characteristic $p \geq 0$. For $S\mathcal{P}(\mathbf{a}, n, v)$, the following are equivalent.*

(6.14.1) *There is an $m = m(\mathbf{a}, n, v)$ such that $m(K_X + \Delta)$ is very ample for every $(X, \Delta) \in S\mathcal{P}(\mathbf{a}, n, v)(k)$ where $\operatorname{char} k = p$.*

(6.14.2) *There are $N = N(\mathbf{a}, n, v)$ and $D = D(\mathbf{a}, n, v)$ such that every $(X, \Delta) \in S\mathcal{P}(\mathbf{a}, n, v)(k)$ is isomorphic to an embedded pair (X, Δ) in $\mathbb{P}^N$ satisfying $\deg X \leq D$ and $\deg \Delta \leq D$.*

(6.14.3) *Then there is a $\mathbb{Q}$-scheme (resp. $\mathbb{F}_p$-scheme) of finite type S and a stable morphism $\pi\colon (X^S, \Delta^S) \to S$ such that every $(X, \Delta) \in S\mathcal{P}(\mathbf{a}, n, v)(k)$ is obtained from π be base change.*

Proof Assume (1). Then $\dim |mK_X + m\Delta| \leq m^n v + n =: N$ by Matsusaka's inequality (11.52.3). Hence all pairs in $S\mathcal{P}(\mathbf{a}, n, v)$ are isomorphic to an embedded pair (X, Δ) in $\mathbb{P}^N$ such that $\deg X = m^n v$. We also know that $\deg(K_X + \Delta) =$

$m^{n-1}v$. A lower bound for deg K_X can be obtained by looking at a general curve section. This gives an upper bound for deg Δ.

(2) $\Rightarrow$ (3) was treated in (6.13).

Finally assume (3) and let $g \in S$ be a generic point. By assumption there is an m_g such that $m_g(K_{X_g} + \Delta_g)$ is very ample. Then the same holds over an open neighborhood $g \in S^\circ \subset S$. We finish by Noetherian induction. □

In the next three sections, we give various stability notions and then check that they all give a good moduli theory as in (6.10).

6.2 Kollár–Shepherd-Barron Stability

This notion of stability is obtained by imposing the strongest possible properties that are satisfied by one-parameter stable families. For surfaces, this was accomplished in Kollár and Shepherd-Barron (1988). There were two reasons why the original paper dealt only with surfaces. First, the existence of stable limits relies on the minimal model program, which was only available for families of surfaces at that time. It was, however, clear that this part should work in all dimensions. Second, the proof of the representability (6.18) relied on detailed properties of lc singularities of surfaces. The theory of hulls and husks, to be discussed in Chapter 9, was developed to prove representability.

We discuss three versions. First, the classical setting of stable varieties without boundary divisors, then a generalization where we allow standard coefficients, and finally arbitrary coefficients in $(\frac{1}{2}, 1]$.

Kollár–Shepherd-Barron Stability without Boundary

6.15 (Stable objects) The stable objects are geometrically reduced, proper k-schemes X with slc singularities such that K_X is ample.

6.16 (Stable families) A family $f : X \to S$ is *KSB-stable* if

(6.16.1) f is flat with slc fibers,

(6.16.2) $\omega_{X/S}^{[m]}$ is a flat family of divisorial sheaves (3.25) for $m \in \mathbb{Z}$,

(6.16.3) f is proper and $\omega_{X/S}^{[M]}$ is an f-ample line bundle for some $M > 0$.

The first two of these conditions define *locally KSB-stable* families.

6.17 (Explanation) This definition restates (3.40). It imposes the strongest restrictions on stable families, thus it gives the smallest scheme structure on the moduli space of stable varieties.

We see in Section 6.3 that assumption (6.16.2) can be weakened, leading to a moduli space with the same underlying reduced space, but with a larger nilpotent structure. The difference between the two versions is explored in Section 6.6.

Theorem 6.18 *KSB-stability, as in (6.15–6.16) is a good moduli theory (6.10)*

Proof As we already noted, only the conditions (6.10.2–4) need checking. For valuative-properness, the stable extension exists by (2.51), and (2.79.2) shows that it satisfies (6.10.3). The existence of embedded moduli spaces is a trivial special case of (6.13). Representability is a restatement of (3.3).

The coarse moduli space exists by (6.6). □

Let us also note another good property of this case.

Proposition 6.19 *For KSB-stable families as in (6.15–6.16), the Hilbert function $\chi(X, \omega_X^{[m]})$ and the plurigenera $h^0(X, \omega_X^{[m]})$ are deformation invariant.*

Proof For the Hilbert function, this follows from the assumption (6.16.2).

If $m \geq 2$ then the higher cohomologies of $\omega_X^{[m]}$ vanish by (11.34). For $m = 1$ we use (2.69). □

Kollár–Shepherd-Barron stability with standard coefficients

Definition 6.20 Let Δ be an effective $\mathbb{R}$-divisor such that coeff $\Delta \subset (\frac{1}{2}, 1]$, that is, $\frac{1}{2} < \text{coeff}_D \Delta \leq 1$ for every irreducible $D \subset \text{Supp}\,\Delta$. There is a unique way of writing $\Delta = \sum_i a_i D_i$ where the D_i are effective $\mathbb{Z}$-divisors, $a_i > \frac{1}{2}$ for every i and $a_i \neq a_j$ for $i \neq j$. We call this the *reduced normal form* of Δ.

6.21 (Stable objects) We parametrize pairs $(X, \Delta = \sum_i a_i D_i)$ in reduced normal form such that

(6.21.1) (X, Δ) is slc,

(6.21.2) $a_i \in \{1 - \frac{1}{3}, 1 - \frac{1}{4}, \ldots, 1\}$ (diminished standard coefficient set),

(6.21.3) X is projective and $K_X + \Delta$ is ample.

6.22 (Stable families) A family $f \colon (X, \Delta = \sum_i a_i D_i) \to S$ is *KSB-stable* if

(6.22.1) $f \colon (X, \Delta) \to S$ is a flat family of pairs as in (6.11),

(6.22.2) the fibers (X_s, Δ_s) satisfy (6.21.1–2),

(6.22.3) the $\omega_{X/S}^{[m]}(\lfloor m\Delta \rfloor - B)$ are flat families of divisorial sheaves (3.25) for every $m \in \mathbb{Z}$ and for every $B = \sum_{j \in J} D_j$ where $a_j = 1$ for $j \in J$, and

(6.22.4) f is proper and $\omega_{X/S}^{[M]}(M\Delta)$ is an f-ample line bundle for some $M > 0$.

The first three of these conditions define *locally KSB-stable* families.

6.23 (Explanation) These conditions are rather straightforward generalizations of (6.16.1–3), but why the restriction on the coefficients?
It follows from (2.79.5) and (4.33) that the $B = 0$ parts of condition (6.22.3) are satisfied if the coefficients of Δ are all $1 - \frac{1}{m}$ and S is reduced. For the $B \neq 0$ cases we use (2.79.8) and (4.33). Note that the conditions on B imply that $B \leq \lfloor \Delta \rfloor$. If S is unibranch, then we could have required (6.22.3) to hold for every $B \leq \lfloor \Delta \rfloor$. However, B has to be a generically Cartier divisor; this is assured if B is a sum of some of the D_i. This is the reason of the somewhat awkward formulation of (6.22.3).
We proved in (2.82) that, if the coefficients are $> \frac{1}{2}$, then the scheme-theoretic specializations of the boundary divisors are reduced and the different $(D_i)_s$ have no common irreducible components. In particular, $\lfloor m\Delta \rfloor_s = \lfloor m\Delta_s \rfloor$ for every $s \in S$. That is, valuative-properness holds. Imposing both of these restrictions gives the coefficient set $\{1 - \frac{1}{3}, 1 - \frac{1}{4}, \ldots, 1\}$.
Pairs satisfying $\frac{1}{2} < a_i \leq 1$ are studied in (6.26–6.27).

Theorem 6.24 *KSB-stability with standard coefficients, as defined in (6.21–6.22) is a good moduli theory (6.10).*

Proof As before, only (6.10.2–4) need checking. We already noted that valuative-properness holds. The existence of embedded moduli spaces follows from (6.13). For representability, the proof of (3.3) – given in (3.42) – carries over with minor changes.
We apply (3.31) with $N_i := \omega_{X/S}^{[i]}(\lfloor i\Delta \rfloor)$ for $1 \leq i < M$ and $L_1 := \omega_{X/S}^{[M]}$. We get $S^{NL} \to S$ such that all the $\omega_{X^{NL}/S^{NL}}^{[m]}(\lfloor m\Delta^{NL} \rfloor)$ are flat families of divisorial sheaves and $\omega_{X^{NL}/S^{NL}}^{[M]}(\lfloor M\Delta^{NL} \rfloor)$ is invertible.
Then (4.45) shows that S^{KSB} is an open subscheme of S^{NL}. □

Proposition 6.25 (Kollár, 2018a, Cor.3) *For KSB-stable families with standard coefficients as in (6.21–6.22), the Hilbert function $\chi(X, \omega_X^{[m]}(\lfloor m\Delta \rfloor))$ and the plurigenera $h^0(X, \omega_X^{[m]}(\lfloor m\Delta \rfloor))$ are deformation invariant.*

Proof For the Hilbert function, this follows from (6.22.3). For the plurigenera, write $mK_X + \lfloor m\Delta \rfloor = K_X + (\lfloor m\Delta \rfloor - (m-1)\Delta) + (m-1)(K_X + \Delta)$. Since the coefficients are standard, $0 \leq \lfloor m\Delta \rfloor - (m-1)\Delta \leq \Delta$, hence (11.34) applies, so the higher cohomologies vanish for $m \geq 2$. For $m = 1$ we use (2.69). □

Kollár–Shepherd-Barron stability with major coefficients

6.26 (Stable objects) We parametrize pairs $(X, \Delta = \sum_i a_i D_i)$ in reduced normal form (6.20) such that

(6.26.1) (X, Δ) is slc,

(6.26.2) $a_i \in (\frac{1}{2}, 1]$,

(6.26.3) X is projective and $K_X + \Delta$ is ample.

6.27 (Stable families) A family $f\colon (X, \Delta = \sum_i a_i D_i) \to S$ is *KSB-stable* if

(6.27.1) $f\colon (X, \Delta) \to S$ is a flat family of pairs as in (6.11),

(6.27.2) the fibers (X_s, Δ_s) satisfy (6.26.1–2),

(6.27.3) the $\omega_{X/S}^{[m]}(\lfloor m\Delta \rfloor - B)$ are flat families of divisorial sheaves (3.25) for every $m \in \mathbb{Z}$ and for every $B = \sum_{j \in J} D_j$ where $a_j = 1$ for $j \in J$, and

(6.27.4) f is proper and $K_{X/S} + \Delta$ is an f-ample $\mathbb{R}$-divisor.

The first three of these conditions define *locally KSB-stable* families.

For technical reasons we introduce a weakening of (3):

(6.27.3') The $\omega_{X/S}^{[m]}(\lfloor m\Delta \rfloor)$ are flat families of divisorial sheaves over S for $m \in M(a_1, \ldots, a_r, n) \subset \mathbb{Z}$; a set of positive density defined in (11.49).

6.28 (Explanation) The restriction that the coefficients be in $(\frac{1}{2}, 1]$ is dictated by (2.82). Example (2.41) shows that flatness of the divisorial part fails with coefficient $= \frac{1}{2}$. The requirement (6.27.3) is dictated by (2.83). The choice of B is discussed in (6.23).

We conjecture that (6.27.3) is always the right assumption. However, (2.83) is known only if the general fiber is normal, so we cannot guarantee that (6.27.3) holds for all families of relative dimension ≥ 3.

Theorem 6.29 *KSB-stability with major coefficients, as defined in (6.26–6.27) is a good moduli theory (6.10), satisfying (6.27.3').*

Furthermore, (6.27.3) is satisfied in relative dimension 2 and on those irreducible components that generically parametrize normal varieties.

Proof The proof closely follows (6.24). We proved in (2.82) that, if the coefficients are $> \frac{1}{2}$, then the scheme-theoretic specializations of the boundary divisors are reduced, so assuming that the D_i are flat divisorial sheaves is correct. Following the proofs in (3.42) and (6.24), we can guarantee the requirements (6.27.1–2) and (6.27.4).

The difficulty is with proving that (6.27.3) holds. Following Kollár (2018a), we outlined a proof in (2.83) when the general fibers are normal. Kollár

(2018b) treats all families of surfaces. Thus (6.29) holds for surfaces and for those irreducible components that generically parametrize normal varieties.

For the version (6.27.3') we use (11.50).

The construction of the moduli space works as before if the a_i are rational. We leave the irrational case to the general theory in Chapter 8; see (8.15). □

Complement 6.29.1 The Hilbert function $\chi(X, \omega_X^{[m]}(\lfloor m\Delta \rfloor))$ is deformation invariant if (6.27.3) holds. Unlike in the earlier cases, the plurigenera need not be deformation invariant; see (Kollár, 2018a, 40–43).

6.3 Strict Viehweg Stability

6.30 (Stable objects) The same as in (6.15): reduced, proper k-schemes X with slc singularities such that K_X is ample.

6.31 (Stable families) A family $f : X \to S$ is V^+-*stable* if the following hold:

(6.31.1) f is flat with slc fibers.

(6.31.2) For every $m \in \mathbb{Z}$ and $x \in X$, $\omega_{X/S}^{[m]}$ is locally free at x iff $\omega_{X_s}^{[m]}$ is locally free at x, where $s = f(x)$.

(6.31.3) f is proper and $\omega_{X/S}^{[M]}$ is an f-ample line bundle for some $M > 0$.

The first two of these conditions define *locally* V^+-*stable* families.

6.32 (Explanation) The original version in Viehweg (1995) assumes (6.31.2) only for some $m > 0$. By (4.37) the latter is equivalent to V^+-stability in characteristic 0, but not in positive characteristic, see Section 8.8.

Already for families of surfaces with quotient singularities this definition gives a large nilpotent structure on the moduli space of stable varieties, even when KSB-stability gives a smooth moduli space, see Section 6.6.

Strict Viehweg Stability with Major Coefficients

6.33 (Stable objects) We parametrize pairs $(X, \Delta = \sum_i a_i D_i)$ in reduced normal form such that

(6.33.1) (X, Δ) is slc,

(6.33.2) $a_i \in (\frac{1}{2}, 1] \cap \mathbb{Q}$ for every i,

(6.33.3) X is projective and $K_X + \Delta$ is ample.

The first two of these conditions define *locally stable* pairs.

6.34 (Stable families) A family $f : (X, \Delta = \sum_i a_i D_i) \to S$ is V^+-*stable* if the following hold:

(6.34.1) $f: X \to S$ is flat and the fibers of $f|_{D_i}: D_i \to S$ are reduced subschemes of pure codimension 1 for every i.

(6.34.2) The fibers (X_s, Δ_s) are stable as in (6.33).

(6.34.3) $\omega_{X/S}^{[m]}(m\Delta)$ is locally free along X_s iff $\omega_{X_s}^{[m]}(m\Delta_s)$ is locally free.

(6.34.4) f is proper and $\omega_{X/S}^{[M]}(M\Delta)$ is an f-ample line bundle for some $M > 0$.

The first three of these conditions define *locally V^+-stable* families.

6.35 (Explanation) These conditions are rather straightforward generalizations of (6.31) and (6.27).

Theorem 6.36 *V^+-stability with major coefficients, as defined in (6.33–6.34) is a good moduli theory (6.10).*

Proof The arguments given in (6.29) work since we no longer require the condition (6.27.3) that gave us trouble there.

Representability is actually simpler, since we work only with the locally free $\omega_{X/S}^{[M]}(M\Delta)$ and ignore the other $\omega_{X/S}^{[m]}(\lfloor m\Delta \rfloor)$.   □

6.4 Alexeev Stability

6.37 (Stable objects) We parametrize pairs $(X, \Delta = \sum_i a_i D_i)$ in reduced normal form (6.20) such that

(6.37.1) (X, Δ) is slc,

(6.37.2) $1, a_1, \ldots, a_r$ are $\mathbb{Q}$-linearly independent,

(6.37.3) X is projective and $K_X + \Delta$ is ample.

6.38 (Stable families) A family $f: (X, \Delta = \sum_i a_i D_i) \to S$ is *A-stable* if the following hold:

(6.38.1) $f: (X, \Delta) \to S$ is a flat family of pairs as in (6.11).

(6.38.2) The fibers (X_s, Δ_s) are stable as in (6.37).

(6.38.3) The $\omega_{X/S}^{[m_0]}(\sum m_i D_i)$ are flat families of divisorial sheaves (3.25) over S for every $m_i \in \mathbb{Z}$.

(6.38.4) f is proper and $K_{X/S} + \Delta$ is an f-ample $\mathbb{R}$-divisor.

The first three of these conditions define *locally A-stable* families.

6.39 (Explanation) The two new features are the $\mathbb{Q}$-linear independence in (6.37) and (6.38.3).

Let us start with $\mathbb{Q}$-linear independence. As a simple example, let X be a smooth, projective variety and $\sum D_i$ an snc divisor with index set $\{i \in I\}$. Then $(X, \sum a_i D_i)$ is an lc pair for every $a_i \in [0, 1]$. So we can ask how the answers to various questions – for example the ampleness of $K_X + \sum a_i D_i$, or the steps of the MMP – depend on the a_i.

In many cases, the answer is that $[0, 1]^I$ admits a rational chamber decomposition such that the answers depend only on the chamber we are in, not the particular choice of the $\{a_i : i \in I\}$ inside the chamber. There is reason to expect that if a point $\{a'_i : i \in I\}$ lies in an open chamber, then $K_X + \sum a'_i D_i$ exhibits generic – hence simplest – behavior.

Since the chambers are polyhedra with rational vertices, a point $\{a'_i : i \in I\}$ whose coordinates are $\mathbb{Q}$-linearly independent, must lie in an open chamber. Thus assumption (6.37.2) is a convenient way to guarantee that we encounter the generic behavior.

By (6.38.4), $K_{X/S} + \sum_i a_i D_i$ is $\mathbb{R}$-Cartier. By (11.43), the $\mathbb{Q}$-linear independence assumption implies that $K_{X/S}$ and the D_i are $\mathbb{Q}$-Cartier. Thus all the $m_0 K_{X/S} + \sum m_i D_i$ are $\mathbb{Q}$-Cartier $\mathbb{Z}$-divisors. Therefore all the sheaves in (6.38.3) should be flat over S with S_2 fibers by (2.79.1). This gives a moduli space with many flat universal sheaves, and, as we see next, it also helps with the proof of existence.

Finally note that, since the D_i are not assumed irreducible, $\lfloor m \sum_i a_i D_i \rfloor$ may not be a linear combination of the D_i, so we do not assume anything about the sheaves $\omega^{[m]}_{X/S}(\lfloor m\Delta \rfloor)$. If the $a_i < \frac{1}{2}$, then these frequently do not have S_2 fibers (2.41–2.44). Although (11.50) shows that infinitely many of them do, it is not clear how to predict which ones.

Theorem 6.40 *A-stability, as in (6.37–6.38) is a good moduli theory (6.10).*

Proof As before, separatedness and valuative-properness holds. The idea of the proof of the existence of embedded moduli spaces is the following. The chamber structure mentioned in (6.39) suggests that, if we pick a rational point $(a'_1, \ldots, a'_r)$ in the interior of the chamber, then the pairs $(X, \sum a_i D_i)$ and $(X, \sum a'_i D_i)$ have the same moduli theory. We can thus work with the rational-coefficient pairs $(X, \sum a'_i D_i)$ as in (6.13). This is basically what we do, but the details are more complicated. See (8.21) for a full treatment.

Representability needs a somewhat different proof. The set of slc fibers is constructible by (4.44), hence there are $M_i > 0$ such that $M_0 K_{X_s}$ and the $M_i D_i|_{X_s}$ are Cartier whenever (X_s, Δ_s) is slc.

We apply (3.31) where the set $\{N\}$ consists of the sheaves $\omega^{[m_0]}_{X/S}(\sum m_i D_i)$ for $0 \le m_i \le M_i$ and the set $\{L\}$ of the sheaves $\omega^{[m_0]}_{X/S}, \mathscr{O}_X(M_1 D_1), \ldots, \mathscr{O}_X(M_r D_r)$.

We get $S^{NL} \to S$ such that the $\omega^{[m_0]}_{X^{NL}/S^{NL}}(\sum m_i D_i^{NL})$ are flat families of divisorial sheaves for all $m_i \in \mathbb{Z}$ and $\omega^{[M_0]}_{X^{NL}/S^{NL}}, \mathscr{O}_{X^{NL}}(M_1 D_1^{NL}), \ldots, \mathscr{O}_{X^{NL}}(M_r D_r^{NL})$ are all invertible. Then (4.45) shows that S^A is an open subscheme of S^{NL}.   □

6.5 First Order Deformations

In this section, we study first order infinitesimal deformations of normal varieties. We describe the deformations of the smooth locus and then try to understand when a deformation of the smooth locus extends to a deformation of the whole variety. The final aim is to get an explicit obstruction theory for lifting sections of powers of the dualizing sheaf. This turns out to be given by the classical notion of divergence.

6.41 (First order thickening) Let k be a field and R a k-algebra. Consider the algebra $R[\varepsilon]$ where ε is a new variable satisfying $\varepsilon^2 = 0$. It is flat over $k[\varepsilon]$ and $R[\varepsilon] \otimes_{k[\varepsilon]} k \simeq R$. We think of $R[\varepsilon]$ as the trivial first order deformation of R.

Let $v \colon R \to R$ be a k-linear derivation. Then

$$\alpha_v \colon r_1 + \varepsilon r_2 \mapsto r_1 + \varepsilon(v(r_1) + r_2) \qquad (6.41.1)$$

defines an automorphism of $R[\varepsilon]$ that is trivial modulo (ε). Conversely, every automorphism of $R[\varepsilon]$ that is trivial modulo (ε) arises this way. (The product (or Leibniz) rule for v is equivalent to the multiplicativity of α_v.)

Let X be a k-scheme. The trivial first order deformation of X is

$$X[\varepsilon] := X \times_k \operatorname{Spec}_k k[\varepsilon]. \qquad (6.41.2)$$

As in (6.41.1), every derivation $v \colon \mathscr{O}_X \to \mathscr{O}_X$ defines an automorphism α_v of $X[\varepsilon]$ that is trivial modulo (ε). This gives an exact sequence

$$0 \to \operatorname{Hom}(\Omega^1_X, \mathscr{O}_X) \to \operatorname{Aut}(X[\varepsilon]) \to \operatorname{Aut}(X) \to 1. \qquad (6.41.3)$$

If X is smooth, or at least normal, then $\mathcal{H}om(\Omega^1_X, \mathscr{O}_X)$ is the tangent sheaf T_X of X, hence we can rewrite the sequence as

$$0 \to H^0(X, T_X) \xrightarrow{\alpha} \operatorname{Aut}(X[\varepsilon]) \to \operatorname{Aut}(X) \to 1. \qquad (6.41.4)$$

Aside On a differentiable manifold M one can identify the Lie algebra of all vector fields with the Lie algebra of the automorphism group. If X is a smooth variety, then this identification works if X is proper, but not otherwise. For instance, an affine curve C of genus ≥ 1 has only finitely many automorphisms, but $H^0(C, T_C)$ is infinite dimensional. Infinitesimal thickenings restore the connection between vector fields and automorphisms.

6.42 (Locally trivial first order deformations) Let k be a field and X a k-scheme. A *deformation* of X over $A := \operatorname{Spec}_k k[\varepsilon]$ is a flat A-scheme X' together with an isomorphism $X' \times_A \operatorname{Spec} k \simeq X$. The set of isomorphism classes of first order deformations is denoted by $T^1(X)$. It is easy to see that $T^1(X)$ is naturally a k-vector space whose zero is the trivial deformation $X[\varepsilon]$, but this is not very important for us now. See Artin (1976) or Hartshorne (2010) for detailed discussions.

We say that X' is *locally trivial* if there is an affine cover $X = \cup_i X_i$ such that each X_i' is a trivial deformation of X_i. We aim to classify all locally trivial first order deformations of arbitrary k-schemes X, but our main interest is in cases when X is smooth and quasi-projective.

Let $X = \cup_i X_i$ be an affine cover. This gives an affine cover $X' = \cup_i X_i'$ and we assume that each X_i' is a trivial deformation of X_i. Fix trivializations $\phi_i \colon X_i' \simeq X_i[\varepsilon]$. Over $X_{ij}' := X_i' \cap X_j'$ we have two trivializations, these differ by an automorphism

$$\alpha_{ij} := \phi_j^{-1} \circ \phi_i \colon X_{ij}' \to X_{ij}', \qquad (6.42.1)$$

which is the identity on X_{ij}. By (6.41.1), the automorphisms α_{ij} correspond to $v_{ij} \in \operatorname{Hom}(\Omega^1_{X_{ij}}, \mathcal{O}_{X_{ij}})$ and these form a 1-cocycle $D := \{v_{ij}\}$. Changing the trivializations changes the cocyle by a coboundary. Thus we get a well defined

$$D = D(X') \in H^1(X, \mathcal{H}om(\Omega^1_X, \mathcal{O}_X)). \qquad (6.42.2)$$

The construction can be reversed. It is left to the reader to check that $D(X')$ is independent of the choices we made. The final outcome is the following.

Claim 6.42.3 Let X be a k-scheme. There is a one-to-one correspondence, denoted by $D \mapsto X_D$, between
  (a) elements of $H^1(X, \mathcal{H}om(\Omega^1_X, \mathcal{O}_X))$, and
  (b) locally trivial deformations of X over $\operatorname{Spec}_k k[\varepsilon]$, up-to isomorphism.
Furthermore, if X is normal then $H^1(X, \mathcal{H}om(\Omega^1_X, \mathcal{O}_X)) = H^1(X, T_X)$. ☐

Next we check that every first order deformation of a smooth variety Y is locally trivial. To see this, we may assume that Y is affine. Then Y' is also affine and we can fix a vector space isomorphism $k[Y'] \simeq k[Y] \otimes k[\varepsilon]$. Pick a point $p \in Y$, local coordinates $y_1, \ldots, y_n$. Then $k(Y)$ is separable over $k(y_1, \ldots, y_n)$. Choose arbitrary lifts $y_1', \ldots, y_n' \in k[Y']$. Any other $z \in k[Y]$ satisfies a monic, separable equation $F(z, \mathbf{y}) = 0$. We claim that z has a unique lift $z' \in k(Y')$ such that $F(z', \mathbf{y}') = 0$. To see this pick any lift z^*. Then $F(z^*, \mathbf{y}') = \varepsilon G(z)$ for some $G(z) \in k[Y]$. We are looking for z' in the form $z' = z^* + \varepsilon g$ where $g \in k[Y]$. Since $F(z^* + \varepsilon g, \mathbf{y}') = \varepsilon G(z) + \varepsilon g \cdot \partial F(z, \mathbf{y})/\partial z$, we see that $g = -G(z)(\partial F(z, \mathbf{y})/\partial z)^{-1}$

is the unique solution. We do this for a finite set of generators $\{z_i\}$ of $k[Y]$ to get a trivialization in a neighborhood where all the $\partial F_i(z, \mathbf{y})/\partial z$ are invertible.

Combining with (6.42.3), this shows that every deformation of a smooth, affine variety over $k[\varepsilon]$ is trivial. (See Hartshorne (1977, exc.II.8.6) for a slightly different proof.)

6.43 (Arbitrary first order deformations) Let k be a field and X a normal k-variety. Let $U \subset X$ be the smooth locus, $Z \subset X$ the singular locus, and $j : U \hookrightarrow X$ the natural injection.

Let $X' \to \mathrm{Spec}_k\, k[\varepsilon]$ be a flat deformation of X. By restriction, it induces a flat deformation U' of U. Note that U' uniquely determines X'. Indeed, $\mathrm{depth}_Z \mathscr{O}_X \geq 2$ since X is normal, hence $\mathrm{depth}_Z \mathscr{O}_{X'} \geq 2$ since $\mathscr{O}_{X'}$ is an extension of two copies of $\mathscr{O}_X$. Therefore $\mathscr{O}_{X'} = j_*\mathscr{O}_{U'}$ by (10.6). Thus we have an injection $T^1(X) \hookrightarrow T^1(U) = H^1(U, T_U)$.

Following Schlessinger (1971), our plan is to study $T^1(X)$ by first describing $T^1(U)$ and then understanding which $D \in H^1(U, T_U)$ correspond to a deformation of X; see also von Essen (1990). The second step is in (6.46).

Definition 6.44 Let X be a k-scheme. Given $v \in \mathrm{Hom}(\Omega^1_X, \mathscr{O}_X)$, *differentiation by v* is defined as the composite

$$v(\ ) : \mathscr{O}_X \xrightarrow{d} \Omega^1_X \xrightarrow{v} \mathscr{O}_X. \tag{6.44.1}$$

Let $x_1, \ldots, x_n$ be (analytic or étale) local coordinates at a smooth point of X and write $v = \sum_i v_i \frac{\partial}{\partial x_i}$. Then the maps are

$$v : f \mapsto \sum_i \frac{\partial f}{\partial x_i} dx_i \mapsto \sum_i v_i \frac{\partial f}{\partial x_i}.$$

Thus if X is smooth and v is identified with a section of T_X, then (6.44.1) agrees with the usual definition.

Next let $D \in H^1(X, \mathcal{H}om(\Omega^1_X, \mathscr{O}_X))$ and choose a representative 1-cocyle $D = \{v_{ij}\}$ using an affine cover $X = \cup X_i$. For any $s \in H^0(X, \mathscr{O}_X)$ the derivatives $\{v_{ij}(s|_{X_{ij}})\}$ form a 1-cocycle with values in $\mathscr{O}_X$. This defines $D(s) \in H^1(X, \mathscr{O}_X)$. We think of it either as a *cohomological differentiation* map

$$D : H^0(X, \mathscr{O}_X) \to H^1(X, \mathscr{O}_X), \tag{6.44.2}$$

or as a k-bilinear map

$$H^1(X, \mathcal{H}om(\Omega^1_X, \mathscr{O}_X)) \times H^0(X, \mathscr{O}_X) \to H^1(X, \mathscr{O}_X). \tag{6.44.3}$$

If X is normal, then we can rewrite this as

$$H^1(X, T_X) \times H^0(X, \mathscr{O}_X) \to H^1(X, \mathscr{O}_X). \tag{6.44.4}$$

Let X_D be the deformation of X corresponding to D. Its structure sheaf sits in an exact sequence

$$0 \to \varepsilon \mathscr{O}_X \to \mathscr{O}_{X_D} \to \mathscr{O}_X \to 0. \tag{6.44.5}$$

Taking cohomology, we see that D in (6.44.2) is the connecting map

$$H^0(X_D, \mathscr{O}_{X_D}) \to H^0(X, \mathscr{O}_X) \xrightarrow{D} H^1(X, \mathscr{O}_X). \tag{6.44.6}$$

Warning 6.44.7 Since the constant $1_X \in H^0(X, \mathscr{O}_X)$ always lifts, $D(1_X) = 0$. Thus D is an $H^0(X, \mathscr{O}_X)$-module homomorphism iff it is identically 0.

We can summarize the above considerations as follows.

Lemma 6.45 *Let X be a k-scheme, $D \in H^1(X, \mathcal{H}om(\Omega^1_X, \mathscr{O}_X))$ and X_D the corresponding deformation of X. Then a global section $s \in H^0(X, \mathscr{O}_X)$ lifts to $s_D \in H^0(X_D, \mathscr{O}_{X_D})$ iff $D(s) \in H^1(X, \mathscr{O}_X)$ is 0.* □

Corollary 6.46 *Let X be a normal, affine variety and $U \subset X$ its smooth locus. Let U_D be the deformation of U corresponding to $D \in H^1(U, T_U)$. Then*
(6.46.1) *U_D extends to a flat deformation X_D of X iff D (as in (6.44.2)) is identically 0.*
(6.46.2) *$T^1(X)$ is the left kernel of $H^1(U, T_U) \times H^0(U, \mathscr{O}_U) \to H^1(U, \mathscr{O}_U)$.*

Proof Assume that U_D extends to a flat deformation X_D of X. Since X is affine, so is X_D and so $H^0(X_D, \mathscr{O}_{X_D}) \to H^0(X, \mathscr{O}_X)$ is surjective. Thus D is identically 0 by (6.45).

Conversely, if D is identically 0, then $H^0(U_D, \mathscr{O}_{U_D}) \to H^0(U, \mathscr{O}_U)$ is surjective and $H^0(U, \mathscr{O}_U) = H^0(X, \mathscr{O}_X)$ since X is normal. We can then take $X_D := \mathrm{Spec}_k H^0(U_D, \mathscr{O}_{U_D})$. This proves the first claim and the second is a reformulation of it. □

Remark 6.47 If X is not affine, then $D \in H^1(U, T_U)$ gives a k-linear map $D \colon \mathscr{O}_X = j_* \mathscr{O}_U \to R^1 j_* \mathscr{O}_U \simeq \mathcal{H}^2_Z(\mathscr{O}_X)$ where $Z := X \setminus U$ is the singular locus. Then U_D extends to a flat deformation X_D of X iff $D \colon \mathscr{O}_X \to \mathcal{H}^2_Z(\mathscr{O}_X)$ is identically 0.

6.48 (Lie derivative) Let M be a smooth, real manifold and v a vector field on M. By integrating v we get a 1-parameter family of diffeomorphisms ϕ_t of M. The *Lie derivative* of a covariant tensor field S is defined as

$$L_v S := \tfrac{d}{dt}(\phi_t^* S)_{t=0}. \tag{6.48.1}$$

In local coordinates $\{y_i\}$, write $v = \sum_i v_i \frac{\partial}{\partial y_i}$. The Lie derivatives of a function s and of a 1-form dy_j are given by the formulas

$$L_v s = v(s) = \sum_i v_i \frac{\partial s}{\partial y_i} \quad \text{and} \quad L_v(dy_j) = dv_j. \qquad (6.48.2)$$

Since functions and 1-forms generate the algebra of covariant tensors, the Lie derivative is uniquely determined by the formulas (6.48.2). One can extend the definition to all tensors by duality.

We can transplant this definition to algebraic geometry as follows.

Let Y be a smooth variety over a field k and $v \in H^0(Y, T_Y)$ a vector field. By (6.41.4) v can be identified with an automorphism α_v of $Y[\varepsilon]$. We write Ω_Y for the module of derivations (frequently denoted by Ω_Y^1). The covariant tensors are sections of the algebra $\sum_{m \geq 0} \Omega_Y^{\otimes m}$.

Let $S \in H^0(Y, \sum_{m \geq 0} \Omega_Y^{\otimes m})$ be a covariant tensor on Y. It has a trivial extension to $Y[\varepsilon]$; denote it by $S[\varepsilon]$. Thus $\alpha_v^*(S[\varepsilon])$ is a global section of $\sum_{m \geq 0} \Omega_{Y[\varepsilon]}^{\otimes m}$. Since α_v is the identity on Y, $\alpha_v^*(S[\varepsilon]) - S[\varepsilon]$ is divisible by ε and we can define the *Lie derivative* of S by the formula

$$\alpha_v^*(S[\varepsilon]) = S[\varepsilon] + \varepsilon L_v S. \qquad (6.48.3)$$

Expanding the identity $\alpha_v^*(S_1[\varepsilon] \otimes S_2[\varepsilon]) = \alpha_v^*(S_1[\varepsilon]) \otimes \alpha_v^*(S_2[\varepsilon])$ shows that the Lie derivative is a k-linear derivation of the tensor algebra

$$L_v : \bigoplus_{m \geq 0} \Omega_Y^{\otimes m} \to \bigoplus_{m \geq 0} \Omega_Y^{\otimes m}. \qquad (6.48.4)$$

The Lie derivative preserves natural quotient bundles of $\Omega_Y^{\otimes m}$. Thus we get similar maps L_v for symmetric and skew-symmetric tensors. Our main interest is in powers of ω_Y. The corresponding map

$$L_v : \omega_Y^m \to \omega_Y^m \qquad (6.48.5)$$

is obtained using the identification $\Omega_Y^{\otimes n} \twoheadrightarrow \Omega_Y^n = \omega_Y$ where $n = \dim Y$. From (6.41.1), we see that

$$\alpha_v^*(s[\varepsilon]) = s[\varepsilon] + \varepsilon v(s) \quad \text{and} \quad \alpha_v^*(dy_j) = d(\alpha_v^*(y_j)) = dy_j + \varepsilon dv_j. \qquad (6.48.6)$$

Comparing with (6.48.2), we see that the algebraic definition coincides with the differential geometry definition.

6.49 (Cartan formula) This is an identity which holds for exterior forms S

$$L_v(S) = d(v \lrcorner S) + v \lrcorner dS, \qquad (6.49.1)$$

where $\lrcorner$ denotes *contraction* or *inner product* by a vector field $v \in H^0(Y, T_Y)$ obtained as follows. We have the contraction map $T_Y \otimes \Omega_Y^m \to \Omega_Y^{m-1}$, thus every $v \in H^0(Y, T_Y)$ gives the $\mathcal{O}_Y$-linear map

$$v \lrcorner : \Omega_Y^m \to \Omega_Y^{m-1}. \tag{6.49.2}$$

In (analytic or étale) local coordinates $y_1, \ldots, y_n$, write $v = \sum_i v_i \frac{\partial}{\partial y_i}$. Then

$$v \lrcorner (dy_1 \wedge \cdots \wedge dy_m) = \sum_r (-1)^{r-1} v_r \cdot dy_1 \wedge \cdots \wedge \widehat{dy_r} \wedge \cdots \wedge dy_m, \tag{6.49.3}$$

where the hat indicates that we omit that term.

To prove (6.49.1), one first checks that $S \mapsto d(v \lrcorner S) + v \lrcorner dS$ is also a derivation. Thus it is sufficient to verify (6.49.1) for a generating set of exterior forms. For functions and for dy_j we recover the identities (6.48.2).

6.50 As in (6.44), let Y be a smooth k-variety. Pick $D \in H^1(Y, T_Y)$ and choose a representative 1-cocycle $D = \{v_{ij}\}$ using an affine cover $Y = \cup Y_i$. For any $S \in H^0(Y, \Omega_Y^{\otimes m})$ the Lie derivatives $\{L_{v_{ij}}(S|_{Y_{ij}})\}$ form a 1-cocycle with values in $\Omega_Y^{\otimes m}$. This defines

$$L_D(S) \in H^1(Y, \Omega_Y^{\otimes m}), \tag{6.50.1}$$

which we view as a *cohomological differentiation* map

$$L_D : \oplus H^0(Y, \Omega_Y^{\otimes m}) \to \oplus H^1(Y, \Omega_Y^{\otimes m}). \tag{6.50.2}$$

As we noted in (6.48), the map L_D respects natural quotient bundles of $\Omega_Y^{\otimes m}$. Thus we get similar maps for symmetric and skew-symmetric tensors and for powers of ω_Y

$$L_D : \oplus H^0(Y, \omega_Y^m) \to \oplus H^1(Y, \omega_Y^m). \tag{6.50.3}$$

For $m = 0$, the map L_D agrees with the map D defined in (6.44.2).

As in (6.44.7), L_D is a k-linear differentiation which is usually not $H^0(Y, \mathscr{O}_Y)$-linear. However, if $D : H^0(Y, \mathscr{O}_Y) \to H^1(Y, \mathscr{O}_Y)$ is 0, then L_D is $H^0(Y, \mathscr{O}_Y)$-linear; this holds both for (6.50.2) and (6.50.3).

Arguing as in (6.45), we obtain the following lifting criterion.

Lemma 6.51 *Let Y be a smooth k-variety and Y_D a first order deformation of Y. Then $S \in H^0(Y, \Omega_Y^{\otimes m})$ lifts to $S_D \in H^0(Y_D, \Omega_{Y_D}^{\otimes m})$ iff $L_D(S) \in H^1(Y, \Omega_Y^{\otimes m})$ is 0. The same holds for all natural quotient bundles of $\Omega_Y^{\otimes m}$.* □

Next we consider what the previous method gives for ω_Y and its powers.

On $M := \mathbb{R}^n$ with coordinates y_i, the divergence of a vector field $v = \sum v_i \frac{\partial}{\partial y_i}$ is $\nabla \cdot v := \sum \frac{\partial v_i}{\partial y_i}$. Note that the y_i give an n-form $\sigma = dy_1 \wedge \cdots \wedge dy_n$, which gives isomorphisms $T_M \simeq \mathcal{H}om(\Omega_M^n, \Omega_M^{n-1}) \simeq \Omega_M^{n-1}$. This identifies the divergence with exterior derivation $d : \Omega_M^{n-1} \to \Omega_M^n$.

6.52 (Divergence) More generally, let Y be a smooth k-variety, $\sigma \in H^0(Y, \omega_Y^m)$ and $v \in H^0(Y, T_Y)$. Then σ and $L_v\sigma$ (6.48.5) are both sections of the line bundle ω_Y^m, hence their quotient is a rational function, called the *divergence* of v with respect to σ,

$$\nabla_\sigma v := \frac{L_v\sigma}{\sigma}. \tag{6.52.1}$$

(Most books seem to use this terminology only when σ is a nowhere 0 section of ω_Y, and σ is frequently suppressed in the notation.)

In order to compute this, start with a section σ of ω_Y. Since $d\sigma = 0$, Cartan's formula (6.49) shows that $L_v : \omega_Y \to \omega_Y$ is the composite map

$$L_v : \omega_Y = \Omega_Y^n \xrightarrow{v \lrcorner} \Omega_Y^{n-1} \xrightarrow{d} \Omega_Y^n = \omega_Y. \tag{6.52.2}$$

In local coordinates $y_1, \dots, y_n$, assume that $\sigma = dy_1 \wedge \cdots \wedge dy_n$ and $v = \sum_i v_i \frac{\partial}{\partial y_i}$. Contraction by v sends σ to

$$\sum_i (-1)^{i-1} v_i \, dy_1 \wedge \cdots \wedge \widehat{dy_i} \wedge \cdots \wedge dy_n. \tag{6.52.3}$$

Exterior differentiation now gives that

$$L_v\sigma = d(v \lrcorner \sigma) = \sum_i \frac{\partial v_i}{\partial y_i} \cdot \sigma, \tag{6.52.4}$$

which is the usual formula for the divergence. Thus, if σ is a nowhere 0 section of ω_Y^m, then we get the divergence as a k-linear map $\nabla_\sigma : T_Y \to \mathscr{O}_Y$. Thus it induces a map on cohomologies; we are especially interested in

$$\nabla_\sigma : H^1(Y, T_Y) \to H^1(Y, \mathscr{O}_Y). \tag{6.52.5}$$

For powers of ω_Y, we get the next formula.

Lemma 6.53 *Let Y be a smooth k-variety of dimension n. Let $v \in H^0(Y, T_Y)$ be a vector field, $s \in H^0(Y, \mathscr{O}_Y)$ a function, and $\sigma \in H^0(Y, \omega_Y)$ an n-form. Then*

$$\nabla_{(s\sigma^m)} v = \frac{v(s)}{s} + m\nabla_\sigma v. \tag{6.53.1}$$

Proof This is really just the assertion that the Lie derivative is a derivation, but it is instructive to do the local computations.

The claimed identities are local, so we use local coordinates $y_1, \dots, y_n$ and assume that $\sigma = dy_1 \wedge \cdots \wedge dy_n$. Write $v = \sum_i v_i \frac{\partial}{\partial y_i}$. We need to compute how the isomorphism α_v acts on $s\sigma^m$. It sends y_i to $y_i + \varepsilon v(y_i) = y_i + \varepsilon v_i$, thus

$$\alpha_v^*(dy_i) = (1 + \varepsilon \frac{\partial v_i}{\partial y_i}) dy_i + \varepsilon (\sum_{j \neq i} \frac{\partial v_i}{\partial y_j} dy_j). \tag{6.53.2}$$

Next we wedge these together. Any two epsilon terms wedge to 0 since $\varepsilon^2 = 0$. Thus $\varepsilon(\sum_{j \neq i} \frac{\partial v_i}{\partial y_j} dy_j)$ gets killed unless it is wedged with all the other dy_j, but the result is then 0 in the exterior algebra. The only term that survives is

$$
\begin{aligned}
\prod_i (1 + \varepsilon \tfrac{\partial v_i}{\partial y_i}) \cdot dy_1 \wedge \cdots \wedge dy_n &= (1 + \varepsilon \sum_i \tfrac{\partial v_i}{\partial y_i}) \cdot dy_1 \wedge \cdots \wedge dy_n \\
&= (1 + \varepsilon \nabla_{\mathbf{y}} v) \cdot dy_1 \wedge \cdots \wedge dy_n.
\end{aligned} \tag{6.53.3}
$$

Thus we get that $s\sigma^m$ is mapped to

$$
\begin{aligned}
(s + \varepsilon v(s))(1 + m\varepsilon \nabla_{\mathbf{y}} v) \cdot \sigma^m &= (s + \varepsilon v(s) + m\varepsilon s \nabla_{\mathbf{y}} v) \cdot \sigma^m \\
&= s\sigma^m + \varepsilon \cdot (\tfrac{v(s)}{s} + m \nabla_{\mathbf{y}} v) \cdot s\sigma^m. \qquad \square
\end{aligned}
$$

Notation 6.54 Let X be a normal, affine k-variety and X_D a flat deformation of X over $k[\varepsilon]$ corresponding to $D \in T^1(X)$. Let $U \subset X$ be the smooth locus. By (6.43), we can think of D as a cohomology class $D \in H^1(U, T_U)$. By (6.44.2), D induces a map

$$
D \colon H^0(U, \mathscr{O}_U) \to H^1(U, \mathscr{O}_U) \tag{6.54.1}
$$

which is identically 0 by (6.46.2). There is a natural exact sequence

$$
0 \to \varepsilon \cdot \omega_U^m \to \omega_{U_D}^m \to \omega_U^m \to 0. \tag{6.54.2}
$$

Taking cohomologies gives an exact sequence

$$
H^0(U_D, \omega_{U_D}^m) \to H^0(U, \omega_U^m) \overset{\delta_m}{\to} H^1(U, \omega_U^m). \tag{6.54.3}
$$

As we noted in (6.50), δ_m is $H^0(U, \mathscr{O}_U)$-linear since D in (6.54.1) is 0.

It was observed in Stevens (1988) that, for cyclic quotients, the deformation obstruction equals the divergence. The next result shows that this is a general phenomenon.

Theorem 6.55 *Let X, $U \subset X$, $D = \{v_{ij}\} \in H^1(U, T_U)$ and X_D be as in (6.54). Assume that ω_U^m has a nowhere 0 section σ_m for some $m > 0$ such that* char $k \nmid m$. *As in (6.52.5), we get $\nabla_{\sigma_m} D := \{\nabla_{\sigma_m}(v_{ij})\} \in H^1(U, \mathscr{O}_U)$. Then*
(6.55.1) $\nabla D := \frac{1}{m} \nabla_{\sigma_m} D \in H^1(U, \mathscr{O}_U)$ *is independent of m and σ_m.*
(6.55.2) *The boundary map $\delta_m \colon H^0(U, \omega_U^m) \to H^1(U, \omega_U^m)$ defined in (6.54.3) is multiplication by $m \nabla D$.*
(6.55.3) $\omega_{U_D}^m$ *is free $\Leftrightarrow \nabla D = 0$ in $H^1(U, \mathscr{O}_U)$.*

Proof Choose affine charts $\{U_i\}$ on U such that $D = \{v_{ij}\}$ and $\sigma_m|_{U_{ij}} = s_{ij} \sigma_{ij}^m$ for some $\sigma_{ij} \in H^0(U_{ij}, \omega_{U_{ij}})$. Any other section of ω_U^m can be written as $g\sigma_m$ where $g \in H^0(U, \mathscr{O}_U)$. Using (6.53), we obtain that

$$\nabla_{\sigma_m} D = \{\nabla_{\sigma_m}(v_{ij})\} = \left\{ \frac{v_{ij}(s_{ij})}{s_{ij}} + m\nabla_{\sigma_{ij}}(v_{ij}) \right\}. \tag{6.55.4}$$

Similarly, we get that

$$\nabla_{g\sigma_m} D = \left\{ \frac{v_{ij}(gs_{ij})}{gs_{ij}} + m\nabla_{\sigma_{ij}}(v_{ij}) \right\}. \tag{6.55.5}$$

Since

$$\frac{v_{ij}(gs_{ij})}{gs_{ij}} = \frac{v_{ij}(g)}{g} + \frac{v_{ij}(s_{ij})}{s_{ij}}, \tag{6.55.6}$$

subtracting (6.55.4) from (6.55.5) yields

$$\nabla_{g\sigma_m} D - \nabla_{\sigma_m} D = \tfrac{1}{g} D(g) \in H^1(U, \mathscr{O}_U). \tag{6.55.7}$$

As we noted in (6.54), $D(g) = 0$ in $H^1(U, \mathscr{O}_U)$. Thus $\nabla_{g\sigma_m} D = \nabla_{\sigma_m} D$ (as classes in $H^1(U, \mathscr{O}_U)$). Independence of the choice of m is shown by the formula

$$\nabla_{(\sigma_m^r)} D = \left\{ \frac{v_{ij}(s_{ij}^r)}{s_{ij}^r} + rm\nabla_{\sigma_{ij}}(v_{ij}) \right\} = r \cdot \left\{ \frac{v_{ij}(s_{ij})}{s_{ij}} + m\nabla_{\sigma_{ij}}(v_{ij}) \right\}. \tag{6.55.8}$$

Thus ∇D is well defined and this proves (1–2).

Finally, $\omega_{U_D}^m$ is free iff σ_m lifts to a section of $\omega_{X_D}^m$, and $\nabla D \cdot \sigma_m$ is the lifting obstruction. This implies (3). □

Remark 6.56 Let $x \in X$ be an isolated normal singularity and $U := X \setminus \{x\}$. Then $H^1(U, \mathscr{O}_U) = H_x^2(X, \mathscr{O}_X)$ and $H^1(U, T_U) = H_x^2(X, T_X)$. Thus if $\omega_U^m \simeq \mathscr{O}_U$ for some $m > 0$ then the divergence in (6.52.5) becomes a map

$$\nabla \colon T^1(X) \to H_x^2(X, \mathscr{O}_X).$$

If $\mathrm{depth}_x \mathscr{O}_X \geq 3$, then $H_x^2(X, \mathscr{O}_X) = 0$ by Grothendieck's vanishing theorem (10.29.5), thus in this case the divergence vanishes and sections of ω_U^m lift to all first order deformations. This, however, already follows from (6.54.3) since $H^1(U, \omega_U^m) = H^1(U, \mathscr{O}_U) = H_x^2(X, \mathscr{O}_X) = 0$.

If X is lc and ω_X is locally free, then sections of ω_X lift to any deformation by Kollár and Kovács (2020); see also (2.67). By (6.55), this implies that $\nabla \colon T^1(X) \to H^1(U, \mathscr{O}_U)$ is the zero map.

This should either have a direct proof or some interesting consequences.

Next we give explicit forms of the maps in the general theory for $X := \mathbb{A}^2$ and $U := \mathbb{A}^2 \setminus \{(0,0)\}$. At first this seems quite foolish to do since we already know that a smooth affine variety has only trivial infinitesimal deformations. However, we will be able to use these computations to get very detailed information about deformations of two-dimensional cyclic quotient singularities.

Notation 6.57 Let k be a field, $X = \mathbb{A}^2_{xy}$ and $U := X \setminus \{(0,0)\}$. Using the affine charts $U_0 := U \setminus (x = 0)$, $U_1 := U \setminus (y = 0)$ and $U_{01} := U \setminus (xy = 0)$, we compute that

$$H^1(U, \mathscr{O}_U) = \left\langle \tfrac{1}{x^i y^j} : i, j \geq 1 \right\rangle \tag{6.57.1}$$

and also that

$$H^1(U, T_U) = \left\langle \tfrac{1}{x^i y^j} \cdot \tfrac{\partial}{\partial x},\ \tfrac{1}{x^i y^j} \cdot \tfrac{\partial}{\partial y} : i, j \geq 1 \right\rangle.$$

Note that $H^1(U, \mathscr{O}_U)$ is naturally a quotient of

$$H^0(U_{01}, \mathscr{O}_{U_{01}}) = k[x^{-i} y^{-j} : i, j \in \mathbb{Z}];$$

but the basis in (6.57.1) depends on the choice of coordinates x, y. Similarly, $H^1(U, T_U)$ is naturally a quotient of $H^0(U_{01}, T_{U_{01}})$.

It is very convenient computationally that the diagonal subgroup $\mathbb{G}_m^2 \subset \mathrm{GL}_2$ acts on these cohomology groups and subsequent constructions are $\mathbb{G}_m^2$-equivariant. In order to keep track of this action, it is better to use the $\mathbb{G}_m^2$-invariant differential operators

$$\partial_x := x \tfrac{\partial}{\partial x} \quad \text{and} \quad \partial_y := y \tfrac{\partial}{\partial y}. \tag{6.57.2}$$

Thus $\partial_x(x^r y^s) = r x^r y^s$, $\partial_y(x^r y^s) = s x^r y^s$ and

$$H^1(U, T_U) = \left\langle \tfrac{\partial_x}{x^i y^j} : i \geq 2, j \geq 1 \right\rangle \bigoplus \left\langle \tfrac{\partial_y}{x^i y^j} : i \geq 1, j \geq 2 \right\rangle. \tag{6.57.3}$$

The $\mathbb{G}_m^2$-eigenspaces in $H^1(U, T_U)$ are usually two-dimensional

$$\left\langle \tfrac{\partial_x}{x^i y^j}, \tfrac{\partial_y}{x^i y^j} \right\rangle \quad \text{for} \quad i, j \geq 2. \tag{6.57.4.a}$$

The one-dimensional eigenspaces are

$$\left\langle \tfrac{\partial_x}{x^i y} \right\rangle \quad \text{and} \quad \left\langle \tfrac{\partial_y}{xy^j} \right\rangle \quad \text{for} \quad i, j \geq 2. \tag{6.57.4.b}$$

The pairing $H^1(U, T_U) \times H^0(U, \mathscr{O}_U) \to H^1(U, \mathscr{O}_U)$ defined in (6.44.3) is especially transparent using the bases (6.57.1–4), since

$$\tfrac{a\partial_x - b\partial_y}{x^i y^j}(x^r y^s) = (ar - bs) \cdot x^{r-i} y^{s-j}, \tag{6.57.5}$$

where $a, b \in k$ and $i, j \geq 1$. This is identically 0 as an element of $H^0(U_{01}, \mathscr{O}_{U_{01}})$ iff $ar - bs = 0$. It is more important to know when this is 0 as an element of $H^1(U, \mathscr{O}_U)$. The latter holds iff

(6.a) either $ar - bs = 0$, or

(6.b) $r \geq i$, or $s \geq j$.

This easily implies that the left kernel of $H^1(U, T_U) \times H^0(U, \mathcal{O}_U) \to H^1(U, \mathcal{O}_U)$ is trivial, hence $T^1(\mathbb{A}^2) = 0$ by (6.46.2); but this we already knew.

Combining (6.51) and (6.53) gives the following.

Lemma 6.58 *Using the notation of (6.57), let $D \in H^1(U, T_U)$ and U_D the corresponding deformation. Then $f(dx \wedge dy)^m$ lifts to a section of $\omega_{U_D}^m$ iff*

$$D(f) + mf\nabla D \in H^1(U, \mathcal{O}_U) \quad \text{vanishes}. \qquad \square$$

We are thus interested in computing the kernels of the operators

$$(D, f) \mapsto D(f) + mf\nabla D.$$

We start by describing the kernel of ∇.

6.59 (Computing the divergence) Set $D := (a\partial_x - b\partial_y)x^{-i}y^{-j}$. By explicit computation,

$$\nabla\left(\frac{a\partial_x - b\partial_y}{x^i y^j}\right) = -\frac{a(i-1) - b(j-1)}{x^i y^j}. \tag{6.59.1}$$

Thus ∇D is identically 0 iff $a(i - 1) - b(j - 1) = 0$. If D is a nonzero element of $H^1(U, T_U)$ then $i, j > 0$ and then ∇D is 0 as an element of $H^1(U, \mathcal{O}_U)$ iff it is identically 0.

If $(i, j) = (1, 1)$, then $\nabla D = 0$, but then D vanishes in $H^1(U, T_U)$. If $\nabla D = 0$ and $i = 1, j > 1$ then $b = 0$ and again D vanishes in $H^1(U, T_U)$. Thus we conclude that

$$\ker[H^1(U, T_U) \xrightarrow{\nabla} H^1(U, \mathcal{O}_U)] = \left\langle \frac{(j-1)\partial_x - (i-1)\partial_y}{x^i y^j} : i, j \geq 2 \right\rangle. \tag{6.59.2}$$

Corollary 6.60 *Let $D \in H^1(U, T_U)$. Then $D(xy), \nabla D \in H^1(U, \mathcal{O}_U)$ are both 0 iff D is contained in the subspace*

$$K_{VW} := \left\langle \frac{\partial_x - \partial_y}{(xy)^i} : i \geq 2 \right\rangle \subset H^1(U, T_U).$$

Proof Corresponding to the two cases in (6.57.6.a–b), the kernel of the map $D \mapsto D(xy) \in H^1(U, \mathcal{O}_U)$ is a direct sum of two subspaces

$$K_1 := \left\langle \frac{\partial_x - \partial_y}{x^i y^j} : i, j \geq 2 \right\rangle \quad \text{and} \quad K_2 := \left\langle \frac{\partial_y}{xy}, \frac{\partial_x}{x^i y} : i, j \geq 2 \right\rangle. \tag{6.60.1}$$

Combining this with (6.59.2) gives the claim. $\qquad \square$

6.6 Deformations of Cyclic Quotient Singularities

We use the methods of the previous section to understand first order deformations of cyclic quotient singularities. It is based on Altmann and Kollár (2019), which uses toric geometry. For cyclic quotients the two approaches are equivalent, but they suggest different generalizations.

Notation 6.61 X is a pure dimensional, S_2 scheme over a field k such that ω_X is locally free outside a closed subset $Z \subset X$ of codimension ≥ 2 and $\omega_X^{[m]}$ is locally free for some $m > 0$. The smallest such $m > 0$ is called the *index* of ω_X. Both of these conditions are satisfied by schemes with slc singularities.

Let $(0, T)$ be a local scheme such that $k(0) \simeq k$ and $p: X_T \to T$ a flat deformation of $X \simeq X_0$. As in (2.5), for every $r \in \mathbb{Z}$ we have maps

$$\mathcal{R}^{[r]}: \omega_{X_T/T}^{[r]}|_{X_0} \to \omega_{X_0}^{[r]}. \tag{6.61.1}$$

These maps are isomorphisms over $X \backslash Z$ and we are interested in understanding those cases when $\mathcal{R}^{[r]}$ is an isomorphisms over X.

By (9.17), if T is Artinian, then $\mathcal{R}^{[r]}$ is an isomorphism $\Leftrightarrow \mathcal{R}^{[r]}$ is surjective $\Leftrightarrow \omega_{X_T/T}^{[r]}$ is flat over T.

Definition 6.62 Let $p: X_T \to T$ be a flat deformation as in (6.61).

(6.62.1) We call $p: X_T \to T$ a *KSB-deformation* if $\mathcal{R}^{[r]}$ is an isomorphism for every r. It is enough to check these for $r = 1, \ldots, \text{index}(\omega_X)$. (These are also called qG-deformations. The letter are short for "quotient of Gorenstein," but this is misleading if $\dim X \geq 3$.) These appear on KSB-stable families (6.16).

(6.62.2) We call $p: X_T \to T$ a *Viehweg-type deformation* (or V-deformation) if $\mathcal{R}^{[r]}$ is an isomorphism for every r divisible by $\text{index}(\omega_X)$. It is enough to check this for $r = \text{index}(\omega_X)$. These appear on V^+-stable families (6.31).

(6.62.3) We call $p: X_T \to T$ a *Wahl-type deformation* (or W-deformation) if $\mathcal{R}^{[r]}$ is an isomorphism for $r = -1$. These deformations were considered in Wahl (1980, 1981) and called ω^*-constant deformations there.

(6.62.4) We call $p: X_T \to T$ a VW-deformation if it is both a V-deformation and a W-deformation.

It is clear that every KSB-deformation is also a VW-deformation. Understanding the precise relationship between these four classes has been a long-standing open problem, especially for quotient singularities of surfaces. For reduced base spaces we have the following, which is a combination of (4.33) and (3.1).

Theorem 6.63 *A flat deformation of an slc variety over a reduced, local scheme of characteristic 0 is a V-deformation iff it is a KSB-deformation.*

This raised the possibility that every V-deformation of an slc singularity is also a KSB-deformation over arbitrary base schemes. It would be enough to check this for Artinian bases. Here we focus on first order deformations and prove that these two classes are quite different from each other.

Definition 6.64 Let X be a scheme satisfying the assumptions of (6.61). Let $T^1(X)$ be the set of isomorphism classes of deformations of X over $k[\varepsilon]$. This is a (possibly infinite dimensional) k-vector space. Let $T^1_{KSB}(X) \subset T^1(X)$ denote the space of first order KSB-deformations, $T^1_V(X)$ the space of first order V-deformations, $T^1_W(X)$ the space of first order W-deformations and $T^1_{VW}(X)$ the space of first order VW-deformations. We have obvious inclusions

$$T^1_{KSB}(X) \subset T^1_{VW}(X) \subset T^1_V(X), T^1_W(X) \subset T^1(X),$$

but the relationship between $T^1_V(X)$ and $T^1_W(X)$ is not clear.

These $T^1_*(X)$ are the tangent spaces to the corresponding miniversal deformation spaces; we denote these by $\mathrm{Def}_{KSB}(X), \mathrm{Def}_V(X)$ and so on. See Artin (1976) or Looijenga (1984) for precise definitions and introductions, or (2.25–2.29) for details on surface quotient singularities.

6.65 (Cyclic quotient singularities) Let $\frac{1}{n}(1, q)$ denote the cyclic group action $g\colon (x, y) \mapsto (\eta x, \eta^q y)$, where η is a primitive nth root of unity. We always assume that char $k \nmid n$ and $(n, q) = 1$; then the action is free outside the origin on $\mathbb{A}^2 = \mathrm{Spec}\, k[x, y]$. The ring of invariants is

$$R_{nq} := k[x, y]^G = k[x^i y^j : i, j \geq 0, \ i + q j \equiv 0 \mod n], \qquad (6.65.1)$$

and the corresponding quotient singularity is

$$S_{n,q} := \mathbb{A}^2 / \tfrac{1}{n}(1, q) = \mathrm{Spec}_k R_{nq}. \qquad (6.65.2)$$

While we work with this affine model, all the results apply to its localization, Henselisation, or completion at the origin.

We can also choose $\eta' = \eta^q$ as our primitive nth root of unity. This shows the isomorphism $S_{n,q} \simeq S_{n,q'}$ if $q q' \equiv 1 \mod n$.

Various ways of studying such singularities go back a long time. The first relevant work might be Jung (1908). See also Brieskorn (1967/1968).

In (6.70), we give an algorithm that yields an explicit, minimal generating set of R_{nq}. The number of generators is the embedding dimension.

For us, the embedding dimension is the most natural invariant, but traditionally the multiplicity is considered the basic one. For cyclic quotients, more generally, for rational surface singularities, these are related by the formula

$$\text{embdim}(S_{n,q}) = \text{mult}(S_{n,q}) + 1. \tag{6.65.3}$$

We completely describe first order KSB-, V- and W-deformations of cyclic quotient singularities. The main conclusion is that KSB-deformations and V-deformations are quite different over Artinian bases; see (6.82).

The A_{n-1}-singularity $\mathbb{A}^2/\frac{1}{n}(1, n - 1)$ has embedding dimension 3, and all of its deformations are KSB. In the other cases, we have the following.

Theorem 6.66 *Let* $S_{n,q} := \mathbb{A}^2/\frac{1}{n}(1, q)$ *be as in (6.65) with* $q \neq n - 1$. *Then*

$$\dim T_V^1(S_{n,q}) - \dim T_{VW}^1(S_{n,q}) = \text{embdim}(S_{n,q}) - 4 \quad or \quad \text{embdim}(S_{n,q}) - 5.$$

In particular, if $\text{embdim}(S_{n,q}) \geq 6$ *then* $S_{n,q}$ *has V-deformations that are not VW-deformations, hence also not KSB-deformations.*

Complement 6.66.1 In (6.85) we list all $S_{n,q}$ for which every V-deformation is a KSB-deformation.

By contrast, KSB-deformations and VW-deformations are quite close to each other, as shown by the next result, proved in (6.84).

Theorem 6.67 *Let* $S_{n,q} := \mathbb{A}^2/\frac{1}{n}(1, q)$ *be as in (6.65).*
(6.67.1) If $(n, q + 1) = 1$, *then* $\text{Def}_{KSB}(S_{n,q}) = \text{Def}_{VW}(S_{n,q}) = \{0\}$.
(6.67.2) If $S_{n,q}$ *admits a KSB-smoothing, then* $\text{Def}_{KSB}(S_{n,q}) = \text{Def}_{VW}(S_{n,q})$.
(6.67.3) In general, $\dim T_{KSB}^1(S_{n,q}) \leq \dim T_{VW}^1(S_{n,q}) \leq \dim T_{KSB}^1(S_{n,q}) + 1$.

Next we discuss what the general theory of the previous section says about deformations of two-dimensional quotient singularities.

6.68 (Deformation of quotients) Let k be a field, X an affine k-scheme that is S_2, $x \in X$ a closed point and $U := X \setminus \{x\}$. Let G be a finite group acting on X such that x is a G-fixed point and the action is free on U. The quotient map $\pi_U \colon U \to U/G$ is finite and étale. This extends to a finite map $\pi_X \colon X \to X/G$ which is ramified at x.

$\mathcal{O}_{U/G}$ is identified with the G-invariant subsheaf $(\pi_*\mathcal{O}_U)^G$ and similarly $\omega_{U/G}$ is identified with $(\pi_*\omega_U)^G$. (For the latter we need that the action is free). Thus

$$H^0(U/G, \mathcal{O}_{U/G}) = H^0(U, \mathcal{O}_U)^G = H^0(X, \mathcal{O}_X)^G, \quad \text{and}$$
$$H^0(U/G, \omega_{U/G}^{[m]}) = H^0(U, \omega_U^{[m]})^G = H^0(X, \omega_X^{[m]})^G. \tag{6.68.1}$$

If char $k \nmid |G|$ then the G-invariant subsheaf is a direct summand, hence by taking cohomologies we similarly see that

$$H^1(U/G, \mathscr{O}_{U/G}) = H^1(U, \mathscr{O}_U)^G \quad \text{and} \quad H^1(U/G, T_{U/G}) = H^1(U, T_U)^G.$$

If $D \in H^1(U, T_U)$ is G-invariant, then U_D descends to a deformation $(U/G)_D$ of U/G; these give all first order deformations. If $H^0(U/G, \mathscr{O}_{U/G})$ is flat over $k[\varepsilon]$, then its spectrum gives a flat deformation of X/G and every flat deformation that is locally trivial on U/G arises this way.

Thus, using (6.46) we get the following fundamental observation.

Theorem 6.69 Schlessinger (1971) *Let k be a field, X a smooth, affine k-variety, $x \in X$ a closed point and $U := X \setminus \{x\}$. Let G be a finite group acting on X such that x is a G-fixed point, the action is free on U and* char $k \nmid |G|$. *Then $T^1(X/G)$ is the left kernel of the pairing*

$$H^1(U, T_U)^G \times H^0(U, \mathscr{O}_U)^G \to H^1(U, \mathscr{O}_U)^G \tag{6.69.1}$$

defined in (6.44). More generally, if X is normal, the left kernel corresponds to those flat deformations of X/G that are locally trivial on U/G. □

Next we compute the terms in (6.69.1) for cyclic quotient singularities.

Notation 6.70 Our aim is to describe the generators of $R_{n,q}$ as in (6.65.1). We assume that char $k \nmid n$ and $(n, q) = 1$.

Most of the following formulas can be found in Riemenschneider (1974); see Stevens (2013) for an introduction and many examples.

The group action preserves the monomials, hence R_{nq} has a generating set consisting of monomials. A nonminimal generating set can be constructed as follows. For any $0 < j < n$ let $0 < \gamma_j < n$ be the unique integer such that $\gamma_j + q\,j \equiv 0 \mod n$. Then

$$x^n, x^{\gamma_1} y, x^{\gamma_2} y^2, \dots, x^{\gamma_{n-1}} y^{n-1}, y^n$$

is a generating set of R_{nq}. We know that $\gamma_1 = n - q$ and $\gamma_{n-1} = q$. This is a minimal generating set of R_{nq} as a $k[x^n, y^n]$-module, but usually not as a k-algebra. Indeed, $x^{\gamma_i} y^i$ divides $x^{\gamma_j} y^j$ if $\gamma_i < \gamma_j$ and $i < j$. In any concrete case one can use this observation to get a minimal set of algebra generators.

We label the monomials of the minimal algebra generators as $M_i = x^{a_i} y^{b_i}$, ordered by increasing y-powers

$$M_0 = x^n, M_1 = x^{n-q} y = x^{a_1} y^{b_1}, M_2 = x^{a_2} y^{b_2}, \dots, M_r = y^n. \tag{6.70.1}$$

At the same time, the a_i form a decreasing sequence. Indeed, if $b_i < b_j$ and $a_i \le a_j$, then M_i divides M_j so the sequence would not be minimal.

From (6.71.2), we obtain that there are relations of the form

$$M_i^{c_i} = M_{i-1}M_{i+1} \quad \text{for} \quad i = 1, \ldots, r-1. \tag{6.70.2}$$

This tells us that the a_i and the c_i are recursively defined by

$$a_0 = n, a_1 = n - q, c_i = \lceil a_{i-1}/a_i \rceil, a_{i+1} = c_i a_i - a_{i-1}. \tag{6.70.3}$$

Similarly, $b_0 = 0, b_1 = 1$ and $b_{i+1} = c_i b_i - b_{i-1}$. These imply that $(a_i, a_{i+1}) = (b_i, b_{i+1}) = 1$ for every i and that the $c_i \geq 2$ are computed by the modified continued fraction expansion

$$\frac{n}{n-q} = c_1 - \cfrac{1}{c_2 - \cfrac{1}{c_3 - \cdots}}. \tag{6.70.4}$$

The following observations about the a_i, b_i, c_i are quite useful. The first two follow from the original construction of the M_i, the third from (6.70.5) and the last one is equivalent to (6.71.3).

$a_{i-1} = \min\{\alpha > 0 : \exists x^\alpha y^\beta \in R_{nq} \text{ such that } \beta < b_i\}$ for $i > 0$.

$b_{i+1} = \min\{\beta > 0 : \exists x^\alpha y^\beta \in R_{nq} \text{ such that } \alpha < a_i\}$ for $i < r$.

$c_i - 1 = \lfloor \frac{a_{i-1}}{a_i} \rfloor = \lfloor \frac{b_{i+1}}{b_i} \rfloor$ for $0 < i < r$.

$a_i b_{i+1} - a_{i+1} b_i = n$ for $0 \leq i < r$.

Note that $r + 1$ is the embedding dimension of S_{nq} and r is its multiplicity. Thus $r = 2$ iff $M_1 = M_{r-1} = xy$ and hence we have the A_{n-1}-singularity $\mathbb{A}^2/\frac{1}{n}(1, -1)$. These are exceptional for many of the subsequent formulas, so we assume from now on that $r \geq 3$.

6.71 (Cones and semigroups) Let $v_0, v_1 \in \mathbb{Z}^2$ be primitive vectors and $C := \mathbb{R}_{\geq 0}v_0 + \mathbb{R}_{\geq 0}v_1 \subset \mathbb{R}^2$ the closed cone spanned by them. Let $\bar{C}(\mathbb{Z})$ be the closed, convex hull of $(\mathbb{Z}^2 \cap C) \setminus \{(0,0)\}$ and $N(C)$ the part of the boundary of $\bar{C}(\mathbb{Z})$ that connects v_0 and v_1. Let $m_0 = v_0, m_1, \ldots, m_{r-1}, m_r = v_1$ be the integral points in $N(C)$ as we move from v_0 to v_1. We leave it to the reader to prove that

(6.71.1) the m_i generate the semigroup $\mathbb{Z}^2 \cap C$,

(6.71.2) there are $c_1, \ldots, c_{r-1} \geq 2$ such that $c_i m_i = m_{i-1} + m_{i+1}$, and

(6.71.3) the triangles with vertices $\{(0,0), m_i, m_{i+1}\}$ all have the same area.

Thus $R(C)$, the semigroup algebra of $\mathbb{Z}^2 \cap C$, is generated by $m_0, \ldots, m_s$.

For $1 \leq q < n$ and $(n, q) = 1$, consider the cone C_{nq} spanned by $v_0 = (1, 0)$ and $v_1 = (q, n)$. Then

$$\mathbb{Z}^2 \cap C_{nq} = \langle \tfrac{i}{n}v_0 + \tfrac{j}{n}v_1 : i, j \geq 0, i + qj \equiv 0 \mod n \rangle.$$

Thus we see that the semigroup algebra $R(C_{nq})$ is isomorphic to the algebra of invariants R_{nq} defined in (6.65). (It is not hard to see that, up to the action of $SL(2, \mathbb{Z})$, every rational cone in $\mathbb{R}^2$ is of the form C_{nq}.)

6.72 (Computing $T^1(S_{nq})$) Continuing with the notation of (6.68–6.70), we see that $D \in H^1(U, T_U)^G$ is in $T^1(S_{nq})$ iff $D(M_i) = 0 \in H^1(U, \mathcal{O}_U)$ for every i.

Since the pairing (6.69.1) is $\mathbb{G}_m^2$-equivariant, it is sufficient to consider one eigenspace at a time. As in (6.57.4.a–b), the eigenspaces in $H^1(U, T_U)^G$ are usually two-dimensional and of the form

$$\left\langle \tfrac{\partial_x}{M}, \tfrac{\partial_y}{M} \right\rangle, \tag{6.72.1}$$

where M is a monomial in the M_i-s involving both x, y. The exceptions are one-dimensional subspaces. For every $s \geq 0$ we have two of them

$$\left\langle \tfrac{\partial_x}{M_0^s M_1} \right\rangle \quad \text{and} \quad \left\langle \tfrac{\partial_y}{M_{r-1} M_r^s} \right\rangle. \tag{6.72.2}$$

Thus we can write $D = (\alpha \partial_x - \beta \partial_y)/M$. Note that

$$D(x^a y^b) = (\alpha a - \beta b)\tfrac{x^a y^b}{M}, \tag{6.72.3}$$

hence if $a < \mathrm{ord}_x M$ and $b < \mathrm{ord}_y M$, then this is 0 in $H^1(U, \mathcal{O}_U)$ iff $\beta/\alpha = a/b$. Thus if M is divisible by at least two different monomials M_i, M_j for $0 < i, j < r$ then $D(M_i) = 0$ and $D(M_j) = 0$ imply that we need to satisfy both of the equations $\beta/\alpha = a_i/b_i$ and $\beta/\alpha = a_j/b_j$, a contradiction. We get a similar contradiction for the eigenspaces (6.72.2) if $s > 0$. We are left with the cases when $M = M_i^s$ for some $0 < i < r$. If $s \geq 2$ then $D(M_i) = 0$ implies that $D = (b_i \partial_x - a_i \partial_y)/M_i^s$. Then $b_i a_j - a_i b_j \neq 0$ for $j \neq i$ hence $D(M_j) = (b_i a_j - a_i b_j)(M_j/M_i^s)$ vanishes in $H^1(U, \mathcal{O}_U)$ iff $sa_i \leq a_j$ or $sb_i \leq b_j$. If $j < i$ then $b_j < b_i$, hence $sa_i \leq a_j$ must hold. Since the a_j form a decreasing sequence, we need $sa_i \leq a_{i-1}$. Similarly, $sb_j \leq b_{j+1}$. By (6.70.5.c), these are equivalent to $s \leq c_i - 1$. We have thus proved the following result of Riemenschneider (1974) and Pinkham (1977).

Proposition 6.73 *Let $M_i = x^{a_i} y^{b_i}$ for $i = 0, \ldots, r$ be the generators of R_{nq} as in (6.70.1). Then $T^1(S_{nq}) \subset H^1(U, T_U)$ has a basis consisting of*

$$\left\{ \tfrac{\partial_x}{M_1}, \tfrac{\partial_y}{M_{r-1}} \right\} \quad and \quad \left\{ \tfrac{\partial_x}{M_i}, \tfrac{\partial_y}{M_i} : 2 \leq i \leq r - 2 \right\}, \tag{6.73.1}$$

plus the possibly empty set

$$\left\{ \tfrac{b_i \partial_x - a_i \partial_y}{M_i^s} : 1 \leq i \leq r - 1, \ 2 \leq s \leq c_i - 1 \right\} \tag{6.73.2}$$

where $c_i = \lceil \tfrac{a_{i-1}}{a_i} \rceil = \lceil \tfrac{b_{i+1}}{b_i} \rceil$ is defined in (6.70.2).

6.74 (Powers of ω) Fix any $m \in \mathbb{Z}$. Then $H^0(U, \omega_U^m)$ has a basis consisting of $M(dx \wedge dy)^m$ where M is any monomial. Thus $H^0(S_{nq}, \omega_{S_{nq}}^{[m]}) = H^0(U/G, \omega_{U/G}^m)$ has a basis consisting of

$$\{x^a y^b (dx \wedge dy)^m : a + qb \equiv -m(1 + q) \mod n\}. \tag{6.74.1}$$

For $D \in T^1(S_{nq})$ let S_D denote the corresponding deformation. By (6.58) $x^a y^b (dx \wedge dy)^m f$ lifts to a section of $\omega_{S_D}^{[m]}$ iff

$$D(x^a y^b) + m x^a y^b \nabla D = 0 \in H^1(U, \mathscr{O}_U). \tag{6.74.2}$$

It is enough to check (6.74.2) for a minimal generating set of $H^0(S_{nq}, \omega_{S_{nq}}^{[m]})$ as an R_{nq}-module. In any given case, this can be worked out by hand, but there are two instances where the answer is simple.
(6.74.3) If $n \mid (q + 1)m$ then $H^0(S_{nq}, \omega_{S_{nq}}^m)$ is cyclic with generator $(dx \wedge dy)^m$.
(6.74.4) If $m = -1$ then $xy(dx \wedge dy)^{-1}$ is G-invariant and every $x^a y^b (dx \wedge dy)^{-1}$ is a multiple of it, save for powers of x or y. Thus $\omega_{S_{nq}}^{-1}$ has 3 generating sections:

$$\frac{xy}{dx \wedge dy}, \quad \frac{x^{q+1}}{dx \wedge dy}, \quad \frac{y^{q'+1}}{dx \wedge dy}.$$

6.75 (V-deformations) If $n \mid (q+1)m$, then $(dx \wedge dy)^m$ is a generator by (6.74.3), thus the condition (6.74.2) is equivalent to $\nabla D = 0$.

Therefore $T_V^1(S_{nq})$ equals the intersection of $T^1(S_{nq})$ with the kernel of ∇. The former was computed in (6.73), the latter in (6.59.2). Thus we see that a basis of $T_V^1(S_{nq})$ is

$$\left\{ \frac{(b_i - 1)\partial_x - (a_i - 1)\partial_y}{M_i} : 2 \leq i \leq r - 2 \right\} \tag{6.75.1.a}$$

and, if M_i is a power of xy for some i, then we have to add

$$\left\{ \frac{\partial_x - \partial_y}{M_i^s} : 2 \leq s \leq c_i - 1 \right\}. \tag{6.75.1.b}$$

6.76 (W-deformations) By (6.74.4), $\omega_{X/G}^{-1}$ has three generating sections. Thus, by (6.74.2), D corresponds to a W-deformation iff $D(xy) - xy\nabla D = 0$, $D(x^{q+1}) - x^{q+1}\nabla D = 0$, and $D(y^{q'+1}) - y^{q'+1}\nabla D = 0$.

The first of these conditions is especially strong. We do not compute it here, rather go directly to the next case where the answer is simpler.

6.77 (VW-deformations) Combining (6.75) and (6.76) we get the description of VW-deformations. These satisfy the conditions
(6.77.1) $D(xy) = 0$, $D(x^{q+1}) = 0$ and $D(y^{q'+1}) = 0$.

We computed the subspace K_{VW} where then first two hold in (6.60). It is spanned by the derivations $(\partial_x - \partial_y)(xy)^{-i}$ for $i \geq 2$. Comparing this with (6.73) we get the following.

Claim 6.77.2 If $T^1_{VW}(S_{nq}) \neq 0$, then R_{nq} has a minimal generator of the form $M_i = (xy)^a$.   □

In order to put this into a cleaner form, assume that $(xy)^s$ is the smallest G-invariant power of xy. Note that $(xy)^n = M_0 M_r$ is G-invariant, but it is not one of the M_i. We have $s(q + 1) \equiv 0 \mod n$, thus if $s < n$ then $b := (n, q + 1) > 1$. We have thus shown the following.

Claim 6.77.3 Assume that $(n, q + 1) = 1$. Then $T^1_{KSB}(S_{nq}) = T^1_{VW}(S_{nq}) = 0$ and $\dim T^1_V(S_{nq}) = r - 3$.   □

Claim 6.77.4 Assume that $M_i = (xy)^a$ for some i (so $a_i = b_i = a$). Then the space of VW-deformations is spanned by

$$\left\{ \frac{\partial_x - \partial_y}{M_i^s} : 1 \leq s \leq \min\left\{c_i - 1, \tfrac{q+1}{a}, \tfrac{q'+1}{a}\right\}\right\}.$$

Proof The first restriction on s we get from (6.73.2). Then $D(x^{q+1}) = 0$ is equivalent to $sa \leq q + 1$ and $D(y^{q'+1}) = 0$ is equivalent to $sa \leq q' + 1$. These give the last 2 restrictions.   □

We thus need to compare the two upper bounds occurring in (6.75.1.b) and (6.77.4). The key is the following general estimate.

Lemma 6.78 *Using the notation of (6.70) we have*

$$\frac{n}{a_ib_i} \leq \frac{a_{i-1}}{a_i}, \frac{b_{i+1}}{b_i} < \frac{n}{a_ib_i} + 1.$$

Proof Note that $n = a_ib_{i+1} - a_{i+1}b_i$ by (6.70.5.d). Dividing by a_ib_i we get that

$$\frac{n}{a_ib_i} = \frac{b_{i+1}}{b_i} - \frac{a_{i+1}}{a_i}.$$

Since the a_i form a decreasing sequence, $\frac{a_{i+1}}{a_i} < 1$.   □

The final estimate connecting (6.75.1.b) and (6.77.4) is easier to state using a different system of indexing the singularities.

Notation 6.79 Set $b = (n, q + 1)$ and write $n = ab$, $q + 1 = bc$ where $(a, c) = 1$. The inverse (modulo ab) of $bc - 1$ is written as $bc' - 1$. We thus have the singularity

$$S_{abc} := S_{nq} = \mathbb{A}^2/\tfrac{1}{ab}(1, bc - 1) \simeq \mathbb{A}^2/\tfrac{1}{ab}(1, bc' - 1). \qquad (6.79.1)$$

Note that $(xy)^a$ is the smallest G-invariant power of xy, but it need not be among the generators M_i; see (6.81).

Corollary 6.80 *Assume in addition that $M_i = (xy)^a$ for some i. Then*

$$\lfloor \tfrac{b}{a} \rfloor \le \min\{c_i - 1, \tfrac{q+1}{a}, \tfrac{q'+1}{a}\} \le c_i - 1 \le \lfloor \tfrac{b}{a} \rfloor + 1. \tag{6.80.1}$$

Proof First, we claim that

$$\tfrac{b}{a} \le \min\{\tfrac{a_{i-1}}{a_i}, \tfrac{b_{i+1}}{b_i}, \tfrac{q+1}{a}, \tfrac{q'+1}{a}\} \le \min\{\tfrac{a_{i-1}}{a_i}, \tfrac{b_{i+1}}{b_i}\} < \tfrac{b}{a} + 1. \tag{6.80.2}$$

To see this, note that $q = bc - 1, q' = bc' - 1$. Thus $b \le q + 1, q' + 1$, so it is enough to show that

$$\tfrac{b}{a} \le \min\{\tfrac{a_{i-1}}{a_i}, \tfrac{b_{i+1}}{b_i}\} < \tfrac{b}{a} + 1.$$

Since $n = ab$ and $a = a_i = b_i$, the latter is equivalent to (6.78). Taking the round-down gives (1) using (6.70.5.c). □

Example 6.81 Assume that $x^\alpha y^\beta$ is G-invariant. From $\alpha + \beta(bc - 1) \equiv 0$ mod ab, we see that $\alpha \equiv \beta$ mod b. Thus if $0 < \alpha, \beta \le 2b$ then either $\alpha = \beta$ or $\alpha = \beta \pm b$.

It turns out that if $a \le b$ then we can write down these invariants explicitly. Corresponding to the first case we have $(xy)^a$ (and its square). In order to get the other cases, let $0 < e < a$ (resp. $0 < e' < a$) be the unique solution of $ec \equiv -1$ mod a (resp. $e'c' \equiv -1$ mod a). Then $(b + e) + e(bc - 1) = b(ec + 1) \equiv 0$ mod ab and $e'(bc' - 1) + (b + e') = b(e'c' + 1) \equiv 0$ mod ab. Thus we get the minimal generators

$$M_{i-1} = x^{b+e} y^e, \quad M_i = x^a y^a, \quad M_{i+1} = x^{e'} y^{b+e'}.$$

This gives that $c_i - 1 = \lfloor \tfrac{b+e}{a} \rfloor = \lfloor \tfrac{b+e'}{a} \rfloor$.

Fixing a, b we can choose any $0 < e < a$ such that $(a, e) = 1$ and then solve for c. Thus we see that if $b \equiv 0$ mod a then $\lfloor \tfrac{b}{a} \rfloor = c_i - 1$ for every e and if $b \equiv -1$ mod a then $\lfloor \tfrac{b}{a} \rfloor = c_i - 2$ for every e, but otherwise both are possible for suitable choice of e.

We see in (6.83) that the condition $a \le b$ holds iff S_{abc} has a nontrivial KSB-deformation, so this is a natural class to consider.

6.82 (Proof of 6.66) Comparing (6.75) and (6.77) we see that the derivations listed in (6.75.1) give V-deformations, but not W-deformations. The only possible exception occurs if $M_i = (xy)^a$ for some i. Thus we have two cases.

If $M_i = (xy)^a$ does not occur, then $\dim T^1_V(S_{nq}) = \dim T^1_{VW}(S_{nq}) + r - 3$.

If $M_i = (xy)^a$ for some i, then (6.75.1) gives $r - 4$ basis vectors that give V-deformations, but not W-deformations. By (6.80), there is at most one derivation as in (6.75.2) that gives a V-deformation that is not a W-deformation. $\qquad\square$

6.83 (KSB-deformations) From (6.58) and (6.70.2) we see that D corresponds to a KSB-deformation iff $D(x^i y^j) + m x^i y^j \nabla D = 0$ whenever $i + j(bc-1) \equiv -mbc$ mod ab.

First, we use this for $(dx \wedge dy)^{ab}$ to conclude that $\nabla D = 0$. Second, we note that since $(a, c) = 1$, the congruence $i + j(bc - 1) \equiv -mbc$ mod ab holds for some m iff $i \equiv j$ mod b. The ring of such monomials is generated by x^b, xy, y^b. Thus D gives a first order KSB-deformation iff

(6.83.1) $\quad \nabla D = 0, D(xy) = 0, D(x^b) = 0$ and $D(y^b) = 0$.

We thus get that $T^1_{KSB}(S_{abc})$ is spanned by the derivations

$$\left\{ \frac{\partial_x - \partial_y}{(xy)^{as}} : \ 1 \le s \le \lfloor b/a \rfloor \right\}. \tag{6.83.2}$$

The corresponding deformations were written down in Wahl (1980, 2.7):

$$(uv - w^b - t_1 w^{b-a} - \cdots - t_r w^{b-ra} = 0)/\tfrac{1}{a}(1, bc - 1, c). \tag{6.83.3}$$

To make this $\mathbb{G}_m^2$-equivariant, the $\mathbb{G}_m^2$-action on t_i should be the same as on $(xy)^{ai}$. Thus (6.83.5) describes a smooth subscheme T of $\mathrm{Def}_{KSB}(S_{abc})$ and $\dim T = \lfloor b/a \rfloor$. By (6.83.2), the tangent space of $\mathrm{Def}_{KSB}(S_{abc})$ has dimension $\lfloor b/a \rfloor$, so $T = \mathrm{Def}_{KSB}(S_{abc})$ and $\mathrm{Def}_{KSB}(S_{abc})$ is smooth.

In particular, there is a nontrivial 1-parameter KSB-deformation iff $a \le b$ and there is a KSB-smoothing iff $a|b$. Note that $a \le b$ is equivalent to $ab \le b^2$ and we have proved the following.

Claim 6.83.6 The singularity S_{nq} has

(a) a KSB-smoothing iff $n|(q + 1)^2$, and

(b) a nontrivial KSB-deformation iff $n \le (n, q + 1)^2$. Furthermore,

(c) $\dim T^1_{KSB}(S_{nq}) = \lfloor b/a \rfloor = \lfloor (n, q + 1)^2/n \rfloor$. $\qquad\square$

If $a|b$ then write $b = ad$. We get the singularities

$$W_{adc} := \tfrac{1}{a^2 d}(1, adc - 1) \simeq (uv - w^{ad} = 0)/\tfrac{1}{a}(1, -1, c). \tag{6.83.7}$$

In this case, $b/a = c_i - 1$ hence the arguments give the following.

Claim 6.83.8 For the singularities $W_{adc} = \mathbb{A}^2/\tfrac{1}{a^2 d}(1, adc - 1)$ every VW-deformation is a KSB-deformation. $\qquad\square$

6.84 (Proof of 6.67) Note that (6.67.1) follows from (6.77.3) and (6.67.2) from (6.83.8) for first order deformations. Since $\mathrm{Def}_{KSB}(S_{n,q})$ is smooth by (2.29) or by the explicit description (6.83.5), equality of the tangent spaces $T^1_{KSB}(S_{n,q}) = T^1_{VW}(S_{n,q})$ implies that $\mathrm{Def}_{KSB}(S_{n,q}) = \mathrm{Def}_{VW}(S_{n,q})$.

In order to prove (6.67.3) we consider two cases. If R_{nq} does not have a minimal generator of the form $M_i = (xy)^a$, then $T^1_{VW}(S_{nq}) = T^1_{KSB}(S_{nq}) = \{0\}$ by (6.77.4).

Otherwise, we have proved in (6.83) that

$$\dim T^1_{KSB}(\mathbb{A}^2/\tfrac{1}{ab}(1, bc-1)) = \lfloor \tfrac{b}{a} \rfloor$$

and (6.80) shows that

$$\dim T^1_{VW}(\mathbb{A}^2/\tfrac{1}{ab}(1, bc-1)) = \min\{c_i - 1, \tfrac{q+1}{a}, \tfrac{q'+1}{a}\} \le \lfloor \tfrac{b}{a} \rfloor + 1. \quad \square$$

Examples 6.85 We work out (6.66.1), that is, list those cyclic quotients singularities for which every V-deformation is a KSB-deformation.

6.85.1 (*Double points*) These are the A_n singularities; every deformation is a KSB-deformation.

6.85.2 (*Triple points*) For cyclic quotient triple points the minimal generators of its coordinate ring are $x^n, x^{n-q}y, xy^{n-q'}, y^n$. Thus $\frac{n}{n-q}$ has a two-step continued fraction expansion involving c_1, c_2. Setting $c_1 = e, c_2 = d$ we have the singularities $\mathbb{A}^2/\frac{1}{ed-1}(1, ed-d-1)$, with invariants $x^{ed-1}, x^dy, xy^e, y^{ed-1}$. By (6.75) we have $T^1_V = T^1_{KSB} = 0$.

6.85.3 (*Quadruple points*) By (6.66) and (6.82), every cyclic quotient singularity of multiplicity 4 has a V-deformation that is not a KSB-deformation, unless M_2 (6.70.1) is a power of xy. Thus in this case the minimal generators of its coordinate ring are $x^n, x^{n-q}y, x^ay^a, xy^{n-q'}, y^n$.

The equation $M_2^{c_2} = M_1M_3$ now implies that $q = q'$. Thus $\frac{n}{n-q}$ has a three-step continued fraction expansion involving $c_1, c_2, c_3 = c_1$. By expanding it we see that $c_1 = a$. Setting $c_2 = d$, the singularity is $\mathbb{A}^2/\frac{1}{a(ad-2)}(1, (ad-2)(a-1)-1)$, and the ring of invariants is $k[x^{a(ad-2)}, x^{ad-1}y, x^ay^a, xy^{ad-1}, y^{a(ad-2)}]$.

Thus $\lfloor (ad-2)/a \rfloor = d-1 = c_2 - 1$ and hence, by (6.75) and (6.83), $T^1_V = T^1_{KSB}$ is spanned by $\left\{\frac{\partial_x - \partial_y}{(xy)^{as}} : 1 \le s \le d-1\right\}$. These singularities admit a KSB-smoothing iff $a = 2$. Then, after replacing $d-1$ by d, the normal form becomes $\mathbb{A}^2/\frac{1}{4d}(1, 2d-1)$. Together with the A_n-series, these are the only cyclic quotient singularities with a KSB-smoothing for which every V-deformation is a KSB-deformation.

6.85.4 (*Higher multiplicity points*) By (6.66), every cyclic quotient singularity of multiplicity ≥ 5 has V-deformations that are not KSB-deformations.

7

Cayley Flatness

There are two traditional notions of what a "family of varieties" is: the older Cayley–Chow variant (3.5) and the currently ubiquitous Hilbert–Grothendieck variant (3.6), which puts flatness at the center.

For stable varieties, the Hilbert–Grothendieck approach gives the correct moduli theory. That is, a stable morphism $X \to S$ is a flat morphism with additional properties, as in Section 6.2.

A major problem in the moduli theory of stable pairs is that, while the underlying varieties X form flat families, the divisorial parts Δ do not. Neither of the two main traditional methods of parametrizing varieties or schemes gives the right answer for the divisorial part.

- Cayley–Chow theory works only over reduced base schemes.
- Hilbert–Grothendieck theory works only when the coefficients of Δ satisfy various restrictions, as in Sections 6.2 and 6.4.

In this chapter we develop a theory – called K-flatness – that interpolates between these two, managing to keep from both of them the properties that we need. The objects that we parametrize are divisors – so the strong geometric flavor of Cayley–Chow theory is preserved – but one can work over Artinian base schemes. The latter is one of the key advantages of the theory of Hilbert schemes. Quite unexpectedly, the new theory behaves better than either of the classical approaches in several aspects; see especially (7.4–7.5).

One might say that the main new result is Definition 7.1; we discuss its origin and relationship to the classical theory of Chow varieties in (7.2). The rest of this chapter is then devoted to proving that it has all the hoped-for properties. (Actually, we end up with several variants, but we conjecture them to be equivalent; see Section 7.4.)

The definition of K-flatness and its main properties are discussed in Section 7.1, while Section 7.2 reviews divisor theory over Artinian schemes. The key notion of divisorial support is introduced and studied in Section 7.3.

Several versions of K-flatness are investigated in Section 7.4. For our treatment, technically the most important is C-flatness, which is treated in detail in Section 7.5. The main results are proved in Section 7.6. Sections 7.7–7.9 are devoted to examples. First, we show that a K-flat deformation of a normal variety is flat. Then we describe first order K-flat deformations of plane curves in Section 7.8 and of seminormal curves in Section 7.9. While the computations are somewhat lengthy, the answers are quite nice.

Assumptions In this Chapter we work over an arbitrary field k.

7.1 K-flatness

We eventually introduce several closely related (possibly equivalent) notions in (7.37). The most natural one is C-flatness, which is closest to the ideas of Cayley. Aiming to create a notion that is independent of projective embeddings led to K-flatness. Conveniently, K is also the first syllable of Cayley.

Definition 7.1 (K-flatness) Let $f: X \to S$ be a projective morphism of pure relative dimension n. A relative Mumford divisor $D \subset X$ as in (4.68) is *K-flat* over S iff one of the following–increasingly more general–conditions hold.

(7.1.1) (S local with infinite residue field) For every finite morphism $\pi: X \to \mathbb{P}^n_S$, $\pi_* D \subset \mathbb{P}^n_S$ is a relative Cartier divisor.

(7.1.2) (S local) For some (equivalently every) flat, local morphism $q: S' \to S$, where S' has infinite residue field, the pull-back $q^* D$ is K-flat over S'.

(7.1.3) (S arbitrary) D is K-flat over every localization of S.

Let us start with some comments on the definition.

(7.1.4) The definition of $\pi_* D$ is not always obvious; in essence Section 7.3 is mainly devoted to establishing it. However, $\pi_* D$ equals the scheme-theoretic image of D if red $D \to \text{red}(\pi(D))$ is birational and π is étale at every generic point of the closed fiber D_s (7.28.2). It is sufficient to check condition (7.1.1) for such morphisms $\pi: X \to \mathbb{P}^n_S$.

(7.1.5) If S is not local, then there may not be any finite morphisms $\pi: X \to \mathbb{P}^n_S$; see (7.7.2) for an example. This is one reason for the three-step definition.

(7.1.6) The residue field extension in (7.1.2) is necessary in some cases; see for example (7.80.9).

(7.1.7) The definition of K-flatness is global in nature, but we show that it is in fact local on X (7.52).

(7.1.8) We eventually define K-flatness also for families of coherent sheaves in (7.37). This turns out to be quite convenient technically. However, while the images $\pi_* D$ carry a lot of information about a Mumford divisor D, much of the sheaf information is lost. Thus it is unlikely that K-flatness can be useful for studying the moduli of sheaves.

7.2 (Why this definition?) The idea in the papers Cayley (1860, 1862) is to associate to a subvariety $Y^{n-1} \subset \mathbb{P}^N_k$ a hypersurface

$$\mathrm{Ch}(Y) := \{L \in \mathrm{Gr}(N-n, \mathbb{P}^N_k): Y \cap L \neq \emptyset\} \subset \mathrm{Gr}(N-n, \mathbb{P}^N_k),$$

we call it the Cayley–Chow hypersurface. In modern terminology, the end result is that, over weakly normal bases, there is a one-to-one correspondence

$$\left\{ \begin{array}{c} \text{well-defined families} \\ \text{of subvarieties} \end{array} \right\} \leftrightarrow \left\{ \begin{array}{c} \text{flat families of} \\ \text{Cayley–Chow hypersurfaces} \end{array} \right\}; \quad (7.2.1)$$

see Section 4.8 or Kollár (1996, sec.I.3) for details.

The correspondence (7.2.1) works well for geometrically reduced, pure dimensional subschemes, but for an arbitrary subscheme $Z \subset \mathbb{P}^N$, its Cayley–Chow hypersurface $\mathrm{Ch}(Z)$ detects only red Z and the multiplicities of Z at the maximal dimensional generic points. This is where the role of X and the Mumford condition become crucial: a Mumford divisor $D \subset X$ is uniquely determined by red D and the multiplicities.

We know how to define flatness in general, so we try to make the equivalence into a definition over an arbitrary base scheme. So let $f: X \to S$ be a flat, projective morphism, say with reduced fibers of pure dimension n. Fix an embedding $X \hookrightarrow \mathbb{P}^N_S$ and let $D \subset X$ be a Mumford divisor. We say that D is C-flat over S iff $\mathrm{Ch}(D/S)$ is flat over S. (This needs a suitable extension of the definition of $\mathrm{Ch}(D/S)$ to allow for multiple fibers; see (7.37) for details.)

There are two immediate disadvantages of C-flatness. Cayley–Chow hypersurfaces are unwieldy objects and the resulting notion is very much tied to the choice of an embedding $X^n \hookrightarrow \mathbb{P}^N_k$.

One can think of a Cayley–Chow hypersurface $\mathrm{Ch}(D/S)$ as encoding the images $\pi(D)$ for all linear projections $\pi: \mathbb{P}^N_S \dashrightarrow \mathbb{P}^n_S$. (This also goes back to Cayley; it is worked out in Catanese (1992), Dalbec and Sturmfels (1995), and Kollár (1999).) One can show that the Cayley–Chow hypersurface $\mathrm{Ch}(D/S)$ is flat over S iff $\pi(D) \subset \mathbb{P}^n_S$ is flat over S, for all *linear* projections $\pi: \mathbb{P}^N_S \dashrightarrow \mathbb{P}^n_S$ that are finite on Supp D; see (7.47). (In fact, by (7.47), it is enough to check this for a dense set of projections. We need S to be local with infinite residue field to ensure that there are enough projections.)

This suggests three different generalizations of C-flatness. We can work with

- projective morphisms $f: X \to S$ and all finite $\pi: X \to \mathbb{P}^n_S$,
- affine morphisms $f: U \to S$ and all finite $\pi: U \to \mathbb{A}^n_S$, or

- morphisms of complete, local schemes $f\colon \widehat{X} \to \widehat{S}$ and all finite $\pi\colon \widehat{X} \to \widehat{\mathbb{A}}^n_S$. The affine version has the problem that, even if S is local, there might not be any finite morphisms $\pi\colon U \to \mathbb{A}^n_S$; see (7.38.4) for more on this. Working with complete, local schemes would be the best theoretically, but several of the technical problems remain unresolved. This leaves us with projective morphisms, which is our definition of K-flatness.

The key technical result (7.40) shows that K-flatness is equivalent to C-flatness for every Veronese embedding $X \hookrightarrow \mathbb{P}^N_S \overset{v_m}{\hookrightarrow} \mathbb{P}^M_S$ (where $M = \binom{N+m}{m} - 1$); we call the resulting notion stable C-flatness.

We conjecture that stable C-flatness, K-flatness, local K-flatness, and formal K-flatness are equivalent, giving a very robust concept. This would show that our notion is truly about the singularities in families of divisors. The equivalence of C-flatness and K-flatness would be very helpful computationally, but does not seem to be theoretically significant.

Good Properties of K-flatness

K-flat families have several good properties. Some of them are needed for the moduli theory of stable pairs, but others, for example (7.5), come as a bonus.

The functoriality of K-flatness is not obvious. Indeed, let $T \subset S$ be a closed subscheme. Then a finite morphism $\pi_T\colon X_T \to \mathbb{P}^n_T$ need not extend to a finite morphism $\pi_S\colon X_S \to \mathbb{P}^n_S$. Thus flatness of all $\pi_S(X_S)$ does not directly imply that $\pi_T(X_T)$ is also flat.

Nonetheless, we prove in (7.40) and (7.50) that being K-flat is preserved by arbitrary base changes and it descends from faithfully flat base changes. Thus we get the functor $\mathcal{KDiv}(X/S)$ of K-flat, relative Mumford divisors on X/S. If we have a fixed relatively ample divisor H on X, then $\mathcal{KDiv}_d(X/S)$ denotes the functor of K-flat, relative Mumford divisors of degree d.

We have a disjoint union decomposition $\mathcal{KDiv}(X/S) = \cup_d \mathcal{KDiv}_d(X/S)$. The main result is the following, to be proved in (7.66).

Theorem 7.3 *Let $f\colon X \to S$ be a projective morphism of pure relative dimension n. Then the functor $\mathcal{KDiv}_d(X/S)$ of K-flat, relative Mumford divisors of degree d is representable by a separated S-scheme of finite type $\mathrm{KDiv}_d(X/S)$.*

Complement 7.3.1 If f is flat with normal fibers, then $\mathrm{KDiv}_d(X/S)$ is proper over S, but otherwise usually not. This is not a problem for us.

7.4 (Properties of K-flatness) We list a series of good properties of K-flatness. Let $f\colon X \to S$ be a projective morphism of pure relative dimension n and D or D_i relative Mumford divisors.

7.4.1 (Comparison with flatness) K-flatness is a generalization of flatness and it is equivalent to it for smooth morphisms and for normal divisors.

- If $f|_D : D \to S$ is flat, then D is K-flat; see (7.54).
- If $f : X \to S$ is smooth, then D is K-flat $\Leftrightarrow D$ is flat over $S \Leftrightarrow D$ is a relative Cartier divisor; see (7.53).
- Assume that D is K-flat, $D_s \subset X_s$ has multiplicity 1, and red(D_s) is normal for some $s \in S$. Then $f|_D : D \to S$ is flat along D_s by (7.67).

These properties also hold locally on X. Hence, the notion of K-flatness gives something new only at the points where f is not smooth and $f|_D$ is not flat.

7.4.2 (Reduced base schemes) If S is reduced then every relative Mumford divisor is K-flat; see (7.29). In retrospect, this is the reason why the moduli theory of pairs could be developed over reduced base schemes without the notion of K-flatness in Chapter 4.

7.4.3 (Artinian base schemes) A divisor $D \subset X$ is K-flat over S iff $D_A \subset X_A$ is K-flat over A for every Artinian subscheme $A \subset S$; see (7.44). Thus one can fully understand K-flatness by studying it over reduced bases (as in Chapter 4) and over Artinian base schemes.

7.4.4 (Push-forward) Let $g : Y \to S$ be another projective morphisms of pure relative dimension n, and $\tau : X \to Y$ a finite morphism. Assume that $D \subset X$ is K-flat and $\tau_* D$ is also a relative Mumford divisor. (That is, g is smooth at generic points of $\tau(D_s)$ for every s.) Then $\tau_* D$ is also K-flat, see (7.45). (See Section 7.3 for the definition of $\tau_* D$.) A similar property fails for flatness; combine (7.7.3) and (7.45).

7.4.5 (Additivity) If D_1, D_2 are K-flat, then so is $D_1 + D_2$, see (7.45). This again fails for flatness; see (7.7.3).

7.4.6 (Multiplicativity) Let $m > 0$ be relatively prime to the residue characteristics. Then D is K-flat iff mD is K-flat; see (7.45).

   By contrast, if A is Artinian, nonreduced, with residue field k of characteristic $p > 0$, then the divisors D on $\mathbb{P}_A^2$ such that pD is K-flat (= relative Cartier), but D is not K-flat, span an infinite dimensional k-vectorspace; see (7.10.4–5). This is an extra difficulty in positive characteristic, see Section 8.8.

7.4.7 (Linear equivalence) K-flatness is preserved by linear equivalence; see (7.33). (Note that flatness is not preserved by linear equivalence (7.7.4).)

7.4.8 (K-flatness depends only on the divisor) It is well understood that in the theory of pairs (X, Δ) one cannot separate the underlying variety X from the divisorial part Δ. For example, if X is a surface with quotient singularities only and $D \subset X$ is a smooth curve, then the pair (X, D) is plt if $D \cap \text{Sing } X = \emptyset$, but not even lc in some other cases. It really matters how exactly D sits inside X.

Thus it is unexpected that K-flatness depends only on the divisor D, not on the ambient variety X, though maybe this is less surprising if one thinks of K-flatness as a variant of flatness.

On the other hand, not all K-flat deformations (7.37) of D are realized on deformations of a given X. For example, for deformations of the pair $(\mathbb{A}^2, D_1 := (xy = 0))$, K-flatness is equivalent to flatness by (7.4.2). However, there are deformations of the pair $((xy = z^2), (z = 0))$ that induce a K-flat, but non-flat deformation of $D_2 := (xy - z^2 = z = 0) \simeq D_1$. A typical example is

$$((xy = z^2 - t^2), (x = z + t = 0) \cup (y = z - t) = 0) \subset \mathbb{A}^3_{xyz} \times \mathbb{A}^1_t.$$

Now we come to a property that is quite unexpected, but makes the whole theory much easier to use: K-flatness is essentially a property of surface pairs (S, D). Thus K-flatness is mostly about families of singular curves.

Theorem 7.5 (Bertini theorems, up and down) *Let $f: X \to S$ be a projective morphism of pure relative dimension n, and D a Mumford divisor on X. Assume that $n \geq 3$, and let $|H|$ be a linear system on X that is base point free in characteristic 0 and very ample in general. Then D is K-flat iff $D|_{H_\lambda}$ is K-flat for general $H_\lambda \in |H|$.*

This is established by combining (7.57–7.59) with (7.40). As a consequence, K-flatness is really a question about families of surfaces and curves on them.

This reduction to surfaces is very helpful both conceptually and computationally, since we have rather complete lists of singularities of log canonical surface pairs (X, Δ), at least when the coefficients of Δ are not too small.

Another variant of the phenomenon, that higher codimension points sometimes do not matter much, is the Hironaka-type flatness theorem (10.72).

7.6 (Problems and questions about K-flatness) There are also some difficulties with K-flatness. We believe that they do not effect the general moduli theory of stable pairs, but they make some of the proofs convoluted and explicit computations lengthy.

7.6.1 (The definition is not formal-local) One expects K-flatness to be a formal-local property on X, but there are some (hopefully only technical) problems with this. See (7.41) and (7.60) for partial results. This is probably the main open foundational question.

7.6.2 (Hard to compute) The definition of K-flatness is quite hard to check, since for $X \subset \mathbb{P}^N$ we need to check not just linear projections $\mathbb{P}^N_S \dashrightarrow \mathbb{P}^n_S$ (7.36), but all morphisms $X \to \mathbb{P}^n_S$ involving all linear systems on X.

It is, however, possible that checking general linear projections is in fact sufficient; see (7.47) and (7.42) for a precise formulation.

In the examples in Sections 7.8–7.9, the computation of the restrictions imposed by general linear projections is the hard part. From the resulting answers it is then easy to read off what happens for all morphisms $X \to \mathbb{P}^n_S$. It would be good to work out more examples of space curves $C \subset \mathbb{A}^3$.

7.6.3 (Tangent space and obstruction theory) We do not know how to write down the tangent space of $\mathrm{KDiv}(X/S)$. A handful of examples are computed in Sections 7.8–7.9, but they do not seem to suggest any general pattern. The obstruction theory of K-flatness is completely open.

7.6.4 (Universal deformations) Let D be a reduced, projective scheme over a field k. Is there a universal deformation space for its K-flat deformations?

Examples 7.7 The first example shows that the space of first order deformations of the smooth divisor $(x = 0) \subset \mathbb{A}^2$, that are Cartier away from the origin, is infinite dimensional. Thus working with generically flat divisors (3.26) does not give a sensible moduli space.

(7.7.1) Start with $X := \mathrm{Spec}\, k[x, y, \varepsilon]_{(x,y)}$ over $\mathrm{Spec}\, k[\varepsilon]$ and set $X^\circ := X \setminus (x = y = 0)$. Let $g(y^{-1}) \in y^{-1}k[y^{-1}]$ be a polynomial of degree n. Then

$$x + g(y^{-1})\varepsilon \in k[x, y, y^{-1}, \varepsilon]_{(x,y)}$$

defines a relative Cartier divisor D°_g, whose restriction to the closed fiber is $(x = 0)$. One can check (7.14) that, if $g_1 \neq g_2$, then $D^\circ_{g_1}$ and $D^\circ_{g_2}$ give different elements of $\mathrm{Pic}(X^\circ)$. Set

$$I_g := (x^2, xy^n + y^n g(y^{-1})\varepsilon, \varepsilon x) \subset k[x, y, \varepsilon]_{(x,y)}, \quad \text{and} \quad D_g := \mathrm{Spec}\, k[x, y, \varepsilon]/I_g.$$

Note that $y^n g(y^{-1})$ is invertible in $k[x, y, \varepsilon]_{(x,y)}$, hence

$$k[x, y, \varepsilon]_{(x,y)}/(x^2, xy^n + y^n g(y^{-1})\varepsilon, \varepsilon x) \simeq k[x, y]_{(x,y)}/(x^2).$$

Thus D_g is the scheme-theoretic closure of D°_g, $(I_g, \varepsilon)/(\varepsilon) = (x^2, xy^n)$, D_g has no embedded points, and $D_{g_1} \sim D_{g_2}$ iff $g_1 = g_2$. More general computations are done in (7.20).

(7.7.2) To illustrate (7.1.5), let C be a smooth projective curve and E a vector bundle over C of rank $n + 1 \geq 2$ and of degree 0. We claim that usually there is no finite morphism $\pi \colon \mathbb{P}_C(E) \to \mathbb{P}^n \times C$.

Indeed, let $p_0, \ldots, p_{n+1} \in \mathbb{P}^n$ be the coordinate vertices plus $(1:\cdots:1)$. Then $C_i := \pi^{-1}(\{p_i\}\times C)$ are $n+2$ disjoint multi-sections of $\mathbb{P}_C(E) \to C$. Pick $p\colon D \to C$ that factors through all of the $C_i \to C$. Then $\mathbb{P}_D(p^*E)$ has $n+2$ disjoint sections in linearly general position, hence $\mathbb{P}_D(p^*E) \simeq \mathbb{P}^n \times D$. Equivalently, $p^*D \simeq L \otimes \mathcal{O}_D^{n+1}$ for some line bundle L of degree 0.

This cannot happen for most line bundles. The simplest example is $E = \mathcal{O}_{\mathbb{P}^1}(1)\oplus\mathcal{O}_{\mathbb{P}^1}(-1)$. More generally, such a line bundle has to be semi-stable. If E is stable, hence comes from a representation $\pi_1(C) \to U(n+1)$, then its image in $PU(n+1)$ must be finite.

(7.7.3) As an example for (7.4.5), set $X := (xy = uv)$ and let $\pi\colon X \to \mathbb{A}_t^1$ be given by $t = x + y$. Then $D_1 := (x = u = 0)$ and $D_2 := (y = v = 0)$ are both flat over $\mathbb{A}_t^1$, but $D_1 \cup D_2$ is not flat.

(7.7.4) As an example for (7.4.7), let $A \subset \mathbb{P}^n$ be a projectively normal abelian variety of dimension ≥ 2 and $C_A \subset \mathbb{P}^{n+1}$ the cone over it. Let $\pi\colon \mathbb{P}^{n+1} \dashrightarrow \mathbb{P}^2$ be a general projection. Let $H \subset C_A$ be a hyperplane section. If H does not pass through the vertex then $H \simeq A$ is smooth and $\pi|_H\colon H \dashrightarrow \mathbb{P}^2$ is flat.

If H does pass through the vertex v, then $\mathrm{depth}_v H = 1$ by (2.35), hence $\pi|_H\colon H \dashrightarrow \mathbb{P}^2$ is not flat at v.

7.2 Infinitesimal Study of Mumford Divisors

In this section, we review the divisor theory of nonreduced schemes. The standard reference books treat Cartier divisors in detail, but for us the interesting cases are precisely when the divisors fail to be Cartier. We start with the general theory, and at the end give explicit formulas for some cases.

Definition 7.8 (Mumford class group) Let S be a scheme and $f\colon X \to S$ a morphism of pure relative dimension n. Two relative Mumford divisors (4.68) $D_1, D_2 \subset X$ are *linearly equivalent over* S if $\mathcal{O}_X(-D_1) \simeq \mathcal{O}_X(-D_2) \otimes f^*L$ for some line bundle L on S. The linear equivalence classes generate the *relative Mumford class group* $\mathrm{MCl}(X/S)$.

This is a higher dimensional version of the generalized Jacobians, worked out in Severi (1947), Rosenlicht (1954), and Serre (1959). It is slightly different from the theory of almost-Cartier divisors of Hartshorne (1986) and Hartshorne and Polini (2015).

By definition, if D is a Mumford divisor then there is a closed subset $Z \subset X$ such that $D|_{X\setminus Z}$ is Cartier and Z/S has relative dimension $\leq n-2$. This gives a natural identification

$$\mathrm{MCl}(X/S) = \lim_Z \mathrm{Pic}((X \setminus Z)/S), \qquad (7.8.1)$$

where the limit is over all closed subsets $Z \subset X$ such that Z/S has relative dimension $\leq n - 2$.

As with the Picard group, it may be better to sheafify $\mathrm{MCl}(X/S)$ in the étale topology as in Bosch et al. (1990, chap.8). However, we use this notion mostly when S is local, so this is not important for our current purposes.

7.9 The infinitesimal method to study families of objects in algebraic geometry posits that we should proceed in three broad steps:

- Study families over Artinian schemes.
- Inverse limits then give families over complete local schemes.
- For arbitrary local schemes, descend properties from the completion.

This approach has been very successful for proper varieties and for coherent sheaves. One of the problems with general (possibly nonflat) families of divisors is that the global and the infinitesimal computations do not match up; in fact they say the opposite in some cases. We discuss two instances of this:

- Relative Cartier divisors on non-proper varieties.
- Generically flat families of divisors on surfaces.

The surprising feature is that the two behave quite differently. We state two cases where the contrast between Artinian and DVR bases is striking.

Claim 7.9.1 Let $\pi \colon X \to (s, S)$ be a smooth, affine morphism, S local.

(a) If S is Artinian, then the restriction map $\mathrm{Pic}(X) \to \mathrm{Pic}(X_s)$ is an isomorphism by (7.10.2).

(b) If $S = \mathrm{Spec}\, k[[t]]$, then $\mathrm{Pic}(X)$ can be infinite dimensional by (7.13.3).

That is, there can be many nontrivial line bundles on X over $\mathrm{Spec}\, k[[t]]$, but we do not see them when working over $\mathrm{Spec}\, k[[t]]/(t^m)$.

The opposite happens for the Mumford class group of projective surfaces.

Claim 7.9.2 $\mathrm{MCl}(\mathbb{P}^2_{k[[t]]/(t^m)}) \simeq \mathbb{Z} + k^{\infty}$ for $m \geq 2$, but $\mathrm{MCl}(\mathbb{P}^2_{k[[t]]}) \simeq \mathbb{Z}$.

Proof Here $\mathbb{P}^2_{k[[t]]}$ is regular, so every Weil divisor on X is Cartier. The first part follows from (7.8.1) and (7.10.3), since $H^1(\mathbb{P}^2 \setminus Z, \mathscr{O}_{\mathbb{P}^2 \setminus Z}) \simeq H^2_Z(\mathbb{P}^2, \mathscr{O}_{\mathbb{P}^2})$ is infinite dimensional. □

7.10 (Picard group over Artinian schemes) Let (A, m, k) be a local Artinian ring and $X_A \to \mathrm{Spec}\, A$ a flat morphism. Let $(\varepsilon) \subset A$ be an ideal such that $I \simeq k$ and set $B = A/(\varepsilon)$. We have an exact sequence

$$0 \longrightarrow \mathscr{O}_{X_k} \xrightarrow{\ e\ } \mathscr{O}^*_{X_A} \longrightarrow \mathscr{O}^*_{X_B} \longrightarrow 1, \qquad (7.10.1)$$

where $e(h) = 1 + h\varepsilon$ is the exponential map. We use its long exact cohomology sequence and induction on length A to compute $\text{Pic}(X_A)$. There are three cases that are especially interesting for us.

Claim 7.10.2 Let $X_A \to \text{Spec } A$ be a flat, affine morphism. Then the restriction map $\text{Pic}(X_A) \to \text{Pic}(X_k)$ is an isomorphism.

Proof We use the exact sequence

$$H^1(X_k, \mathcal{O}_{X_k}) \to \text{Pic}(X_A) \to \text{Pic}(X_B) \to H^2(X_k, \mathcal{O}_{X_k}). \tag{7.10.2.a}$$

Since X is affine, the two groups at the ends vanish, hence we get an isomorphism in the middle. Induction completes the proof. ☐

Claim 7.10.3 Let $X_A \to \text{Spec } A$ be a flat, proper morphism. If $H^0(X_k, \mathcal{O}_{X_k}) = k$, then the kernel of the restriction map $\text{Pic}(X_A) \to \text{Pic}(X_k)$ is a unipotent group scheme of dimension $\leq h^1(X_k, \mathcal{O}_{X_k}) \cdot (\text{length } A - 1)$ and equality holds if $H^2(X_k, \mathcal{O}_{X_k}) = 0$. (In fact, if char $k = 0$, then the kernel is a k-vector space and equality holds even if $H^2(X_k, \mathcal{O}_{X_k}) \neq 0$; see Bosch et al. (1990, chap.8).)

Proof By Hartshorne (1977, III.12.11), $H^0(X_A, \mathcal{O}_{X_A}) \to H^0(X_B, \mathcal{O}_{X_B})$ is surjective and so is $H^0(X_A, \mathcal{O}_{X_A}^*) \to H^0(X_B, \mathcal{O}_{X_B}^*)$. Thus we get the exactness of

$$0 \to H^1(X_k, \mathcal{O}_{X_k}) \to \text{Pic}(X_A) \to \text{Pic}(X_B) \to H^2(X_k, \mathcal{O}_{X_k}). \qquad ☐$$

Claim 7.10.4 Let $X_A \to \text{Spec } A$ be a flat morphism and $Z \subset X_A$ a closed subset of codimension ≥ 2. Set $X_A^\circ := X_A \setminus Z$. Assume that X_k is S_2. Then the kernel of the restriction map $\text{Pic}(X_A^\circ) \to \text{Pic}(X_k^\circ)$ is a unipotent group scheme of dimension $\leq h^1(X_k^\circ, \mathcal{O}_{X_k^\circ}) \cdot (\text{length } A - 1)$.

Proof Since X_k is S_2, $H^0(X_k^\circ, \mathcal{O}_{X_k^\circ}) \simeq H^0(X_k, \mathcal{O}_{X_k})$ and similarly for X_A. Thus $H^0(X_A^\circ, \mathcal{O}_{X_A^\circ}^*) \to H^0(X_B^\circ, \mathcal{O}_{X_B^\circ}^*)$ is surjective and the rest of the argument works as in (7.10.3). ☐

Remark 7.10.5. Although (7.10.4) is very similar to (7.10.3), a key difference is that in (7.10.4) the group $H^1(X_k^\circ, \mathcal{O}_{X_k^\circ})$ can be infinite dimensional. Indeed, $H^1(X_k^\circ, \mathcal{O}_{X_k^\circ}) \simeq H_Z^2(X_k, \mathcal{O}_{X_k})$ and it is
  (a) infinite dimensional if $\dim X_k = 2$,
  (b) finite dimensional if X_k is S_2 and $\text{codim}_{X_k} Z \geq 3$, and
  (c) 0 if X_k is S_3 and $\text{codim}_{X_k} Z \geq 3$.
See, for example, Section 10.3 for these claims.

The following immediate consequence of (7.10.5.c) is especially useful.

Corollary 7.10.6 Let $X \to S$ be a smooth morphism, $D \subset X$ a closed subscheme, and $Z \subset X$ a closed subset. Assume that D is a relative Cartier divisor on $X \setminus Z$, D has no embedded points in Z, and $\mathrm{codim}_{X_s} Z_s \geq 3$ for every $s \in S$. Then D is a relative Cartier divisor. $\qquad\qquad\qquad\qquad\qquad\qquad\qquad\square$

The following is a special case of (4.28).

Lemma 7.11 *Let* $X \to S$ *be a flat morphism with* S_2 *fibers and* D *a divisorial subscheme. Let* $U \subset X$ *be an open subscheme such that* $D|_U$ *is relatively Cartier and* $\mathrm{codim}_{X_s}(X_s \setminus U_s) \geq 2$ *for every* $s \in S$.

Then D *is relatively Cartier iff the generically Cartier pull-back* $\tau^{[*]}D$ *(4.2.7) is relatively Cartier for every Artinian subscheme* $\tau \colon A \hookrightarrow S$. $\qquad\square$

Relative Cartier divisors also have some unexpected properties over nonreduced base schemes. These do not cause theoretical problems, but it is good to keep them in mind.

Example 7.12 (Cartier divisors over $k[\varepsilon]$) Let R be an integral domain over a field k. Relative principal ideals in $R[\varepsilon]$ over $k[\varepsilon]$ are given as $(f + g\varepsilon)$ where $f, g \in R$ and $f \neq 0$. We list some properties of such principal ideals that hold for any integral domain R:

(7.12.1) $(f + g_1\varepsilon) = (f + g_2\varepsilon)$ iff $g_1 - g_2 \in (f)$.

(7.12.2) If $u \in R$ is a unit then so is $u + g\varepsilon$ since $(u + g\varepsilon)(u^{-1} - u^{-2}g\varepsilon) = 1$.

(7.12.3) If f is irreducible then so is $f + g\varepsilon$ for every g.

(7.12.4) $(f + g\varepsilon)(f - g\varepsilon) = f^2$ shows that there is no unique factorization.

(7.12.5) If R is a UFD and the f_i are pairwise relatively prime, then

$$\textstyle\prod_i(f_i + g_i\varepsilon) = \prod_i(f_i + g_i'\varepsilon) \quad \text{iff} \quad g_i - g_i' \in (f_i) \quad \forall i.$$

The following concrete example illustrates several of the above features.

Example 7.13 (Picard group of a constant elliptic curve) Let $(0, E)$ be a smooth, projective elliptic curve. Over any base S we have the constant family $\pi \colon E \times S \to S$ with the constant section $s_0 \colon S \simeq \{0\} \times S$. Let L be a line bundle on $E \times S$. Then $L \otimes \pi^* s_0^* L^{-1}$ has a canonical trivialization along $\{0\} \times S$, hence it defines a morphism $S \to \mathrm{Pic}(E)$. Thus

$$\mathrm{Pic}(E \times S / S) \simeq \mathrm{Mor}(S, \mathrm{Pic}(E)). \qquad\qquad (7.13.1)$$

Corollary 7.13.2 Let (R, m) be a complete local ring. Set $S = \mathrm{Spec}\, R$ and $S_n = \mathrm{Spec}\, R/m^n$. Then $\mathrm{Pic}(E \times S / S) = \varprojlim \mathrm{Pic}(E \times S_n / S_n)$. $\qquad\square$

Corollary 7.13.3 Let $S = \operatorname{Spec} k[t]_{(t)}$ be the local ring of the affine line at the origin and $\widehat{S} = \operatorname{Spec} k[[t]]$ its completion. Then $\operatorname{Pic}(E \times S/S) \simeq \operatorname{Pic}(E)$, but $\operatorname{Pic}(E \times \widehat{S}/\widehat{S})$ is infinite dimensional.   □

Next consider the affine elliptic curve $E^\circ = E \setminus \{0\}$ and the constant affine family $E^\circ \times S \to S$. Note that $\operatorname{Pic}(E^\circ) \simeq \operatorname{Pic}^\circ(E)$.

If S is smooth and D° is a Cartier divisor on $E^\circ \times S$ then its closure $D \subset E \times S$ is also Cartier. More generally, this also holds if S is normal, using (4.4). Thus (7.13.1–2) give the following.

Corollary 7.13.3 If S is normal then $\operatorname{Pic}(E^\circ \times S/S) \simeq \operatorname{Mor}(S, \operatorname{Pic}^\circ(E))$.   □

Corollary 7.13.4 If $S = \operatorname{Spec} A$ is Artinian then $\operatorname{Pic}(E^\circ \times S/S) \simeq \operatorname{Pic}^\circ(E)$. So $\operatorname{Pic}(E^\circ \times S/S)$ has dimension 1, but $\dim_k \operatorname{Mor}(S, \operatorname{Pic}^\circ(E)) = \operatorname{length} A$.   □

For the rest of the section we make some explicit computations about Mumford divisors on schemes that are smooth over an Artinian ring.

Proposition 7.14 *Let (A, k) be a local Artinian ring, $k \simeq (\varepsilon) \subset A$ an ideal, and $B = A/(\varepsilon)$. Let (R_A, m) be a flat, local, S_2, A-algebra and set $X_A := \operatorname{Spec}_A R_A$. Let $f_B \in R_B$ be a non-zerodivisor and set $D_B := (f_B = 0) \subset X_B$.*

Then the set of relative Mumford divisors $D_A \subset X_A$ such that $\operatorname{pure}((D_A)|_B) = D_B$, is a torsor under the k-vector space $H^1_m(D_k, \mathscr{O}_{D_k})$.

Proof We can lift f_B to $f_A \in R_A$. Choose $y \in m$ that is not a zerodivisor on D_B and such that D_A is a principal divisor on $X_A \setminus (y = 0)$. After inverting y, we can write the ideal of D_A as

$$(I, y^{-1}) = (f_A + \varepsilon y^{-r} g_k), \quad \text{where} \quad g_k \in R_k, r \in \mathbb{N}. \tag{7.14.1}$$

We can multiply $f_A + \varepsilon y^{-r} g_k$ by $1 + \varepsilon y^{-s} v$. This changes $y^{-r} g_k$ to $y^{-r} g_k + v y^{-s} f_A$. By (7.15) the relevant information is carried by the residue class

$$\overline{y^{-r} g_k} \in H^0(D_k^\circ, \mathscr{O}_{D_k^\circ}), \tag{7.14.2}$$

where $D_k^\circ \subset D_k$ denotes the complement of the closed point.

If the residue class is in $H^0(D_k, \mathscr{O}_{D_k})$, then we get a Cartier divisor. Thus the non-Cartier divisors are parametrized by

$$H^0(D_k^\circ, \mathscr{O}_{D_k^\circ})/H^0(D_k, \mathscr{O}_{D_k}) \simeq H^1_m(D_k, \mathscr{O}_{D_k}). \tag{7.14.3}$$

We get distinct divisors by (7.17.2).   □

Lemma 7.15 *Let (A, k) be a local Artinian ring, $k \simeq (\varepsilon) \subset A$ an ideal, and $B = A/(\varepsilon)$. Let (R_A, m) be a flat, local, S_2, A-algebra. Let $f_A \in R_A$ and $g_k \in R_k$ be non-zerodivisors and y a non-zerodivisor modulo both f_A and g_k.*

For $I := R_A \cap (f_A + \varepsilon y^{-r} g_k) R_A[y^{-1}]$ the following are equivalent:
(7.15.1) *I is a principal ideal.*
(7.15.2) *The residue class $\overline{y^{-r} g_k}$ lies in $R_k/(f_k)$.*
(7.15.3) $g_k \in (f_k, y^r)$.

Note that we can change $f_A + \varepsilon y^{-r} g_k$ to $(f_A + \varepsilon h_k) + \varepsilon y^{-r}(g_k - y^r h_k)$ for any $h_k \in R_k$, but $g_k \in (f_k, y^r)$ iff $g_k - y^r h_k \in (f_k, y^r)$.

Proof I is a principal ideal iff it has a generator of the form $f_A + \varepsilon h_k$ where $h_k \in R_k$. This holds iff

$$f_A + \varepsilon y^{-r} g_k = (1 + \varepsilon y^{-s} b_k)(f_A + \varepsilon h_k) \quad \text{for some} \quad b_k \in R_A.$$

Equivalently, iff $y^{-r} g_k = h_k + y^{-s} b_k f_k$. If $r > s$ then $g_k = y^r h_k + y^{r-s} b_k f_k$, which is impossible since y is not a zerodivisor modulo g_k. If $r < s$ then $y^{s-r} g_k = y^s h_k + b_k f_k$, which is impossible since y is not a zerodivisor modulo f_k. Thus $r = s$ and then $g_k = y^r h_k + b_k f_k$ is equivalent to $g_k \in (f_k, y^r)$. □

The next will be crucial in the proof of (7.60). To state it, let $\mathrm{nil}(n_A)$ denote the smallest $r \geq 0$ such that $n_A^r = 0$, and for $f \in R_A[y^{-1}]$, let ord_y denote *pole order* in y, that is, the smallest $r \geq 0$ such that $y^r f \in R_A$.

Proposition 7.16 *Let (A, n_A, k) be a local Artinian ring and (R_A, m_R) a flat, local, S_2, A-algebra of dimension ≥ 2. Let $f_k \in m_k$ be a non-zerodivisor and $y \in m_R$ a non-zerodivisor modulo f_k. Let $f_A, f_A' \in R_A[y^{-1}]$ be two liftings of f_k. Assume that $f_A - f_A' \in y^N R_A$, where $N = \mathrm{nil}(n_A) \cdot \mathrm{ord}_y f_A$.*
Then $(f_A) \cap R_A$ is a principal ideal iff $(f_A') \cap R_A$ is.

Proof Note first that $N \geq 0$, so $f_A - f_A' \in y^N R_A$ implies that $\mathrm{ord}_y f_A = \mathrm{ord}_y f_A'$, so the assumption is symmetric in f_A, f_A'. It is thus enough to prove that if $(f_A) \cap R_A$ is a principal ideal, then so is $(f_A') \cap R_A$.
Assume that $(f_A) \cap R_A = (F_A)$. Then there is unit u_A in $R_A[y^{-1}]$ such that $f_A = u_A F_A$. Since $f_k = F_k$, we see that u_k is a unit in R_k.
We claim that $\mathrm{ord}_y f_A = \mathrm{ord}_y u_A$. Indeed, if $\mathrm{ord}_y u_A = r$ then we get a nonzero remainder $\bar{u}_A \in y^{-r} R_A/y^{1-r} R_A \simeq R_A/y R_A$. Multiplication by F_A preserves the pole-order filtration, so

$$\overline{F_A u_A} = F_A \bar{u}_A \in y^{-r} R_A/y^{1-r} R_A \simeq R_A/y R_A.$$

Here $R_A/y R_A$ has a filtration whose successive quotients are $R_k/y R_k$ and F_A acts by multiplication by f_k an each graded piece. Since f_k is a non-zerodivisor modulo y, we see that $\overline{F_A u_A} \neq 0$. So $\mathrm{ord}_y f_A = \mathrm{ord}_y u_A$. Taylor expansion of the inverse shows that $\mathrm{ord}_y(u_A^{-1}) \leq \mathrm{nil}(n_A) \cdot \mathrm{ord}_y f_A =: N$. Thus

$$u_A^{-1} f_A' = u_A^{-1} f_A + u_A^{-1}(f_A' - f_A) = F_A + (y^N u_A^{-1})(y^{-N}(f_A' - f_A)) \in R_A. \qquad \square$$

The connection between (7.14) and (7.10) is given by the following.

7.17 Let X be an affine, S_2 scheme and $D := (s = 0) \subset X$ a Cartier divisor. Let $Z \subset D$ be a closed subset that has codimension ≥ 2 in X. Set $X^\circ := X \setminus Z$ and $D^\circ := D \setminus Z$. Restricting the exact sequence

$$0 \to \mathscr{O}_X \xrightarrow{s} \mathscr{O}_X \to \mathscr{O}_D \to 0$$

to X° and taking cohomologies we get

$$0 \to H^0(X^\circ, \mathscr{O}_{X^\circ}) \xrightarrow{s} H^0(X^\circ, \mathscr{O}_{X^\circ}) \to H^0(D^\circ, \mathscr{O}_{D^\circ}) \xrightarrow{\partial} H^1(X^\circ, \mathscr{O}_{X^\circ}).$$

Note that $H^0(X^\circ, \mathscr{O}_{X^\circ}) = H^0(X, \mathscr{O}_X)$ since X is S_2 and its image in $H^0(D^\circ, \mathscr{O}_{D^\circ})$ is $H^0(D, \mathscr{O}_D)$. Thus ∂ becomes the injection

$$\partial \colon H_Z^1(D, \mathscr{O}_D) \simeq H^0(D^\circ, \mathscr{O}_{D^\circ})/H^0(D, \mathscr{O}_D) \hookrightarrow H_Z^2(X, \mathscr{O}_X). \qquad (7.17.1)$$

We are especially interested in the case when (x, X) is local, two-dimensional and $Z = \{x\}$. In this case (7.17.1) becomes

$$\partial \colon H_x^1(D, \mathscr{O}_D) \hookrightarrow H_x^2(X, \mathscr{O}_X). \qquad (7.17.2)$$

The left side describes first order deformations of D by (7.14) and the right side the Picard group of the first order deformation of $X \setminus \{x\}$ by (7.10.4).

We can be especially explicit about first order deformations in the smooth case. Let us start with the description as in (7.14).

7.18 (Mumford divisors in $k[[u, v]][\varepsilon]$) Set $X = \operatorname{Spec} k[[u, v]][\varepsilon]$ with closed point $x \in X$. By (7.10), the Picard group of the punctured spectrum $X \setminus \{x\}$ is

$$H_x^2(X, \mathscr{O}_X) \simeq \bigoplus_{i,j>0} \frac{1}{u^i v^j} \cdot k.$$

An ideal corresponding to $cu^{-i}v^{-j}$ (where $c \in k^\times$) can be given as

$$I(cu^{-i}v^{-j}) := (u^{2i}, u^i v^j + c\varepsilon, u^i \varepsilon);$$

a more systematic derivation of this is given in (7.20.1).

This is explicit, but we are more interested in the point of view of (7.10).

Lemma 7.19 *Let $f \in k[[u]][v]$ be a monic polynomial in v of degree n defining a curve $C_k \subset \widehat{\mathbb{A}}_{uv}^2$. Let $C \subset \widehat{\mathbb{A}}_{k[\varepsilon]}^2$ be a relative Mumford divisor such that* pure$((C)_k) = C_k$. *Then the restriction of C to the complement of $(u = 0)$ can be uniquely written as*

$$(f + \varepsilon \sum_{i=0}^{n-1} v^i \phi_i(u) = 0) \quad \text{where} \quad \phi_i(u) \in u^{-1} k[u^{-1}].$$

Thus the set of all such C is naturally isomorphic to the infinite dimensional k-vector space $H_m^1(C_k, \mathscr{O}_{C_k}) \simeq \bigoplus_{i=0}^{n-1} u^{-1} k[u^{-1}]$.

Note that, by the Weierstrass preparation theorem, almost every curve in $\widehat{\mathbb{A}}_{uv}$ is defined by a monic polynomial in v.

Proof Note that $k[[u]][v]/(f) \simeq \bigoplus_{i=0}^{n-1} v^i k[[u]]$ as a $k[[u]]$-module, so

$$H^0(C_k, \mathscr{O}_{C_k}) \simeq \bigoplus_{i=0}^{n-1} v^i k[[u]] \quad \text{and} \quad H^0(C_k^\circ, \mathscr{O}_{C_k^\circ}) \simeq \bigoplus_{i=0}^{n-1} v^i k((u)). \quad (7.19.1)$$

That is, if $g \in k((u))[v]$ is a polynomial of degree $< n$ in v, then $g|_{C^\circ}$ extends to a regular function on C iff $g \in k[[u]][v]$.   □

We can also restate (7.19.1) as

$$H_m^1(C_k, \mathscr{O}_{C_k}) \simeq \bigoplus_{i=0}^{n-1} v^i k((u))/k[[u]] \simeq \bigoplus_{i=0}^{n-1} v^i u^{-1} k[u^{-1}]. \quad (7.19.2)$$

Example 7.20 Consider next the special case of (7.19) when $f = v$. We can then write the restriction of C as $(v + \phi(u)\varepsilon = 0)$ where $\phi \in u^{-1} k[u^{-1}]$. Let r denote the pole-order of ϕ and set $q(u) := u^r \phi(u)$. By (7.7.1), the ideal of C is

$$I_C = (v^2, vu^r + q(u)\varepsilon, v\varepsilon). \quad (7.20.1)$$

Thus the fiber over the closed point is $k[[u, v]]/(v^2, vu^r)$. Its torsion submodule is isomorphic to $k[[u, v]]/(v, u^r) \simeq k[u]/(u^r)$.

The ideals of relative Mumford divisors in $k[[u, v]][\varepsilon]$ are likely to be more complicated in general. At least the direct generalization of (7.20.1) does not always give the correct generators.

For example, let $f = v^2 - u^3$ and consider the ideal $I \subset k[[u, v]][\varepsilon]$ extended from $((v^2 - u^3) + u^{-3} v\varepsilon)$. The formula (7.20.1) suggests the elements

$$(v^2 - u^3)^2, \; u^3(v^2 - u^3) + v\varepsilon, \; (v^2 - u^3)\varepsilon \in I.$$

However, $u^3(v^2 - u^3) + v\varepsilon = v^2(v^2 - u^3) + v\varepsilon$, giving that

$$I = ((v^2 - u^3)^2, v(v^2 - u^3) + \varepsilon, (v^2 - u^3)\varepsilon). \quad (7.20.2)$$

Using the isomorphism $R[\varepsilon]/(f^2, fg + \varepsilon, f\varepsilon) \simeq R/(f^2, -f^2 g) \simeq R/(f^2)$, the examples can be generalized to the nonsmooth case as follows.

Claim 7.20.3 Let (R, m) be a local, S_2, k-algebra of dimension 2, and $f, g \in m$ a system of parameters. Then $J_{f,g} = (f^2, fg + \varepsilon, f\varepsilon)$ is (the ideal of) a relative Mumford divisor in $R[\varepsilon]$ whose central fiber is $R/(f^2, fg)$, with embedded subsheaf isomorphic to $R/(f, g)$. $\qquad\square$

7.3 Divisorial Support

There are at least three ways to associate a divisor to a sheaf (7.22), but only one of them – the divisorial support – behaves well in flat families. In this Section we develop this notion and a method to compute it. The latter is especially important for the applications. First, we recall the definition of the Fitting ideal sheaf.

7.21 (Fitting ideal) Let R be a Noetherian ring, M a finite R-module, and

$$R^s \xrightarrow{A} R^r \to M \to 0$$

a presentation of M, where A is given by an $s \times r$-matrix with entries in R. The *Fitting ideal*, or, more precisely, the $0th$ *Fitting ideal* of M, denoted by $\mathrm{Fitt}_R(M)$, is the ideal generated by the determinants of $r \times r$-minors of A. For the following basic properties, see Fitting (1936) or (Eisenbud, 1995, Sec.20.2).

(7.21.1) $\mathrm{Fitt}_R(M)$ is independent of the presentation chosen.

(7.21.2) If R is regular and $M \simeq \oplus_i R/(g_i^{m_i})$ then $\mathrm{Fitt}_R(M) = (\prod g_i^{m_i})$.

(7.21.3) The Fitting ideal commutes with base change. That is, if S is an R-algebra then $\mathrm{Fitt}_S(M \otimes_R S)$ is generated by $\mathrm{Fitt}_R(M) \otimes_R S$.

The following is a special case of Lipman (1969, lem.1).

(7.21.4) Let M be a torsion module. Then $\mathrm{Fitt}_R(M)$ is a principal ideal generated by a non-zerodivisor iff the projective dimension of M is 1.

One direction is easy. If the projective dimension of M is 1, then M has a presentation

$$0 \to R^s \xrightarrow{A} R^r \to M \to 0.$$

Here $r = s$ since M is torsion, thus $\det(A)$ generates $\mathrm{Fitt}_R(M)$.

We prove the converse only in the following special case that we use later, which, however, captures the essence of the general proof.

(7.21.5) Let X be a smooth variety of dimension n and F a coherent sheaf of generic rank 0 on X. Then $\mathrm{Fitt}_X(F)$ is a principal ideal iff F is CM of pure dimension $n - 1$.

Proof This can be checked after localization and completion. Thus we have a module M over $S := k[[x_1, \ldots, x_n]]$, and, after a coordinate change, we may assume that it is finite over $R := k[[x_1, \ldots, x_{n-1}]]$ of generic rank say r. Using first (7.21.2) and then (7.21.3) we get that

$$\dim_k M \otimes_S k[[x_n]] = \dim_k k[[x_n]]/\operatorname{Fitt}_{k[[x_n]]}(M \otimes_S k[[x_n]])$$
$$= \dim_k(S/\operatorname{Fitt}_S(M)) \otimes_S k[[x_n]]. \tag{7.21.6}$$

Next note that M is CM $\Leftrightarrow M$ is free over $R \Leftrightarrow \dim M \otimes_S k[[x_n]] = r$. Using (7.21.1) and the previous equivalences for $S/\operatorname{Fitt}_S(M)$, we get that these are equivalent to $S/\operatorname{Fitt}_S(M)$ being CM. This holds iff $\operatorname{Fitt}_S(M)$ is a height 1 unmixed ideal, hence principal.   □

The following explicit formula is quite useful.

Computation 7.21.7 Let S be a smooth R-algebra and $v \in S$ such that $S/(v) \simeq R$. (The examples we use are $S = R[v]$ and $S = R[[v]]$.) Let M be an S-module that is free of finite rank as an R-module. Write $M = \oplus_{i=1}^r Rm_i$ and $vm_i = \sum_{i=1}^r a_{ij}m_j$ for $a_{ij} \in R$. Then $\operatorname{Fitt}_S(M)$ is generated by $\det(v\mathbf{1}_r - (a_{ij}))$.

Proof A presentation of M as an S-module is given by

$$\oplus_{i=1}^r S e_i \xrightarrow{\phi} \oplus_{i=1}^r S f_i \xrightarrow{\psi} M \to 0,$$

where $\psi(f_i) = m_i$ and $\phi(e_i) = vf_i - \sum_{j=1}^r a_{ij}f_j$. Thus $\phi = v\mathbf{1}_r - (a_{ij})$ and so $\det(v\mathbf{1}_r - (a_{ij}))$ generates $\operatorname{Fitt}_S(M)$.   □

Computation 7.21.8 Let T be a free S-algebra and $t \in T$ a non-zerodivisor. Then $\operatorname{Fitt}_S(T/tT)$ is generated by $\operatorname{norm}_{T/S}(t)$.

Proof We use $0 \to T \xrightarrow{t} T \to T/tT \to 0$ and the definition of the norm.   □

Definition 7.22 (Divisorial support I) Let X be a scheme and F a coherent sheaf on X. One usually defines its *support* $\operatorname{Supp} F$ and its *scheme-theoretic support* $\operatorname{SSupp} F := \operatorname{Spec}_X(\mathscr{O}_X/\operatorname{Ann} F)$.

Assume next that $\operatorname{Supp} F$ is nowhere dense and X is regular at every generic point $x_i \in \operatorname{Supp} F$ that has codimension 1 in X. Then there is a unique divisorial sheaf (3.25) associated to the Weil divisor $\sum \operatorname{length}(F_{x_i}) \cdot [\bar{x}_i]$. We call it the *divisorial support* of F and denote it by $\operatorname{DSupp} F$. Equivalently,

$$\operatorname{DSupp}(F) = \operatorname{Spec}\left(\mathscr{O}_X/\operatorname{Fitt}_X(F)\right), \tag{7.22.1}$$

where pure denotes the pure codimension 1 part (10.1).

If every associated point of F has codimension 1 in X, then we have inclusions of subschemes

$$\operatorname{Supp} F \subset \operatorname{SSupp} F \subset \operatorname{DSupp} F. \tag{7.22.2}$$

In general, all three subschemes are different, though with the same support.

Our aim is to develop a relative version of this notion and some ways of computing it in families. Let $X \to S$ be a morphism and F a coherent sheaf on X. Informally, we would like the relative divisorial support of F, denoted by $\mathrm{DSupp}_S F$, to be a scheme over S whose fibers are $\mathrm{DSupp}(F_s)$ for all $s \in S$. If S is reduced, this requirement uniquely determines $\mathrm{DSupp}_S F$, but in general there are two problems.

- Even in nice situations, this requirement may be impossible to meet.
- For nonreduced base schemes, the fibers do not determine $\mathrm{DSupp}_S F$.

In our main applications, X is smooth over some base scheme S that may well have nilpotent elements. As in (9.12), we need to allow embedded subsheaves that "come from" S, but not the others.

Definition 7.23 (Divisorial support II) Let $X \to S$ be a smooth morphism of pure relative dimension n. Let F be a coherent sheaf on X that is flat over S with CM fibers of pure dimension $n - 1$. We define its divisorial support as

$$\mathrm{DSupp}_S(F) := \mathrm{Spec}(\mathscr{O}_X / \mathrm{Fitt}_X(F)).$$

Lemma 7.24 *Under the assumptions of (7.23),*

(7.24.1) $\mathrm{DSupp}_S(F)$ *is a relative Cartier divisor, and*

(7.24.2) $\mathrm{DSupp}_S(F)$ *commutes with base change. That is, let $h: S' \to S$ be a morphism. By base change we get $g': X' \to S'$, $h_X: X' \to X$. Then $h_X^*(\mathrm{DSupp}\, F) = \mathrm{DSupp}(h_X^* F)$.*

Proof The first claim can be checked after localization and completion. We may thus assume that $S = \mathrm{Spec}\, B$ where (B, m) is local with residue field k, $X = \mathrm{Spec}\, B[[x_1, \ldots, x_n]]$ and F is the sheafification of M. Since $M \otimes_B k$ has dimension $n - 1$ over $k[[x_1, \ldots, x_n]]$, after a general coordinate change we may assume that $M/(x_1, \ldots, x_{n-1}, m)M$ is finite. Thus M is a finite $R := B[[x_1, \ldots, x_{n-1}]]$-module. Set $R_k = R \otimes_B k \cong k[[x_1, \ldots, x_{n-1}]]$. Since M is flat over B, its generic rank over R equals the generic rank of $M \otimes_B k$ over R_k. By assumption, $M \otimes_B k$ is CM, hence free over R_k. Thus the generic rank of M over R equals $\dim_k M \otimes_R k$ and M is free as an R-module. The rest follows from (7.21.7). The second claim is immediate from (7.21.3). □

The following restriction property is also implied by (7.21.3).

Lemma 7.25 *Continuing with the notation and assumptions of (7.23), let $D \subset X$ be a relative Cartier divisor that is also smooth over S. Assume that D does not contain any generic point of $\mathrm{Supp}\, F_s$ for any $s \in S$. Then*

$$\mathrm{DSupp}(F|_D) = (\mathrm{DSupp}\, F)|_D. \qquad\qquad\square$$

Now we are ready to define the sheaves for which the relative divisorial support makes sense, but first we have to distinguish associated points that come from the base from the other ones.

Definition 7.26 Let $X \to S$ be a morphism and F a coherent sheaf on X. The *flat locus* of F is the largest open subset $U \subset X$ such that $F|_U$ is flat over S. We denote it by $\text{Flat}_S(X, F)$.

It is usually more convenient to work with the *flat-CM locus* of F. It is the largest open subset $U \subset X$ such that $F|_U$ is flat with CM fibers over S. We denote it by $\text{FlatCM}_S(X, F)$. If F is generically flat over S of relative dimension d, then $(\text{Supp } F \setminus \text{FlatCM}_S(X, F)) \to S$ has relative dimension $< d$.

Definition 7.27 Let $X \to S$ be a morphism. A coherent sheaf F is a *generically flat family of pure sheaves* of dimension d over S, if F is generically flat (3.26) and $\text{Supp } F \to S$ has pure relative dimension d. This property is preserved by any base change $S' \to S$.

For our current purposes, we can harmlessly replace F by its vertically pure quotient $\text{vpure}(F)$ (9.12). The generic fibers of $\text{vpure}(F)$ are pure of dimension d, but special fibers may have embedded points outside the flat locus (7.26). Vertically purity is preserved by flat base changes.

Definition–Lemma 7.28 (Divisorial support III) Let $g : X \to S$ be a flat morphism of pure relative dimension n and $g^\circ : X^\circ \to S$ the smooth locus of g.

Let F be a coherent sheaf on X that is generically flat and pure over S of dimension $n - 1$. Assume that for every $s \in S$, every generic point of F_s is contained in X°.

Set $U := \text{FlatCM}_S(X, F) \cap X^\circ$ and $j : U \hookrightarrow X$ the natural injection. We define the *divisorial support* of F over S as

$$\text{DSupp}_S(F) := \overline{\text{DSupp}_S(F|_U)}, \qquad (7.28.1)$$

the scheme-theoretic closure of $\text{DSupp}_S(F|_U)$. This makes sense since the latter is already defined by (7.23).

Note that $\text{Supp DSupp}_S(F) = \text{Supp } F$ and $\text{DSupp}_S(F)$ is a generically flat family of pure subschemes of dimension $n - 1$ over S, whose restriction to U is relatively Cartier.

It is enough to check the following equalities at codimension 1 points, which follow from (7.24) and (7.21.3).

Claim 7.28.2 Let $g_i : X_i \to S$ be flat morphisms of pure relative dimension n and $\pi : X_1 \to X_2$ a finite morphism. Let $D \subset X_1$ be a relative Mumford divisor.

Assume that red $D_s \to \text{red}(\pi(D_s))$ is birational and π is étale at generic points of D_s. Then $\text{DSupp}_S(\pi_* \mathcal{O}_D) = \pi(D)$, the scheme-theoretic image of D. □

Claim 7.28.3 (Divisorial support commutes with push-forward) Let $g_i \colon X_i \to S$ be flat morphisms of pure relative dimension n and $\pi \colon X_1 \to X_2$ a finite morphism. Let F be a coherent sheaf on X_1 that is generically flat and pure over S of relative dimension $n - 1$. Assume that g_1 (resp. g_2) is smooth at every generic point of F_s (resp. $\pi_* F_s$) for every $s \in S$. Then

$$\text{DSupp}_S(\pi_* F) = \text{DSupp}_S(\pi_* \mathcal{O}_{\text{DSupp}_S(F)}).$$ □

Claim 7.28.4 Let $g_i \colon X_i \to S$ be flat morphisms of pure relative dimension n and $\pi_1 \colon X_1 \to X_2$, $\pi_2 \colon X_2 \to X_3$ finite morphisms. Let F be a coherent sheaf on X_1 that is generically flat and pure over S of relative dimension $n - 1$. Assume that g_1 (resp. g_2, g_3) is smooth at every generic point of F_s (resp. $\pi_{1*}F_s$, $(\pi_2 \circ \pi_1)_* F_s$) for every $s \in S$. Then

$$\text{DSupp}_S((\pi_2 \circ \pi_1)_* F) = \text{DSupp}_S(\pi_{2*} \mathcal{O}_{\text{DSupp}_S(\pi_{1*}F)}).$$ □

Lemma 7.29 *Let $X \to S$ be a smooth morphism of pure relative dimension n. Let F be a coherent sheaf on X that is generically flat over S with pure fibers of dimension $n - 1$. Assume that either F is flat over S, or S is reduced.*
Then $\text{DSupp}_S F$ is a relative Cartier divisor.

Proof Assume first that F is flat over S. If $x \in X_s$ is a point of codimension ≤ 2, then F_s is CM at x, hence $\text{DSupp}_S F$ is a relative Cartier divisor at x by (7.23). Since $X \to S$ is smooth, $\text{DSupp}_S F$ is a relative Cartier divisor everywhere by (7.10.6).

For the second claim, our argument gives only that $\text{DSupp}_S F$ is a relative, generically Cartier divisor. By (4.34), it is then enough to check the conclusion after base change $T \to S$, where T is the spectrum of a DVR. Then X_T is regular, so $\text{DSupp}_T F_T$ is Cartier. □

7.30 (Restriction to divisors) Let (s, S) be a local scheme and $g \colon X \to S$ a flat morphism of pure relative dimension n. Let F be a generically flat family of pure sheaves of relative dimension $n - 1$ such that g is smooth at every generic point of $\text{Supp} F_s$. Let $D \subset X$ be a relative Cartier divisor.
(7.30.1) Assume that $g|_D$ is smooth and F is flat with CM fiber, at every generic point of $D \cap \text{Supp} F_s$. Then

$$\text{DSupp}_S(F|_D) = \text{vpure}((\text{DSupp}_S F)|_D).$$

(7.30.2) Assume in addition, that D contains neither a generic point of $\operatorname{Supp} F_s \setminus \operatorname{FlatCM}_S(X, F)$, nor a codimension ≥ 2 point of $\operatorname{Supp} F_s$ where $\operatorname{DSupp}_S F$ is not S_2, then

$$\operatorname{DSupp}_S(F|_D) = (\operatorname{DSupp}_S F)|_D.$$

Corollary 7.31 (Bertini theorem for divisorial support) *Let* $g: X \to S$ *be a flat morphism of pure relative dimension n and F a generically flat family of pure sheaves of dimension* $n - 1$ *over* S. *Fix* $s \in S$ *such that g is smooth at every generic point of* $\operatorname{Supp} F_s$. *Let D be a general member of a linear system on X, that is base point free in characteristic 0 and very ample in general. Then there is an open neighborhood* $s \in S^\circ \subset S$ *such that* $\operatorname{DSupp}_S(F|_D) = (\operatorname{DSupp}_S F)|_D$ *holds over* S°.

Lemma 7.32 (Divisorial support commutes with base change) *Let* $g: X \to S$ *be a flat morphism of pure relative dimension n and F a generically flat family of pure sheaves of dimension* $n - 1$ *over* S. *Assume that g is smooth at every generic point of* $\operatorname{Supp} F_s$, *for every* $s \in S$. *Let* $h: S' \to S$ *be a morphism. By base change, we get* $g': X' \to S'$ *and* $h_X: X' \to X$. *Then*

$$h_X^{[*]}(\operatorname{DSupp}_S F) = \operatorname{DSupp}_{S'}(h_X^* F),$$

where $h_X^{[*]}$ *is the generically Cartier pull-back (4.2.7).*

Proof Set $U := \operatorname{FlatCM}_S(X, F) \subset X$ with injection $j: U \hookrightarrow X$. Set $U' := h_X^{-1}(U)$ and $h_U: U' \to U$ the restriction of h_X. Then (7.24) shows the equality $h_U^*(\operatorname{DSupp}_S F|_U) = \operatorname{DSupp}_{S'}(h_U^*(F|_U))$.

By (7.27.4), $h_X^{[*]}(\operatorname{DSupp}_S F)$ is a generically flat family of pure divisors and it agrees with $\operatorname{DSupp}_{S'}(h_X^* F)$ over U'. Thus the two are equal. □

7.33 (Proof of 7.4.7) Assume that we have $f: X \to (s, S)$ of relative dimension n and relative Mumford divisors $D_1, D_2 \subset X$, where (s, S) is local. Let $\operatorname{FlatCM}_S(X) \subset X$ be the largest open subset where f has CM fibers and $Z = X \setminus \operatorname{FlatCM}_S(X)$. Note that $Z \to S$ has relative dimension $\leq n - 2$.

Let $\pi: X \to \mathbb{P}_S^n$ be a finite morphism. Set $P^\circ := \mathbb{P}_S^n \setminus \pi(Z)$ and $X^\circ := \pi^{-1}(P^\circ)$. Then $\pi: X^\circ \to P^\circ$ is finite and flat. If $(f) = D_1 - D_2$ then, by (7.21.8),

$$(\operatorname{norm}_{X^\circ/X^\circ}(f)) = \operatorname{DSupp}_S(D_1)|_{P^\circ} - \operatorname{DSupp}_S(D_2)|_{P^\circ}.$$

Since $Z \to S$ has relative dimension $\leq n - 2$, this implies that $\operatorname{DSupp}_S(D_1)$ and $\operatorname{DSupp}_S(D_2)$ are linearly equivalent. Thus, if one of them is relatively Cartier, then so is the other. □

7.4 Variants of K-Flatness

We introduce five versions of K-flatness, which may well be equivalent to each other. From the technical point of view, Cayley–Chow-flatness (or C-flatness) is the easiest to use, but a priori it depends on the choice of a projective embedding. Then most of the work in the next two sections goes to proving that a modified version (stable C-flatness) is equivalent to K-flatness, hence independent of the projective embedding.

7.34 (Projections of $\mathbb{P}^n$) Let S be an affine scheme. Projecting $\mathbb{P}^n_S$ from the section $(a_0 : \cdots : a_n)$ (where $a_i \in \mathcal{O}_S$) to the $(x_n = 0)$ hyperplane is given by

$$\pi \colon (x_0 : \cdots : x_n) \to (a_n x_0 - a_0 x_n : \cdots : a_n x_{n-1} - a_{n-1} x_n). \tag{7.34.1}$$

It is convenient to normalize $a_n = 1$ and then we get

$$\pi \colon (x_0 : \cdots : x_n) \to (x_0 - a_0 x_n : \cdots : x_{n-1} - a_{n-1} x_n). \tag{7.34.2}$$

Similarly, a Zariski open set of projections of $\mathbb{P}^n_S$ to $L^r = (x_n = \cdots = x_{r+1} = 0)$ is given by

$$\pi \colon (x_0 : \cdots : x_n) \to (x_0 - \ell_0(x_{r+1}, \ldots, x_n) : \cdots : x_r - \ell_r(x_{r+1}, \ldots, x_n)), \tag{7.34.3}$$

where the ℓ_i are linear forms.

Note that in affine coordinates, when we set $x_0 = 1$, the projections become

$$\pi \colon (x_1, \ldots, x_n) \to \left(\tfrac{x_1 - \ell_1}{1 - \ell_0}, \ldots, \tfrac{x_r - \ell_r}{1 - \ell_0} \right), \tag{7.34.4}$$

where again the ℓ_i are (homogeneous) linear forms in the $x_{r+1}, \ldots, x_n$. If $\ell_0 \equiv 0$, then we recover the linear projections, but in general the coordinate functions have a non-linear expansion

$$\tfrac{x_i - \ell_i}{1 - \ell_0} = (x_i - \ell_i)(1 + \ell_0 + \ell_0^2 + \cdots). \tag{7.34.5}$$

Finally, formal projections are given as

$$\pi \colon (x_1, \ldots, x_n) \to (x_1 - \phi_1(x_1, \ldots, x_n), \ldots, x_r - \phi_r(x_1, \ldots, x_n)), \tag{7.34.6}$$

where ϕ_i are power series such that $\phi_i(x_1, \ldots, x_r, 0, \ldots, 0) \equiv 0$ for every i.

7.35 (Approximation of formal projections) Let $v_m \colon \mathbb{P}^n_S \hookrightarrow \mathbb{P}^N_S$ (where $N = \binom{n+m}{n} - 1$) be the mth Veronese embedding. Pulling back the linear coordinates on $\mathbb{P}^N_S$ we get all the monomials of degree m. In affine coordinates $x_1, \ldots, x_n$ as above, we get all monomials of degree $\leq m$.

In particular, we see that given a formal projection π as in (7.34.6) and $m > 0$, there is a unique linear projection π_m of $\mathbb{P}_S^N$ such that $\pi_m \circ v_m$ is

$$(x_1, \ldots, x_n) \to (x_1 - \psi_1, \ldots, x_r - \psi_r), \quad \text{where}$$
$$\psi_i \equiv \phi_i \mod (x_1, \ldots, x_n)^{m+1}, \quad \text{and} \quad \deg \psi_i \leq m \quad \forall i. \tag{7.35.1}$$

That is, we can approximate formal projections by linear projections composed with a Veronese embedding. Thus it is reasonable to expect that K-flatness is very close to C-flatness for all Veronese images; this leads to the notion of stable C-flatness in (7.37.2).

The uniqueness of this approximation is not always an advantage. In practice we would like π_m to be in general position away from the chosen point. This is easy to achieve if we increase m a little. In particular, we get the following obvious result.

Claim 7.35.2 Let (s, S) be a local scheme and $Y \subset \mathbb{P}_S^n$ a closed subset of pure relative dimension d. Let $p \in Y_s$ be a closed point with maximal ideal $\mathbf{m}_p$ such that $x_0(p) \neq 0$. Fix $m \in \mathbb{N}$ and let $(\widehat{g}_1 : \cdots : \widehat{g}_e) : \widehat{Y}_p \to \widehat{\mathbb{A}}_S^e$ be a finite morphism. Then for every $M \geq m + 1$ there are $g_1, \ldots, g_e \in H^0(\mathbb{P}_S^n, \mathscr{O}_{\mathbb{P}_S^n}(M))$ such that $\pi \colon (x_0^M : g_1 : \cdots : g_e) : Y \to \mathbb{P}_S^e$ is a finite morphism, $\pi^{-1}(\pi(p)) \cap Y = \{p\}$, and $\widehat{g}_i \equiv g_i / x_0^M \mod \mathbf{m}_p^m$ for every i. $\qquad\square$

Despite having good approximations, the equivalence of K-flatness and stable C-flatness is not clear. The problem is the following.

Assume for simplicity that S is the spectrum of an Artinian ring A. For sheaves of dimension d, using the notation of (7.21.7), we can write the equation of $\mathrm{DSupp}(\widehat{\pi}_* \widehat{F})$ in the form $\det(v\mathbf{1}_r - M)) = 0$, where the entries of the matrix M involve rational functions in the power series ϕ_i. The problem is that inverses of power series usually do not have good approximations by rational functions. For example, there is no rational function $g(x_1, x_2)$ such that

$$(x_2 - \sin x_1)^{-1} - g(x_1, x_2) \in k[[x_1, x_2]].$$

The exception is the one-variable case, where truncations of Laurent series give good approximations. This is what we exploit in (7.60) to prove that K-flatness is equivalent to stable C-flatness for curves.

Definition 7.36 Let E be a vector bundle over a scheme S and $F \subset E$ a vector subbundle. This induces a natural *linear projection* map $\pi \colon \mathbb{P}_S(E) \dashrightarrow \mathbb{P}_S(F)$. If S is local, then E, F are free. After choosing bases, π is given by a matrix of constant rank with entries in $\mathscr{O}_S$. We call these $\mathscr{O}_S$-*projections* if we want to emphasize this. If S is over a field k, we can also consider *k-projections,* given by a matrix with entries in k. These, however, only make good sense if we have a canonical trivialization of E; this rarely happens for us.

We can now formulate various versions of K-flatness.

Definition 7.37 Let (s, S) be a local scheme with infinite residue field and F a generically flat family of pure, coherent sheaves of relative dimension d on $\mathbb{P}^n_S$ (7.27), with scheme-theoretic support $Y := \operatorname{SSupp} F$.

(7.37.1) F is *C-flat* over S iff $\operatorname{DSupp}(\pi_* F)$ is Cartier over S for every $\mathcal{O}_S$-projection $\pi\colon \mathbb{P}^n_S \dashrightarrow \mathbb{P}^{d+1}_S$ (7.36) that is finite on Y.

(7.37.2) F is *stably C-flat* iff $(v_m)_* F$ is C-flat for every Veronese embedding $v_m\colon \mathbb{P}^n_S \hookrightarrow \mathbb{P}^N_S$ (where $N = \binom{n+m}{n} - 1$).

(7.37.3) F is *K-flat* over S iff $\operatorname{DSupp}(\varrho_* F)$ is Cartier over S for every finite morphism $\varrho\colon Y \to \mathbb{P}^{d+1}_S$.

(7.37.4) F is *locally K-flat* over S at $y \in Y$ iff $\operatorname{DSupp}(\varrho_* F)$ is Cartier over S at $\varrho(y)$ for every finite $\varrho\colon Y \to \mathbb{P}^{d+1}_S$ for which $\{y\} = \operatorname{Supp} \varrho^{-1}(\varrho(y))$.

(7.37.5) F is *formally K-flat* over S at a closed point $y \in Y$ iff $\operatorname{DSupp}(\varrho_* \widehat{F})$ is Cartier over $\widehat{S}$ for every finite morphism $\varrho\colon \widehat{Y} \to \widehat{\mathbb{A}}^{d+1}_{\underline{S}}$, where $\widehat{S}$ (resp. $\widehat{Y}$) denotes the completion of S at s (resp. Y at y).

7.37.6 (Base change properties) We see in (7.50) that being C-flat is preserved by arbitrary base changes and the property descends from faithfully flat base changes. This then implies the same for stable C-flatness. Once we prove that the latter is equivalent to K-flatness, the latter also has the same base change properties. Most likely the same holds for formal K-flatness.

7.37.7 (General base schemes) We say that any of the above notions (7.37.1–5) holds for a local base scheme (s, S) (with finite residue field) if it holds after some faithfully flat base change $(s', S') \to (s, S)$, where $k(s')$ is infinite. Property (7.37.6) assures that this is independent of the choice of S'.

Finally, we say that any of the notions (7.37.1–5) holds for an arbitrary base scheme S, if it holds for all of its localizations.

Variants 7.38 These definitions each have other versions and relatives. I believe that each of the five are natural and maybe even optimal, though they may not be stated in the cleanest form. Here are some other possibilities and equivalent versions.

(7.38.1) It could have been better to define C-flatness using the Cayley–Chow form; the equivalence is proved in (7.47). The Cayley–Chow form version matches better with the study of Chow varieties; the definition in (7.37.1) emphasizes the similarity with the other four.

(7.38.2) In (7.37.2), it would have been better to say that F is *stably C-flat for* $L := \mathscr{O}_Y(1)$. However, we see in (7.62) that this notion is independent of the choice of an ample line bundle L, so we can eventually drop L from the name. (7.38.3) In (7.37.3), we get an equivalent notion if we allow all finite morphisms $\varrho \colon Y \to W$, where $W \to S$ is any smooth, projective morphism of pure relative dimension $d + 1$ over S. Indeed, let $\pi \colon W \to \mathbb{P}_S^{d+1}$ be a finite morphism. If F is K-flat then $\mathrm{DSupp}((\pi \circ \varrho)_* F)$ is a relative Cartier divisor, hence $\mathrm{DSupp}(\varrho_* F)$ is K-flat by (7.28.3). Since $W \to S$ is smooth, $\mathrm{DSupp}(\varrho_* F)$ is a relative Cartier divisor by (7.53).

(7.38.4) It would be natural to consider an affine version of C-flatness: We start with a coherent sheaf F on $\mathbb{A}_S^n$ and require that $\mathrm{DSupp}(\pi_* F)$ be Cartier over S for every projection $\pi \colon \mathbb{A}_S^n \to \mathbb{A}_S^{d+1}$ that is finite on Y.

The problem is that the relative affine version of Noether's normalization theorem does not hold, thus there may not be any such projections (10.47), though one can try to go around this using (10.46.2). This is why (7.37.4) is stated for projective morphisms only.

Although a more local version is defined in (7.51), we did not find a truly local theory. Nonetheless, the notions (7.37.1–4) are étale local on X, and most likely the following Henselian version of (7.37.5) does work.

(7.38.5) Assume that $f \colon (y, Y) \to (s, S)$ is a local morphism of pure relative dimension d of Henselian local schemes such that $k(y)/k(s)$ is finite. Let F be a coherent sheaf on X that is pure of relative dimension d over S. Then F is *K-flat* over S iff $\mathrm{DSupp}(\varrho_* F)$ is Cartier over S for every finite morphism $\varrho \colon Y \to \mathrm{Spec}\, \mathscr{O}_S \langle x_0, \ldots, x_d \rangle$ (where $R\langle \mathbf{x} \rangle$ denotes the Henselization of $R[\mathbf{x}]$).

It is possible that in fact all five versions (7.37.1–5) are equivalent to each other, but for now we can prove only 13 of the 20 possible implications. Four of them are easy to see.

Proposition 7.39 *Let F be a generically flat family of pure, coherent sheaves of relative dimension d on $\mathbb{P}_S^n$. Then*

$$\textit{formally K-flat} \Rightarrow \textit{K-flat} \Rightarrow \textit{locally K-flat} \Rightarrow \textit{stably C-flat} \Rightarrow \textit{C-flat}.$$

Proof A divisor D on a scheme X is Cartier iff its completion $\widehat{D}$ is Cartier on $\widehat{X}$ for every $x \in X$ by (7.11). Thus formally K-flat $\Rightarrow$ K-flat.

K-flat $\Rightarrow$ locally K-flat is clear, and locally K-flat $\Rightarrow$ stably C-flat follows from (7.52). Finally stably C-flat $\Rightarrow$ C-flat is clear; see also (7.56).   □

A key technical result of the chapter is the following, to be proved in (7.63).

Theorem 7.40 *K-flatness is equivalent to local K-flatness and to stable C-flatness.*

It is quite likely that our methods will prove the following.

Conjecture 7.41 *Formal K-flatness is equivalent to K-flatness.*

We prove the special case of relative dimension 1 in (7.60); this is also a key step in the proof of (7.40).

The remaining question is whether C-flat implies stably C-flat. This holds in the examples computed in Sections 7.8–7.9, but we do not have any conceptual argument why these two notions should be equivalent.

Question 7.42 Is C-flatness equivalent to stable C-flatness and K-flatness?

Next we show that K-flatness is automatic over reduced schemes and can be checked on Artinian subschemes.

Proposition 7.43 *Let S be a reduced scheme and F a generically flat family of pure, coherent sheaves on $\mathbb{P}^n_S$. Then F is K-flat over S.*

Proof This follows from (7.29.2). □

Proposition 7.44 *Let S be a scheme and F a generically flat family of pure, coherent sheaves on $\mathbb{P}^n_S$. Then F satisfies one of the properties (7.37.1–5) iff $\tau^* F$ satisfies the same property for every Artinian subscheme $\tau: A \hookrightarrow S$.*

Proof Set $Y := \mathrm{SSupp}\, F$ and let $\pi: Y \to \mathbb{P}^{d+1}_S$ be a finite morphism. By (7.11), $\mathrm{DSupp}_S(\pi_* F)$ is Cartier iff $\mathrm{DSupp}_A((\pi_A)_* \tau^* F)$ is Cartier for every Artinian subscheme $\tau: A \hookrightarrow S$. Thus the Artinian versions imply the global ones.

To check the converse, we may localize at $\tau(A)$. The claim is clear if every finite morphism $\pi_A: Y_A \to \mathbb{P}^{d+1}_A$ can be extended to $\pi: Y \to \mathbb{P}^{d+1}_S$. This is obvious for C-flatness, stable C-flatness, and formal K-flatness, but it need not hold for K-flatness and local K-flatness.

These cases will be established only after we prove (7.40) in (7.63). Thus we have to be careful not to use this direction in Section 7.5. □

7.45 (Push-forward, additivity and multiplicativity) First, as a generalization of (7.4.4), let $f: X \to S$ and $g: Y \to S$ be projective morphisms of pure relative dimension n and $\tau: X \to Y$ a finite morphism. Let F be a coherent sheaf on X that is generically flat and pure over S of dimension $n - 1$ such that

g is smooth at generic points of $f_*(F_s)$ for every $s \in S$. Let $\pi\colon Y \to \mathbb{P}^n_S$ be any finite morphism. Then

$$\mathrm{DSupp}_S((\pi \circ \tau)_* F) = \mathrm{DSupp}_S(\pi_*(\tau_* F)) = \mathrm{DSupp}_S(\pi_* \mathscr{O}_{\mathrm{DSupp}_S(\tau_* F)}),$$

where the first equality follows from the identity $\pi_*(\tau_* F) = (\pi \circ \tau)_* F$ and for the second we apply (7.28.3) to $\tau_* F$. This proves (7.4.4).

Additivity (7.4.5) is essentially a special case of this. Let $f\colon X \to S$ be a projective morphism of pure relative dimension n and $D_1, D_2 \subset X$ K-flat, relative Mumford divisors. Next take two copies $X' := X_1 \cup X_2$ of X, mapping to X by the identity map $\tau\colon X' \to X$. Let $D' \subset X'$ be the union of the divisors $D_i \subset X_i$. Then $\mathrm{DSupp}_S(\tau_* \mathscr{O}_{D'}) = D_1 + D_2$. Thus if the D_i are K-flat, then so is $D_1 + D_2$.

Finally, consider (7.4.6). If D is K-flat, then so is every mD by additivity, the interesting claim is the converse. Let $\pi\colon Y \to \mathbb{P}^n_S$ be any finite morphism. Set $E := \mathrm{DSupp}_S(\pi_* D)$. Then $mE = \mathrm{DSupp}_S(\pi_*(mD))$, thus we need to show that if mE is Cartier and char $k \nmid m$, then E is Cartier. This was treated in (4.37). $\square$

7.5 Cayley–Chow Flatness

Let $Z \subset \mathbb{P}^n$ be a subvariety of dimension d. Cayley (1860, 1862) associates to it the Cayley–Chow hypersurface

$$\mathrm{Ch}(Z) := \{L \in \mathrm{Gr}(n{-}d{-}1, \mathbb{P}^n)\colon Z \cap L \neq \emptyset\} \subset \mathrm{Gr}(n{-}d{-}1, \mathbb{P}^n).$$

We extend this definition to coherent sheaves on $\mathbb{P}^n_S$ over an arbitrary base scheme. We use two variants, but the proof of (7.47) needs two other versions as well. All of these are defined in the same way, but $\mathrm{Gr}(n{-}d{-}1, \mathbb{P}^n)$ is replaced by other universal varieties.

Definition 7.46 (Cayley–Chow hypersurfaces) Let S be a scheme and F a generically flat family of pure, coherent sheaves of dimension d on $\mathbb{P}^n_S$ (7.27). We define four versions of the Cayley–Chow hypersurface associated to F. In all four versions the left-hand side map σ is a smooth fiber bundle.

7.46.1 (Grassmannian version) Consider the diagram

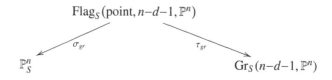

where the flag variety parametrizes pairs (point) $\in L^{n-d-1} \subset \mathbb{P}^n$. Set

$$\mathrm{Ch}_{gr}(F) := \mathrm{DSupp}_S((\tau_{gr})_*\sigma_{gr}^*F).$$

7.46.2 (Product version) Consider the diagram

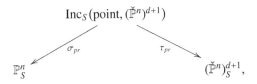

where the incidence variety parametrizes $(d + 2)$-tuples $((\text{point}), H_0, \ldots, H_d)$ satisfying (point) $\in H_i$ for every i. Set

$$\mathrm{Ch}_{pr}(F) := \mathrm{DSupp}_S((\tau_{pr})_*\sigma_{pr}^*F).$$

7.46.3 (Flag version) Consider the diagram

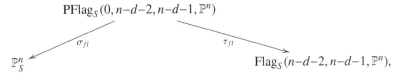

where PFlag parametrizes triples $((\text{point}), L^{n-d-2}, L^{n-d-1})$ such that (point) $\in L^{n-d-1}$ and $L^{n-d-2} \subset L^{n-d-1}$ (but the point need not lie on L^{n-d-2}). Set

$$\mathrm{Ch}_{fl}(F) := \mathrm{DSupp}_S((\tau_{fl})_*\sigma_{fl}^*F).$$

7.46.4 (Incidence version) Consider the diagram

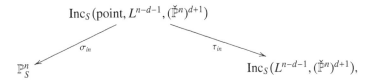

where the $(d+3)$-tuples $((\text{point}), L^{n-d-1}, H_0, \ldots, H_d)$ satisfy (point) $\in L^{n-d-1} \subset H_i$ for every i. Set

$$\mathrm{Ch}_{in}(F) := \mathrm{DSupp}_S((\tau_{in})_*\sigma_{in}^*F).$$

Theorem 7.47 *Let S be a scheme and F a generically flat family of pure, coherent sheaves of dimension d on $\mathbb{P}_S^n$. The following are equivalent:*

(7.47.1) $\mathrm{Ch}_{pr}(F) \subset (\check{\mathbb{P}}^n)_S^{d+1}$ *is Cartier over S.*

(7.47.2) $\mathrm{Ch}_{gr}(F) \subset \mathrm{Gr}_S(n - d - 1, \mathbb{P}^n)$ *is Cartier over S.*

If S is local with infinite residue field, then these are also equivalent to

(7.47.3) $\mathrm{DSupp}(\pi_* F)$ *is Cartier over* S *for every* $\mathcal{O}_S$-*projection* $\pi\colon \mathbb{P}^n_S \dashrightarrow \mathbb{P}^{d+1}_S$ *(7.36) that is finite on* $\mathrm{Supp}\, F$.

(7.47.4) $\mathrm{DSupp}(\pi_* F)$ *is Cartier over* S *for a dense set of* $\mathcal{O}_S$-*projections* $\pi\colon \mathbb{P}^n_S \dashrightarrow \mathbb{P}^{d+1}_S$.

Proof The extreme cases $d = 0$ and $d = n - 1$ are somewhat exceptional, so we deal with them first.

If $d = n-1$, then $\mathrm{Gr}_S(n-d-1, \mathbb{P}^n_S) = \mathrm{Gr}_S(0, \mathbb{P}^n_S) \simeq \mathbb{P}^n_S$ and the only projection is the identity. Furthermore, $\mathrm{Ch}_{gr}(F) = \mathrm{DSupp}_S(F)$ by definition, so (7.47.2) and (7.47.3) are equivalent. If these hold, then $\mathrm{Ch}_{pr}(F) = \mathrm{Ch}_{pr}(\mathrm{DSupp}_S(F))$ is also flat by (7.23). For (7.47.1) $\Rightarrow$ (7.47.2), the argument in (7.48) works.

If $d = 0$, then F is flat over S and (7.47.1–3) hold by (7.29).

We may thus assume from now on that $0 < d < n - 1$. These cases are discussed in (7.48–7.49). $\square$

7.48 (Proof of 7.47.1 $\Leftrightarrow$ 7.47.2) To go between the product and the Grassmannian versions, the basic diagram is the following.

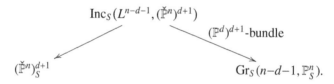

The right-hand side projection

$$\pi_2\colon \mathrm{Inc}_S(L^{n-d-1}, (\check{\mathbb{P}}^n)^{d+1}) \to \mathrm{Gr}_S(n-d-1, \mathbb{P}^n_S)$$

is a $(\mathbb{P}^d)^{d+1}$-bundle. Therefore $\mathrm{Ch}_{in}(F) = \pi_2^* \mathrm{Ch}_{gr}(F)$. Thus $\mathrm{Ch}_{gr}(F)$ is Cartier over S iff $\mathrm{Ch}_{in}(F)$ is Cartier over S. It remains to compare $\mathrm{Ch}_{in}(F)$ and $\mathrm{Ch}_{pr}(F)$.

The left-hand side projection

$$\pi_1\colon \mathrm{Inc}_S(L^{n-d-1}, (\check{\mathbb{P}}^n)^{d+1}) \to (\check{\mathbb{P}}^n)^{d+1}_S$$

is birational. It is an isomorphism over $(H_0, \ldots, H_d) \in (\check{\mathbb{P}}^n)^{d+1}_S$ iff $\dim(H_0 \cap \cdots \cap H_d) = n-d-1$, the smallest possible. That is, when the rank of the matrix formed from the equations of the H_i is $d + 1$. Thus π_1^{-1} is an isomorphism outside a subset of codimension $n + 1 - d$ in each fiber of $(\check{\mathbb{P}}^n)^{d+1}_S \to S$.

Therefore, if $\mathrm{Ch}_{in}(F)$ is Cartier over S then $\mathrm{Ch}_{pr}(F)$ is Cartier over S, outside a subset of codimension $n + 1 - d \ge 3$ on each fiber of $(\check{\mathbb{P}}^n)^{d+1}_S \to S$. Then $\mathrm{Ch}_{pr}(F)$ is Cartier over S everywhere by (7.10.6).

Conversely, let E be the support of the π_1-exceptional divisor. If $\mathrm{Ch}_{pr}(F)$ is a relative Cartier divisor, then so is $\pi_1^* \mathrm{Ch}_{pr}(F)$, which agrees with $\mathrm{Ch}_{in}(F)$ outside E.

Note that E consists of those $(L^{n-d-1}, H_0, \dots, H_d)$ for which $H_0, \dots, H_d$ are linearly dependent. This is easiest to describe using π_2, which is a $(\mathbb{P}^d)^{d+1}$-bundle over $\mathrm{Gr}_S(n-d-1, \mathbb{P}_S^n)$. In a local trivialization, the points in the ith copy of $\mathbb{P}^d$ have coordinates $(a_{i,0}: \cdots : a_{i,d})$. Then the equation of E is $\det(a_{i,j}) = 0$. Thus E is irreducible and the restriction of π_2

$$\mathrm{Inc}_S\left(L^{n-d-1}, (\check{\mathbb{P}}^n)^{d+1}\right) \setminus E \to \mathrm{Gr}_S(n-d-1, \mathbb{P}_S^n)$$

is surjective. Since $\mathrm{Ch}_{in}(F) = \pi_2^* \mathrm{Ch}_{gr}(F)$, this implies that $\mathrm{Ch}_{gr}(F)$ is relative Cartier (2.92.1). $\quad\square$

7.49 (Proof of 7.47.2 $\Rightarrow$ 7.47.3 $\Rightarrow$ 7.47.4 $\Rightarrow$ 7.47.2) To go between the Grassmannian version and the projection versions, the basic diagram is the following:

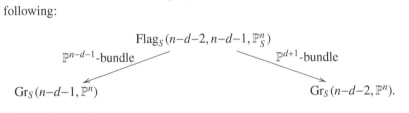

The left-hand side projection

$$\varrho_1 \colon \mathrm{Flag}_S(n-d-2, n-d-1, \mathbb{P}_S^n) \to \mathrm{Gr}_S(n-d-1, \mathbb{P}_S^n)$$

is a $\mathbb{P}^{n-d-1}$-bundle and $\mathrm{Ch}_{fl}(X) = \varrho_1^* \mathrm{Ch}_{gr}(X)$. Thus $\mathrm{Ch}_{gr}(F)$ is Cartier over S iff $\mathrm{Ch}_{fl}(F)$ is Cartier over S.

The right-hand side projection

$$\varrho_2 \colon \mathrm{Flag}_S(n-d-2, n-d-1, \mathbb{P}_S^n) \to \mathrm{Gr}_S(n-d-2, \mathbb{P}_S^n)$$

is a $\mathbb{P}^{d+1}$-bundle, but $\mathrm{Ch}_{fl}(X)$ is not a pull-back from $\mathrm{Gr}_S(n-d-2, \mathbb{P}_S^n)$.

Let $L \subset \mathbb{P}_S^n$ be a flat family of $(n-d-2)$-planes over S. The preimage of $[L]$ is the set of all $n-d-1$-planes that contain L; we can identify this with sections of the target of the projection $\pi_L \colon \mathbb{P}^n \dashrightarrow L^\perp$. Thus the restriction of $\mathrm{Ch}_{fl}(X)$ to the preimage of L is $\mathrm{DSupp}((\pi_L)_*(F))$.

So, if $\mathrm{Ch}_{fl}(F)$ is Cartier over S, then $\mathrm{DSupp}((\pi_L)_*(F)) = \mathrm{Ch}_{fl}(F)|_{L^\perp}$ is also Cartier over S. Thus (7.47.2) $\Rightarrow$ (7.47.3) and (7.47.3) $\Rightarrow$ (7.47.4) is obvious.

Conversely, assume that $\mathrm{DSupp}((\pi_L)_*(F))$ is Cartier over S for general L. By (7.10.6) it is enough to show that $\mathrm{Ch}_{fl}(F)$ is flat over S, outside a subset of codimension ≥ 3.

Let $U_F \subset \mathrm{Gr}_S(n{-}d{-}2, \mathbb{P}^n_S)$ be the open subset consisting of those L^{n-d-2} that are disjoint from $\mathrm{DSupp}(F)$. The restriction of the projection π_f to $\mathrm{Supp}\,\sigma^*_f F$ is finite over $\varrho_2^{-1} U_F$, thus $\mathrm{Ch}_f(F) = \mathrm{DSupp}_S((\pi_f)_* \sigma^*_f F)$ is flat over S, outside a codimension ≥ 2 subset of each fiber of $\varrho_2^{-1} U_F \to U_F$ by (7.29). By assumption, the non-flat locus is disjoint from the generic fiber, hence the non-flat locus has codimension ≥ 3 over U_F.

It remains to understand what happens over $Z_F := \mathrm{Gr}_S(n{-}d{-}2, \mathbb{P}^n_S) \setminus U_F$. Note that $\varrho_2^{-1}(Z_F)$ has codimension 2 in $\mathrm{Flag}_S(n{-}d{-}2, n{-}d{-}1, \mathbb{P}^n_S)$, so it is enough to show that $\mathrm{Ch}_{fl}(F)$ is flat over S at a general point of a general fiber over Z_F.

Thus let L^{n-d-2} be a general point of Z_F. Then $\mathrm{DSupp}(F) \cap L^{n-d-2}$ is a single point p and F is flat over S at p. Furthermore, a general $L^{n-d-1} \supset L^{n-d-2}$ still intersects $\mathrm{DSupp}(F)$ only at p. Thus $\sigma^*_{fl}(F)$ is flat over S at

$$(p, L^{n-d-2}, L^{n-d-1}) \in \mathrm{PFlag}_S(0, n{-}d{-}2, n{-}d{-}1, \mathbb{P}^n),$$

and $\mathrm{Supp}\,\sigma^*_{fl} F$ is finite over $(L^{n-d-2}, L^{n-d-1}) \in \mathrm{Flag}_S(n{-}d{-}2, n{-}d{-}1, \mathbb{P}^n_S)$.

Since $\mathrm{Ch}_{fl}(F) = \mathrm{DSupp}_S((\pi_{fl})_* \sigma^*_{fl} F)$ by (7.46.3), it is flat over S at the point (L^{n-d-2}, L^{n-d-1}) by (7.29). $\square$

Corollary 7.50 *Let S be a scheme and F a generically flat family of pure, coherent sheaves of dimension d on $\mathbb{P}^n_S$. Let $h\colon S' \to S$ be a morphism. By base change, we get $g'\colon X' \to S'$ and $F' = \mathrm{vpure}(h_X^* F)$ (9.12).*
(7.50.1) If F is C-flat, then so is F'.
(7.50.2) If F' is C-flat and h is scheme-theoretically dominant, then F is C-flat.

Proof We may assume that S is local with infinite residue field. Being C-flat is exactly (7.47.3), which is equivalent to (7.47.1). $F \mapsto \mathrm{Ch}_{pr}(F)$ commutes with base change by (7.32) and, if h is scheme-theoretically dominant, then, by (4.28), a divisorial sheaf is Cartier iff its divisorial pull-back is. $\square$

Definition 7.51 Let S be a local scheme with infinite residue field and F a generically flat family of pure, coherent sheaves of dimension d over S (7.27). F is *locally C-flat* over S at $y \in Y := \mathrm{SSupp}\,F$ iff $\mathrm{DSupp}(\pi_* F)$ is Cartier over S at $\pi(y)$ for every $\mathscr{O}_S$-projection $\pi\colon \mathbb{P}^n_S \dashrightarrow \mathbb{P}^{d+1}_S$ that is finite on Y for which $\{y\} = \mathrm{Supp}(\pi^{-1}(\pi(y)) \cap Y)$.

Lemma 7.52 *Let S be a local scheme with infinite residue field and F a generically flat family of pure, coherent sheaves of dimension d on $\mathbb{P}^n_S$. Then F is C-flat iff it is locally C-flat at every point.*

Proof It is clear that C-flat implies locally C-flat. Conversely, assume that F is locally C-flat. Set $Z_s := \text{Supp}(F_s) \setminus \text{FlatCM}_S(X, F)$ and pick points $\{y_i : i \in I\}$, one in each irreducible component of Z_s. If $\pi: \mathbb{P}_S^n \dashrightarrow \mathbb{P}_S^{d+1}$ is a general $\mathcal{O}_S$-projection, then $\{y_i\} = \pi^{-1}(\pi(y_i)) \cap Y$ for all $i \in I$.

Note that $\text{DSupp}(\pi_* F)$ is a relative Cartier divisor along $\mathbb{P}_S^{d+1} \setminus \pi(Z_s)$ by (7.23) and it is also relative Cartier at the points $\pi(y_i)$ for $i \in I$ since F is locally C-flat. Thus $\text{DSupp}(\pi_* F)$ is a relative Cartier divisor outside a codimension ≥ 3 subset of $\mathbb{P}_S^{d+1}$, hence a relative Cartier divisor everywhere by (7.10.6). □

Corollary 7.53 *Let (s, S) be a local scheme and $X \subset \mathbb{P}_S^n$ a closed subscheme that is flat over S of pure relative dimension $d + 1$. Let $D \subset X$ be a relative Mumford divisor. Let $x \in X_s$ be a smooth point. Then $\mathcal{O}_D$ is locally C-flat at x iff D is a relative Cartier divisor at x.*

Proof We may assume that S has infinite residue field. A general linear projection $\pi: X \to \mathbb{P}_S^{d+1}$ is étale at x, and $D \cap \pi^{-1}(\pi(x)) = \{x\}$. Thus $\pi|_D : D \to \pi(D)$ is a local isomorphism at x, hence D is a relative Cartier divisor at x iff $\pi(D)$ is a relative Cartier divisor at $\pi(x)$. By (7.28.2) $\text{DSupp}_S(\pi_* \mathcal{O}_D) = \pi(D)$, thus D is a relative Cartier divisor at x iff $\text{DSupp}_S(\pi_* \mathcal{O}_D)$ is a relative Cartier divisor at $\pi(x)$. That is, iff $\mathcal{O}_D$ is locally C-flat at x. □

Corollary 7.54 *Let S be a scheme and F a generically flat family of pure, coherent sheaves of dimension d over S. If F is flat at $y \in Y := \text{SSupp} F$ then it is also locally C-flat at y.*

Proof We may assume that (s, S) is local. By (10.17), F_s is CM outside a subset $Z_s \subset Y_s$ of dimension $\leq d - 2$. Let $W_s \subset Y_s$ be the set of points where F is not flat. Let $\pi: Y \to \mathbb{P}_S^{d+1}$ be a general linear projection. By (7.23), $\text{DSupp}(\pi_* F)$ is a relative Cartier divisor outside $\pi(Z_s \cup W_s)$, so we may assume that $\pi(y) \notin \pi(W_s)$. Thus, in a neighborhood of $\pi(y)$, $\text{DSupp}(\pi_* F)$ is a relative Cartier divisor outside $\pi(Z_s)$, which has dimension $\leq d - 2$. Thus $\text{DSupp}(\pi_* F)$ is a relative Cartier divisor at y by (7.10.6). □

Lemma 7.55 *Let S be a scheme and F a generically flat family of pure, coherent sheaves of dimension d on $\mathbb{P}_S^n$. Let $g_m: Y \hookrightarrow \mathbb{P}_S^N$ be an embedding such that $g_m^* \mathcal{O}_{\mathbb{P}_S^N}(1) \simeq \pi^* \mathcal{O}_{\mathbb{P}_S^{d+1}}(m)$. If $(g_m)_* F$ is C-flat then F is C-flat.*

Proof We may assume that S is local with infinite residue field. Let $\pi: \mathbb{P}_S^n \dashrightarrow \mathbb{P}_S^{d+1}$ be a general linear projection. We need to show that $\text{DSupp}(\pi_* F)$ is a relative Cartier divisor.

Choosing $d + 2$ general sections of $\mathcal{O}_{\mathbb{P}_S^{d+1}}(m)$ gives a morphism $w_m \colon \mathbb{P}_S^{d+1} \to$
$\mathbb{P}_S^{d+1}$. There is a linear projection $\varrho \colon \mathbb{P}_S^N \dashrightarrow \mathbb{P}_S^{d+1}$ such that $w_m \circ \pi = \varrho \circ g_m$. By
assumption $\mathrm{DSupp}((\varrho \circ g_m)_* F)$ is a relative Cartier divisor, hence so is

$$\mathrm{DSupp}((w_m \circ \pi)_* F) = \mathrm{DSupp}((w_m)_* \mathcal{O}_{\mathrm{DSupp}(\pi_* F)}),$$

where the equality follows from (7.30.2).

Pick a point $x \in \mathrm{DSupp}(\pi_* F)$. Then a general w_m is étale at x and also $\{x\} =$
$w_m^{-1}(w_m(x)) \cap \mathrm{DSupp}(\pi_* F)$. Thus $w_m \colon \mathrm{DSupp}(\pi_* F) \to \mathrm{DSupp}((w_m \circ \pi)_* F)$ is
étale at x. Thus $\mathrm{DSupp}(\pi_* F)$ is Cartier at x.   □

Corollary 7.56 *Let S be a scheme and F a generically flat family of pure,
coherent sheaves of dimension d on $\mathbb{P}_S^n$. Let $v_m \colon \mathbb{P}_S^n \hookrightarrow \mathbb{P}_S^N$ be the mth Veronese
embedding. If $(v_m)_* F$ is C-flat then so is F.*   □

There are very useful Bertini theorems for C-flatness. The going-down
versions are straightforward.

Lemma 7.57 *Let (s, S) be a local scheme and F a C-flat family of pure, coherent sheaves of dimension $d \geq 1$ on $\mathbb{P}_S^n$ (7.27). Then there is a finite set of points
$\Sigma \subset \mathrm{Supp}\, F_s$ with the following property.*

*Let $H \subset \mathbb{P}_S^n$ be a hyperplane that does not contain any point in Σ and H_s is
smooth at generic points of $H \cap \mathrm{Supp}\, F_s$. Then $F|_H$ is C-flat.*

Proof We may assume that the residue field is infinite. Every projection $H \dashrightarrow$
$\mathbb{P}_S^d$ is obtained as the restriction of a projection $\mathbb{P}_S^n \dashrightarrow \mathbb{P}_S^{d+1}$. The rest follows
from (7.30.2).   □

Corollary 7.58 *Let (s, S) be a local scheme and F a stably C-flat family of
pure, coherent sheaves of dimension $d \geq 1$ on $\mathbb{P}_S^n$. Set $Y := \mathrm{SSupp}\, F$. Let
$D \subset Y$ be a relative Cartier divisor that does not contain any point in Σ (7.57)
and D_s is smooth at generic points of $D \cap \mathrm{Supp}\, F_s$. Then $F|_D$ is also stably
C-flat.*

Proof We may assume that the residue field is infinite. By (7.52) it is sufficient
to prove that $F|_D$ is locally C-flat. Pick a point $y \in D$ and let $H \supset \mathbb{P}_S^n$ be a
general hypersurface such that $H \cap Y$ equals D in a neighborhood of y. After a
Veronese embedding, H becomes a hyperplane section, and then (7.57) implies
that $F|_H$ is stably C-flat. Hence $F|_H$ is locally C-flat by (7.52) and so $F|_D$ also
locally C-flat at y.   □

The going-up version needs a little more care.

Lemma 7.59 *Let* (s, S) *be a local Artinian scheme with infinite residue field and F a generically flat family of pure, coherent sheaves of dimension* $d \geq 2$ *on* $\mathbb{P}^n_S$. *Then F is C-flat iff* $F|_H$ *is C-flat for a dense set of hyperplanes* $H \subset \check{\mathbb{P}}^n_S$.

Proof The hyperplanes are parametrized by $H^0(\mathbb{P}^n_S, \mathcal{O}_{\mathbb{P}^n_S}(1)) \simeq \mathcal{O}^{n+1}_S$. Since $\mathcal{O}_S$ is Artinian, it makes sense to talk about a dense set of hyperplanes. (This is the only reason why the lemma is stated for Artinian schemes.)

One direction follows from (7.57). Conversely, if $F|_H$ is C-flat for a dense set of hyperplanes H, then there is a dense set of projections $\pi \colon \mathbb{P}^n_S \dashrightarrow \mathbb{P}^{d+1}_S$ such that, for a dense set of hyperplanes $L \subset \mathbb{P}^{d+1}_S$, the restriction of F to $\pi^{-1}(L)$ is C-flat. Thus $\mathrm{DSupp}(\pi_* F)$ is a relative Cartier divisor in an open neighborhood of such an L by (7.31). Since $d \geq 2$, this implies that $\mathrm{DSupp}(\pi_* F)$ is a relative Cartier divisor everywhere by (7.10.6). Thus F is C-flat by (7.47). □

Now we come to the key result.

Proposition 7.60 *Let* (s, S) *be a local scheme and F a generically flat family of pure, coherent sheaves of dimension* 1 *on* $\mathbb{P}^n_S$. *Then F is stably C-flat* $\Leftrightarrow$ *K-flat* $\Leftrightarrow$ *formally K-flat.*

Proof By (7.39) formally K-flat $\Rightarrow$ K-flat $\Rightarrow$ stably C-flat.

Thus assume that F is stably C-flat. Set $Y := \mathrm{SSupp}\, F$ and pick a closed point $p \in Y$. We need to show that F is formally K-flat at p. By the already proved parts of (7.44), it is enough to prove this for Artinian base schemes with infinite residue field. We may thus assume that $S = \mathrm{Spec}\, A$ for a local Artinian ring (A, n_A, k) with k infinite, and $p \in Y(k)$ is the origin $(1{:}0{:}\cdots{:}0)$.

Let $\pi \colon \widehat{Y} \to \widehat{\mathbb{A}}^2_S = \mathrm{Spec}\, A[[u, v]]$ be a finite morphism. We need to show that $\mathrm{DSupp}(\pi_* \widehat{F})$ is Cartier.

Let m_0 be as in (7.61). By (7.35.2), for $m \gg m_0$ we can choose homogeneous polynomials $g_1, g_2 \in H^0(\mathbb{P}^n_A, \mathcal{O}_{\mathbb{P}^n_A}(m))$ such that

$$\tau \colon Y \to \mathbb{P}^2_S \quad \text{given by} \quad (x_0^m {:} g_1 {:} g_2) \tag{7.60.1}$$

is a finite morphism, p is the only point of Y that maps to $(1{:}0{:}0)$,

$$g_1/x_0^m \equiv \pi^* u \mod n_R^{m_0}, \quad \text{and} \quad g_2/x_0^m \equiv \pi^* v \mod n_R^{m_0}, \tag{7.60.2}$$

where n_R is the ideal sheaf of $p \in Y$.

Since F is stably C-flat, $\mathrm{DSupp}(\tau_* F)$ is a Cartier divisor and so is its completion at the image of p. Then $\mathrm{DSupp}(\pi_* \widehat{F})$ is Cartier by (7.61). □

Proposition 7.61 *Let (A, n_A, k) be an Artinian k-algebra, (R, n_R) a local, S_1, generically flat A-algebra of dimension 1, and F a generically free, finite R-module. Let $\pi \colon \operatorname{Spec} R \to \operatorname{Spec} A[[u, v]]$ be a projection such that R is finite over $A[[u]]$ and $\pi^* u, \pi^* v$ are non-zerodivisors. Then there is an m_0 such that*

(7.61.1) *if $\tau \colon \operatorname{Spec} R \to \operatorname{Spec} A[[u, v]]$ satisfies $\tau^* u \equiv \pi^* u \mod n_R^{m_0}$ and $\tau^* v \equiv \pi^* v \mod n_R^{m_0}$, then $\operatorname{DSupp}(\pi_* F)$ is Cartier iff $\operatorname{DSupp}(\tau_* F)$ is.*

Proof We follow the computation of $\operatorname{DSupp}(\pi_* F)$ as in (7.21) and show that the formula for $\operatorname{DSupp}(\tau_* F)$ is very similar. Then we finish using (7.16).

Set $s := \pi^* u$. Since R is finite over $A[[u]]$, (s) is n_R-primary, hence $n_R^e \subset (s)$ for some $e \geq 1$. Since F is generically free over $A[[s]]$, it contains a free $A[[s]]$-module $G = \oplus_j A[[s]]e_j$ of the same generic rank $= r$. Since R is a finite $A[[s]]$-algebra, $RG \subset s^{-c}G$ for some $c \geq 0$. Hence $\operatorname{DSupp}(\pi_* F)$ agrees with $\operatorname{DSupp}(\pi_* G)$ on the open set $(u \neq 0)$.

We can thus compute $\operatorname{DSupp}(\pi_* F)$ using multiplication by $\pi^* v$ on G, which is given by a meromorphic matrix

$$M_\pi(s) : \oplus_j A[[s]]e_j \simeq G \xrightarrow{\pi^* v} s^{-d}G \simeq \oplus_j s^{-d}A[[s]]e_j$$

for some $d \geq 0$. Our bound on m_0 depends on r, c, d, e, and $\operatorname{nil}(n_A)$.

Claim 7.61.2 If $s_1 \equiv s \mod (s^m)$ and $m \geq c + 1$, then $s_1^r G = s^r G$ for $r \geq 0$.

Proof Note that $s_1 G \subset sG + s^{m-c}(s^c RG) \subset sG + s^{m-c}G \subset sG$. Also, $s_1^c RG = Rs_1^c G \subset Rs^c G = s^c RG \subset G$, thus we can interchange s, s_1 in the previous argument to get that $s_1 G = sG$. □

In particular, if $t := \tau^* u \equiv \pi^* u \mod (s^m)$ and $m \geq c + 1$, then $G = \oplus_j A[[t]]e_j$. Thus we can use the same G for computing the divisorial support of $\tau_* F$. Multiplication by $\tau^* v$ is given by another meromorphic matrix $M_\tau(t) : G \to t^{-d}G$. Next we compare M_π and M_τ.

Claim 7.61.3 Assume that $\tau^* v \equiv \pi^* v \mod (s^{m+c+d})$ and $t \equiv s \mod (s^{m+c})$. Then $M_\pi(u) \equiv M_\tau(u) \mod u^m A[[u]]$.

Proof The assumptions imply that $G/s^m G = G/t^m G$, $s^{-d}G/s^m G = t^{-d}G/s^m G$, and $\tau^* v, \pi^* v$ induce the same map $G/s^m G \to s^{-d}G/s^m G$. □

Claim 7.61.4 Assume that $M_\pi(u) \equiv M_\tau(u) \mod u^{m+rd-d}A[[u]]$. Then

$$\det(v \mathbf{1}_r - M_\pi) \equiv \det(v \mathbf{1}_r - M_\tau) \mod u^m A[[u]].$$

Proof The difference of the two sides involves terms that contain at most $r - 1$ entries of M_π and at least one entry of $M_\pi - M_\tau$. □

Putting these together, we get that if (7.61.1) holds and m_0 is large enough, then $\det(v\mathbf{1}_r - M_\pi) \equiv \det(v\mathbf{1}_r - M_\tau) \mod u^m A[[u]]$ and $m \geq \mathrm{nil}(n_A) \cdot d$. The proposition now follows from (7.16). $\qquad\square$

Corollary 7.62 *Let (s, S) be a local scheme and F a generically flat family of pure, coherent sheaves of dimension $d \geq 1$ on $\mathbb{P}^n_S$. Let L, M be relatively ample line bundles on $Y := \mathrm{SSupp}\, F$. Then F is stably C-flat for L (as in (7.38.2)) iff it is stably C-flat for M.*

Proof We already proved (7.44) for stable C-flatness, thus it is enough to prove our claim when S is Artinian with infinite residue field.

Assume that F is stably C-flat for M. By (7.56), we may assume that L is very ample. Repeatedly using (7.58) we get that, for general $L_i \in |L|$, the restriction of F to the complete intersection curve $L_1 \cap \cdots \cap L_{d-1} \cap Y$ is stably C-flat for M. Thus the restriction of F to $L_1 \cap \cdots \cap L_{d-1} \cap Y$ is formally K-flat by (7.60). Using (7.60) in the other direction for L, we get that the restriction of F to $L_1 \cap \cdots \cap L_{d-1} \cap Y$ is stably C-flat for L. Now we can use (7.59) to conclude that F is stably C-flat for L. $\qquad\square$

7.63 (Proof of 7.40 and 7.44) We already noted in (7.39) that K-flat $\Rightarrow$ stably C-flat.

To see the converse, assume that F is stably C-flat. We aim to prove that it is K-flat. By the already established directions of (7.44), it is enough to prove this over Artinian rings. Thus assume that S is the spectrum of an Artinian ring and let $\pi\colon X \to \mathbb{P}^{d+1}_S$ be a finite projection. Set $L := \pi^* \mathscr{O}_{\mathbb{P}^{d+1}_S}(1)$. By (7.62) F is stably C-flat for L, hence $\mathrm{DSupp}(\pi_* F)$ is a relative Cartier divisor by (7.55). This proves (7.40).

We already proved (7.44) for stable C-flatness. By the just established (7.40), stable C-flatness is equivalent to K-flatness and local C-flatness, hence (7.44) also holds for these. $\qquad\square$

7.6 Representability Theorems

Definition 7.64 Let S be a scheme and F a generically flat family of pure, coherent sheaves on $\mathbb{P}^n_S$. As in (3.16.1), the *functor of K-flat pull-backs* is

$$\mathcal{K}\!flat_F(q\colon T \to S) = \begin{cases} \{\emptyset\} & \text{if } q_{\mathbb{P}}^{[*]} F \to T \text{ is K-flat, and} \\ \emptyset & \text{otherwise,} \end{cases}$$

where $q_{\mathbb{P}} \colon \mathbb{P}^n_T \to \mathbb{P}^n_S$ is the induced morphism and $q_{\mathbb{P}}^{[*]} F := \mathrm{vpure}(q_{\mathbb{P}}^* F)$ is the divisorial pull-back as in (4.2.7) or (9.12). If $Y \subset \mathbb{P}^n_S$ is a generically flat family of pure subschemes then we write $\mathcal{K}\mathit{flat}_Y$ instead of $\mathcal{K}\mathit{flat}_{\mathcal{O}_Y}$.

If $\mathcal{K}\mathit{flat}_F$ is representable by a morphism, we denote it by $j^{\mathrm{kflat}}_F \colon S^{\mathrm{kflat}}_F \to S$. Note that j^{kflat}_F is necessarily a monomorphism.

One defines analogously the *functor of C-flat pull-backs* $C\mathit{flat}_F$ and the *functor of stably C-flat pull-backs* $SC\mathit{flat}_F$. The monomorphisms representing them are denoted by $j^{\mathrm{cflat}}_F \colon S^{\mathrm{cflat}}_F \to S$ and $j^{\mathrm{scflat}}_F \colon S^{\mathrm{scflat}}_F \to S$.

In our cases, several of the monomorphisms are subschemes $S^* \hookrightarrow S$ such that $\mathrm{red}\, S = \mathrm{red}\, S^*$. (In particular, $S^* \subset S$ is both open and closed.) We call such a subscheme *full*.

Proposition 7.65 *Let S be a scheme and F a generically flat family of pure, coherent sheaves of dimension d on $\mathbb{P}^n_S$. Then the functors of C-flat, stably C-flat or K-flat pull-backs of F are represented by full subschemes*

$$S^{\mathrm{kflat}}_F = S^{\mathrm{scflat}}_F \subset S^{\mathrm{cflat}}_F \subset S.$$

Proof By (7.47), $j^{\mathrm{cflat}}_F \colon S^{\mathrm{cflat}}_F \to S$ is the same as $j^{\mathrm{car}}_{\mathrm{Ch}_{pr}(F)} \colon S^{\mathrm{car}}_{\mathrm{Ch}_{pr}(F)} \to S$, with the Cayley–Chow hypersurface $\mathrm{Ch}_{pr}(F)$ as defined in (7.46.2). Thus (4.28) gives $S^{\mathrm{cflat}}_F \subset S$.

We can apply this to each Veronese embedding $v_m \colon \mathbb{P}^n_S \hookrightarrow \mathbb{P}^N_S$, to get full subschemes $S^{\mathrm{cflat}}_{v_m(F)} \subset S$. Their intersection gives $S^{\mathrm{scflat}}_F \subset S$. (An intersection of closed subschemes is a subscheme.) Finally $S^{\mathrm{kflat}}_F = S^{\mathrm{scflat}}_F$ by (7.40). $\square$

7.66 (Proof of 7.3) Fix an embedding $X \hookrightarrow \mathbf{P}_S$. By (4.76), there is a universal family of generically flat Mumford divisors $\mathrm{Univ}^{\mathrm{md}}_d \to \mathrm{MDiv}_d(X \subset \mathbf{P}_S)$. By (7.65), we get $\mathrm{KDiv}_d(X)$ as a full subscheme

$$j^{\mathrm{kflat}} \colon \mathrm{KDiv}_d(X) = \mathrm{MDiv}_d(X \subset \mathbf{P}_S)^{\mathrm{kflat}} \hookrightarrow \mathrm{MDiv}_d(X \subset \mathbf{P}_S). \qquad \square$$

7.7 Normal Varieties

In the next three sections, we aim to give explicit descriptions of K-flat deformations of certain varieties. First, we show that every K-flat deformation of a normal variety is flat. Then we consider K-flat deformations of planar curves and of seminormal curves. In both cases, we give a complete answer for first order deformations only.

Theorem 7.67 *Let* $g\colon Y \to (s, S)$ *be a projective morphism. Assume that* red(Y_s) *is normal, g is K-flat, g is smooth at the generic points of Y_s, and $\mathcal{O}_Y$ is vertically pure. Then g is flat along Y_s.*

Proof If $\dim Y_s = 1$, then the claim follows from (7.68). In general, there is a smallest, closed subset $Z \subset Y_s$ such that g is flat along $Y_s \setminus Z$. Using the Bertini-type theorem (7.5), we see that the codimension of Z is ≥ 2. In this case, flatness holds even without K-flatness by (10.71). □

Lemma 7.68 *Let* $g\colon (y, Y) \to (s, S)$ *be a local morphism of pure relative dimension 1, that is, essentially of finite type. Assume that g is smooth along $Y \setminus \{y\}$, g is formally K-flat at y, and* pure(Y_s) *is smooth at y. Then g is smooth at y.*

Proof By (7.44), we may assume that S is Artinian. Then we can reduce it further to the case when Y is complete and $k(y) = k(s) =: k$; see (10.57) and (7.50). Write $Y = \operatorname{Spec} R_A$.

By induction on the length of A, we may assume that there is an ideal $A \supset (\varepsilon) \simeq k$ such that pure($R_A / \varepsilon R_A) \simeq (A/\varepsilon)[[\bar{x}]]$.

Let $x \in R_A$ be a lifting of $\bar{x}$. Set $J := \ker[R_A \to \operatorname{pure}(R_A/\varepsilon R_A)]$. Then J is a rank 1 R_k-module, hence free; let $y \in J$ be a generator. We have $x^r y = \varepsilon g_k(x)$, where $g_k \in k[[x]]$ is a unit and $r = \dim_k(J/\varepsilon R_A)$. These determine a projection of R_A whose image in $\operatorname{Spec} A[[x, y]]$ is given by the ideal

$$A[[x, y]] \cap (y - \varepsilon x^{-r} g_k(x)) A[[x, x^{-1}, y]].$$

By (7.15), this is a principal ideal iff $g_k(x) \in (y, x^r)$, that is, when $r = 0$. Thus $R_A = A[[x]]$. □

7.8 Hypersurface Singularities

In this section we give a detailed description of K-flat deformations of hypersurface singularities over $k[\varepsilon]$.

7.69 (Non-flat deformations) Let $X \subset \mathbb{A}^n$ be a reduced subscheme of pure dimension d. We aim to describe nonflat deformations of X that are flat outside a subset $W \subset X$. Choose equations $g_1, \ldots, g_{n-d}$ such that

$$(g_1 = \cdots = g_{n-d} = 0) = X \cup X',$$

where $Z := X \cap X'$ has dimension $< d$. Let h be an equation of $X' \cup W$ that does not vanish on any irreducible component of X. Thus X is a complete intersection in $\mathbb{A}^n \setminus (h = 0)$ with equation $g_1 = \cdots = g_{n-d} = 0$. Its flat deformations over an Artinian ring (A, m, k) are then given by

$$g_i(\mathbf{x}) = \Psi_i(\mathbf{x}), \quad \text{where} \quad \Psi_i \in m[x_1, \ldots, x_n, h^{-1}]. \tag{7.69.1}$$

Note that we can freely change the Ψ_i by an element of the ideal $(g_i - \Psi_i)$. For $A = k[\varepsilon]$ the equations can be written as

$$g_i(\mathbf{x}) = \Phi_i(\mathbf{x})\varepsilon, \quad \text{where} \quad \Phi_i \in k[x_1, \ldots, x_n, h^{-1}]. \tag{7.69.2}$$

Now we can freely change the Φ_i by any element of the ideal $\varepsilon(g_1, \ldots, g_{n-d})$. Thus the relevant information is carried by $\phi_i := \Phi_i|_X$. So, generically, first order flat deformations can be given in the form

$$g_i = \phi_i \varepsilon, \quad \text{where} \quad \phi_i \in H^0(X, \mathscr{O}_X)[h^{-1}]. \tag{7.69.3}$$

Set $X^\circ := X \setminus (Z \cup W)$. By varying h, we see that in fact

$$g_i = \phi_i \varepsilon, \quad \text{where} \quad \phi_i \in H^0(X^\circ, \mathscr{O}_{X^\circ}). \tag{7.69.4}$$

This shows that the choice of h is largely irrelevant.

If the deformation is flat then the equations defining X lift, that is, $\phi_i \in H^0(X, \mathscr{O}_X)$. In some simple cases, for example if X is a complete intersection, this is equivalent to flatness. In the examples that we compute, the most important information is carried by the polar parts

$$\bar{\phi}_i \in H^0(X^\circ, \mathscr{O}_{X^\circ})/H^0(X, \mathscr{O}_X). \tag{7.69.5}$$

We study first order K-flat deformations of hypersurface singularities. Plane curves turn out to be the most interesting ones.

7.70 Consider a hypersurface singularity $X := (f = 0) \subset \mathbb{A}^n_{\mathbf{x}}$ and a generically flat deformation of it

$$\mathbf{X} \subset \mathbb{A}^{n+r}_{\mathbf{x},\mathbf{z}}[\varepsilon] \to \operatorname{Spec} k[\varepsilon]. \tag{7.70.1}$$

Aiming to work inductively, we assume that the deformation is flat outside the origin. Choose coordinates such that the x_i do not divide f.

As in (7.69.3), any such deformation can be given as

$$f(\mathbf{x}) = \psi(\mathbf{x})\varepsilon \quad \text{and} \quad z_j = \phi_j(\mathbf{x})\varepsilon, \tag{7.70.2}$$

where $\psi, \phi_j \in \cap_i H^0(X, \mathscr{O}_X)[x_n^{-1}]$. If $n \geq 3$, then $\cap_i \mathscr{O}_X[x_i^{-1}] = \mathscr{O}_X$ and we get the following special case of (10.73).

Claim 7.70.3 Let $X := (f = 0) \subset \mathbb{A}^n$ be a hypersurface singularity and $\mathbf{X} \subset \mathbb{A}^{n+r}[\varepsilon]$ a first order deformation of X that is flat outside the origin. If $n \geq 3$ then $\mathbf{X}$ is flat over $k[\varepsilon]$. $\square$

For $n = 2$, we use the following:

Notation 7.70.4 Let $B = (f(x,y) = 0) \subset \mathbb{A}^2$ be a reduced curve singularity. Set $B^\circ := B \setminus \{(0,0)\}$. A nonflat deformation **B** over $k[\varepsilon]$ is written as

$$f(x,y) = \Psi(x,y)\varepsilon \quad \text{and} \quad z_j = \Phi_j(x,y)\varepsilon.$$

As in (7.69), we set $\psi := \Psi|_B, \phi_j := \Phi_j|_B$ and $\bar{\psi}, \bar{\phi}_j \in H^0(B^\circ, \mathscr{O}_{B^\circ})/H^0(B, \mathscr{O}_B)$ denote their polar parts.

We say that a (flat, resp. generically flat) deformation over $k[\varepsilon]$ *globalizes* if it is induced from a (flat, resp. generically flat) deformation over $k[[t]]$.

Theorem 7.71 *Consider a generically flat deformation* **B** *of the plane curve singularity* $B := (f = 0) \subset \mathbb{A}^2_{xy}$ *given in (7.70.4).*
(7.71.1) If **B** *is C-flat, then* $\psi \in H^0(B, \mathscr{O}_B)$.
(7.71.2) If $\psi \in H^0(B, \mathscr{O}_B)$, *then the deformation is*
  (a) flat iff $\phi_j \in H^0(B, \mathscr{O}_B)$ *and*
  (b) C-flat iff $f_x\phi_j, f_y\phi_j \in H^0(B, \mathscr{O}_B)$.
(7.71.3) If B *is reduced and* $\psi = 0$, *then the deformation globalizes iff* $\phi_j \in H^0(\bar{B}, \mathscr{O}_{\bar{B}})$, *where* $\bar{B} \to B$ *is the normalization.*

Remark 7.71.4 Note that Ω^1_B is generated by $dx|_B, dy|_B$, while ω_B is generated by $f_y^{-1}dx = -f_x^{-1}dy$.

If B is reduced, then Ω^1_B and ω_B are naturally isomorphic over the smooth locus B°. This gives a natural inclusion $\mathrm{Hom}(\Omega^1_B, \omega_B) \hookrightarrow \mathscr{O}_{B^\circ}$. Then (7.71.2.b) says that $\bar{\phi}_j \in \mathrm{Hom}(\Omega^1_B, \omega_B)/\mathscr{O}_B$. See (7.72) for monomial curves.

Proof For simplicity, we compute with one z coordinate. If $\psi, \phi \in H^0(B, \mathscr{O}_B)$ then we can assume that Ψ, Φ are regular, so the deformation is flat. The converse in (7.71.2.a) is clear.

As for (7.71.2.b), we write down the equation of image of the projection

$$(x,y,z) \mapsto (\bar{x}, \bar{y}) = (x - \alpha(x,y,z)z, y - \gamma(x,y,z)z),$$

where α, γ are constants for linear projections and power series that are non-zero at the origin in general. Since $z^2 = \phi^2\varepsilon^2 = 0$, Taylor expansion gives that

$$f(\bar{x},\bar{y}) = f(x,y) - \alpha(x,y,z)f_x(x,y)z - \gamma(x,y,z)f_y(x,y)z.$$

Similarly, for any polynomial $F(x,y)$, we get that $F(\bar{x},\bar{y}) \equiv F(x,y) \mod \varepsilon\mathscr{O}_\mathbf{B}$, hence $F(\bar{x},\bar{y})z = F(x,y)z$ in $\mathscr{O}_\mathbf{B}$ since $z\varepsilon = 0$. Thus the equation is

$$f(\bar{x},\bar{y}) - (\psi(\bar{x},\bar{y}) - \alpha(\bar{x},\bar{y},0)f_x(\bar{x},\bar{y})\phi - \gamma(\bar{x},\bar{y},0)f_y(\bar{x},\bar{y})\phi) \cdot \varepsilon = 0. \quad (7.71.5)$$

By (7.15.2), this defines a relative Cartier divisor for every α, γ iff $\psi, f_x\phi, f_y\phi \in \mathscr{O}_B$, proving (7.71.2.b). (Thus linear and formal projections give the same restrictions, hence C-flatness implies formal K-flatness in this case.)

If **B** globalizes then $\phi \in H^0(\bar{B}, \mathscr{O}_{\bar{B}})$, this is the $n = 1$ case of (7.73.1). To prove the converse assertion in (7.71.3), we would like to write the global deformation as

$$(f(x, y) = 0, z = \phi(x, y)s) \subset \mathbb{A}^4_{xyzs}.$$

The problem with this is that ϕ has a pole at the origin. Thus we write $\phi = \phi_1 h^{-r}$ where ϕ_1 is regular at the origin and h is a general linear form in x, y. Then the correct equations are

$$(f(x, y) = 0, zh^r = \phi_1(x, y)s) \subset \mathbb{A}^4_{xyzs}.$$

Note that typically $\phi_1(0,0) = 0$, hence the two-plane $(x = y = 0) \subset \mathbb{A}^4_{xyzs}$ appears as an extra irreducible component. We need one more equation to eliminate it.

If $\phi \in H^0(\bar{B}, \mathscr{O}_{\bar{B}})$, then it satisfies an equation

$$\phi^m + \sum_{j=0}^{m-1} r_j\phi^j = 0, \quad \text{where} \quad r_j \in H^0(B, \mathscr{O}_B).$$

Thus $z = \phi s$ satisfies the equation $z^m + \sum_{j=0}^{m-1} r_j z^j s^{m-j} = 0$. Now the three equations

$$f(x, y) = zh^r - \phi_1(x, y)s = z^m + \sum_{j=0}^{m-1} r_j z^j s^{m-j} = 0$$

define the required globalization of the infinitesimal deformation.   □

7.71.6 (Nonreduced curves) Consider $B = (y^2 = 0)$ with deformations

$$y^2 = (y\psi_1(x) + \psi_0(x))\varepsilon \quad \text{and} \quad z = (y\phi_1(x) + \phi_0(x))\varepsilon,$$

where $\psi_i, \phi_i \in k[x, x^{-1}]$. If this is C-flat, then $\psi_i \in k[x]$ by (7.71.1). Since $f_x \equiv 0$, (7.71.2.b) gives only one condition, that $y(y\phi_1(x) + \phi_0(x))$ be regular. Since $y^2 = 0$, we get that $\phi_0 \in k[x]$, but no condition on ϕ_1. So it can have a pole of arbitrary high order. Note that if ϕ_1 has a pole of order m, then regularizing the second equation we get $zx^m = y\varepsilon + (\text{other terms})$. This suggests that if these deformations lie on a family of surfaces, the total space must have more and more complicated singularity at the origin as $m \to \infty$.

Example 7.72 (Monomial curves) We can be quite explicit if B is the irreducible monomial curve $B := (x^a = y^c) \subset \mathbb{A}^2$ where $(a, c) = 1$. Its miniversal space of flat deformations is given as

$$x^a - y^c + \sum_{i=0}^{a-2} \sum_{j=0}^{c-2} s_{ij} x^i y^j = 0.$$

Its dimension is $(a - 1)(c - 1)$.

In order to compute C-flat deformations, we parametrize B as $t \mapsto (t^c, t^a)$. Thus $\mathcal{O}_B = k[t^c, t^a]$. Let $E_B = \mathbb{N}a + \mathbb{N}c \subset \mathbb{N}$ denote the semigroup of exponents. Then the condition (7.71.2.b) becomes

$$t^{ac-c}\phi(t), t^{ac-a}\phi(t) \in k[t^a, t^c]. \tag{7.72.1}$$

This needs to be checked one monomial at a time.

For $\phi = t^m$ and $m \geq 0$ the conditions (7.72.1) are automatic, and the deformation is nonflat iff $m \notin E_B$. These give a space of dimension $\frac{1}{2}(a-1)(c-1)$. (This is an integer since one of a, c must be odd.)

For $\phi = t^{-m}$ and $m \geq 0$, we get the conditions $ac-c-m \in E_B$ and $ac-a-m \in E_B$. By (7.72.4), these are equivalent to $ac - a - c - m \in E_B$. The largest value of m satisfying this gives the deformation

$$(x^a - y^c = z - t^{-ac+a+c}\varepsilon = 0) \quad \text{over} \quad k[\varepsilon]. \tag{7.72.2}$$

Note also that for $0 \leq m \leq ac - a - c$, we have that $ac - a - c - m \in E_B$ iff $m \notin E_B$. These again have $\frac{1}{2}(a-1)(c-1)$ solutions.

Thus we see that the space of C-flat deformations that are nonflat has $(a-1)(c-1)$ extra dimensions; the same as the space of flat deformations. This looks very promising, but the next example shows that we get different answers for non-monomial curve singularities.

7.72.3 (Non-monomial example) Consider the curve singularity $B = (x^4 + y^5 + x^2 y^3 = 0)$. Blowing up the origin, we get $(x/y)^4 + y + (x/y)^2 y = 0$. Thus B is irreducible, it can be parametrized as $x = t^5 + \cdots, y = t^4 + \cdots$, and it is an equisingular deformation of the monomial curve $(x^4 + y^5 = 0)$.

In the monomial case we have the deformation (7.72.2) where $z - t^{-11}\varepsilon = 0$. We claim that B does not have a C-flat deformation $z - \phi\varepsilon = 0$ where $\phi = t^{-11} + \cdots$. Indeed, such a deformation would satisfy

$$f_x\phi = y \cdot (\text{local unit}) \quad \text{and} \quad f_y\phi = x \cdot (\text{local unit}).$$

Eliminating ϕ gives that $(xf_x)/(yf_y) = (\text{local unit})$. We can compute the left-hand side as

$$\frac{4x^4 + 2x^2 y^3}{5y^5 + 3x^2 y^3} = \frac{-4y^5 - 4x^2 y^3 + 2x^2 y^3}{5y^5 + 3x^2 y^3} = -\frac{4}{5} \cdot \frac{1 + (1/2)(x/y)^2}{1 + (3/5)(x/y)^2}.$$

This is invertible at the origin of the normalization of B, but it is not regular on B since $\frac{x}{y} = t + \cdots$. □

The following is left as an exercise.

Claim 7.72.4 For $(a, c) = 1$, set $E = \mathbb{N}a + \mathbb{N}c \subset \mathbb{N}$. Then

(a) If $0 \leq m \leq \min\{ac - a, ac - c\}$ then $ac - a - m, ac - c - m \in E$ iff $ac - a - c - m \in E$.

(b) If $0 \le m \le ac - a - c$ then $ac - a - c - m \in E$ iff $m \notin E$.   □

7.73 (Normalization of a deformation) Let T be the spectrum of a DVR with maximal ideal (t) and residue field k. Let $g\colon X \to T$ be a flat morphism of pure relative dimension d with generically reduced fibers. Set $Z\colon = \operatorname{Supp} \operatorname{tors}(X_0)$ and let $\pi\colon \bar{X} \to X$ be the normalization.

By composition, we get $\bar{g}\colon \bar{X} \to T$. Note that $\pi_0\colon \bar{X}_0 \to X_0$ is an isomorphism over $X_0 \setminus Z$ and $\bar{X}_0$ is S_1. In particular, $\bar{X}_0$ is dominated by the normalization X_0^{nor} of X_0.

Note that $t^n \mathcal{O}_X$ usually has some embedded primes contained in Z. The intersection of its height 1 primary ideals (also called the nth symbolic power of $t\mathcal{O}_X$) is $(t\mathcal{O}_X)^{(n)} = \mathcal{O}_X \cap t^n \mathcal{O}_{\bar{X}}$. In particular, we have injections

$$(t\mathcal{O}_X)^{(n)}/(t\mathcal{O}_X)^{(n+1)} \hookrightarrow t^n \mathcal{O}_{\bar{X}}/t^{n+1} \mathcal{O}_{\bar{X}} \simeq \mathcal{O}_{\bar{X}_k}. \tag{7.73.1}$$

A closely related computation is the following.

Example 7.74 Kollár (1999, 4.8) Using (7.34.1), we see that the ideal of Chow equations of the codimension 2 subvariety $(x_{n+1} = f(x_0, \ldots, x_n) = 0) \subset \mathbb{P}^{n+1}$ is generated by the forms

$$f(x_0 - a_0 x_{n+1} : \cdots : x_n - a_n x_{n+1}) \quad \text{for all} \quad a_0, \ldots, a_n. \tag{7.74.1}$$

If the characteristic is 0, then Taylor's theorem gives that

$$f(x_0 - a_0 x_{n+1} : \cdots : x_n - a_n x_{n+1}) = \sum_I \frac{(-1)^I}{I!} a^I \frac{\partial^I f}{\partial x^I} x_{n+1}^{|I|}, \tag{7.74.2}$$

where $I = (i_0, \ldots, i_n) \in \mathbb{N}^{n+1}$. The $a^{|I|}$ are linearly independent, hence we get that the ideal of Chow equations is

$$I^{\mathrm{ch}}\big(f(x_0, \ldots, x_n), x_{n+1}\big) = \big(f, x_{n+1} D(f), \ldots, x_{n+1}^m D^m(f)\big), \tag{7.74.3}$$

where we can stop at $m = \deg f$. Here we use the usual notation

$$D(f) := \Big(f, \tfrac{\partial f}{\partial x_0}, \ldots, \tfrac{\partial f}{\partial x_n}\Big) \tag{7.74.4}$$

for derivative ideals.

If we want to work locally at the point $p = (x_1 = \cdots = x_n = 0)$, then we can set $x_0 = 1$ to get the local version

$$I^{\mathrm{ch}}\big(f(1, x_1, \ldots, x_n), x_{n+1}\big) = \big(f, x_{n+1} D(f), \ldots, x_{n+1}^m D^m(f)\big), \tag{7.74.5}$$

where we can now stop at $m = \operatorname{mult}_p f$. This also holds if f is an analytic function, though this needs to be worked out using the more complicated formulas (7.34.6) that for us become

$$\pi\colon (x_1, \ldots, x_{n+1}) \to (x_1 - x_{n+1}\psi_1, \ldots, x_n - x_{n+1}\psi_n), \tag{7.74.6}$$

where $\psi_i = \psi_i(x_0, \ldots, x_{n+1})$ are analytic functions. Expanding as in (7.74.2) we see that

$$f(x_1 - x_{n+1}\psi_1, \ldots, x_n - x_{n+1}\psi_n) \in I^{\mathrm{ch}}(f(x_1, \ldots, x_n), x_{n+1}). \qquad (7.74.7)$$

Thus we get the same ideal if we compute I^{ch} using analytic projections.

7.9 Seminormal Curves

Over an algebraically closed field k, every seminormal curve singularity is formally isomorphic to

$$B_n := \operatorname{Spec} k[x_1, \ldots, x_n]/(x_i x_j : i \neq j) \subset \mathbb{A}_{\mathbf{x}}^n,$$

formed by the union of the n coordinate axes. In this section, we study deformations of B_n over $k[\varepsilon]$ that are flat outside the origin.

A normal form is worked out in (7.75.4), which shows that the space of these deformations is infinite dimensional. Then we describe the flat deformations (7.76) and their relationship to smoothings (7.77).

We compute C-flat and K-flat deformations in (7.79); these turn out to be quite close to flat deformations.

The ideal of Chow equations of B_n is computed in Kollár (1999, 4.11). For $n = 3$, these are close to C-flat deformations, but the difference between the two classes increases rapidly with n.

7.75 (Generically flat deformations of B_n) Let $\mathbf{B}_n \subset \mathbb{A}_{\mathbf{x}}^m[\varepsilon]$ be a generically flat deformation of $B_n \subset \mathbb{A}_{\mathbf{x}}^m$ over $k[\varepsilon]$.

If $\mathbf{B}_n$ is flat over $k[\varepsilon]$, then we can assume that $n = m$, but a priori we only know that $n \leq m$. Following (7.69), we can describe $\mathbf{B}_n$ as follows.

Along the x_j-axis and away from the origin, the deformation is flat. Thus, in the $(x_j \neq 0)$ open set, $\mathbf{B}_n$ can be given as

$$x_i = \Phi_{ij}(x_1, \ldots, x_m)\varepsilon, \quad \text{where } i \neq j \text{ and } \Phi_{ij} \in k[x_1, \ldots, x_m, x_j^{-1}]. \qquad (7.75.1)$$

Note that $(x_1, \ldots, \widehat{x_j}, \ldots, x_m, \varepsilon)^2$ is identically 0 on $\mathbf{B}_n \cap (x_j \neq 0)$, so the terms in this ideal can be ignored. Thus along the x_j-axis we can change (7.75.1) to the simpler form

$$x_i = \phi_{ij}(x_j)\varepsilon, \quad \text{where } i \neq j \text{ and } \phi_{ij} \in k[x_j, x_j^{-1}]. \qquad (7.75.2)$$

There is one more simplification that we can make. Write

$$\phi_{ij} = \phi'_{ij} + \gamma_{ij} \quad \text{where} \quad \phi'_{ij} \in k[x_j^{-1}], \gamma_{ij} \in (x_j) \subset k[x_j],$$

and set $x_i' = x_i - \sum_{j \neq i} \gamma_{ij}(x_j)$. Then we get the description

$$x_i' = \phi_{ij}'(x_j')\varepsilon \quad \text{where} \quad i \neq j \quad \text{and} \quad \phi_{ij}' \in k[x_j'^{-1}]. \tag{7.75.3}$$

For most of our computations, the latter coordinate change is not very important. Thus we write our deformations as

$$\mathbf{B}_n: \{x_i = \phi_{ij}(x_j)\varepsilon \quad \text{along the } x_j\text{-axis}\}, \tag{7.75.4}$$

where $\phi_{ij}(x_j) \in k[x_j, x_j^{-1}]$, but we keep in mind that we can choose $\phi_{ij}(x_j) \in k[x_j^{-1}]$ if it is convenient. Writing $\mathbf{B}_n$ as in (7.75.4) is almost unique; see (7.76.3) for one more coordinate change that leads to a unique normal form.

Writing $x_i x_j$ in two ways using (7.75.4) we get that

$$x_i x_j = \left(x_i \phi_{ji}(x_i) 1_i + x_j \phi_{ij}(x_i) 1_j\right)\varepsilon, \tag{7.75.5}$$

where 1_ℓ denotes the function that is 1 on the x_ℓ-axis and 0 on the others.

In order to deal with the cases when $m > n$, we make the following:

Convention 7.75.6 We set $\phi_{ij} \equiv 0$ for $j > n$.

We get the same result (7.75.4) if we work with the analytic or formal local scheme of B_n: we still end up with $\phi_{ij}(x_j) \in k[x_j^{-1}]$.

Proposition 7.76 *For $n \geq 3$, the generically flat deformation* $\mathbf{B}_n \subset \mathbb{A}_x^n[\varepsilon]$ *as in (7.75.4) is flat iff*
(7.76.1) *either $n \geq 3$ and the ϕ_{ij} have no poles,*
(7.76.2) *or $n = 2$ and ϕ_{12}, ϕ_{21} have only simple poles with the same residue.*

Proof $\mathbf{B}_n$ is flat iff the equations $x_i x_j = 0$ of B_n lift to equations of $\mathbf{B}_n$. We computed in (7.75.5) that $x_i x_j = (x_i \phi_{ji}(x_i)1_i + x_j \phi_{ij}(x_i)1_j)\varepsilon$, thus $x_i x_j$ lifts to an equation iff $x_i \phi_{ji}(x_i)1_i + x_j \phi_{ij}(x_i)1_j$ is regular. Thus the ϕ_{ij} have only simple poles and the residues must agree along all the axes. $x_i \phi_{ji}(x_i)1_i + x_j \phi_{ij}(x_i)1_j$ vanishes along the other $n - 2$ axes for $n \geq 3$, so the residues must be 0. $\square$

Corollary 7.76.3 The first order flat deformation space $T_{B_n}^1$ has dimension $n(n-1) - n = n(n-2)$.

Proof By (7.75.3) and (7.76), flat deformations can be given as

$$\mathbf{B}_n: \{x_i = e_{ij}\varepsilon \quad \text{along the } x_j\text{-axis, where } e_{ij} \in k\}.$$

The constants e_{ij} are not yet unique, $x_i \mapsto x_i - a_i$ changes $e_{ij} \mapsto e_{ij} - a_j$. $\square$

Strangely, (7.76.3) says that every flat first order deformation of B_n is obtained by translating the axes independently of each other. These deformations all globalize in the obvious way, but the globalization is not a flat deformation of B_n unless the translated axes all pass through the same point. If this point is $(a_1\varepsilon, \ldots, a_n\varepsilon)$, then $e_{ij} = a_j$ and applying (7.76.3) we get the trivial deformation. See (7.77) for smoothings of B_n.

If $n = 2$, then the universal deformation is $x_1 x_2 + \varepsilon = 0$. One may ask why this deformation does not lift to a deformation of B_3: smooth two of the axes to a hyperbola and just move the third axis along. If we use $x_1 x_2 + t = 0$, then the x_3-axis should move to the line $(x_1 - \sqrt{t} = x_2 - \sqrt{t} = 0)$. This gives the flat deformation given by equations

$$x_1 x_2 + t = x_3(x_1 - \sqrt{t}) = x_3(x_2 - \sqrt{t}) = 0.$$

Of course this only makes sense if t is a square. Thus setting $\varepsilon = \sqrt{t} \mod t$ the $t = \varepsilon^2 \mod t$ term becomes 0 and we get

$$x_1 x_2 = x_3 x_1 - x_3 \varepsilon = x_3 x_2 - x_3 \varepsilon = 0,$$

which is of the form given in (7.76.1).

Example 7.77 (Smoothing B_n) Rational normal curves $R_n \subset \mathbb{P}^n$ have a moduli space of dimension $(n + 1)(n + 1) - 1 - 3 = n^2 + 2n - 3$. The $B_n \subset \mathbb{P}^n$ have a moduli space of dimension $n + n(n - 1) = n^2$. Thus the smoothings of B_n have a moduli space of dimension $n^2 + 2n - 3 - n^2 = 2n - 3$. We can construct these smoothings explicitly as follows.

Fix distinct $p_1, \ldots, p_n \in k$ and consider the map

$$(t, z) \mapsto \left(\tfrac{t}{z - p_1}, \ldots, \tfrac{t}{z - p_n} \right).$$

Eliminating z gives the equations

$$(p_i - p_j)x_i x_j + (x_i - x_j)t = 0 \colon 1 \le i \ne j \le n \tag{7.77.1}$$

for the closure of the image, which is an affine cone over a degree n rational normal curve $R_n \subset \mathbb{P}^n_{t,\mathbf{x}}$. So far this is an $(n - 1)$-dimensional space.

Applying the torus action $x_i \mapsto \lambda_i^{-1} x_i$, we get new smoothings given by

$$(p_i - p_j)x_i x_j + (\lambda_j x_i - \lambda_i x_j)t = 0 \colon 1 \le i \ne j \le n. \tag{7.77.2}$$

Writing it in the form (7.75.4), we get

$$x_i = \tfrac{\lambda_i}{p_i - p_j}\varepsilon \quad \text{along the } x_j\text{-axis.} \tag{7.77.3}$$

This looks like a $2n$-dimensional family, but $\mathbf{Aut}(\mathbb{P}^1)$ acts on it, reducing the dimension to the expected $2n - 3$. The action is clear for $z \mapsto \alpha z + \beta$, but $z \mapsto z^{-1}$ also works out since

$$\frac{\lambda_i}{p_i^{-1} - p_j^{-1}} = \frac{-\lambda_i p_i^2}{p_i - p_j} + \lambda_i p_i.$$

Claim 7.77.4 For distinct $p_i \in k$ and $\lambda_j \in k^*$, the vectors

$$\left(\frac{\lambda_j}{p_i - p_j} : i \neq j\right) \quad \text{span} \quad T^1_{B_n} \simeq k^{\binom{n}{2}}.$$

So the flat infinitesimal deformations determined in (7.76.3) form the Zariski tangent space of the smoothings.

Proof Assume that there is a linear relation $\sum_{ij} m_{ij} \frac{\lambda_j}{p_i - p_j} = 0$. If we let $p_i \to p_j$ and keep the others fixed, we get that $m_{ij} = 0$. □

Remark 7.77.5 If $n = 3$, then the Hilbert scheme of degree 3 reduced space curves with $p_a = 0$ is smooth; see Piene and Schlessinger (1985).

Example 7.78 (Simple poles) Among nonflat deformations, the simplest ones are given by $\phi_{ij}(x_j) = c_{ij}x_j^{-1} + e_{ij}$. By (7.75.5), $x_i x_j = (c_{ji}1_i + c_{ij}1_j)\varepsilon$. For $n \geq 3$ and general choices of the c_{ij}, the rational functions $c_{ji}1_i + c_{ij}1_j$ span $\mathcal{O}_{\tilde{B}_n}/\mathcal{O}_{B_n}$. Thus we get an exact sequence

$$0 \to \varepsilon \cdot \mathcal{O}_{\tilde{B}_n} \to \mathcal{O}_{\mathbf{B}_n} \to \mathcal{O}_{B_n} \to 0. \tag{7.78.1}$$

The main result is the following.

Theorem 7.79 *For a first order deformation of $B_n \subset \mathbb{A}^m$ specified by*

$$\mathbf{B}_n : \{x_i = \phi_{ij}(x_j)\varepsilon \quad \text{along the } x_j\text{-axis}\}, \tag{7.79.1}$$

the following are equivalent:
(7.79.2) $\mathbf{B}_n$ *is C-flat.*
(7.79.3) $\mathbf{B}_n$ *is K-flat.*
(7.79.4) *The ϕ_{ij} have only simple poles and ϕ_{ij}, ϕ_{ji} have the same residue.*

Recall that $\phi_{ij} \equiv 0$ for $j > n$ by (7.75.6), hence (4) implies that ϕ_{ij} has no poles for $i > n$.

Proof The proof consist of two parts. First, we show in (7.80) that (7.79.2) and (7.79.4) are equivalent by explicitly computing linear projections.

We see in (7.81) that if the ϕ_{ij} have only simple poles, then there is only one term of the equation of a nonlinear projection that could have a pole. This term is the same for the linearization of the projection. Hence it vanishes iff it vanishes for linear projections. This shows that (7.79.4) $\Rightarrow$ (7.79.3). □

Remark 7.79.5 If $j > n$ then $\phi_{ij} \equiv 0$ by (7.75.6), so ϕ_{ji} is regular by (7.79.4). Evaluating them at the origin gives the vector $\mathbf{v}_j \in k^n$. If $\sum_{j>n} \lambda_j \mathbf{v}_j = 0$ then

$$\sum_{j>n} \lambda_j \left(x_j - \sum_{i=1}^n \phi_{ji} x_\ell \varepsilon\right)$$

is regular and identically 0 on $\mathbf{B}_n$. We can thus eliminate some of the x_j for $j > n$ and obtain that every K-flat deformation of $\mathbf{B}_n$ lives in $\mathbb{A}^{2n-1}$.

7.80 (Linear projections) Recall that by our convention (7.75.6), $\phi_{ij} \equiv 0$ for $j > n$. Extending this, in the following proof all sums/products involving i go from 1 to m and sums/products involving j go from 1 to n.

With $\mathbf{B}_n$, as in (7.79.1), consider the special projections

$$\pi_\mathbf{a} \colon \mathbb{A}^n_\mathbf{x}[\varepsilon] \to \mathbb{A}^2_{uv}[\varepsilon] \quad \text{given by} \quad u = \textstyle\sum x_i, v = \textstyle\sum a_i x_i, \tag{7.80.1}$$

where $a_i \in k[\varepsilon]$. Write $a_i = \bar{a}_i + a'_i \varepsilon$. (One should think that $a'_i = \partial a_i / \partial \varepsilon$.)

In order to compute the projection, we follow the method of (7.21.7). Since we compute over $k[u, u^{-1}, \varepsilon]$, we may as well work with the $k[u, \varepsilon]$-module $M := \oplus_j k[x_j, \varepsilon]$ and write $1_j \in k[x_j, \varepsilon]$ for the jth unit. Then multiplication by u and v are given by

$$\begin{aligned} u \cdot 1_j &= (\textstyle\sum_i x_i) 1_j = x_j + \textstyle\sum_i \phi_{ij}\varepsilon, \quad \text{and} \\ v \cdot 1_j &= (\textstyle\sum_i a_i x_i) 1_j = a_j x_j + \textstyle\sum_i a_i \phi_{ij}\varepsilon. \end{aligned} \tag{7.80.2}$$

Thus $v \cdot 1_j = (a_j u + \sum_i (a_i - a_j)\phi_{ij}(u)\varepsilon) \cdot 1_j$ and the v-action on M is given by the diagonal matrix

$$\mathrm{diag}(a_j u + \textstyle\sum_i (a_i - a_j)\phi_{ij}(u)\varepsilon).$$

By (7.21.7), the equation of the projection is its characteristic polynomial

$$\textstyle\prod_j (v - a_j u - \sum_i (a_i - a_j)\phi_{ij}(u)\varepsilon) = 0. \tag{7.80.3}$$

Expanding it, we get an equation of the form

$$\begin{aligned} &\textstyle\prod_j (v - \bar{a}_j u) - E(u, v, a, \phi)\varepsilon = 0, \quad \text{where} \\ &E(u, v, a, \phi) = \textstyle\sum_j (\prod_{i \neq j} (v - \bar{a}_j u)) \cdot (a'_j u + \sum_i (\bar{a}_i - \bar{a}_j)\phi_{ij}(u)). \end{aligned} \tag{7.80.4}$$

This is a polynomial of degree $\leq n - 1$ in v, hence by (7.19) its restriction to the curve $(\prod_j (v - \bar{a}_j u) = 0)$ is regular iff $E(u, v, a, \phi)$ is a polynomial in u as well. Let r be the highest pole order of the ϕ_{ij} and write

$$\phi_{ij}(u) = c_{ij} u^{-r} + \text{(higher terms)}. \tag{7.80.5}$$

Then the leading part of the coefficient of v^{n-1} in $E(u, v, a, \phi)$ is

$$\textstyle\sum_j \sum_i (\bar{a}_i - \bar{a}_j) c_{ij} u^{-r} = u^{-r} \sum_i \bar{a}_i (\sum_j (c_{ij} - c_{ji})). \tag{7.80.6}$$

Since the $\bar{a}_i$ are arbitrary, we get that

$$\textstyle\sum_j (c_{ij} - c_{ji}) = 0 \quad \text{for every } i. \tag{7.80.7}$$

Next we use a linear reparametrization of the lines $x_i = \lambda_i^{-1} y_i$ and then apply a projection $\pi_{\mathbf{a}}$ as in (7.80.1). The equations $x_i = \phi_{ij}(x_j)\varepsilon$ become

$$y_i = \lambda_i \phi_{ij}(\lambda_j^{-1} y_j)\varepsilon$$

and c_{ij} changes to $\lambda_i \lambda_j^r c_{ij}$. Thus the equations (7.80.7) become

$$\sum_j (\lambda_i \lambda_j^r c_{ij} - \lambda_j \lambda_i^r c_{ji}) = 0 \quad \forall i. \tag{7.80.8}$$

If $r \geq 2$, this implies that $c_{ij} = 0$ and if $r = 1$ then we get that $c_{ij} = c_{ji}$. This completes the proof of (7.79.2) $\Leftrightarrow$ (7.79.4).

Remark 7.80.9 Note that if we work over $\mathbb{F}_2$, then necessarily $\lambda_i = 1$, hence (7.80.8) does not exclude the $r \geq 2$ cases.

7.81 (Non-linear projections) Consider a general non-linear projection

$$(x_1, \ldots, x_n) \mapsto (\Phi_1(x_1, \ldots, x_n), \Phi_2(x_1, \ldots, x_n)).$$

After a formal coordinate change, we may assume that $\Phi_1 = \sum_i x_i$. Note that the monomials of the form $x_i x_j x_k$, $x_i^2 x_j^2$, $x_i x_j \varepsilon$ vanish on $\mathbf{B}_n$, so we can discard these terms from Φ_2. Thus, in suitable local coordinates, a general nonlinear projection can be written as

$$u = \sum_i x_i, \quad v = \sum_i \alpha_i(x_i) + \sum_{i \neq j} x_i \beta_{ij}(x_j), \tag{7.81.1}$$

where $\alpha_i(0) = \beta_{ij}(0) = 0$. Note that $\alpha_i'(0) = a_i$ in the notation of (7.80). Now

$$u \cdot 1_j = x_j + \sum_i \phi_{ij}(x_j)\varepsilon, \quad \text{and}$$
$$v \cdot 1_j = \alpha_j(x_j) + \sum_{i \neq j} \alpha_i(\phi_{ij}(x_j)\varepsilon) + \sum_{i \neq j} \phi_{ij}(x_j)\beta_{ij}(x_j)\varepsilon. \tag{7.81.2}$$

Note further that $\alpha_i(\phi_{ij}(x_j)\varepsilon) = \alpha_i'(0)\phi_{ij}(x_j)\varepsilon$ and

$$\alpha_j(x_j) = \alpha_j(u - \sum_i \phi_{ij}(x_j)\varepsilon) = \alpha_j(u) - \alpha_j'(u)\sum_i \phi_{ij}(x_j)\varepsilon.$$

Thus, as in (7.80.4), the projection is defined by the vanishing of

$$\prod_j \left(v - \alpha_j(u) - \sum_i (\beta_{ij}(u) + \alpha_i'(0) - \alpha_j'(u))\phi_{ij}(u)\varepsilon\right)$$
$$=: \prod_j (v - \bar{\alpha}_j(u)) - E(u, v, \alpha, \beta, \phi)\varepsilon. \tag{7.81.3}$$

Let $\bar{\beta}_{ij}, \bar{\alpha}_j'$ denote the residue of β_{ij}, α_j' modulo ε and write $\alpha_j(u) = \bar{\alpha}_j(u) + \partial_\varepsilon \alpha_j(u)\varepsilon$. As in (7.80.5), expanding the product gives that $E(u, v, \alpha, \beta, \phi)$ equals

$$\sum_j (\prod_{i \neq j}(v - \bar{\alpha}_i(u))) \cdot (\partial_\varepsilon \alpha_j(u) + \sum_i (\bar{\beta}_{ij}(u) + \bar{\alpha}_i'(0) - \bar{\alpha}_j'(u))\phi_{ij}). \tag{7.81.4}$$

We already know that $\phi_{ij}(u) = c_{ij} u^{-1} + \text{(higher terms)}$, hence $E(u, v, \alpha, \beta, \phi)$ has at most simple pole along $(u = 0)$. Computing its residue gives that

$$v^{n-1}\sum_j \sum_i (\bar{\beta}_{ij}(0) + \bar{\alpha}_i'(0) - \bar{\alpha}_j'(0))c_{ij} = v^{n-1}\sum_{ij}(\bar{a}_i - \bar{a}_j)c_{ij}. \tag{7.81.5}$$

These are the same as in (7.80.6). Thus $E(u, v, \alpha, \beta, \phi)$ is regular iff it is regular for the linearization. This completes the proof of (7.79.4) $\Rightarrow$ (7.79.3).

Example 7.82 The image of a general linear projection of $B_n \subset \mathbb{A}^n$ to $\mathbb{A}^2$ is n distinct lines through the origin. A general nonlinear projection to $\mathbb{A}^2$ gives n smooth curve germs with distinct tangent lines through the origin.

As a typical example, the miniversal deformation of $(x^n + y^n = 0)$ is

$$\left(x^n + y^n + \sum_{i,j \leq n-2} t_{ij} x^i y^j = 0\right) \subset \mathbb{A}^2_{xy} \times \mathbb{A}^{(n-1)^2}_\mathbf{t}. \tag{7.82.1}$$

Deformations with tangent cone $(x^n + y^n = 0)$ form the subfamily

$$\left(x^n + y^n + \sum_{i+j>n} t_{ij} x^i y^j = 0\right) \subset \mathbb{A}^2_{xy} \times \mathbb{A}^{\binom{n-3}{2}}_\mathbf{t}. \tag{7.82.2}$$

For $n \leq 4$, there is no such pair (i, j), thus, for $n \leq 4$, every analytic projection $\widehat{B}_n \to \widehat{\mathbb{A}}^2$ is obtained as the composite of an automorphism of $\widehat{B}_n$, followed by a linear projection and an automorphism of $\widehat{\mathbb{A}}^2$.

For $n = 5$, we get the deformations $(x^5 + y^5 + t x^3 y^3 = 0) \subset \mathbb{A}^2_{xy} \times \mathbb{A}_t$. For $t \neq 0$, these give curve germs that are images of $\widehat{B}_n$ by a nonlinear projection, but cannot be obtained as the image of a linear projection, up to automorphisms.

8

Moduli of Stable Pairs

We bring together the moduli theory of Chapter 6 with K-flatness of Chapter 7 to obtain the moduli theory of stable pairs in full generality. The basic definitions originate in the papers Kollár and Shepherd-Barron (1988) and Alexeev (1996); the resulting moduli spaces are usually called KSBA moduli spaces.

In Section 8.1 we discuss a bookkeeping device called marking: we need to know not only what the boundary divisor Δ is, but also how it is written as a linear combination of effective $\mathbb{Z}$-divisors. In the cases considered in Chapter 6, there was always a unique, obvious marking; this is why the notion was not introduced before. Simple examples show that, without marking, we get infinite dimensional moduli spaces, already for pointed curves (8.2).

The notion of Kollár–Shepherd-Barron–Alexeev stability is introduced in Section 8.2. The proof that we get a good moduli theory, as defined in (6.10), follows the methods of Chapter 6 if the coefficients are rational (8.9), but a few more steps are need if they are irrational (8.15).

The end result is the following consequence of (8.9) and (8.15).

Theorem 8.1 *Fix a base scheme S of characteristic 0, a coefficient vector* $\mathbf{a} = (a_1, \ldots, a_r) \in [0, 1]^r$, *an integer n, and a real number v. Let $\mathcal{SP}(\mathbf{a}, n, v)$ denote the functor of marked, stable pairs of dimension n and volume v. Then $\mathcal{SP}(\mathbf{a}, n, v)$ is good moduli theory (6.10) and it has a coarse moduli space* $\mathrm{SP}(\mathbf{a}, n, v)$, *which is projective over S.*

A variant with floating coefficients is treated in Section 8.3 and the moduli theory of more general polarized pairs is discussed in Sections 8.4–8.5.

The construction of moduli spaces as quotients by group actions is treated in Section 8.6, and a short overview of descent is in Section 8.7.

In Section 8.8, we discuss several unexpected problems that appear in positive characteristic. Quite likely, these necessitate substantial changes in the moduli theory of varieties of dimension ≥ 3 in positive and mixed characteristics.

Further Results

An early difficulty of KSBA theory was that good examples were not easy to write down. The first notable successes were Alexeev (2002); Hacking (2004). By now there is a rapidly growing body of fully understood cases.

Various moduli spaces are worked out in the papers Abramovich and Vistoli (2000); van Opstall (2005, 2006b,a); Hacking (2012); Alexeev (2015); Franciosi et al. (2015b, 2017, 2018); Alexeev (2016); Ascher and Gallardo (2018); Ascher and Bejleri (2019, 2021a,b); Ascher et al. (2020); Alexeev and Thompson (2021); Bejleri and Inchiostro (2021).

Examples of stable degenerations and their relations to other invariants are exhibited in Hassett (1999, 2000, 2001); Alexeev (2008); Tziolas (2009, 2010); Hacking and Prokhorov (2010); Hacking (2013, 2016); Urzúa (2016b,a); Rana (2017); Hacking et al. (2017); Rana and Urzúa (2019); Franciosi et al. (2022).

Computations of invariants of stable surfaces are given in Liu and Rollenske (2014); Franciosi et al. (2015a); Stern and Urzúa (2016); Tziolas (2017, 2022).

Special examples are computed in detail in Hacking et al. (2006, 2009); Thompson (2014); Ascher and Molcho (2016); Alexeev and Liu (2019a,b); Donaldson (2020).

Other approaches to the moduli spaces are discussed in Abramovich and Vistoli (2002); Alexeev and Knutson (2010); Abramovich and Hassett (2011); Abramovich et al. (2013, 2017); Abramovich and Chen (2014); Abramovich and Fantechi (2017).

Assumptions In this Chapter we work over a $\mathbb{Q}$-scheme. The definitions are set up in full generality, but some of the theorems fail in positive characteristic; see Section 8.8 for a discussion.

8.1 Marked Stable Pairs

So far, we have studied slc pairs (X, Δ), but usually did not worry too much about how Δ was written as a sum of divisors. As long as we look at a single variety, we can write Δ uniquely as $\sum a_i D_i$ where the D_i are prime divisors, and there is usually not much reason to do anything else. However, the situation changes when we look at families.

8.2 (Is $D = \frac{1}{n}(nD)$?) Assume that we have an slc family over an irreducible base $f\colon (X, \Delta) \to S$ with generic point $g \in S$. Then the natural approach is to write $\Delta_g = \sum a_i D_g^i$, where the D_g^i are prime divisors on the generic fiber X_g. For any other point $s \in S$ this gives a decomposition $\Delta_s = \sum a_i D_s^i$, where D_s^i is the specialization of D_g^i. Note that the D_s^i need not be prime divisors. They can have several irreducible components with different multiplicities and two different D_s^i, D_s^j can have common irreducible components. Thus $\Delta_s = \sum a_i D_s^i$ is not the "standard" way to write Δ_s.

Let us now turn this around. We fix a proper slc pair (X_0, Δ_0) and aim to understand all deformations of it. A first suggestion could be the following:

8.2.1 (*Naive definition*) An slc deformation of (X_0, Δ_0) over a local scheme $(0 \in S)$ is a proper slc morphism $f\colon (X, \Delta) \to S$ whose central fiber $(X, \Delta)_0$ is isomorphic to (X_0, Δ_0).

As an example, start with $(\mathbb{P}^1_{xy}, (x = 0))$. Pick $n \geq 1$ and variables t_i. Then

$$\left(\mathbb{P}^1_{xy} \times \mathbb{A}^n_{\mathbf{t}}, \tfrac{1}{n}(x^n + t_{n-1}x^{n-1}y + \cdots + t_0 y^n = 0)\right) \tag{8.2.2}$$

is a deformation of $(\mathbb{P}^1_{xy}, (x = 0))$ over $\mathbb{A}^n$ by the naive definition (8.2.1). We get a deformation space of dimension $n - 2$ using $\mathbf{Aut}(\mathbb{P}^1, (0{:}1))$. Letting n vary results in an infinite dimensional deformation space.

The polynomial in (8.2.2) is irreducible over $k(t_0, \ldots, t_{n-1})$, thus our recipe says that we should write $\Delta = \frac{1}{n}D_g$ (where D_g is irreducible). Then the special fiber is written as $(x = 0) = \frac{1}{n}(x^n = 0)$.

The situation becomes even less clear if we take two deformations as in (8.2.2) for two different values n, m and glue them together over the origin. The family is locally stable. One side says that the fiber over the origin should be $\frac{1}{n}(x^n = 0)$, the other side that it should be $\frac{1}{m}(x^m = 0)$.

As (8.2) suggests, some bookkeeping is necessary to control the multiplicities of the divisorial part of a pair (X, Δ) in families. This is the role of the marking we introduce next.

Once we control how a given $\mathbb{R}$-divisor Δ is written as a linear combination of $\mathbb{Z}$-divisors, we obtain finite dimensional moduli spaces.

Definition 8.3 (Marked pairs) A *marking* of an effective Weil $\mathbb{R}$-divisor Δ is a way of writing $\Delta = \sum a_i D_i$, where the D_i are effective $\mathbb{Z}$-divisors and $0 < a_i \in \mathbb{R}$. We call $\mathbf{a} = (a_1, \ldots, a_r)$ the *coefficient vector.*

A *marked pair* is a pair (X, Δ), plus a marking $\Delta = \sum a_i D_i$.

We allow the D_i to be empty; this has the advantage that the restriction of a marking to an open subset is again a marking. However, in other contexts, this is not natural and we will probably sometimes disregard empty divisors.

Observe that $\Delta = \sum a_i D_i$ and $\Delta = \sum (\frac{1}{2} a_i)(2 D_i)$ are different as markings. This seems rather pointless for one pair but, as we observed in (8.2), it is a meaningful distinction when we consider deformations of a pair.

Note that, for a given (X, Δ), markings are combinatorial objects that are not constrained by the geometry of X. If $\Delta = \sum_i b_i B_i$ and the B_i are distinct prime divisors, then the markings correspond to ways of writing the vector $(b_1, \ldots, b_r)$ as a positive linear combination of nonnegative integral vectors.

Comments Working with such markings is a rather natural thing to do. For example, plane, curves C of degree d can be studied using the log CY pair $(\mathbb{P}^2, \Delta_C := \frac{3}{d} C)$ as in Hacking (2004). Thus, even if C is reducible, we want to think of the $\mathbb{Q}$-divisor Δ_C as $\frac{3}{d} C$; hence as a marked divisor with $\mathbf{a} = (\frac{3}{d})$. Similarly, in most cases when we choose the boundary divisor Δ, it has a natural marking.

However, when a part of Δ is forced upon us, for instance coming from the exceptional divisor of a resolution, there is frequently no "natural" marking, though usually it is possible to choose a marking that works well enough.

If (X, Δ) is slc and $a_i > \frac{1}{2}$ for every i, then the marking is almost determined by Δ. For example, if the a_i are distinct then the obvious marking of $\Delta = \sum a_i D_i$ is the unique one. If all the $a_i = 1$, then the markings of $\sum_{i \in I} D_i$ correspond to partitions of I.

If we allow $a_i = \frac{1}{2}$, then an irreducible divisor D can have three different markings: $[D]$, $\frac{1}{2}[2D]$, or $\frac{1}{2}[D] + \frac{1}{2}[D]$. The smaller the a_i, the more markings are possible.

Definition 8.4 (Families of marked pairs) Fix a real vector $\mathbf{a} = (a_1, \ldots, a_r)$. A *family of marked pairs* with coefficient vector $\mathbf{a} = (a_1, \ldots, a_r)$ consists of

(8.4.1) a flat morphism $f : X \to S$ with demi-normal fibers (11.36),

(8.4.2) an effective, relative, Mumford $\mathbb{R}$-divisor Δ, plus

(8.4.3) a marking $\Delta = \sum a_i D_i$, where the D_i are effective, relative, Mumford $\mathbb{Z}$-divisors (4.68).

As we discussed in Section 4.1, the relative Mumford assumption on the D_i assures that markings can be pulled back by base-change morphisms $W \to S$.

However, being relative Mumford is not automatic. This means that not all markings of Δ give a family of marked pairs.

Examples 8.5 (Marking and stability) Given a family of pairs $f : (X, \Delta) \to S$, it can happen that it is KSBA-stable for one choice of the marking, but not for other markings. Although we define KSBA-stability only in the next section, these examples influenced the precise definitions of KSBA-stability, especially (8.13), so this is their right place.

(8.5.1) If S is normal then, by (4.4), every marking of Δ yields a family of marked pairs $f : (X, \Delta) \to S$.

(8.5.2) Assume that Δ is a $\mathbb{Q}$-divisor and S is reduced. There is a smallest $N \in \mathbb{N}^{>0}$ such that $N\Delta$ is a $\mathbb{Z}$-divisor. In characteristic 0, $D := N\Delta$ is a relative Mumford divisor by (4.39), thus $\Delta = \frac{1}{N}D$ gives a marking of (X, Δ).

(8.5.3) For $\mathbb{R}$-divisors, there are markings $\Delta = \sum_i \lambda_i \Delta_i$ where the λ_i are $\mathbb{Q}$-linearly independent; see (11.47). In characteristic 0 we get the same stable families, independent of the choice of the λ_i by (11.43.4) and (4.39).

(8.5.4) The simplest case is when $\Delta = cD$ with a single irreducible D. The only possible markings are $\Delta = \sum a_i(m_i D)$ for some $m_i \in \mathbb{N}$ and $c = \sum a_i m_i$. In characteristic 0 we get the same stable families, independent of the choices by (4.39), but not in positive characteristic, see (8.76).

(8.5.5) Let B be a curve with a single node b and $B^\circ := B \setminus \{b\}$. Let $b_1, b_2 \in \bar{B}$ be the preimages of the node in the normalization $\bar{B}$. Set $\bar{S} = \mathbb{P}^1 \times \bar{B}$.

Let $\bar{E}_i := \{p_i\} \times \bar{B}$ sections for $i = 1, 2, 3$ and $\bar{\Delta}$ their sum. If we use the marking with only 1 divisor $D_1^\circ := \sum_i E_i^\circ$, then we can use any of the six automorphisms of $\mathbb{P}^1$ that preserve $\{p_1, p_2, p_3\}$ to descend $(\bar{S}, \bar{\Delta})$ to a family of marked pairs over B. If we use the marking with three divisors $D_i^\circ := E_i^\circ$, then the identity gives the only descent.

8.2 Kollár–Shepherd-Barron–Alexeev Stability

Now we come to the main theorem of the book, the existence of a good moduli theory for all marked stable pairs (X, Δ) in characteristic 0.

The principle is that, once we have K-flatness to replace flatness in Section 6.2, the rest of the arguments should go through with small changes. This is indeed true for rational coefficients, so we start with that case.

For irrational coefficients, it is less clear how to cook up ample line bundles, so the existence of embedded moduli spaces needs more work.

KSBA Stability with Rational Coefficients

Fix a rational coefficient vector $\mathbf{a} = (a_1, \ldots, a_r)$ and let $\mathrm{lcd}(\mathbf{a})$ denote the least common denominator of the a_i.

8.6 (Stable objects) These are marked pairs $(X, \Delta = \sum_i a_i D_i)$ with coefficient vector $\mathbf{a}$ such that

(8.6.1) (X, Δ) is slc,

(8.6.2) X is projective and $K_X + \Delta$ is ample.

8.7 (Stable families) A family $f \colon (X, \Delta = \sum_i a_i D_i) \to S$ is *KSBA-stable* if the following hold:

(8.7.1) $f \colon X \to S$ is flat, finite type, pure dimensional.

(8.7.2) The D_i are K-flat families of relative, Mumford, $\mathbb{Z}$-divisors (7.1).

(8.7.3) The fibers (X_s, Δ_s) are slc.

(8.7.4) $\omega_{X/S}^{[m]}(m\Delta - B)$ is a flat family of divisorial sheaves, provided $\mathrm{lcd}(\mathbf{a}) \mid m$ and $B = \sum_{j \in J} D_j$ with $a_j = 1$ for $j \in J$.

(8.7.5) f is proper and $K_{X/S} + \Delta$ is f-ample.

The first four of these conditions define *locally KSBA-stable* families.

8.8 (Explanation) These conditions are mostly straightforward generalizations of (6.16.1–3). We discussed K-flatness in Chapter 7.

The main question is (8.7.4). We should think of it as the minimal assumption, which should be made more stronger whenever possible, without changing the reduced structure of the moduli space.

The main case is $B = 0$. For $\omega_{X/S}^{[m]}(m\Delta)$ to make sense, $m\Delta$ must be a $\mathbb{Z}$-divisor. If the D_i have no multiple or common irreducible components, this holds only if m is a multiple of $\mathrm{lcd}(\mathbf{a})$. The nonzero choices of B in (8.7.4) are discussed in (6.23).

We could also ask about the sheaves $\omega_{X/S}^{[m]}(\sum \lfloor ma_i \rfloor D_i)$, as in (6.22.3). As we saw in (2.41), they are not flat families of divisorial sheaves in general, but (2.79) discusses various examples where they are. Thus, on a case-by-case basis, a strengthening of (8.6.4) is possible and useful. This was one of the themes of Chapter 6.

Theorem 8.9 *KSBA-stability with rational coefficients, as defined in (8.6–8.7), is a good moduli theory (6.10).*

Proof We need to check the conditions (6.10.1–5).

Separatedness (6.10.1) follows from (2.50); valuative-properness (6.10.2) is proved in (2.51) and (7.4.2). Assumption (8.7.4) follows from (2.79.1) if $B = 0$

and from (2.79.8) when $B \neq 0$. Representability is proved in (7.65) and (3.31). Boundedness holds by (6.8.1) and (6.14).

Once we know that $m(K_X + \Delta)$ is very ample for every $(X, \Delta) \in S\mathcal{P}(\mathbf{a}, n, v)$ for some fixed m, embedded moduli spaces (6.10.3) are constructed in (8.52). However, the universal family over $C^m ESP(\mathbf{a}, n, \mathbb{P}^N_{\mathbb{Q}})$ satisfies (8.7.3) only for multiples of m. We can then handle the other values as in the proof of (6.24). The coarse moduli space exists by (6.6). □

As in (6.25), we get the following from (8.7.3).

Proposition 8.10 *For KSBA-stable families as in (8.6–8.7), let m be a multiple of* $\mathrm{lcd}(\mathbf{a})$. *Then* $\chi(X, \omega_X^{[m]}(m\Delta))$ *and* $h^0(X, \omega_X^{[m]}(m\Delta))$ *are both deformation invariant.* □

KSBA Stability with Arbitrary Coefficients

Fix a coefficient vector $\mathbf{a} = (a_1, \ldots, a_r)$ where $a_i \in [0, 1]$ are arbitrary real numbers. By (11.43.1), if $K_X + \sum_i a_i D_i$ is $\mathbb{R}$-Cartier, then we can get many $\mathbb{Q}$-Cartier divisors. We start by listing them.

Definition 8.11 Fix $\mathbf{a} = (a_1, \ldots, a_r)$ with linear $\mathbb{Q}$-envelope $\mathrm{LEnv}_{\mathbb{Q}}(1, \mathbf{a}) \subset \mathbb{Q}^{r+1}$ as in (11.44). For $\Delta = \sum_{i=1}^r a_i D_i$, set

$$\mathrm{LEnv}_{\mathbb{Z}}(K_X + \Delta) := \{m_0 K_X + \sum m_i D_i : (m_0, \ldots, m_r) \in \mathrm{LEnv}_{\mathbb{Q}}(1, \mathbf{a}) \cap \mathbb{Z}^{r+1}\}.$$

Let us mention two extreme cases.

(8.11.1) If all $a_i \in \mathbb{Q}$, then $\mathrm{LEnv}_{\mathbb{Z}}(K_X + \Delta)$ consists of all $\mathbb{Z}$-multiples of $\mathrm{lcd}(\mathbf{a})(K_X + \Delta)$.

(8.11.2) If $\{1, a_1, \ldots, a_r\}$ is a $\mathbb{Q}$-linearly independent set, then $\mathrm{LEnv}_{\mathbb{Z}}(K_X + \Delta)$ consist of all $\mathbb{Z}$-linear combinations $m_0 K_X + \sum m_i D_i$.

It is very important that, by (11.44) and (11.43.1), if $K_X + \Delta$ is $\mathbb{R}$-Cartier, then all elements of $\mathrm{LEnv}_{\mathbb{Z}}(K_X + \Delta)$ are $\mathbb{Q}$-Cartier $\mathbb{Z}$-divisors. (There may be other linear combinations that are $\mathbb{Q}$-Cartier $\mathbb{Z}$-divisors.)

The stable objects are the same as before, but the definition of stable families again looks different.

8.12 (Stable objects) We parametrize marked pairs $(X, \Delta = \sum_i a_i D_i)$ with coefficient vector $\mathbf{a}$ such that

(8.12.1) (X, Δ) is slc,

(8.12.2) X is projective and $K_X + \Delta$ is ample.

8.13 (Stable families) A family $f \colon (X, \Delta = \sum_i a_i D_i) \to S$ is *KSBA-stable* if the following hold

(8.13.1) $f \colon X \to S$ is flat, finite type, pure dimensional.

(8.13.2) The D_i are K-flat families of relative, Mumford, $\mathbb{Z}$-divisors (7.1).

(8.13.3) The fibers (X_s, Δ_s) are slc.

(8.13.4) $\omega_{X/S}^{[m_0]}(\sum m_i D_i - B)$ is a flat family of divisorial sheaves, whenever $(m_0, \ldots, m_r) \in \mathrm{LEnv}_{\mathbb{Z}}(K_X + \Delta)$ and $B = \sum_{j \in J} D_j$, where $a_j = 1$ for $j \in J$.

(8.13.5) f is proper and $K_{X/S} + \Delta$ is f-ample.

The first four of these conditions define *locally KSBA-stable* families.

8.14 (Explanation) These conditions are mostly straightforward generalizations of (8.7), again the main question is assumption (8.13.4).

If the a_i are rational, then, by (8.11.1), $\mathrm{LEnv}_{\mathbb{Z}}(K_X + \Delta)$ consists of the integer multiples of $\mathrm{lcd}(\mathbf{a})(K_X + \Delta)$, so (8.13.4) specializes to (8.7.4). If $1, a_1, \ldots, a_r$ are $\mathbb{Q}$-linearly independent, then, by (8.11.2), we specialize to (6.38).

For the intermediate cases we follow the philosophy behind KSB stability as in Section 6.2. Whenever we can prove to have a flat family of divisorial sheaves over DVR's, we require this property over all schemes.

Working with all of $\mathrm{LEnv}_{\mathbb{Z}}(K_X + \Delta)$ is (almost) necessary for our proof. We are using several rational perturbations of $K_X + \Delta$ to get enough ample $\mathbb{Q}$-divisors. These span $\mathrm{LEnv}_{\mathbb{Z}}(K_X + \Delta)$ (at least with $\mathbb{Q}$-coefficients).

The choice of B in (8.13.4) is discussed in (6.23).

The sheaves $\omega_{X/S}^{[m]}(\sum \lfloor m a_i \rfloor D_i)$ are not easy to understand. As we already noted in (8.8), they are not always flat families of divisorial sheaves, though the latter holds for infinitely many m, depending on the coefficient vector $\mathbf{a}$. Unfortunately, the method of (11.50) is ineffective, it is not at all clear how to produce such values m.

Theorem 8.15 *KSBA-stability, as defined in (8.12–8.13), is a good moduli theory (6.10).*

Proof We need to check the conditions (6.10.1–5).

Separatedness and valuative-properness (6.10.1–2) are as for (8.9). Embedded moduli spaces (6.10.3) are worked out in (8.21). Representability holds by (7.65) and (3.31). Boundedness is discussed in (6.8.2). □

Let us note the following strengthening of (2.65) and (2.69).

Theorem 8.16 *Let $f \colon (X, \Delta = \sum_{i \in I} a_i D_i) \to S$ be a KSBA stable family. Let $B = \sum_{j \in J} D_j$ be a divisor, where $a_j = 1$ for $j \in J$ and L an f-semi-ample*

divisorial sheaf (3.25) on X. Then $R^i f_*(L^{[-1]}(-B))$ and $R^i f_*(\omega_{X/S} \boxtimes L(B))$ are locally free and compatible with base change for every i.

Proof As in the proof of (2.65), a suitable cyclic cover reduces the first part to the case when $L = \mathcal{O}_X$, which follows from Kovács and Schwede (2016, 5.1). (Note that the latter is stated over a smooth base, but that is not used in the proof. Also, $\mathcal{O}_X(-B)$ is a flat family of divisorial sheaves by (8.13.4), thus (X_s, B_s) is a Du Bois pair for every $s \in S$.) This implies the second part as in (2.69). □

KSBA Stability, Stronger Version

K-flatness is designed to work for all boundary divisors $\Delta = \sum_i a_i D_i$, thus it cannot capture the stronger properties of those D_i that have coefficient $> \frac{1}{2}$. The notion of strong KSBA-stability takes care of this. The resulting moduli space has the same underlying reduced subscheme, but a smaller nilpotent structure.

8.17 (Stable objects) $(X, \Delta = \sum_{i \in I} a_i D_i)$, same as in (8.12).

8.18 (Stable families) Families $f: (X, \Delta = \sum_{i \in I} a_i D_i) \to S$ as in (8.13), with the following additional assumption, taken from (2.82).
(8.18.1) Let $J \subset I$ be any subset such that $a_j > \frac{1}{2}$ for every $j \in J$ and set $D_J := \cup_{j \in J} D_j$. Then $f|_{D_J}: D_J \to S$ is flat with reduced fibers.
Note It is possible that some variant of (6.27.3) could be added, but (2.83) does not seem strong enough for this.

The proof of (8.15) carries over without changes to give the following.

Theorem 8.19 *Strong KSBA-stability, as defined in (8.17–8.18), is a good moduli theory (6.10).* □

Example 8.20 To see that we do get a smaller scheme structure, even for surfaces, start with the A_1 singularity $S_0 := (y^2 - x^2 + z^2 = 0)$ and the nodal curve $C_0 := (z = y^2 - x^2 = 0)$. Then (S_0, C_0) is lc. Over $k[\varepsilon]$, consider the trivial deformation $S := (y^2 - x^2 + z^2 = 0)$. For C_0 we take the simplest K-flat, but nonflat deformation. Using the notation of (7.70.5), it is given by $y^2 - x^2 = 0$ and $z = \frac{y}{x}\varepsilon$ in the chart $(x \neq 0)$. The closure is given by

$$C = (y^2 - x^2 = zx - y\varepsilon = zy - x\varepsilon = z^2 = z\varepsilon = 0).$$

Then $(S, C) \to \operatorname{Spec} k[\varepsilon]$ is locally stable as in (8.13), but C is not flat over $\operatorname{Spec} k[\varepsilon]$. Hence (8.18.1) is not satisfied.

8.21 (Construction of embedded moduli spaces) A way of approximating an $\mathbb{R}$-Cartier pair with $\mathbb{Q}$-Cartier pairs is given in (11.47).

Depending on the vector $\mathbf{a}$, we have $\mathbb{Q}$-linear maps $\sigma_j^m \colon \mathbb{R} \to \mathbb{Q}$, extended to divisors by $\sigma_j^m(\sum a_i D_i) := \sum \sigma_j^m(a_i) D_i$, with the following properties.

(8.21.1) If $K_{X/S} + \Delta$ is $\mathbb{R}$-Cartier then the $K_{X/S} + \sigma_j^m(\Delta)$ are $\mathbb{Q}$-Cartier.

(8.21.2) $\lim_{m \to \infty} \sigma_j^m(\Delta) = \Delta$.

(8.21.3) Δ is a convex $\mathbb{R}$-linear combination of the $\sigma_j^m(\Delta)$ for every fixed m.

(8.21.4) If $(X, \Delta) \to S$ is stable then so are the $(X, \sigma_j^m(\Delta)) \to S$ for $m \gg 1$.

Since $\mathcal{SP}(\mathbf{a}, n, v)$ is bounded by (6.8.2), there is a fixed M such that

(8.21.5) the $(X, \sigma_j^M(\Delta))$ are stable for every j and every $(X, \Delta) \in \mathcal{SP}(\mathbf{a}, n, v)$.

The volume of $(X, \sigma_j^M(\Delta))$ may depend on (X, Δ), but it is locally constant in families, so only finitely many values can occur for $\mathcal{SP}(\mathbf{a}, n, v)$. Denote this set by $V \subset \mathbb{R}$.

Let $\mathcal{SP}(\sigma_*^M(\mathbf{a}), n, V)$ be the moduli functor of all pairs (X, Δ) of dimension n, for which all the $(X, \sigma_j^M(\Delta))$ are stable and $\mathrm{vol}(X, \sigma_1^M(\Delta)) \in V$. We claim that this is a good moduli theory.

Indeed, first $\mathcal{SP}(\sigma_1^M(\mathbf{a}), n, V)$ is a good moduli theory by (8.9). Then we have to add the conditions that the $K_{X/S} + \sigma_j^M(\Delta)$ are $\mathbb{Q}$-Cartier for $j \neq 1$; these are representable by (4.29). Finally, once the $K_{X/S} + \sigma_j^M(\Delta)$ are $\mathbb{Q}$-Cartier, ampleness of these is an open condition. Thus we have the moduli space $\mathrm{SP}(\sigma_*^M(\mathbf{a}), n, V)$.

Since Δ is a convex $\mathbb{R}$-linear combination of the $\sigma_j^m(\Delta)$, $\mathrm{SP}(\mathbf{a}, n, v)$ is a closed subspace of $\mathrm{SP}(\sigma_*^M(\mathbf{a}), n, V)$ by (11.4.4). □

8.3 Stability with Floating Coefficients

Much of the technical subtlety of the KSBA approach is caused by the presence of boundary divisors that are not $\mathbb{Q}$-Cartier. As we discussed in Section 6.4, one way to avoid these is to work with marked pairs $(X, \sum_{i \in I} a_i D_i)$ as in (8.3), where the $a_i \in \mathbb{R}$ are $\mathbb{Q}$-linearly independent. However, in many important cases, the a_i are dictated by geometric considerations and they are rational.

By working on a $\mathbb{Q}$-factorialization $\pi \colon X' \to X$, we can achieve that the D_i' are $\mathbb{Q}$-Cartier. The price we pay is that $K_X + \sum a_i D_i'$ is only nef, giving a non-separated moduli space. We can restore separatedness if we know which linear combinations $- \sum_i c_i D_i'$ are π-ample. (The negative sign works better later.)

If we fix c_i, then $- \sum_i (c_i + \eta_i) D_i'$ is also π-ample for all $|\eta_i| \ll |c_i|$. Thus we can choose the η_i such that the $(a_i - c_i - \eta_i)$ are $\mathbb{Q}$-linearly independent; we are then back to the situation of Alexeev stability, as in Section 6.4. However, we do not wish to fix the c_i.

Using floating coefficients was considered early on by Alexeev, Hassett, and Kovács. There is a short discussion in Alexeev (2006), but the first significant example of it is treated in Alexeev (2015). The general type case with a floating coefficient is worked out in Filipazzi and Inchiostro (2021).

Keeping in mind the chambers discussed in (6.39), it is clear that one cannot float several coefficients independently. A solution is to fix an ordering of the index set I; this is natural in many cases, but not always.

The key observation is that, for a normal pair $(X, \sum_{i=1}^{r} a_i D_i)$, there is at most one small modification $\pi\colon X' \to X$ such that

• $-\sum_{i=1}^{r} \varepsilon_i D_i'$ is π-ample for all $0 < \varepsilon_1 \ll \cdots \ll \varepsilon_r$.

(The notation means that there is a $\delta > 0$ such that ampleness holds whenever $\varepsilon_i \le \delta \varepsilon_{i+1}$ for every i. By (11.43), then the $D_i' := \pi_*^{-1} D_i$ are $\mathbb{Q}$-Cartier.)

To get a good moduli theory out of this, we need to allow certain nonsmall birational maps $X' \to X$. There is a further issue that going freely between X and X' seems to need the Abundance Conjecture to hold (Kollár and Mori, 1998, 3.12). Thus the working definition is more complicated.

8.22 (Canonical contractions and models of nef slc pairs) By Kollár (2011c), there are projective surfaces with normal crossing singularities whose canonical ring is not finitely generated. Thus it is not possible to define the canonical model of a proper slc pair (Y, Δ) in general.

There are problems even if we assume that $K_Y + \Delta$ is semiample. As a typical example, let $S \subset \mathbb{P}^3$ be a surface of degree ≥ 5 with a single singular point $s \in S$ that is simple elliptic (2.21.4.1). Let $S' \to S$ be the minimal resolution with exceptional curve $E' \subset S'$. Next take two copies of (S_i', E_i') and glue them along $E_1' \simeq E_2'$ to get a surface T' with normal crossing singularities. Note that $\omega_{T'}$ is generated by global sections and it maps T' to the surface T obtained by gluing two copies (S_i, s_i) at the points $s_1 \simeq s_2$. Thus T is not S_2. Here the problem is that T' is singular along the exceptional divisor of $T' \to T$. It is easy to see that this is the only obstacle in general.

Claim 8.22.1 Let $g\colon Y \to X$ be a proper, birational morphism of pure dimensional, reduced schemes. Assume that $g_* \mathcal{O}_Y = \mathcal{O}_X$, Y is S_2 and none of the g-exceptional divisors is contained in Sing Y. Then X is S_2. □

Corollary 8.22.2 Let (Y, Δ) be an slc pair such that $K_Y + \Delta$ is semiample, inducing a proper morphism $g\colon Y \to X$. Assume that g is birational and none of the g-exceptional divisors is contained in Sing Y. Then $(X, g_* \Delta)$ is slc, $K_X + g_* \Delta$ is ample and g is a *crepant* contraction. That is, $K_Y + \Delta$ is numerically g-trivial. □

We call $(X, g_* \Delta)$ the *canonical model* of (Y, Δ) and denote it by (Y^c, Δ^c).

We stress that here we are considering only those cases for which $(Y, \Delta) \to (Y^c, \Delta^c)$ is a crepant contraction.

Lemma 8.23 *Let* (Y, Δ) *be an slc pair and* $g \colon Y \to X$ *a proper morphism such that* $g_* \mathcal{O}_Y = \mathcal{O}_X$ *and* $K_Y + \Delta$ *is numerically g-trivial. Let* Θ_1, Θ_2 *be effective divisors such that* $\operatorname{Supp} \Theta_i \subset \operatorname{Supp} \Delta$. *Assume that* $-\Theta_1$ *is g-ample and* $-\Theta_2$ *is g-nef. Then the following hold.*

(8.23.1) $(X, g_* \Delta)$ *is slc and g is birational.*

(8.23.2) $g_* \mathcal{O}_Y(mK_Y + \lfloor m\Delta \rfloor) = \mathcal{O}_X(mK_X + \lfloor mg_* \Delta \rfloor)$ *for every* $m \geq 1$.

(8.23.3) $-\Theta_2$ *is g-semiample.*

Proof Since $-\Theta_1$ is g-ample, $\operatorname{Ex}(g) \subset \operatorname{Supp} \Theta_1 \subset \operatorname{Supp} \Delta$. In particular, g is birational and Y is smooth at every generic point of $\operatorname{Ex}(g)$ that has codimension 1 in Y. Thus X is S_2 by (8.22.1), hence demi-normal, so (1) holds by (4.50). Next (2) follows from (11.61).

For (3), assume first that Y is normal. We apply (11.28.2) to $(Y, \Delta - \varepsilon\Theta_2)$. Set $Z = g(\Theta_2)$. Then $(X \setminus Z, g_* \Delta)$ is the canonical model of $(Y \setminus g^{-1}(Z), \Delta - \varepsilon\Theta_2)$. Since $-\Theta_2$ is g-nef, $\operatorname{Supp} \Theta_2 = g^{-1}(Z)$, hence none of the lc centers of $(Y, \Delta - \varepsilon\Theta_2)$ is contained in $g^{-1}(Z)$. Thus $K_Y + \Delta - \varepsilon\Theta_2$ is g-semiample and so is $-\Theta_2$.

In general, we can apply the above to the normalization $\bar{Y} \to Y$, get a canonical model of $(\bar{Y}, \bar{\Delta} - \varepsilon\bar{\Theta}_2)$ and then use (11.38) to conclude. □

8.24 (Stable objects) Alexeev–Filipazzi–Inchiostro stability parametrizes projective, marked, slc pairs

$$(X, \Delta = \textstyle\sum_{j \in J} b_j B^j + \sum_{i \in I} a_i D^i), \tag{8.24.1}$$

where the divisors are indexed by the disjoint union $J \cup I$. We write $\Delta_0 := \sum_{j \in J} b_j B^j$; this divisor will be treated as in KSBA stability. The new aspect is the treatment of the divisors D^i. The index set I is ordered, so we identify it with $\{1, \ldots, r\}$.

The sole assumption that we would like to have is the following.

(8.24.2) $K_X + \Delta - \sum_i \varepsilon_i D^i$ is ample for all $0 < \varepsilon_1 \ll \cdots \ll \varepsilon_r \ll 1$.

Since we can choose the $a_i - \varepsilon_i$ to be $\mathbb{Q}$-linearly independent of the b_j, we see that $K_X + \Delta_0$ and the $D^1, \ldots, D^r$ are necessarily $\mathbb{R}$-Cartier by (11.43).

Fixing m and $0 < \varepsilon_m \ll \cdots \ll \varepsilon_r$, letting the others go to 0 gives that $K_X + \Delta - \sum_{i=m+1}^r \varepsilon_i D^i$ is nef for $0 \leq m \leq r$. If the Abundance Conjecture holds, then these divisors are semiample, but this is not known. So for now we have to add the assumption:

(8.24.3) $K_X + \Delta$ is semiample.

An slc pair (X, Δ) as in (1) is *AFI-stable* if it satisfies assumptions (2–3).

If (3) holds, then we have a crepant contraction to the canonical model $\pi\colon (X, \Delta) \to (X^c, \Delta^c)$. Then (2) is equivalent to the following condition.

(8.24.4) There are $\eta_{im} > 0$ such that $-\sum_{i=m}^{r} \eta_{im} D^i$ is π-nef for $1 < m < r$ and π-ample for $m = 1$.

8.25 (Explanation) By (8.23.3), the assumptions (8.24.2–3) imply that

(8.25.1) $K_X + \Delta - \sum_{i=m+1}^{r} \varepsilon_i D^i$ is semiample for $0 \leq m \leq r$.

Thus for each m we get a morphism $\pi_m\colon X \to X_m$. Then (8.23.1) shows that π_m is birational. For the rest of the section, we use a subscript m to denote the image of a divisor on X_m.

Using (8.23), we get that, for $0 < \varepsilon_m \ll \cdots \ll \varepsilon_r \ll 1$,

(8.25.2) (X_m, Δ_m) is slc and $K_{X_m} + \Delta_m - \sum_{i=m+1}^{r} \varepsilon_i D_m^i$ is ample.

In particular, we have

(8.25.3) a tower of morphisms $X =: X_0 \to X_1 \to \cdots \to X_r$ such that

(8.25.4) $K_{X_r} + \Delta_r$ is ample on X_r, and

(8.25.5) $-D_{m-1}^m$ is $(X_{m-1} \to X_m)$-ample for every m.

Repeatedly using Hartshorne (1977, exc.II.7.14) we get that (3–5) are equivalent to (8.24.2–3).

In (8.30), we show how to transform the conditions (8.24.2) involving variable $\varepsilon_i > 0$ into a set of conditions with fixed $\delta_i > 0$. The result is, however, ineffective, and it would be good to find a more constructive approach.

Note that (X_r, Δ_r) is the canonical model (X^c, Δ^c) of (X, Δ), it is thus independent of the ordering of I. By contrast, the intermediate $X \to X_m \to X^c$ do depend on the ordering of I.

8.26 (Stable families) A family $f\colon (X, \Delta = \sum_{j \in J} b_j B^j + \sum_i a_i D^i) \to S$ is *AFI-stable* if the following hold:

(8.26.1) $f\colon (X, \Delta) \to S$ is locally stable.

(8.26.2) $K_{X/S} + \sum_{j \in J} b_j B^j$ and the $D^1, \ldots, D^r$ are relatively $\mathbb{R}$-Cartier.

(8.26.3) The fibers (X_s, Δ_s) are AFI-stable as in (8.24).

(8.26.4) $\omega_{X/S}^{[m_0]}(\sum_{j \in J} m_j B^j + \sum_{i \in I} n_i D^i)$ is a flat family of divisorial sheaves if $n_i \in \mathbb{Z}$ and $(m_j\colon j \in \{0\} \cup J) \in \mathrm{LEnv}_{\mathbb{Z}}(K_X + \sum_{j \in J} b_j B^j)$.

The next assumption may be redundant; we discuss it in (8.27) and (8.34).

(8.26.5) $f\colon (X, \Delta) \to S$ has a crepant contraction to its simultaneous canonical model. That is, to a stable morphism $f^c\colon (X^c, \Delta^c) \to S$ whose fibers are the canonical models $((X_s)^c, (\Delta_s)^c)$ of the fibers.

8.27 (Explanation) Assumptions (8.26.1–3) closely follow (8.7). Assumption (8.26.4) is modeled on (8.11) and (8.13).

The D^i are $\mathbb{Q}$-Cartier by (11.43.2), hence K-flat by (7.4.6), at least in characteristic 0. Thus if $\Delta_0 := \sum_{j \in J} b_j B^j$ is the 0 divisor, then we avoid using K-flatness entirely.

Since ampleness is an open condition (11.54), we see using (8.24.4) that, if S is Noetherian, then $K_X + \Delta - \sum_i \varepsilon_i D^i$ is f-ample for all $0 < \varepsilon_1 \ll \cdots \ll \varepsilon_r \ll 1$. Thus one can view (8.26) as picking one of the chambers discussed in (6.39).

By (8.25), each fiber (X_s, Δ_s) has a canonical model $((X_s)^c, (\Delta_s)^c)$. More generally, we prove in (8.35) that, if S is reduced, then (8.26.1–4) imply (8.26.5). This implication is not known over arbitrary bases, but we prove in (8.33) that $(X, \Delta) \to S$ uniquely determines its simultaneous canonical model.

Assumption (8.26.5) guarantees that taking the relative canonical model is a natural transformation from AFI-stable families to KSBA-stable families.

A stronger variant of (8.26.5) would be to require that the towers (8.25.3) of the fibers form a flat family. The latter might be equivalent to (8.26.5).

Theorem 8.28 *AFI-stability, as defined in (8.24–8.26) is a good moduli theory (6.10).*

We start with a general discussion on ample perturbations, followed by results on simultaneous canonical models and boundedness. With these preliminaries in place, the proof of (8.28) given in (8.37) is quite short.

Definition 8.29 A pair $(X, \Delta = \sum_{j \in J} b_j B^j + \sum_{i=1}^r a_i D^i)$ as in (8.24) has an intersection form

$$\mathcal{I}(t_0, \ldots, t_r) := (t_0(K_X + \Delta) + t_1 D^1 + \cdots + t_r D^r)^n. \qquad (8.29.1)$$

For a family as in (8.26), the intersection form $\mathcal{I}(t_0, \ldots, t_r)$ is a locally constant function on the base by (8.26.2). We can thus decompose the functor of AFI-stable pairs into open and closed subfunctors

$$\mathcal{AFI}(\mathbf{b}, \mathbf{a}, n, \mathcal{I}(t_0, \ldots, t_r)). \qquad (8.29.2)$$

Next we see that, for each of these subfunctors, there is a uniform choice of ample divisors.

Proposition 8.30 *Fix $\mathbf{b}, \mathbf{a}, n$ and $\mathcal{I}(t_0, \ldots, t_r)$. Then there are $\delta_i > 0$ such that $(X, \Delta) \in \mathcal{AFI}(\mathbf{b}, \mathbf{a}, n, \mathcal{I}(t_0, \ldots, t_r))$ iff the following hold.*
(8.30.1) (X, Δ) *is slc and the D^i are $\mathbb{R}$-Cartier,*
(8.30.2) $(t_0(K_X + \Delta) + t_1 D^1 + \cdots + t_r D^r)^n = \mathcal{I}(t_0, \ldots, t_r),$
(8.30.3) $K_X + \Delta - \sum_{i=1}^r \delta_i D^i$ *is ample, and*
(8.30.4) $K_X + \Delta - \sum_{i=m+1}^r \delta_i D^i$ *is semiample for $m = 1, \ldots, r$.*

Proof If $(X, \Delta) \in \mathcal{AFI}(\mathbf{b}, \mathbf{a}, n, \mathcal{I}(t_0, \ldots, t_r))$ then (1–2) hold by assumption. The key point is that one can find δ_i that do not depend on (X, Δ). Start with the canonical model $(X_r, \Delta_r) = (X^c, \Delta^c)$. Since

$$(K_{X^c} + \Delta^c)^n = (K_X + \Delta)^n = \mathcal{I}(1, 0, \ldots, 0),$$

the canonical models form a bounded family by (6.8.1). In particular, there is an $\eta > 0$ such that $((K_{X_r} + \Delta_r) \cdot C) \geq \eta$ for every nonzero effective curve $C \subset X_r$ for every X_r. Thus $((K_X + \Delta) \cdot C)$ is either 0 or $\geq \eta$ for every nonzero effective curve $C \subset X$.

Choose $0 < c_1 \ll \cdots \ll c_{r-1} \ll c_r = a_r$ such that $K_X + \Delta - \varepsilon_r \sum c_i D^i$ is ample for some $\varepsilon_r > 0$. Applying (8.31) with $\Delta_2 := \sum c_i D^i$ and $\Delta_1 := \Delta - \Delta_2$, we get a fixed δ_r such that

$$K_X + \Delta_1 + (1 - \delta_r)\Delta_2 = K_X + (\Delta - \delta_r D^r) - \sum_{i=1}^{r-1} c_i \delta_r D^i$$

is ample. This holds for all $0 < c_1 \delta_r \ll \cdots \ll c_{r-1} \delta_r \ll 1$, so $K_X + \Delta - \delta_r D^r$ is nef, so semiample by (8.23.3). By induction on r, we get the other δ_i and the divisors in (4) are nef. $K_X + \Delta$ is semiample by assumption (8.24.3). This implies the rest of (4) by (8.23.3).

Conversely, convex linear combinations of the divisors $K_X + \Delta - \sum_{i=m+1}^{r} \delta_i D^i$ for $m = 0, \ldots, r$ show that (8.24.2) holds. $\square$

Lemma 8.31 (Filipazzi and Inchiostro, 2021, 2.15) *Let $(X, \Delta_1 + \Delta_2)$ be a proper slc pair of dimension n. Assume that there is an $\eta > 0$ such that*

(8.31.1) *$((K_X + \Delta_1 + \Delta_2) \cdot C)$ is either 0, or $\geq \eta$ for every nonzero effective curve C, and*

(8.31.2) *$K_X + \Delta_1 + (1 - \varepsilon_0)\Delta_2$ is ample for some $\varepsilon_0 > 0$.*

Then $K_X + \Delta_1 + (1 - \varepsilon)\Delta_2$ is ample for every $\eta/(2n + \eta) > \varepsilon > 0$.

Proof We may assume that X is normal. If $\varepsilon_0 > \varepsilon > 0$, then $K_X + \Delta_1 + (1 - \varepsilon)\Delta_2$ is a convex linear combination of $K_X + \Delta_1 + (1 - \varepsilon_0)\Delta_2$ and of $K_X + \Delta_1 + \Delta_2$, hence ample.

Thus consider the case when $\varepsilon_0 < \varepsilon$. We check Kleiman's ampleness criterion. For $Z \in \overline{NE}(X)$ we have that

$$\begin{aligned}
&((K_X + \Delta_1 + (1 - \varepsilon)\Delta_2) \cdot Z) \\
&= \tfrac{\varepsilon - \varepsilon_0}{1 - \varepsilon_0}((K_X + \Delta_1) \cdot Z) + \tfrac{1 - \varepsilon}{1 - \varepsilon_0}((K_X + \Delta_1 + (1 - \varepsilon_0)\Delta_2) \cdot Z).
\end{aligned} \tag{8.31.3}$$

So the criterion holds on the part where $((K_X + \Delta_1) \cdot Z) \geq 0$.

By the Cone theorem of Fujino (2017, 4.6.2), the rest of $\overline{NE}(X)$ is generated by curves C_i for which $-2n \leq ((K_X + \Delta_1) \cdot C_i) < 0$. If C_i is such a curve, then, applying (8.31.3) with $\varepsilon_0 = 0$, we get that

$$((K_X+\Delta_1+(1-\varepsilon)\Delta_2)\cdot C_i) = \varepsilon((K_X+\Delta_1)\cdot C_i)+(1-\varepsilon)((K_X+\Delta_1+\Delta_2)\cdot C_i). \quad (8.31.4)$$

Now set $\varepsilon = \varepsilon_0$. Then the left-hand side is positive, hence so is $((K_X + \Delta_1 + \Delta_2) \cdot C_i)$. Thus $((K_X + \Delta_1 + \Delta_2) \cdot C_i) \geq \eta$ by assumption. Thus (8.31.4) gives that, for every $\varepsilon > 0$, $((K_X + \Delta_1 + (1 - \varepsilon)\Delta_2) \cdot C_i) \geq -2n\varepsilon + (1 - \varepsilon)\eta$. The latter is positive if $\varepsilon < \eta/(2n + \eta)$. $\square$

Definition 8.32 Let $\pi_X : X \to S$ be a flat, proper morphism with S_2 fibers. A *simultaneous contraction* is a factorization $\pi_X : X \xrightarrow{\tau} Y \to S$ where

(8.32.1) $\pi_Y : Y \to S$ is flat, proper with S_2 fibers, and

(8.32.2) $\tau_* \mathscr{O}_X = \mathscr{O}_Y$.

This implies that $(\tau_s)_* \mathscr{O}_{X_s} = \mathscr{O}_{Y_s}$ for every $s \in S$.

If the τ_s are birational, then $X \to S$ and the τ_s uniquely determine Y. When S is Artinian, this is (8.33), which in turn implies the general case.

Lemma 8.33 *Let A be a local, Artinian ring with residue field k. Let $g_k : X_k \to Y_k$ be a birational morphism between proper, pure dimensional, S_2-schemes such that $(g_k)_* \mathscr{O}_{X_k} = \mathscr{O}_{Y_k}$. Let $X_A \to \mathrm{Spec}\, A$ be a flat, proper morphism.*

(8.33.1) *There is at most one flat $Y_A \to \mathrm{Spec}\, A$ such that g_k lifts to $g_A : X_A \to Y_A$.*

(8.33.2) *If Y_A exists, then $\mathscr{O}_{Y_A} = (g_A)_* \mathscr{O}_{X_A}$.*

Proof Note that X_A, Y_A are S_2 since $X_A \to \mathrm{Spec}\, A$ and $Y_A \to \mathrm{Spec}\, A$ are flat.

Let $U_k \subset Y_k$ be the largest open set over which g_k is an isomorphism. Thus we get open sets $V_k \subset X_k$, $V_A \subset X_A$ and $U_A \subset Y_A$. Note that $Y_A \setminus U_A$ has codimension ≥ 2. Thus $\mathscr{O}_{Y_A}$ is the push-forward of $\mathscr{O}_{U_A}$ by the injection $j_A : U_A \hookrightarrow Y_A$ (10.6).

Since $g_A : V_A \to U_A$ is an isomorphism, we see that $\mathscr{O}_{Y_A} = (j_A)_*(g_A)_* \mathscr{O}_{V_A}$ is determined by X_A and g_k. This also implies that $(g_A)_* \mathscr{O}_{X_A} = \mathscr{O}_{Y_A}$. $\square$

Definition 8.34 Let $h : (X, \Delta) \to S$ be proper and locally stable such that $K_{X/S} + \Delta$ is h-nef. A *simultaneous, canonical, crepant, birational contraction* is a simultaneous contraction $\pi_X : X \xrightarrow{\tau} Y \to S$ such that

(8.34.1) $\pi_Y : (Y, \tau_*\Delta) \to S$ is stable, and

(8.34.2) $\tau_s : (X_s, \Delta_s) \to (Y_s, g_*\Delta_s)$ is the crepant, birational contraction to its canonical model as in (8.22) for every $s \in S$.

By (8.32), $\pi_Y : (Y, \tau_*\Delta) \to S$ is uniquely determined by $\pi_X : (X, \Delta) \to S$, even when S is nonreduced.

Using the rational approximations $\Delta_j^n \to \Delta$ as in (11.47), we see that
(8.34.3) $X \to Y \to S$ is a simultaneous canonical contraction for Δ iff it is a simultaneous canonical contraction for Δ_j^n for $n \gg 1$ for every j.

Proposition 8.35 *Let* $g\colon (X, \Delta) \to S$ *be proper and locally stable, S reduced. Assume that (X_s, Δ_s) has a crepant, birational contraction to its canonical model for every $s \in S$. Then $g\colon (X, \Delta) \to S$ has a simultaneous, canonical, crepant, birational contraction.*

Proof By (8.34.3) it is enough to deal with the case when Δ is a $\mathbb{Q}$-divisor.

Next we prove that $g_*(\mathscr{O}_X(mK_{X/S} + m\Delta))$ is locally free and commutes with base change for m sufficiently divisible. By Grauert's theorem (as stated in Hartshorne (1977, III.12.9)) it is enough to prove this when S is a smooth curve. In this case (11.28) and (11.38) show that the relative canonical model exists, τ_s is an isomorphism for the generic point $s \in S$ and a finite, universal homeomorphism (10.78) for closed points $s \in S$. However, (5.4) then implies that in fact τ_s is an isomorphism for every $s \in S$. Thus

$$h^0(X_s, \mathscr{O}_{X_s}(mK_{X_s} + m\Delta_s)) = h^0((X_s)^c, \mathscr{O}_{X_s^c}(mK_{X_s^c} + m\Delta_s^c))$$
$$= h^0((X^c)_s, \mathscr{O}_{X_s^c}(m(K_{X^c/S} + \Delta^c)_s))$$

is independent of $s \in S$ for m sufficiently divisible, since $X^c \to S$ is flat and $K_{X^c/S} + \Delta^c$ is relatively ample.

With arbitrary S, we get the simultaneous canonical model

$$X^c = \mathrm{Proj}_S \oplus_{r \in \mathbb{N}} g_*(\mathscr{O}_X(rmK_{X/S} + rm\Delta)). \qquad \square$$

For representability, the key step is the following.

Proposition 8.36 *Let* $g\colon (X, \Delta) \to S$ *be proper and locally stable. Then there is a locally closed partial decomposition $S^{\mathrm{sccc}} \to S$ such that, for any $T \to S$, the base change $g_T\colon (X_T, \Delta_T) \to T$ has a simultaneous, canonical, crepant, birational contraction iff T factors through S^{sccc}.*

Proof As before, using (8.34.3), it is enough to deal with the case when Δ is a $\mathbb{Q}$-divisor. We may assume that S is connected.

Assume that (X_s, Δ_s) has a crepant, birational contraction to its canonical model (X_s^c, Δ_s^c). The self-intersection of $K_{X_s^c} + \Delta_s^c$ equals the self-intersection of $K_{X_s} + \Delta_s$, which is independent of $s \in S$. Thus the pairs (X_s^c, Δ_s^c) are in a bounded family by (6.8.1). In particular, there is an $m > 0$, independent of s, such that $rmK_{X_s^c} + rm\Delta_s^c$ is Cartier, very ample, and has no higher cohomologies for $r > 0$. Moreover, we get only finitely many possible Hilbert functions.

Thus $\mathscr{O}_{X_s}(rmK_{X_s} + rm\Delta_s)$ is locally free, globally generated, and maps (X_s, Δ_s) to its canonical model. This implies that if $\pi_T \colon (X_T, \Delta_T) \to T$ has a simultaneous canonical, crepant contraction, then, for every $r > 0$,

(8.36.1) $\mathscr{O}_{X_T}(rmK_{X_T/T} + rm\Delta_T)$ is relatively globally generated, and

(8.36.2) $(\pi_T)_* \mathscr{O}_{X_T}(rmK_{X_T/T} + rm\Delta_T)$ is locally free and commutes with base change.

For each Hilbert function, these conditions are representable by a locally closed subscheme by (3.21). □

Remark 8.36.3 If the Abundance Conjecture holds, then red S^{sccc} is an open subset of red S. The scheme-theoretic situation is not clear.

Example 8.36.4 Being semiample and big is not a constructible condition for families of line bundles. As a simple example, let $E \subset \mathbb{P}^2$ be an elliptic curve, $\pi \colon X := \mathrm{Proj}_E(\mathscr{O}_E + \mathscr{O}_E(3)) \to E$ the resolution of the cone over E. Consider the line bundles $L_X := \mathscr{O}_X(1) \otimes \pi^* L$ where $L \in \mathbf{Pic}^\circ(E)$.

Then L_X is nef and big for every $L \in \mathbf{Pic}^\circ(E)$, but semiample only if L is a torsion point of $\mathbf{Pic}^\circ(E)$. Thus the set $\{L \colon L_X \text{ is big and semiample}\}$ is not constructible.

A much subtler example of Lesieutre (2014) shows that being nef and big is also not a constructible condition in families of smooth surfaces.

8.37 (Proof of 8.28) First, we show that $\mathscr{AFI}(\mathbf{b}, \mathbf{a}, n, \mathcal{I})$ is bounded and representable. By (8.30), there are fixed $\delta_i > 0$ such that $K_X + \Delta - \sum_{i=1}^r \delta_i D^i$ is ample. The self-intersection of this divisor is $\mathcal{I}(1, a_1 - \delta_1, \ldots, a_r - \delta_r)$, hence all such stable pairs form a bounded family by (6.8.1). We can choose the $a_i - \delta_i$ to be $\mathbb{Q}$-linearly independent of the b_i. Then the D^i are $\mathbb{R}$-Cartier by (11.43).

Note that (6.8.1) gives boundedness for pairs such that $(X, \Delta - \sum_{i=1}^r \delta_i D^i)$ is slc, but we want (X, Δ) to be slc. By (7.65), local stability of (X, Δ) is a representable condition. (Actually, (11.48) shows that, for a suitable choice of the δ_i, the (X, Δ) are in fact slc.)

Semiampleness of the divisors $K_X + \Delta - \sum_{i=m+1}^r \delta_i D^i$ is a representable condition by (8.36). Representability of (8.26.4) is handled as in (6.40). (8.26.5) was treated in (8.36).

Separatedness follows from (11.40) as usual, applied to $(X, \Delta - \sum_{i=m+1}^r \delta_i D^i)$. To see valuative-properness, assume that we have $f^\circ \colon (X^\circ, \Delta^\circ = \sum_{j \in J} b_j B^{j\circ} + \sum_{i=1}^r a_i D^{i\circ}) \to C^\circ$ over an open subset of a smooth curve $C^\circ \subset C$.

Applying (8.35) to the divisors $K_X + \Delta - \sum_{i=m+1}^r \delta_i D^i$ we also have a tower

$$f^\circ \colon (X^\circ, \Delta^\circ) \to (X_1^\circ, \Delta_1^\circ) \to \cdots \to (X_r^\circ, \Delta_r^\circ) \to C^\circ,$$

where $(X_r^\circ, \Delta_r^\circ)$ is the relative canonical model of (X°, Δ°).

First, we use (2.51) to get that, after a base change (which we suppress in the notation), $(X_r^\circ, \Delta_r^\circ) \to C^\circ$ extends to a stable morphism $(X_r, \Delta_r) \to C$. Next we extend $(X_{r-1}^\circ, \Delta_{r-1}^\circ) \to C^\circ$. By construction, $K_{X_{r-1}^\circ} + \Delta_{r-1}^\circ - \varepsilon_r D_{r-1}^{r,\circ}$ is relatively ample on $X_{r-1}^\circ \to X_r^\circ$ for $0 < \varepsilon_r \ll 1$ and relatively semiample for $\varepsilon_r = 0$. Thus, by Hacon and Xu (2013, 1.5), after a base change (which we again suppress in the notation), it extends to a model $(X_{r-1}, \Delta_{r-1} - \varepsilon_r D_{r-1}^r) \to C$ with the same properties (with a possibly smaller upper bound for ε_r). This gives $(X_{r-1}, \Delta_{r-1}) \to C$. We can continue this until we get the tower

$$f \colon (X, \Delta) \to (X_1, \Delta_1) \to \cdots \to (X_r, \Delta_r) \to C,$$

proving valuative properness. □

8.4 Polarized Varieties

Assumptions In this section, we work with arbitrary schemes. Because of functoriality, the situation over $\operatorname{Spec} \mathbb{Z}$ determines everything.

8.38 (Ampleness conditions) Let X be a proper scheme over a field k and L a line bundle on X. The most important positivity notion is *ampleness,* but in connection with projective geometry the notion of *very ampleness* seems more relevant. If L is ample then L^r is very ample for $r \gg 1$ and there are numerous Matsusaka-type theorems that give effective control over r; see Matsusaka (1972); Lieberman and Mumford (1975); Kollár and Matsusaka (1983). In practice, this will not be a major difficulty for us.

A problem with very ampleness is that it is not open in flat families (X_s, L_s). Thus one needs to consider stronger variants. The two most frequently needed additional conditions are the following.

(8.38.1) $H^i(X, L) = 0$ for $i > 0$.

(8.38.2) $H^0(X, L)$ generates the ring $\sum_{r \geq 0} H^0(X, L^r)$.

These are connected by the notion of Castelnuovo–Mumford regularity; see Lazarsfeld (2004, sec.I.8) for details.

For our purposes the relevant issue is (1). Thus we say that a line bundle L is *strongly ample* if it is very ample and $H^i(X, L^m) = 0$ for $i, m > 0$. By Lazarsfeld (2004, I.8.3), if this holds for all $m \leq \dim X + 1$ then it holds for all m. Thus strong ampleness is an open condition in flat families.

Let $f \colon X \to S$ be a proper, flat morphism and L a line bundle on X. We say that L is *strongly f-ample* or *strongly ample over S*, if L is strongly ample on the fibers. Equivalently, if $R^i f_* L^m = 0$ for $i, m > 0$ and L is f-very ample. Thus $f_* L$ is locally free and we get an embedding $X \hookrightarrow \mathbb{P}_S(f_* L)$.

The main case for us is when $f: (X, \Delta) \to S$ is stable and $L = \omega_{X/S}^{[r]}(r\Delta)$ for some $r > 0$. If $r > 1$ then $R^i f_* L^m = 0$ for $i, m > 0$ by (11.34).

Definition 8.39 (Polarization) A *polarized scheme* is a pair (X, L) consisting of a projective scheme X plus an ample line bundle L on X.

In the most basic version of the definition, a *polarized family of schemes* over a scheme S consists of a flat, projective morphism $f: X \to S$, plus a relatively ample line bundle L on X. (See (8.40) for other variants.)

We are interested only in the relative behavior of L, thus two families (X, L) and (X, L') are considered equivalent if there is a line bundle M on S such that $L \simeq L' \otimes f^* M$. There are some quite subtle issues with this in general Raynaud (1970), but if S is reduced and $H^0(X_s, \mathcal{O}_{X_s}) \simeq k(s)$ for every $s \in S$, then $L \simeq L' \otimes f^* M$ for some M iff $L|_{X_s} \simeq L'|_{X_s}$ for every $s \in S$ by Grauert's theorem, as in Hartshorne (1977, III.12.9). See also (8.40) for further comments on this.

For technical reasons, it is more convenient to deal with the cases when, in addition, L is strongly f-ample (8.38). We call such an L a *strong polarization*. Thus the "naive" functor of strongly polarized schemes

$$S \mapsto \mathcal{P}^s\mathcal{S}ch(n, N)(S) \tag{8.39.1}$$

associates to a scheme S the equivalence classes of all $f: (X, L) \to S$ such that

(8.39.2) f is flat, proper, of pure relative dimension n,

(8.39.3) X_s is pure and $H^0(X_s, \mathcal{O}_{X_s}) \simeq k(s)$ for every $s \in S$,

(8.39.4) L is strongly f-ample (8.38), and

(8.39.5) $f_* L$ is locally free of rank $N + 1$.

Since L is flat over S, strong f-ampleness implies that $f_* L$ is locally free.

(8.39.6) If we fix the whole Hilbert polynomial $\chi(X, r) := \chi(X, L^r)$, we get the functor $S \mapsto \mathcal{P}^s\mathcal{S}ch(\chi)(S)$.

Let $f: X \to S$ be a flat, proper morphism and L a line bundle on X. Having pure fibers is an open condition (10.11) and then pure dimensionality is an open condition. Thus there is a maximal open subscheme $S° \subset S$ such that $f°: (X°, L°) \to S°$ satisfies the assumptions (2–5).

Definition 8.40 (Pre-polarization) The definition in (8.39) is geometrically clear, but it does not have the sheaf property. In analogy with the notion of a presheaf, we could define a *pre-polarization* of a projective morphism $f: X \to S$ to consist of

(8.40.1) an open cover $\cup_i U_i \to S$, and

(8.40.2) relatively ample line bundles L_i on $X_i := X \times_S U_i$ such that,

(8.40.3) for every i, j, the restrictions of L_i and L_j to $X_{ij} := X \times_S U_i \times_S U_j$ are identified as in (8.39).

(That is, there are line bundles M_{ij} on $U_i \times_S U_j$ such that $L_i|_{X_{ij}} \simeq L_j|_{X_{ij}} \otimes f_{ij}^* M_{ij}$.)

Pre-polarizations form a presheaf, hence the "right" notion of polarization should be a global section of the corresponding sheaf.

If $\cup_i U_i \to S$ is a cover by Zariski open subsets, the resulting notion is very similar to what we have in (8.39). The only difference is in property (8.39.5) since $f_* L$ need not exists globally. However, $\mathbb{P}_S(f_* L)$ does exist as a Zariski locally trivial $\mathbb{P}^N$-bundle over S and we usually use $\mathbb{P}_S(f_* L)$ anyhow.

If the $U_i \to S$ are étale, then we still get an object $\mathbb{P}_S(f_* L) \to S$, but this is a Severi–Brauer scheme, that is, an étale locally trivial $\mathbb{P}^N$-bundle over S. (See (8.40.5) for an example with $N = 1$.) From the theoretical point of view, it is most natural to use the étale topology for the moduli theory of varieties. Pre-polarizations define a pre-sheaf in the étale topology and sheafifying gives the functors

$$S \mapsto \mathcal{P}^s\!\mathcal{S}\!ch^{\text{et}}(n, N)(S) \quad \text{and} \quad S \mapsto \mathcal{P}^s\!\mathcal{S}\!ch^{\text{et}}(\chi)(S). \tag{8.40.4}$$

(For arbitrary schemes one needs finer topologies; see Raynaud (1970).)

For the difference between $\mathcal{P}^s\!\mathcal{S}\!ch^{\text{et}}$ and $\mathcal{P}^s\!\mathcal{S}\!ch$, a simple example to keep in mind is the following. Consider

$$X := (x^2 + sy^2 + tz^2 = 0) \subset \mathbb{P}^2_{xyz} \times (\mathbb{A}^2_{st} \setminus (st = 0)), \tag{8.40.5}$$

with coordinate projection to $S := \mathbb{A}^2_{st} \setminus (st = 0)$. The fibers are all smooth conics. In the analytic or étale topology, there is a pre-polarization whose restriction to each fiber is a degree 1 line bundle, but there is no such line bundle on X. However, $\mathcal{O}_{\mathbb{P}^2}(1)$ gives a line bundle on X whose restriction to each fiber has degree 2.

We will, however, stick to the naive versions for several reasons.

- Stable families come with preferred polarizing line bundles $\omega_{X/S}^{[m]}(m\Delta)$.
- $\mathcal{P}^s\!\mathcal{S}\!ch^{\text{et}}$ and $\mathcal{P}^s\!\mathcal{S}\!ch$ have the same coarse moduli spaces (8.56.1).
- A suitable power of any pre-polarization naturally gives an actual polarization using (8.66.6).

So, at the end, the distinction between the functors $\mathcal{P}^s\!\mathcal{S}\!ch^{\text{et}}$ and $\mathcal{P}^s\!\mathcal{S}\!ch$ does not matter much for us. There is, however, another related notion that does lead to different coarse moduli spaces.

8.40.7 (*Numerical polarization*) Given $f: X \to S$, two relatively ample line bundles L and L' on X are considered equivalent if $L_s \equiv L'_s$ (p.xv) for every geometric point $s \to S$. This is the original definition used by Matsusaka

(1972) and it may be the most natural notion for general polarized pairs. Stable varieties come with an ample divisor, not just with an ample numerical equivalence class, which simplifies our task.

8.41 (Strongly embedded schemes) Fix a projective space $\mathbb{P}^N_{\mathbb{Z}}$. Over the Hilbert scheme there is a universal family, hence we get

$$\mathrm{Univ}(\mathbb{P}^N_{\mathbb{Z}}) \subset \mathbb{P}^N_{\mathbb{Z}} \times \mathrm{Hilb}(\mathbb{P}^N_{\mathbb{Z}}), \tag{8.41.1}$$

and $\mathscr{O}_{\mathbb{P}^N}(1)$ gives a polarization of $\mathrm{Univ}(\mathbb{P}^N_{\mathbb{Z}}) \to \mathrm{Hilb}(\mathbb{P}^N_{\mathbb{Z}})$. As in (8.39) there is a largest open subset

$$\mathrm{Hilb}^{\mathrm{str}}_n(\mathbb{P}^N_{\mathbb{Z}}) \subset \mathrm{Hilb}_n(\mathbb{P}^N_{\mathbb{Z}}), \tag{8.41.2}$$

over which the polarization is strong (8.39.2–5). One should think of this as pairs (X, L) that "naturally live" in $\mathbb{P}^N$. The universal family restricts to

$$\mathrm{Univ}^{\mathrm{str}}_n(\mathbb{P}^N_{\mathbb{Z}}) \to \mathrm{Hilb}^{\mathrm{str}}_n(\mathbb{P}^N_{\mathbb{Z}}). \tag{8.41.3}$$

The corresponding functor associates to a scheme S the set of all flat families of closed subschemes of pure dimension n of $\mathbb{P}^N_S$

$$f \colon (X \subset \mathbb{P}^N_S; \mathscr{O}_X(1)) \to S, \tag{8.41.4}$$

where $\mathscr{O}_X(1)$ is strongly f-ample. Equivalently, we parametrize objects

$$\left(f \colon (X; L) \to S; \phi \in \mathrm{Isom}_S\left(\mathbb{P}_S(f_*L), \mathbb{P}^N_S\right) \right), \tag{8.41.5}$$

consisting of strongly polarized, flat families of purely n-dimensional schemes, plus an isomorphism $\phi \colon \mathbb{P}_S(f_*L) \simeq \mathbb{P}^N_S$. We call the latter a *projective framing* of f_*L or of L. We can also fix the Hilbert polynomial χ of X and, for $N :=$ $\chi(1) - 1$ consider the subschemes

$$\mathrm{Univ}^{\mathrm{str}}_\chi(\mathbb{P}^N_{\mathbb{Z}}) \to \mathrm{Hilb}^{\mathrm{str}}_\chi(\mathbb{P}^N_{\mathbb{Z}}) \subset \mathrm{Hilb}^{\mathrm{str}}_n(\mathbb{P}^N_{\mathbb{Z}}). \tag{8.41.6}$$

By the theory of Hilbert schemes, the spaces $\mathrm{Hilb}^{\mathrm{str}}_\chi(\mathbb{P}^N_{\mathbb{Z}})$ are quasi-projective, though usually non-projective, reducible and disconnected; see Grothendieck (1962), Kollár (1996, chap.I), or Sernesi (2006).

We can summarize these discussions as follows.

Proposition 8.42 *Fix a polynomial $\chi(t)$. Then*

$$\mathrm{Univ}^{str}_\chi(\mathbb{P}^N_{\mathbb{Z}}) \to \mathrm{Hilb}^{str}_\chi(\mathbb{P}^N_{\mathbb{Z}})$$

constructed in (8.41) represents the functor of strongly polarized schemes with Hilbert polynomial χ and a projective framing. That is, for every scheme S, pull-back gives a one-to-one correspondence between

(8.42.1) $\mathrm{Mor}_{\mathbb{Z}}(S, \mathrm{Hilb}_{\chi}^{str}(\mathbb{P}_{\mathbb{Z}}^N))$, *as in (8.63), and*

(8.42.2) *flat, projective families of purely n-dimensional schemes $f: X \to S$ with a strong polarization L of Hilbert polynomial χ, plus an isomorphism $\mathbb{P}_S(f_*L) \simeq \mathbb{P}_S^N$, where $N + 1 = \chi(1)$.* $\square$

The general correspondence between the moduli of polarized varieties and the moduli of embedded varieties (8.56.1) gives now the following.

Corollary 8.43 *Fix a Hilbert polynomial χ with $N + 1 = \chi(1)$. Then the stack $[\mathrm{Hilb}_{\chi}^{str}(\mathbb{P}^N)/\mathrm{PGL}_{N+1}]$ represents the functor $\mathcal{P}^s\mathcal{S}ch^{et}(\chi)$ defined in (8.40.3).* $\square$

8.44 (Marking points) So far we have studied varieties with marked divisors on them. It is sometimes useful to also mark some points. For curves, the points are also divisors and they interact with the log canonical structure. By contrast, in dimension ≥ 2, the points and the log canonical structure are independent of each other. This makes the resulting notion much less interesting theoretically, but it gives a quick way to rigidify slc pairs, which was quite useful in Section 5.9.

A flat family of *r-pointed* schemes is a flat morphism $f: X \to S$ plus r sections $\sigma_i: S \to X$. This gives a functor of r-pointed schemes.

Consider the Hilbert scheme with its universal family $\mathrm{Univ}(\mathbb{P}^N) \to \mathrm{Hilb}(\mathbb{P}^N)$. Then the r-fold fiber product

$$\mathrm{Univ}(\mathbb{P}^N) \times_{\mathrm{Hilb}(\mathbb{P}^N)} \mathrm{Univ}(\mathbb{P}^N) \cdots \times_{\mathrm{Hilb}(\mathbb{P}^N)} \mathrm{Univ}(\mathbb{P}^N)$$

represents the functor of r-pointed subschemes of $\mathbb{P}^N$. More generally, for any functor that is representable by a flat universal family $\mathrm{Univ}_M \to M$, its r-pointed version is representable by the r-fold fiber product of Univ_M over M.

In particular, we get MpSP, the moduli of pointed stable pairs.

8.5 Canonically Embedded Pairs

Assumptions In this section, we work with arbitrary schemes. As before, the situation over $\mathrm{Spec}\,\mathbb{Z}$ determines everything.

Definition 8.45 A *strongly polarized family of schemes marked with K-flat divisors* is written as

(8.45.1) $f: (X; D^1, \ldots, D^r; L) \to S$, where

(8.45.2) $f: X \to S$ satisfies (8.39.2–5),

(8.45.3) the D^i are K-flat families of relative Mumford divisors (7.1), and

(8.45.4) L is strongly f-ample (8.39).

If we fix the relative dimension and the rank of f_*L, then, as in (8.39.6), we get the functor

$$\mathcal{P}^s\mathcal{MSch}(r, n, N). \tag{8.45.5}$$

We write $\mathcal{P}^s\mathcal{MSch}(r, \chi)$ if the Hilbert polynomial $\chi = \chi(X_s, L_s^m)$ of L is also fixed. These can also be sheafified in the étale topology as in (8.40.3). (The notation does not indicate K-flatness; but it has enough letters in it already.)

The embedded version is denoted by

$$\mathcal{E}^s\mathcal{MSch}(r, n, \mathbb{P}^N). \tag{8.45.6}$$

These functors associate to a scheme S the set of all families of closed subschemes of a given $\mathbb{P}_S^N$ (where $N = \chi(1) - 1$) marked with K-flat divisors

$$f\colon (X \subset \mathbb{P}_S^N; D^1, \dots, D^r; \mathcal{O}_X(1)) \to S, \tag{8.45.7}$$

where $\mathcal{O}_X(1)$ is strongly ample.

Equivalently, we can view $\mathcal{E}^s\mathcal{MSch}(r, n, \mathbb{P}^N)$ as parametrizing objects

$$\left(f\colon (X; D^1, \dots, D^m; L) \to S; \phi \in \mathrm{Isom}_S\big(\mathbb{P}_S(f_*L), \mathbb{P}_S^N\big) \right) \tag{8.45.8}$$

consisting of a strongly polarized family of schemes marked with K-flat divisors, plus a projective framing $\phi\colon \mathbb{P}_S(f_*L) \simeq \mathbb{P}_S^N$ as in (8.41.5).

8.46 (Universal family of strongly embedded, marked schemes) Fix a projective space $\mathbb{P}_{\mathbb{Z}}^N$ and integers $n \geq 1$ and $r \geq 0$. By (8.41) we have a universal family of strongly embedded schemes

$$\mathrm{Univ}_n^{\mathrm{str}}(\mathbb{P}_{\mathbb{Z}}^N) \to \mathrm{Hilb}_n^{\mathrm{str}}(\mathbb{P}_{\mathbb{Z}}^N) \tag{8.46.1}$$

satisfying (8.39.2–5). The universal family of K-flat, Mumford divisors

$$\mathrm{KDiv}(\mathrm{Univ}_n^{\mathrm{str}}(\mathbb{P}_{\mathbb{Z}}^N) / \mathrm{Hilb}_n^{\mathrm{str}}(\mathbb{P}_{\mathbb{Z}}^N)) \to \mathrm{Hilb}_n^{\mathrm{str}}(\mathbb{P}_{\mathbb{Z}}^N)$$

was constructed in (7.3). If we need r such divisors, the base of the universal family we want is the r-fold fiber product

$$\mathrm{E}^s\mathrm{MSch}(r, n, \mathbb{P}_{\mathbb{Z}}^N) := \times_{\mathrm{Hilb}_n(\mathbb{P}_{\mathbb{Z}}^N)}^r \mathrm{KDiv}(\mathrm{Univ}_n^{\mathrm{str}}(\mathbb{P}_{\mathbb{Z}}^N) / \mathrm{Hilb}_n^{\mathrm{str}}(\mathbb{P}_{\mathbb{Z}}^N)). \tag{8.46.2}$$

We denote the universal family by

$$\mathbf{F}\colon (\mathbf{X}, \mathbf{D}^1, \dots, \mathbf{D}^r; \mathbf{L}) \to \mathrm{E}^s\mathrm{MSch}(r, n, \mathbb{P}_{\mathbb{Z}}^N), \tag{8.46.3}$$

where we really should have written the rather cumbersome

$$\left(\mathbf{X}(r, n, \mathbb{P}_{\mathbb{Z}}^N), \mathbf{D}^1(r, n, \mathbb{P}_{\mathbb{Z}}^N), \dots, \mathbf{D}^r(r, n, \mathbb{P}_{\mathbb{Z}}^N); \mathbf{L}(r, n, \mathbb{P}_{\mathbb{Z}}^N) \right).$$

It is clear from the construction that the spaces $E^s\mathrm{MSch}(r, n, \mathbb{P}^N_{\mathbb{Z}})$ parametrize polarized families of schemes marked with divisors, equipped with an extra framing.

Proposition 8.47 *Fix* r, n, N. *Then the scheme of embedded, marked schemes* $E^s\mathrm{MSch}(r, n, \mathbb{P}^N_{\mathbb{Z}})$ *constructed in (8.46.3) represents* $\mathcal{E}^s\mathcal{MSch}(r, n, \mathbb{P}^N_{\mathbb{Z}})$, *defined in (8.45). That is, for every $\mathbb{Z}$-scheme S, pulling back the family (8.46.3) gives a one-to-one correspondence between*

(8.47.1) $\mathrm{Mor}_{\mathbb{Z}}(S, E^s\mathrm{MSch}(r, n, \mathbb{P}^N_{\mathbb{Z}}))$, *and*

(8.47.2) *families* $f : (X; D^1, \ldots, D^r; L) \to S$ *of n-dimensional schemes, with a strong polarization and marked with K-flat Mumford divisors, plus a projective framing* $\mathbb{P}_S(f_*L) \simeq \mathbb{P}^N_S$. □

As in (8.43) and (8.56.1), this implies the following.

Corollary 8.48 *Fix* n, m, N. *Then the stack* $[E^s\mathrm{MSch}(r, n, \mathbb{P}^N_{\mathbb{Z}})/\mathrm{PGL}_{N+1}]$ *represents the functor* $\mathcal{P}^s\mathcal{MSch}(r, n, N)$, *defined in (8.45).* □

8.49 (Boundedness conditions) The schemes $E^s\mathrm{MSch}(r, n, \mathbb{P}^N)$ have infinitely many irreducible components since we have not fixed the degrees of X and of the divisors D^i. Set

$$\deg_L(X; D^1, \ldots, D^r) := (\deg_L X, \deg_L D^1, \ldots, \deg_L D^r) \in \mathbb{N}^{r+1}. \qquad (8.49.1)$$

This multidegree is a locally constant function on $E^s\mathrm{MSch}(r, n, \mathbb{P}^N)$, hence its level sets give a decomposition

$$E^s\mathrm{MSch}(r, n, \mathbb{P}^N) = \amalg_{\mathbf{d} \in \mathbb{N}^{r+1}} E^s\mathrm{MSch}(r, n, \mathbf{d}, \mathbb{P}^N). \qquad (8.49.2)$$

The schemes $E^s\mathrm{MSch}(r, n, \mathbf{d}, \mathbb{P}^N)$ are still not of finite type since the fibers are allowed to be nonreduced. However, the subscheme

$$E^s\mathrm{MV}(r, n, \mathbf{d}, \mathbb{P}^N) \subset E^s\mathrm{MSch}(r, n, \mathbf{d}, \mathbb{P}^N), \qquad (8.49.3)$$

which parametrizes geometrically reduced fibers, is quasi-projective, though usually non-projective, reducible, and disconnected.

Definition 8.50 A family of marked pairs $f : (X, \Delta) \to S$ as in (8.4) is *m-canonically strongly polarized* if

(8.50.1) $\omega_{X/S}$ is locally free outside a codimension ≥ 2 subset of each fiber,

(8.50.2) $\omega_{X/S}^{[m]}(m\Delta)$ is a line bundle, and

(8.50.3) $\omega_{X/S}^{[m]}(m\Delta)$ is strongly f-ample.

If $X \subset \mathbb{P}_S^N$ then $f\colon (X, \Delta) \to S$ is *m-canonically strongly embedded* if, in addition,

(8.50.4) $\omega_{X/S}^{[m]}(m\Delta) \simeq \mathscr{O}_{\mathbb{P}^N}(1) \otimes f^* M_S$ for some line bundle M_S on S.

These define the functors $C^m \mathcal{P}^s \mathcal{MSch}$ and $C^m \mathcal{E}^s \mathcal{MSch}$.

Theorem 8.51 *Fix* $m, n, N \in \mathbb{N}$ *and a rational vector* $\mathbf{a} = (a_1, \dots, a_r)$. *Then the functor* $C^m \mathcal{E}^s \mathcal{MSch}(\mathbf{a}, n, \mathbb{P}^N)$ *is represented by a monomorphism*

$$C^m E^s MSch(\mathbf{a}, n, \mathbb{P}_{\mathbb{Z}}^N) \to E^s MSch(r, n, \mathbb{P}_{\mathbb{Z}}^N)$$

Proof Start with the universal family, as in (8.46.3),

$$\mathbf{F} \colon (\mathbf{X}, \mathbf{D}^1, \dots, \mathbf{D}^r; \mathbf{L}) \to E^s MSch(r, n, \mathbb{P}_{\mathbb{Z}}^N).$$

Note that (8.50.1) is an open condition and it holds iff ω_{X_s} is locally free outside a closed subset of codimension ≥ 2 of X_s for every $s \in S$. Being a line bundle is representable by (3.30) and, once it holds, being a strong polarization is an open condition. Applying (3.22) to $\omega_{X/S}^{[m]}(m\Delta)(-1)$ shows that condition (8.50.4) is representable.   □

By (4.45), if $K_{X/S} + \Delta$ is $\mathbb{Q}$-Cartier, then the stable fibers are parametrized by an open subset, at least in characteristic 0. Thus we get the following.

Corollary 8.52 *Fix* $m, n, N \in \mathbb{N}$ *and a rational vector* $\mathbf{a} = (a_1, \dots, a_r)$. *Then, over* $\operatorname{Spec} \mathbb{Q}$, *there is an open subscheme*

$$C^m ESP(\mathbf{a}, n, \mathbb{P}_{\mathbb{Q}}^N) \subset C^m E^s MSch(\mathbf{a}, n, \mathbb{P}_{\mathbb{Q}}^N),$$

representing the functor of m-canonically, strongly embedded, stable families.

Warning 8.52.1 The reduced subspace of $C^m ESP$ is the correct one, but its scheme structure is still a little too large. The reason is that (8.7.3) imposes restrictions on $\omega_{X/S}^{[r]}(r\Delta)$ for various values of r, but we took care only of our chosen m (and its multiples).

We dropped the superscript from E^s since, as we noted in (8.38), an m-canonical polarization is automatically strong.

8.6 Moduli Spaces as Quotients by Group Actions

Notation 8.53 For a scheme S, we use $\operatorname{PGL}_n(S)$ to denote the group scheme PGL_n over S. We will formulate definitions and results for general algebraic group schemes whenever possible, but in the applications we use only PGL_n, which is smooth and geometrically reductive.

Keep in mind that, if k is a field, then, in the literature, $\text{PGL}_n(k)$ usually denotes the k-points of the group scheme PGL_n, not $\text{PGL}_n(\text{Spec } k)$. It is customary to use PGL_n to denote $\text{PGL}_n(\text{Spec } \mathbb{Z})$ if we work with arbitrary schemes and $\text{PGL}_n(\text{Spec } \mathbb{Q})$ if we work in characteristic 0.

8.54 (Comment on algebraic spaces) We will consider quotients of schemes by algebraic groups, primarily PGL_n. It turns out that in many cases such quotients are not schemes, but algebraic spaces. For this reason, it is natural to formulate the basic definitions using algebraic spaces.

In our cases, these quotients turn out to be schemes, even projective, but this is not easy to prove.

In any case, this means that the reader can substitute "scheme" for "algebraic space" in the sequel, without affecting the final theorems.

Definition 8.55 An action of an algebraic group scheme G on an algebraic space X is a morphism $\sigma \colon G \times X \to X$ that satisfies the scheme-theoretic version of the condition $g_1(g_2(x)) = (g_1 g_2)(x)$. That is, the diagram

$$
\begin{array}{ccc}
G \times G \times X & \xrightarrow{\ 1_G \times \sigma\ } & G \times X \\
{\scriptstyle m \times 1_X} \downarrow & & \downarrow {\scriptstyle \sigma} \\
G \times X & \xrightarrow{\quad \sigma \quad} & X
\end{array}
$$

commutes. If G acts on X_1, X_2 then $\pi \colon X_1 \to X_2$ is a G-morphism if the following diagram commutes:

$$
\begin{array}{ccc}
G \times X_1 & \xrightarrow{\ \sigma_1\ } & X_1 \\
{\scriptstyle 1_G \times \pi} \downarrow & & \downarrow {\scriptstyle \pi} \\
G \times X_2 & \xrightarrow{\ \sigma_2\ } & X_2.
\end{array}
$$

The *categorical quotient* is a G-morphism $q \colon X \to Y$ such that the G-action is trivial on Y and q is universal among such.

Fix N and consider the functor $\mathcal{P}^s\mathcal{S}ch(N)$ of strongly polarized schemes of embedding dimension N. By (8.42), its embedded version has a moduli space with a universal family $\text{Univ}^{\text{str}}(\mathbb{P}^N) \to \text{Hilb}^{\text{str}}(\mathbb{P}^N)$. The connection between the two versions is the following impressive sounding, but quite simple claim.

Theorem 8.56 *The categorical quotient* $\text{Hilb}^{\text{str}}(\mathbb{P}^N)/\text{PGL}_{N+1}$ *is also the categorical moduli space* $\text{P}^s\text{Sch}(n, N)$.

Proof We have a universal family over $\text{Hilb}^{\text{str}}(\mathbb{P}^N)$, so we get $\text{Hilb}^{\text{str}}(\mathbb{P}^N) \to$ $\mathcal{P}^s\text{Sch}(n, N)$ which is PGL_{N+1}-equivariant.

Conversely, let $f\colon (X, L) \to S$ be a family in $\mathcal{P}^s\text{Sch}(n, N)$. Then f_*L is locally free of rank $N + 1$ on S, hence S has an open cover $S = \cup S_i$ such that each $f_*L|_{S_i}$ is free. Choosing a trivialization gives embedded families, hence morphisms $\phi_i\colon S_i \to \text{Hilb}^{\text{str}}(\mathbb{P}^N)$. Over $S_i \cap S_j$ we have two different trivializations, these differ by a section of $g_{ij} \in H^0(S_i \cap S_j, \text{PGL}_{N+1})$. Thus, composing with the quotient map $q\colon \text{Hilb}^{\text{str}}(\mathbb{P}^N) \to \text{Hilb}^{\text{str}}(\mathbb{P}^N)/\text{PGL}_{N+1}$ we get that $q \circ (\phi_i|_{S_i \cap S_j}) = q \circ (g_{ij}(\phi_j|_{S_i \cap S_j})) = q \circ (\phi_j|_{S_i \cap S_j})$, since q is PGL_{N+1}-equivariant. Thus the $q \circ \phi_i$ glue to a morphism $\phi\colon S \to \text{Hilb}^{\text{str}}(\mathbb{P}^N)/\text{PGL}_{N+1}$.   □

Remark 8.56.1 Since one can glue a morphism from étale charts, we see that $\mathcal{P}^s\text{Sch}^{\text{et}}$ and $\mathcal{P}^s\text{Sch}$ have the same categorical moduli spaces (8.40.5). For those conversant with stacks, this argument proves (8.43) and (8.48).

The same proof applies to pairs and we get the following.

Corollary 8.57 *Fix $m, n, N \in \mathbb{N}$ and a rational vector $\mathbf{a} = (a_1, \ldots, a_r)$. Then the categorical quotient $C^m\text{ESP}(\mathbf{a}, n, \mathbb{P}^N_\mathbb{Q})/\text{PGL}_{N+1}$ is also the categorical moduli space of $\mathcal{SP}(\mathbf{a}, n, *, m, N)$, the functor of stable families that have an m-canonical, strong embedding into $\mathbb{P}^N$.*   □

Existence of Quotients

Let G be an algebraic group acting on an algebraic space X. Under very mild conditions, the categorical quotient X/G exists, but it may be very degenerate. For example, consider $\mathbb{A}^n_k$ with the scalar $\mathbb{G}_m$-action $x_i \mapsto \lambda x_i$. Then $\mathbb{A}^n/\mathbb{G}_m = \text{Spec}\, k$, but $(\mathbb{A}^n \setminus \{\mathbf{0}\})/\mathbb{G}_m = \mathbb{P}^{n-1}$. Note that here the stabilizer is $\mathbb{G}_m$ for the origin, but trivial for every other point. This and many other examples suggest that points with infinite stabilizers cause problems.

With PGL_{N+1} acting on the Hilbert scheme, the stabilizer of the point $[X]$ corresponding to a strongly embedded $X \subset \mathbb{P}^N$ is the automorphism group of the polarized scheme $(X, \mathcal{O}_X(1))$. As we saw in Section 1.8, infinite automorphism groups cause many problems.

We get the best results if all automorphism groups are trivial; we discuss these in Section 8.7. For stable pairs the automorphism groups are finite, but we need a scheme-theoretic version of this.

Definition 8.58 (Proper action) Let $\sigma\colon G \times X \to X$ be an algebraic group scheme acting on an algebraic space X. Combining σ with the coordinate projection to X gives $(\sigma, \pi_X)\colon G \times X \longrightarrow X \times X$. The action is called *proper* if (σ, π_X) is proper and called *free* if (σ, π_X) is a closed embedding. Note that the preimage of a diagonal point (x, x) is the stabilizer of x. Thus free implies that all stabilizers are trivial and, if G is affine (for example PGL), then proper implies that all stabilizers are finite. (The converses are, however, not true; see Mumford (1965, p.11).)

Assume that $X \subset C^m \text{ESP}(\mathbf{a}, n, \mathbb{P}^N)$ parametrizes pluricanonically embedded stable subvarieties in $\mathbb{P}^N$ and $G = \text{PGL}_{N+1}$. We claim that the properness of the PGL_{N+1}-action is equivalent to the uniqueness of stable extensions considered in (2.50) (and called separatedness there).[1]

Over X, we have a universal family $Y \to X$. Let T be the spectrum of a DVR with generic point η and $(q_1, q_2) \colon T \to X \times X$ a morphism. Thus the $q_i^* Y \to T$ give families of stable varieties over T. The generic point η lifts to $G \times T$ iff there is a $g(\eta) \in G_\eta$ such that $q_1(\eta) = \sigma(g(\eta), q_2(\eta))$. Equivalently, if the generic fibers $(q_1^* Y)_\eta$ and $(q_2^* Y)_\eta$ are isomorphic, (2.50) then says that the families $q_1^* Y$ and $q_2^* Y$ are isomorphic. This isomorphism gives $q_G \colon T \to G$ and $(q_G, q_2) \colon T \to G \times X$ shows that the valuative criterion of properness holds for $G \times X \to X \times X$.

Now we come to the definition of the right class of quotients.

Definition 8.59 (Mumford, 1965, p.4) Let G be an algebraic group scheme acting on an algebraic space X with categorical quotient $q \colon X \to X/G$ (8.55). It is called a *geometric quotient* if

(8.59.1) $q(K) \colon X(K)/G(K) \to (X/G)(K)$ is a bijection of sets, whenever K is algebraically closed,

(8.59.2) q is of finite type and universally surjective, and

(8.59.3) $\mathcal{O}_{X/G} = (q_* \mathcal{O}_X)^G$.

The geometric quotient is denoted by $X /\!/ G$.

The fundamental theorem for the existence of geometric quotients is the following. Seshadri (1962/1963, 1972) came close to proving it. His ideas were developed in Kollár (1997) to settle many cases, including PGL that we need. The general case was treated in Keel and Mori (1997); see Olsson (2016) for a thorough treatment.

Theorem 8.60 *Let G be a flat group scheme acting properly on an algebraic space X. Then the geometric quotient $X /\!/ G$ exists.* □

For free actions, the quotient map is especially simple. Over fields, this is proved in Mumford (1965, prop.0.9). The general case follows from Stacks (2022, tag 0CQJ).

---

[1] This clash of terminologies is, unfortunately, well entrenched.

Complement 8.61 *Assume in addition that the G-action is free on X. Then* $X \to X/\!/G$ *is a principal G-bundle.* □

For us the main application is the following.

Theorem 8.62 *Fix* $m, n, N \in \mathbb{N}$ *and a rational vector* $\mathbf{a} = (a_1, \ldots, a_r)$. *Then the* PGL_{N+1}-*action on* $\mathrm{C}^m\mathrm{ESP}(\mathbf{a}, n, \mathbb{P}^N_\mathbb{Q})$ *(8.52) is proper.*

Thus the geometric quotient $\mathrm{C}^m\mathrm{ESP}(\mathbf{a}, n, \mathbb{P}^N_\mathbb{Q})/\!/\mathrm{PGL}_{N+1}$ *exists and it is the coarse moduli space of* $\mathcal{SP}(\mathbf{a}, n, *, m, N)$, *the functor of stable families that have an m-canonical, strong embedding into* $\mathbb{P}^N$.

Proof For simplicity, write $\mathrm{Univ} \to \mathrm{C}^m\mathrm{ESP}$ for the universal family over $\mathrm{C}^m\mathrm{ESP}(\mathbf{a}, n, \mathbb{P}^N_\mathbb{Q})$. Following (8.58), we need to show that

$$\mathrm{PGL}_{N+1} \times \mathrm{C}^m\mathrm{ESP} \longrightarrow \mathrm{C}^m\mathrm{ESP} \times \mathrm{C}^m\mathrm{ESP} \qquad (8.62.1)$$

is proper. First, we claim that (8.62.1) is isomorphic to

$$\mathbf{Isom}(\pi_1^* \mathrm{Univ}, \pi_2^* \mathrm{Univ}) \longrightarrow \mathrm{C}^m\mathrm{ESP} \times \mathrm{C}^m\mathrm{ESP}, \qquad (8.62.2)$$

where the $\pi_i \colon \mathrm{C}^m\mathrm{ESP} \to \mathrm{C}^m\mathrm{ESP} \times \mathrm{C}^m\mathrm{ESP}$ are the coordinate projections. This is simply the statement that giving a stable pair (X, Δ) plus two m-canonical embeddings into $\mathbb{P}^N$ is the same as giving one m-canonical embedding into $\mathbb{P}^N$ plus an element of PGL_{N+1}.

The properness of (8.62.2) follows from (8.64). The rest then follow from (8.60) and (8.57). □

8.63 (Morphism schemes) For S-schemes X, Y let $\mathrm{Mor}_S(X, Y)$ be the set of morphisms that commute with projections to S. We get the functor of morphisms on S-schemes $T \mapsto \mathrm{Mor}_T(X_T, Y_T)$.

Claim 8.63.1 Assume that $X \to S$ is flat, proper and $Y \to S$ is of finite type. Then the functor of morphisms is representable by a scheme $\mathbf{Mor}_S(X, Y)$.

Proof We can identify a morphism with its graph, which is in $\mathrm{Hilb}_S(X \times_S Y)$ since $X \to S$ is flat. Conversely, a subscheme $Z \subset X \times_S Y$ is the graph of a morphism iff the first projection $\pi_X \colon Z \to X$ is finite and $(\pi_X)_* \mathcal{O}_Z \cong \mathcal{O}_X$. The first of these is always an open condition, for the second we need the flatness of $Z \to S$ (10.54). □

We also get sets $\mathrm{Isom}_S(X, Y)$, $\mathrm{Aut}_S(X)$ and schemes $\mathbf{Isom}_S(X, Y)$ $\mathbf{Aut}_S(X)$. that represent the functor of isomorphisms (resp. automorphisms). The identity is always in automorphism, thus we have the identity section $S \subset \mathbf{Aut}_S(X)$. We say that X is *rigid* (over S) if $S = \mathbf{Aut}_S(X)$.

The definitions of Mor, Isom, Aut and **Mor, Isom, Aut** also apply to pairs.

With the definition of stable families in place, we get the following consequence of (11.40) about isomorphism schemes.

Proposition 8.64 *Let* f_i: $(X_i, \Delta_i) \to S$ *be stable morphisms. Then the structure map* $\mathbf{Isom}_S((X_1, \Delta_1), (X_2, \Delta_2)) \to S$ *is finite.*

Proof Choose m such that the divisors $m(K_{X_i/S} + \Delta_i)$ are very f_i-ample. Set $F_i := (f_i)_* \mathcal{O}_{X_i}(mK_{X_i/S} + m\Delta_i)$. Then

$$\mathbf{Isom}_S((X_1, \Delta_1), (X_2, \Delta_2)) \subset \mathbf{Isom}_S(\mathbb{P}_S(F_1), \mathbb{P}_S(F_2))$$

is closed, hence affine over S.

Let T be the spectrum of a DVR over k with generic point t_g and $\phi_g \colon t_g \to \mathbf{Isom}_S((X_1, \Delta_1), (X_2, \Delta_2))$ a morphism. We can view it as an isomorphism of the generic fibers $\phi_g \colon (X_1, \Delta_1) \times_S \{t_g\} \simeq (X_2, \Delta_2) \times_S \{t_g\}$. By (2.50), ϕ_g extends uniquely to an isomorphism $\Phi \colon (X_1, \Delta_1) \times_S T \simeq (X_2, \Delta_2) \times_S T$. This is the valuative criterion of properness for $\mathbf{Isom}_S((X_1, \Delta_1), (X_2, \Delta_2))$, which is thus both affine and proper, hence finite over S. □

Next we verify (1.77.1) for stable pairs.

Corollary 8.65 *Let* $f \colon (X, \Delta) \to S$ *be a stable morphism. Then the structure map* $\pi \colon \mathbf{Aut}_S(X, \Delta) \to S$ *is finite, the subset* $S° \subset S$ *of rigid fibers is open and* $\mathbf{Aut}_S(X, \Delta) = S$ *iff* $\mathrm{Aut}(X_s, \Delta_s)$ *is trivial for every geometric point* $s \to S$.

Proof Finiteness follows from (8.64). The identity section gives that $\mathcal{O}_S$ is a direct summand of $\pi_* \mathcal{O}_{\mathbf{Aut}_S(X,\Delta)}$. Thus $S°$ is the complement of the support of $\pi_* \mathcal{O}_{\mathbf{Aut}_S(X,\Delta)}/\mathcal{O}_S$. The fibers of $\mathbf{Aut}_S(X, \Delta) \to S$ are the $\mathrm{Aut}(X_s, \Delta_s)$. □

8.7 Descent

Let $q \colon S' \to S$ be a morphism of schemes and assume that we have an object over S'. We say that the object *descends* to S if it is isomorphic to the pull-back of an object on S. Typical examples are

- a (quasi)coherent sheaf F', in which case we want to get a (quasi)coherent sheaf F on S such that $F' \simeq q^* F$, or
- a morphism $X' \to S'$, in which case we want to get a morphism $X \to S$ such that $X' \simeq X \times_S S'$.

A systematic theory was developed in Grothendieck (1962, lec.1), treating the case when $S' \to S$ is faithfully flat; see also Grothendieck (1971, chap. VIII), Bosch et al. (1990, chap.6) or Stacks (2022, tag 0306) for more detailed treatments. We explain the basic idea during the proof of (8.69).

Here we discuss the consequences of descent theory for the moduli of stable pairs; the main one is (8.71). We also prove some special cases that are representative of the general theory, yet can be obtained by simpler methods.

8.66 (Functorial polarization) Kollár (1990) Let $\mathcal{F}$ be a subfunctor of $\mathcal{P}^s\mathcal{S}ch$. A *functorial polarization* (of level r) of $\mathcal{F}$ assigns

(8.66.1) to any $(f\colon (X, L) \to S) \in \mathcal{F}(S)$ another $(f\colon (X, \bar{L}) \to S) \in \mathcal{F}(S)$ such that $\bar{L}$ is equivalent to L^r, and

(8.66.2) to every $q\colon S' \to S$ an isomorphism $\sigma(q)\colon q_X^*(\bar{L}) \simeq \overline{(q_X^*L)}$ such that

(8.66.3) $\sigma(q \circ q') = \sigma(q') \circ (q_X')^*\sigma(q)$ for every $q'\colon S'' \to S'$ and $q\colon S' \to S$.

Note that in (2) we need to fix an isomorphism, it is not enough to say that the two sides are isomorphic.

If the choice of $\bar{L}$ is specified, then we say that $\mathcal{F}$ is *functorially polarized.*

The following are examples of functorial polarizations.

(8.66.4) If $L_s \simeq \omega_{X_s}$ for $s \in S$, then $\bar{L} := \omega_{X/S}$ is a functorial polarization.

(8.66.5) If every family in $\mathcal{F}$ has a natural section $\sigma\colon S \to X$, then we can take $\bar{L} := L \otimes f^*(\sigma^*L)^{-1}$. This applies, for instance, to pointed varieties and (depending on our definition) to polarized abelian varieties.

(8.66.6) Assume that $r := \chi(X_s, L_s)$ is constant and positive for every (X_s, L_s) in $\mathcal{F}$. Then, using the notation of (3.24.3), $\bar{L} := L^r \otimes f^*(\det R^\bullet f_*L)^{-1}$ is a level r functorial polarization.

(8.66.7) $(\mathbb{P}^1, \mathscr{O}_{\mathbb{P}^1}(1))$ does not have a functorial polarization of level 1, since that would lead to a nontrivial representation of $\mathrm{Aut}(\mathbb{P}^1)$ on $H^0(\mathbb{P}^1, \mathscr{O}_{\mathbb{P}^1}(1)) \simeq k^2$. On the other hand, $(\mathbb{P}^1, \omega_{\mathbb{P}^1}^{-1})$ gives a functorial polarization of level 2.

Functorial polarizations also give natural line bundles on the base spaces of families. Let $\mathcal{F}$ be a functorially polarized subfunctor of $\mathcal{P}\mathcal{S}ch$. For any $(f\colon (X, \bar{L}) \to S) \in \mathcal{F}(S)$ we get the line bundle $\det R^\bullet f_*(\bar{L}^{\otimes k})$ as in (3.24.3). For $k \gg 1$ it is given by the simpler formula $\det f_*(\bar{L}^{\otimes k})$.

These line bundles are functorial for base changes, thus they give line bundles on the moduli stack of $\mathcal{F}$.

Uniqueness of descent now follows easily.

Proposition 8.67 *Let $S' \to S$ be a faithfully flat morphism and $X' \to S'$ a flat, proper morphism such that X' is rigid over S'. Then there is at most one scheme $X \to S$ such that $X' \simeq X \times_S S'$.*

Proof Assume that we have $X_1 \to S$ and $X_2 \to S$. Since the $X_i \times_S S' \simeq X'$ are flat and proper, so are $X_i \to S$. We aim to prove that $\mathbf{Isom}_S(X_1, X_2) \simeq S$. To see this, take any $T \to S'$, and note that

$$\mathrm{Isom}_T(X'_T, X'_T) = \mathrm{Isom}_T(X_1 \times_S T, X_2 \times_S T) = \mathrm{Mor}_S(T, \mathbf{Isom}_S(X_1, X_2)).$$

If X' is rigid over S' then $\mathrm{Isom}_T(X'_T, X'_T)$ has only 1 element, so $\mathrm{Mor}_S(T, S) = \mathrm{Mor}_S(T, \mathbf{Isom}_S(X_1, X_2))$ for every T. Thus $S = \mathbf{Isom}_S(X_1, X_2)$. □

The simplest descent result is the following; see (1.73).

Lemma 8.68 *Let K/k be a finite, separable field extension and (X, L) a rigid, functorially polarized, projective variety defined over K. Then (X, L) descends to k iff $(X, L) \simeq (X^\sigma, L^\sigma)$ for every $\sigma \in \mathrm{Gal}(\bar{k}/k)$.*

Proof We may assume that K/k is Galois. Then only the $\sigma \in \mathrm{Gal}(K/k)$ matter. We get an action of $\mathrm{Gal}(K/k)$ on $H^0(X, L)$ by

$$H^0(X, L) \overset{\sigma-lin}{\longrightarrow} H^0(X^\sigma, L^\sigma) \overset{K-isom}{\longrightarrow} H^0(X, L).$$

This is well defined since the K-isomorphism is unique, even on L. By the fundamental lemma on quasi-linear maps (see Shafarevich (1974, sec.A.3)) there is a unique k-subspace $V(X, L) \subset H^0(X, L)$ such that $V(X, L) \otimes_k K = H^0(X, L)$. Since $X = \mathrm{Proj}_K \sum H^0(X, L^m)$, we see that $X_k := \mathrm{Proj}_k \sum V(X, L^m)$ defines the descent. □

Theorem 8.69 *Let $S' \to S$ be a faithfully flat morphism and $f' : (X', L') \to S'$ a flat, functorially polarized projective morphism that is rigid over S'. The following are equivalent.*
(8.69.1) $f' : (X', L') \to S'$ descends to $f : (X, L) \to S$.
(8.69.2) For every Artinian scheme $\tau : A \to S$, the pull-back $f'_A : (X'_A, L'_A) \to A$ is independent of the lifting $\tau' : A \to S'$.
If S is normal and $S' \to S$ is smooth, then it is enough to check (2) for spectra of fields.

Proof We just explain how this fits in the framework of faithfully flat descent, for which we refer to Stacks (2022, tag 0306).

Let $\pi_i : S' \times_S S' \to S'$ denote the coordinate projections for $i = 1, 2$. Pulling back $f' : (X', L') \to S'$ to $S' \times_S S'$ by the π_i, we get two families

$$f'_i : (X'_i, L'_i) \to S' \times_S S'.$$

If $f : (X, L) \to S$ exists then these are both isomorphic to the pull-back of $f : (X, L) \to S$, hence to each other $\sigma_{12} : (X'_1, L'_1) \simeq (X'_2, L'_2)$. The existence of σ_{12} is a necessary condition for descent. The key observation is that it is not sufficient, one also needs certain compatibility conditions over the triple product $S' \times_S S' \times_S S'$. However, if (X', L') is rigid over S', then σ_{12} is unique and the compatibility conditions are automatic.

To prove that σ_{12} exists, consider

$$\pi \colon \operatorname{Isom}_{S' \times_S S'}((X'_1, L'_1), (X'_2, L'_2)) \to S' \times_S S'.$$

Since (X', L') is rigid over S', π is a monomorphism. Assumption (2) implies that it is scheme-theoretically surjective, hence an isomorphism.

If $S' \to S$ is smooth then $S' \times_S S' \to S$ is also smooth, hence $S' \times_S S'$ is normal if S is normal. In that case, surjectivity is a set-theoretic question. □

Corollary 8.70 *Let G be a flat group scheme over S and $S' \to S$ a principal G-bundle. Let $f' \colon (X', L') \to S'$ be a flat, functorially polarized projective morphism that is rigid over S'. Assume that the G actions lifts to (X', L'). Then $f' \colon (X', L') \to S'$ descends to $f \colon (X, L) \to S$.*

Proof We need to check assumption (8.69.2). So fix $\tau \colon A \to S$ and liftings $\tau_i \colon A \to S'$. Then S'_A is a principal G-bundle with two sections τ_i. Thus $\tau_2 = g_{12} \circ \tau_1$ for some section g_{12} of G_A. Since the G-action lifts to (X', L'), the corresponding pull-backs are isomorphic. □

Now we come to the main theorem.

Theorem 8.71 *Let $\mathrm{SP}^{rigid} \subset \mathrm{SP}$ be the open subset parametrizing stable pairs without automorphisms. Then there is a universal family over SP^{rigid}.*

Proof First, note that SP^{rigid} is indeed open by (8.65).

For rigid families the existence is a local question. We may thus fix the dimension n, the number of marked divisors r, the coefficient vector $(a_1, \ldots, a_r)$, the volume v and the intended embedding dimension N.

First, consider the case when the a_i are rational and also fix $m > 1$, a multiple of $\mathrm{lcd}(a_1, \ldots, a_r)$. Set $\mathbf{d} := (n, r, a_1, \ldots, a_r, m, v, N)$.

Let $\mathcal{SP}(\mathbf{d})(S)$ denote the set of marked families $f \colon (X, \Delta) \to S$ with these numerical data, for which $m(K_{X/S} + \Delta)$ is a Cartier $\mathbb{Z}$-divisor and a strong polarization, and such that $f_*\mathcal{O}_X(m(K_{X/S} + \Delta))$ has rank $N + 1$. Similarly, let $\mathcal{EMSP}(\mathbf{d})(S)$ denote the set of these objects together with a strong embedding into $\mathbb{P}^N_S$.

By (8.52), we have the moduli spaces $\mathrm{EMSP}^{rigid}(\mathbf{d}) \subset \mathrm{EMSP}(\mathbf{d})$, with universal families. By (8.61), $\mathrm{EMSP}^{rigid}(\mathbf{d}) \to \mathrm{SP}^{rigid}(\mathbf{d})$ is a principal PGL_{N+1}-bundle. Hence the universal family over $\mathrm{EMSP}^{rigid}(\mathbf{d})$ descends to $\mathrm{SP}^{rigid}(\mathbf{d})$ by (8.70).

The case of irrational coefficients is very similar. We need to work with the rational approximations $(X, \sigma_j^m(\Delta)) \to S$ as in (8.21). □

Complement 8.71.1 The same proof works for other variants of the moduli of stable pairs, in particular we get universal families over the moduli space MpSP^{rigid} of rigid, pointed, stable pairs (8.44).

8.8 Positive Characteristic

We discuss, mostly through examples, two types of problems that complicate the moduli theory of pairs in positive characteristic.

The first problem is that, as we already noted in (2.4), the four versions of the definition of local stability in (2.3) are not equivalent in positive characteristic. The first such examples are in Kollár (2022); these are families of 3-folds. In (8.73) we discuss a series of higher dimensional examples that have very mild singularities.

The second is due to p-torsion in local class groups, visible most clearly in (4.39). As we see starting with (8.75), this issue appears already for the moduli of 4 points on $\mathbb{P}^1$. This difficulty can be avoided either by working only over weakly normal bases, or by a strong reliance on markings.

Theorem 8.72 *Kollár (2022) Let k be an algebraically closed field of characteristic $\neq 0$. There are flat, projective morphisms $f : (X, \Delta) \to \mathbb{A}_k^1$ of relative dimension 3 such that*

(8.72.1) $(X, X_t + \Delta)$ *is lc for every $t \in \mathbb{A}^1$,*

(8.72.2) $(\bar{X}_t, \text{Diff}_{\bar{X}_t} \Delta)$ *is lc for every $t \in \mathbb{A}^1$,*

(8.72.3) $(\bar{X}_0, \text{Diff}_{\bar{X}_0} \Delta)$ *lifts to characteristic 0, yet*

(8.72.4) X_0 *is not weakly normal, Sing X_0 is 1-dimensional, and $\bar{X}_0 \to X_0$ is purely inseparable over Sing X_0.*

The singularities of the 3-folds in Kollár (2022) are rather complicated. We discuss here instead another series of examples, arising from cones over homogeneous spaces. These are higher dimensional, but similar to the various examples discussed in Section 2.3.

Example 8.73 (Kovács–Totaro–Bernasconi examples) Let $X = G/P$ be a projective, homogeneous space. If P is reduced, then G/P is Fano and Kodaira vanishing holds on X in any characteristic by the Bott–Kempf theorem.

The cases when P is non-reduced were studied in Haboush and Lauritzen (1993). For some of these, $X = G/P$ is Fano, but Kodaira vanishing fails for a multiple of the canonical class. The first example was identified by Kovács (2018); giving a seven-dimensional canonical singularity in characteristic 2, that is not CM. A large series of examples is exhibited in Totaro (2019), leading to terminal singularities in any characteristic $p > 0$, that are not CM. These were further studied by Bernasconi (2018).

Kollár (2022) observed that they can be used to construct stable degenerations, where the generic fibers are smooth with ample canonical class and the special fibers have isolated, nonnormal singularities.

Assume that $X = G/P$ as above and $-K_X = mH$ for some ample divisor H for some $m \geq 1$. $|H|$ is very ample by Lauritzen (1996), so it gives an embedding $X \hookrightarrow \mathbb{P}^N$, where $N = \dim |H|$. Let $Y := C(X, H) \subset \mathbb{P}^{N+1}$ be the projective cone over X with vertex v.

Let $D \in |H|$ be a smooth divisor and $D_Y \subset Y$ its preimage. Since $K_X + D \sim (m-1)H$, (2.35) shows that (Y, D_Y) is a log canonical pair if $m = 1$, a canonical pair if $m > 1$.

$D_Y \subset Y$ is a Cartier divisor that is smooth outside v. Thus D_Y is normal $\Leftrightarrow \mathrm{depth}_v D_Y \geq 2 \Leftrightarrow \mathrm{depth}_v Y \geq 3$; see (2.36). Since $H_v^{i+1}(Y, \mathcal{O}_Y) \simeq \sum_{m \in \mathbb{Z}} H^i(X, \mathcal{O}_X(mH))$ by (2.35.1), D_Y is normal iff $H^1(X, \mathcal{O}_X(mH)) = 0$ for all $m \in \mathbb{Z}$ by (10.29.5).

Therefore, if $H^1(X, \mathcal{O}_X(H)) \neq 0$, then D_Y is not normal. Intersecting Y with a pencil of hyperplanes with base locus $Z \not\ni v$, we get a locally stable morphism $\pi : B_Z Y \to \mathbb{P}^1$. It has one fiber isomorphic to D_Y, the others are isomorphic to X.

Taking a suitable cyclic cover (2.13), we get a series of examples of stable families, where the generic fibers are smooth varieties with ample canonical class and the special fibers have isolated nonnormal singularities.

The cases described in Totaro (2019) have $m = 2$. Then the normalization of D_Y has canonical singularities, hence these families occur in what is usually considered the "interior" of the moduli space.

Aside 8.73.1 Another class of non-CM, cyclic, quotient singularities is described in Yasuda (2019). These all have depth ≥ 3 by Ellingsrud and Skjelbred (1980), so they do not lead to families as in (8.72).

8.74 (Cartier or $\mathbb{Q}$-Cartier?) One of the early key conceptual steps of the minimal model program was the realization that, starting with dimension 3, minimal models can be singular. Moreover, their canonical class need not be Cartier. It was gradually understood that the more general $\mathbb{Q}$-Cartier condition is the important one.

In moduli theory, we frequently start with pairs (X, B) where X is smooth and B is Cartier, but in compactifying their moduli space we encounter pairs

(X', B') where X' is singular and $K_{X'}$ is only $\mathbb{Q}$-Cartier. Thus the usual approach is to work with pairs (X, B) where $K_X + B$ is $\mathbb{Q}$-Cartier.

Next we discuss various problems that arise when the denominators involve the characteristic.

8.75 (Moduli of points on $\mathbb{P}^1$) We consider the moduli problem of $n = 2r + 1 \geq 3$ unordered, distinct points in $\mathbb{P}^1$. Fix an index set I of n elements. There is only one natural way of defining the objects of this theory.

(8.75.1) (Geometric objects) $(\mathbb{P}^1, \sum_{i \in I} [p_i])$ where the p_i are distinct points.

(8.75.2) (Objects over a field) $(\mathbb{P}^1, Z)$ where $Z \subset \mathbb{P}^1$ is a geometrically reduced, 0-dimensional subscheme of degree n.

The question becomes more subtle when families are considered.

(8.75.3) (Families) $(P_S \to S, D)$ where $P_S \to S$ is a locally trivial $\mathbb{P}^1$-bundle and $D \subset P_S$ is a divisor over S of degree n. For ordered points the traditional choice is to take D to be a union of sections of $P_S \to S$, but for unordered points we have two natural choices.

(3.a) (Cartier) D is a relative Cartier divisor over S.

(3.b) ($\mathbb{Q}$-Cartier) D is a relative $\mathbb{Q}$-Cartier divisor over S.

The first is closest to the traditional choice of union of sections, the second is more in the spirit of the higher dimensional theory.

(8.75.4) (Base spaces) Ideally we should work over arbitrary base schemes, but it turns out that unexpected things happen even when the base is quite nice. We consider three classes of base schemes.

(4.a) (Reduced)

(4.b) (Seminormal)

(4.c) (Weakly normal)

The cases (3.a–b) and (4.a–c) are in principle independent, thus we have six different settings for the moduli problem. We might expect that, for all of them, $M_{0,n}/S_n \simeq (\mathrm{Sym}^n \mathbb{P}^1 \setminus (\text{diagonal}))/\mathrm{PGL}_2$ is a fine moduli space.

Theorem 8.76 *Consider the above six settings of the moduli problem of $n \geq 3$ unordered points in $\mathbb{P}^1$ over a field k.*

(8.76.1) *If $\mathrm{char}\, k = 0$ then $M_{0,n}/S_n$ is a fine moduli space in all six settings.*

(8.76.2) *If $\mathrm{char}\, k > 0$ then $M_{0,n}/S_n$ is a fine moduli space, provided either (8.75.3.a) or (8.75.4.c) holds.*

(8.76.3) *If $\mathrm{char}\, k > 0$ and we are in (8.75.3.b+4.a) or (8.75.3.b+4.b), then $M_{0,n}/S_n$ is not even a coarse moduli space. In fact the categorical moduli space (1.9) is $\mathrm{Spec}\, k$.*

Proof Let $(P_S \to S, D)$ be as in (8.75.3). If D is flat over S, then choosing an open cover $S = \cup_j U_j$ and isomorphisms $P_{U_j} \simeq \mathbb{P}^1 \times U_j$ gives morphisms $\phi_j \colon U_j \to \mathrm{Hilb}_n(\mathbb{P}^1)$. Changing the local trivialization changes the ϕ_j by an element of $\mathrm{Aut}(\mathbb{P}^1)$. Thus the ϕ_j glue to give a global morphism $\phi \colon S \to M_{0,n}/S_n$.

Since $P_S \to S$ is smooth, a relatively $\mathbb{Q}$-Cartier divisor D is Cartier by (4.39) if char $k = 0$. The same holds in any characteristic if the base is weakly normal by (4.41). In both cases D is flat over S, showing (1) and (2).

The proof of (3) relies on the following construction.

Let k be a field of characteristic $p > 0$, B a smooth projective curve over k and S a k-variety, for example a smooth curve. Let Δ be an effective, relative Cartier divisor on $B \times S \to S$. Any universal homeomorphism $\tau \colon S \to T$ (10.78) factors through a power of the Frobenius (for some $q = p^m$) as

$$F_q \colon S \xrightarrow{\ \tau\ } T \xrightarrow{\ \tau'\ } S.$$

Taking product with B we get $\tau_B \colon B \times S \to B \times T$ and $\tau'_B \colon B \times T \to B \times S$. Set $\Delta_T := (\tau_B)_* \Delta$ on $B \times T$. If τ is birational, the coefficients of Δ_T are the same as the coefficients of Δ. Also, Δ_T is $\mathbb{Q}$-Cartier since $q\Delta_T = (\tau'_B)^*\Delta$. However, the Cartier index may get multiplied by q. We have thus proved the following.

Claim 8.76.4 If $(B \times S, \Delta) \to S$ is in our moduli problem using (8.75.3.b), then so is $(B \times T, \Delta_T) \to T$. $\qquad\qquad\square$

A typical example with concrete equations is in (4.12).

Assume now that we work in the settings (8.75.3.b+4.a) or (8.75.3.b+4.b). Let $\mathbf{M}_n$ be the categorical moduli space. If $(\mathbb{P}^1 \times S, D)$ is a family of n points on $\mathbb{P}^1$, then we get a moduli map $\phi \colon S \to \mathbf{M}_n$. By the above construction, for any $\tau \colon S \to T$ we get a factorization $\phi \colon S \xrightarrow{\tau} T \to \mathbf{M}_n$.

Corollary 8.76.5 If the universal push-out of all the above $\tau \colon S \to T$ is $S \to$ Spec k, then the moduli map $\phi \colon S \to \mathbf{M}_n$ is constant.

Instead of proving this in general, we work out some typical examples.

Example 8.76.6 The map Spec $k[x] \to$ Spec $k[(x - c)^r, (x - c)^s]$ is a birational, universal homeomorphism for any $(r, s) = 1$ and $c \in k$. The universal push-out of all of them is Spec $k[x] \to$ Spec k; cf. (10.87).

Indeed, if $f(x) \in k[(x - c)^r, (x - c)^s]$ vanishes at c then it has a zero of multiplicity $\geq \min\{r, s\}$. Thus only the constants are contained in the intersection of all of them.

This settles (8.75.3.b+4.a), but the curves Spec $k[(x - c)^r, (x - c)^s]$ are not seminormal if $r, s > 1$. Over an algebraically closed field k, there are two-dimensional seminormal examples.

Example 8.76.7 Set $R_q := k[x] + (y^q - x)k[x, y] \subset k[x, y]$. R_q is seminormal, but not weakly normal and its normalization is $k[x, y]$. The conductor ideal is $(y^q - x)k[x, y]$. It is a principal ideal in $k[x, y]$, but not in R_q.

The map $\operatorname{Spec} k[x, y] \to \operatorname{Spec} R_q$ is birational. It is again easy to check that the universal push-out of all of them is $\operatorname{Spec} k[x, y] \to \operatorname{Spec} k[x]$. Thus if we combine the maps $\operatorname{Spec} k[x, y] \to \operatorname{Spec} R_q$ with all linear coordinate changes, then the universal push-out is $\operatorname{Spec} k[x, y] \to \operatorname{Spec} k$.   □

9

Hulls and Husks

Given a coherent sheaf F over a proper scheme, the quot-scheme parametrizes all quotients $F \twoheadrightarrow Q$. In many applications, it is necessary to understand not only surjections $F \twoheadrightarrow Q$ but also "almost surjections" $F \to G$. Such objects are called *quotient husks*. Special cases appeared in Kollár (2008a); Pandharipande and Thomas (2009); Alexeev and Knutson (2010); and Kollár (2011b). In this chapter, we study quotient husks, prove that they have a fine moduli space QHusk(F), and then apply this to families of hulls.

The notion of the *hull* of a coherent sheaf F is the generalization of the concept of reflexive hull of a module over a normal domain. In Section 9.1 we discuss the absolute case, denoted usually by $F^{[**]}$, and in Section 9.2 the relative case, denoted by F^H. For many applications, the key is the following.

Question 9.1 Let $f : X \to S$ be a proper morphism and F a coherent sheaf on X. Do the hulls $F_s^{[**]}$ of the fibers F_s form a coherent sheaf that is flat over S?

If the answer is yes, the resulting sheaf is called the *universal hull* of F over S. Local criteria for its existence are studied in Section 9.3.

In order to get global criteria, husks and quotient husks are defined in Section 9.4. In Section 9.5, the first main result of the Chapter proves that if $X \to S$ is projective and F is a coherent sheaf on X then the functor of all quotient husks with a given Hilbert polynomial has a fine moduli space $\text{QHusk}_p(X)$, which is a proper algebraic space over S. The proof closely follows the arguments given in Kollár (2008a).

This is used in a global study of hulls in Section 9.6. A third answer to our question is given in Section 9.7 in terms of a decomposition of S into locally closed subschemes. Local versions of these results are studied in Section 9.8.

Assumptions In this chapter we are mostly interested in schemes of finite type over an arbitrary base scheme.

347

However, the results of Section 9.1 work for Noetherian schemes that have a dimension function dim( ) such that closed points have dimension 0, and if $W_1 \subsetneq W_2$ is a maximal (with respect to inclusion) irreducible subscheme of an irreducible $W_2 \subset X$, then dim W_1 = dim $W_2 - 1$. (That is, X is catenary (Stacks, 2022, tag 02I0).) This holds for schemes of finite type over a local CM scheme; see Stacks (2022, tags 00NM and 02JT).

9.1 Hulls of Coherent Sheaves

We use the results on S_2 sheaves, to be discussed in Section 10.1.

Let X be an integral, normal scheme and F a coherent sheaf on X. The reflexive hull of F is the double dual $F^{**} := \mathcal{H}om_X(\mathcal{H}om_X(F, \mathcal{O}_X), \mathcal{O}_X)$. We would like to extend this notion to arbitrary schemes and arbitrary coherent sheaves. For this, the key properties of the reflexive hull are the following:

- F^{**} is S_2, and
- F^{**} is the smallest S_2 sheaf containing $F/(\text{torsion})$.

These are the properties that we use to define the hull of a sheaf. Note, however, that for this, we need to agree what the "torsion subsheaf" of a sheaf should be. Two natural candidates, emb(F) and tors(F), are discussed in (10.1).

Here we work with tors(F), the largest subsheaf whose support has dimension < dim F. An advantage is that pure(F) := $F/\text{tors}(F)$ is pure dimensional; but one needs the dimension function to be reasonable. A theory of hulls using emb(F) is developed in Kollár (2017).

A useful property of pure sheaves is the following.

Lemma 9.2 *Let $p: X \to Y$ be a finite morphism and F a coherent sheaf on X. Then F is pure and S_m iff p_*F is pure and S_m.*

Proof The last remark of (10.2) implies that the depth is preserved by push-forward. Thus the only question is whether (co)dimension is preserved or not; this is where our assumptions on the dimension function come in. □

Definition 9.3 (Hull of a sheaf) Let X be a scheme and F a coherent sheaf on X. Set $n = \dim F$. The *hull* of F is a coherent sheaf $F^{[**]}$ together with a map $q: F \to F^{[**]}$ such that

(9.3.1) Supp(ker q) has dimension $\leq n - 1$,

(9.3.2) Supp(coker q) has dimension $\leq n - 2$, and

(9.3.3) $F^{[**]}$ is pure and S_2.

We sometimes say S_2-*hull* or *pure hull* if we want to emphasize these properties. We see below that a hull is unique and it exists if X is excellent.

By definition, $F^{[**]} = (F/\operatorname{tors}(F))^{[**]}$, hence it is enough to construct hulls of pure, coherent sheaves.

The notation $F^{[**]}$ is chosen to emphasize the close connection between the hull and the reflexive hull F^{**}; see (9.4). We introduce a relative version, denoted by F^H in (9.8).

The following property is clear from the definition.

(9.3.4) Let G be a pure, coherent, S_2 sheaf and $F \subset G$ a subsheaf. Then $G = F^{[**]}$ iff $\dim(G/F) \leq \dim G - 2$.

From (9.2) and (10.10) we obtain the following base change properties of hulls.

(9.3.5) Let $p: X \to Y$ be a finite morphism. Then $p_*(F^{[**]}) = (p_*F)^{[**]}$.

(9.3.6) Let $g: Z \to X$ be flat, pure dimensional, with S_2 fibers. Then there is a natural isomorphism $g^*(F^{[**]}) = (g^*F)^{[**]}$.

Proposition 9.4 *Let X be an irreducible, normal scheme and F a torsion free coherent sheaf on X. Then $F^{[**]} = F^{**} := \mathcal{H}om_X(\mathcal{H}om_X(F, \mathscr{O}_X), \mathscr{O}_X)$.*

Proof F is locally free outside a codimension ≥ 2 subset $Z \subset X$. Thus the natural map $F \to F^{**}$ is an isomorphism over $X \setminus Z$. Since F^{**} is S_2 by (10.8), it satisfies the assumptions of (9.3). □

This can be used to construct the hull over schemes of finite type over a field. Indeed, we may assume that X is affine and $X = \operatorname{Supp} F$. By Noether normalization, there is a finite surjection $p: X \to \mathbb{A}^n$. Thus, by (9.3.5) and (9.4), $F^{[**]}$ can be identified with $(p_*F)^{**}$, as a $p_*\mathscr{O}_X$-module. Hulls also exist over excellent schemes; see Kollár (2017) for a more general result.

Proposition 9.5 *Let F be a pure, coherent sheaf on an excellent scheme X.*

(9.5.1) *There is a closed subset $Z \subset \operatorname{Supp} F$ of dimension $\leq \dim F - 2$ such that F is S_2 over $X \setminus Z$.*

(9.5.2) *Let $Z \subset \operatorname{Supp} F$ be any closed subset of dimension $\leq \dim F - 2$ such that F is S_2 over $U := X \setminus Z$. Then $F^{[**]} = j_*(F|_U)$, and, for every coherent sheaf G, every morphism $G|_U \to F|_U$ uniquely extends to $G \to F^{[**]}$.*

Proof The first claim follows from (10.27). To see (2), note that $j_*(F|_U)$ is coherent by (10.26), S_2 over U by assumption, and $\operatorname{depth}_Z j_*(F|_U) \geq 2$ by (10.6). Thus $j_*(F|_U)$ is a hull of F and we get $\tau: G \to j_*(G|_U) \to j_*(F|_U)$.

Let $F^{[**]}$ be any hull of F. Then $F^{[**]}|_U$ is a hull of $F|_U$; let $W \subset U$ be the support of their quotient. Then $\operatorname{codim}_X W \geq 2$ hence $F^{[**]}|_U = F|_U$ by (10.6.2). Thus we get a map $F^{[**]} \to j_*(F|_U)$. Applying (10.6) again gives that $F^{[**]} = j_*(F|_U)$. □

Corollary 9.6 *Let* $0 \to F_1 \to F_2 \to F_3$ *be an exact sequence of coherent sheaves of the same dimension. Then the hulls also form an exact sequence* $0 \to F_1^{[**]} \to F_2^{[**]} \to F_3^{[**]}.$ □

9.7 (Quasi-coherent hulls) Following (9.5.2), one should define the hull of a torsion-free, quasi-coherent sheaf F as $F^{[**]} := \varinjlim (j_Z)_*(F|_{X \setminus Z})$, where Z runs through all codimension ≥ 2 closed subsets of $\operatorname{Supp} F$. It is easy to see that $F^{[**]}$ is S_2, as defined in Grothendieck (1968, exp.III).

9.2 Relative Hulls

Next we develop a relative version of the notion of hull for coherent sheaves on a scheme X over a base scheme S.

In the absolute case, the hull is an S_2 sheaf that we can associate to any coherent sheaf on X, in particular, the hull does not have embedded points.

In the relative case, assume for simplicity that $f: X \to S$ is smooth; then $\mathcal{O}_X$ should be its own "relative hull." Note, however, that the structure sheaf $\mathcal{O}_X$ has no embedded points if and only if the base scheme S has no embedded points. Thus if we want to say that $\mathcal{O}_X$ is its own relative hull then we have to distinguish embedded points that are caused by S (these are allowed) from other embedded points (these are forbidden).

The distinction between these two types of embedded points seems to be meaningful only if F is generically flat (3.26).

Definition 9.8 (Relative hull) Let $f: X \to S$ be a morphism of finite type and F a coherent sheaf on X. Let n be the relative dimension of $\operatorname{Supp} F \to S$. A *hull* (or *relative hull*) of F over S is a coherent sheaf F^H together with a morphism $q: F \to F^H$ such that[1]
(9.8.1) $\operatorname{Supp}(\ker q) \to S$ has fiber dimension $\leq n - 1$,
(9.8.2) $\operatorname{Supp}(\operatorname{coker} q) \to S$ has fiber dimension $\leq n - 2$,
(9.8.3) there is a closed subset $Z \subset X$ with complement $U := X \setminus Z$ such that $Z \to S$ has fiber dimension $\leq n - 2$, $(F/\ker q) \to F^H$ is an isomorphism over U, $F^H|_U$ is flat over S with pure, S_2 fibers, and $\operatorname{depth}_Z F^H \geq 2$.
Note that $\operatorname{Supp}(\operatorname{coker} q) \subset Z$ by (3), hence in fact (3) implies (2). We state the latter separately to emphasize the parallels with (9.3).

Note that, while the hull always exists, the relative hull frequently does not; see (9.13) for a criterion. We have the following obvious comparisons.

---

[1] F^h would have been more consistent, but it is frequently used to denote the Henselization.

Claim 9.8.4 Assume that F^H exists and S is reduced. Then $(F^H)_g = (F_g)^{[**]}$ for every generic point $g \in S$. □

Claim 9.8.5 Assume that F^H exists and S is S_2. Then $F^H = F^{[**]}$. □

The converse fails. As an example, let $f: X := \mathbb{A}^2_{st} \to S := \mathbb{A}^1_t$ be the projection and $F \subset \mathscr{O}_X$ the ideal sheaf of the point $(0, 0)$. Then $F^{[**]} = \mathscr{O}_X$, but $F \to \mathscr{O}_X$ is not a relative hull since coker($F \to \mathscr{O}_X$) has codimension 1 on X_0.

Lemma 9.9 *Let $(0, T)$ be the spectrum of a DVR, $f: X \to T$ a morphism of finite type, and $q: F \to G$ a map between pure, coherent sheaves on X that are flat over T. Then G is a relative hull of F iff G_g is the hull of F_g, G_0 is S_1, and $q_0: F_0 \to G_0$ is an isomorphism outside a subset $Z_0 \subset \operatorname{Supp} G_0$ of codimension ≥ 2.*

Proof Assume that $G = F^H$ and let $Z \subset X$ be as in (9.8). By assumption, $G|_{X \setminus Z}$ has S_2 fibers thus $G|_{X \setminus Z}$ is S_2. Hence G is S_2 since $\operatorname{depth}_Z G \geq 2$ and so G_0 is S_1 and $q_0: F_0 \to G_0$ is an isomorphism outside $X_0 \cap Z$.

Conversely, if (1–3) hold then G is S_2 by (1–2). By (9.5) there is a closed subset $Z_1 \subset X_0$ of codimension ≥ 2 such that F_0 is S_2 over $X_0 \setminus Z_1$. Thus $q: F \to G$ satisfies the conditions (9.8.1–3) where Z is the union of three closed sets: Z_0, Z_1 and the closure of $\operatorname{Supp}(\operatorname{coker} q_g)$. □

Corollary 9.10 *Let $(0, T)$ be the spectrum of a DVR, $f: X \to T$ a morphism of finite type and F a pure, coherent sheaf on X that is flat over T. Then $F = F^H$ $\Leftrightarrow F$ is $S_2 \Leftrightarrow F_g$ is S_2 and F_0 is S_1.* □

Corollary 9.11 (Bertini theorem for relative hulls) *Let $(0, T)$ be the spectrum of a DVR, $X \subset \mathbb{P}^n_T$ a quasi-projective scheme and F a coherent sheaf on X with relative hull $q: F \to F^H$. Then $q|_L: F|_L \to F^H|_L$ is the relative hull of $F|_L$ for a general hyperplane $L \subset \mathbb{P}^n_T$.*

Proof We use (10.18) and (10.19) both for the special fiber X_0 and the generic fiber X_g. We get open subsets $U_0 \subset \check{\mathbb{P}}^n_0$ and $U_g \subset \check{\mathbb{P}}^n_g$ such that $F^H|_{L_0}$ is S_1 for $L_0 \in U_0$, $(F/\operatorname{tors}(F))|_{L_0} = (F|_{L_0})/\operatorname{tors}(F|_{L_0})$ for $L_0 \in U_0$, the natural map $(F|_{L_0})/\operatorname{tors}(F|_{L_0}) \to G_{L_0}$ is an isomorphism outside a subset of codimension ≥ 2 for $L_0 \in U_0$, and $F^H|_{L_g}$ is the hull of $F|_{L_g}$ for $L_g \in U_g$.

Let $W_T \subset \check{\mathbb{P}}^n_T$ denote the closure of $\check{\mathbb{P}}^n_g \setminus U_g$. For dimension reasons, W_T does not contain $\check{\mathbb{P}}^n_0$. Thus any hyperplane corresponding to a section through a point of $U_0 \setminus W_T$ works. □

Definition 9.12 (Vertical purity) Let $g: X \to S$ be a finite type morphism and G a coherent sheaf on X. We say that G is *vertically* pure of dimenion n, if for every $W \in \text{Ass}(G)$, every fiber of $g|_W: W \to S$ is either empty or has pure dimension n,

Let F be a coherent sheaf on X such that $\text{Supp}\, F \to S$ has relative dimension n. Let $\{W_i: i \in I\} \subset \text{Ass}(F)$ be those associated subschemes for which the generic fiber of $g|_{W_i}: W_i \to S$ has dimension $< n$. Set $Z := \cup_{i \in I} W_i$. The *vertically pure quotient* of F is $\text{vpure}(F) := F/\text{tors}_Z(F)$, using the notation of (10.1). Note that if $q: F \to F^H$ is a relative hull, then $\text{vpure}(F) = \text{im}\, q$.

Next we state the precise conditions needed for the existence of relative hulls. Then we show that a relative hull is unique, generalizing (9.5).

Lemma 9.13 *Let* $f: X \to S$ *be a morphism of finite type and* F *a coherent sheaf on* X. *Let* n *denote the maximum fiber dimension of* $\text{Supp}\, F \to S$. *Then* F *has a relative hull iff*

(9.13.1) F *is generically flat (3.26), and*

(9.13.2) *there is an open* $j: U \hookrightarrow X$ *such that* $\text{vpure}(F)|_U$ *is a flat family of* S_2 *sheaves and* $(\text{Supp}\, F \setminus U) \to S$ *has fiber dimension* $\leq n - 2$.

If this holds, then

(9.13.3) $F^H = j_*(\text{vpure}(F)|_U)$ *is the unique relative hull of* F *over* S, *and*

(9.13.4) *any* $\tau_U: G|_U \to F|_U$ *uniquely extends to* $\tau: G \to F^H$.

Proof If $q: F \to F^H$ is a relative hull, then $\text{vpure}(F) = \text{im}\, q$, so the conditions (9.13.1–2) are satisfied.

Conversely, if the conditions (9.13.1–2) are satisfied, then we can harmlessly replace F by $\text{vpure}(F)$. Then $j_*(F|_U)$ is coherent by (10.26), $F \to j_*(F|_U)$ is an isomorphism over U by construction, and $\text{depth}_Z\, j_*(F|_U) \geq 2$ by (10.6).

The last claim follows from the universal property of the push-forward and it implies that F^H is independent of the choice of U. $\square$

Corollary 9.14 *Let* $f: X \to S$ *be a morphism of finite type and* G *a coherent sheaf on* X *that is flat over* S *with pure,* S_2 *fibers of dimension* n. *Let* $F \subset G$ *be a subsheaf. Then* $G = F^H$ *iff the fiber dimension of* $\text{Supp}(G/F) \to S$ *is* $\leq n - 2$. $\square$

9.3 Universal Hulls

For many applications, a key question is to understand the behavior of relative hulls under a base change.

Notation 9.15 Let $f \colon X \to S$ be a morphism of finite type and F a coherent sheaf satisfying (9.13.1–2). As in (3.18.1), for any $g \colon T \to S$ we get

$$X \xleftarrow{g_X} X_T := X \times_S T \xrightarrow{f_T} T.$$

Set $U_T := g_X^{-1}(U)$ and $F_T := g_X^* F$. The relative hulls F^H and $(F_T)^H$ exists, and, as in (3.27.2), we have *restriction maps*

$$r_T^S \colon (F^H)_T \to (F_T)^H. \tag{9.15.1}$$

Definition 9.16 Let $f \colon X \to S$ be a morphism of finite type and F a coherent sheaf on X satisfying (9.13.1–2).

We say that F^H is a *universal hull* of F at $x \in X$ if the restriction map r_T^S (9.15.2) is an isomorphism along $g_X^{-1}(x)$ for every $g \colon T \to S$. F^H is a *universal hull* of F if this holds at every $x \in X$. Equivalently, iff the functor $F \mapsto F^H$ *commutes with base change.*

We say that $F \mapsto F^H$ is *universally flat* if $(F_T)^H$ is flat over T for every $g \colon T \to S$.

The following theorem gives several characterizations of universal hulls.

Theorem 9.17 *Let $f \colon X \to S$ be a morphism of finite type and F a coherent sheaf on X that has a relative hull F^H over S. The following are equivalent.*
(9.17.1) F^H *is a universal hull of F.*
(9.17.2) $F \mapsto F^H$ *is universally flat.*
(9.17.3) F^H *is flat over S with pure, S_2 fibers.*
(9.17.4) F^H *is flat over S with pure, S_2 fibers over closed points of S.*
(9.17.5) $r_s^S \colon F^H \to (F_s)^H$ *is surjective for every closed point $s \in S$.*
(9.17.6) $(F_A)^H$ *is a universal hull of F_A for every Artinian scheme $A \to S$.*

Proof The only obvious implications are $(3) \Rightarrow (4)$ and $(1) \Rightarrow (5)$, but $(4) \Rightarrow (3)$ directly follows from the openness of the S_2-condition (10.11).

Note that the properties in (3) are preserved by base change, thus $(F^H)_T$ is flat over T and $((F^H)_T)_t$ is S_2 for every point $t \in T$. By (9.14) this implies that $(F^H)_T$ is the relative hull of F_T. Therefore $(F^H)_T = (F_T)^H$, so $F \mapsto F^H$ is universally flat and commutes with base change. That is, $(3) \Rightarrow (2)$ and $(3) \Rightarrow (1)$ both hold.

If (4) holds, then $(F^H)_s = (F_s)^H$ by (9.3.4), thus $(4) \Rightarrow (5)$. Applying (10.71) to every localization of S at closed points shows that $(5) \Rightarrow (4)$.

Next we show that $(2) \Rightarrow (6)$. We may assume that $S = \operatorname{Spec} A$, where (A, m) is a local, Artinian ring. Choose the smallest $r \geq 0$ such that $m^{r+1} = 0$;

so $m^r \simeq \oplus_i A/m$, the sum of a certain number of copies of A/m. This gives an injection $j_r \colon \oplus_i F_s \hookrightarrow F$ which then extends to $j_r^H \colon \oplus_i (F_s)^H \hookrightarrow F^H$.

Since F^H is flat over A, the image $j_r^H(\oplus_i (F_s)^H)$ is also isomorphic to $(m^r) \otimes_A F^H$ which is the same as $\oplus_i (F^H)_s$. Thus $(F_s)^H = (F^H)_s$ and, by the above arguments, (2) implies the properties (1–5) for local, Artinian base schemes.

In order to see (6) $\Rightarrow$ (5), we may replace S by its completion at s. For $r \in \mathbb{N}$ set $A_r := \mathrm{Spec}_S \mathscr{O}_S/m_s^r$. By base change we get $f_r \colon X_r \to A_r$ and $F_r := F|_{X_r}$. By assumption, $(F_r)^H$ is flat over A_r and we have proved that $F \mapsto F^H$ commutes with base change over Artinian schemes. Set $\tilde{F} := \varprojlim(F_r)^H$. Then $\tilde{F}$ is flat over S, coherent by Hartshorne (1977, II.9.3.A), agrees with F over U, and $\tilde{F} \to F_S^H$ is surjective. Thus $\tilde{F} = F^H$ by (9.14), giving (5). □

We can restate the characterization (9.17.3) as follows.

Corollary 9.18 *Let* $f \colon X \to S$ *be a morphism of finite type,* $q \colon F \to G$ *a map of coherent sheaves on* X. *Let* n *denote the maximum fiber dimension of* $\mathrm{Supp}(F) \to S$. *Then* G *is the universal hull of* F *over* S *iff the following hold.*

(9.18.1) $q_s \colon F_s \to G_s$ *is an isomorphism at all* n-*dimensional points of* X_s *for every* $s \in S$.

(9.18.2) G *is flat with purely* n-*dimensional,* S_2 *fibers over* S, *and*

(9.18.3) $\mathrm{Supp}(\mathrm{coker}(q)) \to S$ *has fiber dimension* $\leq n - 2$. □

Combining (9.18) and (10.12) shows that a relative hull is a universal hull over a dense open subset of the base. Thus Noetherian induction gives the following. A much more precise form will be proved in (9.40).

Corollary 9.19 *Let* $f \colon X \to S$ *be a proper morphism and* F *a coherent sheaf on* X. *Then there is a locally closed decomposition* $j \colon S' \to S$ *such that* $j_X^* F$ *has a universal hull.* □

The following example illustrates several aspects of (9.17).

Example 9.20 Let $g \colon X \to S$ be a flat family of projective varieties, S reduced and connected, with g-ample line bundle L. As in (2.35), we get the relative affine cone $C_S(X) := \mathrm{Spec}_S \oplus_{m \in \mathbb{N}} g_* \mathscr{O}_X(m)$, with vertex $V \simeq S$. Note that $C_S(X) \setminus V$ is a $\mathbb{G}_m$-bundle over X, so flat over S. By contrast, $C_S(X)$ is flat over S iff $h^0(X_s, L_s^m)$ is independent of $s \in S$ for all $m \in \mathbb{N}$.

The simplest examples where h^0 jumps are given by taking $X = C \times \mathrm{Jac}(C)$ for some smooth curve C of genus ≥ 2 and L a universal line bundle of relative degree $0 < d < 2g - 2$.

In these cases, the structure sheaf of $C_S(X)$ is its own relative hull, but it is not a universal hull.

9.4 Husks of Coherent Sheaves

Definition 9.21 Let X be a scheme and F a coherent sheaf on X. An n-dimensional *quotient husk* of F is a quasi-coherent sheaf G together with a homomorphism $q \colon F \to G$ such that

(9.21.1) G is pure of dimension n and

(9.21.2) $q \colon F \to G$ is surjective at all generic points of $\operatorname{Supp} G$.

A quotient husk is called a *husk,* if $n = \dim F$ and

(9.21.3) $q \colon F \to G$ is an isomorphism at all n-dimensional points of X.

Note 9.21.4 If $h \in \operatorname{Ann}(F)$, then $hG \subset G$ is supported in dimension $< n$, thus it is 0. Therefore G is also an $\mathscr{O}_X / \operatorname{Ann}(F)$ sheaf, so we get the same husks if we replace X with any subscheme containing $\operatorname{Spec}_X(\mathscr{O}_X / \operatorname{Ann}(F))$.

Any coherent sheaf F has a maximal husk $M(F) := \varinjlim \, (j_Z)_*(F|_{X \setminus Z})$, where Z runs through all closed subsets of $\operatorname{Supp} F$ such that $\dim Z < \dim F$. If $\dim F \geq 1$ then $M(F)$ is never coherent, but it is the union of coherent husks.

Lemma 9.22 *Let F be a coherent sheaf on X and $q \colon F \to G$ an n-dimensional (quotient) husk of F.*

(9.22.1) *Let $g \colon X \to Z$ be a finite morphism. Then g_*G is an n-dimensional (quotient) husk of g_*F.*

(9.22.2) *Let $h \colon Y \to X$ be a flat morphism of pure relative dimension r with S_1 fibers. Then h^*G is an $(n + r)$-dimensional (quotient) husk of h^*F.*

Proof If g is a finite morphism and M is a sheaf then the associated primes of g_*M are the images of the associated primes of M. This implies (1). Similarly, if h is flat then the associated primes of h^*M are the preimages of the associated primes of M. Since h^*G is S_1 by (10.10), we get (2). □

9.23 (Bertini theorem for (quotient) husks) Let F be a coherent sheaf on a quasi-projective variety $X \subset \mathbb{P}^n$ and $q \colon F \to G$ a coherent (quotient) husk. Let $H \subset \mathbb{P}^n$ be a general hyperplane. Then $G|_H$ is pure by (10.18). If, in addition, H does not contain any of the associated primes of $\operatorname{coker} q$ then $q|_H \colon F|_H \to G|_H$ is also a (quotient) husk.

Definition 9.24 Let X be a scheme and F a coherent sheaf on X. Set $n := \dim F$. A husk $q \colon F \to G$ is called *tight* if $q \colon F / \operatorname{tors}(F) \hookrightarrow G$ is an isomorphism at all $(n - 1)$-dimensional points of X.

Thus the hull $q \colon F \to F^{[**]}$ defined in (9.3) is a tight husk of F. We see below that the hull is the maximal tight husk.

Lemma 9.25 *Let X be a scheme and F a coherent sheaf on X with hull $q: F \to F^{[**]}$. Let $r: F \to G$ be any tight husk. Then q extends uniquely to an injection $q_G: G \hookrightarrow F^{[**]}$. Therefore $F^{[**]}$ is the unique tight husk that is S_2.*

Proof After replacing F with $F/\operatorname{tors}(F)$ we may assume that F is pure. Set $Z := \operatorname{Supp}(\operatorname{coker} r) \cup \operatorname{Supp}(F^{[**]}/F)$. Then Z has codimension ≥ 2 and F is S_2 on $X \setminus Z$. Using (9.5.2) we get that $G \subset j_*(G|_{X\setminus Z}) = j_*(F|_{X\setminus Z}) = F^{[**]}$. If G is also S_2, then, (9.5.2) gives that $G = F^{[**]}$.   □

Lemma 9.26 *Let X be a projective scheme, F a coherent sheaf of pure dimension n and $F \to G$ a quotient husk. The following are equivalent.*
(9.26.1) $G = F^{[**]}$.
(9.26.2) G is S_2 and $\chi(X, F(t)) - \chi(X, G(t))$ has degree $\leq n - 2$.
(9.26.3) $\chi(X, F^{[**]}(t)) \equiv \chi(X, G(t))$ *(identical as polynomials).*

Proof The exact sequence $0 \to K \to F \to G \to Q \to 0$ defines K, Q and

$$\chi(X, F(t)) - \chi(X, G(t)) \equiv \chi(X, K(t)) - \chi(X, Q(t)).$$

Note that K has pure dimension n and $\dim Q \leq n - 1$. If $G = F^{[**]}$ then $K = 0$ and $\dim Q \leq n - 2$ which implies (2) and (1) $\Rightarrow$ (3) is obvious.

Conversely, assume that $\chi(X, F(t)) - \chi(X, G(t))$ has degree $\leq n - 2$. Since $\deg\chi(X, Q(t)) \leq n - 1$, we see that $\deg\chi(X, K(t)) \leq n - 1$. However, K has pure dimension n, thus in fact $K = 0$ and so G is a tight husk of F. If G is S_2 then (9.25) implies that $G = F^{[**]}$, hence (2) $\Rightarrow$ (1).

Finally, if (3) holds, then $\chi(X, F(t)) - \chi(X, G(t))$ has degree $\leq n - 2$, hence, as we proved, G is a tight husk of F. By (9.25.1) G is a subsheaf of $F^{[**]}$. Thus $G = F^{[**]}$ since they have the same Hilbert polynomials.   □

Definition 9.27 (Husks over a base scheme) Let $f: X \to S$ be a morphism and F a coherent sheaf on X. A *quotient husk* of F over S is a quasi-coherent sheaf G on X, together with a homomorphism $q: F \to G$ such that
(9.27.1) G is flat and pure over S, and
(9.27.2) $q_s: F_s \to G_s$ is a quotient husk for every $s \in S$.
A quotient husk is called a *husk* if
(9.27.3) $q_s: F_s \to G_s$ is a husk for every $s \in S$.
We sometimes omit "over S" if our choice of S is clear from the context. The following properties are useful.
(9.27.4) Husks are preserved by base change. That is, let $q: F \to G$ be a (quotient) husk over S and $g: T \to S$ a morphism. Set $X_T := X \times_S T$ and let

$g_X \colon X_T \to X$ be the first projection. Then $g_X^* q \colon g_X^* F \to g_X^* G$ is a (quotient) husk over T.

(9.27.5) Assume that f is proper and we have $q \colon F \to G$ where G is flat and pure over S. By (10.54.1) there is a largest open S° such that $q^\circ \colon F^\circ \to G^\circ$ is a quotient husk over $S^\circ \subset S$.

9.5 Moduli Space of Quotient Husks

Definition 9.28 Let $f \colon X \to S$ be a proper morphism and F a coherent sheaf on X. Let $\mathcal{QHusk}(F/S)(*)$ (resp. $\mathcal{Husk}(F/S)(*)$) be the functor that to an S-scheme $g \colon T \to S$ associates the set of all coherent quotient husks (resp. husks) of $g_X^* F$, where $g_X \colon T \times_S X \to X$ is the projection.

We write $\mathcal{QHusk}(F)$ and $\mathcal{Husk}(F)$ if the choice of S is clear.

By (10.54.1) $\mathcal{Husk}(F/S)(*)$ is an open subfunctor of $\mathcal{QHusk}(F/S)(*)$.

If f is projective, H is an f-ample divisor class and $p(t)$ is a polynomial, then $\mathcal{QHusk}_p(F/S)(*)$ (resp. $\mathcal{Husk}_p(F/S)(*)$) denote the subfunctors of all coherent quotient husks (resp. husks) of $g_X^* F$ with Hilbert polynomial $p(t)$.

The main existence theorem of this section is the following.

Theorem 9.29 *Let $f \colon X \to S$ be a projective morphism and F a coherent sheaf on X. Let H be an f-ample divisor class and $p(t)$ a polynomial. Then $\mathcal{QHusk}_p(F/S)$ has a fine moduli space $\mathrm{QHusk}_p(F/S) \to S$, which is a proper algebraic space over S.*

When S is a point, the projectivity of $\mathrm{QHusk}_p(F)$ is proved in Lin (2015), see also Wandel (2015).

As we noted, $\mathcal{Husk}_p(F/S)$ is represented by an open subspace $\mathrm{Husk}_p(F/S) \subset \mathrm{QHusk}_p(F/S)$, which is usually not closed. There are, however, many important cases when $\mathrm{Husk}_p(F/S)$ is also proper over S.

Corollary 9.30 *Let $f \colon X \to S$ be a projective morphism and F a coherent sheaf that is generically flat over S (3.26). Let H be an f-ample divisor class and $p(t)$ a polynomial. Then $\mathcal{Husk}_p(F/S)$ has a fine moduli space $\mathrm{Husk}_p(F/S) \to S$ which is a proper algebraic space over S.*

The implication (9.29) $\Rightarrow$ (9.30) is proved in (9.31), where we also establish the valuative criteria of properness and separatedness for $\mathcal{QHusk}(F/S)$.

As a preliminary step, note that the problem is local on S, thus we may assume that S is affine. Then f, X, F are defined over a finitely generated subalgebra of $\mathcal{O}_S$, hence we may assume in the sequel that S is of finite type.

9.31 (The valuative criteria of separatedness and properness) More generally, we show that $Q\mathcal{H}usk(F/S)$ satisfies the valuative criteria of separatedness and properness whenever f is proper.

Let T be the spectrum of an excellent DVR with closed point $0 \in T$ and generic point $t \in T$. Given $g\colon T \to S$, let $g_X\colon T \times_S X \to X$ denote the projection. We have the coherent sheaf $g_X^* F$ and, over the generic point, a quotient husk $q_t\colon g_X^* F_t \to g_X^* G_t$. We aim to extend it to a quotient husk $\tilde{q}\colon g_X^* F \to \tilde{G}$.

Let $K \subset g_X^* F$ be the largest subsheaf that agrees with $\ker q_t$ over the generic fiber. Then $g_X^* F/K$ is a coherent sheaf on X_T and none of its associated primes is contained in X_0. Thus $g_X^* F/K$ is flat over T. Let $Z_0 \subset X_0$ be the union of the embedded primes of $(g_X^* F/K)_0$.

By construction q_t descends to a morphism $q_t'\colon (g_X^* F/K)_t \hookrightarrow g_X^* G_t$. Let $Z_t \subset \mathrm{Supp}(g_X^* F/K)_t$ be the closed subset where q_t' is not an isomorphism and $Z_T \subset X_T$ its closure. Finally set $Z = Z_0 \cup (Z_T \cap X_0)$.

The restriction of the sheaf $g_X^* F/K$ to $X_T \backslash (Z_0 \cup Z_T)$ is flat and pure over T and $g_X^* G_t$ is pure on $X_t = X_T \backslash X_0$. Furthermore, when restricted to $X_T \backslash (X_0 \cup Z_T)$, both of these sheaves are naturally isomorphic to $g_X^* F/K$. Thus we can glue them to get a single sheaf G' defined on $X_T \backslash Z$ that is is flat and pure over T.

Let $j\colon X_T \backslash Z \hookrightarrow X_T$ be the injection. By (10.6.6), $\tilde{G} := j_* G'$ is the unique extension that is flat and pure over T, hence $\tilde{q}\colon g_X^* F \to g_X^* F/K \to \tilde{G}$ is the unique quotient husk extending $q_t\colon F_t \to G_t$. Thus $Q\mathcal{H}usk(F/S)$ satisfies the valuative criteria of separatedness and properness.

If f is projective then $\tilde{G}_0$ has the same Hilbert polynomial as G_t.

Finally note that if F is generically flat over S and $q_t\colon g_X^* F_t \to g_X^* G_t$ is a husk then $K \subset g_X^* F$ is zero at the generic points of $X_0 \cap \mathrm{Supp}\, g_X^* F$, thus $\tilde{q}\colon g_X^* F \to g_X^* F/K \to \tilde{G}$ is a husk.

This shows that if F is generically flat over S then $\mathrm{Husk}(F/S)$ is closed in $\mathrm{QHusk}(F/S)$ hence (9.30) follows from (9.29).

9.32 (Construction of $\mathrm{QHusk}_p(F/S)$) We may assume that $X = \mathbb{P}_S^N$ for some N; the only consequence we actually need is that $f_* \mathcal{O}_X = \mathcal{O}_S$, and this holds after any base change.

We use the existence and basic properties of quot-schemes (9.33) and hom-schemes (9.34). Also, as we discuss in (9.35), there is fixed m such that $G_s(m)$ is generated by global sections and its higher cohomologies vanish for all quotient husks of $F_s \to G_s$ with Hilbert polynomial $p(t)$. Thus each $G_s(m)$ can be written as a quotient of $\mathscr{O}_{X_s}^{p(m)}$. Let

$$Q_{p(t)} := \mathrm{Quot}^\circ_{p(t)}(\mathscr{O}_X^{p(m)}) \subset \mathrm{Quot}(\mathscr{O}_X^{p(m)})$$

be the universal family of quotients $q_s \colon \mathscr{O}_{X_s}^{p(m)} \twoheadrightarrow M_s$ that have Hilbert polynomial $p(t)$, are pure, have no higher cohomologies and the induced map

$$q_s \colon H^0(X_s, \mathscr{O}_{X_s}^{p(m)}) \to H^0(X_s, M_s)$$

is an isomorphism. Openness of purity is the $m = 1$ case of (10.12), the other two properties are discussed in (9.35).

Let $\pi \colon Q_{p(t)} \to S$ be the structure map, $\pi_X \colon Q_{p(t)} \times_S X \to X$ the second projection and M the universal sheaf on $Q_{p(t)} \times_S X$.

By (10.54.1) the hom-scheme $\mathbf{Hom}(\pi_X^* F, M)$ (9.34) has an open subscheme $W_{p(t)}$ parametrizing maps from F to M that are surjective outside a subset of dimension $\leq n - 1$. Let $\sigma \colon W_{p(t)} \to Q_{p(t)}$ be the structure map and $\sigma_X \colon W_{p(t)} \times_S X \to Q_{p(t)} \times_S X$ the fiber product.

Note that $W_{p(t)}$ parametrizes triples

$$w := [F_w \xrightarrow{r_w} G_w \xleftarrow{q_w} \mathscr{O}_{X_w}^{p(m)}(-m)],$$

where $r_w \colon F_w \to G_w$ is a quotient husk with Hilbert polynomial $p(t)$ and $q_w(m) \colon \mathscr{O}_{X_w}^{p(m)} \to G_w(m)$ is a surjection that induces an isomorphism on the spaces of global sections.

Let $w' \in W_{p(t)}$ be another point corresponding to the triple

$$[F_{w'} \xrightarrow{r_{w'}} G_{w'} \xleftarrow{q_{w'}} \mathscr{O}_{X_{w'}}^{p(m)}(-m)]. \quad \text{such that} \quad [F_w \xrightarrow{r_w} G_w] \simeq [F_{w'} \xrightarrow{r_{w'}} G_{w'}].$$

The difference between w and w' comes from the different ways that we can write $G_w \simeq G_{w'}$ as quotients of $\mathscr{O}_{X_w}(-m)^{\oplus p(t)}$. Since we assume that the induced maps

$$q_w(m), q_{w'}(m) \colon H^0(X_w, \mathscr{O}_{X_w}^{p(m)}) \rightrightarrows H^0(X_w, G_w(m)) = H^0(X_w, G_{w'}(m))$$

are isomorphisms, the different choices of q_w and $q_{w'}$ correspond to different bases in $H^0(X_w, G_w(mH))$. Thus the fiber of $\mathrm{Mor}(*, W_{p(t)}) \to \mathcal{Q}Husk_p(F/S)(*)$ over $\pi \circ \sigma(w) = \pi \circ \sigma(w') =: s \in S$ is a principal homogeneous space under the algebraic group $\mathrm{GL}(p(m), k(s)) = \mathrm{Aut}(H^0(X_s, G_s(m)))$.

Thus the group scheme $\mathrm{GL}(p(m), S)$ acts on $W_{p(t)}$ and, arguing as in (8.56),

$$\mathrm{QHusk}_p(F/S) = W_{p(t)}/\mathrm{GL}(p(m), S).$$

9.33 (Quot-schemes) Let $f \colon X \to S$ be a morphism and F a coherent sheaf on X. $\mathcal{Q}uot(F/S)(*)$ denotes the functor that to a scheme $g \colon T \to S$ associates the set of all quotients of $g_X^* F$ that are flat over T with proper support, where $g_X : T \times_S X \to X$ is the projection.

If $F = \mathscr{O}_X$, then a quotient can be identified with a subscheme of X, thus $Quot(\mathscr{O}_X/S) = \mathcal{H}ilb(X/S)$, the Hilbert functor.

If H is an f-ample divisor class and $p(t)$ a polynomial, then $Quot_p(F/S)(*)$ denotes those flat quotients that have Hilbert polynomial $p(t)$.

By Grothendieck (1962, lect.IV), $Quot_p(F/S)$ is bounded, proper, separated and it has a fine moduli space $\mathrm{Quot}_p(F/S)$. See Sernesi (2006, sec.4.4) for a detailed proof.

Note that one can write F as a quotient of $\mathscr{O}_{\mathbb{P}^n}(-m)^r$ for some m, r, thus $Quot_p(F/S)$ can be viewed as a subfunctor of $Quot(\mathscr{O}_{\mathbb{P}^n}^r/S)$. The theory of the latter is essentially the same as the study of the Hilbert functor.

9.34 (Hom-schemes) Let $f : X \to S$ be a morphism and F, G quasi-coherent sheaves on X. Let $\mathrm{Hom}_S(F, G)$ be the set of $\mathscr{O}_X$-linear maps of F to G.

For $q : T \to S$, we have $\mathrm{Hom}_S(F_T, G_T)$, where $g_X : T \times_S X \to X$ is the projection and $F_T = g_X^* F$, $G_T = g_X^* G$.

As a special case of Grothendieck (1960, III.7.7.8–9), if f is proper, F, G are coherent and G is flat over S, then this functor is represented by an S-scheme $\mathbf{Hom}_S(F, G)$. That is, for any $g : T \to S$, there is a natural isomorphism

$$\mathrm{Hom}_T(F_T, G_T) \simeq \mathrm{Mor}_S(T, \mathbf{Hom}_S(F, G)).$$

To see this, note first that there is a natural identification between

(9.34.1) homomorphisms $\phi : F \to G$, and

(9.34.2) quotients $\Phi : (F + G) \twoheadrightarrow Q$ that induce an isomorphism $\Phi|_G : G \simeq Q$.

Next let $\pi : \mathrm{Quot}_S(F + G) \to S$ denote the quot-scheme parametrizing quotients of $F + G$ with universal quotient $u : \pi_X^*(F + G) \to Q$, where π_X denotes the induced map $\pi_X : \mathrm{Quot}_S(F + G) \times_S X \to X$.

Consider now the restriction of u to $u_G : \pi_X^* G \to Q$. By (10.54) there is an open subset

$$\mathrm{Quot}_S^\circ(F + G) \subset \mathrm{Quot}_S(F + G)$$

that parametrizes those quotients $v : F + G \to Q$ that induce an isomorphism $v_G : G \simeq Q$. Thus $\mathbf{Hom}_S(F, G) = \mathrm{Quot}_S^\circ(F + G)$. $\square$

9.35 (Boundedness of quotient husks) Let us say that a set of sheaves $\{F_\lambda : \lambda \in \Lambda\}$ is *bounded* if there is fixed m such that, $F_\lambda(m)$ is generated by global sections and its higher cohomologies vanish for all $\lambda \in \Lambda$.

By an argument going back to Mumford (1966, lec.14), a set of pure sheaves $\{F_\lambda : \lambda \in \Lambda\}$ on $\mathbb{P}^N$ with given Hilbert polynomial is bounded iff their restrictions to general linear subspaces of codimension $d - 1$ are bounded; see Huybrechts and Lehn (1997, 3.3.7) for a stronger result.

Since being a quotient husk commutes with restriction to general linear subspaces (9.23), after replacing S by the Grassmannian $\mathrm{Gr}_S(\mathbb{P}^{N-d+1}, \mathbb{P}^N)$, it is sufficient to prove boundedness in relative dimension 1.

If $\dim F_s = 1$, then we can choose m such that $F_s(m)$ is generated by global sections and its H^1 vanishes for all $s \in S$. Since $\mathrm{coker}(F_s \to G_s)$ has dimension 0, we get that $G_s(m)$ is also generated by global sections and its H^1 vanishes.

9.6 Hulls and Hilbert Polynomials

Recall that we use $\leq$ (resp. $\equiv$) to denote the lexicographic ordering (resp. identity) of polynomials, see (5.14).

Let $f : X \to S$ be a projective morphism with relatively ample line bundle $\mathcal{O}_X(1)$. For a coherent sheaf F on X we aim to understand flatness of F and of its hull F^H in terms of the Hilbert polynomials $\chi(X_s, F_s(t))$ of the fibers F_s. Note that the $\chi(X_s, F_s(t))$ carry no information about the nilpotents in $\mathcal{O}_S$, so we assume that S is reduced.

As we noted in (3.20), $s \mapsto \chi(X_s, F_s(*))$ is an upper semi-continuous function on S and F is flat over S iff this function is locally constant.

The next result says that the same holds for $s \mapsto \chi(X_s, F_s^{[**]}(*))$. This does not follow directly from (3.20), since in general there is no sheaf on X whose fibers are $F_s^{[**]}$.

Theorem 9.36 *Let $f : X \to S$ be a projective morphism with relatively ample line bundle $\mathcal{O}_X(1)$ and F a mostly flat family of coherent, S_2 sheaves (3.26). Assume that S is reduced. Then $s \mapsto \chi(X_s, F_s^{[**]}(*))$ is an upper semi-continuous function and the following are equivalent.*

(9.36.1) *$s \mapsto \chi(X_s, F_s^{[**]}(*))$ is locally constant on S.*

(9.36.2) *$r_s^S : F_s \to F_s^{[**]}$ is an isomorphism for $s \in S$.*

(9.36.3) *F is flat over S with S_2 fibers (9.17).*

Proof We follow the method of (5.30). By generic flatness (Eisenbud, 1995, 14.4), there is a dense open subset $S^\circ \subset S$ such that F^H is flat with S_2 fibers $(F^H)_s = F_s^{[**]}$ over S°. Thus the function $s \mapsto \chi(X_s, F_s^{[**]}(t))$ is locally constant on S°, hence constructible on S by Noetherian induction. Thus it is enough to prove upper semicontinuity when $(0 \in S)$ is the spectrum of a DVR with generic point g.

Then F is S_2 and flat over S. Thus $\chi(X_0, F_0(t)) \equiv \chi(X_g, F_g(t))$ and F_0 is S_1, hence the restriction map (9.15) $r_0^S : F_0 \to F_0^H$ is an injection. The exact sequence

$$0 \to F_0 \to F_0^H \to Q_0 \to 0$$

defines Q_0 and $\chi(X_0, F_0^H(t)) \equiv \chi(X_0, F_0(t)) + \chi(X_0, Q_0(t))$. This gives that

$$\chi(X_0, F_0^H(t)) \geq \chi(X_0, F_0(t)) \equiv \chi(X_g, F_g(t)).$$

Equality holds iff $r_0^S : F_0 \to F_0^H$ is an isomorphism, that is, when F_0 is S_2.

We have thus proved that if $s \mapsto \chi(X_s, F_s^{[**]}(t))$ is locally constant and S is regular, one-dimensional, then F^H is flat over S with S_2 fibers. We show in (9.41) that this implies the general case. $\qquad\square$

Complement 9.36.4 If $\dim Q_0 = 0$, then we get that $\chi(X_0, F_0^H) \geq \chi(X_g, F_g)$ and equality holds iff r_0^S is an isomorphism.

Proposition 9.37 *Let* $f : X \to S$ *be a projective morphism with relatively ample line bundle* $\mathscr{O}_X(1)$ *and* F *a mostly flat family of coherent,* S_2 *sheaves. Then* F^H *is a universal hull iff for every local, Artinian ring* (A, m_A) *with residue field* $k = A/m_A$ *and every morphism* $\operatorname{Spec} A \to S$, *we have*

$$\chi(X_A, (F_A)^H(t)) \equiv \chi(X_k, (F_k)^H(t)) \cdot \operatorname{length} A.$$

Proof We show that the condition holds iff $(F_A)^H$ is flat over A and then conclude using (9.17.6).

Let $U \subset X$ be the largest open set where F is flat with S_2 fibers. Pick a maximum length filtration of A and lift it to a filtration

$$0 = G_0^U \subset G_1^U \subset \cdots \subset G_r^U = F_A|_{U_A}$$

such that $G_{i+1}^U/G_i^U \simeq F_k|_{U_k}$ and $r = \operatorname{length} A$. By pushing it forward to X_A we get a filtration

$$0 = G_0 \subset G_1 \subset \cdots \subset G_r = (F_A)^H$$

such that $G_{i+1}/G_i \subset (F_k)^H$. Therefore

$$\chi(X_A, (F_A)^H(t)) \leq \chi(X_k, (F_k)^H(t)) \cdot \operatorname{length} A.$$

Equality holds iff $G_{i+1}/G_i = (F_k)^H$ for every i, that is, iff F_A^H is flat over A. $\qquad\square$

The next result roughly says that local constancy of H^0 implies flatness for globally generated shaves. It is similar to Grauert's theorem on direct images; the key difference is that we do not have a flat sheaf to start with.

Proposition 9.38 *Let* $f : X \to S$ *a proper morphism to a reduced scheme and* F *a mostly flat family of coherent,* S_2 *sheaves on* X. *Assume that*

(9.38.1) $s \mapsto h^0(X_s, F_s^H)$ *is a locally constant function on* S, *and*

(9.38.2) F_s^H *is generated by its global sections for every* $s \in S$.

Then F^H *is a universal hull and* $f_*(F^H)$ *is locally free.*

Proof Assume first that S is the spectrum of a DVR. We may replace F by $F^{[**]}$, hence assume that F is flat over S. Then $F_s \hookrightarrow F_s^H$ is an injection and we have inequalities

$$h^0(X_g, F_g) \le h^0(X_s, F_s) \le h^0(X_s, F_s^H). \qquad (9.38.3)$$

By (1) these are equalities. Since F_s^H is generated by its global sections, this implies that $F_s = F_s^H$. As we explain in (9.41), this implies that F^H is a universal hull for every S. The last claim then follows from Grauert's theorem. $\qquad\qquad\square$

9.7 Moduli Space of Universal Hulls

Definition 9.39 Let $f : X \to S$ be a morphism and F a coherent sheaf on X. As in (3.16.1)), for a scheme $g : T \to S$ set $\mathcal{H}ull(F/S)(T) = \{\emptyset\}$ if $g_X^* F$ has a universal hull, and $\mathcal{H}ull(F/S)(T) = \emptyset$ otherwise, where $g_X : T \times_S X \to X$ is the projection.

If f is projective and p is a polynomial we set $\mathcal{H}ull_p(F/S)(T) = 1$ if $g_X^* F$ has a universal hull with Hilbert polynomial p.

The following result is the key to many applications of the theory.

Theorem 9.40 (Flattening decomposition for universal hulls) *Let* $f : X \to S$ *be a projective morphism and* F *a coherent sheaf on* X. *Then*

(9.40.1) $\mathcal{H}ull_p(F/S)$ *has a fine moduli space* $\mathrm{Hull}_p(F/S)$.

(9.40.2) $\mathrm{Hull}_p(F/S) \to S$ *is a locally closed embedding (10.83).*

(9.40.3) *The structure map* $\mathrm{Hull}(F/S) = \amalg_p \mathrm{Hull}_p(F/S) \to S$ *is a locally closed decomposition (10.83).*

Proof Let n be the relative dimension of $\mathrm{Supp}\, F/S$ and $S_n \subset S$ the closed subscheme parametrizing n-dimensional fibers. We construct $\mathrm{Hull}_n(F/S)$, the fine moduli space of n-dimensional universal hulls. Then repeat the argument for $S \setminus S_n$.

Let $\pi\colon \mathrm{Husk}(F/S) \to S$ be the structure map, $\pi_X\colon \mathrm{Husk}(F/S) \times_S X \to X$ the second projection and $q_{\mathrm{univ}}\colon \pi_X^* F \to G_{\mathrm{univ}}$ the universal husk. The set of points $y \in \mathrm{Husk}(F/S)$ such that $(G_{\mathrm{univ}})_y$ is S_2 and has pure dimension n is open by (10.12). The fiber dimension of

$$\mathrm{Supp}\,\mathrm{coker}[\pi_X^* F \to G_{\mathrm{univ}}] \to \mathrm{Husk}(F/S)$$

is upper semi-continuous. Thus there is a largest open set $W_n \subset \mathrm{Husk}(F/S)$ parametrizing husks $F_s \to G_s$ such that G_s is S_2, has pure dimension n and $\dim \mathrm{Supp}\, G_s/F_s \leq n - 2$. By (9.18), $\mathrm{Hull}_n(F/S) = W_n$.

Since hulls are unique (9.13), $\mathrm{Hull}(F/S) \to S$ is a monomorphism (10.82). In order to prove that each $\mathrm{Hull}_p(F/S) \to S$ is a locally closed embedding, we check the valuative criterion (10.84).

Let $(0, T)$ be the spectrum of a DVR with generic point g and $p\colon T \to S$ a morphism such that the hulls of F_g and of F_0 have the same Hilbert polynomials. Let G_g denote the hull of F_g and extend G_g to a husk $F_T \to G_T$. By assumption and by flatness

$$\chi(X_0, (G_T)_0(t)) \equiv \chi(X_g, (G_T)_g(t)) \equiv \chi(X_g, (F_g)^H(t)) \equiv \chi(X_0, (F_0)^H(t)).$$

Hence $(G_T)_0 = (F_0)^H$ by (9.26) and so G_T is the relative hull of F_T. Thus G_T defines the lifting $T \to \mathrm{Hull}_p(F/S)$. $\square$

9.41 (End of the proof of 9.36 and 9.38) By definition, F has a universal hull over $\mathrm{Hull}(F/S)$, thus we need to show that $\tau\colon \mathrm{Hull}(F/S) \to S$ is an isomorphism.

By (9.40), τ is a locally closed decomposition, and, by (10.83.2), a proper, locally closed decomposition is an isomorphism if S is reduced.

To check properness, let T be the spectrum of a DVR and $p\colon T \to S$ a morphism. We already proved for both (9.36) and (9.38) that $(p^*F)^H$ is a universal hull. Thus $p\colon T \to S$ lifts to $\tilde{p}\colon T \to \mathrm{Hull}(F/S)$, so $\mathrm{Hull}(F/S) \to S$ is proper. $\square$

Let $f\colon X \to S$ be a morphism. Two coherent sheaves F, G on X are called *relatively isomorphic* or f-*isomorphic* if there is a line bundle L_S on S such that $F \simeq G \otimes f^* L_S$. We are interested in understanding all morphisms $q\colon T \to S$ such that the hulls of $q_X^* F$ and $q_X^* G$ are relatively isomorphic, that is, there is a line bundle L_T on T such that $(q_X^* F)^H \simeq (q_X^* G)^H \otimes f_T^* L_T$.

Proposition 9.42 *Let* $f\colon X \to S$ *be a flat, projective morphism with* S_2 *fibers such that* $H^0(X_s, \mathscr{O}_{X_s}) \simeq k(s)$ *for every* $s \in S$. *Let* M_1, M_2 *be mostly flat families of divisorial sheaves on* X. *Then there is a locally closed subscheme* $i\colon S^{riso} \hookrightarrow$

S such that, for any $q\colon T \to S$, the pull-backs $q_X^* M_1$ and $q_X^* M_2$ have relatively isomorphic hulls iff q factors as $q\colon T \to S^{riso} \hookrightarrow S$.

Proof Set $L := \mathcal{H}om_X(M_1, M_2)$. Then $q_X^* M_1$ and $q_X^* M_2$ have relatively isomorphic hulls iff L is relatively isomorphic to $\mathcal{O}_X$.

We may assume that S is connected. Then $p(*) := \chi(X_s, \mathcal{O}_{X_s}(*))$ is independent of $s \in S$. Thus $i\colon S^{riso} \hookrightarrow S$ factors through $\mathrm{Hull}_p(L/S) \to S$. After replacing S by $\mathrm{Hull}_p(L/S)$ it remains to prove the special case when L is flat over S. The latter follows from (3.21). □

9.43 (Pure quotients) We get a similar flattening decomposition for pure quotients. The proofs are essentially the same as for hulls, so we just state the results.

Let $f\colon X \to S$ be a morphism of finite type and F a coherent sheaf on X. We say that F is f-*pure* or *relatively pure*, if F is flat over S and has pure fibers (10.1). We say that $q\colon F \to G$ is an f-*pure quotient* or *relatively pure quotient* of F if G is f-pure and $G_s = \mathrm{pure}(F_s)$ for every $s \in S$. Note that $\ker q$ is then the largest subsheaf $K \subset F$ such that $\dim(\mathrm{Supp}\, K_s) < \dim(\mathrm{Supp}\, F_s)$ for every $s \in S$. In particular, a relatively pure quotient is unique.

This gives the functor of relatively pure quotients $\mathcal{P}ureq(F/S)$. If f is projective, it can be decomposed $\mathcal{P}ureq(F/S) = \amalg_p \mathcal{P}ureq_p(F/S)$ using Hilbert polynomials. As in (9.40), we get the following.

Claim 9.43.1 Let $f\colon X \to S$ be a projective morphism and F a coherent sheaf on X. The functor of pure quotients is represented by a locally closed decomposition $\mathrm{Pureq}(F/S) \to S$. □

Arguing as in (9.36) gives the following.

Corollary 9.43.2 Let S be a reduced scheme, $g\colon X \to S$ a projective morphism and F a coherent sheaf on X. Then F has a g-pure quotient $F \twoheadrightarrow G$ iff $s \mapsto \chi(\mathrm{pure}(F_s)(*))$ is locally constant on S. □

9.8 Non-projective Versions

The proofs in Section 9.7 used in an essential way the projectivity of $X \to S$. Here we consider similar questions for non-projective morphisms in two cases. If $X \to S$ is affine then a good theory seems possible only if S is local and complete. Then we study the case when $X \to S$ is proper.

For affine morphisms we have the following variant of (9.40).

Theorem 9.44 *Let* (S, m_S) *be a complete local ring, R a finite type S-algebra and F a finite R-module that is mostly flat with S_2 fibers over S (3.28). Then there is a quotient $S \twoheadrightarrow S^u$ that represents* $\mathrm{Hull}(F/S)$ *for local, Artinian S-algebras.*

Since universal hulls commute with completion (9.17.6), (9.44) implies the same statement for complete, local S-algebras. That is, for every local morphism $h: (S, m_S) \to (T, m_T)$, the hull $(F_T)^H$ is universal iff there is a factorization $h: S \twoheadrightarrow S^u \to T$.

Note that, compared with (9.40), we only identify the stratum containing the closed point of $\mathrm{Spec}\, S$.

Proof We follow the usual method of deformation theory; Artin (1976); Seshadri (1975); Hartshorne (2010). As a first step we construct S^u.

For an ideal $I \subset S$ set $F_I := F \otimes (R/IR)$. First, we claim that if $(F_I)^H$ and $(F_J)^H$ are universal hulls then so is $(F_{I \cap J})^H$. Start with the exact sequence

$$0 \to S/(I \cap J) \to S/I + S/J \to S/(I + J) \to 0. \tag{9.44.1}$$

F is mostly flat over S, thus (9.44.1) stays left exact after tensoring by F and taking the hull. Thus we obtain the exact sequence

$$0 \to (F_{I \cap J})^H \to (F_I)^H + (F_J)^H \to (F_{I+J})^H. \tag{9.44.2}$$

$(F_J)^H \to (F_{I+J})^H$ is surjective since $(F_J)^H$ is a universal hull, hence (9.44.2) is also right exact.

Set $k := S/m_S$. Since $(F_I)^H$ is a universal hull, $(F_I)^H \otimes k \simeq (F_m)^H$, and the same holds for J and $I + J$. Thus tensoring (9.44.2) with k yields

$$(F_{I \cap J})^H \otimes k \to (F_m)^H + (F_m)^H \xrightarrow{p} (F_m)^H \to 0. \tag{9.44.3}$$

Since $\ker p \simeq (F_m)^H$ we see that $(F_{I \cap J})^H \otimes k \to (F_m)^H$ is surjective. By (9.17) this implies that $(F_{I \cap J})^H$ is a universal hull.

Let $I^u \subset S$ be the intersection of all those ideals I such that $(F_I)^H$ is a universal hull and $S^u := S/I^u$. By (9.17.6) $(F_{S^u})^H$ is a universal hull.

By construction, if $h: S \twoheadrightarrow W := S/I_W$ is a quotient such that $(F_W)^H$ is a universal hull then $I^u \subset I_W$. We still need to prove that if (A, m_A) is a local Artinian S-algebra such that $(F_A)^H$ is a universal hull then $h: S \to A$ factors through S^u.

Let $K := A/m_A$ denote the residue field. $F/m_S F$ has a hull by (9.5), so $I^u \subset m_S$. Thus $S \to A \to K$ factors through S^u. Working inductively we may assume that there is an ideal $J_A \subset A$ such that $J_A \simeq K$ and $h': S \to A' := A/J$ factors through S^u. Therefore $h: S \to A$ factors through $S \to S/m_S I^u$. Note that $I^u/m_S I^u$ is a finite dimensional k-vector space, call it V_k, and we have a commutative diagram

$$
\begin{array}{ccccccccc}
0 & \longrightarrow & V_k & \longrightarrow & S/m_S I^u & \longrightarrow & S^u & \longrightarrow & 0 \\
& & \downarrow{\scriptstyle\lambda} & & \downarrow{\scriptstyle h} & & \downarrow{\scriptstyle h'} & & \\
0 & \longrightarrow & K & \longrightarrow & A & \longrightarrow & A & \longrightarrow & 0
\end{array}
\qquad (9.44.4)
$$

for some k-linear map $\lambda\colon V_k \to K$. If $\lambda = 0$ then h factors through S^u, this is what we want. If $\lambda \neq 0$ then we show that there is an ideal $J^u \subsetneq I^u$ such that F has a universal hull over S/J^u. This contradicts our choice of I^u and proves the theorem.

It is easier to write down the obstruction map in scheme-theoretic language. To simplify notation, we may assume that $m_S I^u = 0$. Thus set $X := \mathrm{Spec}_S R$ and let $i\colon U \hookrightarrow X$ be the largest open set over which $\tilde{F}$ (the sheaf associated to F) is flat over S. For any $S \to T$ by base change we get $i\colon U_T \hookrightarrow X_T$. Let $\mathcal{F}_T$ denote the restriction of $\tilde{F}_T$ to U_T. Then $i_*\mathcal{F}_T$ is the sheaf associated to $(F_T)^H$ and we have a commutative diagram

$$
\begin{array}{ccccccc}
V_k \otimes_k i_*\mathcal{F}_k & \longrightarrow & i_*\mathcal{F}_S & \longrightarrow & i_*\mathcal{F}_{S^u} & \overset{\delta}{\longrightarrow} & V_k \otimes_k R^1 i_*\mathcal{F}_k \\
\downarrow{\scriptstyle\lambda} & & \downarrow{\scriptstyle h} & & \downarrow{\scriptstyle h'} & & \downarrow{\scriptstyle\lambda} \\
i_*\mathcal{F}_K & \longrightarrow & i_*\mathcal{F}_A & \longrightarrow & i_*\mathcal{F}_{A'} & \overset{\Delta}{\longrightarrow} & R^1 i_*\mathcal{F}_K.
\end{array}
\qquad (9.44.5)
$$

Here $\Delta = 0$ since $i_*\mathcal{F}_A$ is a universal hull. The right-hand square factors as

$$
\begin{array}{ccccc}
\delta\colon i_*\mathcal{F}_{S^u} & \longrightarrow & i_*\mathcal{F}_k & \overset{\delta_k}{\longrightarrow} & V_k \otimes_k R^1 i_*\mathcal{F}_k \\
\downarrow{\scriptstyle h'} & & \downarrow{\scriptstyle h_k} & & \downarrow{\scriptstyle\lambda\otimes 1} \\
\Delta\colon i_*\mathcal{F}_{A'} & \longrightarrow & i_*\mathcal{F}_K & \overset{\Delta_K}{\longrightarrow} & K \otimes_k R^1 i_*\mathcal{F}_k.
\end{array}
\qquad (9.44.6)
$$

By assumption $\Delta_K = 0$. Choosing a basis $\{v_j\}$ of V_k, this means that the components $\delta_{k,j}\colon i_*\mathcal{F}_k \to R^1 i_*\mathcal{F}_k$ are linearly dependent over K. So they are linearly dependent over k, that is, there is a nonzero $\mu\colon V_k \to k$ such that $\mu \circ \delta_k = 0$. Set $J^u := m_S I^u + \ker\mu$ and $S' := S/J^u$. Note that $I^u/J^u \simeq k$. The extension $I^u/J^u \to S' \to S^u$ gives the exact sequence

$$
(I^u/J^u) \otimes_k i_*\mathcal{F}_k \hookrightarrow i_*\mathcal{F}_{S'} \to i_*\mathcal{F}_{S^u} \overset{\mu\circ\delta}{\longrightarrow} (I^u/J^u) \otimes_k R^1 i_*\mathcal{F}_k.
\qquad (9.44.7)
$$

Since $\mu \circ \delta = 0$ the map $i_*\mathcal{F}_{S'} \to i_*\mathcal{F}_{S^u}$ is surjective and so is the composite $i_*\mathcal{F}_{S'} \to i_*\mathcal{F}_{S^u} \to i_*\mathcal{F}_k$. Thus $i_*\mathcal{F}_{S'}$ is a universal hull by (9.17). This contradicts the choice of S^u. $\qquad\qquad\square$

One can see that (9.44) does not hold for arbitrary local schemes S, but the following consequence was pointed out by E. Szabó.

Corollary 9.45 *The conclusion of (9.44) remains true if S is a Henselian local ring, that is the localization of an algebra of finite type over a field or over an excellent DVR.*

Proof There is a general theorem (Artin, 1969, 1.6) about representing functors over Henselian local rings; we check that its conditions are satisfied.

Let $\hat{S}$ denote the completion of S. As in (3.16.1), define a functor on local S-algebras by setting $\mathcal{F}(T) = \{\emptyset\}$ if $(F_T)^H$ is a universal hull and $\mathcal{F}(T) = \emptyset$ otherwise.

It is easy to see that if $\mathcal{F}(T) = \{\emptyset\}$, then there is a factorization $S \to T' \to T$ such that T' is of finite type over S and $\mathcal{F}(T') = \{\emptyset\}$. So $\mathcal{F}$ is locally of finite presentation over S, as in Artin (1969, 1.5). The universal family over $(\hat{S})^u$ gives an effective versal deformation of the fiber over m_S. The existence of S^u now follows from Artin (1969, 1.6).   □

Next we present an alternative approach to hulls and husks that does not use projectivity, works for proper algebraic spaces, but leaves properness of $\mathrm{Husk}(F/S)$ unresolved. The proofs were worked out jointly with M. Lieblich.

Theorem 9.46 *Let S be a Noetherian algebraic space and $p\colon X \to S$ a proper morphism of algebraic spaces. Let F be a coherent sheaf on X. Then $Q\mathcal{H}usk(F/S)$ is separated and it has a fine moduli space $\mathrm{QHusk}(F/S)$.*

Proof Let $f\colon X \to S$ be a proper morphism. The functor of flat families of coherent sheaves $\mathcal{F}lat(X/S)$ is represented by an algebraic stack $\mathrm{Flat}(X/S)$ which is locally of finite type, but very non-separated; see Laumon and Moret-Bailly (2000, 4.6.2.1).

Let $\sigma\colon \mathrm{Flat}(X/S) \to S$ be the structure morphism and $U_{X/S}$ the universal family. By (10.12), there is an open substack $\mathrm{Flat}^n(X/S) \subset \mathrm{Flat}(X/S)$ parametrizing pure sheaves of dimension n. Let $U^n_{X/S}$ be the corresponding universal family. Consider $X \times_S \mathrm{Flat}^n(X/S)$ with coordinate projections π_1, π_2. The stack $\mathbf{Hom}(\pi_1^* F, \pi_2^* U^n_{X/S})$ parametrizes all maps from the sheaves F_s to pure, n-dimensional sheaves N_s.

We claim that $\mathrm{QHusk}(F/S)$ is an open substack of $\mathbf{Hom}(\pi_1^* F, \pi_2^* U^n_{X/S})$. Indeed, by (10.54), for a map of sheaves $M \to N$ with N flat over S, it is an open condition to be an isomorphism at the generic points of the support.

As we discussed in (9.31), $\mathrm{QHusk}(F/S)$ satisfies the valuative criteria of separatedness and properness, so the diagonal of $\mathrm{QHusk}(F/S)$ is a monomorphism. Every algebraic stack with this property is an algebraic space; see Laumon and Moret-Bailly (2000, sec.8).   □

The connected components of QHusk(F/S) are not proper over S, this fails even for the quot-scheme, but the following should be true.

Conjecture 9.47 *Every irreducible component of* QHusk(F/S) *is proper.*

The construction of Hull(F/S) given in (9.40) applies to algebraic spaces as well, but it does not give boundedness. Nonetheless, we claim that Hull(F/S) is of finite type. First, it is locally of finite type since QHusk(F/S) is. Second, we claim that red Hull(F/S) is dominated by an algebraic space of finite type. In order to see this, consider the (reduced) structure map red Hull(F/S) → red S. It is an isomorphism at the generic points, hence there is an open dense $S^\circ \subset$ red S such that S° is isomorphic to an open subspace of red Hull(F/S). Repeating this for red $S \setminus S^\circ$, by Noetherian induction we eventually write red Hull(F/S) as a disjoint union of finitely many locally closed subspaces of red S. These imply that Hull(F/S) is of finite type. Using (9.13.4), we get the following.

Theorem 9.48 (Flattening decomposition for hulls) *Let $f: X \to S$ be a proper morphism of algebraic spaces and F a coherent sheaf on X. Then*

(9.48.1) *Hull(F/S) is separated and it has a fine moduli space* Hull(F/S),

(9.48.2) *Hull(F/S) is an algebraic space of finite type over S, and*

(9.48.3) *the structure map* Hull(F/S) → S *is a surjective monomorphism.* □

Example 9.49 Let C, D be two smooth projective curves. Pick points $p, q \in C$ and $r \in D$. Let X be the surface obtained from the blow-up $B_{(p,r)}(C \times D)$ by identifying $\{q\} \times D$ with the birational transform of $\{p\} \times D$. Note that X is a proper, but non-projective scheme and there is a natural proper morphism $\pi: X \to C'$ where C' is the nodal curve obtained from C by identifying the points p, q. Then Hull($\mathscr{O}_X/S$) = $C \setminus \{q\}$. The natural map $C \setminus \{q\} \to C'$ is a surjective monomorphism, but not a locally closed embedding.

10

Ancillary Results

In this chapter, we discuss various results that were used earlier and for which good references are scarce or scattered. We work over an arbitrary base scheme, whenever possible.

10.1 S_2 Sheaves

Definition 10.1 Let F be a quasi-coherent sheaf on a scheme X. Its *annihilator*, denoted by $\mathrm{Ann}(F)$, is the largest ideal sheaf $I \subset \mathscr{O}_X$ such that $I \cdot F = 0$. The *support* of F is the zero set $Z(I) \subset X$, denoted by $\mathrm{Supp}\, F$.

The *dimension* of F at a point x, denoted by $\dim_x F$, is the dimension of its support at x. The dimension of F is $\dim F := \dim \mathrm{Supp}\, F$.

The set of all associated points (or primes) of a quasi-coherent sheaf F is denoted by $\mathrm{Ass}(F)$. An associated point of F is called *embedded* if it is contained in the closure of another associated point of F. Let $\mathrm{emb}(F) \subset F$ denote the largest subsheaf whose associated points are all embedded points of F. Thus $F/\mathrm{emb}(F)$ has no embedded points, hence it is S_1 (10.5). Informally speaking, $F \mapsto F/\mathrm{emb}(F)$ is the best way to associate an S_1 sheaf to another sheaf.

If F is coherent then it has only finitely many associated points and $\mathrm{Supp}\, F$ is the union of their closures.

Let $Z \subset X$ be a closed subscheme. Then $\mathrm{tors}_Z(F) \subset F$ denotes the Z-torsion subsheaf, consisting of all local sections whose support is contained in Z. There is a natural isomorphism $\mathrm{tors}_Z(F) \simeq \mathcal{H}^0_Z(X, F)$.

If X has a dimension function (see the Assumptions on p.347), then we use $\mathrm{tors}(F) \subset F$ to denote the *torsion* subsheaf, consisting of all local sections whose support has dimension $< \dim \mathrm{Supp}\, F$. A coherent sheaf F is called *pure* (of dimension n) if (the closure of) every associated point of F has dimension n. Thus $\text{pure}(F) := F/\text{tors}(F)$ is the maximal *pure quotient* of F. A scheme is pure iff its structure sheaf is.

If $\text{Supp}\, F$ is pure dimensional, then $\text{emb}(F) = \text{tors}(F)$.

Let $f: X \to S$ be of finite type and F a coherent sheaf on X such that F_s is pure for every $s \in S$. Then the same holds after any base change $S' \to S$.

Warning. If X is pure dimensional, F is coherent and $\dim F = \dim X$, then our terminology agrees with every usage of "torsion" that we know of. However, the above distinction between $\text{emb}(F)$ and $\text{tors}(F)$ is not standard.

10.2 (Regular sequences and depth) Let A be a ring and M an A-module. Recall that $x \in A$ is M-*regular* if it is not a zero divisor on M, that is, if $m \in M$ and $xm = 0$ implies that $m = 0$. Equivalently, if x is not contained in any of the associated primes of M.

A sequence $x_1, \ldots, x_r \in A$ is an M-*regular sequence* if x_1 is not a zero divisor on M and x_i is not a zero divisor on $M/(x_1, \ldots, x_{i-1})M$ for all i.

Let $\text{rad}\, A$ denote the *radical* (or Jacobson radical) of A, that is, the intersection of all maximal ideals. Let $I \subset \text{rad}\, A$ be an ideal. The *depth* of M along I is the maximum length of an M-regular sequence $x_1, \ldots, x_r \in I$. It is denoted by $\text{depth}_I M$. If A is Noetherian, M is finite over A and $I \subset \text{rad}\, A$, then all maximal M-regular sequences $x_1, \ldots, x_r \in I$ have the same length; see Matsumura (1986, p.127) or Eisenbud (1995, sec.17).

Warning The literature is not fully consistent on the depth if $M = 0$ or if $I \not\subset \text{rad}\, A$. While the definition of depth makes sense for arbitrary rings and ideals, it can give unexpected results.

10.3 (Comments on depth and S_m) Let F be a coherent sheaf on X. The *depth of F at x*, denoted by $\text{depth}_x F$, is defined as the depth of its localization F_x along $m_{x,X}$ (as an $\mathscr{O}_{x,X}$-module). For a closed subscheme $Z \subset X$ we set

$$\text{depth}_Z F := \inf\{\text{depth}_z F : z \in Z\}. \tag{10.3.1}$$

If $X = \text{Spec}\, A$ is affine, $Z = V(I)$ for some ideal $I \subset \text{rad}\, A$ and $M = H^0(X, F)$ then $\text{depth}_Z F = \text{depth}_I M$. (This definition is for coherent sheaves only. See Grothendieck (1968, exp.III) for quasi-coherent sheaves.)

A coherent sheaf F on a scheme X satisfies *Serre's condition S_m* if

$$\text{depth}_x F \geq \min\{m, \text{codim}(x, \text{Supp}\, F)\} \quad \text{for every } x \in X; \tag{10.3.2}$$

see Stacks (2022, tag 033P) for details.

It is important to note that over a local scheme (x, X), being S_m is *not* the same as $\text{depth}_x F \geq m$; neither implies the other.

Definition 10.4 F is *Cohen–Macaulay* or *CM* if

$$\text{depth}_x F = \dim_x F \quad \text{for every } x \in X. \tag{10.4.1}$$

It is easy to see that if F is CM then the local rings of $\text{Supp}\, F$ are pure dimensional (Stacks, 2022, tag 00N2). In the literature, the definition of CM frequently includes the assumption that $\text{Supp}\, F$ be pure dimensional; we will most likely lapse into this habit too.

In contrast with the S_m situation (10.3), if (10.4.1) holds at closed points, then it holds at every point of $\text{Supp}\, F$; see Matsumura (1986, 17.4).

Condition S_1 can be described in terms of embedded points.

Lemma 10.5 *Let F be a coherent sheaf on a scheme X and $Z \subset X$ a closed subscheme. Then $\text{depth}_Z F \geq 1$ iff none of the associated points of F is contained in Z. In particular, F is S_1 iff it has no embedded associated points.* □

The following lemma gives several characterizations of S_2 sheaves.

Lemma 10.6 *Let F be a coherent sheaf and $Z \subset \text{Supp}\, F$ a nowhere dense subscheme. The following are equivalent.*

(10.6.1) $\text{depth}_Z F \geq 2$.

(10.6.2) $\text{depth}_Z F \geq 1$ *and* $\text{depth}_Z(F|_D) \geq 1$ *whenever D is a Cartier divisor in an open subset of X that does not contain any associated prime of F.*

(10.6.3) $\text{tors}_Z(F) = 0$ *and* $\text{tors}_Z(F|_D) = 0$ *whenever D is as above.*

(10.6.4) *An exact sequence $0 \to F \to F' \to Q \to 0$ splits if $\text{Supp}\, Q \subset Z$.*

(10.6.5) $\text{depth}_Z F \geq 1$ *and for any exact sequence $0 \to F \to F' \to Q \to 0$ such that $\emptyset \neq \text{Supp}\, Q \subset Z$, F' has an associated point in $\text{Supp}\, Q$.*

(10.6.6) $F = j_*(F|_{X \setminus Z})$ *where $j \colon X \setminus Z \hookrightarrow X$ is the natural injection.*

(10.6.7) $\mathscr{H}^0_Z(X, F) = \mathscr{H}^1_Z(X, F) = 0$.

(10.6.8) *Let $z \in Z$ be any point. Then $H^0_z(X_z, F_z) = H^1_z(X_z, F_z) = 0$.*

Proof All but (4) are clearly local conditions on X. By assumption, $\text{tors}_Z(F) = 0$. Thus, if in (4) there is a splitting locally then the unique splitting is given by $\text{tors}_Z(F') \subset F'$. Thus (4) is also local, so we can assume that X is affine.

Conditions (2) and (3) are just restatements of the inductive definition of depth. Assume (1) and consider an extension $0 \to F \to F' \to Q \to 0$ where $\text{Supp}\, Q \subset Z$. If $\text{tors}_Z(F') \to Q$ is surjective then it gives a splitting. If not, then after quotienting out by $\text{tors}_Z(F')$ and taking a coherent subsheaf $F'' \subset F'/\text{tors}_Z(F')$, we get an extension $0 \to F \to F'' \to Q'' \to 0$ where tors$_Z(F'')$ $=$ 0. Pick s $\in$ I_Z that is not a zero divisor on F and F'', but $s \cdot (F''/F)$ $=$ 0. Then sF'' is a nonzero submodule of F/sF supported on Z. This proves (1) $\Rightarrow$ (4).

Assuming (4), we claim that tors$_Z(F)$ $=$ 0. After localizing at a generic point of tors$_Z(F)$, we may assume that tors$_Z(F)$ is supported at $z \in Z$. Since the injective hull of $k(z)$ over $\mathscr{O}_X$ has infinite length, there is a nonsplit extension j: tors$_Z(F)$ $\hookrightarrow$ G. Then the cokernel of $(1, j)$: tors$_Z(F)$ $\to$ $F + G$ gives a non-split extension of F. The rest of (5) is clear.

If depth$_Z$ F $\geq$ 1, then the natural map F $\to$ $j_*(F|_{X\setminus Z})$ is an injection. The quotient is supported on Z, thus (5) $\Rightarrow$ (6).

Assume (6). Then F $\to$ $j_*(F|_{X\setminus Z})$ is an injection, so depth$_Z$ F $\geq$ 1. If depth$_Z$ F $<$ 2, then we can pick s $\in$ I_Z such that F/sF has a subsheaf Q supported on Z. Let $F' \subset F$ be the preimage of Q. Then $s^{-1}F' \subset j_*(F|_{X\setminus Z})$ shows that (6) $\Rightarrow$ (1). We discuss (7) and (8) in (10.29). □

Corollary 10.7 *Let F be a coherent, S_2 sheaf and $G \subset F$ a subsheaf. Then G is S_2 iff every associated point of F/G has codimension $\leq$ 1 in* Supp F.

Proof Let $Z \subset$ Supp F be a closed subset of codimension $\geq$ 2 and j: U := $X \setminus Z \hookrightarrow X$ the injection. Then $j_*(G|_U) \subset F$ and depth$_Z$ $G < 2 \Leftrightarrow G \neq j_*(G|_U)$ $\Leftrightarrow$ $j_*(G|_U)/G \subset F/G$ is a nonzero subsheaf supported on Z. □

Corollary 10.8 *Let F be a coherent, S_2 sheaf and G any coherent sheaf. Then $\mathcal{H}om_X(G, F)$ is also S_2.*

Proof It is clear that every irreducible component of Supp $\mathcal{H}om_X(G, F)$ is also an irreducible component of Supp F.

Let $Z \subset$ Supp F be a closed subset of codimension $\geq$ 2 and j: $X \setminus Z \hookrightarrow X$ the injection. Any homomorphism ϕ: $G|_{X\setminus Z}$ $\to$ $F|_{X\setminus Z}$ uniquely extends to $j_*\phi$: $j_*(G|_{X\setminus Z})$ $\to$ $j_*(F|_{X\setminus Z})$. Since F is S_2, the target equals F. We have a natural map G $\to$ $j_*(G|_{X\setminus Z})$, whose kernel is tors$_Z(G)$. Thus $\mathcal{H}om_X(G, F)$ $=$ $j_*(\mathcal{H}om_X(G, F)|_{X\setminus Z})$, hence $\mathcal{H}om_X(G, F)$ is S_2. □

An important property of S_2 sheaves is the following, which can be obtained by combining Hartshorne (1977, III.7.3 and III.12.11).

Proposition 10.9 (Enriques–Severi–Zariski lemma) *Let f: $X \to S$ be a projective morphism and F a coherent sheaf on X that is flat over S, with S_2 fibers of pure dimension $\geq$ 2. Then $f_*F(-m) = R^1f_*F(-m) = 0$ for $m \gg 1$.*

Therefore, if $H \in |\mathscr{O}_X(m)|$ does not contain any of the associated points of F, then the restriction map $f_*F \to (f|_H)_*(F|_H)$ is an isomorphism. □

10.10 (Depth and flatness) Let $p\colon Y \to X$ be a morphism and G a coherent sheaf on Y that is flat over X. It is easy to see that for any point $y \in Y$ we have

$$\operatorname{depth}_y G = \operatorname{depth}_{p(y)} X + \operatorname{depth}_y G_{p(y)}. \tag{10.10.1}$$

Similarly, if $p\colon Y \to X$ is flat and F is a coherent sheaf on X, then

$$\operatorname{depth}_y p^*F = \operatorname{depth}_{p(y)} F + \operatorname{depth}_y Y_{p(y)}. \tag{10.10.2}$$

In particular, if $p\colon Y \to X$ is flat with S_m fibers and F is a quasi-coherent S_m sheaf on X then p^*F is also S_m. The converse also holds if p is faithfully flat.

The assumption on the fibers is necessary and a flat pull-back of an S_m sheaf need not be S_m; not even for products. Let X_1, X_2 be k-schemes. Then $X_1 \times X_2$ is S_m iff both of the X_i are S_m.

10.2 Flat Families of S_m Sheaves

We consider how the S_m property (2.72) varies in flat families.

Theorem 10.11 (Grothendieck, 1960, IV.12.1.6) *Let $\pi\colon X \to S$ be a morphism of finite type and F a coherent sheaf on X that is flat over S. Fix $m \in \mathbb{N}$. Then the set of points $\{x \in X\colon F_{\pi(x)}$ is pure and S_m at $x\}$ is open in X.*

This immediately implies the following variant for proper morphisms.

Corollary 10.12 *5 Let $\pi\colon X \to S$ be a proper morphism and F a coherent sheaf on X that is flat over S. Fix $m \in \mathbb{N}$. Then the set of points $\{s \in S\colon F_s$ is pure and $S_m\}$ is open in S.* $\qquad\square$

For nonproper morphisms we get the following.

Corollary 10.13 *Let S be an integral scheme, $\pi\colon X \to S$ a morphism of finite type, and F a coherent sheaf on X. Assume that F is pure and S_m. Then there is a dense open subset $S^\circ \subset S$ such that F_s is pure and S_m for every $s \in S^\circ$.*

Proof Let $Z \subset X$ denote the set of points $x \in X$ such that either F is not flat at x or $F_{\pi(x)}$ is not pure and S_m at x. Note that Z is closed in X by (10.11) and by generic flatness (Eisenbud, 1995, 14.4).

The local rings of the generic fiber of π are also local rings of X, hence the restriction of F to the generic fiber is pure and S_m. Thus Z is disjoint from the generic fiber of π. Therefore $\pi(Z) \subset S$ is a constructible subset that does not contain the generic point, hence $S \setminus \pi(Z)$ contains a dense open subset $S^\circ \subset S$. □

10.14 (Nagata's openness criterion) In many cases, one can check openness of a subset of a scheme using the following easy to prove test, which is sometimes called the *Nagata openness criterion*.

Let X be a Noetherian topological space and $U \subset X$ an arbitrary subset. Then U is open iff the following conditions are satisfied.

(10.14.1) If $x_1 \in \bar{x}_2$ and $x_1 \in U$ then $x_2 \in U$.

(10.14.2) If $x \in U$ then there is a nonempty open $V \subset \bar{x}$ such that $V \subset U$.

Assume now that we want to use this to check openness of a fiber-wise property $\mathcal{P}$ for a morphism $\pi \colon X \to S$.

We start with condition (10.14.1). Pick points $x_1, x_2 \in X$ such that $x_1 \in \bar{x}_2$. Let T be the spectrum of a DVR with closed point $0 \in T$, generic point $t_g \in T$, and $q \colon T \to X$ a morphism such that $q(0) = x_1$ and $q(t_g) = x_2$. After base change using $\pi \circ q$ we get $Y \to T$. Usually one cannot guarantee that the residue fields are unchanged under q. However, if property $\mathcal{P}$ is invariant under field extensions, then it is enough to check (10.14.1) for $Y \to T$. Thus we may assume that S is the spectrum of a DVR.

As for (10.14.2), we can replace S by the closure of $\pi(x)$. Then $\pi(x)$ is the generic point of S and then we may assume that S is regular.

We can summarize these considerations in the following form.

Proposition 10.15 (Openness criterion) *Let $\mathcal{P}$ be a property defined for coherent sheaves on schemes over fields. Assume that $\mathcal{P}$ is invariant under base field extensions. The following are equivalent.*

(10.15.1) *Let $\pi \colon X \to S$ be a morphism of finite type and F a coherent sheaf on X that is flat over S. Then $\{x \in X \colon F_{\pi(x)} \text{ satisfies property } \mathcal{P} \text{ at } x\}$ is open in X.*

(10.15.2) *The following hold, where $\sigma \colon S \to X$ denotes a section.*

  (a) If S is the spectrum of a DVR with closed point 0, generic point g and $\mathcal{P}$ holds for $\sigma(0) \in X_0$, then $\mathcal{P}$ holds for $\sigma(g) \in X_g$.

  (b) If S is the spectrum of a regular ring with generic point g and $\mathcal{P}$ holds for $\sigma(g) \in X_g$, then $\mathcal{P}$ holds in a nonempty open $U \subset \sigma(S)$. □

10.16 (Proof of 10.11) By (10.15), we may assume that S is affine and regular. We may also assume that π is affine and $X = \text{Supp } F$.

First, we check (10.15.2.a) for $m = 1$. (Note that pure and S_1 is equivalent to pure (10.1).) Let $W \subset X$ be the closure of an associated prime of F. Then the irreducible components of $W \cap X_0$ are associated primes of F_0 by (10.22).

Since F_0 is pure, $W \cap X_0$ is an irreducible component of Supp F_0. Hence W is an irreducible component of Supp F. Thus F_g is also pure.

Next we check (10.15.2.a) for $m > 1$. Since S_m implies S_1, we already know that every fiber of F is pure. By (10.17) there is a subset $Z \subset X$ of codimension ≥ 2 such that F is CM over $X \setminus Z$. Let $Z \subset H \subset X$ be a Cartier divisor that does not contain any of the associated primes of F_0. Then $F|_H$ is flat over S and $(F|_H)_0 = F_0|_H$ is pure and S_{m-1}. Thus, by induction, $F|_H$ is pure and S_{m-1} on the generic fiber, hence F_{s_g} is pure and S_m along H. It is even CM on $X \setminus H$, hence F_{s_g} is pure and S_m.

For (10.15.2.b) we start with $m = 1$. We may assume that F_{s_g} is pure. By Noether normalization, after passing to some open subset of S, there is a finite surjection $p: X \to \mathbb{A}^n_S$ for some n. Note that $p_* F$ is flat over S and it is pure on the generic fiber by (9.2), hence torsion-free. Using (9.2) in the reverse direction for the other fibers, we are reduced to the case when $X = \mathbb{A}^n_S$ and F is torsion-free at $x := \sigma(g)$ on the generic fiber. Thus there is an injection of the localizations $F_x \hookrightarrow \mathscr{O}^m_{x,X}$. By generic flatness (Eisenbud, 1995, 14.4), the quotient $\mathscr{O}^m_{x,X}/F_x$ is flat over an open, dense subset $S^\circ \subset S$. Thus if $s \in S^\circ$ then we have an injection $F|_U \hookrightarrow \mathscr{O}^m_U$. Thus every fiber F_s is torsion-free over $U \cap \pi^{-1}(S^\circ)$. For $m > 1$, we follow the same argument as above using $Z \subset H \subset X$ and induction.   □

Lemma 10.17 *Let $\pi: X \to S$ be a morphism of finite type and F a coherent sheaf on X that is flat over S. Assume that Supp F is pure-dimensional over S. As in (7.26), let* $\mathrm{FlatCM}_S(X, F) \subset X$ *be the set of points x such that $F_{\pi(x)}$ is CM at x. Then, for every $s \in S$,*

(10.17.1) Supp $F_s \cap \mathrm{FlatCM}_S(X, F)$ *is dense in* Supp F_s, *and,*

(10.17.2) *if F_s is pure, then its complement has codimension ≥ 2 in* Supp F_s.

Proof We may assume that π is affine and $X = $ Supp F. By (10.49), after replacing X with an étale neighborhood of x, there is a finite surjection $g: X \to Y$ where $\tau: Y \to S$ is smooth.

Since $g_* F$ is flat over S, it is locally free at a point $y \in Y$ iff the restriction of $g_* F$ to the fiber $Y_{\tau(y)}$ is locally free at y. The latter holds outside a codimension ≥ 1 subset of each fiber Y_s. If F is pure then $g_* F$ is torsion-free on each fiber, so local freeness holds outside a subset of codimension ≥ 2.   □

Let F be a coherent, S_m sheaf on $\mathbb{P}^n$. If a hyperplane $H \subset \mathbb{P}^n$ does not contain any of the irreducible components of Supp F then $F|_H$ is S_{m-1}, essentially by definition. The following result says that $F|_H$ is even S_m for general hyperplanes, though we cannot be very explicit about the meaning of "general."

Corollary 10.18 (Bertini theorem for S_m) *Let F be a coherent, pure, S_m sheaf on a finite type k-scheme and $|V|$ a base point free linear system on X. Then there is a dense, open $U \subset |V|$ such that $F|_H$ is also pure and S_m for $H \in U$.*

Proof Let $Y \subset X \times |V|$ be the incidence correspondence (that is, the set of pairs (point $\in H$) with projections π and $\check{\pi}$). Note that π is a $\mathbb{P}^{n-1}$-bundle for $n = \dim |V|$, thus $\pi^* F$ is also pure and S_m by (10.10).

By (10.13) there is a dense open subset $U \subset |V|$ such that $F|_H$ is also pure and S_m for $H \in U$. For a divisor H, the restriction $F|_H$ is isomorphic to the restriction of $\pi^* F$ to the fiber of $\check{\pi}$ over $H \in |V|$. □

Corollary 10.19 (Bertini theorem for hulls) *Let $|V|$ be a base point free linear system on a finite type k-scheme X. Let F be a coherent sheaf on X with hull $q \colon F \to F^{[**]}$. Then there is a dense, open subset $U \subset |V|$ such that*

$$(F^{[**]})|_H = (F|_H)^{[**]} \quad \text{for } H \in U.$$

Proof By definition we have an exact sequence

$$0 \to K \to F \to F^{[**]} \to Q \to 0,$$

where $\dim K \le n - 1$ and $\dim Q \le n - 2$. If $H \in |V|$ is general, then the restriction stays exact

$$0 \to K|_H \to F|_H \to (F^{[**]})|_H \to Q|_H \to 0,$$

$\dim K|_H \le n - 2$ and $\dim Q|_H \le n - 3$. Thus $(F^{[**]})|_H = (F|_H)^{[**]})$. □

Corollary 10.20 (Bertini theorem for S_m in families) *Let T be the spectrum of a local ring, $X \subset \mathbb{P}^n_T$ a quasi-projective scheme and F a coherent sheaf on X that is flat over T with pure, S_m fibers.*

Assume that either X is projective over T or $\dim T \le 1$. Then $F|_{H \cap X}$ is also flat over T with pure and S_m fibers for a general hyperplane $H \subset \mathbb{P}^n_T$.

Proof The hyperplanes correspond to sections of $\check{\mathbb{P}}^n_T \to T$. If X is projective over T then we use (10.18) for the special fiber X_0 and conclude using (10.12).

If $\dim T = 1$ then we use (10.18) both for the special fiber X_0 and the generic fibers X_{g_i}. We get open subsets $U_0 \subset \check{\mathbb{P}}^n_0$ and $U_{g_i} \subset \check{\mathbb{P}}^n_{g_i}$. Let $W_i \subset \check{\mathbb{P}}^n_T$ denote the closure of $\check{\mathbb{P}}^n_{g_i} \setminus U_{g_i}$. For dimension reasons, W_i does not contain $\check{\mathbb{P}}^n_0$. Thus any hyperplane corresponding to a section through a point of $U_0 \setminus (\cup_i W_i)$ works. □

Example 10.21 If dim $T \geq 2$ then (10.20) does not hold for nonproper maps. Here is a similar example for the classical Bertini theorem on smoothness. Set

$$X := (x^2 + y^2 + z^2 = s) \setminus (x = y = z = s = 0) \subset \mathbb{A}^3_{xyz} \times \mathbb{A}^2_{st}$$

with smooth second projection $f \colon X \to \mathbb{A}^2_{st}$. Over the origin we start with the hyperplane $H_{00} := (x = 0)$, it is a typical member of the base point free linear system $|ax + by + cz = 0|$.

A general deformation of it is given by $H_{st} := x + b(s, t)y + c(s, t)z = d(s, t)$. It is easy to compute that the intersection $H_{st} \cap X_{st}$ is singular iff $s(1 + b^2 + c^2) = d^2$. This equation describes a curve in $\mathbb{A}^2_{st}$ that passes through the origin.

10.22 (Associated points of restrictions) Let X be a scheme, $D \subset X$ a Cartier divisor and F a coherent sheaf on X. We aim to compare $\mathrm{Ass}(F)$ and $\mathrm{Ass}(F|_D)$. If D does not contain any of the associated points of G then $\mathrm{Tor}^1(G, \mathscr{O}_D) = 0$. Thus if $0 = F_0 \subset \cdots \subset F_r = F$ is a filtration of F by subsheaves and D does not contain any of the associated points of F_i/F_{i-1} then $0 = F_0|_D \subset \cdots \subset F_r|_D = F|_D$ is a filtration of $F|_D$ and $F_i|_D/F_{i-1}|_D \simeq (F_i/F_{i-1})|_D$. We can also choose any of the associated points of F to be an associated point of F_1, proving the following.

Claim 10.22.1 If D does not contain any of the associated points of F, then
   (a) $\mathrm{Ass}(F|_D) \subset \cup_i \mathrm{Ass}((F_i/F_{i-1})|_D)$ and
   (b) for every $x \in \mathrm{Ass}(F)$, every generic point of $D \cap \bar{x}$ is in $\mathrm{Ass}(F|_D)$.   □

By (10.25), we can choose the F_i such that $\mathrm{Ass}(F_i/F_{i-1})$ is a single associated point of F for every i. Thus it remains to understand $\mathrm{Ass}(G|_D)$ when G is pure. Let $G^{[**]} \supset G$ denote the hull of G and set $Q := G^{[**]}/G$. As we have noted, if D does not contain any of the associated points of Q then $G^{[**]}|_D \supset G|_D$, thus $\mathrm{Ass}(G^{[**]}|_D) = \mathrm{Ass}(G|_D)$. Finally, since $G^{[**]}$ is S_2, the restriction $G^{[**]}|_D$ is S_1, hence its associated points are exactly the generic points of $D \cap \mathrm{Supp}\, G$. We have thus proved the following.

Claim 10.22.2 Let $D \subset X$ be a Cartier divisor that contains neither an associated point of F nor an associated point of $(F_i/F_{i-1})^{[**]}/(F_i/F_{i-1})$. Then
   (a) the associated points of $F|_D$ are exactly the generic points of $D \cap \bar{x}$ for all $x \in \mathrm{Ass}(F)$, and
   (b) $(F/\mathrm{emb}(F))|_D \simeq (F|_D)/(\mathrm{emb}(F|_D))$.   □

Note that the associated points of $(F_i/F_{i-1})^{[**]}/(F_i/F_{i-1})$ depend on the choice of the F_i, they are not determined by F. For the Claim to hold, it is enough to take the intersection of all possible sets. This set is still hard to determine, but in many applications the key point is that, as long as X is excellent, we need D to avoid only a finite set of points.

The next result describes how the associated points of fibers of a flat sheaf fit together. The proof is a refinement of the arguments used in (10.16).

Theorem 10.23 *Let $f: X \to S$ be a morphism of finite type and F a coherent sheaf on X. Then the following hold.*

(10.23.1) *There are finitely many locally closed $W_i \subset X$ such that $\mathrm{Ass}(F_s)$ equals the set of generic points of the $(W_i)_s$ for every $s \in S$.*

(10.23.2) *If F is flat over S then we can choose the W_i to be closed and such that each $f|_{W_i}: W_i \to f(W_i)$ is equidimensional.*

Proof Using Noetherian induction it is enough to prove that (1) holds over a non-empty open subset of red S. We may thus assume that S is integral with generic point $g \in S$.

Assume first that X is integral and F is torsion-free. By Noether normalization, after again passing to some non-empty open subset of S there is a finite surjection $p: X \to \mathbb{A}_S^m$. Then p_*F is torsion-free of generic rank say r, hence there is an injection $j: p_*F \hookrightarrow \mathscr{O}_{\mathbb{A}_S^m}^r$. After again passing to some non-empty open subset we may assume that coker(j) is flat over S, thus

$$j_s: p_*(F_s) = (p_*F)_s \hookrightarrow \mathscr{O}_{\mathbb{A}_S^m}^r$$

is an injection for every $s \in S$. Thus each F_s is torsion-free and its associated points are exactly the generic points of the fiber X_s.

In general, we use (10.25) for the generic fiber and then extend the resulting filtration to X. Thus, after replacing S by a nonempty open subset if necessary, we may assume that there is a filtration $0 = F^0 \subset \cdots \subset F^n = F$ such that each F^{m+1}/F^m is a coherent, torsion-free sheaf over some integral subscheme $W_m \subset X$ and $W_{m_1} \not\subset W_{m_2}$ for $m_1 > m_2$. As we proved, we may assume that the associated points of each $(F^{m+1}/F^m)_s$ are exactly the generic points of the fiber $(W_m)_s$. Using generic flatness, we may also assume that each F^{m+1}/F^m is flat over S and, after further shrinking S, none of the generic points of $(W_{m_1})_s$ are contained in $(W_{m_2})_s$ for $m_1 > m_2$. Then the associated points of each F_s are exactly the generic points of the fibers $(W_m)_s$ for every m. This proves (1).

In order to see (2), consider first the case when the base $(0 \in T)$ is the spectrum of a DVR. The filtration given by (10.25) for the generic fiber extends to a filtration $0 = F^0 \subset \cdots \subset F^n = F$ over X giving closed integral subschemes $W_m \subset X$. Since T is the spectrum of a DVR, the F^{m+1}/F^m are flat over T, hence the associated points of F_0 are exactly the generic points of the fibers $(W_m)_0$ for every m.

To prove (2) in general, we take the $W_i \subset X$ obtained in (1) and replace them by their closures. A possible problem arises if $f|_{W_i} \colon W_i \to f(W_i)$ is not equidimensional. Assume that $W_i \to f(W_i)$ has generic fiber dimension d and let $(W_i)_s$ be a special fiber. Pick any closed point $x \in (W_i)_s$ and the spectrum of a DVR ($0 \in T$) mapping to W_i such that the special point of T maps to $f(x)$ and the generic point of T to the generic point of $f(W_i)$. After base change to T we see that F_s has a d-dimensional associated subscheme containing x. Thus $(W_i)_s$ is covered by d-dimensional associated subschemes of F_s. Since F_s is coherent, this is only possible if $\dim(W_i)_s = d$ and every generic point of the $(W_i)_s$ is an associated point of F_s.   □

10.24 (Semicontinuity and depth) Let X be a scheme and F a coherent sheaf on X. As we noted in (10.3), the function $x \mapsto \operatorname{depth}_x F$ is not lower semicontinuous. This is, however, caused by the non-closed points. A quick way to see this is the following.

Assume that X is regular and let $0 \in X$ be a closed point. By the Auslander–Buchsbaum formula as in Eisenbud (1995, 19.9), F_0 has a projective resolution of length $\dim X - \operatorname{depth}_0 F$. Thus there is an open subset $0 \in U \subset X$ such that $F|_U$ has a projective resolution of length $\dim X - \operatorname{depth}_0 F$. This shows that

$$\operatorname{depth}_x F \geq \operatorname{depth}_0 F - \dim \bar{x} \quad \forall x \in U. \tag{10.24.1}$$

That is, $x \mapsto \operatorname{depth}_x F$ is lower semicontinuous for closed points. In general, we have the following analog of (10.11).

Proposition 10.24.2 Let $\pi \colon X \to S$ be a morphism of finite type and F a coherent sheaf on X that is flat over S with pure fibers. Let $0 \in X$ be a closed point. Then there is an open subset $0 \in U \subset X$ such that

$$\operatorname{depth}_x F_{\pi(x)} \geq \operatorname{depth}_0 F_{\pi(0)} - \operatorname{tr-deg}_{k(\pi(x))} k(x) \quad \forall x \in U,$$

where $F_{\pi(x)}$ is the restriction of F to the fiber $X_{\pi(x)}$ and tr-deg denotes the transcendence degree. Hence $x \mapsto \operatorname{depth}_x F_{\pi(x)}$ is lower semicontinuous on closed points.

Proof Using Noether normalization and (10.17.1) as in (10.16), we can reduce to the case when $X = \mathbb{A}^n_S$ for some n. Next we take a projective resolution of the fiber $F_{\pi(0)}$ and lift it to a suitable neighborhood $0 \in U \subset X$ using the flatness of F.   □

Dévissage is a method that writes a coherent sheaf as an extension of simpler coherent sheaves and uses these to prove various theorems. There are many ways to do this, and different ones are useful in different contexts; see Stacks (2022, tag 07UN) for some of them.

Recall that $\text{Ass}(F)$ denotes the set of associated points of a sheaf F (10.1) and that a sheaf is S_1 iff it has no embedded points (10.5). As in (10.1), $\text{tors}_Z(F) \subset F$ is the largest subsheaf whose support is contained in Z.

Lemma 10.25 (Dévissage) *Let X be a Noetherian scheme, F a coherent sheaf on X. Write $\text{Ass}(F) = \{w_i : i = 1, \ldots, m\}$ in some fixed order and let W_i be the closure of w_i. Assume that $W_j \not\subset W_i$ for $i < j$. Then the following hold.*

(10.25.1) *There is a unique filtration $0 = G_0 \subset G_1 \subset \cdots \subset G_m = F$ such that each G_i/G_{i-1} is a torsion-free sheaf supported on W_i. Moreover, the natural map $\text{tors}_{W_i}(F) \to G_i/G_{i-1}$ is an isomorphism at w_i.*

(10.25.2) *There is a non-unique refinement $G_i = G_{i,0} \subset G_{i,1} \subset \cdots \subset G_{i,r_i} = G_{i+1}$ such that each $G_{i,j+1}/G_{i,j}$ is a rank 1, torsion-free sheaf over $\text{red } W_i$.*

Proof It is easy to see that we must set $G_1 = \text{tors}_{W_1}(F)$. Then pass to F/G_1 and use induction on the number of associated points to get (1).

For (2), any filtration of $(G_{i+1}/G_i)_{w_i}$ whose successive quotients are $k(w_i)$ extends uniquely to the required $G_{i,j}$. □

10.3 Cohomology over Non-proper Schemes

The cohomology theory of coherent sheaves is trivial over affine schemes and well understood over proper schemes. If X is a scheme and $j : U \hookrightarrow X$ is an open subscheme then one can study the cohomology theory of coherent sheaves on U by understanding the cohomology theory of quasi-coherent sheaves on X and the higher direct image functors $R^i j_*$. The key results are (10.26) and (10.30); see Grothendieck (1960, IV.5.11.1).

Proposition 10.26 *Let X be an excellent scheme, $Z \subset X$ a closed subscheme and $U := X \setminus Z$ with injection $j : U \hookrightarrow X$. Let G be a coherent sheaf on U. Then j_*G is coherent iff $\text{codim}_W(Z \cap W) \geq 2$, whenever $W \subset X$ is the closure of an associated point w of G.*

The case of arbitrary Noetherian schemes is discussed in Kollár (2017).

Proof This is a local question, hence we may assume that X is affine. By (10.25) G has a filtration $0 = G_0 \subset \cdots \subset G_r = G$ such that each G_{m+1}/G_m is isomorphic to a subsheaf of some $\mathcal{O}_{W \cap U}$ where w is an associated prime of G. Since j_* is left exact, it is enough to show that each $j_*\mathcal{O}_{W \cap U}$ is coherent.

Ancillary Results

Let $p \colon V \to W$ be the normalization. Since X is excellent, p is finite. $\mathscr{O}_V$ is S_2 (by Serre's criterion) and so is $p_* \mathscr{O}_V$ by (9.2). Thus

$$j_* \mathscr{O}_{W \cap U} \subset j_*(p_* \mathscr{O}_V|_U) = p_* \mathscr{O}_V,$$

where the equality follows from (10.6) using $\mathrm{codim}_W(Z \cap W) \geq 2$. Thus $j_* \mathscr{O}_{W \cap U}$ is coherent. $\qquad\square$

It is frequently quite useful to know that coherent sheaves are "nice" over large open subsets. For finite type schemes this was established in (10.17).

Proposition 10.27 *Let X be a Noetherian scheme. Assume that every integral subscheme $W \subset X$ has an open dense subscheme $W^\circ \subset W$ that is regular, or at least CM. (For example, X is excellent.) Let F be a coherent sheaf on X.*

(10.27.1) *There is a closed subset $Z_1 \subset \mathrm{Supp}\, F$ of codimension ≥ 1 such that F is CM on $X \setminus Z_1$.*

(10.27.2) *If F is S_1 then there is a closed subset $Z_2 \subset \mathrm{Supp}\, F$ of codimension ≥ 2 such that F is CM on $X \setminus Z_2$.*

Proof We put the intersections of different irreducible components of $\mathrm{Supp}\, F$ into Z_1. Since (1) is a local question, we may thus assume that $\mathrm{Supp}\, F$ is irreducible. Since an extension of CM sheaves of the same dimensional support is CM (10.28), using (10.25) we may assume that F is torsion-free over an integral subscheme $W \subset X$. Then F is locally free over a dense open subset $W^\circ \subset W$ and we can take $Z_1 := W \setminus W^*$, where W^* is the regular locus of W°.

In order to prove (2), we may assume that X is affine. Let $s = 0$ be a local equation of Z_1. We apply the first part to F/sF to obtain a closed subset $Z_2 \subset \mathrm{Supp}(F/sF)$ of codimension ≥ 1 such that F/sF is CM on $X \setminus Z_2$. Thus F is CM on $X \setminus Z_2$. $\qquad\square$

The next lemma is quite straightforward; see Kollár (2013b, 2.60).

Lemma 10.28 *Let X be a scheme and $0 \to F' \to F \to F'' \to 0$ a sequence of coherent sheaves on X that is exact at $x \in X$.*

(10.28.1) *If $\mathrm{depth}_x F \geq r$ and $\mathrm{depth}_x F'' \geq r - 1$ then $\mathrm{depth}_x F' \geq r$.*

(10.28.2) *If $\mathrm{depth}_x F \geq r$ and $\mathrm{depth}_x F' \geq r - 1$ then $\mathrm{depth}_x F'' \geq r - 1$.* $\qquad\square$

10.29 (Cohomology over quasi-affine schemes) Grothendieck (1967)

Let X be an affine scheme, $Z \subset X$ a closed subscheme and $U := X \setminus Z$. Here our primary interest is in the case when $Z = \{x\}$ is a closed point.

For a quasi-coherent sheaf F on X, let $H_Z^0(X, F)$ denote the space of global sections whose support is in Z. There is a natural exact sequence

$$0 \to H_Z^0(X, F) \to H^0(X, F) \to H^0(U, F|_U).$$

This induces a long exact sequence of the corresponding higher cohomology groups. Since X is affine, $H^i(X, F) = 0$ for $i > 0$, hence the long exact sequence breaks up into a shorter exact sequence

$$0 \to H^0_Z(X, F) \to H^0(X, F) \to H^0(U, F|_U) \to H^1_Z(X, F) \to 0 \qquad (10.29.1)$$

and a collection of isomorphisms

$$H^i(U, F|_U) \simeq H^{i+1}_Z(X, F) \quad \text{for } i \geq 1. \qquad (10.29.2)$$

The vanishing of the local cohomology groups is closely related to the depth of the sheaf F. Two instances of this follow from already established results. First, for coherent sheaves (10.5) can be restated as

$$H^0_Z(X, F) = 0 \iff \operatorname{depth}_Z F \geq 1. \qquad (10.29.3)$$

Second, (10.6) tells us when the map $H^0(X, F) \to H^0(U, F|_U)$ in (10.29.1) is an isomorphism. This implies that, for coherent sheaves,

$$H^0_Z(X, F) = H^1_Z(X, F) = 0 \iff \operatorname{depth}_Z F \geq 2. \qquad (10.29.4)$$

More generally, Grothendieck's vanishing theorem (see Grothendieck (1967, sec.3) or Bruns and Herzog (1993, 3.5.7)) says that

$$\operatorname{depth}_Z F = \min\{i : H^i_Z(X, F) \neq 0\}. \qquad (10.29.5)$$

Combined with (10.29.2–3), this shows that

$$H^i(U, F|_U) = 0 \quad \text{for } 1 \leq i \leq \operatorname{depth}_Z F - 2. \qquad (10.29.6)$$

All these cohomology groups are naturally modules over $H^0(X, \mathcal{O}_X)$ and we need to understand when they are finitely generated.

More generally, let G be a coherent sheaf on U. When is the group $H^i(U, G)$ a finite $H^0(X, \mathcal{O}_X)$-module? Since X is affine, $H^i(U, G) = H^0(X, R^i j_* G)$, where $j \colon U \hookrightarrow X$ denotes the natural open embedding. Thus $H^i(U, G)$ is a finite $H^0(X, \mathcal{O}_X)$-module iff $R^i j_* G$ is a coherent sheaf. For $i \geq 1$, the sheaves $R^i j_* G$ are supported on Z, which implies the following.

Claim 10.29.7 Assume that $i \geq 1$. Then every associated prime of $H^i(U, G)$ (viewed as an $H^0(X, \mathcal{O}_X)$-module) is contained in Z, and, if $Z = \{x\}$, then $H^i(U, G)$ is a finite $H^0(X, \mathcal{O}_X)$-module iff $H^i(U, G)$ has finite length. □

The general finiteness condition is stated in (10.30); but first we work out the special cases that we use. We start with $H^0(U, G)$; here we have the following restatement of (10.26).

Claim 10.29.8 Let X be an excellent, affine scheme, $Z \subset X$ a closed subscheme, $U := X \setminus Z$, and G a coherent sheaf on U. Assume, in addition, that $Z \cap \bar{W}_i$

has codimension ≥ 2 in $\bar{W}_i$ for every associated prime $W_i \subset U$ of G. Then $H^0(U,G)$ is a finite $H^0(X, \mathcal{O}_X)$-module. □

It is considerably harder to understand finiteness for $H^1(U,G)$. The following special case is used in Section 5.4.

Claim 10.29.9 Let X be an excellent scheme, $Z \subset X$ a closed subscheme, $U := X \setminus Z$, and G a coherent sheaf on U. Assume in addition that G is S_2, there is a coherent CM sheaf F on X and an injection $G \hookrightarrow F|_U$, and Z has codimension ≥ 3 in $\mathrm{Supp}\,F$. Then $R^1 j_* G$ is coherent.

Proof Set $Q = F|_U / G$. Since G is S_2, it has no extensions with a sheaf whose support has codimension ≥ 2 by (10.6), thus every associated prime of Q has codimension ≤ 1 in $\mathrm{Supp}\,F$. Thus Q satisfies the assumptions of (10.26) and so $j_* Q$ is coherent. By (10.29.4) $R^1 j_*(F|_U) = 0$, hence the exact sequence

$$0 \to j_* G \to j_*(F|_U) \to j_* Q \to R^1 j_* G \to R^1 j_*(F|_U) = 0$$

shows that $R^1 j_* G$ is coherent. □

Not every S_2-sheaf can be realized as a subsheaf of a CM sheaf, but this can be arranged in some important cases.

Claim 10.29.10 Assume in addition that X is embeddable into a regular, affine scheme R as a closed subscheme, $\mathrm{Supp}\,G$ has pure dimension $n \geq 3$, $Z = \{x\}$ is a closed point, and G is S_2.

Then $H^1(U,G)$ has finite length. Thus, if X is of finite type over a field k, then $H^1(U,G)$ is a finite dimensional k-vector space.

Outline of proof X plays essentially no role. Let $Y \subset R$ be a complete intersection subscheme defined by $\dim R - n$ elements of $\mathrm{Ann}\,G$. Then Y is Gorenstein, we can view G as a coherent sheaf on $Y \setminus \{x\}$, and $H^i(X \setminus \{x\}, G) = H^i(Y \setminus \{x\}, G)$. Thus it is enough to prove vanishing of the latter for $i = 1$. By (10.29.11) there is an embedding $G \hookrightarrow \mathcal{O}^m_{Y \setminus \{x\}}$, hence (10.29.9) applies. □

Claim 10.29.11 Let U be a quasi-affine scheme of pure dimension n and G a pure, coherent sheaf on U of dimension n. Assume that either U is reduced, or U is Gorenstein at its generic points.

Then G is isomorphic to a subsheaf of $\mathcal{O}^m_U$ for some m.

Outline of proof Assume that such an embedding exists at the generic points. Then we have an embedding $G \hookrightarrow \mathcal{O}^m_U$ over some dense open set $U^\circ \subset U$. Pick $s \in \mathcal{O}_U$ invertible at the generic points and vanishing along $U \setminus U^\circ$. Multiplying by s^r for $r \gg 1$ gives the embedding $G \hookrightarrow \mathcal{O}^m_U$.

The remaining question is, what happens at the generic point. The existence of the embedding is clear if U is reduced.

In general, we are reduced to the following algebra question: given an Artinian ring A, is every finite A-module M a submodule of A^m for some m? Usually, the answer is no. However, local duality theory (see, for instance, Eisenbud (1995, secs.21.1–2)) shows that every finite A-module is a submodule of ω_A^m for some m. Finally, A is Gorenstein iff $A \simeq \omega_A$. □

Much of the next result can be proved using these methods, but local duality theory works better, as in Grothendieck (1968, VIII.2.3).

Theorem 10.30 *Let X be an excellent scheme, $Z \subset X$ a closed subscheme, $U := X \setminus Z$, and $j\colon U \hookrightarrow X$ the open embedding. Assume in addition that X is locally embeddable into a regular scheme. For a coherent sheaf G on U and $n \in \mathbb{N}$ the following are equivalent.*

(10.30.1) *$R^i j_* G$ is coherent for $i < n$.*

(10.30.2) *$\operatorname{depth}_u G \geq n$ for every point $u \in U$ such that $\operatorname{codim}_{\bar{u}}(Z \cap \bar{u}) = 1$.* □

10.4 Volumes and Intersection Numbers

We have used several general results that compare intersection numbers and volumes under birational morphisms.

Definition 10.31 (Lazarsfeld, 2004, sec.2.2.C) Let X be a proper scheme of dimension n over a field and D a Mumford $\mathbb{R}$-divisor on X. Its *volume* is

$$\operatorname{vol}(D) := \lim_{m \to \infty} \frac{h^0(X, \mathscr{O}_X(\lfloor mD \rfloor))}{m^n/n!}.$$

Numerically equivalent divisors have the same volume, and, for $D = \sum d_i D_i$, the volume is a continuous function of the d_i; see Lazarsfeld (2004, 2.2.41–44). If D is nef then $\operatorname{vol}(D) = (D^n)$ (11.52).

Proposition 10.32 *Let $p\colon Y \to X$ be a birational morphism of normal, proper varieties of dimension n. Let D_Y be a p-nef $\mathbb{R}$-Cartier $\mathbb{R}$-divisor such that $D_X := p_*(D_Y)$ is also $\mathbb{R}$-Cartier. Then*

(10.32.1) *$\operatorname{vol}(D_X) \geq \operatorname{vol}(D_Y)$, and*

(10.32.2) *if D_X is ample then equality holds iff $D_Y \sim_{\mathbb{R}} p^* D_X$.*

Furthermore, let H be an ample divisor on X. Then

(10.32.3) *$I(H, D_X) \geq I(p^*H, D_Y)$ (with $I(*, *)$ as in (5.13)), and*

(10.32.4) *equality holds iff $D_Y \sim_{\mathbb{R}} p^* D_X$.*

Proof Write $D_Y = p^* D_X - E$ where E is p-exceptional. By assumption $-E$ is p-nef, hence E is effective by (11.60). Thus $\mathrm{vol}(D_X) = \mathrm{vol}(p^* D_X) \geq \mathrm{vol}(D_Y)$, proving (1). Parts (2) and (4) are special cases of (10.39), but here is a more direct argument.

Set $r = \dim(p(\mathrm{Supp}\, E))$. For any $\mathbb{R}$-Cartier divisors A_i on X, the intersection number $(p^* A_1 \cdots p^* A_j \cdot E)$ vanishes whenever $j > r$. Thus, if $j > r$ then

$$(p^* H^j \cdot D_Y^{n-j}) = (p^* H^j \cdot (p^* D_X - E)^{n-j}) = (p^* H^j \cdot p^* D_X^{n-j}) = (H^j \cdot D_X^{n-j}),$$

and for $j = r$ we get that

$$(p^* H^r \cdot D_Y^{n-r}) = (H^r \cdot D_X^{n-r}) + (p^* H^r \cdot (-E)^{n-r}).$$

Thus we need to understand $(p^* H^r \cdot (-E)^{n-r})$. We may assume that H is very ample. Intersecting with $p^* H$ is then equivalent to restricting to the preimage of a general member of $|H|$. Using this r-times (and normalizing if necessary), we get a birational morphism $p' : Y' \to X'$ between normal varieties of dimension $n - r$ and an effective, nonzero, p-exceptional $\mathbb{R}$-Cartier $\mathbb{R}$-divisor E' such that $-E'$ is p'-nef and $p'(E_i')$ is 0-dimensional. Thus, by (10.33), $(p^* H^r \cdot (-E)^{n-r}) = (-E')^{n-r} < 0$ which proves (3–4).

If D_X is ample then we can use this for $H := D_X$. Then $(H^r \cdot D_X^{n-r}) = (D_X^n)$ and we get (2). □

Lemma 10.33 *Let* $p \colon Y \to X$ *be a proper, birational morphism of normal schemes. Let* E *be an effective, nonzero, p-exceptional $\mathbb{R}$-Cartier $\mathbb{R}$-divisor such that* $p(E)$ *is 0-dimensional and* $-E$ *is p-nef. Set* $n = \dim E$.
Then $-(-E)^{n+1} = (-E|_E)^n > 0$.

Proof Assume that there is an effective, nonzero, p-exceptional $\mathbb{R}$-Cartier $\mathbb{R}$-divisor F such that $p(F) = p(E)$, $-F$ is p-nef and $-(-F)^{n+1} > 0$. Note that E, F have the same support, namely $p^{-1}(p(E))$, thus $E - \varepsilon F$ is effective for $0 < \varepsilon \ll 1$. Thus $-(-E)^n \geq -(-\varepsilon F)^n$ by (10.34) applied to $N_2 = -E, N_1 = -\varepsilon F$.

Such a divisor F exists on the normalization of the blow-up $B_{p(E)} X$. Let $Z \to X$ be a proper, birational morphism that dominates both Y and $B_{p(E)} X$. We can apply the above observation to the pull-backs of E and F to Z. □

Lemma 10.34 *Let* N_1, N_2 *be $\mathbb{R}$-Cartier divisors with proper support on an* $n + 1$*-dimensional scheme. Assume that there exists an effective divisor with proper support D such that $D \sim_{\mathbb{R}} N_1 - N_2$ and the $N_i|_D$ are both nef. Then* $(N_1^{n+1}) \geq (N_2^{n+1})$.

Proof $(N_1^{n+1}) - (N_2^{n+1}) = D \cdot \sum_{i=0}^{n} N_1^i N_2^{n-i} = \sum_{i=0}^{n} (N_1|_D)^i (N_2|_D)^{n-i}$. $\qquad\square$

The next results compare the volumes of different perturbations of the canonical divisor.

Lemma 10.35 *Let X be a normal, proper variety of dimension n, and D an effective $\mathbb{R}$-divisor such that $K_X + D$ is $\mathbb{R}$-Cartier, nef and big. Let Y be a smooth, proper variety birational to X. Then*

(10.35.1) $\mathrm{vol}(K_Y) \leq (K_X + D)^n$, *and*

(10.35.2) *equality holds iff $D = 0$ and X has canonical singularities.*

Proof Let Z be a normal, proper variety birational to X such that there are morphisms $q: Z \to Y$ and $p: Z \to X$. Write

$$K_Z \sim_{\mathbb{R}} q^* K_Y + E \quad \text{and} \quad K_Z \sim_{\mathbb{R}} p^*(K_X + D) - p_*^{-1} D + F, \qquad (10.35.3)$$

where E is effective, q-exceptional and F is p-exceptional (not necessarily effective). Thus

$$q^* K_Y \sim_{\mathbb{R}} p^*(K_X + D) - p_*^{-1} D + F - E. \qquad (10.35.4)$$

Write $F - E = G^+ - G^-$ where G^+, G^- are effective and without common irreducible components. Note that G^+ is p-exceptional, therefore

$$H^0(Z, \mathscr{O}_Z(\lfloor mp^*(K_X + D) + mG^+ \rfloor)) = H^0(Z, \mathscr{O}_Z(\lfloor mp^*(K_X + D)\rfloor)), \quad \text{so}$$
$$H^0(Z, \mathscr{O}_Z(\lfloor mp^*(K_X + D) - p_*^{-1}(mD) + mG^+ - mG^- \rfloor))$$
$$= H^0(Z, \mathscr{O}_Z(\lfloor mp^*(K_X + D) - p_*^{-1}(mD) - mG^- \rfloor)).$$

This implies that

$$\mathrm{vol}(K_Y) = \mathrm{vol}(p^*(K_X + D) - p_*^{-1} D + G^+ - G^-)$$
$$= \mathrm{vol}(p^*(K_X + D) - p_*^{-1} D - G^-)$$
$$\leq \mathrm{vol}(p^*(K_X + D)) = \mathrm{vol}(K_X + D) = (K_X + D)^n.$$

Furthermore, by (10.39) equality holds iff $p_*^{-1} D + G^- = 0$, that is, when $D = 0$ and $G^- = 0$. In such a case (10.35.4) becomes $q^* K_Y \sim_{\mathbb{R}} p^* K_X + G^+$ and G^+ is effective. Thus $a(E, X) \geq a(E, Y)$ for every divisor E by (11.4.3), hence X has canonical singularities. $\qquad\square$

A similar birational statement does not hold for pairs in general, but a variant holds if Y is a resolution of X. We can also add some other auxiliary divisors; these are needed in our applications.

Lemma 10.36 *Let X be a normal, proper variety of dimension n and Δ a reduced, effective $\mathbb{R}$-divisor on X. Let A be an $\mathbb{R}$-Cartier $\mathbb{R}$-divisor and D an effective $\mathbb{R}$-divisor such that $K_X + \Delta + A + D$ is $\mathbb{R}$-Cartier, nef and big. Let $p\colon Y \to X$ be any log resolution of (X, Δ). Then*

(10.36.1) $\mathrm{vol}(K_Y + p_*^{-1}\Delta + p^*A) \leq (K_X + \Delta + A + D)^n$ *and*

(10.36.2) *equality holds iff $D = 0$ and (X, Δ) is canonical.*

Proof There are p-exceptional, effective divisors F_i such that

$$K_Y + p_*^{-1}\Delta \sim_{\mathbb{R}} p^*(K_X + \Delta + D) - p_*^{-1}D - F_1 + F_2. \qquad (10.36.3)$$

As in (10.35), we get that

$$H^0(Y, \mathscr{O}_Y(\lfloor mp^*(K_X + \Delta + A + D) - p_*^{-1}(mD) - mF_1 + mF_2 \rfloor))$$
$$= H^0(Y, \mathscr{O}_Y(\lfloor mp^*(K_X + \Delta + A + D) - p_*^{-1}(mD) - mF_1 \rfloor)), \quad \text{and}$$

$$\mathrm{vol}(K_Y + p_*^{-1}\Delta + p^*A) = \mathrm{vol}(p^*(K_X + \Delta + A + D) - p_*^{-1}D + F_2 - F_1)$$
$$= \mathrm{vol}(p^*(K_X + \Delta + A + D) - p_*^{-1}D - F_1) \leq \mathrm{vol}(p^*(K_X + \Delta + A + D))$$
$$= \mathrm{vol}(K_X + \Delta + A + D) = (K_X + \Delta + A + D)^n.$$

Furthermore, by (10.39), equality holds iff $p_*^{-1}D + F_1 = 0$, that is, when $D = 0$ and $F_1 = 0$. Thus (10.36.3) becomes $K_Z + p_*^{-1}\Delta \sim_{\mathbb{R}} p^*(K_X + \Delta) + F_2$, where F_2 is effective. This says that (X, Δ) is canonical. $\quad\square$

Essentially the same argument gives the following log canonical version.

Lemma 10.37 *Let X be a normal, proper variety of dimension n, Δ a reduced, effective $\mathbb{R}$-divisor on X and A an $\mathbb{R}$-Cartier $\mathbb{R}$-divisor on X. Let $q\colon \bar{X} \to X$ be a proper birational morphism, $\bar{E}$ the reduced q-exceptional divisor, $\bar{\Delta} := q_*^{-1}\Delta$, and $\bar{D}$ an effective $\mathbb{R}$-divisor on $\bar{X}$ such that $K_{\bar{X}} + \bar{\Delta} + \bar{E} + D + q^*A$ is $\mathbb{R}$-Cartier, nef and big. Let $p\colon Y \to X$ be any log resolution of singularities with reduced exceptional divisor E. Then*

(10.37.1) $\mathrm{vol}(K_Y + p_*^{-1}\Delta + E + p^*A) \leq (K_{\bar{X}} + \bar{\Delta} + \bar{E} + \bar{D} + q^*A)^n$ *and*

(10.37.2) *equality holds iff $\bar{D} = 0$ and $(\bar{X}, \bar{\Delta} + \bar{E})$ is log canonical.* $\quad\square$

We have also used the following elementary estimate.

Lemma 10.38 *Let $p\colon Y \to X$ be a separable, generically finite morphism between smooth, proper varieties. Then $\mathrm{vol}(K_Y) \geq \deg(Y/X) \cdot \mathrm{vol}(K_X)$.*

Proof This is obvious if $\mathrm{vol}(K_X) = 0$, hence we may assume that K_X is big. Pulling back differential forms gives a natural map $p^*\omega_X \to \omega_Y$. This gives an injection $\omega_X^r \otimes p_* \omega_Y \hookrightarrow p_*(\omega_Y^{r+1})$. Since $p_* \omega_Y$ has rank $\deg(Y/X)$ and K_X is big, $H^0(X, \omega_X^r \otimes p_* \omega_Y)$ grows at least as fast as $\deg(Y/X) \cdot H^0(X, \omega_X^r)$. □

The following result describes the variation of the volume near a nef and big divisor. The assertions are special cases of Fulger et al. (2016, thms.A–B).

Theorem 10.39 *Let X be a proper variety, L a big $\mathbb{R}$-Cartier divisor, and E an effective divisor. The following are equivalent.*

(10.39.1) $\mathrm{vol}(L - E) = \mathrm{vol}(L)$*, and*

(10.39.2) $H^0(\mathscr{O}_X(\lfloor mL - mE \rfloor)) = H^0(\mathscr{O}_X(\lfloor mL \rfloor))$ *for every $m \geq 0$.*

If L is nef then these are further equivalent to

(10.39.3) $E = 0$.

Note that (3) $\Rightarrow$ (2) $\Rightarrow$ (1) are clear, but the converse is somewhat surprising. It says that although the volume measures only the asymptotic growth of the Hilbert function, one cannot change the Hilbert function without changing the volume. For proofs, see Fulger et al. (2016, thms.A–B).

10.5 Double Points

We used a variety of results about hypersurface double points. For the rest of the section, we work with rings R that contain $\frac{1}{2}$. In this case, all the definitions that we have seen are equivalent to the ones given below. If $\frac{1}{2} \notin R$, there are differing conventions, especially if $\mathrm{char}\, R = 2$.

The following results on normal forms, deformations, and resolutions of double points are well known, but not easy to find in one place.

Definition 10.40 A *quadratic form* over a field k is a degree 2 homogeneous polynomial $q(x_1, \ldots, x_n) \in k[x_1, \ldots, x_n]$. The *rank* of q is defined either as the dimension of the space spanned by the derivatives $\left\langle \frac{\partial q}{\partial x_1}, \ldots, \frac{\partial q}{\partial x_n} \right\rangle$, or as the rank of $\mathrm{Hess}(q) := \left(\frac{\partial^2 q}{\partial x_i \partial x_j} \right)$, or as the number of variables in any diagonalized form $q = a_1 y_1^2 + \cdots + a_r y_r^2$ where $a_i \in k^\times$. More abstractly, if V is a k-vector space, we can think of q as an element of the symmetric square of its dual $S^2(V^*)$.

Definition 10.41 Let (S, m) be a regular local ring with residue field k such that $\mathrm{char}\, k \neq 2$. We can identify m^2/m^3 with $S^2(m/m^2)$. Thus, for any $g \in m^2$, we can view its image in m^2/m^3 as a quadratic form.

Let Y be a smooth variety over a field of characteristic $\neq 2$ and $X = (g = 0) \subset Y$ a hypersurface. Given a point $p \in X$, we let $\mathrm{rank}_p X$ denote the rank of the image of g in m_p^2/m_p^3.

We say that $p \in X$ is a *double point* if $\text{rank}_p X \geq 1$, a *cA point* if $\text{rank}_p X \geq 2$, and an *ordinary double point* if $\text{rank}_p X = \dim_p X$. An ordinary double point is also called a *node*, especially if $\dim S = 2$.

If $y_1, \ldots, y_n$ are étale coordinates on Y then $\text{Hess}_y(g) = \left(\frac{\partial^2 g}{\partial y_i \partial y_j} \right)$. Since the rank is lower semicontinuous, $\{ p \in \text{Sing } X \colon \text{rank}_p X \geq r \}$ is open in $\text{Sing } X$ for every r. For us the most interesting case is $r = 2$. The relative version is then the following.

Claim 10.41.1 Let $f \colon Y \to S$ be smooth and $X \subset Y$ a relative Cartier divisor. Then $\{ p \in X \colon p \text{ is } cA \text{ (or smooth) on } X_{f(p)} \} \subset X$ is open.　　　　□

This implies that if $X \to S$ is proper and X_s has only cA-singularities (and smooth points) outside a closed subset $Z_s \subset X_s$ of codimension $\geq m$ for some $s \in S$ then the same holds in an open neighborhood $s \in S^\circ \subset S$.

Corollary 10.42 *Let $\pi \colon X \to S$ be a flat and pure dimensional morphism. Then the set of points $\{ x \colon X_{\pi(x)} \text{ is demi-normal at } x \}$ is open in X.*

Proof Being S_2 is an open condition by (10.12). An S_1 scheme is geometrically reduced iff it is generically smooth and smoothness is an open condition. Thus being S_2 and geometrically reduced is an open condition.

It remains to show that having only nodes in codimension 1 is also an open condition. If all residue characteristics are $\neq 2$, this follows from (10.41.3) since having only cA-singularities in codimension 1 is an open condition. See Kollár (2013b, 1.41) for nodes in characteristic 2.　　　　□

Let f be a function on $\mathbb{R}^n$ that has an ordinary critical point at the origin. The Morse lemma says that in suitable local coordinates $y_1, \ldots, y_n$ we can write f as $\pm y_1^2 \pm \cdots \pm y_n^2$; see Milnor (1963, p.6) and Arnol'd et al. (1985, vol.I.sec.6.2) for differentiable and analytic versions. Algebraically, the best is to work with formal power series. We prove a form that also works if $\text{char}(R/m) = 2$.

Lemma 10.43 (Formal Morse lemma with parameters) *Let (R, m) be a complete local ring and $G \in R[[x_1, \ldots, x_n]]$. Assume that $G = q + H$, where q is a quadratic form with reduction modulo m denoted by $\bar{q}$ such that*

(10.43.1) $\dim \langle \partial \bar{q} / \partial x_1, \ldots, \partial \bar{q} / \partial x_n \rangle = n$, *and*

(10.43.2) $H \in (x_1, \ldots, x_n)^3 + mR[[x_1, \ldots, x_n]]$.

Then there are local coordinates $y_1, \ldots, y_n$ such that

(10.43.3) $y_i \equiv x_i \mod (x_1, \ldots, x_n)^2 + mR[[x_1, \ldots, x_n]]$, *and*

(10.43.4) $G = q(y_1, \ldots, y_n) + b$ *for some $b \in m$.*

Proof Let us start with the case when $R = k$ is a field. Set $x_{2,i} := x_i$. Assume inductively (starting with $r = 2$) that there are local coordinate systems $(x_{s,1}, \ldots, x_{s,n})$ for $3 \le s \le r$ such that

$$x_{s,i} \equiv x_{s-1,i} \mod (x_1, \ldots, x_n)^{s-1} \quad \text{and}$$
$$G \equiv q(x_{r,1}, \ldots, x_{r,n}) \mod (x_1, \ldots, x_n)^{r+1}.$$

Next we choose $x_{r+1,i} := x_{r,i} + h_{r,i}$ for suitable $h_{r,i} \in (x_1, \ldots, x_n)^r$. Note that

$$q(x_{r+1,1}, \ldots, x_{r+1,n}) = q(x_{r,1}, \ldots, x_{r,n}) + \sum_i h_{r,i} \tfrac{\partial q}{\partial x_i} \mod (x_1, \ldots, x_n)^{2r}.$$

(We use this only modulo $(x_1, \ldots, x_n)^{r+2}$.) Since q is nondegenerate,

$$\sum_i \tfrac{\partial q}{x_i}(x_1, \ldots, x_n)^r = (x_1, \ldots, x_n)^{r+1}.$$

Thus we can choose the $h_{r,i}$ such that

$$G - q(x_{r+1,1}, \ldots, x_{r+1,n}) \in (x_1, \ldots, x_n)^{r+2}.$$

In the limit we get $(x_{\infty,1}, \ldots, x_{\infty,n})$ as required.

Applying this to $k = R/m$, we can assume from now on that

$$G - q(x_1, \ldots, x_n) \in mR[[x_1, \ldots, x_n]].$$

Working inductively (starting with $r = 1$), assume that there are local coordinate systems $(y_{s,1}, \ldots, y_{s,n})$ for $3 \le s \le r$ such that

$$y_{s,i} \equiv y_{s-1,i} \mod m^{s-1} R[[x_1, \ldots, x_n]] \quad \text{and}$$
$$G \equiv q(y_{r,1}, \ldots, y_{r,n}) \mod m + m^r R[[x_1, \ldots, x_n]].$$

Next we choose $y_{r+1,i} := y_{r,i} + c_{r,i}$ for suitable $c_{r,i} \in m^r R[[x_1, \ldots, x_n]]$. Note that

$$q(y_{r+1,1}, \ldots, y_{r+1,n}) = q(y_{r,1}, \ldots, y_{r,n}) + \sum_i c_{r,i} \tfrac{\partial q}{\partial x_i} \mod m^{2r} R[[x_1, \ldots, x_n]].$$

(We use this only modulo $m^{r+1} R[[x_1, \ldots, x_n]]$.) Since q is nondegenerate,

$$\sum_i \tfrac{\partial q}{\partial x_i} m^r R[[x_1, \ldots, x_n]] = (x_1, \ldots, x_n) m^r R[[x_1, \ldots, x_n]].$$

Thus we can choose the $c_{r,i}$ such that

$$G - q(y_{r+1,1}, \ldots, y_{r+1,n}) \in m + m^{r+1} R[[x_1, \ldots, x_n]].$$

In the limit we get $(y_{\infty,1}, \ldots, y_{\infty,n})$ as required. $\square$

In (1.27), we used various results on resolutions of double points of surfaces that contain a pair of lines and double points of 3–folds that contain a pair of planes. The normal forms can be obtained using the method of (10.43), but we did not follow how linear subvarieties transform under the (non-linear) coordinate changes used there. However, in the next examples, one can be quite explicit about the coordinate changes and the resolutions.

10.44 (Ordinary double points of surfaces) Let $S := (h(x_1, x_2, x_3) = 0) \subset \mathbb{A}^3$ be a surface with an ordinary double point at the origin that contains the pair of lines $(x_1 x_2 = x_3 = 0)$. Then h can be written as

$$h = f(x_1, x_2, x_3)x_1 x_2 - g(x_1, x_2, x_3)x_3.$$

Since the quadratic part has rank 3, then $f(0, 0, 0) \neq 0$ and we can write $g = x_1 g_1 + x_2 g_2 + x_3 g_3$ for some polynomials g_i. Thus

$$h = f(x_1 - f^{-1}g_1 x_3)(x_2 - f^{-1}g_2 x_3) - (g_3 + f^{-1}g_1 g_2)x_3^2.$$

Here $g_3 + f^{-1}g_1 g_2$ is nonzero at $(0, 0, 0)$ and we can set

$$y_1 := x_1 - f^{-1}g_1 x_3, \quad y_2 := f(x_2 - f^{-1}g_2 x_3)(g_3 + f^{-1}g_1 g_2)^{-1} \quad \text{and} \quad y_3 := x_3$$

to bring the equation to the normal form $S = (y_1 y_2 - y_3^2 = 0)$. The pair of lines is $(y_1 y_2 = y_3 = 0)$.

Now we consider three ways of resolving the singularity of X. First, one can blow up the origin $0 \in \mathbb{A}^3$. We get $B_0 \mathbb{A}^3 \subset \mathbb{A}_y^3 \times \mathbb{P}_s^2$ defined by the equations $\{y_i s_j = y_j s_i : 1 \leq i, j \leq 3\}$. Besides these equations, $B_0 S$ is defined by $y_1 y_2 - y_3^2 = s_1 s_2 - s_3^2 = y_1 s_2 - y_3 s_3 = s_1 y_2 - y_3 s_3 = 0$.

One can also blow up (y_1, y_3). We get $B_{(y_1, y_3)}\mathbb{A}^3 \subset \mathbb{A}_y^3 \times \mathbb{P}_{u_1 u_3}^1$ defined by the equation $y_1 u_3 = y_3 u_1$. Besides this equation, $B_{(y_1, y_3)}S$ is defined by $y_1 y_2 - y_3^2 = u_1 y_2 - u_3 y_3 = 0$.

These two blow-ups are actually isomorphic, as shown by the embedding

$$\mathbb{A}_y^3 \times \mathbb{P}_{u_1 u_3}^1 \hookrightarrow \mathbb{A}_y^3 \times \mathbb{P}_s^2 \quad : \quad ((y_1, y_2, y_3), (u_1 : u_3)) \mapsto ((y_1, y_2, y_3), (u_1^2 : u_3^2 : u_1 u_3))$$

restricted to $B_{(y_1, y_3)}S$. The same things happen if we blow up (y_2, y_3).

10.45 (Ordinary double points of 3-folds) Let $X := (h(x_1, \ldots, x_4) = 0) \subset \mathbb{A}^4$ be a hypersurface with an ordinary double point at the origin that contains the pair of planes $(x_1 x_2 = x_3 = 0)$. Then h can be written as

$$h = f(x_1, \ldots, x_4)x_1 x_2 - g(x_1, \ldots, x_4)x_3.$$

The quadratic part has rank 4 iff $f(0, \ldots, 0) \neq 0$ and x_4 appears in g with nonzero coefficient. In this case, we can set $y_i := x_i$ for $i = 1, 2, 3$ and $y_4 := f^{-1}g$ to bring the equation to the normal form $X = (y_1 y_2 - y_3 y_4 = 0)$. The original pair of planes is $(y_1 y_2 = y_3 = 0)$.

Now we consider three ways of resolving the singularity of X. First, one can blow up the origin $0 \in \mathbb{A}^4$. We get $B_0 \mathbb{A}^4 \subset \mathbb{A}^4_\mathbf{y} \times \mathbb{P}^3_\mathbf{s}$, defined by the equations $\{y_i s_j = y_j s_i : 1 \le i, j \le 4\}$, and $p \colon B_0 X \to X$ by the additional equations

$$y_1 y_2 - y_3 y_4 = s_1 s_2 - s_3 s_4 = y_i s_{3-i} - y_j s_{7-j} = 0 : i \in \{1, 2\}, j \in \{3, 4\}.$$

The exceptional set is the smooth quadric $(s_1 s_2 = s_3 s_4) \subset \mathbb{P}^3$ lying over the origin $0 \in \mathbb{A}^4$.

One can also blow up (y_1, y_3). Then $B_{(y_1, y_3)} \mathbb{A}^4 \subset \mathbb{A}^4_\mathbf{y} \times \mathbb{P}^1_{u_1 u_3}$ is defined by the equation $y_1 u_3 = y_3 u_1$. Besides this equation, $B_{(y_1, y_3)} X$ is defined by $y_1 y_2 - y_3 y_4 = u_1 y_2 - u_3 y_4 = 0$. The exceptional set is the smooth rational curve $E \simeq \mathbb{P}^1_{u_1 u_3}$ lying over the origin $0 \in \mathbb{A}^4$.

Note furthermore that the birational transform P^*_{24} of the plane $P_{24} := (y_2 = y_4 = 0)$ is the blown-up plane $B_0 P_{24}$, but the birational transform P^*_{14} of the plane $P_{14} := (y_1 = y_4 = 0)$ is the plane $(y_1 = u_1 = 0)$. The latter intersects E at the point $(u_1 = 0) \in E$, thus $(P^*_{14} \cdot E) = 1$. Since $P^*_{14} + P^*_{24}$ is the pullback of the Cartier divisor $(y_4 = 0)$, it has 0 intersection number with E. Thus $(P^*_{24} \cdot E) = -1$.

By direct computation, the rational map $p \colon \mathbb{A}^4_\mathbf{y} \times \mathbb{P}^3_\mathbf{s} \dashrightarrow \mathbb{A}^4_\mathbf{y} \times \mathbb{P}^1_\mathbf{u}$ given by $p_1 \colon (y_1, \dots, y_4, s_1 : \cdots : s_4) \mapsto (y_1, \dots, y_4, s_1 : s_3)$ gives a morphism $p_1 \colon B_0 X \to B_{(y_1, y_3)} X$. Similarly, we obtain $p_2 \colon B_0 X \to B_{(y_2, y_3)} X$ and an isomorphism

$$p_1 \times p_2 \colon B_0 X \simeq B_{(y_1, y_3)} X \times_X B_{(y_2, y_3)} X.$$

Finally, set $S := (y_3 = y_4) \subset X$. By the computations of (10.44), the p_i restrict to isomorphisms $p_i \colon B_0 S \simeq B_{(y_i, y_3)} S$. Thus $p^{-1} S = B_0 S \cup E$ and $B_0 S$ is the graph of the isomorphism $p_2 \circ p_1^{-1} \colon B_{(y_1, y_3)} S \simeq B_{(y_2, y_3)} S$.

10.6 Noether Normalization

10.46 (Classical versions) Noether's normalization theorem says that if X is an affine (resp. projective) k-variety of dimension n then it admits a finite morphism to $\mathbb{A}^n_k$ (resp. $\mathbb{P}^n_k$).

We aim to generalize this to arbitrary morphisms. For the projective case, let $X \subset \mathbb{P}^N_S$ be projective over S and $n = \dim X_s$ for some $s \in S$. Choose a linear subspace $L_s \subset \mathbb{P}^N_s$ of dimension $N - n - 1$ that is disjoint from X_s. (This is always possible if $k(s)$ is infinite, otherwise we may need to take a high enough Veronese embedding first.) Lifting L_s to $\mathbb{P}^N_S$ and projecting from it gives the following.

Claim 10.46.1 Let $p \colon X \to S$ be a projective morphism and $n = \dim X_s$ for some $s \in S$. Then there is an open neighborhood $s \in S^\circ \subset S$ such that $p|_{X^\circ}$ can be factored as

$$p|_{X^\circ} : X^\circ \xrightarrow{\text{finite}} \mathbb{P}^n_{S^\circ} \longrightarrow S^\circ. \qquad\qquad \square$$

In general, we have the following weaker local version.

Claim 10.46.2 Let $p : X \to S$ be a finite type morphism and $x \in X$ a closed point. Then there is an open neighborhood $x \in X^\circ \subset X$ and an open embedding $X^\circ \hookrightarrow X^*$, where $p^* : X^* \to S$ is projective of relative dimension $\leq \dim X_s$.

Proof Set $d := \dim X_s$ and pick $g_1, \ldots, g_d \in \mathcal{O}_{x,X}$ that generate an $m_{x,X}$-primary ideal. They give a rational map $X \dashrightarrow \mathbb{A}^d_S \hookrightarrow \mathbb{P}^d_S$ that is quasi-finite on some $x \in X^\circ \subset X$. We then take $X^\circ \subset X^*$ such that $X^* \to \mathbb{P}^d_S$ is finite. $\quad\square$

Next we give two examples showing that in (10.46.2) one cannot choose X° such that $X^\circ \to \mathbb{A}^d_S$ is finite, not even when S is local. After that we discuss an étale local version for finite type morphisms due to Raynaud and Gruson (1971). Arbitrary morphisms are discussed in (10.52); these results work best for morphisms of complete local schemes.

Example 10.47 We give an example of a morphism of pure relative dimension one $p: X \to S$ from an affine 3-fold X to a smooth, pointed surface $s \in S$ that cannot be factored as

$$p: X \xrightarrow{\text{finite}} \mathbb{A}^1 \times S \longrightarrow S,$$

not even over a formal neighborhood of s. Such examples are quite typical and there does not seem to be any affine version of Noether normalization over base schemes of dimension ≥ 2.

Let S denote the localization (or completion) of $\mathbb{A}^2_{st}$ at the origin and consider the affine scheme

$$X := ((x^3 + y^3 + 1)(1 + tx) + sy = 0) \subset \mathbb{A}^2_{xy} \times S.$$

Then $\pi: X \to S$ is a flat family of curves. We claim that there is no finite morphism of it onto $\mathbb{A}^1 \times S$.

Assume to the contrary that such a map $g: X \to \mathbb{A}^1 \times S$ exists. Then g can be extended to a finite morphism $\bar{g}: \bar{X} \to \mathbb{P}^1 \times S$.

Here $\bar{X}_{(0,0)}$ is a compactification of $X_{(0,0)}$, hence a curve of genus 1.

For $t \neq 0$, the line $(1 + tx = s = 0)$ gives an irreducible component of $\bar{X}_{(0,t)}$ that is a rational curve. As $t \to 0$, the limit of these rational curves is a union of rational, irreducible, geometric components of $\bar{X}_{(0,0)}$, a contradiction.

Example 10.48 In $\mathbb{A}^4_{xyst}$ consider the surface $X := (x - sy^2 = y - tx^2 = 0)$. Projection to $\mathbb{A}^2_{xy}$ is birational with inverse $(x, y) \mapsto (s, t) = (x/y^2, y/x^2)$. The projection to $\mathbb{A}^2_{st}$ is quasi-finite.

Consider the projection $\pi: \mathbb{A}^4_{xyst} \to \mathbb{A}^3_{zst}$ given by $z = x + y$. We claim that the closure of its image contains the z-axis. Indeed, for any c, the curve

$$t \mapsto \left(t, c - t, \tfrac{t}{(c-t)^2}, \tfrac{c-t}{t^2}\right)$$

lies on X and its projection converges to $(c, 0, 0)$ as $t \to \infty$.

It is easy to see that the same happens for every perturbation of π. In fact, given $(x, y) \mapsto (a(s, t)x + b(s, t)y + c(s, t))$, the closure of the image of X contains the z-axis whenever $a(0, 0) \neq 0 \neq b(0, 0)$.

The next result of Raynaud and Gruson (1971) shows that Noether normalization works étale locally. The version given in Stacks (2022, tag 052D) states the first part, but following the proof gives the additional information about the choices.

Theorem 10.49 *Let* $f: X \to S$ *be a finite type morphism. Pick* $s \in S$, *a closed point* $x \in X_s$ *and set* $n = \dim_x X_s$. *Then there is an elementary étale neighborhood (2.18)* $\pi: (x', X') \to (x, X)$ *such that* $f \circ \pi$ *factors as*

$$(x', X') \xrightarrow{g} (y, Y) \xrightarrow{\tau} (s, S), \tag{10.49.1}$$

where g *is finite,* $g^{-1}(y) = \{x'\}$ *(as sets),* τ *is smooth of relative dimension* n, *and* $k(y) = k(s)$.

Moreover, pick $c \in \mathbb{N}$ *and* $x_1, \ldots, x_n \in m_{x,X_s}$ *that generate an* m_{x,X_s}-*primary ideal. Then we can choose (10.49.1) such that there are* $y_1, \ldots, y_n \in m_{y,Y_s}$ *satisfying* $g_s^* y_i \equiv \pi_s^* x_i \mod m_{x',X_s'}^c$ *for every i.* □

If X_s is generically geometrically reduced, then we can choose $y_{n+1} \in m_{x,X_s}$ with specified residue modulo $m_{x',X_s'}^c$ and which embeds the generic fiber of $X_s \to Y_s$ into $\mathbb{A}^1_{k(Y_s)}$. Lifting it to X' and setting $Y' := \mathbb{A}^1_Y$ gives the following birational version of Noether normalization.

Corollary 10.50 *Let* $f: X \to S$ *be a finite type morphism. Pick* $s \in S$ *and a closed point* $x \in X_s$. *Assume that* X_s *is generically geometrically reduced and of pure dimension* n. *Then there is an elementary étale neighborhood* $\pi: (x', X') \to (x, X)$ *such that* $f \circ \pi$ *factors as*

$$(x', X') \xrightarrow{g'} (y', Y') \xrightarrow{\tau'} (s, S), \tag{10.50.1}$$

where g' is finite, $(g')^{-1}(y') = \{x'\}$ (as sets), $g'_s \colon X'_s \to Y_s \times \mathbb{A}^1$ is birational, τ' is smooth of relative dimension $n + 1$, and $k(y') = k(s)$.

Moreover, pick $c \in \mathbb{N}$ and $x_1, \ldots, x_{n+1} \in m_{x,X_s}$ that generate an m_{x,X_s}-primary ideal. Then we can arrange that there are $y_1, \ldots, y_{n+1} \in m_{y,Y_s}$ satisfying $g'^*_s y_i \equiv \pi^*_s x_i \mod m^c_{x',X'_s}$ for every i.   □

Corollary 10.51 *Let $f \colon X \to S$ be a finite type morphism of pure relative dimension n. Pick $s \in S$ and a closed point $x \in X_s$ such that $k(x) = k(s)$. Assume that S is normal and f is flat at the generic points of X_s. Assume also that* embdim$_x$ pure$(X_s) \le n+1$. *Then there is an elementary étale $\pi \colon (x', X') \to (x, X)$ such that (10.50.1) further factors as*

$$(x', X') \xrightarrow{g'} (y', D') \hookrightarrow (y', Y') \xrightarrow{\tau'} (s, S), \qquad (10.51.1)$$

where, $D' \subset Y'$ is a relative Cartier divisor, g' is birational, $g'_s \colon X'_s \to D'_s$ is birational and induces a local isomorphism pure$(X'_s) \to D'_s$ *at x'.*

Proof Since embdim(pure$(X_s)) \le n+1$, we can choose $x_1, \ldots, x_{n+1} \in m_{x,X_s}$ that generate the ideal of $x \in$ pure(X_s). Applying (10.50) with $c = 2$ guarantees that pure$(X'_s) \to D'_s$ is a local isomorphism at x'.

D' is a relative Cartier divisor by (4.4) and then (10.54) implies that g' is a local isomorphism at the generic points of X'_s. Thus g' is birational.   □

Informally speaking, (10.51) says that partial normalizations of flat deformations of hypersurfaces describe all deformations over normal base schemes. For double points this approach leads to a complete answer (10.68). More substantial applications are in de Jong and van Straten (1991).

Next we turn to local morphisms of Noetherian local schemes

10.52 (Noether normalization, local version) Let $f \colon (x, X) \to (s, S)$ be a morphism of local, Noetherian schemes. We would like to factor f as

$$f \colon (x, X) \xrightarrow{p} (s', S') \xrightarrow{q} (s, S), \qquad (10.52.1)$$

where p has "finiteness" properties and q has "smoothness" properties. Let us start with the case when $k(x) \supset k(s)$ is a finitely generated field extension. Pick any transcendence basis $\bar{y}_1, \ldots, \bar{y}_n$ of $k(x)/k(s)$ and lift these back to $y_1, \ldots, y_n \in \mathcal{O}_X$. We can then take S' to be the localization of $\mathbb{A}^n_S$ at the generic point of the fiber over $s \in S$. Thus we have proved the following.

Claim 10.52.2 Let $f \colon (x, X) \to (s, S)$ be a local morphism of local, Noetherian schemes such that $k(x) \supset k(s)$ is a finitely generated field extension. Then we can factor f as $f \colon (x, X) \xrightarrow{p} (s', S') \xrightarrow{q} (s, S)$, where $k(x)/k(s')$ is a finite field extension, q is the localization of a smooth morphism and $q^{-1}(s) = s'$ (as schemes). $\qquad\square$

For Henselian schemes, we can do better. Pick $\bar{y} \in k(x)$ that is separable over $k(s')$ with separable, monic equation $\bar{g}(\bar{y}) = 0$. If $\mathscr{O}_X$ is Henselian then we can lift $\bar{y}$ to $y \in \mathscr{O}_X$ such that y satisfies a separable, monic equation $g(y) = 0$. We can now replace S' with the Henselization of $\mathscr{O}_{S'}[y]/(g(y))$ at the generic point of the central fiber, and obtain the following.

Claim 10.52.3 Let $f: (x, X) \to (s, S)$ be a local morphism of local, Henselian, Noetherian schemes such that $k(x)/k(s)$ is a finitely generated field extension. Then we can factor f as $f: (x, X) \overset{p}{\to} (s', S') \overset{q}{\to} (s, S)$, where p is finite, $k(x)/k(s')$ is a purely inseparable field extension, q is the localization of a smooth morphism and $q^{-1}(s) = s'$ (as schemes). $\qquad\square$

Combining these with (10.53) gives the following.

Claim 10.52.4 Let $f: (x, X) \to (s, S)$ be a local morphism of local, complete, Noetherian schemes. Then we can factor f as $f: (x, X) \overset{p}{\to} (s', S') \overset{q}{\to} (s, S)$, where $k(x)/k(s')$ is a purely inseparable field extension, q is formally smooth, faithfully flat, regular and $q^{-1}(s) = s'$ (as schemes). $\qquad\square$

Putting these together we get the following.

Claim 10.52.5 Let $f: (x, X) \to (s, S)$ be a local morphism of local, complete, Noetherian schemes such that $k(x)/k(s)$ is separable. Set $n := \dim X_s$.
Then we can factor f as

$$f: (x, X) \overset{p}{\to} ((s', 0), \hat{\mathbb{A}}^n_{S'}) \overset{\pi}{\to} (s', S') \overset{q}{\to} (s, S),$$

where p is finite, $k(x) = k(s', 0) = k(s')$, π is the coordinate projection, $q^{-1}(s) = s'$ (as schemes), q is the localization of a smooth morphism if $k(x)/k(s)$ is finitely generated and formally smooth, faithfully flat and regular in general.

Proof By (10.52.4), we have $q: (s', S') \to (s, S)$ such that $k(x) = k(s')$. Since $\mathscr{O}_{X_s}$ has dimension n, there are $\bar{t}_1, \ldots, \bar{t}_n \in \mathscr{O}_{X_s}$ that generate an ideal that is primary to the maximal ideal. Lift these back to $t_1, \ldots, t_n \in \mathscr{O}_X$. These define $p: (x, X) \to ((s', 0), \hat{\mathbb{A}}^n_{S'})$. By construction, $\mathscr{O}_X/(m_S, t_1, \ldots, t_n) \cong \mathscr{O}_{X_s}/(\bar{t}_1, \ldots, \bar{t}_n)$ is finite over $k(s')$. Thus p is finite. $\qquad\square$

Notation 10.52.6 Let R be a complete, local ring and $Y = \operatorname{Spec} R$. We write $\hat{\mathbb{A}}^n_Y := \operatorname{Spec} R[[x_1, \ldots, x_n]]$. Note that $\hat{\mathbb{A}}^n_Y$ is *not* the product of $\hat{\mathbb{A}}^n$ with Y in any sense. If $X \to Y$ is a finite morphism then $\hat{\mathbb{A}}^n_X \cong X \times_Y \hat{\mathbb{A}}^n_Y$.

10.53 (Residue field extensions) Let (s, S) be a Noetherian, local scheme and $K/k(s)$ a field extension. By Grothendieck (1960, $0_{III}.10.3.1$), there is a Noetherian, local scheme (x, X) and a flat morphism $g \colon (x, X) \to (s, S)$ such that $g^* m_{s,S} = m_{x,X}$ (that is, the scheme fiber $g^{-1}(s)$ is the reduced point $\{x\}$) and $k(x) \simeq K$.

If $K/k(s)$ is a finitely generated separable extension then we can choose $g \colon (x, X) \to (s, S)$ to be the localization of a smooth morphism. In particular, if S is normal then so is X.

Combining Grothendieck (1960, $0_{III}.10.3.1$) and Stacks (2022, tag 07PK) shows that if $K/k(s)$ is an arbitrary separable extension, then we can choose $g \colon (x, X) \to (s, S)$ to be formally smooth. If S is complete then g is also regular. In particular, if S is normal then so is X.

Note that infinite inseparable extensions do cause problems in the above arguments. One difficulty is that they can lead to non-excellent schemes; see Nagata (1962, p.206).

10.54 (Openness for isomorphism) Let $g \colon (x, X) \to (s, S)$ be a local morphism of local, Noetherian schemes and $g \colon G \to F$ a map of coherent sheaves on X. Assume that F is flat over S. Then g is an isomorphism iff g_s is an isomorphism, and g_s is injective iff g is injective and coker g is flat over S. See Matsumura (1986, 22.5) or Kollár (1996, I.7.4.1) for proofs. Applying this to the structure sheaf of a scheme and its image, we get the following.

Claim 10.54.1 Let $\pi \colon X \to Y$ be a finite morphism of S-schemes. Assume that X is flat over S. Then π is an isomorphism (resp. closed embedding) in a neighborhood of a fiber X_s iff $\pi_s \colon X_s \to Y_s$ is an isomorphism (resp. closed embedding).

10.7 Flatness Criteria

Let $g \colon X \to S$ be a morphism and F a coherent sheaf on X. We are mainly interested in those cases when F is flat over S with pure fibers of dimension d for some d. In practice, we already know that $F|_U$ is flat for some dense open subset $U \subset X$ and we aim to find conditions that guarantee flatness.

Note that such a result is possible only if $F|_U$ determines F. Thus we at least need to assume that none of the associated point of F are contained in Z.

10.55 (Flatness and associated points) Let $f \colon X \to S$ be a morphism of Noetherian schemes and F a coherent sheaf on X.

Claim 10.55.1 If F is flat over S then $f(\mathrm{Ass}(F)) \subset \mathrm{Ass}(S)$.

Proof Let $x \in X$ be an associated point of F and $s := f(x)$. Assume that s is not an associated point of S. Then there is an $r \in m_{s,S}$ such that $r: \mathcal{O}_S \to \mathcal{O}_S$ is injective near s. Tensoring with F shows that $r: F \to F$ is injective near X_s. Thus none of the points of X_s is in $\mathrm{Ass}(F)$. $\square$

Claim 10.55.2 Assume that F is flat over S and $x \in \mathrm{Ass}(F)$. Then every generic point of $\mathrm{Supp}(\bar{x} \cap X_s)$ is an associated point of F_s. In particular, if F is flat with pure fibers then every $x \in \mathrm{Ass}(F)$ is a generic point of $\mathrm{Supp}(F_{f(x)})$.

Proof Let $G \subset F$ be the largest subsheaf supported on $\bar{x}$. After localizing at a generic point of $\mathrm{Supp}(\bar{x} \cap X_s)$, we have $\mathrm{Supp}(\bar{x} \cap X_s) = \{w\}$, a single closed point. There is an $n \geq 0$ such that $G \subset m_{s,S}^n F$, but $G \not\subset m_{s,S}^{n+1} F$. Thus $m_{s,S}^n F / m_{s,S}^{n+1} F \simeq (m_{s,S}^n / m_{s,S}^{n+1}) \otimes F_s$ has a nonzero subsheaf supported on w. $\square$

Note that flatness is needed for (10.55.2) as illustrated by the restriction of either of the coordinate projections to the union of the axes $(xy = 0)$.

Claim 10.55.3 Assume f is of finite type, F is flat over S, and $x \in \mathrm{Ass}(F)$. Then every fiber of $\bar{x} \to f(\bar{x})$ has the same dimension.

Proof We may assume that $f(x)$ is a minimal associated point of S. Assume that we have $s \in f(\bar{x})$ such that $\dim(X_s \cap \bar{x})$ is larger than the expected dimension d. By restricting to a general relative Cartier divisor $H \subset X$, $F|_H$ is flat along H_s by (10.56) and $H_s \cap \bar{x}$ is a union of associated points of $F|_H$ by (10.22.1). Repeating this $d + 1$ times we get Cartier divisors $H^1, \ldots, H^{d+1} \subset X$ and a complete intersection $Z := H^1 \cap \cdots \cap H^{d+1}$ such that $F|_Z$ is flat along Z_s, the generic points of $Z \cap \bar{x}$ are associated points of $F|_Z$ yet they do not dominate $f(\bar{x})$. This is impossible by (10.55.1). $\square$

Next we discuss some basic reduction steps.

Let $f: X \to S$ be a morphism that we would like to prove to be flat. We can usually harmlessly assume that S is local.

If f is of finite type, then flatness is an open property. Let $U \subset X$ denote the largest open set over which f is flat and set $Z := X \setminus U$. The situation is technically simpler if Z is a single closed point. To achieve this, one can use (10.56) to pass to a general hyperplane section of X and repeat if necessary, until Z becomes zero-dimensional. A potential drawback is that, while we can choose general hyperplanes, some fibers are nongeneral complete intersections, so may be harder to control.

Alternatively, we can localize at a generic point of Z. Then f is no longer of finite type, which can cause problems.

Once S and X are both local, we can take their completions. Now we have a local morphism of complete, local, Noetherian schemes. Note, however, that some of our results hold only over base schemes that are normal, seminormal, or reduced. These properties are preserved by completion for excellent schemes, but not in general.

Proposition 10.56 (Bertini theorem for flatness) (Matsumura, 1986, p.177) *Let* $(x, X) \to (s, S)$ *be a local morphism of local schemes,* $r \in m_{x,X}$ *and* F *a coherent sheaf on X. The following are equivalent.*

(10.56.1) *r is a non-zerodivisor on F and F/rF is flat over S.*

(10.56.2) *r is a non-zerodivisor on F_s and F is flat over S.* □

10.57 (Flatness and residue field extension) The following simple trick reduces most flatness questions for local morphisms $f : (x, X) \to (s, S)$ with finitely generated residue field extension $k(x)/k(s)$ to the special case when $k(x) = k(s)$ and they are infinite. (See 10.52–10.53 for other versions.)

If $k(x)/k(s)$ is a generated by n elements then there is a point $s' \in \mathbb{A}^n_{k(s)}$ such that $k(x) \subset k(s')$ and $k(s')$ is infinite.

Consider next the trivial lifting $f' : X' := \mathbb{A}^n_X \to S' := \mathbb{A}^n_S$. Set $s' \in \mathbb{A}^n_{k(s)} \subset S'$ and $x' := (s', x) \in X'$ projecting to x. Thus we have a commutative diagram of pointed schemes

$$
\begin{array}{ccc}
(x' \in X') & \xrightarrow{\ \pi_X\ } & (x \in X) \\
{\scriptstyle f'}\downarrow & & \downarrow{\scriptstyle f} \\
(s' \in S') & \xrightarrow{\ \pi_S\ } & (s \in S)
\end{array}
\qquad (10.57.1)
$$

where π_X, π_S are smooth, $k(x') = k(s')$ and f is flat at x iff f' is flat at x'.

Many properties of schemes and morphisms are preserved by composing with smooth morphisms; see Matsumura (1986, sec.23) for a series of such results. Thus the properties of (s, S) are inherited by (s', S'). Once we prove a result about (x', X') it descends to (x, X).

Over reduced bases, flatness is usually easy to check if we know all the fibers. For projective morphisms there are criteria using the Hilbert function (3.20). In the local case, we have the following.

Lemma 10.58 *Let S be a reduced scheme and $f : X \to S$ a morphism that is of finite type, pure dimensional and with geometrically reduced fibers. Then f is flat.*

Proof By (4.38), it is enough to show this when (s, S) is the spectrum of a DVR. In this case f is flat iff none of the associated points of X is contained in X_s. By assumption X_s is reduced, so only generic points of X_s could occur. Then the corresponding irreducible component of X_s is also an irreducible component of X, but we also assumed that f has pure relative dimension. □

10.59 (Format of flatness criteria) In many cases we have some information about the fibers of a morphism, but we do not fully understand them. So we are looking for results of the following type.

Let (s, S) be a local scheme, $f : X \to S$ a morphism and F a vertically pure coherent sheaf on X. Let $Z \subset X$ be a closed subset such that $F_{X \setminus Z}$ is flat over S. We make various assumptions on pure(F_s) (involving Z_s) and on S. The conclusion should be that F is flat and F_s is pure.

The natural way to organize the results is by the *relative codimension;* in the local case this equals $\operatorname{codim}_{X_s}(Z_s)$. The starting case is when $Z = X$, so the codimension is 0.

The main theorems are (10.60), (10.63), (10.67), (10.71) and (10.73).

Flatness in Relative Codimension 0

The basic result is the following, proved in Grothendieck (1971, II.2.3).

Theorem 10.60 *Let* $f : (x, X) \to (s, S)$ *be a local morphism of local, Noetherian schemes of the same dimension such that* $f^{-1}(s) = x$ *as schemes, that is,* $m_{x,X} = m_{s,S} \mathscr{O}_X$. *Assume that* $k(x) \supset k(s)$ *is separable and* $\hat{S}$, *the completion of* S, *is normal. (Note that if* S *is normal and excellent, then* $\hat{S}$ *is normal.)*

Then f *is flat at* x.

Proof We may replace S and X by their completions. As in (10.52.4), we can factor f as

$$f : (x, X) \xrightarrow{p} (y, Y) \xrightarrow{q} (s, S)$$

where (y, Y) is also complete, local, Noetherian, $k(x) = k(y)$, $m_{x,X} = m_{y,Y} \mathscr{O}_X$ and q is flat.

Thus $p^* : m_{y,Y}/m_{y,Y}^2 \to m_{x,X}/m_{x,X}^2$ is surjective, hence $p^* : \mathscr{O}_Y \to \mathscr{O}_X$ is surjective by the Nakayama lemma. Equivalently, $p : X \to Y$ is a closed embedding. It is thus an isomorphism, provided Y is integral.

In order to ensure these properties of Y we need to know more about q. If $k(x)/k(s)$ is finitely generated then q is the localization of a smooth morphism (10.52.3). Thus Y is normal and $\dim Y = \dim S$, as required. The general case is technically harder. We use that q is formally smooth and geometrically regular (10.52.4) to reach the same conclusions as before.

Thus p is an isomorphism, so $f = q$ and f is flat.   □

Examples 10.61 These examples show that the assumptions in (10.60) and (10.63) are necessary.

(10.61.1) Assume that char $k \neq 2$ and set $C := (y^2 = ax^2 + x^3)$ where $a \in k$ is not a square. Let $f: \bar{C} \to C$ denote the normalization. Then the fiber over the origin is the spectrum of $k(\sqrt{a})$, which is a separable extension of k. Here C is not normal and f is not flat.

(10.61.2) The extension $\mathbb{C}[x, y] \subset \mathbb{C}[\frac{x}{y}, y]_{(y)}$ is not flat yet $(x, y) \cdot \mathbb{C}[\frac{x}{y}, y]_{(y)}$ is the maximal ideal and the residue field extension is purely transcendental. However, the dimension of the larger ring is 1.

A similar thing happens with $\mathbb{C}[x, y] \hookrightarrow \mathbb{C}[[t]]$ given by $(x, y) \mapsto (t, \sin t)$. The fiber over the origin is the origin with reduced scheme structure.

(10.61.3) On $\mathbb{C}[x, y]$ consider the involution $\tau(x) = -x, \tau(y) = -y$. The invariant ring is $\mathbb{C}[x^2, xy, y^2] \subset \mathbb{C}[x, y]$. The fiber over the origin is the spectrum of $\mathbb{C}[x, y]/(x^2, xy, y^2)$; it has length 3 and embedding dimension 2. The fiber over any other point has length 2. Thus the extension is not flat.

(10.61.4) As in Kollár (1995a, 15.2), on $S := k[x_1, x_2, y_1, y_2]$ consider the involution $\tau(x_1, x_2, y_1, y_2) = (x_2, x_1, y_2, y_1)$. The ring of invariants is

$$R := k[x_1 + x_2, x_1 x_2, y_1 + y_2, y_1 y_2, x_1 y_1 + x_2 y_2].$$

The resulting extension is not flat along $(x_1 - x_2 = y_1 - y_2 = 0)$.

If char $k = 2$, then $x_1 - x_2, y_1 - y_2$ are invariants. Set $P := (x_1 - x_2, y_1 - y_2)R$. Then $S/PS = S/(x_1 - x_2, y_1 - y_2)S \simeq k[x_1, y_1]$ and $R/P \simeq k[x_1^2, y_1^2]$.

Thus $S_P \supset R_P$ is a finite extension whose fiber over P is $k(x_1, y_1) \supset k(x_1^2, y_1^2)$. This is an inseparable field extension, generated by 2 elements.

(10.61.5) Set $X := (z = 0) \cup (z - x = z - y = 0) \subset \mathbb{A}^3$ with coordinate projection $\pi: X \to \mathbb{A}^2_{xy}$. Then π is finite, has curvilinear fibers, but not flat.

These examples leave open only one question: What happens with curvilinear fibers?

Definition 10.62 (Curvilinear schemes) Let k be a field and (A, m) a local, Artinian k-algebra. We say that $\text{Spec}_k A$ is *curvilinear* if A is cyclic as a $k[t]$-module for some t. That is, if A can be written as a quotient of $k[t]$. It is easy to see that this holds if either A/m is a finite, separable extension of k and m is a principal ideal, or A is a field extension of k of degree $= \text{char } k$.

Let B be an Artinian k-algebra. Then $\mathrm{Spec}_k\, B$ is called *curvilinear* if all of its irreducible components are curvilinear. If k is an infinite field, this holds iff B can be written as a quotient of $k[t]$. If K/k is a field extension and $\mathrm{Spec}_k\, B$ is curvilinear then so is $\mathrm{Spec}_K(B \otimes_k K)$.

Let $\pi: X \to S$ be a finite-type morphism. The embedding dimension of fibers is upper semicontinuous, thus the set $\{x \in X: X_{\pi(x)}$ is curvilinear at $x\}$ is open.

Theorem 10.63 *Let* $f: X \to S$ *be a finite type morphism with curvilinear fibers such that every associated point of X dominates S. Assume that either S is normal, or there is a closed $W \subset S$ such that* $\mathrm{depth}_W\, S \geq 2$ *and f is flat over* $S \setminus W$. *Then f is flat.*

Proof We start with the classical case when X, S are complex analytic, S is normal, f is finite, and $X \subset S \times \mathbb{C}$. Let $s \in S$ be a smooth point. Then $S \times \mathbb{C}$ is smooth along $\{s\} \times \mathbb{C}$ thus X is a Cartier divisor near X_s. In particular, f is flat over the smooth locus $S^{\mathrm{ns}} \subset S$. Set $d := \deg f$. For each $s \in S^{\mathrm{ns}}$ there is a unique monic polynomial $t^d + a_{d-1}(s)t^{d-1} + \cdots + a_0(s)$ of degree d whose zero set is precisely $X_s \subset \mathbb{C}$. As in the proof of the analytic form of the Weierstrass preparation theorem (see, for instance, Griffiths and Harris (1978, p.8) or Gunning and Rossi (1965, sec.II.B)) we see that the $a_i(s)$ are analytic functions on S^{ns}. By Hartogs's theorem, they extend to analytic functions on the whole of S; we denote these still by $a_i(s)$. Thus

$$X = (t^d + a_{d-1}(s)t^{d-1} + \cdots + a_0(s) = 0) \subset S \times \mathbb{C}$$

is a Cartier divisor and f is flat. This completes the complex analytic case.

In general, we argue similarly, but replace the polynomial $t^d + a_{d-1}(s)t^{d-1} + \cdots + a_0(s)$ by the point in the Hilbert scheme corresponding to X_s.

Assume first that f is finite. Again set $d := \deg f$ and let $S^\circ \subset S$ denote a dense open subset over which f is flat. Since f is finite, it is (locally) projective, thus we have

$$
\begin{array}{ccc}
\mathrm{Univ}_d(X/S) & \xrightarrow{\;p\;} & X \\[4pt]
{\scriptstyle u}\big\downarrow & & \big\downarrow{\scriptstyle f} \\[4pt]
\mathrm{Hilb}_d(X/S) & \xrightarrow{\;\pi\;} & S
\end{array}
\qquad (10.63.1)
$$

parametrizing length d quotients of the fibers of f. If $s \in S^\circ$ then $\mathscr{O}_{X_s}$ has length d, hence its sole length d quotient is itself. Thus π is an isomorphism over S°.

Let $s \to S$ be a geometric point. Then $X_s \cong \mathrm{Spec}\, k(s)[t]/(\prod_i(t - a_i)^{m_i})$ for some $a_i \in k(s)$ and $m_i \in \mathbb{N}$. Thus the fiber of p over s is a finite set corresponding to length d quotients of $k(s)[t]/(\prod_i(t - a_i)^{m_i})$, equivalently, to solutions of the equation $\sum_i m_i' = d$ where $0 \le m_i' \le m_i$. We have not yet proved that $\text{Hilb}_d(X/S)$ has no embedded points over $\text{Sing}\,S$, but we obtain that $\text{pure}(\text{Hilb}_d(X/S)) \to S$ is finite and birational, hence an isomorphism if S is normal or if $S^\circ \supset S \setminus W$ in case (2) by (10.6.4). The natural map

$$\text{pure}(p)\colon \ \text{Univ}_d(X/S) \times_{\text{Hilb}_d(X/S)} \text{pure}(\text{Hilb}_d(X/S)) \to X$$

is a closed immersion whose image is isomorphic to X over S°. Thus $\text{pure}(p)$ is an isomorphism, so f is flat and $\text{Hilb}_d(X/S) \simeq S$.

Finally, (10.49) reduces the general case to the finite one. (Note that any finite type, quasi-finite morphism can be extended to a finite morphism, but there is no reason to believe that the extension still has curvilinear fibers. So we need to use the more difficult (10.49).)   □

Over a nonnormal base there does not seem to be any simple analog of (10.63), but the following is quite useful.

Proposition 10.64 *Let* $f\colon X \to (s, S)$ *be a finite morphism with curvilinear fibers. Assume that*

(10.64.1) *the pair* $(s \in S)$ *is weakly normal (10.74),*

(10.64.2) f *is flat of constant degree* d *over* $S \setminus \{s\}$,

(10.64.3) X *has no associated points in* $\text{Supp}\,f^{-1}(s)$, *and*

(10.64.4) *either* $x := \text{Supp}\,f^{-1}(s)$ *is a single point and* $k(x)/k(s)$ *is purely inseparable, or* f *has well-defined specializations (4.2.9).*

Then f *is flat.*

Proof Again consider the diagram (10.63.1). By (2), p and π are isomorphisms over $S \setminus \{s\}$. We claim that π is an isomorphism. Since $(s \in S)$ is weakly normal, this holds if $\pi^{-1}(s)$ has a unique geometric point. If f has well-defined specializations, this holds by definition.

Otherwise, let $s' \to s$ be a geometric point. Since $k(x)/k(s)$ is purely inseparable, $X_{s'} \simeq \text{Spec}\,k(s')[t]/(t^r)$ for some r, which has a unique subscheme of length d. Thus π is an isomorphism. As in the proof of (10.63), we conclude that p is also an isomorphism.   □

The proof given in (4.21) applies with minor modifications to give the following result of Ramanujam (1963) and Samuel (1962); see also Grothendieck (1960, IV.21.14.1).

Theorem 10.65 (Principal ideals in power series rings) *Let* (R, m) *be a normal, complete, local ring and* $P \subset R[[x_1, \ldots, x_n]]$ *a height 1 prime ideal that is not contained in* $mR[[x_1, \ldots, x_n]]$. *Then* P *is principal.*   □

Corollary 10.66 (Unique factorization in power series rings) *Let (R, m) be a normal, complete, local ring and $g \in R[[x_1, \ldots, x_n]]$ a power series not contained in $mR[[x_1, \ldots, x_n]]$. Then g has a unique factorization as $g = \prod_i p_i$ where each (p_i) is a prime ideal.*

Proof Let P_i be a height 1 associated prime ideal of (g). Then P_i is not contained in $mR[[x_1, \ldots, x_n]]$ thus it is principal by (10.65). □

Example 10.66.1 A lemma of Gauss says that if R is a UFD then $R[t]$ is also a UFD. More generally, if Y is a normal scheme then $\mathrm{Cl}(Y \times \mathbb{A}^n) \simeq \mathrm{Cl}(Y)$. If $\mathbb{A}^n$ is replaced by a smooth variety X then there is an obvious inclusion

$$\mathrm{Cl}(Y) \times \mathrm{Cl}(X) \hookrightarrow \mathrm{Cl}(Y \times X),$$

but, as the next example shows, this map is not surjective, not even if $\mathrm{Cl}(Y) = \mathrm{Cl}(X) = 0$.

Let $E \subset \mathbb{P}^2$ be a cubic defined over $\mathbb{Q}$ such that $\mathrm{Pic}(E)$ is generated by a degree 3 point $P := E \cap L$ for some line $L \subset \mathbb{P}^2$. Let $S \subset \mathbb{A}^3$ be the affine cone over E and $E^\circ := E \setminus P$. Then $\mathrm{Cl}(S) = 0$ and $\mathrm{Cl}(E^\circ) = 0$. However, we claim that $\mathrm{Cl}(S \times E^\circ)$ is infinite.

To see this, pick any $\phi \in \mathrm{End}(E)$. (For example, for any m we have multiplication by $3m + 1$ which sends $p \in E(\bar{\mathbb{Q}})$ to the unique point $\phi(p) \sim (3m + 1)p - mP$.) The lines $\{\ell_p \times \{\phi(p)\} : p \in E\}$ sweep out a divisor in $S \times E$, where $\ell_p \subset S$ denotes the line over $p \in E$. It is not hard to see that this gives an isomorphism $\mathrm{End}(E) \simeq \mathrm{Cl}(S \times E^\circ)$.

As another application, let R denote the complete local ring of S at its vertex. The above considerations also show that R is a UFD, but $R[[t]]$ is not.

Flatness in Relative Codimension 1

The following result is stated in all dimensions, but we will have stronger theorems when the codimension is ≥ 2.

Theorem 10.67 *Let $f : X \to S$ be a finite type morphism of Noetherian schemes, $s \in S$ a closed point, and $Z \subset X_s$ a nowhere dense closed subset such that f is flat along $X_s \setminus Z$. Assume that*
(10.67.1) $\mathrm{pure}_Z(X_s)$ is smooth,
(10.67.2) $\dim S \geq 1$ and S has no embedded points, and
(10.67.3) X has no embedded points.
Then f is smooth.

Proof Pick a closed point $x \in Z$. By (10.57) we may assume that $k(x) = k(s)$. Choose local coordinates $x_1, \ldots, x_n \in m_{x,X_s}$ and apply (10.50). Then there is an elementary étale $\pi \colon (x', X') \to (x, X)$ such that $f \circ \pi$ factors as

$$(x', X') \xrightarrow{g} (y, Y) \xrightarrow{\tau} (s, S),$$

where g is finite, $g^{-1}(y') = \{x'\}$ (as sets), τ is smooth of relative dimension n, and $k(y) = k(s)$. We also know that g_s induces an embedding pure$(X'_s) \to Y_s$.

We claim that g is an isomorphism. To see this, note first that, since $X' \to S$ is flat along $X'_s \setminus Z'$, (10.54) implies that there is a smallest closed subset $W \subset Y$ such that $g^{-1}(X'_s \cap W) \subset Z'$ and g is an isomorphism over $Y \setminus W$. Since $Y \to S$ is smooth, we are done if $W = \emptyset$.

To see this, pick a generic point $w \in W$ with projections $p_Y \in Y$ and $p \in S$. Since Y_p is smooth and $X'_p \to Y_p$ is an isomorphism outside W, we see that pure$_W(X_p) \simeq Y_p$. Thus X'_p has an embedded point in $g^{-1}(W \cap Y_p)$. Therefore p is not a generic point of S by (3). Then

$$\text{depth}_{p_Y} Y = \text{depth}_{p_Y} Y_p + \text{depth}_p S \geq 1 + 1 = 2,$$

and X' has no associated points contained in $g^{-1}(W)$ (3). Hence g is an isomorphism by (10.6).   □

In codimension 1, an slc pair is either smooth or has nodes. Next we show that a close analog of (10.67) holds for nodal fibers if the base scheme is normal; the latter assumption is necessary by (10.70.1).

Corollary 10.68 *Let* (s, S) *be a normal, local scheme and* $f \colon X \to S$ *a finite type morphism of pure relative dimension 1. Assume that f is generically flat along X_s and* pure(X_s) *has a single singular point x, which is a node. Then, in a neighborhood of x, one of the following holds:*

(10.68.1) f *is flat and its fibers have only nodes.*

(10.68.2) f *is not flat, X is not S_2 and the normalization* $\bar{f} \colon \bar{X} \to S$ *is smooth.*

Proof By (10.51), after étale coordinate changes, we may assume that X is a partial normalization of a relative hypersurface $H = (h = 0) \subset \mathbb{A}^2_S$ such that h_s has a single node.

If the generic fiber H_g is smooth, then H is normal and so $X = H$. Otherwise, $\partial h / \partial x = \partial h / \partial y = 0$ is an étale section. After an étale base change, we may assume that the fibers are singular along the zero section $Z \subset \mathbb{A}^2_S \to S$. Blowing it up gives the normalization $\tau \colon \bar{H} \to H$, which is smooth over S. Furthermore, we have an exact sequence

$$0 \to \mathscr{O}_H \to \tau_* \mathscr{O}_{\bar{H}} \to \mathscr{O}_Z \to 0.$$

Since X lies between $\bar{H}$ and H, there is an ideal sheaf $J \subset \mathcal{O}_Z$ such that $\mathcal{O}_X / \mathcal{O}_H \simeq J$.

If $J = 0$ then $X \simeq H$. If $J = \mathcal{O}_Z$, then $X \simeq \bar{H}$. The projection to S is flat in both cases. Otherwise $\mathrm{Supp}(\mathcal{O}_{\bar{H}} / \mathcal{O}_X) = \mathrm{Supp}(\mathcal{O}_Z / J)$ has codimension ≥ 2 in H, thus X is not S_2 by (10.6). □

With different methods, the following generalization of (10.68) is proved in Kollár (2011b). The projectivity assumption should not be necessary.

Theorem 10.69 *Let (s, S) be a normal, local scheme and $f\colon X \to S$ a projective morphism of pure relative dimension 1. Assume that X is S_2 and $\mathrm{pure}(X_s)$ is seminormal (resp. has only simple, planar singularities).*

Then f is flat with reduced fibers that are seminormal (resp. have only simple, planar singularities). □

See Arnol'd et al. (1985, I.p.245) for the conceptual definition of simple, planar singularities. For us, it is quickest to note that a plane curve singularity $(f(x, y) = 0)$ is simple iff $(z^2 + f(x, y) = 0)$ is a Du Val surface singularity.

Examples 10.70 The next examples show that (10.68–10.69) do not generalize to nonnormal bases or to other curve singularities.

10.70.1 (*Deformations of ordinary double points*) Let $C \subset \mathbb{P}^2$ be a nodal cubic with normalization $p\colon \mathbb{P}^1 \to C$. Over the coordinate axes $S := (xy = 0) \subset \mathbb{A}^2$ consider the family X that is obtained as follows.

Over the x-axis take a smoothing of C, over the y-axis take $\mathbb{P}^1 \times \mathbb{A}^1_y$ and glue them over the origin using $p\colon \mathbb{P}^1 \to C$ to get $f\colon X \to S$.

Then X is seminormal and S_2, the central fiber is C with an embedded point, yet f is not flat.

10.70.2 (*Deformations of ordinary triple points*) Consider the family of plane cubic curves

$$\mathbf{C} := ((x^2 - y^2)(x + t) + t(x^3 + y^3) = 0) \subset \mathbb{A}^2_{xy} \times \mathbb{A}^1_t.$$

For every t, the origin is a singular point, but it has multiplicity 3 for $t = 0$ and multiplicity 2 for $t \neq 0$. Thus blowing up the line $(x = y = 0)$ gives the normalization for $t \neq 0$, but it introduces an extra exceptional curve over $t = 0$. The normalization of $\mathbf{C}$ is obtained by contracting this extra curve. The fiber over $t = 0$ is then isomorphic to three lines though the origin in $\mathbb{A}^3$.

10.70.3 (*Deformations of ordinary quadruple points*) Let $\mathbf{C}_4 \to \mathbb{P}^{14}$ be the universal family of degree 4 plane curves and $\mathbf{C}_{4,1} \to S^{12}$ the 12-dimensional subfamily whose general members are elliptic curves with two nodes. We normalize both the base and the total space to get $\bar{\pi} \colon \bar{C}_{4,1} \to \bar{S}^{12}$.

We claim that the fiber of $\bar{\pi}$ over the plane quartic with an ordinary quadruple point $C_0 \colon = (x^3 y - x y^3 = 0)$ is C_0 with at least two embedded points. Most likely, the family is not even flat.

We prove this by showing that in different families of curves through $[C_0] \in S^{12}$ we get different flat limits.

To see this, note that the seminormalization C_0^{sn} of C_0 can be thought of as four general lines through a point in $\mathbb{P}^4$. In suitable affine coordinates, we can write it as $k[x, y]/(x^3 y - x y^3) \hookrightarrow k[u_1, \ldots, u_4]/(u_i u_j \colon i \neq j)$ using the map $(x, y) \mapsto (u_1 + u_3 + u_4, u_2 + u_3 - u_4)$. Any three-dimensional linear subspace $\langle u_1, \ldots, u_4 \rangle \supset W_\lambda \supset \langle u_1 + u_3 + u_4, u_2 + u_3 - u_4 \rangle$. corresponds to a projection of C_0^{sn} to $\mathbb{P}^3$; call the image $C_\lambda \subset \mathbb{P}^3$. Then C_λ is four general lines through a point in $\mathbb{P}^3$; thus it is a $(2, 2)$-complete intersection curve of arithmetic genus 1. (Note that the C_λ are isomorphic to each other, but the isomorphism will not commute with the map to C_0 in general.) Every C_λ can be realized as the special fiber in a family $S_\lambda \to B_\lambda$ of $(2, 2)$-complete intersection curves in $\mathbb{P}^3$ whose general fiber is a smooth elliptic curve.

By projecting these families to $\mathbb{P}^2$, we get a one-parameter family $S'_\lambda \to B_\lambda$ of curves in S^{12} whose special fiber is C_0.

Let $\bar{S}'_\lambda \subset \bar{C}_{4,1}$ be the preimage of this family in the normalization. Then $\bar{S}'_\lambda$ is dominated by the surface S_λ. In particular, the preimage of C_0 in $\bar{C}_{4,1}$ is connected.

There are two possibilities. First, if $\bar{S}'_\lambda$ is isomorphic to S_λ, then the fiber of $\bar{C}_{4,1} \to \bar{S}^{12}$ over $[C_0]$ is C_λ. This, however, depends on λ, a contradiction. Second, if $\bar{S}'_\lambda$ is not isomorphic to S_λ, then the fiber of $\bar{S}'_\lambda \to B_\lambda$ over the origin is C_0 with some embedded points. Since C_0 has arithmetic genus 3, we must have at least two embedded points.

Flatness in Relative Codimension ≥ 2

Once we know flatness at codimension 1 points of the fibers, the following general result, valid for coherent sheaves, can be used to prove flatness everywhere. We no longer need any restrictions on the base scheme S.

Theorem 10.71 *Let* $f \colon X \to S$ *be a finite type morphism of Noetherian schemes,* (s, S) *local. Let* F *be a vertically pure coherent sheaf on* X *and* $Z \subset \operatorname{Supp} F_s$ *a nowhere dense closed subset. Assume that*

(10.71.1) $\operatorname{depth}_Z \operatorname{pure}_Z(F_s) \geq 2$, *and*

(10.71.2) F *is flat over* S *along* $X \setminus Z$.

Then F is flat over S and $\text{tors}_Z(F_s) = 0$.

Proof Set $m := m_{s,S}$ and $X_n := \text{Spec}_X(\mathscr{O}_X/m_{s,S}^n \mathscr{O}_X)$ and $F_n := F|_{X_n}$. We may assume that S is m-adically complete. There are natural complexes

$$0 \to (m^n/m^{n+1}) \cdot F_0 \to F_{n+1} \xrightarrow{r_n} F_n \to 0, \qquad (10.71.3)$$

which are exact on $X \setminus Z$, but not (yet) known to be exact along Z, except that r_n is surjective. We also know that

$$(m^n/m^{n+1}) \cdot \text{pure}_Z(F_0) \to \text{pure}_Z(\ker r_n) \qquad (10.71.4)$$

is an isomorphism on $X \setminus Z$. Since $\text{depth}_Z \text{pure}_Z(F_0) \geq 2$, this implies that (10.71.4) is an isomorphism on X by (10.6). Next we show that the induced

$$r_n: \text{tors}_Z(F_{n+1}) \to \text{tors}_Z(F_n) \quad \text{is surjective.} \qquad (10.71.5)$$

Set $K_{n+1} := r_n^{-1}(\text{tors}_Z(F_n))$. We have an exact sequence

$$0 \to \text{pure}_Z(\ker r_n) \to K_{n+1}/\text{tors}_Z(\ker r_n) \to \text{tors}_Z(F_n) \to 0. \qquad (10.71.6)$$

Using that (10.71.4) is an isomorphism, we have $\text{depth}_Z \text{pure}_Z(\ker r_n) \geq 2$, hence the sequence (10.71.6) splits by (10.6).

Thus $T := \varprojlim \text{tors}_Z(F_n)$ is a subsheaf of F and $X_s \cap \text{Supp}\, T \subset Z$. Thus $T = 0$ since F is vertically pure, and $\text{tors}_Z(F_n) = 0$ for every n by (10.71.5).

Now (10.71.4) says that $(m^n/m^{n+1}) \cdot F_0 \simeq \ker r_n$. Therefore the sequences (10.71.3) are exact, F is flat and $\text{tors}_Z(F_0) = 0$. $\quad\square$

Putting together the flatness criteria (10.60), (10.68), (10.69.1) and (10.71) gives the following strengthening of Hironaka (1958).

Theorem 10.72 *Let* (s, S) *be a normal, local, excellent scheme,* X *an* S_2 *scheme, and* $f: X \to S$ *a finite type morphism of pure relative dimension* n. *Assume that* $\text{pure}(X_s)$ *is*

(10.72.1) *either geometrically normal*

(10.72.2) *or geometrically seminormal and* S_2.

Then f *is flat with reduced fibers that are normal in case (1) and seminormal and* S_2 *in case (2).* $\quad\square$

Flatness in Relative Codimension ≥ 3

The following gives an even stronger result in codimension ≥ 3; see Kollár (1995a, thm.12). Lee and Nakayama (2018) pointed out that the purity assumption in (2) is also necessary.

Theorem 10.73 *Let* $f: X \to S$ *be a finite type morphism of Noetherian schemes,* (s, S) *local. Let* F *a coherent sheaf on* X *and* $Z \subset \operatorname{Supp} F$ *a closed subset such that* $X_s \cap Z \subset \operatorname{Supp} F_s$ *has codimension* ≥ 3. *Let* $j: X_s \setminus Z \hookrightarrow X_s$ *be the natural injection. Assume that*

(10.73.1) $\operatorname{depth}_{X_s \cap Z}(j_*(F_s|_{X_s \setminus Z})) \geq 3$,

(10.73.2) $F|_{X \setminus Z}$ *is flat over* S *with pure,* S_2 *fibers, and*

(10.73.3) $\operatorname{depth}_Z F \geq 2$.

Then F *is flat over* S *and* $F_s = j_*(F_s|_{X_s \setminus Z})$.

Proof Set $m := m_{s,S}$, $X_n := \operatorname{Spec}_X(\mathscr{O}_X / m^n \mathscr{O}_X)$ and $F_n := F|_{X_n}$. We may assume that $\mathscr{O}_S$ and $\mathscr{O}_X$ are m-adically complete. Set $G_n := F_n|_{X_n \setminus Z}$ and let j denote any of the injections $X_n \setminus Z \hookrightarrow X_n$. By assumption (2) we have exact sequences

$$0 \to (m^n/m^{n+1}) \cdot G_0 \to G_{n+1} \longrightarrow G_n \to 0. \tag{10.73.4}$$

Pushing it forward we get the exact sequences

$$0 \to (m^n/m^{n+1}) \otimes j_*G_0 \to j_*G_{n+1} \overset{r_n}{\to} j_*G_n \\ \to (m^n/m^{n+1}) \otimes R^1 j_*G_0. \tag{10.73.5}$$

Here j_*G_0 is coherent and assumption (1) implies (in fact is equivalent to) $R^1 j_*G_0 = 0$ by Grothendieck (1968, III.3.3, II.6 and I.2.9) or (10.29).

Thus the r_n are surjective. This shows that $G := \varprojlim j_*G_n$ is a coherent sheaf on X that is flat over S with S_2 fibers. Furthermore, the natural map $\varrho: F \to G$ is an isomorphism along $X_s \setminus Z$. Thus (10.6) implies that it is an isomorphism. So $F \simeq G$ is flat with central fiber $j_*G_0 = j_*(F_s|_{X_s \setminus Z})$. □

10.8 Seminormality and Weak Normality

Normalization is a very useful operation that can be used to "improve" a scheme X. However, the normalization $X^n \to X$ usually creates new points, and this makes it harder to relate X and X^n. The notions of semi and weak normalization intend to do as much of the normalization as possible, without creating new points.

Definition 10.74 Let X be a Noetherian scheme and $Z \subset X$ a closed, nowhere dense subset. A *finite modification* of X *centered* at Z is a finite morphism $p: Y \to X$ such that the restriction $p: Y \setminus p^{-1}(Z) \to X \setminus Z$ is an isomorphism and none of the associated primes of Y is contained in $p^{-1}(Z)$.

A pair $(Z \subset X)$ is called *normal* if every finite modification $p\colon Y \to X$ centered at Z is an isomorphism. It is called *seminormal* (resp. *weakly normal*) if such a p is an isomorphism, provided $k(x) \hookrightarrow k(\operatorname{red} p^{-1}(x))$ is an isomorphism (resp. purely inseparable) for every $x \in X$.

A reduced scheme X is *normal* (resp. *seminormal* or *weakly normal*) if every pair $(Z \subset X)$ is normal (resp. seminormal or weakly normal).

Let X be a reduced scheme with normalization X^n. There are unique

$$X^n \longrightarrow X^{sn} \overset{\pi_{sn}}{\longrightarrow} X \qquad \text{and} \qquad X^n \longrightarrow X^{wn} \overset{\pi_{wn}}{\longrightarrow} X,$$

where X^{sn} is seminormal, X^{wn} is weakly normal, and $k(x) \hookrightarrow k(\operatorname{red} \pi_{sn}^{-1}(x))$ (resp. $k(x) \hookrightarrow k(\operatorname{red} \pi_{wn}^{-1}(x))$) is an isomorphism (resp. purely inseparable) for every $x \in X$. Note that $X^{wn} = X^{sn}$ in characteristic 0.

For more details, see Kollár (1996, sec.I.7.2) and Kollár (2013b, sec.10.2).

Examples 10.75 The curve examples led to the general definition of seminormalization, but they do not adequately show how complicated seminormal schemes are in higher dimensions.

(10.75.1) The normalization of the higher cusps $C_{2m+1} := (x^2 = y^{2m+1})$ is

$$\pi_{2m+1}\colon \mathbb{A}_t^1 \to C_{2m+1} \quad \text{given by} \quad t \mapsto (t^{2m+1}, t^2).$$

The map π_{2m+1} is a homeomorphism, so it is also the seminormalization. By contrast, the normalization of the higher tacnode $C_{2m} := (x^2 = y^{2m})$ is

$$\pi_{2m}\colon \mathbb{A}_t^1 \times \{\pm 1\} \to C_{2m} \quad \text{given by} \quad (t, \pm 1) \mapsto (\pm t^m, t).$$

The map π_{2m} is not a homeomorphism since $(0,0) \in C_{2m}$ has two preimages, $(0,1)$ and $(0,-1)$. The seminormalization of C_{2m} is

$$\tau_{2m}\colon C_2 \simeq (s^2 = t^2) \to C_{2m} \quad \text{given by} \quad (s,t) \mapsto (s^m, t).$$

(10.75.2) Let $g(t) \in k[t]$ be a polynomial without multiple factors and set $C_g := \operatorname{Spec}_k(k + g \cdot k[t])$. We can think of C_g as obtained from $\mathbb{A}^1$ by identifying all roots of g. It is an integral curve whose normalization is $\mathbb{A}^1$. It has a unique singular point $c_g \in C_g$ and $k(c_g) = k$.

If g is separable then C_g is seminormal and weakly normal. If g is irreducible and purely inseparable then C_g is seminormal, but not weakly normal; the weak normalization is $\mathbb{A}^1$.

(10.75.3) If B is a seminormal curve, then every irreducible component of B is also seminormal, but an irreducible component of a seminormal scheme need not be seminormal. In fact, every reduced and irreducible affine variety that is smooth in codimension 1, occurs as an irreducible component of a seminormal complete intersection scheme, see Kollár (2013b, 10.12).

(10.75.4) If X is S_2 (but possibly nonreduced) and Z has codimension ≥ 2, then $(Z \subset X)$ is a normal pair by (10.6).

The following properties are proved in Kollár (2016c). The last equivalence is surprising since the completion of a normal local ring is not always normal.

Proposition 10.76 *For a Noetherian scheme X without isolated points, the following are equivalent.*

(10.76.1) *X is normal (resp. seminormal, weakly normal).*

(10.76.2) *$Z \subset X$ is a normal (resp. seminormal, weakly normal) pair for every closed, nowhere dense subset $Z \subset X$.*

(10.76.3) *$\{x\} \subset \operatorname{Spec} \mathscr{O}_{x,X}$ is a normal (resp. seminormal, weakly normal) pair for every nongeneric point $x \in X$.*

(10.76.4) *$\{x\} \subset \operatorname{Spec} \widehat{\mathscr{O}}_{x,X}$ is a normal (resp. seminormal, weakly normal) pair for every nongeneric point $x \in X$.*   □

The next results show that many questions about schemes can be settled using points and specializations only, up to homeomorphisms.

Definition 10.77 Let $f \colon X \to Y$ be a morphism, R a DVR and $q \colon \operatorname{Spec} R \to Y$ a morphism. We say that q *lifts after a finite extension* if there is a DVR $R' \supset R$ that is the localization of a finite extension of R such that $q' \colon \operatorname{Spec} R' \to \operatorname{Spec} R \to Y$ lifts to $q'_X \colon \operatorname{Spec} R' \to X$.

10.78 (Universal homeomorphism) A morphism $f \colon U \to V$ of S-schemes is a *universal homeomorphism* if $f \times_S 1_W \colon U \times_S W \to V \times_S W$ is a homeomorphism for every S-scheme W; see Stacks (2022, tag 04DC). Equivalently, if f is integral, surjective and geometrically injective, see Stacks (2022, tag 04DF).

The following characterization for local schemes is simple, but useful.

Claim 10.78.1 Let (s, S) be a local scheme and $f \colon U \to S$ a finite type morphism that is geometrically injective. Then f is a finite, universal homeomorphism iff every local, component-wise dominant (4.30) morphism from the spectrum of a DVR to S, lifts to U, after a finite extension.

Proof For any generic point $g_S \in S$ there is a $q \colon (t, T) \to (s, S)$ such that $q(t_g) = s_g$ and $q(t) = s$ where T is the spectrum of a DVR. Thus every irreducible component of S is dominated by a unique irreducible component of U. Let $V \subset U$ be their union. Extend $f|_V$ to a finite $h \colon \bar{V} \to S$.

Pick a point $\bar{v} \in g^{-1}(s)$. There is a $q \colon T \to \bar{V}$ such that $q(t) = \bar{v}$ and $q(t_g)$ is a generic point of $\bar{V}$. Then q is the only possible lifting of $h \circ q$, hence $\bar{v} \in V$. Thus $V = \bar{V}$ and h is a universal homeomorphism. Since f is geometrically injective we must have $V = U$. □

The following is a special case of Stacks (2022, tag 0CNF).

Claim 10.78.2 A finite morphism $Y \to X$ of $\mathbb{F}_p$-schemes is a universal homeomorphism iff it factors a power of the Frobenius $F_q \colon X_q \to Y \to X$. □

Definition 10.79 For a scheme X let $|X|$ denote its underlying point set. Let X, Y be reduced schemes and $\phi \colon |X| \to |Y|$ a set-map of the underlying sets. We say that ϕ is a *morphism* if there is a morphism $\Phi \colon X \to Y$ inducing ϕ. Note that such a Φ is unique since its graph is determined by its points.

Our aim is to find simple conditions that guarantee that a subset is Zariski closed or that a set-map is a morphism.

We say that ϕ is a *morphism on points* if the natural inclusion $k(x) \hookrightarrow k(x, \phi(x))$ is an isomorphism for every $x \in X$, where we view $(x, \phi(x))$ as a point in $X \times Y$. (This in effect says that there is a natural injection $k(\phi(x)) \hookrightarrow k(x)$.)

We say that ϕ is a *morphism on DVRs* (resp. component-wise dominant DVRs) if the composite $\phi \circ h$ is a morphism whenever $h \colon T \to X$ is a morphism (resp. a component-wise dominant morphism (4.30)) from the spectrum of a DVR to X.

Lemma 10.80 (Valuative criterion of being a section) *Let $h \colon X \to S$ be a separated morphism of finite type and $B \subset |X|$ a subset. Then there is a Zariski closed $Z \subset X$ such that $B = |Z|$ and $h|_Z \colon Z \to S$ is a finite, universal homeomorphism (10.78) iff every point $s \in S$ has a unique preimage $b_s \in B$, $k(b_s)/k(s)$ is purely inseparable, and the following holds.*

Let R be an excellent DVR and $q \colon \operatorname{Spec} R \to S$ a component-wise dominant morphism. Then q lifts after a finite extension (10.77) to $q' \colon \operatorname{Spec} R' \to X$ whose image is in B.

Proof By assumption, $h|_B \colon B \to S$ is a universal bijection. Let $s_g \in S$ be a generic point and $b_g \in B$ its preimage. We claim that $\bar{b}_g \subset B$. For any $b_0 \in \bar{b}_g$, there is a component-wise dominant morphism $\tau \colon (t, T) \to S$ that maps the generic point to $h(b_g)$ and the closed point to $h(b_0)$, where T is the spectrum of a DVR. Lifting it shows that $b_0 \in B$.

Thus Z is the union of all $\bar{b}_g$, hence Zariski closed and $h|_Z \colon Z \to S$ is a finite, universal bijection, hence a homeomorphism. □

Lemma 10.81 (Valuative criterion of being a morphism) *Let X, Y be schemes of finite type, X seminormal, and Y separated. Then a set-map $\phi \colon |X| \to |Y|$ is a morphism iff it is a morphism on points and on component-wise dominant DVRs.*

Proof Let $Z \subset X \times Y$ be the graph of ϕ and $h \colon X \times Y \to X$ the projection. By (10.80) $h|_Z \colon Z \to X$ is a finite, universal homeomorphism that is residue field preserving since ϕ is a morphism on points. Thus $h|_Z \colon Z \to X$ is an isomorphism since X is seminormal.   □

Definition 10.82 A morphism $p \colon X \to Y$ is *geometrically injective* if for every geometric point $\bar{y} \to Y$ the fiber $X \times_Y \bar{y}$ consists of at most one point.

Equivalently, for every point $y \in Y$, its preimage $p^{-1}(y)$ is either empty or a single point and $k(p^{-1}(y))$ is a purely inseparable extension of $k(y)$.

If, furthermore, $k(p^{-1}(y))$ equals $k(y)$ then we say that p *preserves residue fields*. The two notions are equivalent in characteristic 0.

A morphism of schemes $f \colon X \to Y$ is a *monomorphism* if for every scheme Z, the induced map of sets $\mathrm{Mor}(Z, X) \to \mathrm{Mor}(Z, Y)$ is an injection.

A monomorphism is geometrically injective. The normalization of the cusp $\pi \colon \mathrm{Spec}\, k[t] \to \mathrm{Spec}\, k[t^2, t^3]$ is geometrically injective, but not a monomorphism. The problem is with the fiber over the origin, which is $\mathrm{Spec}\, k[t]/(t^2) \simeq \mathrm{Spec}\, k[\varepsilon]$ (where $\varepsilon^2 = 0$). The two maps $g_i \colon \mathrm{Spec}\, k[\varepsilon] \to \mathrm{Spec}\, k[t]$ given by $g_0^*(t) = 0$ and $g_1^*(t) = \varepsilon$ are different, but $\pi \circ g_0 = \pi \circ g_1$. A similar argument shows that a morphism is a monomorphism iff it is geometrically injective and unramified; see Grothendieck (1960, IV.17.2.6).

As this example shows, in order to understand when a map between moduli spaces is a monomorphism, the key is to study the corresponding functors over $\mathrm{Spec}\, k[\varepsilon]$ for all fields k.

See (1.64) for an example that is geometrically bijective but, unexpectedly, not a monomorphism.

A closed, open or locally closed embedding is a monomorphism. A typical example of a monomorphism that is not a locally closed embedding is the normalization of the node with a point missing, that is $\mathbb{A}^1 \setminus \{-1\} \to (y^2 = x^3 + x^2)$ given by $(t \mapsto (t^2 - 1, t^3 - t))$.

Claim 10.82.1 (Stacks, 2022, tag 04XV) A proper monomorphism $f \colon X \to Y$ is a closed embedding.   □

Definition 10.83 A morphism $g \colon X \to Y$ is a *locally closed embedding* if it can be factored as $g \colon X \to Y^\circ \hookrightarrow Y$ where $X \to Y^\circ$ is a closed embedding and $Y^\circ \hookrightarrow Y$ is an open embedding.

A monomorphism $g: X \to Y$ is called a *locally closed partial decomposition* of Y if the restriction of g to every connected component $X_i \subset X$ is a locally closed embedding.

If g is also surjective, it is called a *locally closed decomposition* of Y. For reduced schemes, the key example is the following.

Claim 10.83.1 Let $h: Y \to \mathbb{Z}$ be a constructible, upper semi-continuous function and set $Y_i := \{y \in Y: h(y) = i\}$. Then $\amalg_i Y_i \to Y$ is a locally closed decomposition. □

The following direct consequence of (10.82.1) is quite useful.

Claim 10.83.2 A proper, locally closed partial decomposition $g: X \to Y$ is a closed embedding. If Y is reduced, then a proper, locally closed decomposition $g: X \to Y$ is an isomorphism. □

Proposition 10.84 (Valuative criterion of locally closed embedding) *For a geometrically injective morphism of finite type $f: X \to Y$, the following are equivalent.*

(10.84.1) $f(X) \subset Y$ *is locally closed and $X \to f(X)$ is finite.*

(10.84.2) *Every component-wise dominant morphism, from the spectrum of an excellent DVR to $f(X)$, lifts to X, after a finite extension (10.77).*

If f is a monomorphism, then these are further equivalent to

(10.84.3) *f is a locally closed embedding.*

Proof It is clear that $(1) \Rightarrow (2)$. Next assume (2). A geometrically injective morphism of finite type is quasi-finite, hence, by Zariski's main theorem, there is a finite morphism $\bar{f}: \bar{X} \to Y$ extending f. Set $Z := \bar{X} \setminus X$.

If $Z \neq \bar{f}^{-1}\bar{f}(Z)$ then there are points $z \in Z$ and $x \in X$ such that $\bar{f}(z) = \bar{f}(x)$. Let T be the spectrum of a DVR and $h: T \to \bar{X}$ a component-wise dominant morphism. Set $g := \bar{f} \circ h$. Then $g(T) \subset f(X)$ and the only lifting of g to $T \to \bar{X}$ is h, but $h(T) \not\subset X$.

Thus $Z = \bar{f}^{-1}\bar{f}(Z)$ hence $X \to Y \setminus \bar{f}(Z)$ is proper, proving (1). A proper monomorphism is a closed embedding by (10.82.1), showing the equivalence with (3). □

A major advantage of seminormality over normality is that seminormalization $X \mapsto X^{\mathrm{sn}}$ is a functor from the category of excellent schemes to the category of excellent seminormal schemes. (The injection $\mathrm{Sing}\, X \hookrightarrow X$ rarely lifts to the normalizations.) It is thus reasonable to expect that taking the coarse moduli space commutes with seminormalization. This is indeed the case for coarse moduli spaces satisfying the following mild condition.

Definition 10.85 A functor $\mathcal{M}$: (schemes) → (sets) with coarse moduli space M has *enough one-parameter families* if the following holds.

(10.85.1) Let R be a DVR and $\operatorname{Spec} R \to M$ a morphism. Then there is a DVR $R' \supset R$ that is the localization of a finite extension of R and $F' \in \mathcal{M}(\operatorname{Spec} R')$ such that $\operatorname{Spec} R' \to \operatorname{Spec} R \to M$ is the moduli map of F'.

This condition holds if M is obtained as a quotient $M = E/G$, where G is an algebraic group acting properly on E and there is a universal family over E. Thus it is satisfied by all moduli spaces considered in this book.

Proposition 10.86 *Let $\mathcal{M}$: (schemes) → (sets) be a functor defined on finite type schemes over a field of characteristic 0. Assume that $\mathcal{M}$ has a finite type coarse moduli space M and enough one-parameter families.*

Then M^{sn} is the coarse moduli space for $\mathcal{M}^{sn}$, the restriction of $\mathcal{M}$ to the category Sch^{sn} of finite type, seminormal schemes.

Proof Since seminormalization is a functor, every morphism $W \to M$ lifts to $W^{sn} \to M^{sn}$. Thus we have a natural transformation $\Phi\colon \mathcal{M}^{sn} \to \operatorname{Mor}(*, M^{sn})$.

Assume that M' is a finite type, seminormal scheme and we have another natural transformation $\Psi\colon \mathcal{M}^{sn} \to \operatorname{Mor}(*, M')$. Every geometric point $s \mapsto M^{sn}$ comes from a scheme X_s. Let $Z \subset M^{sn} \times M'$ denote the union of the points $(s, \Phi[X_s])$. Since M is a coarse moduli space and $M^{sn} \to M$ is geometrically bijective, the coordinate projection $Z \to M^{sn}$ is also geometrically bijective. Since $\mathcal{M}$ has enough one-parameter families, $Z \to M^{sn}$ is a universal homeomorphism by (10.80). Thus $Z \to M^{sn}$ is an isomorphism since M^{sn} is seminormal and the characteristic is 0.

Thus we get a morphism $M^{sn} \to M'$ and Ψ factors through Φ.   □

The next examples show that the characteristic 0 assumption is likely necessary in (10.86) and that the analogous claim for the underlying reduced subscheme is likely to be false.

Examples 10.87 Let $\mathcal{D}$ be any diagram of schemes with direct limit $\lim \mathcal{D}$. Since seminormalization is a functor, we get a diagram $\mathcal{D}^{sn}$ and a natural morphism $\lim(\mathcal{D}^{sn}) \to (\lim \mathcal{D})^{sn}$. However, this need not be an isomorphism.

(10.87.1) Consider the diagram of all maps $\phi_a\colon \operatorname{Spec} k[x] \to \operatorname{Spec} k[(x - a)^2, (x - a)^3]$ for $a \in k$ where k is an infinite field.

If $\operatorname{char} k = 0$ then the direct limit is $\operatorname{Spec} k$. After seminormalization, the maps ϕ_a become isomorphisms $\phi_a^{sn}\colon \operatorname{Spec} k[x] \simeq \operatorname{Spec} k[x]$. Now the direct limit is $\operatorname{Spec} k[x]$.

(10.87.2) If char $k = p > 0$ then $x^p - a^p = (x - a)^p \in k[(x - a)^2, (x - a)^3]$ shows that the direct limit is Spec $k[x^p]$. After seminormalization, the direct limit is again Spec $k[x]$. Here Spec $k[x^p]$ behaves like a coarse moduli space.

(10.87.3) Consider the maps $\sigma_i \colon k[x] \to k[x, \varepsilon]$ given by $\sigma_0(g(x)) = g(x)$ and $\sigma_1(g(x)) = g(x) + g'(0)\varepsilon$. We get a universal push-out diagram

$$
\begin{array}{ccc}
\text{Spec } k[x, \varepsilon] & \xrightarrow{\ \sigma_0\ } & \text{Spec } k[x] \\
\ \downarrow{\sigma_1} & & \downarrow \\
\text{Spec } k[x] & \longrightarrow & \text{Spec } k[x^2, x^3].
\end{array}
$$

If we pass to the underlying reduced subspaces, the push-out is Spec $k[x]$.

11

Minimal Models and Their Singularities

We review the definitions and results of the minimal model program that we used repeatedly.

Assumptions The theorems of Sections 11.1–11.3 are currently known in characteristic 0. See Kollár and Mori (1998) or Kollár (2013b) for varieties; Lyu and Murayama (2022) and Fujino (2022) in general.

Most of the older literature works with $\mathbb{Q}$-divisors. We treat $\mathbb{R}$-divisors on arbitrary schemes in Section 11.4.

11.1 Singularities of Pairs

Singularities of pairs are treated thoroughly in Kollár (2013b). Here we aim to be concise, discussing all that is necessary for the main results in this book, but leaving many details untouched.

Definition 11.1 (Pairs) We are primarily interested in pairs (X, Δ) where X is a normal variety over a field and $\Delta = \sum a_i D_i$ a formal linear combination of prime divisors with rational or real coefficients. More generally, X can be a reduced scheme and $\Delta = \sum a_i D_i$ a formal linear combination of prime, Mumford divisors (4.16.4), that is, none of the D_i are contained in Sing X.

For a prime divisor E, we use $\mathrm{coeff}_E(\Delta)$ to denote the *coefficient* of E in Δ. That is, $E \not\subset \mathrm{Supp}(\Delta - \mathrm{coeff}_E(\Delta) \cdot E)$. We use $\mathrm{coeff}(\Delta)$ to denote the set of all nonzero coefficients in Δ.

If Δ is $\mathbb{R}$-Cartier, $\pi \colon X' \to X$ is birational and E' is a prime divisor on X', then $\mathrm{coeff}_{E'}(\Delta) := \mathrm{coeff}_{E'}(\pi^*\Delta)$ defines the coefficient of every prime divisor over X in Δ.

For any $c \in \mathbb{R}$ we set $\Delta^{>c} := \sum_{i:\, a_i > c} a_i D_i$, and similarly for $\Delta^{=c}, \Delta^{<c}$.

Definition 11.2 (Canonical or dualizing sheaf) A pure dimensional, projective scheme over a field has a dualizing sheaf as in Hartshorne (1977, III.7), but for arbitrary schemes the existence of a dualizing sheaf is a complicated issue. The following quite general setting is sufficient for our purposes.

Let $g \colon X \to S$ be a finite type morphism. As in Stacks (2022, tag 0E9M), there is a *relative dualizing complex*. If X is pure dimensional, the lowest non-zero cohomology of it is the *relative dualizing sheaf*, or *relative canonical sheaf*, denoted by $\omega_{X/S}$.

We are interested in cases where $\omega_{X/S}$ depends very little on S. This happens when $\mathscr{O}_S$ is a dualizing complex on S Stacks (2022, tag 0AWV). We only need to know that this occurs in four important cases:

- S is the spectrum of a field,
- S is smooth over a field,
- S is regular and of dimension 1, or
- S is the spectrum of a regular, local ring.

We declare $\omega_{X/S}$ to be a *canonical sheaf* of X and denote it by ω_X.

Note that we do not need $X \to S$ to be surjective. So if we want to work over a quasi-projective scheme S, we choose an embedding $S \hookrightarrow \mathbb{P}^N$ and work over $\mathbb{P}^N$. Similarly, if S is the spectrum of a complete local ring, we can embed it into the spectrum of a regular, complete local ring. However, ω_X is well defined only up to tensoring with pull-backs by line bundles from S. Thus one should use it only for properties of X that are local on S.

Definition 11.3 (Canonical class II) Let X be a scheme that has a canonical sheaf ω_X. If ω_X is invertible outside a subset of codimension ≥ 2 – for example, X is normal or demi-normal – then it corresponds to a linear equivalence class of Mumford divisors K_X, called the *canonical class* of X.

Assumptions In Sections 11.1–11.3, we work with pairs that have a canonical class.

Definition 11.4 (Discrepancy of divisors) Let $(X, \Delta = \sum a_i D_i)$ be a pair as in (11.1) that has a canonical class (11.3). We are looking at cases when the pull-back of $K_X + \Delta$ by birational morphisms makes sense. If Δ is a $\mathbb{Q}$-divisor, the natural assumption is that $K_X + \Delta$ is $\mathbb{Q}$-Cartier, that is, $m(K_X + \Delta)$ is Cartier for some $m > 0$.

If Δ is an $\mathbb{R}$-divisor, we need to assume that $K_X + \Delta$ is $\mathbb{R}$-*Cartier*, we discuss this notion in detail in Section 11.4. (See (4.48) for the even more general notion of numerically $\mathbb{R}$-Cartier divisors).

Let $f \colon Y \to X$ be a proper, birational morphism from a demi-normal scheme Y (11.36), $\mathrm{Ex}(f) \subset Y$ the exceptional locus, and $E_i \subset \mathrm{Ex}(f)$ the irreducible exceptional divisors. Assume that $\mathrm{Ex}(f) \cap \mathrm{Sing}\, Y$ and $f(\mathrm{Ex}(f))$ have codimension ≥ 2 in Y and X; these are automatic if X and Y are normal. Let $f_*^{-1}\Delta := \sum a_i f_*^{-1} D_i$ denote the birational transform of Δ. Fix any canonical divisor K^Y in the linear equivalence class K_Y and set $K^X := f_*(K^Y)$.

Assume next that $K_X + \Delta$ is $\mathbb{R}$-Cartier. Then $K^Y + f_*^{-1}\Delta - f^*(K^X + \Delta)$ makes sense and it is exceptional, hence we can write

$$K^Y + f_*^{-1}\Delta = f^*(K^X + \Delta) + \sum_i a(E_i, X, \Delta)E_i. \tag{11.4.1}$$

The $a(E_i, X, \Delta) \in \mathbb{R}$ are independent of the choice of K^Y. This defines $a(E, X, \Delta)$ for exceptional divisors. Set $a(E, X, \Delta) := -\mathrm{coeff}_E \Delta$ for non-exceptional divisors $E \subset X$.

The real number $a(E, X, \Delta)$ is called the *discrepancy* of E with respect to (X, Δ); it depends only on the valuation defined by E, not on the choice of f. (See Kollár and Mori (1998, 2.22) for a more canonical definition.)

Warning 11.4.2 For most cases of interest to us, $a(E, X, \Delta) \geq -1$, so some authors use *log discrepancies*, $a_\ell(E, X, \Delta) := 1 + a(E, X, \Delta)$. Unfortunately, some people use $a(E, X, \Delta)$ to denote the log discrepancy, leading to confusion.

The discrepancies have the following obvious monotonicity and linearity properties; see Kollár and Mori (1998, 2.27).

Claim 11.4.3 Let Δ' be an effective, $\mathbb{R}$-Cartier divisor and E a divisor over X. Then $a(E, X, \Delta + \Delta') = a(E, X, \Delta) - \mathrm{coeff}_E \Delta'$. In particular, $a(E, X, \Delta + \Delta') \leq a(E, X, \Delta)$, and $a(E, X, \Delta + \Delta') < a(E, X, \Delta)$ iff $\mathrm{center}_X E \subset \mathrm{Supp}\, \Delta'$. □

Claim 11.4.4 Assume that $K_X + \Delta_i$ are $\mathbb{R}$-Cartier. Fix $\lambda_i \geq 0$ such that $\sum \lambda_i = 1$ and set $\Delta := \sum \lambda_i \Delta_i$. Then $K_X + \Delta$ is $\mathbb{R}$-Cartier and $a(E, X, \Delta) = \sum \lambda_i a(E, X, \Delta_i)$ for every divisor E over X. In particular, using the next definition, if the (X, Δ_i) are lc (resp. dlt, klt, canonical, terminal) then so is (X, Δ). □

Definition 11.5 Let X be a normal scheme of dimension ≥ 2 and $\Delta = \sum a_i D_i$ an $\mathbb{R}$-divisor such that $K_X + \Delta$ is $\mathbb{R}$-Cartier. We say that (X, Δ) is

terminal			> 0	for every exceptional E,
canonical			≥ 0	for every exceptional E,
klt	if $a(E, X, \Delta)$ is		> -1	for every E,
plt			> -1	for every exceptional E,
dlt			> -1	if $\mathrm{center}_X E \subset \mathrm{non\text{-}snc}(X, \Delta)$,
lc			≥ -1	for every E.

Here klt is short for *Kawamata log terminal*, plt for *purely log terminal, dlt* for *divisorial log terminal,* lc for *log canonical,* and non-snc(X, Δ) denotes the set of points where (X, Δ) is not a simple normal crossing pair (p.xvi). We define semi-log-canonical or slc pairs in (11.37).

Claim 11.5.1 If (X, Δ) is in any of these 6 classes, $0 \leq \Delta' \leq \Delta$ and $K_X + \Delta'$ is $\mathbb{R}$-Cartier, then (X, Δ') is also in the same class. □

Claim 11.5.2 Assume that (X, Δ) is terminal (resp. klt) and Θ is an effective $\mathbb{R}$-Cartier divisor. If (X, Δ) has a log resolution (p.xvi), then $(X, \Delta + \varepsilon\Theta)$ is also terminal (resp. klt) for $0 \leq \varepsilon \ll 1$. (See (11.10.6) for the other cases.)

We gave some examples in (1.33) and (1.40); see also Section 2.2 for such surfaces, (2.35) for cones, and Kollár (2013b) for a detailed treatment.

For computing discrepancies, the following are useful; see also (Kollár and Mori, 1998, 2.29–30).

Lemma 11.6 *Let* $(X, \Delta - \Theta)$ *be an snc pair, where* $\Delta = \sum(1 - a_i)D_i$ *and* Θ *are effective. Let* E *be a divisor over* X *such that* $a(E, X, \Delta - \Theta) < 0$. *Then* $a(E, X, \lceil\Delta\rceil) = -1$ *and* $a(E, X, \Delta - \Theta) \geq a(E, X, \Delta) = -1 + \sum a_i \cdot \mathrm{coeff}_E D_i < 0$.

Proof $(X, \lceil\Delta\rceil)$ is lc by Kollár and Mori (1998, 2.31), so $a(E, X, \Delta - \Theta) \geq a(E, X, \lceil\Delta\rceil) = -1$ by (11.4.3). The rest follows from (11.4.3.a). □

Corollary 11.7 *Using the notation of (11.6), for every* $\varepsilon > 0$ *there is* $\eta > 0$ *such that the following holds.*

Let $(X, \Delta' - \Theta')$ *be a pair, where* $\mathrm{Supp}\,\Theta = \mathrm{Supp}\,\Theta'$ *and* $\Delta' = \sum(1 - a_i')D_i$ *such that* $|a_i - a_i'| < \eta$ *for every* i *and* $a_i' = 0$ *iff* $a_i = 0$. *Then, for every* E,

$$\left| a(E, X, \Delta - \Theta) - a(E, X, \Delta' - \Theta') \right| < \varepsilon,$$

whenever one of the discrepancies is < 0. □

Definition 11.8 Let (X, Δ) be an lc or slc (11.37) pair and $W \subset X$ an irreducible, closed subset. The *minimal log discrepancy* of W is defined as the infimum of the numbers $1 + a(E, X, \Delta)$ where E runs through all divisors over X such that $\mathrm{center}_X(E) = W$. It is denoted by

$$\mathrm{mld}(W, X, \Delta) \quad \text{or by} \quad \mathrm{mld}(W) \tag{11.8.1}$$

if the choice of (X, Δ) is clear. Note that if W is an irreducible divisor on X and $W \not\subset \mathrm{Sing}\,X$ then $\mathrm{mld}(W, X, \Delta) = 1 - \mathrm{coeff}_W \Delta$. If $W \subset X$ is a closed subset with irreducible components W_i, then we set $\mathrm{mld}(W, X, \Delta) = \max_i\{\mathrm{mld}(W_i, X, \Delta)\}$.

If (X, Δ) is slc then, by definition, $\mathrm{mld}(W, X, \Delta) \geq 0$ for every W. The subvarieties with $\mathrm{mld}(W, X, \Delta) = 0$ play a key role in understanding (X, Δ).

Definition 11.9 Let (X, Δ) be an slc pair. An irreducible subset $W \subset X$ is a *log canonical center* or *lc center* of (X, Δ) if $\mathrm{mld}(W, X, \Delta) = 0$. If (X, Δ) has a log resolution, then there is a divisor E over X such that $a(E, X, \Delta) = -1$ and $\mathrm{center}_X E = W$.

11.10 (Properties of log canonical centers) Let (X, Δ) be an slc pair over a field of characteristic 0. (11.10.1) There are only finitely many lc centers.
(11.10.2) Any union of lc centers is seminormal and Du Bois (11.12.1–2).
(11.10.3) Any intersection of lc centers is also a union of lc centers; see Ambro (2003, 2011), Fujino (2017), or (11.12.4).
(11.10.4) If (X, Δ) is snc then the lc centers of (X, Δ) are exactly the *strata* of $\Delta^{=1}$, that is, the irreducible components of the various intersections $D_{i_1} \cap \cdots \cap D_{i_s}$ where the $\mathrm{coeff}_{D_{i_k}} \Delta = 1$; see Kollár (2013b, 2.11). More generally, this also holds if (X, Δ) is dlt; see Fujino (2007, sec.3.9) or Kollár (2013b, 4.16).
(11.10.5) At codimension 2 normal points, the union of lc centers is either smooth or has a node; see Kollár (2013b, 2.31).
(11.10.6) Let (X, Δ) be slc and Θ effective, $\mathbb{R}$-Cartier. Then $(X, \Delta + \varepsilon \Theta)$ is slc for $0 < \varepsilon \ll 1$ iff $\mathrm{Supp}\,\Theta$ does not contain any lc center of (X, Δ).
(11.10.7) Assume that (X, Δ) is slc and $\varepsilon \Theta \leq \Delta$ is an effective $\mathbb{Q}$-Cartier divisor. Then $\mathrm{Supp}\,\Theta$ does not contain any lc center of $(X, \Delta - \varepsilon \Theta)$ by (11.4.3).

Definition 11.11 Let (X, Δ) be an slc pair. An irreducible subset $W \subset X$ is a *log center* of (X, Δ) if $\mathrm{mld}(W, X, \Delta) < 1$. (It is frequently convenient to consider every irreducible component of X a log center.)

Building on earlier results of Ambro (2003, 2011), and Fujino (2017), part 1 of the following theorem is proved in Kollár and Kovács (2010). The rest in Kollár (2014); see also Kollár (2013b, chap.7).

Theorem 11.12 *Let (X, Δ) be an slc pair over a field of characteristic 0 and $Z, W \subset X$ closed, reduced subschemes.*
(11.12.1) *If $\mathrm{mld}(Z, X, \Delta) = 0$, then Z is Du Bois.*
(11.12.2) *If $\mathrm{mld}(Z, X, \Delta) < \frac{1}{6}$, then Z is seminormal (10.74).*
(11.12.3) *If $\mathrm{mld}(Z, X, \Delta) + \mathrm{mld}(W, X, \Delta) < \frac{1}{2}$, then $Z \cap W$ is reduced.*
(11.12.4) $\mathrm{mld}(Z \cap W, X, \Delta) \leq \mathrm{mld}(Z, X, \Delta) + \mathrm{mld}(W, X, \Delta)$.   □

Adjunction is a classical method that allows induction on the dimension by lifting information from divisors to the ambient scheme.

Definition 11.13 (Poincaré residue map) Let X be a (pure dimensional) CM scheme and $S \subset X$ a divisorial subscheme. Then $\omega_S = \mathcal{E}xt^1(\mathcal{O}_S, \omega_X)$ and $\mathcal{E}xt^1(\mathcal{O}_X, \omega_X) = 0$. Thus, applying $\mathcal{H}om(\ , \omega_X)$ to the exact sequence

$$0 \to \mathcal{O}_X(-S) \to \mathcal{O}_X \to \mathcal{O}_S \to 0,$$

we get the short exact sequence

$$0 \to \omega_X \to \omega_X(S) \overset{\mathcal{R}_S}{\to} \omega_S \to 0. \tag{11.13.1}$$

The map $\mathcal{R}_S : \omega_X(S) \to \omega_S$ is called the *Poincaré residue map*. By taking tensor powers, we get maps

$$\mathcal{R}_S^{\otimes m} : (\omega_X(S))^{\otimes m} \to \omega_S^{\otimes m},$$

but, if $m(K_X + S)$ and mK_S are Cartier for some $m > 0$ then we really would like to get a corresponding map between the locally free sheaves

$$\omega_X^{[m]}(mS)|_S \overset{???}{\dashrightarrow} \omega_S^{[m]}. \tag{11.13.2}$$

There is no such map in general; one needs a correction term.

Definition 11.14 (Different) Let X be a demi-normal scheme (11.36), S a reduced divisor (p.xv) on X, and Δ an $\mathbb{R}$-divisor on X. We assume that there are no coincidences, that is, the irreducible components of $\operatorname{Supp} S$, $\operatorname{Supp} \Delta$ and $\operatorname{Sing} X$ are all different from each other.

Let $\pi \colon \bar{S} \to S$ denote the normalization. There is a closed subscheme $Z \subset S$ of codimension 1 such that $S \setminus Z$ and $X \setminus Z$ are both smooth along $S \setminus Z$, the restriction $\pi \colon (\bar{S} \setminus \pi^{-1}Z) \to (S \setminus Z)$ is an isomorphism and $\operatorname{Supp} \Delta \cap S \subset Z$.

Assume first that Δ is a $\mathbb{Q}$-divisor and $m(K_X + S + \Delta)$ is Cartier for some $m > 0$. Then the Poincaré residue map (11.13) gives an isomorphism

$$\mathcal{R}_{S \setminus Z}^m \colon \pi^* \omega_X^{[m]}(mS + m\Delta)|_{(\bar{S} \setminus \pi^{-1}Z)} \simeq \omega_{\bar{S}}^{[m]}|_{(\bar{S} \setminus \pi^{-1}Z)}.$$

Hence there is a unique (not necessarily effective) divisor $\Delta_{\bar{S}}$ on $\bar{S}$ supported on $\pi^{-1}Z$ such that $\mathcal{R}_{S \setminus Z}^m$ extends to an isomorphism

$$\mathcal{R}_{\bar{S}}^m \colon \pi^* \omega_X^{[m]}(mS + m\Delta)|_{\bar{S}} \simeq \omega_{\bar{S}}^{[m]}(\Delta_{\bar{S}}). \tag{11.14.1}$$

We formally divide by m and define the *different* of Δ on $\bar{S}$ as the $\mathbb{Q}$-divisor

$$\operatorname{Diff}_{\bar{S}}(\Delta) := \tfrac{1}{m}\Delta_{\bar{S}}. \tag{11.14.2}$$

We can write (11.14.1) in terms of $\mathbb{Q}$-divisors as

$$(K_X + S + \Delta)|_{\bar{S}} \sim_{\mathbb{Q}} K_{\bar{S}} + \operatorname{Diff}_{\bar{S}}(\Delta). \tag{11.14.3}$$

Note that (11.14.3) has the disadvantage that it indicates only that the two sides are $\mathbb{Q}$-linearly equivalent, whereas (11.14.1) is a *canonical* isomorphism.

If $K_X + S + \Delta$ is $\mathbb{R}$-Cartier, then, by (11.43.4), we can write $\Delta = \Delta' + \Delta''$ where $K_X + S + \Delta'$ is $\mathbb{Q}$-Cartier and Δ'' is $\mathbb{R}$-Cartier. Then we set

$$\mathrm{Diff}_{\bar{S}}(\Delta) := \mathrm{Diff}_{\bar{S}}(\Delta') + \pi^* \Delta''. \tag{11.14.4}$$

If X, S are smooth than $K_S = (K_X + S)|_S$, hence in this case $\mathrm{Diff}_{\bar{S}}(\Delta) = \pi^* \Delta$. Let $f \colon Y \to X$ be a proper birational morphism, $S_Y := f_*^{-1} S$ and write $K_Y + S_Y + \Delta_Y \sim_{\mathbb{R}} f^*(K_X + S + \Delta)$. Then

$$\mathrm{Diff}_{\bar{S}}(\Delta) = (f|_{\bar{S}_Y})_* \mathrm{Diff}_{\bar{S}_Y}(\Delta_Y). \tag{11.14.5}$$

Proposition 11.15 (Kollár, 2013b, 4.4–8) *Using the notation of (11.14) write* $\mathrm{Diff}_{\bar{S}}(\Delta) = \sum d_i V_i$ *where $V_i \subset \bar{S}$ are prime divisors. Then the following hold.*

(11.15.1) *If $(X, S + \Delta)$ is lc (or slc) then $(\bar{S}, \mathrm{Diff}_{\bar{S}}(\Delta))$ is lc.*

(11.15.2) *If $\mathrm{coeff}(\Delta) \subset \{1, \frac{1}{2}, \frac{2}{3}, \frac{3}{4}, \dots\}$, then the same holds for $\mathrm{Diff}_{\bar{S}}(\Delta)$.*

(11.15.3) *If S is Cartier, then $\mathrm{Diff}_{\bar{S}}(\Delta) = \pi^* \Delta$.*

(11.15.4) *If $K_X + S$ and D are both Cartier, then $\mathrm{Diff}_{\bar{S}} D$ is a $\mathbb{Z}$-divisor and $(K_X + S + D)|_{\bar{S}} \sim K_{\bar{S}} + \mathrm{Diff}_{\bar{S}} D$.* □

The following facts about codimension 1 behavior of the different can be proved by elementary computations; see Kollár (2013b, 2.31, 2.36).

Lemma 11.16 *Let S be a normal surface, $E \subset S$ a reduced curve and $\Delta = \sum d_i D_i$ an effective $\mathbb{R}$-divisor. Assume that $0 \le d_i \le 1$ and $D_i \not\subset \mathrm{Supp}\, E$ for every i. Let $\pi \colon \bar{E} \to E$ be the normalization and $x \in \bar{E}$ a point.*

(11.16.1) *If E is singular at $\pi(x)$, then $\mathrm{coeff}_x \mathrm{Diff}_{\bar{E}}(\Delta) \ge 1$, and equality holds iff E has a node at $\pi(x)$, E is Cartier at $\pi(x)$ and $\pi(x) \notin \mathrm{Supp}\, \Delta$.*

(11.16.2) *If $\pi(x) \in D_i$, then $\mathrm{coeff}_x \mathrm{Diff}_{\bar{E}}(\Delta) \ge d_i$.* □

The next theorem – proved in Kollár (1992b, 17.4) and Kawakita (2007) – is frequently referred to as *adjunction* if we assume something about X and obtain conclusions about S, or *inversion of adjunction* if we assume something about S and obtain conclusions about X. See Kollár (2013b, 4.8–9) for a proof of a more precise version.

Theorem 11.17 *Let X be a normal scheme over a field of characteristic 0 and S a reduced divisor on X with normalization $\pi_S \colon \bar{S} \to S$. Let Δ be an effective $\mathbb{R}$-divisor that has no irreducible components in common with S and such that $K_X + S + \Delta$ is $\mathbb{R}$-Cartier. Then*

(11.17.1) *$(\bar{S}, \mathrm{Diff}_{\bar{S}}(\Delta))$ is klt iff $(X, S + \Delta)$ is plt in a neighborhood of S, and*

(11.17.2) *$(\bar{S}, \mathrm{Diff}_{\bar{S}}(\Delta))$ is lc iff $(X, S + \Delta)$ is lc in a neighborhood of S.*

(11.17.3) $\mathrm{mld}(Z, \bar{S}, \mathrm{Diff}_{\bar{S}}(\Delta)) = \mathrm{mld}(\pi_S(Z), X, S + \Delta)$ *for any irreducible and closed subset* $Z \subsetneq \bar{S}$, *provided one of them is* ≤ 1.
(11.17.4) *The claims also hold for slc pairs by (11.37).* □

Many divisorial sheaves on an lc pair are Cohen–Macaulay (CM for short). The following variant is due to Kollár and Mori (1998, 5.25) and Fujino (2017, 4.14); see also Kollár (2013b, 2.88).

Theorem 11.18 *Let* (X, Δ) *be a dlt pair over a field of characteristic 0, L a $\mathbb{Q}$-Cartier $\mathbb{Z}$-divisor, and $D \leq \lfloor \Delta \rfloor$ an effective $\mathbb{Z}$-divisor. Then the sheaves $\mathcal{O}_X$, $\mathcal{O}_X(-D - L)$ and $\omega_X(D + L)$ are CM.*
If $D + L$ is effective, then $\mathcal{O}_{D+L}$ is also CM. □

We also need the following; see Kollár (2011a) or Kollár (2013b, 7.31).

Theorem 11.19 *Let* (X, Δ) *be dlt over a field of characteristic 0, D a (not necessarily effective) $\mathbb{Z}$-divisor and $\Delta' \leq \Delta$ an effective $\mathbb{R}$-divisor on X such that $D \sim_{\mathbb{R}} \Delta'$. Then $\mathcal{O}_X(-D)$ is CM.* □

If (X, Δ) is lc then frequently $\mathcal{O}_X$ is not CM. The following variant of the above theorems, while much weaker, is quite useful. In increasing generality it was proved by Alexeev (2008), Kollár (2011a), and Fujino (2017); see Kollár (2013b, 7.20) for the slc case and Kovács (2011) and Alexeev and Hacon (2012) for other versions. The main applications are in (2.79) and (4.33).

Theorem 11.20 *Let* (X, Δ) *be slc over a field of characteristic 0 and $x \in X$ a point that is not an lc center (11.10). Let D be a Mumford $\mathbb{Z}$-divisor. Assume that there is an effective $\mathbb{R}$-divisor $\Delta' \leq \Delta$ such that $D \sim_{\mathbb{R}} \Delta'$. Then*
(11.20.1) $\mathrm{depth}_x \mathcal{O}_X(-D) \geq \min\{3, \mathrm{codim}_X x\}$, *and*
(11.20.2) $\mathrm{depth}_x \omega_X(D) \geq \min\{3, \mathrm{codim}_X x\}$.

Proof The first claim is proved in Kollár (2013b, 7.20). To get the second, note that, working locally, $K_X + \Delta \sim_{\mathbb{R}} 0$, thus $-(K_X + D) \sim_{\mathbb{R}} \Delta - \Delta'$ and $\Delta - \Delta' \leq \Delta$ is effective. Thus, by the first part, $\omega_X(D) \simeq \mathcal{O}_X(-(-(K_X + D)))$ has depth $\geq \min\{3, \mathrm{codim}_X x\}$. □

Corollary 11.21 Alexeev (2008) *Let* (X, Δ) *be slc. If x is not an lc center and $\mathrm{codim}_X x \geq 3$, then $\mathrm{depth}_x \mathcal{O}_X \geq 3$ and $\mathrm{depth}_x \omega_X \geq 3$.* □

11.22 (Hurwitz formula) The main example is when $\pi: Y \to X$ is a finite, separable morphism between normal varieties of the same dimension, but we also need the case when $\pi\colon Y \to X$ is a finite, separable morphism between demi-normal schemes such that π is étale over the nodes of X. Working over the closure of the open set where K_X is Cartier, we get that

$$K_Y \sim_{\mathbb{Q}} R + \pi^* K_X, \qquad (11.22.1)$$

where R is the *ramification divisor* of π. If none of the ramification indices is divisible by the characteristic, then $R = \sum_D (e(D) - 1)D$ where $e(D)$ denotes the ramification index of π along the divisor $D \subset Y$.

Note that if π is quasi-étale, that is, étale outside a subset of codimension ≥ 2, then $R = 0$, hence $K_Y \sim_{\mathbb{Q}} \pi^* K_X$.

11.23 Let $\pi\colon Y \to X$ be a finite, separable morphism as in (11.22) and Δ_X an $\mathbb{R}$-divisor on X (not necessarily $\mathbb{R}$-Cartier). Set

$$\Delta_Y := -R + \pi^* \Delta_X. \qquad (11.23.1)$$

With this choice, (11.22.1) gives that

$$K_Y + \Delta_Y \sim_{\mathbb{R}} \pi^* (K_X + \Delta_X). \qquad (11.23.2)$$

Reid's covering lemma compares the discrepancies of divisors over X and Y. For precise forms see Reid (1980), Kollár and Mori (1998, 5.20), or Kollár (2013b, 2.42–43). We need the following special cases.

Claim 11.23.3 Assume also, that Δ_X and Δ_Y are both effective, and, either the characteristic is 0, or π is Galois and $\deg \pi$ is not divisible by the characteristic, or $\deg \pi$ is less than the characteristic. Then (X, Δ_X) is klt (resp. lc or slc) iff (Y, Δ_Y) is klt (resp. lc or slc). □

Special case 11.23.4 If π is quasi-étale, then $\Delta_Y = \pi^* \Delta_X$; thus we compare (X, Δ_X) and $(Y, \pi^* \Delta_X)$.

Special case 11.23.5 Let D_X be a reduced divisor on X such that π is étale over $X \setminus D_X$. Set $D_Y := \operatorname{red} \pi^*(D_X)$. Then $D_Y + R = \pi^*(D_X)$, thus we compare $(X, D_X + \Delta_X)$ and $(Y, D_Y + \pi^* \Delta_X)$.

11.24 (Cyclic covers) See Kollár and Mori (1998, 2.49–52) or Kollár (2013b, sec.2.3) for details.

Let X be an S_2 scheme, L a divisorial sheaf (3.25) and s a section of $L^{[m]}$. These data define a *cyclic cover* or μ_m-*cover* $\pi\colon Y \to X$ such that we have direct sum decompositions into μ_m-eigensheaves

$$\pi_* \mathcal{O}_Y = \bigoplus_{i=0}^{m-1} L^{[-i]}, \quad \text{and}$$
$$\pi_* \omega_{Y/C} \simeq \mathcal{H}om_X(\pi_* \mathcal{O}_Y, \omega_{X/C}) = \bigoplus_{i=0}^{m-1} L^{[i]} [\otimes] \omega_{X/C},$$

where $[\otimes]$ denotes the double dual of the tensor product. The morphism π is étale over $x \in X$ iff L is locally free at x, $s(x) \neq 0$ and char $k(x) \nmid m$. Thus π is quasi-étale iff s is a nowhere zero section and char $k(x) \nmid m$.

One can reduce many questions about $\mathbb{Q}$-Cartier divisors to Cartier divisors.

Proposition 11.25 *Let (x, X) be a local scheme over a field of characteristic 0 and $\{D_i : i \in I\}$ a finite set of $\mathbb{Q}$-Cartier, Mumford $\mathbb{Z}$- divisors. Then there is a finite, abelian, quasi-étale cover $\pi \colon \tilde{X} \to X$ such that the $\pi^* D_i$ are Cartier.*

Furthermore, if (X, Δ) is klt (resp. lc or slc) for some $\mathbb{R}$-divisor Δ, then $(\tilde{X}, \tilde{\Delta} := \pi^* \Delta)$ is also klt (resp. lc or slc). □

11.2 Canonical Models and Modifications

We used many times canonical models in the relative setting.

Definition 11.26 Let (Y, Δ_Y) be an lc pair and $p_Y \colon Y \to S$ a proper morphism. We say that (Y, Δ_Y) is a *canonical model over S*, if $K_Y + \Delta_Y$ is p_Y-ample.

Let (X, Δ) be an lc pair and $p \colon X \to S$ a proper morphism. We say that (X^c, Δ^c) is a *canonical model* of (X, Δ) *over S* if there is a diagram

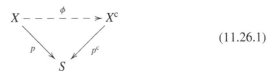

$$(11.26.1)$$

such that

(11.26.2) (X^c, Δ^c) is a canonical model over S,

(11.26.3) ϕ is a birational contraction (p.xiv),

(11.26.4) $\Delta^c = \phi_* \Delta$, and

(11.26.5) $\phi_* \mathcal{O}_X(mK_X + \lfloor m\Delta \rfloor) = \mathcal{O}_{X^c}(mK_{X^c} + \lfloor m\Delta^c \rfloor)$ for every $m \geq 0$.

Comments 11.26.6 Since ϕ is a birational contraction, there are open sets $U \subset X$ and $U^c \subset X^c$ whose complements have codimension ≥ 2 such that the restriction of ϕ is a morphism $\phi_U \colon U \to U^c$. Thus (11.26.5) is equivalent to saying that $\phi_* \mathcal{O}_U(mK_U + \lfloor m\Delta|_U \rfloor) = \mathcal{O}_{U^c}(mK_{U^c} + \lfloor m\Delta^c|_{U^c} \rfloor)$ for every $m \geq 0$. (One needs (11.62.2) to see that this is equivalent to Kollár and Mori (1998, 3.50).)

For $\mathbb{Q}$-divisors we have the following direct generalization of (1.38).

Proposition 11.27 *Let (X, Δ) be an lc pair and $p \colon X \to S$ a proper morphism. Assume that X is irreducible and Δ is a $\mathbb{Q}$-divisor. Then (X, Δ) has a canonical model over S iff the generic fiber is of general type and the canonical algebra*

$\bigoplus_{m \geq 0} p_* \mathscr{O}_X(mK_X + \lfloor m\Delta \rfloor)$ *is finitely generated. If these hold then the canonical model is* $X^c := \mathrm{Proj}_S \bigoplus_{m \geq 0} p_* \mathscr{O}_X(mK_X + \lfloor m\Delta \rfloor)$.

The main conjecture on canonical models says that the relative canonical models always exist if the generic fiber is of general type. The following known cases, due to Birkar et al. (2010) and Hacon and Xu (2013, 2016), and generalized in Lyu and Murayama (2022) are the most important for us.

Theorem 11.28 *Let* (X, Δ) *be an lc pair over a field of characteristic 0 and* $p \colon X \to S$ *a proper morphism, S irreducible. The relative canonical model exists in the following cases.*

(11.28.1) (X, Δ) *is klt and the generic fiber is of general type.*

(11.28.2) (X, Δ) *is dlt, the relative canonical model exists over an open $S^\circ \subset S$, and every lc center intersects $p^{-1}(S^\circ)$.* □

Definition 11.29 (Canonical modification) Let Y be a scheme over a field k. (We allow Y to be reducible and nonreduced, but in applications usually pure dimensional.) Its *canonical modification* is the unique proper, birational morphism $p^{cm} \colon Y^{cm} \to \mathrm{red}\, Y$ such that Y^{cm} has canonical singularities and $K_{Y^{cm}}$ is ample over Y.

Let Δ be an effective divisor on Y. We define the *canonical modification* $p^{cm} \colon (Y^{cm}, \Delta^{cm}) \to (Y, \Delta)$ as the unique proper, birational morphism for which (Y^{cm}, Δ^{cm}) has canonical singularities and $K_{Y^{cm}} + \Delta^{cm}$ is ample over Y; where Δ^{cm} is the birational transform of $\Delta|_{\mathrm{red}\, Y}$; see Kollár (2013b, 1.31).

The *log canonical modification* $p^{lcm} \colon (Y^{lcm}, \Delta^{lcm}) \to (Y, \Delta)$ is defined similarly. The change is that $(Y^{lcm}, \Delta^{lcm} + E^{lcm})$ is log canonical and $K_{Y^{lcm}} + \Delta^{lcm} + E^{lcm}$ is ample over Y, where E^{lcm} denotes the reduced exceptional divisor.

The canonical modification of (X, Δ) is unique. It exist in characteristic 0 if coeff $\Delta \subset [0, 1]$ by (11.28). The lc modification is also unique. As for its existence, we clearly need to assume that coeff $\Delta \subset [0, 1]$. Conjecturally, this is the only necessary condition, but this is known only in some cases. C. Xu pointed out that the arguments in Odaka and Xu (2012) give the following.

Theorem 11.30 *Let X be a normal variety and Δ an $\mathbb{R}$-divisor on X with* coeff$(\Delta) \subset [0, 1]$. *If $K_X + \Delta$ is numerically $\mathbb{R}$-Cartier (4.48), then (X, Δ) has a log canonical modification.* □

Proposition 11.31 (Kollár, 2018a, prop.19) *Let (X, Δ) be a potentially lc pair (11.5.1) over a field of characteristic 0. Then*

(11.31.1) *it has a projective, small, lc modification* $\pi : (X^{lcm}, \Delta^{lcm}) \to (X, \Delta)$,

(11.31.2) π *is a local isomorphism at every lc center of* (X^{lcm}, Δ^{lcm}), *and*

(11.31.3) π *is a local isomorphism over* $x \in X$ *iff* $K_X + \Delta$ *is* $\mathbb{R}$*-Cartier at* x. $\square$

The following typical application of (11.31) reduces some questions about Weil divisors to $\mathbb{Q}$-Cartier Weil divisors.

Proposition 11.32 *Let* (X, Δ) *be an lc pair over a field of characteristic 0 and* Θ *an effective* $\mathbb{R}$*-divisor such that* $\mathrm{Supp}\,\Theta \subset \mathrm{Supp}\,\Delta$. *Let* B *be a Weil* $\mathbb{Z}$*-divisor such that* $B \sim_{\mathbb{R}} -\Theta$. *Then there is a small, lc modification* $\pi\colon X' \to X$ *such that the following hold, where we use* ' *to denote the birational transform of a divisor on* X'.

(11.32.1) B' *is* $\mathbb{Q}$*-Cartier and* π*-ample,*

(11.32.2) $\mathrm{Ex}(\pi) \subset \mathrm{Supp}\,\Theta'$,

(11.32.3) *none of the lc centers of* $(X', \Delta' - \varepsilon\Theta')$ *are contained in* $\mathrm{Ex}(\pi)$,

(11.32.4) $\pi_* \mathcal{O}_{X'}(B') = \mathcal{O}_X(B)$,

(11.32.5) $R^i \pi_* \mathcal{O}_{X'}(B') = 0$ *for* $i > 0$, *and*

(11.32.6) $H^i(X, \mathcal{O}_X(B)) = H^i(X, \mathcal{O}_{X'}(B'))$.

Proof We construct $\pi\colon (X', \Delta') \to (X, \Delta)$ by applying (11.31) to $(X, \Delta - \varepsilon\Theta)$. Then $-\varepsilon\Theta' \sim_{\mathbb{R}} K_{X'} + \Delta' - \varepsilon\Theta'$ is $\mathbb{R}$-Cartier and π-ample, hence (1) holds by (11.43). This gives (2). Then (3) follows from (11.10.7). Next, (4) holds since π is small. We can write $B' \sim_{\mathbb{R}} K_{X'} + (\Delta' - \varepsilon\Theta') + (1 - \varepsilon)(-\Theta')$; then (5) follows from (3) and (11.34). Finally, the Leray spectral sequence shows (6). $\square$

One of the difficulties in dealing with slc pairs is that analogous small modifications need not exist for them; see Kollár (2013b, 1.40).

We use generalizations of Kodaira's vanishing theorem, see Kollár and Mori (1998, secs.2.4–5) for an introductory treatment. The following is proved in Ambro (2003) and (Fujino, 2014, 1.10). See also Fujino (2017, sec.5.7) and Fujino (2017, 6.3.5), where it is called a Reid–Fukuda–type theorem.

Definition 11.33 Let (X, Δ) be an slc pair, $f\colon X \to S$ a proper morphism, and L an $\mathbb{R}$-Cartier, f-nef divisor on X. Then L is called *log* f-*big* if $L|_W$ is big on the generic fiber of $f|_W\colon W \to f(W)$ for every lc center W of (X, Δ) and also for every irreducible component $W \subset X$.

Theorem 11.34 *Let* (X, Δ) *be an slc pair over a field of characteristic 0 and* D *a Mumford* $\mathbb{Z}$*-divisor on* X. *Let* $f\colon X \to S$ *be a proper morphism. Assume that* $D \sim_{\mathbb{R}} K_X + L + \Delta$, *where* L *is* $\mathbb{R}$*-Cartier,* f*-nef and log* f*-big. Then*

$$R^i f_* \mathcal{O}_X(D) = 0 \quad for \quad i > 0.$$

$\square$

11.3 Semi-log-canonical Pairs

Definition 11.35 Let (R, m) be a local ring such that $\operatorname{char}(R/m) \neq 2$. We say that $\operatorname{Spec} R$ has a *node* if there is a regular local ring (S, m_S) of dimension 2, generators $m_S = (x, y)$, a unit $a \in S \setminus m_S$ and $h \in m_S^3$ such that $R \simeq S/(x^2 - ay^2 + h)$. (See Kollár (2013b, 1.41) for characteristic 2.)

If R is complete, then we can arrange that $h = 0$. If R/m is algebraically closed, then we can take $a = 1$. Over an algebraically closed field we get the more familiar form $k[[x, y]]/(xy)$.

As a very simple special case of (2.27) or of (10.43), over a field all deformations of a node can be obtained, étale locally, by pull-back from

$$(x^2 - ay^2 = 0) \subset (x^2 - ay^2 + t = 0) \subset \mathbb{A}^2_{xy} \times \mathbb{A}^1_t. \tag{11.35.1}$$

Definition 11.36 Recall that, by Serre's criterion, a scheme X is normal iff it is S_2 and regular at all codimension 1 points. As a weakening of normality, a scheme is called *demi-normal* if it is S_2 and its codimension 1 points are either regular points or nodes.

A one-dimensional demi-normal variety is a curve C with nodes. It can be thought of as a smooth curve $\bar{C}$ (the normalization of C) together with pairs of points $p_i, p'_i \in \bar{C}$, obtained as the preimages of the nodes. Equivalently, we have the nodal divisor $\bar{D} = \sum_i (p_i + p'_i)$ on $\bar{C}$, plus a fixed point free involution on $\bar{D}$ given by $\tau \colon p_i \leftrightarrow p'_i$.

We aim to get a similar description for any demi-normal scheme X. Let $\pi \colon \bar{X} \to X$ denote the normalization and $D \subset X$ the divisor obtained as the closure of the nodes of X. Set $\bar{D} := \pi^{-1}(D)$ with reduced structure. Then $D, \bar{D}$ are the conductors of π, and the induced map $\bar{D} \to D$ has degree 2 over the generic points. The map between the normalizations $\bar{D}^n \to \bar{D}^n$ has degree 2 over all irreducible components, determining an involution $\tau \colon \bar{D}^n \to \bar{D}^n$, which is not the identity on any irreducible component. We always assume this condition from now on. (Note that τ is only a rational involution on $\bar{D}$.)

It is easy to see (Kollár, 2013b, 5.3) that a demi-normal scheme X is uniquely determined by the triple $(\bar{X}, \bar{D}, \tau)$.

However, it is surprisingly difficult to understand which triples $(\bar{X}, \bar{D}, \tau)$ correspond to demi-normal schemes. The solution of this problem in the log canonical case, given in (11.38), is a key result for us.

Roughly speaking, the concept of semi-log-canonical is obtained by replacing "normal" with "demi-normal" in the definition of log canonical (11.5).

Definition 11.37 Let X be a demi-normal scheme with normalization $\pi\colon \bar{X} \to X$ and with conductors $D \subset X$ and $\bar{D} \subset \bar{X}$. Let Δ be an effective $\mathbb{R}$-divisor whose support does not contain any irreducible component of D, and $\bar{\Delta}$ the divisorial part of $\pi^{-1}(\Delta)$. The pair (X, Δ) is called *semi-log-canonical* or *slc* if

(11.37.1) $K_X + \Delta$ is $\mathbb{R}$-Cartier, and

(11.37.2) $(\bar{X}, \bar{D} + \bar{\Delta})$ is lc.

Alternatively, one can define $a(E, X, \Delta)$ using semi-resolutions (as in Kollár (2013b, Sec.10.4)) and then replace (2) by

(11.37.3) $a(E, X, \Delta) \geq -1$ for every exceptional divisor E over X.

This is now the exact analog of the definition of log canonical given in (11.5); the equivalence is proved in Kollár (2013b, 5.10).

This formula suggests that if $D_i \subset D$ is an irreducible component, then we should declare that $a(D_i, X, \Delta) = -1$.

Warning 11.37.4 It can happen that (2) holds, hence $K_{\bar{X}} + \bar{D} + \bar{\Delta}$ is $\mathbb{R}$-Cartier, but $K_X + \Delta$ is not; see (2.22.1) for an instructive special case of dimension 2. By contrast, this cannot happen in codimensions ≥ 3 by (11.42).

The following theorem, proved in Kollár (2016b) and Kollár (2013b, 5.13), describes slc pairs using their normalizations.

Theorem 11.38 *Let S be a scheme over a field of characteristic 0 as in (11.2). Then normalization gives a one-to-one correspondence:*

$$
\left\{
\begin{array}{l}
\text{Proper, slc pairs} \\
g\colon (X, \Delta) \to S, \\
K_X + \Delta \text{ is } g\text{-ample.}
\end{array}
\right\}
\longleftrightarrow
\left\{
\begin{array}{l}
\text{Proper, lc pairs } \bar{g}\colon (\bar{X}, \bar{D} + \bar{\Delta}) \to S \\
\text{with involution } \tau \curvearrowright (\bar{D}^n, \mathrm{Diff}_{\bar{D}^n}\, \bar{\Delta}), \\
K_{\bar{X}} + \bar{D} + \bar{\Delta} \text{ is } \bar{g}\text{-ample.}
\end{array}
\right\}.
$$

(As in (11.36), τ is not the identity on any irreducible component.) $\square$

In applications, we usually know the codimension 1 points of X and $\bar{X}$. The codimension 1 points of $(\bar{D}^n, \mathrm{Diff}_{\bar{D}^n}\, \bar{\Delta})$ correspond to codimension 2 points of X and $\bar{X}$. Since we understand two-dimensional slc pairs quite well, we frequently have good control over codimension ≤ 2 points of lc and slc pairs. The next theorems show that one can sometimes ignore the higher codimension points.

The first result of this type, due to Matsusaka and Mumford (1964), shows how to extend isomorphisms across subsets of codimension ≥ 2.

Theorem 11.39 *Let S be a Noetherian scheme, $X_i \to S$ projective morphisms and H_i relatively ample $\mathbb{R}$-divisor classes on X_i. Let $Z_i \subset X_i$ be closed subsets such that $\mathrm{depth}_{Z_i} X_i \geq 2$. Let*

$$\tau^\circ : (X_1 \setminus Z_1, H_1|_{X_1 \setminus Z_1}) \simeq (X_2 \setminus Z_2, H_2|_{X_2 \setminus Z_2})$$

be an isomorphism. Then τ° extends to an isomorphism $\tau : X_1 \simeq X_2$.

Proof Let $\Gamma \subset X_1 \times_S X_2$ be the closure of the graph of τ° with projections $\pi_i : \Gamma \to X_i$. Then π_i is an isomorphism over $X_i \setminus Z_i$. By (5.32.4), π_i is an isomorphism iff it is finite. The latter can be checked locally on S after completion. We can now also assume that the X_i are normal and replace Γ with its normalization.

There is a divisor $E \sim_{\mathbb{R}} \pi_1^* H_1 - \pi_2^* H_2$ that is supported on the union of the π_i-exceptional loci. Since $\pi_i(E) \subset Z_i$, we see that $\mathrm{Supp}(E) \subset \mathrm{Ex}(\pi_i)$ for $i = 1, 2$.

Next note that $-E$ is π_1-nef and exceptional, so $E \geq 0$ by (11.60). Also E is π_2-nef and exceptional, so $E \leq 0$. Thus $E = 0$, hence $\pi_1^* H_1 \equiv \pi_2^* H_2$. We now finish by (11.39.1). □

Claim 11.39.1 Let $X_i \to S$ be projective morphisms and H_i relatively ample $\mathbb{R}$-divisor classes on X_i. Let $p : Y \to X_1 \times_S X_2$ be a finite morphism such that $p^* \pi_1^* H_1 \sim_{\mathbb{R}} p^* \pi_2^* H_2$. Then $\pi_i \circ p : Y \to X_i$ are finite.

Proof If a curve $C \subset Y$ is contracted by $\pi_1 \circ p$ then it cannot be contracted by $\pi_2 \circ p$ since p is finite. Thus $(C \cdot p^* \pi_1^* H_1) = 0$, but $(C \cdot p^* \pi_2^* H_2) = (\pi_2 \circ p(C) \cdot H_2) > 0$ since H_2 is ample. □

The $\mathrm{depth}_Z X \geq 2$ assumption in (11.39) holds if X is normal and $Z \subset X$ has codimension ≥ 2; the main case in most applications. If Z has codimension 1, we usually get very little information about X from $X \setminus Z$. Nonetheless, we have the following very useful result about slc pairs.

Theorem 11.40 *Let S be a scheme over a field of characteristic 0 as in (11.2), and let $f_i : (X_i, \Delta_i) \to S$ proper morphisms from slc pairs such that $K_{X_i} + \Delta_i$ is f_i-ample. Let $Z_S \subset S$ be a closed subset and set $Z_i := f_i^{-1}(Z_S)$. Let*

$$\tau^\circ : (X_1 \setminus Z_1, \Delta_1|_{X_1 \setminus Z_1}) \simeq (X_2 \setminus Z_2, \Delta_2|_{X_2 \setminus Z_2}) \tag{11.40.1}$$

be an isomorphism. Assume that none of the log centers (11.11) of (X_i, Δ_i) is contained in Z_i for $i = 1, 2$.

Then τ° extends to an isomorphism $\tau : X_1 \simeq X_2$.

Proof Since every irreducible component of X is a log center, the Z_i are nowhere dense in X_i.

Using (11.38) we may assume that the X_i are normal. Let $\Gamma \to X^1 \times_S X^2$ be the normalization of the closure of the graph of τ° with projections $\pi_i : \Gamma \to X_i$.

As in (1.28), we use the log canonical class to compare the X_i. If F is an irreducible component of Δ_i then $a(F, X_i, \Delta_i) = -\text{coeff}_F \Delta_i < 0$, thus $F \not\subset Z_i$. In particular, $(\pi_1)_*^{-1}\Delta^1 = (\pi_2)_*^{-1}\Delta^2$; let us denote this divisor by Δ_Γ. Write

$$K_\Gamma + \Delta_\Gamma \sim_\mathbb{R} \pi_i^*(K_{X_i} + \Delta_i) + E_i, \tag{11.40.2}$$

where E_i is π_i-exceptional and $\pi_i(\text{Supp}\, E_i) \subset Z_i$. Note that E_i is effective by our assumption on the log centers.

Subtracting the $i = 1, 2$ cases of (11.40.2) from each other, we get that

$$E_1 - E_2 \sim_\mathbb{R} \pi_2^*(K_{X_2} + \Delta_2) - \pi_1^*(K_{X_1} + \Delta_1). \tag{11.40.3}$$

Thus $E_1 - E_2$ is π_1-nef and $-(\pi_1)_*(E_1 - E_2) = (\pi_1)_*(E_2)$ is effective. Thus $E_2 - E_1$ is effective by (11.60). Using π_2 shows that $E_1 - E_2$ is effective, hence $E_1 = E_2$. Thus $\pi_1^*(K_{X_1} + \Delta_1) \sim_\mathbb{R} \pi_2^*(K_{X_2} + \Delta_2)$. We finish by (11.39.1). $\square$

Remark 11.40.4 The assumption on log centers is crucial. To see an example, consider the family of curves

$$X := \left(xyz(x + y + z) + t(x^4 + y^4 + z^4) = 0\right) \subset \mathbb{P}^2_{xyz} \times \mathbb{A}^1_t.$$

It is smooth along the central fiber X_0, which consists of four lines L_i, each with self-intersection -3. We can contract any of them $p_i \colon X \to X_i$, to get $f_i \colon X_i \to \mathbb{A}^1$. Note that $p_j \circ p_i^{-1} \colon X_i \dashrightarrow X_j$ is an isomorphism over $\mathbb{A}^1 \setminus \{0\}$, but not an isomorphism for $i \neq j$. Here X_i has a singularity of type $\mathbb{A}^2/\frac{1}{3}(1, 1)$, which is log terminal and the singularities are log centers of $(X_i, 0)$.

Corollary 11.41 *Let S be a scheme over a field of characteristic 0 as in (11.2) and $S^\circ \subset S$ a dense, open subscheme. Let $g^\circ \colon (X^\circ, \Delta^\circ) \to S^\circ$ be a proper, slc pair with normalization $\pi^\circ \colon (\bar{X}^\circ, \bar{\Delta}^\circ + \bar{D}^\circ) \to (X^\circ, \Delta^\circ)$.*

Assume that there is an slc pair $(\bar{X}, \bar{\Delta} + \bar{D}) \supset (\bar{X}^\circ, \bar{\Delta}^\circ + \bar{D}^\circ)$ that is proper over S such that $K_{\bar{X}} + \bar{\Delta} + \bar{D}$ is ample over S and every codimension ≤ 2 log center of $(\bar{X}, \bar{\Delta} + \bar{D})$ has nonempty intersection with $\bar{X}^\circ$.

Then there is a unique slc pair $(X, \Delta) \supset (X^\circ, \Delta^\circ)$ that is proper over S and whose normalization is $(\bar{X}, \bar{\Delta} + \bar{D})$.

Proof Since every irreducible component of $\bar{X}$ is a log center, $\bar{X}^\circ$ is dense in $\bar{X}$. Let $n \colon \bar{D}^n \to \bar{D}$ denote the normalization. By inversion of adjunction (11.17.2), $(\bar{D}^n, \text{Diff}_{\bar{D}^n}\bar{\Delta})$ is also lc and $K_{\bar{D}^n} + \text{Diff}_{\bar{D}^n}\bar{\Delta}$ is ample over S.

Using (11.17.3), every irreducible component of $\text{Diff}_{\bar{D}^n}$ lies over a codimension 2 log center of $(\bar{X}, \bar{\Delta} + \bar{D})$, hence none of the irreducible components of $\text{Diff}_{\bar{D}^n}$ is disjoint from $\bar{X}^\circ$.

Thus the involution τ° of $(\bar{D}^\circ)^n$ extends to an involution τ on $\bar{D}^n$ by (11.40), and $\text{Diff}_{\bar{D}^n}$ is $\bar{\tau}$-invariant. Hence (11.38) gives the existence of (X, Δ). $\square$

Corollary 11.42 *Let (X, Δ) be demi-normal with lc normalization $(\bar{X}, \bar{\Delta} + \bar{D})$. Assume that there is a closed subset $W \subset X$ of codimension ≥ 3 such that $(X \setminus W, \Delta|_{X \setminus W})$ is slc. Then (X, Δ) is slc.*

Proof Apply (11.41) with $S = X$, $X^\circ = X \setminus W$. □

11.4 $\mathbb{R}$-divisors

It is easy to see that, on a $\mathbb{Q}$-factorial scheme, $\mathbb{R}$-divisors behave very much like $\mathbb{Q}$-divisors. The same holds in general, but it needs a little more work. The basics are discussed in Lazarsfeld (2004, sec.1.3), but most other facts are scattered in the literature; see, for example, Kollár (2013b, 2.21), Birkar (2017), or Fujino and Miyamoto (2021).

11.43 ($\mathbb{R}$-divisors) Let X be a reduced, S_2 scheme and $\Delta = \sum b_i B_i$ a Mumford $\mathbb{R}$-divisor. There is a unique such way of writing Δ where the B_i are irreducible, distinct and $b_i \neq 0$ for every i. The $\mathbb{Q}$-vector space spanned by the coefficients is denoted by $\mathrm{CoSp}(\Delta) = \sum_i \mathbb{Q} \cdot b_i \subset \mathbb{R}$.

We say that Δ is $\mathbb{R}$-*Cartier* if it can be written as an $\mathbb{R}$-linear combination of Cartier $\mathbb{Z}$-divisors $\Delta = \sum r_i D_i$. By (11.46), we can choose the D_i to have the same support as Δ, but we do not assume this to start with. Two $\mathbb{R}$-divisors are $\mathbb{R}$-*linearly equivalent,* denoted by $\Delta_1 \sim_{\mathbb{R}} \Delta_2$, if $\Delta_1 - \Delta_2$ is an $\mathbb{R}$-linear combination of principal divisors. Claim (11.43.2.d) shows that for $\mathbb{Q}$-divisors we do not get anything new.

Let $\sigma \colon \mathbb{R} \to \mathbb{Q}$ be a $\mathbb{Q}$-linear map. It extends to a $\mathbb{Q}$-linear map from $\mathbb{R}$-divisors to $\mathbb{Q}$-divisors as $\sigma(\sum d_i D_i) := \sum \sigma(d_i) D_i$.

Claim 11.43.1 Let $\sigma \colon \mathbb{R} \to \mathbb{Q}$ be a $\mathbb{Q}$-linear map. Then
  (a) $\mathrm{Supp}(\sigma(D)) \subset \mathrm{Supp}(D)$,
  (b) if $D_1 \sim_{\mathbb{R}} D_2$ then $\sigma(D_1) \sim_{\mathbb{Q}} \sigma(D_2)$,
  (c) if D is $\mathbb{R}$-Cartier then $\sigma(D)$ is $\mathbb{Q}$-Cartier, and
  (d) $D \mapsto \sigma(D)$ commutes with pull-back for $\mathbb{R}$-Cartier divisors.

Proof The first claim is clear. If $D_1 - D_2 = \sum c_i(f_i)$ then $\sigma(D_1) - \sigma(D_2) = \sum \sigma(c_i)(f_i)$, showing (b), which in turn implies (c) and (d) is clear. □

Let D be an $\mathbb{R}$-divisor. Choosing a $\mathbb{Q}$-basis $d_i \in \mathrm{CoSp}(D)$, we can write $D = \sum d_i D_i$ where the D_i are $\mathbb{Q}$-divisors (usually reducible). The D_i depend on the choice of the basis. Nonetheless, they inherit many properties of D.

Claim 11.43.2 Let D_i be $\mathbb{Q}$-divisors and $d_i \in \mathbb{R}$ linearly independent over $\mathbb{Q}$. Then

(a) $\sum d_i D_i$ is $\mathbb{R}$-Cartier iff each D_i is $\mathbb{Q}$-Cartier.

(b) $\sum d_i D_i \sim_{\mathbb{R}} 0$ iff $D_i \sim_{\mathbb{Q}} 0$ for every i.

(c) If X is proper then $\sum d_i D_i \equiv 0$ iff $D_i \equiv 0$ for every i.

(d) A $\mathbb{Q}$-divisor D_i is $\mathbb{R}$-Cartier iff it is $\mathbb{Q}$-Cartier.

(e) $D_1 \sim_{\mathbb{R}} D_2$ iff $D_1 \sim_{\mathbb{Q}} D_2$.

(f) $\operatorname{Supp} D_i \subset \operatorname{Supp} D$.

Proof If the $d_i \in \mathbb{R}$ are linearly independent then we can choose σ_i such that $\sigma_i(d_i) = 1$ and $\sigma_i(d_j) = 0$ for $i \neq j$. Then $\sigma_i(D) = D_i$, thus (11.43.1) shows (a) and (b).

For (c), assume that $\sum d_i D_i \equiv 0$ and let $C \subset X$ be a curve. Then $\sum d_i (D_i \cdot C) = 0$. Since $(D_i \cdot C) \in \mathbb{Q}$ and the d_i are linearly independent, we get that $(D_i \cdot C) = 0$ for every i. Applying (a) to D_i gives (d). Applying (b) to $D_1 - D_2$ gives (e). Finally (f) follows from the linear independence over $\mathbb{Q}$. □

Corollary 11.43.3 Let Θ be a Mumford $\mathbb{R}$-divisor and $\{d_i\}$ a basis of $\operatorname{CoSp}(\Theta)$ over $\mathbb{Q}$. Then we get a unique representation $\Theta = \sum d_i D_i$ where the D_i are $\mathbb{Q}$-divisors. If Θ is $\mathbb{R}$-Cartier, then the D_i are $\mathbb{Q}$-Cartier. □

Corollary 11.43.4 Let Δ be a Mumford $\mathbb{R}$-divisor and $\{d_i'\}$ a $\mathbb{Q}$-basis of $\mathbb{Q} + \operatorname{CoSp}(\Delta)$ such that $\sum d_i' = 1$. Then we get a unique representation $\Delta = \sum d_i' D_i$ where the D_i are $\mathbb{Q}$-divisors. If $K_X + \Delta$ is $\mathbb{R}$-Cartier, then $K_X + D_i$ are $\mathbb{Q}$-Cartier.

Proof Note that $K_X + \Delta = \sum d_i'(K_X + D_i)$, so the last assertion follows from (11.43.2.a). □

Next we show that $\mathbb{R}$-divisors can be approximated by $\mathbb{Q}$-divisors in a way that many properties are preserved. We start with some general comments on vector spaces and field extensions. At the end we care only about $\mathbb{R} \supset \mathbb{Q}$.

Definition–Lemma 11.44 Let K/k be a field extension, V a k-vector space and $w \in V \otimes_k K$. The *linear k-envelope* of w, denoted by $\operatorname{LEnv}_k(w) \subset V$, is the smallest vector subspace such that $w \in \operatorname{LEnv}_k(w) \otimes_k K$. Then $\operatorname{LEnv}_k(w)$ is spanned by any of the following three sets, where σ runs through all k-linear maps $K \to k$.

(11.44.1) All $(1_V \otimes \sigma)(w)$.

(11.44.2) All $\sum \sigma(c_i) v_i$, where $v_i \in V$ is a basis and $w = \sum c_i v_i$.

(11.44.3) All $\sum_i a_{ij} v_i$, where $e_j \in K$ is a k-basis and $w = \sum_{ij} a_{ij} e_j v_i$.

The *affine k-envelope* of w, denoted by $\operatorname{AEnv}_k(w) \subset V$, is the smallest affine-linear subspace such that $w \in \operatorname{AEnv}_k(w) \otimes_k K$. Then $\operatorname{AEnv}_k(w)$ is spanned by any of the following three sets, where σ runs through all k-linear maps $K \to k$ such that $\sigma(1) = 1$.

(11.44.4) All $(1_V \otimes \sigma)(w)$.

(11.44.5) All $\sum \sigma(c_i)v_i$, where $v_i \in V$ is a basis and $w = \sum c_i v_i$.

(11.44.6) All $\sum_i a_{ij}v_i$, where $e_j \in K$ is a k-basis such that $e_1 = 1$ and $w = \sum_{ij} a_{ij}e_j v_i$.

11.45 (Approximating by rational simplices) Fix real numbers $d_1, \dots, d_m$ and consider a $\mathbb{Q}$-vector space W with basis $\mathfrak{d}_1, \dots, \mathfrak{d}_m$. Set $\mathfrak{d} := \sum d_i \mathfrak{d}_i \in W_{\mathbb{R}}$ and $V := \mathrm{AEnv}_{\mathbb{Q}}(\mathfrak{d})$. We inductively construct a sequence of simplices

$$V \supset S_1 \supset S_2 \supset \cdots \quad \text{such that} \quad \cap_n S_n = \{\mathfrak{d}\}.$$

Set $S_0 := V$. For each $n \in \mathbb{N}$ the cubes of the lattice $\frac{1}{n}\mathbb{Z}^m$ give a cubical chamber decomposition of $W_{\mathbb{R}}$. There is a smallest chamber C_n that contains $\mathfrak{d}$. Then $\mathfrak{d}$ is an interior point of $C_n \cap S_{n-1}$ (in its affine-linear span). The vertices of $C_n \cap S_{n-1}$ are in $\mathbb{Q}^n$. Thus $\mathfrak{d}$ can be written as a convex linear combination of suitably chosen $\dim V + 1$ vertices of $V_{\mathbb{R}} \cap C_n$; denote them by $\mathfrak{d}_j^n$. These span S_n. By (11.44), there are $\mathbb{Q}$-linear maps $\sigma_j^n \colon \mathbb{R} \to \mathbb{Q}$ such that $\mathfrak{d}_j^n = \sigma_j^n(\mathfrak{d})$. We can thus write

(11.45.1) $\mathfrak{d} = \sum_j \lambda_j^n \mathfrak{d}_j^n$, where

(11.45.2) $\mathfrak{d}_j^n = \sum_i \sigma_j^n(d_i)\mathfrak{d}_i$,

(11.45.3) $\sum_j \lambda_j^n = 1$ and $\sum_j \lambda_j^n \sigma_j^n(d_i) = d_i \quad \forall i$,

(11.45.4) $\lim_{n \to \infty} \mathfrak{d}_j^n = \mathfrak{d} \quad \forall j$, and

(11.45.5) for fixed n, the λ_j^n are linearly independent over $\mathbb{Q}$. (To see this, note that 1 and the d_i are $\mathbb{Q}$-linear combinations of the λ_j^n for fixed n.)

Remark 11.45.6 The choice of the vertices is not unique, but once we choose them, the constants λ_j^n are unique, and so are the restrictions of σ_j^n to $\mathrm{LEnv}_{\mathbb{Q}}(\mathfrak{d})$. Thus, from now on, we view σ_j^n and λ_j^n as depending only on $j, n \in \mathbb{N}$ and $d_1, \dots, d_m \in \mathbb{R}$. Note that these are not continuous functions of the d_i, even the number of the j-indices varies discontinuously with $d_1, \dots, d_m$.

Also, we only care about the restriction of the σ_j^n to $\mathrm{LEnv}_{\mathbb{Q}}(\mathfrak{d})$, so we are really dealing with finite dimensional linear algebra.

Proposition 11.46 (Convex approximation of $\mathbb{R}$-divisors I) *Let X be a reduced, S_2 scheme and $\Theta = \sum_i d_i D_i$ a Mumford $\mathbb{R}$-divisor, where the D_i are $\mathbb{Q}$-divisors. Let σ_j^n and λ_j^n be as in (11.45) and set $\Theta_j^n := \sum \sigma_j^n(d_i)D_i$. Then*

(11.46.1) $\Theta = \sum_j \lambda_j^n \Theta_j^n$ *and the Θ_j^n are $\mathbb{Q}$-divisors.*

(11.46.2) *Let $E \subset X$ be a prime divisor on X. Then $\lim_{n \to \infty} \mathrm{coeff}_E \Theta_j^n = \mathrm{coeff}_E \Theta$ and $\mathrm{coeff}_E \Theta_j^n = \mathrm{coeff}_E \Theta$ if $\mathrm{coeff}_E \Theta \in \mathbb{Q}$.*

(11.46.3) Θ *is effective iff the* Θ_j^n *are effective for every j for* $n \gg 1$ *(then they have the same support).*

Assume next that Θ *is* ℝ*-Cartier. Then the following also hold.*

(11.46.4) *The* Θ_j^n *are* ℚ*-Cartier.*

(11.46.5) *Let* E *be prime divisor over* X *(11.1). Then* $\lim_{n\to\infty} \operatorname{coeff}_E \Theta_j^n = \operatorname{coeff}_E \Theta$ *and* $\operatorname{coeff}_E \Theta_j^n = \operatorname{coeff}_E \Theta$ *if* $\operatorname{coeff}_E \Theta \in ℚ$.

(11.46.6) *Let* C *be a proper curve on* X. *Then* $\lim_{n\to\infty}(C \cdot \Theta_j^n) = (C \cdot \Theta)$ *and* $(C \cdot \Theta_j^n) = (C \cdot \Theta)$ *if* $(C \cdot \Theta) \in ℚ$.

(11.46.7) Θ *is ample (11.51) iff the* Θ_j^n *are ample for every j for* $n \gg 1$.

Proof (1) is a formal consequence of (11.45.2), while the limit in (2) follows from (11.45.3). If $\operatorname{coeff}_E \Theta =: c \in ℚ$, then $\sum_i x_i \operatorname{coeff}_E D_i = c$ defines a rational hyperplane in W (as in (11.45)). It contains $\mathfrak{d}$, hence also V and the other $\mathfrak{d}_j^n$. The Θ_j^n are the images of the $\mathfrak{d}_j^n$.

By (11.45.4) the λ_j^n are linearly independent over ℚ. Thus, if Θ is ℝ-Cartier then the Θ_j^n are ℚ-Cartier by (11.43.2), proving (4). Also, in this case, $\operatorname{coeff}_E \Theta$ makes sense for divisors over X and same for the intersection numbers $(C \cdot \Theta)$. The proofs of (5–7) are now the same as for (2). $\square$

Proposition 11.47 (Convex approximation of ℝ-divisors II) *Let* X *be a demi-normal scheme and* $\Delta = \sum d_i D_i$ *a Mumford* ℝ*-divisor, where the* D_i *are* ℚ*-divisors. Assume that* $K_X + \Delta$ *is* ℝ*-Cartier. Let* σ_j^n *and* λ_j^n *be as in (11.45) and set* $\Delta_j^n := \sum \sigma_j^n(d_i)D_i$. *Then*

(11.47.1) $\Delta = \sum_j \lambda_j^n \Delta_j^n$ *and* $K_X + \Delta = \sum_j \lambda_j^n(K_X + \Delta_j^n)$.

(11.47.2) Δ *is effective iff the* Δ_j^n *are effective for every j for* $n \gg 1$ *(then they have the same support).*

(11.47.3) $K_X + \Delta_j^n$ *are* ℚ*-Cartier.*

(11.47.4) $K_X + \Delta$ *is ample iff the* $K_X + \Delta_j^n$ *are ample for every j for* $n \gg 1$.

(11.47.5) *Let* E *be a prime divisor. Then* $\lim_{n\to\infty} a(E, X, \Delta_j^n) = a(E, X, \Delta)$ *and* $a(E, X, \Delta_j^n) = a(E, X, \Delta)$ *if* $a(E, X, \Delta) \in ℚ$.

(11.47.6) *Let* C *be a proper curve. Then* $\lim_{n\to\infty}(C \cdot (K_X + \Delta_j^n)) = (C \cdot (K_X + \Delta))$ *and* $(C \cdot (K_X + \Delta_j^n)) = (C \cdot (K_X + \Delta))$ *if* $(C \cdot (K_X + \Delta)) \in ℚ$.

Assume next that (X, Δ) *has a log resolution and fix* $\varepsilon > 0$. *Then, for every j and every* $n \gg 1$, *the following hold.*

(11.47.7) $|a(E, X, \Delta) - a(E, X, \Delta_j^n)| < \varepsilon$ *for every divisor* E *over* X, *whenever one of the discrepancies is* < 0.

(11.47.8) (X, Δ) *is lc (resp. dlt or klt) iff* (X, Δ_j^n) *is lc (resp. dlt or klt).*

(11.47.9) (X, Δ) *and* (X, Δ_j^n) *have the same dlt modifications.*

Proof (1–2) follow directly from (11.46) and (3) follows from (11.46.4) and (1). Since ampleness is an open condition, (3) implies (4).

The proofs of (5) and (6) are the same as the proof of (11.46.2). If (X, Δ) has a log resolution then (7) follows from (11.7) and being lc (resp. dlt or klt) can be read off from the discrepancies, hence (7) implies (8) and (9). $\square$

In the slc case, we have the following remarkable sharpening.

Complement 11.48 (Han et al., 2020, 5.6) *In (11.47) assume in addition that* $(X, \Delta = \sum d_i D_i)$ *is slc. Then we can choose the σ_j^n and λ_j^n to depend only on* $(d_1, \dots, d_r)$ *and the dimension.* $\square$

We also get some information about pluricanonical sheaves for $\mathbb{R}$-divisors.

Theorem 11.49 *Fix a finite set $C := \{c_1, \dots, c_r\} \subset [0, 1]$. Then there is a subset $M(C, n) \subset \mathbb{Z}$ of positive density such that, if $(X, \Delta = \sum c_i D_i)$ is an slc pair of dimension n, then $(X, \lfloor \Delta \rfloor + \sum \{mc_i\} D_i)$ is slc for $m \in M(C, n)$, and has the same lc centers as (X, Δ).*

Proof Let $A \subset \mathbb{R}^r$ be the affine envelope of $\mathbf{c} := (c_1, \dots, c_r) \in \mathbb{R}^r$ and $H \subset \mathbb{R}^n$ the closed subgroup generated by A. Then A is a connected component of H and $H/(H \cap \mathbb{Z}^r) \subset \mathbb{R}^r/\mathbb{Z}^r$ is a closed subgroup. Furthermore, by a theorem of Weyl, the multiples of $\mathbf{c}$ are equidistributed in $H/(H \cap \mathbb{Z}^r)$; see, for example, Kuipers and Niederreiter (1974, sec.1.1).

Pick now σ_j^n as in (11.48). Then the convex linear combinations of the $\sigma_j^n(\mathbf{c})$ give an open neighborhood $\mathbf{c} \in U \subset H/(H \cap \mathbb{Z}^r)$. If $(\{mc_1\}, \dots, \{mc_r\}) \in U$ then (1–2) hold. $\square$

Applying (11.20) gives the following.

Corollary 11.50 *Using the notation of (11.49), let $(X, \Delta = \sum c_i D_i)$ be an slc pair of dimension n. Then, for every $m \in M(C, n)$,*

$$\mathrm{depth}_x \, \omega_X^{[m]}(\sum \lfloor mc_i \rfloor D_i) \geq \min\{3, \mathrm{codim}_X x\} \qquad (11.50.1)$$

whenever x is not an lc center of (X, Δ). $\square$

Example 11.50.2 Let $X \subset \mathbb{A}^4$ be the quadric cone and $|A|, |B|$ the two families of planes on X. Fix $r \in \mathbb{N}$ and for $0 < c \leq 1/r$ consider the pair

$$(X, \Delta_c := B + cA_1 + \cdots + cA_r + (1 - rc)A_0).$$

Then (X, Δ_c) is canonical and

$$\mathcal{O}_X(\lfloor m\Delta_c \rfloor) \simeq \begin{cases} \mathcal{O}_X(-A) & \text{if } \{mc\} \leq 1/r, \quad \text{and} \\ \mathcal{O}_X(-dA) & \text{for some } d \geq 2 \text{ otherwise.} \end{cases}$$

An easy computation, as in Kollár (2013b, 3.15.2), shows that $\mathscr{O}_X(\lfloor m\Delta_c \rfloor)$ is CM iff $\{mc\} \le 1/r$. If c is irrational, then the set $\{m: \{mc\} \le 1/r\}$ has no periodic subsets.

Definition 11.51 Let $g: X \to S$ be a proper morphism. An $\mathbb{R}$-Cartier divisor H is g-ample iff it is linearly equivalent to a positive linear combination $H \sim_{\mathbb{R}} \sum c_i H_i$ of g-ample Cartier divisors.

Ampleness is preserved under perturbations. Indeed, let $D_1, \dots, D_r$ be $\mathbb{Q}$-Cartier divisors. There are $m_j > 0$ such that the $m_j H_1 + D_j$ are g-ample. Then

$$H + \sum_j \eta_j D_j \sim_{\mathbb{R}} (c_1 - \sum_j \eta_j m_j)H_1 + \sum_{i \neq 1} c_i H_i + \sum_j \eta_j(m_j H_1 + D_j)$$

shows that $H + \sum_j \eta_j D_j$ is g-ample if $\eta_j \ge 0$ and $\sum_j \eta_j m_j \le c_1$.

This implies that if H is g-ample, $m \gg 1$ and $\lfloor mH \rfloor$ is Cartier, then $\lfloor mH \rfloor$ is very g-ample. However, frequently $\lfloor mH \rfloor$ is not even $\mathbb{Q}$-Cartier for every $m > 0$, making the proofs of the basic ampleness criteria more complicated.

Theorem 11.52 (Asymptotic Riemann–Roch) *Let X be a normal, proper algebraic space of dimension n and D a nef $\mathbb{R}$-Cartier divisor. Then*

$$
\begin{aligned}
h^0(X, \mathscr{O}_X(\lfloor mD \rfloor)) &= \tfrac{m^n}{n!}(D^n) + O(m^{n-1}), \quad \text{and} \\
h^0(X, \mathscr{O}_X(\lceil mD \rceil)) &= \tfrac{m^n}{n!}(D^n) + O(m^{n-1}).
\end{aligned}
\tag{11.52.1}
$$

Proof By Chow's lemma we may assume that X is projective. Write $D = \sum a_i A_i$ where the A_i are effective, ample $\mathbb{Z}$-divisors and $a_i \in \mathbb{R}$. Then

$$\sum \lfloor ma_i \rfloor A_i \le \lfloor mD \rfloor \le mD \le \lceil mD \rceil \le H + \sum \lceil ma_i \rceil A_i \tag{11.52.2}$$

for any H ample and effective. It is thus enough to prove that (11.52.1) holds for the two divisors on the sides of (11.52.2) for suitable H. Note that

$$\sum \lceil ma_i \rceil A_i \sim_{\mathbb{R}} \sum (\lceil ma_i \rceil - ma_i)A_i + mD,$$

thus $\sum \lceil ma_i \rceil A_i$ is nef for every $m \ge 0$, even though some of the $\lceil ma_i \rceil$ may be negative. Next choose H such that (11.52.4) holds (with $F = \mathscr{O}_X$) and $H + \sum A_i$ is linearly equivalent to an irreducible divisor B. Then, by Riemann–Roch,

$$h^0(X, \mathscr{O}_X(H + \sum \lceil ma_i \rceil A_i)) = \chi(X, \mathscr{O}_X(H + \sum \lceil ma_i \rceil A_i)) = \tfrac{m^n}{n!}(D^n) + O(m^{n-1}).$$

Restricting $\mathscr{O}_X(H + \sum \lceil ma_i \rceil A_i)$ to B, the kernel is

$$\mathscr{O}_X(\sum \lceil ma_i \rceil A_i - \sum A_i) \subset \mathscr{O}_X(\sum \lfloor ma_i \rfloor A_i)$$

(the two are equal iff none of the ma_i are integers). Thus

$$h^0(X, \mathscr{O}_X(H + \sum \lceil ma_i \rceil A_i)) - h^0(X, \mathscr{O}_X(\sum \lfloor ma_i \rfloor A_i))$$

is at most $h^0(B, \mathscr{O}_B(H|_B + \sum \lceil ma_i \rceil A_i|_B))$. The latter is bounded by $O(m^{n-1})$ using (11.52.3). □

11.52.3 (Matsusaka inequality) Let X be a proper variety of dimension n, L a nef and big $\mathbb{Z}$-divisor and D a Weil $\mathbb{Z}$-divisor giving a dominant map $|D|: X \dashrightarrow Z$. Then

$$h^0(X, \mathscr{O}_X(D)) \leq \frac{(D \cdot L^{n-1})^{\dim Z}}{(L^n)^{\dim Z - 1}} + \dim Z.$$

See Matsusaka (1972) or Kollár (1996, VI.2.15) for proofs.

11.52.4 (Fujita vanishing) Let X be a projective scheme and F a coherent sheaf on X. Then there is an ample line bundle L such that

$$H^i(X, F \otimes L \otimes M) = 0 \quad \forall i > 0, \ \forall \text{ nef line bundle } M.$$

See Fujita (1983) (or Lazarsfeld (2004, I.4.35) for the characteristic 0 case).

Corollary 11.53 (Kodaira lemma) *Let X be a normal, proper, irreducible algebraic space of dimension n and D a nef $\mathbb{R}$-divisor. Then D is big (p.xvi) $\Leftrightarrow$ $(D^n) > 0 \Leftrightarrow$ one can write $D = cB + E$, where B is a big $\mathbb{Z}$-divisor, $c > 0$, and E is an effective $\mathbb{R}$-divisor. If X is projective, then one can choose B to be ample.*

Proof With (11.52) in place, the arguments in Kollár and Mori (1998, 2.61) or Lazarsfeld (2004, 2.2.6) work. See also Shokurov (1996, 6.17) (for characteristic 0) and Birkar (2017, 1.5) for the original proofs, or Fujino and Miyamoto (2021, 2.3). □

The proof of the Nakai–Moishezon criterion for $\mathbb{R}$-divisors uses induction on all proper schemes, so first we need some basic results about them.

11.54 ($\mathbb{R}$-Cartier divisor classes) Fujino and Miyamoto (2021) On an arbitrary scheme one can define $\mathbb{R}$-*line bundles* or $\mathbb{R}$-*Cartier divisor classes* as elements of $\mathrm{Pic}(X) \otimes_{\mathbb{Z}} \mathbb{R}$. It is better to think of these as coming from line bundles, but writing divisors keeps the additive notation.

Claim 11.54.1 Let X be a proper algebraic space, $p: Y \to X$ its normalization and Θ an $\mathbb{R}$-Cartier divisor class on X. Then Θ is ample iff $p^*\Theta$ is ample.

Proof For Cartier divisors, this is Hartshorne (1977, ex.III.5.7), which implies the $\mathbb{Q}$-Cartier case. Next we reduce the $\mathbb{R}$-Cartier case to it.

By assumption, we can write $\Theta \sim_{\mathbb{R}} \sum_i d_i D_i$ where the D_i are $\mathbb{Q}$-Cartier. By (11.46), there are $c_{ij} \in \mathbb{Q}$ and $0 < \lambda_j \in \mathbb{R}$ such that the $\Theta_j^Y := \sum_i c_{ij} p^* D_i$ are ample and $d_i = \sum_k \lambda_j c_{ij}$ for every i. In particular, $p^*\Theta \sim_{\mathbb{R}} \sum_j \lambda_j \Theta_j^Y$.

Set $\Theta_j := \sum_i c_{ij} D_i$. Then $\Theta \sim_{\mathbb{R}} \sum_j \lambda_j \Theta_j$ and $p^* \Theta_j = \Theta_j^Y$. The Θ_j are $\mathbb{Q}$-Cartier, hence ample, hence so is Θ. □

Corollary 11.54.2 Let $g \colon X \to S$ be a proper morphism of algebraic spaces and Θ an $\mathbb{R}$-Cartier divisor class on X. Then

$$S^{\text{amp}} := \{s \in S : \Theta_s \text{ is ample on } X_s\} \subset S \quad \text{is open.}$$

Proof Write $\Theta = \sum d_i D_i$ and apply (11.46) to its restriction to X_s. Thus we get $\mathbb{Q}$-Cartier divisors $\Theta_j := \Theta_j^n$ (for $n \gg 1$) such that $\Theta = \sum_j \lambda_j \Theta_j$ and each $\Theta_j|_{X_s}$ is ample. The Θ_j are ample over some open $s \in S^\circ \subset S$, hence so is Θ. □

Theorem 11.55 Fujino and Miyamoto (2021) *Let X be a proper algebraic space and D an $\mathbb{R}$-Cartier divisor class on X. Then D is ample iff $(D^{\dim Z} \cdot Z) > 0$ for every integral subscheme $Z \subset X$.*

Proof By (11.54.1), we may assume that X is normal. By (11.52), we may assume that D is an effective $\mathbb{R}$-divisor. By (11.46), we can write $D = \sum \lambda_i D_i$ where the D_i are effective, $\mathbb{Q}$-Cartier. $D - D_i$ can be chosen arbitrarily small.

Let $p \colon Y \to \operatorname{Supp} D \hookrightarrow X$ be the normalization of $\operatorname{Supp} D$. By dimension induction, $p^* D$ is ample, and so are the $p^* D_i$ if the $D - D_i$ are small enough.

Thus the $D_i|_{\operatorname{Supp} D}$ are ample, hence the D_i are semiample by (11.55.1). Since $(D \cdot C) > 0$ for every curve, $\operatorname{Supp} D$ is not disjoint from any curve, hence the same holds for $\operatorname{Supp} D_i = \operatorname{Supp} D$. So the D_i are ample, and the converse is clear. □

Claim 11.55.1 (Lazarsfeld, 2004, p.35) Let X be a proper algebraic space and D an effective $\mathbb{Q}$-Cartier divisor such that $D|_{\operatorname{Supp} D}$ is ample. Then D is semiample. Thus if D is not disjoint from any curve, then D is ample. □

The usual proof of the Seshadri criterion (see Lazarsfeld (2004, 1.4.13)) now gives the following.

Corollary 11.56 (Seshadri criterion) *Let X be a proper algebraic space and D an $\mathbb{R}$-Cartier divisor on X. Then D is ample iff there is an $\varepsilon > 0$ such that $(D \cdot C) \geq \varepsilon \operatorname{mult}_p C$ for every pointed, integral curve $p \in C \subset X$.* □

Next we study a way to pull back Weil divisors.

11.57 (Intersection theory on normal surfaces) Mumford (1961) Let S be a normal, two-dimensional scheme and $p \colon S' \to S$ a resolution with exceptional curves E_i. The intersection matrix $(E_i \cdot E_j)$ is negative definite by the Hodge index theorem (see Kollár (2013b, 10.1)). Let D be an $\mathbb{R}$-divisor on S. Then there is a unique p-exceptional $\mathbb{R}$-divisor E_D such that

$$(E_i \cdot (p_*^{-1}D + E_D)) = 0 \quad \text{for every} \quad i. \qquad (11.57.1)$$

If D is effective, then E_D is effective by Kollár (2013b, 10.3.3) and $(E_i \cdot E_D) \le 0$ for every i.

We call $p^*D := p_*^{-1}D + E_D$ the *numerical pull-back* of D. If D is $\mathbb{R}$-Cartier then this agrees with the usual pull-back.

More generally, the numerical pull-back is also defined if S' is only normal: we first pull-back to a resolution of S' and then push forward to S.

If D_1, D_2 are $\mathbb{R}$-divisors and one of them has proper support, then one can define their intersection cycle as

$$(D_1 \cdot D_2) = p_*(p_*^{-1}D_1 \cdot p^*D_2) = p_*(p_*^{-1}D_2 \cdot p^*D_1). \qquad (11.57.2)$$

If S is proper, we get the usual properties of intersection theory, except that, even if the D_i are $\mathbb{Z}$-divisors, their intersection numbers can be rational.

The following connects the numerical and sheaf-theoretic pull-backs.

Claim 11.57.3 Let $p\colon T \to S$ be a proper, birational morphism between normal surfaces with exceptional curve $E = \cup E_i$. Let B be an $\mathbb{R}$-divisor on T such that $-B$ is p-nef. Then $p_*\mathscr{O}_T(\lfloor B \rfloor) = \mathscr{O}_S(\lfloor p_*B \rfloor)$.

Moreover, if E is connected, then $g_*\mathscr{O}_T(\lfloor B - \varepsilon E \rfloor) = \mathscr{O}_S(\lfloor p_*B \rfloor)$ for $0 \le \varepsilon \ll 1$, save when B is a $\mathbb{Z}$-divisor and $B \sim 0$ in a neighborhood of E.

Proof Write $B = B_v + B_h$ as a sum of its exceptional and nonexceptional parts. We can harmlessly replace B_h with its round down, so we assume that B_h is a $\mathbb{Z}$-divisor. Let ϕ be a local section of $\mathscr{O}_S(p_*B_h)$. Then $\phi \circ p$ is a rational section of $\mathscr{O}_T(B_h)$, with possible poles along the exceptional curves. There is thus a smallest exceptional $\mathbb{Z}$-divisor F such that $\phi \circ p$ is a section of $\mathscr{O}_T(B_h + F)$. In particular, $(E_i \cdot (B_h + F)) \ge 0$ for every i. Thus

$$(E_i \cdot (F - B_v)) = (E_i \cdot (B_h + F - B)) \ge (E_i \cdot (B_h + F)) \ge 0$$

for every i. By the Hodge index theorem (Kollár, 2013b, 10.3.3), this implies that $B_v - F$ is effective, thus $B_h + F \le \lfloor B \rfloor$.

Moreover, $B_v - F - \varepsilon E$ is effective, unless

$$(E_i \cdot (-B)) = 0 \quad \text{and} \quad (E_i \cdot (B_h + F)) = 0$$

for every i. Then $B_h + F \sim 0$ and $B_h + B_v \sim_{\mathbb{Q}} 0$. Thus $F = B_v$, hence $B \sim 0$. $\square$

Corollary 11.57.4 Let $p\colon T \to S$ be a proper, birational morphism between normal surfaces and D an $\mathbb{R}$-divisor on S. Then $p_*\mathscr{O}_T(\lfloor p^*D \rfloor) = \mathscr{O}_S(\lfloor D \rfloor)$. $\square$

Next we propose a higher dimensional version of pull-back, focusing on its numerical properties. A different notion, using sheaf-theoretic properties, is defined in de Fernex and Hacon (2009).

11.58 (Numerical pull-back) Let $g: Y \to X$ be a projective, birational morphism of normal schemes and H a g-ample Cartier divisor. We define the *H-numerical pull-back* of ℝ-divisors

$$g_H^{(*)}: \mathrm{WDiv}_\mathbb{R}(X) \to \mathrm{WDiv}_\mathbb{R}(Y)$$

as follows. Let $D \subset X$ be an ℝ-divisor. We inductively define

$$g_H^{(*)}(D) = g_*^{-1}D + \sum_{i \geq 2} F_i(D), \tag{11.58.1}$$

where $\mathrm{Supp}\, F_i(D)$ consists of g-exceptional divisors $E_{i\ell}$ for which $g(E_{i\ell}) \subset X$ has codimension i.

Assume that we already defined the $F_i(D)$ for $i < j$. Let $x \in X$ be a point of codimension j. After localizing at x, we have $g_x: Y_x \to X_x$. Let $F_j(D)_x$ be the unique divisor supported on $g_x^{-1}(x)$ such that

$$(E_{j\ell} \cdot (g_*^{-1}D + \sum_{i < j} F_i(D) + F_j(D)_x) \cdot H^{j-2}) = 0 \quad \forall \ell. \tag{11.58.2}$$

To make sense of this, we may assume that H is very ample. Let S be a general complete intersection of $j - 2$ members of $|H|$. Then S is a normal surface, so we are working with intersection numbers as in (11.57). Also, if S is general, then the $g_x|_S$-exceptional curves are in one-to-one correspondence with the divisors $E_{j\ell}$, so any linear combination of $g_x|_S$-exceptional curves corresponds to a linear combination of the divisors $E_{j\ell}$.

If we have proper, but non-projective $Y \to X$, we can apply our definition to a projective modification $Y' \to Y \to X$ and then push forward to Y. This defines $g_H^{(*)}$ in general.

Already in simple situations, for example, for cones over cubic surfaces, the divisors $g_H^{(*)}(D)$ do depend on H. However, the notion has several good properties and it is quite convenient in some situations. See, for example, (11.52) or Fulger et al. (2016, 3.3).

Theorem 11.59 *Let $g: Y \to X$ be a projective, birational morphism of normal schemes and H a g-ample Cartier divisor. Then*

(11.59.1) $g_H^{(*)}: \mathrm{WDiv}_\mathbb{R}(X) \to \mathrm{WDiv}_\mathbb{R}(Y)$ *is ℝ-linear,*

(11.59.2) $g_* \circ g_H^{(*)}$ *is the identity,*

(11.59.3) *if D is ℝ-Cartier, then $g_H^{(*)}(D) = g^*(D)$,*

(11.59.4) *if D is effective, then so is $g_H^{(*)}(D)$,*

(11.59.5) $g_H^{(*)}$ *respects ℝ-linear equivalence,*

(11.59.6) $g_* \mathcal{O}_Y(\lfloor g_H^{(*)}(B) \rfloor) = \mathcal{O}_X(\lfloor B \rfloor)$, *and*

(11.59.7) $g_H^{(*)}$ *maps ℚ-divisors to ℚ-divisors.*

Proof Here (1–3) are clear from the definition. (4) follows from its surface case, which we noted after (11.57.1). If $D_1 \sim_{\mathbb{R}} D_2$ then, using first (1) and then (3), we get that

$$g_H^{(*)}(D_1) = g_H^{(*)}(D_2) + g_H^{(*)}(D_1 - D_2) = g_H^{(*)}(D_2) + g^*(D_1 - D_2),$$

giving (5). Finally (6) is a local question. We may thus assume that (6) holds outside a closed point $x \in X$. Assume to the contrary that $\mathscr{O}_Y(\lfloor g_H^{(*)}(B) \rfloor)$ has a rational section that has poles along $g^{-1}(x)$. After restricting to a general complete intersection surface $S \subset Y$ as in (11.58), we would get a contradiction to (11.57.3).   □

The following negativity lemmas are quite useful.

Lemma 11.60 (Kollár and Mori, 1998, 3.39) *Let $h \colon Z \to Y$ be a proper birational morphism between normal schemes. Let $-B$ be an h-nef $\mathbb{R}$-Cartier divisor on Z. Then*

(11.60.1) *B is effective iff $h_* B$ is.*

(11.60.2) *Assume that B is effective. Then for every $y \in Y$, either $h^{-1}(y) \subset$ Supp B or $h^{-1}(y) \cap$ Supp $B = \emptyset$.*   □

Lemma 11.61 Kollár (2018a) *Let $\pi \colon Y \to X$ be a proper, birational contraction of demi-normal schemes such that none of the π-exceptional divisors is contained in* Sing Y. *Let N, B be Mumford $\mathbb{R}$-divisors such that N is π-nef and B is effective and non-exceptional. Then*

$$\pi_* \mathscr{O}_Y(\lfloor -N - B \rfloor) = \mathscr{O}_X(\lfloor \pi_*(-N - B) \rfloor). \qquad (11.61.1)$$

Moreover, fix $x \in X$ and let E_x be the divisorial part of $\pi^{-1}(x)$. Then

$$\pi_* \mathscr{O}_Y(\lfloor -N - B - \varepsilon E_x \rfloor) = \mathscr{O}_X(\lfloor \pi_*(-N - B) \rfloor) \qquad (11.61.2)$$

for $0 \le \varepsilon \ll 1$, save when $N + B$ is a $\mathbb{Z}$-divisor and $N + B \sim 0$ in a neighborhood of $\pi^{-1}(x)$.

Proof If dim $Y = 2$, then B is also π-nef, so the claim follows from (11.57.3). In general, we may assume that π is projective, take the normalization, and reduce to the surface case as in the proof of (11.59.6).   □

11.62 (Divisorial base locus) Let X be a normal scheme and D a $\mathbb{Z}$-divisor. The divisorial part of the base locus of $|D|$ is denoted by $\mathrm{Bs}^{\mathrm{div}}(D)$. Define the *divisorial base locus* of an $\mathbb{R}$-divisor Δ as $\mathrm{Bs}^{\mathrm{div}}(\Delta) := \mathrm{Bs}^{\mathrm{div}}(\lfloor \Delta \rfloor) + \{\Delta\}$. In particular, $H^0(X, \mathscr{O}_X(\lfloor D - \mathrm{Bs}^{\mathrm{div}} \rfloor(D))) = H^0(X, \mathscr{O}_X(\lfloor D \rfloor))$.

Assume now that we can write $\Delta = \sum_j a_j A_j$ where the A_j are Cartier divisors such that $\text{Bs}(A_j) = \emptyset$ and $a_j > 0$ (This is always possible if X is quasi-affine.) Then $\sum_j \lfloor ma_j \rfloor A_j \leq \lfloor m\Delta \rfloor$ for any $m > 0$, which shows that

$$\text{Bs}^{\text{div}}(m\Delta) \leq \textstyle\sum_j A_j. \tag{11.62.1}$$

Claim 11.62.2 Let $g\colon Y \to X$ be a proper, birational morphism of normal schemes and Δ an ℝ-Cartier, ℝ-divisor on X. Let E be a g-exceptional divisor. Then $g_*\mathcal{O}_Y(\lfloor mg^*\Delta + mE \rfloor) = \mathcal{O}_X(\lfloor m\Delta \rfloor)$ for infinitely many $m \geq 1$ iff E is effective.

Proof Use (11.61) with $B = 0$ and $N = -g^*\Delta$ for the if part. For the converse, we may assume that X is affine. Write $\Delta = \sum_j a_j A_j$ as above.

If $g_*\mathcal{O}_Y(\lfloor mg^*\Delta + mE \rfloor) = \mathcal{O}_X(\lfloor m\Delta \rfloor)$ then $-mE \leq \text{Bs}^{\text{div}}(mg^*\Delta) \leq \sum_j g^*A_j$ by (11.62.1). If this holds for infinitely many $m \geq 1$, then E is effective. □

References

Abramovich, Dan, and Chen, Qile. 2014. Stable logarithmic maps to Deligne–Faltings pairs II. *Asian J. Math.*, **18**(3), 465–488. cited on page(s): 307

Abramovich, Dan, and Fantechi, Barbara. 2017. Configurations of points on degenerate varieties and properness of moduli spaces. *Rend. Semin. Mat. Univ. Padova*, **137**, 1–17. cited on page(s): 307

Abramovich, Dan, and Hassett, Brendan. 2011. Stable varieties with a twist. Pages 1–38 of: *Classification of algebraic varieties*. EMS Ser. Congr. Rep. Eur. Math. Soc., Zurich. cited on page(s): 307

Abramovich, Dan, and Karu, Kalle. 2000. Weak semistable reduction in characteristic 0. *Invent. Math.*, **139**(2), 241–273. cited on page(s): 173

Abramovich, Dan, and Vistoli, Angelo. 2000. Complete moduli for fibered surfaces. Pages 1–31 of: *Recent progress in intersection theory (Bologna, 1997)*. Trends Math. Boston, MA: Birkhäuser Boston. cited on page(s): 307

Abramovich, Dan, and Vistoli, Angelo. 2002. Compactifying the space of stable maps. *J. Amer. Math. Soc.*, **15**(1), 27–75. cited on page(s): 307

Abramovich, Dan, Cadman, Charles, Fantechi, Barbara, and Wise, Jonathan. 2013. Expanded degenerations and pairs. *Comm. Algebra*, **41**(6), 2346–2386. cited on page(s): 307

Abramovich, Dan, Chen, Qile, Marcus, Steffen, and Wise, Jonathan. 2017. Boundedness of the space of stable logarithmic maps. *J. Eur. Math. Soc. (JEMS)*, **19**(9), 2783–2809. cited on page(s): 307

A'Campo, Norbert, Ji, Lizhen, and Papadopoulos, Athanase. 2016 (Feb.). *On the early history of moduli and Teichmüller spaces*. arxiv:1602.07208. cited on page(s): 14, 19

Adiprasito, Karim, Liu, Gaku, and Temkin, Michael. 2019. *Semistable reduction in characteristic 0*. *Sém. Lothar. Combin.* 82B, Art. 25. cited on page(s): 173

Alexeev, Valery. 1993. Two two-dimensional terminations. *Duke Math. J.*, **69**(3), 527–545. cited on page(s): 27, 76, 225

Alexeev, Valery. 1996. Moduli spaces $M_{g,n}(W)$ for surfaces. Pages 1–22 of: *Higher-dimensional complex varieties (Trento, 1994)*. Berlin: de Gruyter. cited on page(s): 306

Alexeev, Valery. 2002. Complete moduli in the presence of semiabelian group action. *Ann. of Math. (2)*, **155**(3), 611–708. cited on page(s): 27, 307

446

Alexeev, Valery. 2006. Higher-dimensional analogues of stable curves. Pages 515–536 of: *International Congress of Mathematicians*, vol. II. Eur. Math. Soc., Zurich. cited on page(s): 218, 221, 315

Alexeev, Valery. 2008. Limits of stable pairs. *Pure Appl. Math. Q.*, **4**(3, part 2), 767–783. cited on page(s): 307, 418

Alexeev, Valery. 2015. *Moduli of weighted hyperplane arrangements*. Advanced Courses in Mathematics. CRM Barcelona. Basel: Birkhäuser/Springer. Edited by Gilberto Bini, Martí Lahoz, Emanuele Macrì, and Paolo Stellari. cited on page(s): 218, 221, 307, 315

Alexeev, Valery. 2016. Divisors of Burniat surfaces. Pages 287–302 of: *Development of moduli theory—Kyoto 2013*. Adv. Stud. Pure Math., vol. 69. Math. Soc. Japan [Tokyo]. cited on page(s): 307

Alexeev, Valery, and Hacon, Christopher D. 2012. Non-rational centers of log canonical singularities. *J. Algebra*, **369**, 1–15. cited on page(s): 418

Alexeev, Valery, and Knutson, Allen. 2010. Complete moduli spaces of branchvarieties. *J. Reine Angew. Math.*, **639**, 39–71. cited on page(s): 307, 343

Alexeev, Valery, and Liu, Wenfei. 2019a. Log surfaces of Picard rank one from four lines in the plane. *Eur. J. Math.*, **5**(3), 622–639. cited on page(s): 26, 307

Alexeev, Valery, and Liu, Wenfei. 2019b. Open surfaces of small volume. *Algebr. Geom.*, **6**(3), 312–327. cited on page(s): 307

Alexeev, Valery, and Thompson, Alan. 2021. ADE surfaces and their moduli. *J. Algebraic Geom.*, **30**(2), 331–405. cited on page(s): 307

Altmann, Klaus, and Kollár, János. 2019. The dualizing sheaf on first-order deformations of toric surface singularities. *J. Reine Angew. Math.*, **753**, 137–158. cited on page(s): 24, 219, 221, 247

Ambro, Florin. 2003. Quasi-log varieties. *Tr. Mat. Inst. Steklova*, **240**(Biratsion. Geom. Linein. Sist. Konechno Porozhdennye Algebry), 220–239. cited on page(s): 415, 422

Ambro, Florin. 2011. Basic properties of log canonical centers. Pages 39–48 of: *Classification of algebraic varieties*. EMS Ser. Congr. Rep. Eur. Math. Soc., Zurich. cited on page(s): 415

Ambro, Florin, and Kollár, János. 2019. Minimal models of semi-log-canonical pairs. Pages 1–13 of: *Moduli of K-stable varieties*. Springer INdAM Ser., vol. 31. Cham: Springer. cited on page(s): 212

Arnold V. I., Guseĭn-Zade, S. M., and Varchenko, A. N. 1985. *Singularities of differentiable maps*, vol. I–II. Monographs in Mathematics, vol. 82–83. Boston, MA: Birkhäuser Boston. Translated from the Russian by Ian Porteous and Mark Reynolds. cited on page(s): 84, 384, 401

Artin, Michael. 1969. Algebraization of formal moduli. I. Pages 21–71 of: *Global analysis (Papers in honor of K. Kodaira)*. Tokyo: University of Tokyo Press. cited on page(s): 131, 133, 363

Artin, Michael. 1974. Algebraic construction of Brieskorn's resolutions. *J. Algebra*, **29**, 330–348. cited on page(s): 21, 58

Artin, Michael. 1976. *Deformations of singularities*. Bombay: Tata Institute. cited on page(s): 25, 84, 220, 238, 248, 361

Artin, Michael. 1977. Coverings of the rational double points in characteristic *p*. Pages 11–22 of: *Complex analysis and algebraic geometry*. Tokyo: Iwanami Shoten. cited on page(s): 76

Ascher, Kenneth, and Bejleri, Dori. 2019. Moduli of fibered surface pairs from twisted stable maps. *Math. Ann.*, **374**(1–2), 1007–1032. cited on page(s): 307

Ascher, Kenneth, and Bejleri, Dori. 2021a. Moduli of double covers and degree one del Pezzo surfaces. *Eur. J. Math.*, **7**(2), 557–569. cited on page(s): 307

Ascher, Kenneth, and Bejleri, Dori. 2021b. Moduli of weighted stable elliptic surfaces and invariance of log plurigenera. *Proc. Lond. Math. Soc. (3)*, **122**(5), 617–677. With an appendix by Giovanni Inchiostro. cited on page(s): 307

Ascher, Kenneth, and Gallardo, Patricio. 2018. A generic slice of the moduli space of line arrangements. *Algebra Number Theory*, **12**(4), 751–778. cited on page(s): 307

Ascher, Kenneth, and Molcho, Samouil. 2016. Logarithmic stable toric varieties and their moduli. *Algebr. Geom.*, **3**(3), 296–319. cited on page(s): 307

Ascher, Kenneth, Dubé, Connor, Gershenson, Daniel, and Hou, Elaine. 2020. Enumerating Hassett's wall and chamber decomposition of the moduli space of weighted stable curves. *Exp. Math.*, **29**(1), 36–53. cited on page(s): 307

Ash, Avner, Mumford, David, Rapoport, Michael, and Tai, Yung-Sheng. 1975. *Smooth compactification of locally symmetric varieties*. Brookline, MA: Mathematical Science Press. Lie Groups: History, Frontiers and Applications, vol. IV. cited on page(s): 27

Barlet, Daniel. 1975. Espace analytique réduit des cycles analytiques complexes compacts d'un espace analytique complexe de dimension finie. Pages 1–158. Lecture Notes in Math., vol. 482 of: *Fonctions de plusieurs variables complexes, II (Sém. François Norguet, 1974–1975)*. Berlin: Springer. cited on page(s): 131

Barlet, Daniel, and Magnússon, Jón. 2020. *Cycles Analytiques Complexes II: L'espace des Cycles*. Cours Specialises, vol. 27. Paris: Société Mathématique de France. cited on page(s): 131

Barth, W., Peters, C., and Van de Ven, A. 1984. *Compact complex surfaces*. Ergebnisse der Mathematik und ihrer Grenzgebiete (3), vol. 4. Berlin: Springer. cited on page(s): 31, 55

Bejleri, Dori, and Inchiostro, Giovanni. 2021. Stable pairs with a twist and gluing morphisms for moduli of surfaces. *Selecta Math. (N.S.)*, **27**(3), Paper No. 40, 44. cited on page(s): 307

Bernasconi, Fabio. 2018. *Non-normal purely log terminal centres in characteristic p ≥ 3*. *Eur. J. Math.*, **5**(4), 1242–1251. cited on page(s): 339

Bhatt, Bhargav, and de Jong, Aise Johan. 2014. Lefschetz for local Picard groups. *Ann. Sci. Éc. Norm. Supér. (4)*, **47**(4), 833–849. cited on page(s): 122, 123, 124

Birkar, Caucher. 2017. The augmented base locus of real divisors over arbitrary fields. *Math. Ann.*, **369**, 905–921. cited on page(s): 426, 432

Birkar, Caucher, Cascini, Paolo, Hacon, Christopher D., and McKernan, James. 2010. Existence of minimal models for varieties of log general type. *J. Amer. Math. Soc.*, **23**(2), 405–468. cited on page(s): 420

Bombieri, Enrico. 1973. Canonical models of surfaces of general type. *Inst. Hautes Études Sci. Publ. Math.*, 171–219. cited on page(s): 222

Bosch, Siegfried, Lütkebohmert, Werner, and Raynaud, Michel. 1990. *Néron models.* Ergebnisse der Mathematik und ihrer Grenzgebiete (3), vol. 21. Berlin: Springer. cited on page(s): 220, 265, 267, 335

Boutot, Jean-François. 1978. *Schéma de Picard local.* Lecture Notes in Math., vol. 632. Berlin: Springer. cited on page(s): 168

Brieskorn, Egbert. 1967/1968. Rationale Singularitäten komplexer Flächen. *Invent. Math.,* **4,** 336–358. cited on page(s): 77, 86, 249

Bruns, Winfried, and Herzog, Jürgen. 1993. *Cohen–Macaulay rings.* Cambridge Studies in Advanced Mathematics, vol. 39. Cambridge: Cambridge University Press. cited on page(s): 377

Buchsbaum, David A., and Eisenbud, David. 1977. Algebra structures for finite free resolutions, and some structure theorems for ideals of codimension 3. *Amer. J. Math.,* **99**(3), 447–485. cited on page(s): 84

Campana, Frédéric, Koziarz, Vincent, and Păun, Mihai. 2012. Numerical character of the effectivity of adjoint line bundles. *Ann. Inst. Fourier (Grenoble),* **62**(1), 107–119. cited on page(s): 169

Catanese, Fabrizio. 1992. Chow varieties, Hilbert schemes and moduli spaces of surfaces of general type. *J. Algebraic Geom.,* **1**(4), 561–595. cited on page(s): 177, 260

Cayley, Arthur. 1860. A new analytic representation of curves in space. *Quarterly J. Math.,* **3,** 225–236. cited on page(s): 9, 16, 127, 260, 283

Cayley, Arthur. 1862. A new analytic representation of curves in space. *Quarterly J. Math.,* **5,** 81–86. cited on page(s): 9, 16, 127, 260, 283

Cheltsov, Ivan. 2010. Factorial threefold hypersurfaces. *J. Algebraic Geom.,* **19**(4), 781–791. cited on page(s): 32

Chiang-Hsieh, Hung-Jen, and Lipman, Joseph. 2006. A numerical criterion for simultaneous normalization. *Duke Math. J.,* **133**(2), 347–390. cited on page(s): 133

Chow, Wei-Liang, and van der Waerden, Bartel Leendert. 1937. Zur algebraischen Geometrie. IX. *Math. Ann.,* **113**(1), 692–704. cited on page(s): 16

Conrad, Brian. 2000. *Grothendieck duality and base change.* Lecture Notes in Math., vol. 1750. Berlin: Springer. cited on page(s): 28, 108, 110

Dalbec, John, and Sturmfels, Bernd. 1995. Introduction to Chow forms. Pages 37–58 of: *Invariant methods in discrete and computational geometry (Curaçao, 1994).* Dordrecht: Kluwer. cited on page(s): 177, 260

de Fernex, Tommaso. 2005. Negative curves on very general blow-ups of $\mathbb{P}^2$. Pages 199–207 of: *Projective varieties with unexpected properties.* Berlin: Walter de Gruyter. cited on page(s): 41

de Fernex, Tommaso, and Hacon, Christopher D. 2009. Singularities on normal varieties. *Compositio Mathematica,* **145**(2), 393–414. cited on page(s): 435

de Jong, A. Johan. 1996. Smoothness, semi-stability and alterations. *Inst. Hautes Études Sci. Publ. Math.,* 51–93. cited on page(s): 104

de Jong, Theo, and van Straten, Duco. 1991. On the base space of a semi-universal deformation of rational quadruple points. *Ann. of Math.,* **134**(3), 653–678. cited on page(s): 390

Deligne, P., and Mumford, D. 1969. The irreducibility of the space of curves of given genus. *Inst. Hautes Études Sci. Publ. Math.,* **63,** 75–109. cited on page(s): 20, 21, 219

Dolgachev, Igor. 2003. *Lectures on invariant theory*. London Mathematical Society Lecture Note Series, vol. 296. Cambridge: Cambridge University Press. cited on page(s): 50, 55

Donaldson, Simon. 2020. Fredholm topology and enumerative geometry: reflections on some words of Michael Atiyah. Pages 1–31 of: *Proceedings of the Gökova Geometry-Topology Conferences 2018/2019*. Somerville, MA: International Press. cited on page(s): 307

Du Bois, Philippe. 1981. Complexe de de Rham filtré d'une variété singulière. *Bull. Soc. Math. France*, **109**(1), 41–81. cited on page(s): 105

Du Bois, Philippe, and Jarraud, Pierre. 1974. Une propriété de commutation au changement de base des images directes supérieures du faisceau structural. *C. R. Acad. Sci. Paris Sér. A*, **279**, 745–747. cited on page(s): 105

Du Val, P. 1934. On isolated singularities of surfaces which do not affect the conditions of adjunction I–II. *Proc. Camb. Phil. Soc.*, **30**, 453–465, 483–491. cited on page(s): 75

Durfee, Alan H. 1979. Fifteen characterizations of rational double points and simple critical points. *Enseign. Math. (2)*, **25**(1–2), 131–163. cited on page(s): 76

Earle, Clifford J. 1971. On the moduli of closed Riemann surfaces with symmetries. Pages 119–130. Ann. of Math. Studies, No. 66 of: *Advances in the theory of riemann surfaces (Proc. Conf., Stony Brook, NY, 1969)*. Princeton, NJ: Princeton University Press. cited on page(s): 62

Eisenbud, David. 1995. *Commutative algebra: With a view toward algebraic geometry*. Graduate Texts in Mathematics, vol. 150. New York: Springer. cited on page(s): 273, 357, 366, 369, 371, 374, 379

Eisenbud, David, and Harris, Joe. 1987. On varieties of minimal degree (a centennial account). Pages 3–13 of: *Algebraic geometry (Bowdoin, 1985, Brunswick, ME, 1985)*. Proc. Sympos. Pure Math., vol. 46. Providence, RI: American Mathematics Society. cited on page(s): 39

Ekedahl, Torsten. 1988. Canonical models of surfaces of general type in positive characteristic. *Inst. Hautes Études Sci. Publ. Math.*, 97–144. cited on page(s): 222

Ellingsrud, Geir, and Skjelbred, Tor. 1980. Profondeur d'anneaux d'invariants en caractéristique *p*. *Compositio Math.*, **41**(2), 233–244. cited on page(s): 340

Esser, Louis, Totaro, Burt, and Wang, Chengxi. 2021. Varieties of general type with doubly exponential asymptotics. *arXiv e-prints*, Sept., arXiv:2109.13383. cited on page(s): 26

Farkas, Gavril, and Morrison, Ian (eds). 2013. *Handbook of moduli. vols. I–III*. Advanced Lectures in Mathematics (ALM), vol. 24, 25, 26. Somerville, MA: International Press. cited on page(s): 21

Filipazzi, Stefano, and Inchiostro, Giovanni. 2021. *Moduli of $\mathbb{Q}$-Gorenstein pairs and applications*. cited on page(s): 315, 320

Fitting, Hans. 1936. Die Determinantenideale eines Moduls. *Jahresber. Deutsch. Math.-Verein.*, **46**, 195–228. cited on page(s): 273

Franciosi, Marco, Pardini, Rita, and Rollenske, Sönke. 2015a. Computing invariants of semi-log-canonical surfaces. *Math. Z.*, **280**(3-4), 1107–1123. cited on page(s): 307

Franciosi, Marco, Pardini, Rita, and Rollenske, Sönke. 2015b. Log-canonical pairs and Gorenstein stable surfaces with $K_X^2 = 1$. *Compos. Math.*, **151**(8), 1529–1542. cited on page(s): 307

Franciosi, Marco, Pardini, Rita, and Rollenske, Sönke. 2017. Gorenstein stable surfaces with $K_X^2 = 1$ and $p_g > 0$. *Math. Nachr.*, **290**(5-6), 794–814. cited on page(s): 307

Franciosi, Marco, Pardini, Rita, and Rollenske, Sönke. 2018. Gorenstein stable Godeaux surfaces. *Selecta Math. (N.S.)*, **24**(4), 3349–3379. cited on page(s): 307

Franciosi, Marco, Pardini, Rita, Rana, Julie, and Rollenske, Sönke. 2022. I-surfaces with one T-singularity. *Boll. Unione Mat. Ital.*, **15**(1-2), 173–190. cited on page(s): 307

Fujino, Osamu. 2000. Abundance theorem for semi log canonical threefolds. *Duke Math. J.*, **102**(3), 513–532. cited on page(s): 169

Fujino, Osamu. 2007. What is log terminal? Pages 49–62 of: *Flips for 3-folds and 4-folds*. Oxford Lecture Ser. Math. Appl., vol. 35. Oxford: Oxford University Press. cited on page(s): 415

Fujino, Osamu. 2014. Fundamental theorems for semi log canonical pairs. *Algebr. Geom.*, **1**(2), 194–228. cited on page(s): 169, 422

Fujino, Osamu. 2017. *Foundations of the minimal model program*. MSJ Memoirs, vol. 35. Tokyo: Mathematical Society of Japan. cited on page(s): 320, 415, 417, 418, 422

Fujino, Osamu. 2018. Semipositivity theorems for moduli problems. *Ann. of Math. (2)*, **187**(3), 639–665. cited on page(s): 27, 227

Fujino, Osamu. 2022. *Minimal model program for projective morphisms between complex analytic spaces*. https://arxiv.org/pdf/2201.11315.pdf. cited on page(s): 411

Fujino, Osamu, and Miyamoto, Keisuke. 2021. *Nakai–Moishezon ampleness criterion for real line bundles*. cited on page(s): 426, 432, 433

Fujita, Takao. 1983. Vanishing theorems for semipositive line bundles. Pages 519–528 of: *Algebraic geometry (Tokyo/Kyoto, 1982)*. Lecture Notes in Math., vol. 1016. Berlin: Springer. cited on page(s): 432

Fulger, Mihai, Kollár, János, and Lehmann, Brian. 2016. Volume and Hilbert function of $\mathbb{R}$-divisors. *Michigan Math. J.*, **65**(2), 371–387. cited on page(s): 383, 436

Gieseker, D. 1977. Global moduli for surfaces of general type. *Invent. Math.*, **43**(3), 233–282. cited on page(s): 25

Gongyo, Yoshinori. 2013. Abundance theorem for numerically trivial log canonical divisors of semi-log canonical pairs. *J. Algebraic Geom.*, **22**(3), 549–564. cited on page(s): 169

Griffiths, Phillip, and Harris, Joseph. 1978. *Principles of algebraic geometry*. New York: Wiley-Interscience, John Wiley & Sons. Pure and Applied Mathematics. cited on page(s): 19, 222, 397

Gross, Mark, Hacking, Paul, and Keel, Sean. 2015. Mirror symmetry for log Calabi-Yau surfaces I. *Publ. Math. Inst. Hautes Études Sci.*, **122**, 65–168. cited on page(s): 84

Grothendieck, Alexander. 1960. Éléments de géométrie algébrique. I–IV. *Inst. Hautes Études Sci. Publ. Math.* cited on page(s): 112, 162, 355, 369, 376, 392, 398, 408

Grothendieck, Alexander. 1962. *Fondements de la géométrie algébrique. [Extraits du Séminaire Bourbaki, 1957–1962.]*. Paris: Secrétariat mathématique. cited on page(s): 17, 127, 326, 335, 355

Grothendieck, Alexander. 1967. *Local cohomology*. Lecture Notes in Math., vol. 41. Berlin: Springer. cited on page(s): 377

Grothendieck, Alexander. 1968. *Cohomologie locale des faisceaux cohérents et théorèmes de Lefschetz locaux et globaux (SGA 2)*. Amsterdam: North-Holland Publishing. Augmenté d'un exposé par Michèle Raynaud, Séminaire de Géométrie Algébrique du Bois-Marie, 1962, Advanced Studies in Pure Mathematics, vol. 2. cited on page(s): 124, 346, 366, 379, 404

Grothendieck, Alexander. 1971. *Revêtements étales et groupe fondamental*. Lecture Notes in Math., vol. 224. Heidelberg: Springer. cited on page(s): 163, 335, 395

Gunning, Robert C., and Rossi, Hugo. 1965. *Analytic functions of several complex variables*. Englewood Cliffs, NJ: Prentice-Hall. cited on page(s): 397

Haboush, William, and Lauritzen, Niels. 1993. Varieties of unseparated flags. Pages 35–57 of: *Linear algebraic groups and their representations (Los Angeles, CA, 1992)*. Contemp. Math., vol. 153. Providence, RI: American Mathematics Society. cited on page(s): 339

Hacking, Paul. 2004. Compact moduli of plane curves. *Duke Math. J.*, **124**(2), 213–257. cited on page(s): 307, 309

Hacking, Paul. 2012. Compact moduli spaces of surfaces of general type. Pages 1–18 of: *Compact moduli spaces and vector bundles*. Contemp. Math., vol. 564. Providence, RI: American Mathematics Society. cited on page(s): 307

Hacking, Paul. 2013. Exceptional bundles associated to degenerations of surfaces. *Duke Math. J.*, **162**(6), 1171–1202. cited on page(s): 307

Hacking, Paul. 2016. Compact moduli spaces of surfaces and exceptional vector bundles. Pages 41–67 of: *Compactifying moduli spaces*. Adv. Courses Math. CRM Barcelona. Birkhäuser/Springer. cited on page(s): 307

Hacking, Paul, and Prokhorov, Yuri. 2010. Smoothable del Pezzo surfaces with quotient singularities. *Compos. Math.*, **146**(1), 169–192. cited on page(s): 307

Hacking, Paul, Keel, Sean, and Tevelev, Jenia. 2006. Compactification of the moduli space of hyperplane arrangements. *J. Algebraic Geom.*, **15**(4), 657–680. cited on page(s): 307

Hacking, Paul, Keel, Sean, and Tevelev, Jenia. 2009. Stable pair, tropical, and log canonical compactifications of moduli spaces of del Pezzo surfaces. *Invent. Math.*, **178**(1), 173–227. cited on page(s): 307

Hacking, Paul, Tevelev, Jenia, and Urzúa, Giancarlo. 2017. Flipping surfaces. *J. Algebraic Geom.*, **26**(2), 279–345. cited on page(s): 307

Hacon, Christopher D., and Xu, Chenyang. 2013. Existence of log canonical closures. *Invent. Math.*, **192**(1), 161–195. cited on page(s): 174, 323, 420

Hacon, Christopher D., and Xu, Chenyang. 2016. On finiteness of B-representations and semi-log canonical abundance. Pages 361–377 of: *Minimal models and extremal rays (Kyoto, 2011)*. Adv. Stud. Pure Math., vol. 70. Math. Soc. Japan [Tokyo]. cited on page(s): 169, 420

Hacon, Christopher D., McKernan, James, and Xu, Chenyang. 2014. ACC for log canonical thresholds. *Annals of Math.*, **180**, 523–571. cited on page(s): 226

Hacon, Christopher D., McKernan, James, and Xu, Chenyang. 2018. Boundedness of moduli of varieties of general type. *J. Eur. Math. Soc. (JEMS)*, **20**(4), 865–901. cited on page(s): 27, 225, 226

Han, Jingjun, Liu, Jihao, and Shokurov, V. V. 2020. *ACC for minimal log discrepancies of exceptional singularities*. cited on page(s): 430

Hartshorne, Robin. 1966. *Residues and duality*. Lecture notes of a seminar on the work of A. Grothendieck, given at Harvard 1963/64. With an appendix by P. Deligne. Lecture Notes in Math., No. 20. Berlin: Springer. cited on page(s): 28, 108

Hartshorne, Robin. 1977. *Algebraic geometry*. New York: Springer. Graduate Texts in Mathematics, No. 52. cited on page(s): 10, 13, 15, 28, 108, 109, 127, 130, 134, 135, 239, 267, 317, 321, 324, 349, 368, 411, 433

Hartshorne, Robin. 1986. Generalized divisors on Gorenstein curves and a theorem of Noether. *J. Math. Kyoto Univ.*, **26**(3), 375–386. cited on page(s): 152, 265

Hartshorne, Robin. 2010. *Deformation theory*. Graduate Texts in Mathematics, vol. 257. New York: Springer. cited on page(s): 84, 220, 238, 361

Hartshorne, Robin, and Polini, Claudia. 2015. Divisor class groups of singular surfaces. *Trans. Amer. Math. Soc.*, **367**(9), 6357–6385. cited on page(s): 152, 265

Hassett, Brendan. 1999. Stable log surfaces and limits of quartic plane curves. *Manuscripta Math.*, **100**(4), 469–487. cited on page(s): 307

Hassett, Brendan. 2000. Local stable reduction of plane curve singularities. *J. Reine Angew. Math.*, **520**, 169–194. cited on page(s): 307

Hassett, Brendan. 2001. Stable limits of log surfaces and Cohen-Macaulay singularities. *J. Algebra*, **242**(1), 225–235. cited on page(s): 307

Hassett, Brendan, and Hyeon, Donghoon. 2013. Log minimal model program for the moduli space of stable curves: the first flip. *Ann. of Math. (2)*, **177**(3), 911–968. cited on page(s): 60

Hironaka, Heisuke. 1958. A note on algebraic geometry over ground rings: The invariance of Hilbert characteristic functions under the specialization process. *Illinois J. Math.*, **2**, 355–366. cited on page(s): 130, 403

Hu, Zhi, and Zong, Runhong. 2020. *On Base Change of Local Stability in Positive Characteristics*. https://arxiv.org/pdf/2001.04083.pdf. cited on page(s): 74

Humphreys, J. 1975. *Linear algebraic groups*. New York: Springer. cited on page(s): 85

Huybrechts, Daniel, and Lehn, Manfred. 1997. *The geometry of moduli spaces of sheaves*. Aspects of Mathematics, E31. Braunschweig: Friedr. Vieweg & Sohn. cited on page(s): 356

Iitaka, Shigeru. 1971. On *D*-dimensions of algebraic varieties. *J. Math. Soc. Japan*, **23**, 356–373. cited on page(s): 33

Iitaka, Shigeru. 1972. Genera and classifications of algebraic varieties (in Japanese). *Sugaku*, **24**, 14–27. cited on page(s): 227

Illusie, Luc. 1971. *Complexe cotangent et déformations. I–II*. Lecture Notes in Math., vols. 239, 283. Berlin: Springer. cited on page(s): 25

Jung, Heinrich W. E. 1908. Darstellung der Funktionen eines algebraischen Körpers zweier unabhängigen Veränderlichen x, y in der Umgebung einer Stelle $x = a$, $y = b$. *J. Reine Angew. Math.*, **133**, 289–314. cited on page(s): 249

Karu, Kalle. 2000. Minimal models and boundedness of stable varieties. *J. Algebraic Geom.*, **9**(1), 93–109. cited on page(s): 173

Kawakita, Masayuki. 2007. Inversion of adjunction on log canonicity. *Invent. Math.*, **167**(1), 129–133. cited on page(s): 417

Kawamata, Yujiro. 1985. Minimal models and the Kodaira dimension of algebraic fiber spaces. *J. Reine Angew. Math.*, **363**, 1–46. cited on page(s): 169

Kawamata, Yujiro. 2013. On the abundance theorem in the case of numerical Kodaira dimension zero. *Amer. J. Math.*, **135**(1), 115–124. cited on page(s): 169

Keel, Seán, and Mori, Shigefumi. 1997. Quotients by groupoids. *Ann. of Math. (2)*, **145**(1), 193–213. cited on page(s): 25, 227, 333

Kempf, G., Knudsen, F. F., Mumford, D., and Saint-Donat, B. 1973. *Toroidal embeddings. I*. Lecture Notes in Math., vol. 339. Berlin: Springer. cited on page(s): 27, 104

Klein, Felix, and Fricke, Robert. 1892. *Vorlesungen über die Theorie der elliptischhen Modulfunktionen*. Leipzig: B. G. Teubner. cited on page(s): 19

Knudsen, F. F. 1983. The projectivity of the moduli space of stable curves, II. *Math. Scand.*, **52**, 161–199. cited on page(s): 174

Kodaira, K., and Spencer, D. C. 1958. On deformations of complex analytic structures. I, II. *Ann. of Math. (2)*, **67**, 328–466. cited on page(s): 36

Kollár, János. 1990. Projectivity of complete moduli. *J. Differential Geom.*, **32**(1), 235–268. cited on page(s): 27, 227, 335

Kollár, János. 1992a. Cone theorems and bug-eyed covers. *J. Algebraic Geom.*, **1**(2), 293–323. cited on page(s): 58

Kollár, János (ed). 1992b. *Flips and abundance for algebraic threefolds*. Société Mathématique de France. Papers from the Second Summer Seminar on Algebraic Geometry, University of Utah, Salt Lake City, Utah, August 1991, Astérisque No. 211 (1992). cited on page(s): 417

Kollár, János. 1995a. Flatness criteria. *J. Algebra*, **175**(2), 715–727. cited on page(s): 82, 396, 403

Kollár, János. 1995b. *Shafarevich maps and automorphic forms*. M. B. Porter Lectures. Princeton, NJ: Princeton University Press. cited on page(s): 105

Kollár, János. 1996. *Rational curves on algebraic varieties*. Ergebnisse der Mathematik und ihrer Grenzgebiete. 3. Folge., vol. 32. Berlin: Springer. cited on page(s): 16, 17, 127, 128, 129, 131, 132, 154, 156, 174, 176, 181, 202, 260, 326, 392, 404, 432

Kollár, János. 1997. Quotient spaces modulo algebraic groups. *Ann. of Math. (2)*, **145**(1), 33–79. cited on page(s): 25, 227, 333

Kollár, János. 1999. Effective Nullstellensatz for arbitrary ideals. *J. Eur. Math. Soc. (JEMS)*, **1**(3), 313–337. cited on page(s): 177, 260, 299, 300

Kollár, János. 2006. Non-quasi-projective moduli spaces. *Ann. of Math. (2)*, **164**(3), 1077–1096. cited on page(s): 227

Kollár, János. 2008a. *Hulls and husks*. arXiv:0805.0576. cited on page(s): 24, 343

Kollár, János. 2008b. Is there a topological Bogomolov–Miyaoka–Yau inequality? *Pure Appl. Math. Q.*, **4**(2, part 1), 203–236. cited on page(s): 43

Kollár, János. 2010. Exercises in the birational geometry of algebraic varieties. Pages 495–524 of: *Analytic and algebraic geometry*. IAS/Park City Math. Ser., vol. 17. Providence, RI: American Mathematics Society. cited on page(s): 47

Kollár, János. 2011a. A local version of the Kawamata-Viehweg vanishing theorem. *Pure Appl. Math. Q.*, **7**(4, Special Issue: In memory of Eckart Viehweg), 1477–1494. cited on page(s): 418

Kollár, János. 2011b. Simultaneous normalization and algebra husks. *Asian J. Math.*, **15**(3), 437–449. cited on page(s): 133, 343, 400

Kollár, János. 2011c. Two examples of surfaces with normal crossing singularities. *Sci. China Math.*, **54**(8), 1707–1712. cited on page(s): 46, 315

Kollár, János. 2013a. Grothendieck–Lefschetz type theorems for the local Picard group. *J. Ramanujan Math. Soc.*, **28A**, 267–285. cited on page(s): 71, 123, 124, 188

Kollár, János. 2013b. *Singularities of the minimal model program*. Cambridge Tracts in Mathematics, vol. 200. Cambridge: Cambridge University Press. With the collaboration of Sándor Kovács. cited on page(s): 6, 8, 9, 10, 34, 35, 38, 47, 66, 67, 68, 69, 72, 74, 76, 79, 80, 88, 89, 90, 91, 92, 97, 99, 102, 103, 104, 108, 141, 166, 167, 169, 173, 193, 214, 377, 384, 404, 405, 411, 414, 415, 417, 418, 419, 421, 422, 423, 426, 431, 434, 435

Kollár, János. 2014. Semi-normal log centres and deformations of pairs. *Proc. Edinb. Math. Soc. (2)*, **57**(1), 191–199. cited on page(s): 119, 415

Kollár, János. 2015. *How much of the Hilbert function do we really need to know?* arXiv:1503.08694. cited on page(s): 186

Kollár, János. 2016a. Maps between local Picard groups. *Algebr. Geom.*, **3**(4), 461–495. cited on page(s): 122, 124, 168, 188

Kollár, János. 2016b. Sources of log canonical centers. Pages 29–48 of: *Minimal models and extremal rays (Kyoto, 2011)*. Adv. Stud. Pure Math., vol. 70. Math. Soc. Japan [Tokyo]. cited on page(s): 423

Kollár, János. 2016c. Variants of normality for Noetherian schemes. *Pure Appl. Math. Q.*, **12**(1), 1–31. cited on page(s): 405

Kollár, János. 2017. Coherence of local and global hulls. *Methods Appl. Anal.*, **24**(1), 63–70. cited on page(s): 344, 345, 376

Kollár, János. 2018a. Log-plurigenera in stable families. *Peking Math. J.*, **1**(1), 81–107. cited on page(s): 119, 121, 233, 234, 421, 436

Kollár, János. 2018b. Log-plurigenera in stable families of surfaces. *Peking Math. J.*, **1**(1), 109–124. cited on page(s): 234

Kollár, János. 2019. *Families of divisors*. arXiv:1910.00937. cited on page(s): 24, 221

Kollár, János. 2021a. Deformations of varieties of general type. *Milan J. Math.*, **89**(2), 345–354. cited on page(s): 36, 42

Kollár, János. 2021b. Mumford's influence on the moduli theory of algebraic varieties. *Pure Appl. Math. Q.*, **17**(2), 619–647. cited on page(s): 14, 19

Kollár, János. 2022. *Families of stable 3-folds in positive characteristic*. cited on page(s): 23, 69, 339

Kollár, János, and Kovács, Sándor J. 2010. Log canonical singularities are Du Bois. *J. Amer. Math. Soc.*, **23**(3), 791–813. cited on page(s): 27, 105, 107, 108, 415

Kollár, János, and Kovács, Sándor J. 2020. Deformations of log canonical and F-pure singularities. *Algebr. Geom.*, **7**(6), 758–780. cited on page(s): 105, 107, 108, 245

Kollár, János, and Matsusaka, Teruhisa. 1983. Riemann–Roch type inequalities. *Amer. J. Math.*, **105**(1), 229–252. cited on page(s): 323

Kollár, János, and Mori, Shigefumi. 1992. Classification of three-dimensional flips. *J. Amer. Math. Soc.*, **5**(3), 533–703. cited on page(s): 36, 43, 168

Kollár, János, and Mori, Shigefumi. 1998. *Birational geometry of algebraic varieties*. Cambridge Tracts in Mathematics, vol. 134. Cambridge: Cambridge University Press. With the collaboration of C. H. Clemens and A. Corti; translated from the 1998 Japanese original. cited on page(s): 10, 11, 28, 85, 97, 108, 122, 172, 315, 411, 413, 414, 417, 419, 420, 422, 432, 436

Kollár, János, and Shepherd-Barron, N. I. 1988. Threefolds and deformations of surface singularities. *Invent. Math.*, **91**(2), 299–338. cited on page(s): 8, 22, 24, 141, 191, 206, 207, 218, 219, 230, 306

Kollár, János, Smith, Karen E., and Corti, Alessio. 2004. *Rational and nearly rational varieties*. Cambridge Studies in Advanced Mathematics, vol. 92. Cambridge: Cambridge University Press. cited on page(s): 61

Kontsevich, Maxim, and Tschinkel, Yuri. 2019. Specialization of birational types. *Invent. Math.*, **217**, 415–432. cited on page(s): 31

Kovács, Sándor J. 1999. Rational, log canonical, Du Bois singularities: On the conjectures of Kollár and Steenbrink. *Compositio Math.*, **118**(2), 123–133. cited on page(s): 105

Kovács, Sándor J. 2011. Irrational centers. *Pure Appl. Math. Q.*, **7**(4, Special Issue: In memory of Eckart Viehweg), 1495–1515. cited on page(s): 418

Kovács, Sándor J. 2018. Non-Cohen–Macaulay canonical singularities. Pages 251–259 of: *Local and global methods in algebraic geometry*. Contemp. Math., vol. 712. Providence, RI: American Mathematics Society. cited on page(s): 339

Kovács, Sándor J., and Patakfalvi, Zsolt. 2017. Projectivity of the moduli space of stable log-varieties and subadditivity of log-Kodaira dimension. *J. Amer. Math. Soc.*, **30**(4), 959–1021. cited on page(s): 27, 227

Kovács, Sándor J., and Schwede, Karl. 2016. Inversion of adjunction for rational and Du Bois pairs. *Algebra Number Theory*, **10**(5), 969–1000. cited on page(s): 313

Kuipers, L., and Niederreiter, H. 1974. *Uniform distribution of sequences*. Pure and Applied Mathematics. New York: Wiley-Interscience. cited on page(s): 430

Kunz, Ernst. 1976. On Noetherian rings of characteristic p. *Amer. J. Math.*, **98**(4), 999–1013. cited on page(s): 177

Laufer, Henry B. 1977. On minimally elliptic singularities. *Amer. J. Math.*, **99**(6), 1257–1295. cited on page(s): 83

Laumon, Gérard, and Moret-Bailly, Laurent. 2000. *Champs algébriques*. Ergebnisse der Mathematik und ihrer Grenzgebiete. 3. Folge., vol. 39. Berlin: Springer. cited on page(s): 61, 363, 364

Lauritzen, Niels. 1996. Embeddings of homogeneous spaces in prime characteristics. *Amer. J. Math.*, **118**(2), 377–387. cited on page(s): 339

Lazarsfeld, Robert. 2004. *Positivity in algebraic geometry. I–II*. Ergebnisse der Mathematik und ihrer Grenzgebiete. 3. Folge., vol. 48–49. Berlin: Springer. cited on page(s): 33, 171, 210, 324, 380, 426, 432, 433, 434

Lee, Yongnam, and Nakayama, Noboru. 2018. Grothendieck duality and $\mathbb{Q}$-Gorenstein morphisms. *Publ. Res. Inst. Math. Sci.*, **54**(3), 517–648. cited on page(s): 74, 403

Lee, Yongnam, and Park, Jongil. 2007. A simply connected surface of general type with $p_g = 0$ and $K^2 = 2$. *Invent. Math.*, **170**(3), 483–505. cited on page(s): 42

Lesieutre, John. 2014. The diminished base locus is not always closed. *Compos. Math.*, **150**(10), 1729–1741. cited on page(s): 197, 322

Lieberman, D., and Mumford, D. 1975. Matsusaka's big theorem. Pages 513–530 of: *Algebraic geometry (Humboldt State University, Arcata, CA, 1974)*. Proc. Sympos. Pure Math., vol. 29. Providence, RI: Amer. Math. Soc. cited on page(s): 323

Lin, Yinbang. 2015 (Dec.). *Moduli spaces of stable pairs*. arxiv:1512.03091. cited on page(s): 353

Lipman, Joseph. 1969. On the Jacobian ideal of the module of differentials. *Proc. Amer. Math. Soc.*, **21**, 422–426. cited on page(s): 273

Liu, Wenfei, and Rollenske, Sönke. 2014. Pluricanonical maps of stable log surfaces. *Adv. Math.*, **258**, 69–126. cited on page(s): 307

Looijenga, E. J. N. 1984. *Isolated singular points on complete intersections.* London Mathematical Society Lecture Note Series, vol. 77. Cambridge: Cambridge University Press. cited on page(s): 84, 248

Looijenga, Eduard, and Wahl, Jonathan. 1986. Quadratic functions and smoothing surface singularities. *Topology*, **25**(3), 261–291. cited on page(s): 84

Luo, Zhao Hua. 1987. Kodaira dimension of algebraic function fields. *Amer. J. Math.*, **109**(4), 669–693. cited on page(s): 35

Lyu, Shiji, and Murayama, Takumi. 2022. *The relative minimal model program for excellent algebraic spaces and analytic spaces in equal characteristic zero.* https://arxiv.org/pdf/2209.08732.pdf. cited on page(s): 411, 420

Matsumura, Hideyuki. 1986. *Commutative ring theory.* Cambridge Studies in Advanced Mathematics, vol. 8. Cambridge: Cambridge University Press. Translated from the Japanese by M. Reid. cited on page(s): 366, 367, 392, 394

Matsusaka, Teruhisa. 1964. *Theory of Q-varieties.* Math. Soc. Japan [Tokyo]. cited on page(s): 25

Matsusaka, Teruhisa. 1972. Polarized varieties with a given Hilbert polynomial. *Amer. J. Math.*, **94**, 1027–1077. cited on page(s): 27, 323, 326, 432

Matsusaka, Teruhisa, and Mumford, David. 1964. Two fundamental theorems on deformations of polarized varieties. *Amer. J. Math.*, **86**, 668–684. cited on page(s): 31, 424

Milnor, John Willard. 1963. *Morse theory.* Based on lecture notes by M. Spivak and R. Wells. Annals of Mathematics Studies, No. 51. Princeton, NJ: Princeton University Press. cited on page(s): 384

Mori, Shigefumi. 1975. On a generalization of complete intersections. *J. Math. Kyoto Univ.*, **15**(3), 619–646. cited on page(s): 52

Mori, Shigefumi. 1987. Classification of higher-dimensional varieties. Pages 269–331 of: *Algebraic geometry (Bowdoin, 1985, Brunswick, Maine, 1985).* Proc. Sympos. Pure Math., vol. 46. Providence, RI: Amer. Math. Soc. cited on page(s): 227

Mumford, David. 1961. The topology of normal singularities of an algebraic surface and a criterion for simplicity. *Inst. Hautes Études Sci. Publ. Math.*, 5–22. cited on page(s): 434

Mumford, David. 1965. *Geometric invariant theory.* Ergebnisse der Mathematik und ihrer Grenzgebiete, Neue Folge, Band 34. Berlin: Springer. cited on page(s): 19, 21, 25, 38, 50, 55, 333

Mumford, David. 1966. *Lectures on curves on an algebraic surface.* With a section by G. M. Bergman. Annals of Mathematics Studies, No. 59. Princeton, NJ: Princeton University Press. cited on page(s): 13, 17, 133, 356

Mumford, David. 1970. *Abelian varieties.* Tata Inst. Fund. Res. Studies in Math., No. 5. Bombay: Tata Institute of Fundamental Research. cited on page(s): 135

Mumford, David. 1977. Stability of projective varieties. *Enseignement Math. (2)*, **23**(1-2), 39–110. cited on page(s): 22

Mumford, David. 1978. Some footnotes to the work of C. P. Ramanujam. Pages 247–262 of: *C. P. Ramanujam: a tribute.* Tata Inst. Fund. Res. Studies in Math., vol. 8. Berlin: Springer. cited on page(s): 207

Nagata, Masayoshi. 1955. On the normality of the Chow variety of positive 0-cycles of degree m in an algebraic variety. *Mem. Coll. Sci. Univ. Kyoto. Ser. A. Math.*, **29**, 165–176. cited on page(s): 131

Nagata, Masayoshi. 1962. *Local rings.* Interscience Tracts in Pure and Applied Mathematics, No. 13. London: Wiley Interscience. cited on page(s): 392

Nakayama, Noboru. 1986. Invariance of the plurigenera of algebraic varieties under minimal model conjectures. *Topology*, **25**(2), 237–251. cited on page(s): 190

Nakayama, Noboru. 1987. The lower semicontinuity of the plurigenera of complex varieties. Pages 551–590 of: *Algebraic geometry, Sendai, 1985.* Adv. Stud. Pure Math., vol. 10. Amsterdam: North-Holland. cited on page(s): 190

Nakayama, Noboru. 2004. *Zariski-decomposition and abundance.* MSJ Memoirs, vol. 14. Math. Soc. Japan [Tokyo]. cited on page(s): 36

Odaka, Yuji, and Xu, Chenyang. 2012. Log-canonical models of singular pairs and its applications. *Math. Res. Lett.*, **19**(2), 325–334. cited on page(s): 214, 421

Oguiso, Keiji. 2017. Isomorphic quartic K3 surfaces in the view of Cremona and projective transformations. *Taiwanese J. Math.*, **21**(3), 671–688. cited on page(s): 51

Olsson, Martin. 2016. *Algebraic spaces and stacks.* American Mathematical Society Colloquium Publications, vol. 62. Providence, RI: Amer. Math. Soc. cited on page(s): 333

Ottem, John Christian, and Schreieder, Stefan. 2020. On deformations of quintic and septic hypersurfaces. *J. Math. Pures Appl.*, **135**, 140–158. cited on page(s): 52

Pan, L., and Shen, J. 2013 (Sept.). *An example on volumes jumping over Zariski dense set.* arxiv:1309.7535. cited on page(s): 197

Pandharipande, Rahul, and Thomas, Richard. P. 2009. Curve counting via stable pairs in the derived category. *Invent. Math.*, **178**(2), 407–447. cited on page(s): 343

Pandharipande, Rahul. 2018a. 2018 ICM plenary lecture, Rio de Janeiro. cited on page(s): 21

Pandharipande, Rahul. 2018b. *Geometry of the moduli space of curves.* slides of the 2018ICM lecture, https://people.math.ethz.ch/~rahul/Rio2.pdf. cited on page(s): 21

Park, Heesang, Park, Jongil, and Shin, Dongsoo. 2009a. A simply connected surface of general type with $p_g = 0$ and $K^2 = 3$. *Geom. Topol.*, **13**(2), 743–767. cited on page(s): 42

Park, Heesang, Park, Jongil, and Shin, Dongsoo. 2009b. A simply connected surface of general type with $p_g = 0$ and $K^2 = 4$. *Geom. Topol.*, **13**(3), 1483–1494. cited on page(s): 42

Patakfalvi, Zsolt. 2013. Base change behavior of the relative canonical sheaf related to higher dimensional moduli. *Algebra Number Theory*, **7**(2), 353–378. cited on page(s): 96, 108

Patakfalvi, Zsolt. 2014. Semi-positivity in positive characteristics. *Ann. Sci. Éc. Norm. Supér. (4)*, **47**(5), 991–1025. cited on page(s): 227

Patakfalvi, Zsolt. 2017. *On the projectivity of the moduli space of stable surfaces in characteristic $p > 5$.* arxiv:1710.03818. cited on page(s): 227

Piene, Ragni, and Schlessinger, Michael. 1985. On the Hilbert scheme compactification of the space of twisted cubics. *Amer. J. Math.*, **107**(4), 761–774. cited on page(s): 302

Pinkham, Henry C. 1974. *Deformations of algebraic varieties with G_m action*. Paris: Société Mathématique de France. Astérisque, No. 20. cited on page(s): 84, 87

Pinkham, Henry C. 1977. Deformations of quotient surface singularities. Pages 65–67 of: Several complex variables *Part 1 (Williams Coll., Williamstown, MA, 1975)*. Proc. Sympos. Pure Math., vol. XXX. Providence, RI: American Mathematics Society. cited on page(s): 253

Ramanujam, Chakravarthi Padmanabhan. 1963. Appendix to: Quotient space by an abelian variety, by C. S. Seshadri. *Math. Ann.*, **152**, 185–194. cited on page(s): 147, 398

Rana, Julie. 2017. A boundary divisor in the moduli spaces of stable quintic surfaces. *Internat. J. Math.*, **28**(4), 1750021, 61. cited on page(s): 307

Rana, Julie, and Urzúa, Giancarlo. 2019. Optimal bounds for T-singularities in stable surfaces. *Adv. Math.*, **345**, 814–844. cited on page(s): 307

Raynaud, Michel. 1970. Spécialisation du foncteur de Picard. *Inst. Hautes Études Sci. Publ. Math.*, 27–76. cited on page(s): 324, 325

Raynaud, Michel, and Gruson, Laurent. 1971. Critères de platitude et de projectivité. Techniques de "platification" d'un module. *Invent. Math.*, **13**, 1–89. cited on page(s): 388, 389

Reid, Miles. 1980. Canonical 3-folds. Pages 273–310 of: *Journées de Géometrie Algébrique d'Angers, Juillet 1979/Algebraic Geometry, Angers, 1979*. Alphen aan den Rijn: Sijthoff & Noordhoff. cited on page(s): 34, 35, 419

Riemenschneider, Oswald. 1974. Deformationen von Quotientensingularitäten (nach zyklischen Gruppen). *Math. Ann.*, **209**, 211–248. cited on page(s): 250, 253

Rosenlicht, Maxwell. 1954. Generalized Jacobian varieties. *Ann. of Math. (2)*, **59**, 505–530. cited on page(s): 265

Saito, Kyoji. 1974. Einfach-elliptische Singularitäten. *Invent. Math.*, **23**, 289–325. cited on page(s): 83

Samuel, Pierre. 1962. Sur une conjecture de Grothendieck. *C. R. Acad. Sci. Paris Sér. A-B*, **255**, 3101–3103. https://arxiv.org/pdf/2204.04400.pdf. cited on page(s): 147, 398

Sato, Kenta, and Takagi, Shunsuke. 2022. *Deformations of log terminal and semi log canonical singularities*. cited on page(s): 208

Schlessinger, Michael. 1971. Rigidity of quotient singularities. *Invent. Math.*, **14**, 17–26. cited on page(s): 239, 250

Schubert, David. 1991. A new compactification of the moduli space of curves. *Compositio Math.*, **78**(3), 297–313. cited on page(s): 60

Sernesi, Edoardo. 2006. *Deformations of algebraic schemes*. Grundlehren der Mathematischen Wissenschaften, vol. 334. Berlin: Springer. cited on page(s): 17, 25, 127, 128, 220, 326, 355

Serre, Jean-Pierre. 1955–6. Géométrie algébrique et géométrie analytique. *Ann. Inst. Fourier, Grenoble*, **6**, 1–42. cited on page(s): 105

Serre, Jean-Pierre. 1959. *Groupes algébriques et corps de classes*. Publications de l'institut de mathématique de l'université de Nancago, VII. Paris: Hermann. cited on page(s): 174, 265

Seshadri, C. S. 1962/1963. Some results on the quotient space by an algebraic group of automorphisms. *Math. Ann.*, **149**, 286–301. cited on page(s): 333

Seshadri, C. S. 1972. Quotient spaces modulo reductive algebraic groups. *Ann. of Math. (2)*, **95**, 511–556; errata, ibid. (2) **96** (1972), 599. cited on page(s): 333

Seshadri, C. S. 1975. Theory of moduli. Pages 263–304 of: *Algebraic geometry (Humboldt State University, Arcata, CA, 1974)*. Proc. Sympos. Pure Math., vol. 29. Providence, RI: American Mathematics Society. cited on page(s): 361

Severi, Francesco. 1947. *Funzioni quasi abeliane*. Pontificiae Academiae Scientiarum Scripta Varia, v. 4. Vaitcan City: Publisher unknown. cited on page(s): 265

Shafarevich, Igor R. 1974. *Basic algebraic geometry*. Die Grundlehren der mathematischen Wissenschaften, Band 213. New York: Springer. cited on page(s): 19, 28, 337

Shimada, Ichiro, and Shioda, Tetsuji. 2017. On a smooth quartic surface containing 56 lines which is isomorphic as a $K3$ surface to the Fermat quartic. *Manuscripta Math.*, **153**(1-2), 279–297. cited on page(s): 51

Shimura, Goro. 1972. On the field of rationality for an abelian variety. *Nagoya Math. J.*, **45**, 167–178. cited on page(s): 62

Shokurov, V. V. 1996. 3-fold log models. *J. Math. Sci.*, **81**(3), 2667–2699. cited on page(s): 432

Siegel, Carl Ludwig. 1969. *Topics in complex function theory. I–III*. Translated from the original German. London: Wiley-Interscience. cited on page(s): 19

Simonetti, Angelica. 2022. $\mathbb{Z}/2\mathbb{Z}$-equivariant smoothings of cusp singularities. https://arxiv.org/pdf/2201.02871.pdf. cited on page(s): 86

Simpson, Carlos. 1993. Subspaces of moduli spaces of rank one local systems. *Ann. Sci. École Norm. Sup. (4)*, **26**(3), 361–401. cited on page(s): 169

Siu, Yum-Tong. 1998. Invariance of plurigenera. *Invent. Math.*, **134**(3), 661–673. cited on page(s): 36

Smyth, David Ishii. 2013. Towards a classification of modular compactifications of $M_{g,n}$. *Invent. Math.*, **192**(2), 459–503. cited on page(s): 60

Stacks, Authors. 2022. *Stacks Project*. http://stacks.math.columbia.edu. cited on page(s): 49, 75, 108, 113, 124, 199, 333, 335, 337, 344, 366, 367, 375, 389, 392, 406, 408, 412

Steenbrink, J. H. M. 1983. Mixed Hodge structures associated with isolated singularities. Pages 513–536 of: *Singularities, Part 2 (Humboldt State Univ., Arcata, Calif., 1981)*. Proc. Sympos. Pure Math., vol. 40. Providence, RI: American Mathematics Society. cited on page(s): 105

Stern, Arié, and Urzúa, Giancarlo. 2016. KSBA surfaces with elliptic quotient singularities, $\pi_1 = 1$, $p_g = 0$, and $K^2 = 1, 2$. *Israel J. Math.*, **214**(2), 651–673. cited on page(s): 307

Stevens, Jan. 1988. On canonical singularities as total spaces of deformations. *Abh. Math. Sem. Univ. Hamburg*, **58**, 275–283. cited on page(s): 244

Stevens, Jan. 1998. Degenerations of elliptic curves and equations for cusp singularities. *Math. Ann.*, **311**(2), 199–222. cited on page(s): 84

Stevens, Jan. 2013. The versal deformation of cyclic quotient singularities. Pages 163–201 of: *Deformations of surface singularities*. Bolyai Soc. Math. Stud., vol. 23. Budapest: János Bolyai Math. Soc, Budapest. cited on page(s): 250

Teichmüller, Oswald. 1944. Veränderliche Riemannsche Flächen. *Deutsche Math.*, **7**, 344–359. cited on page(s): 19

Thompson, Alan. 2014. Degenerations of K3 surfaces of degree two. *Trans. Amer. Math. Soc.*, **366**(1), 219–243. cited on page(s): 307

Totaro, Burt. 2019. The failure of Kodaira vanishing for Fano varieties, and terminal singularities that are not Cohen-Macaulay. *J. Algebraic Geom.*, **28**(4), 751–771. cited on page(s): 339, 340

Tziolas, Nikolaos. 2009. Q-Gorenstein deformations of nonnormal surfaces. *Amer. J. Math.*, **131**(1), 171–193. cited on page(s): 307

Tziolas, Nikolaos. 2010. Smoothings of schemes with nonisolated singularities. *Michigan Math. J.*, **59**(1), 25–84. cited on page(s): 307

Tziolas, Nikolaos. 2017. Automorphisms of smooth canonically polarized surfaces in positive characteristic. *Adv. Math.*, **310**, 235–289. cited on page(s): 307

Tziolas, Nikolaos. 2022. Vector fields on canonically polarized surfaces. *Math. Z.*, **300**(3), 2837–2883. cited on page(s): 307

Urzúa, Giancarlo. 2016a. Q-Gorenstein smoothings of surfaces and degenerations of curves. *Rend. Semin. Mat. Univ. Padova*, **136**, 111–136. cited on page(s): 307

Urzúa, Giancarlo. 2016b. Identifying neighbors of stable surfaces. *Ann. Sc. Norm. Super. Pisa Cl. Sci. (5)*, **16**(4), 1093–1122. cited on page(s): 307

Urzúa, Giancarlo, and Yáñez, José Ignacio. 2018. Characterization of Kollár surfaces. *Algebra Number Theory*, **12**(5), 1073–1105. cited on page(s): 43

Vakil, Ravi. 2006. Murphy's law in algebraic geometry: badly-behaved deformation spaces. *Invent. Math.*, **164**(3), 569–590. cited on page(s): 25

van Opstall, Michael A. 2005. Moduli of products of curves. *Arch. Math. (Basel)*, **84**(2), 148–154. cited on page(s): 307

van Opstall, Michael A. 2006a. Stable degenerations of surfaces isogenous to a product of curves. *Proc. Amer. Math. Soc.*, **134**(10), 2801–2806. cited on page(s): 307

van Opstall, Michael A. 2006b. Stable degenerations of symmetric squares of curves. *Manuscripta Math.*, **119**(1), 115–127. cited on page(s): 307

Viehweg, Eckart. 1995. *Quasi-projective moduli for polarized manifolds*. Ergebnisse der Mathematik und ihrer Grenzgebiete (3), vol. 30. Berlin: Springer. cited on page(s): 24, 25, 218, 234

von Essen, Hartwig. 1990. Nonflat deformations of modules and isolated singularities. *Math. Ann.*, **287**(3), 413–427. cited on page(s): 239

Wahl, Jonathan M. 1980. Elliptic deformations of minimally elliptic singularities. *Math. Ann.*, **253**(3), 241–262. cited on page(s): 248, 256

Wahl, Jonathan M. 1981. Smoothings of normal surface singularities. *Topology*, **20**(3), 219–246. cited on page(s): 248

Wandel, Malte. 2015. Moduli spaces of semistable pairs in Donaldson-Thomas theory. *Manuscripta Math.*, **147**(3-4), 477–500. cited on page(s): 353

Wang, Xiaowei, and Xu, Chenyang. 2014. Nonexistence of asymptotic GIT compactification. *Duke Math. J.*, **163**(12), 2217–2241. cited on page(s): 22, 25

Weil, André. 1946. *Foundations of Algebraic Geometry*, vol. 29. New York: American Mathematical Society. cited on page(s): 61

Xu, Chenyang. 2020. *K-stability of Fano varieties: an algebro-geometric approach*. arxiv:2011.10477. cited on page(s): 28

Yasuda, Takehiko. 2019. Discrepancies of *p*-cyclic quotient varieties. *J. Math. Sci. Univ. Tokyo*, **26**(1), 1–14. cited on page(s): 340

Index

$\partial_x, = x \cdot (\partial/\partial x)$, 245
∇, divergence, 242
$\sim$, linear equivalence, xv
$\sim_{\mathbb{Q}}$, $\mathbb{Q}$-linear equivalence, xv
$\sim_{\mathbb{R}}$, $\mathbb{R}$-linear equivalence, 434
$\equiv$, numerical equivalence, xv
$\equiv$, identity of sequences or polynomials, 187
$[\otimes]$, tensor product for divisorial sheaves, 129
$\dashrightarrow$, rational map, xiv
$\rightarrow$, morphism, xiv
$\{\ \}$, fractional part, xv
$\lfloor\ \rfloor, \lceil\ \rceil$, rounding down or up, xv
$\widehat{\ }$, completion, xvii, 397
$\leq$, for sequences, 187
$\leq$, for polynomials, 187

$a(\ ,\ ,\ )$, discrepancy of divisor, 420
$a_\ell(\ ,\ ,\ ) := a(\ ,\ ,\ ) + 1$, log discrepancy, 420
$\mathbb{A}_k^n$ or $\mathbb{A}_x^n$, affine n-space, xiv
$\mathbb{A}^n / \frac{1}{m}(a_1, \ldots, a_n)$, quotient singularity, 27
  two-dimensional, 248
Adjunction, 424
  inversion of, 424
AEnv( ), affine envelope, 435
AFI, Alexeev–Filipazzi–Inchiostro, 320
  functor, $\mathcal{AFI}(\ ,\ )$, 321
Alexeev stable, 234
  good moduli theory, 235
Ample
  fiber-wise, 193
  $\mathbb{R}$-divisor, 439

relatively, xv
strongly, 326
Approximation of $\mathbb{R}$-divisors, 436, 437
Ass( ), associated points or subschemes, 370
Asymptotic Riemann–Roch, 439
$\mathrm{Aut}_S(\ )$, $\mathbf{Aut}_S(\ )$, xvii, 337
  finite for stable families, 338

Base change
  dualizing sheaf, 100, 102
  K-flat, 281
  notation, xvii
  pluricanonical, 112
Bertini theorem
  C-flatness, 290
  divisorial support, 278
  flatness, 400
  generically Cartier family, 153
  hulls, 377
  husks, 355
  inverse, local stability, 64, 184
  K-flatness, 263
  local stability, 64
  relative hull, 351
  S_m, 377
Big, divisor, xvi
Birational
  transform, xv
  fiberwise, 11
  map, xiv
Boundary, xvi
Boundedness, 16
  marked pairs, 332

set of sheaves, 360
strong, 224
weak, 224

$C_a(X, L)$, affine cone, 81
Calabi–Yau pair, xvi
Canonical
    algebra, 427
    class, 17, 419
    class, relative, 61, 133
    divisor, 17
    line bundle, 17
    model, xvi, 23, 427
    model of resolutions, xvi
    model, existence, 428
    model, nef slc case, 318
    model, simult. numerical criterion, 186
    model, simult. of resolutions, 185
    model, simultaneous, 185
    modification, xvi, 428
    modification, simultaneous, 188
    ring, 23, 26
    ring, not finitely generated, 36
    sheaf, 17, 61
    sheaf, absolute, 419
    sheaf, relative, 419
    singularity, 24, 421
    surface singularity, list, 68
Canonically
    embedded family, 333
    polarized family, 332
Cartier
    divisor, 146
    divisor, relative, 150
    divisor, valuative criterion, 156
    generically ~ pull-back, 140, 154
    generically ~, relative, 152
    index, xv
    non-~ locus, 152
Categorical
    moduli as ~ quotient, 334, 335
    quotient by group action, 334
Cayley–Chow
    correspondence, 176
    correspondence, Mumford divisors, 176
    correspondence, over fields, 171
    family, 120
    form, 171
    hypersurface, 171, 284
    hypersurface, flag, Grassmann, incidence,
        product versions, 284, 285

inverse, scheme-theoretic, 177
type hypersurface, 172
$C^m \mathcal{E}^s \mathcal{MSch}(\ )$, $C^m E^s MSch(\ )$, functor and
    moduli of canonically embedded,
    marked schemes, 333
Center
    log ~, 422
    log canonical ~, 422
    of a divisor, xv
$C^m \mathcal{E} \mathcal{SP}(\ )$, $C^m ESP(\ )$, functor and moduli of
    m-canonically embedded, stable pairs,
    222, 333, 337
C-flat, 281
    Bertini theorem, 290
    locally ~, 288
    stably ~, 281
$Ch(\ )$, Cayley–Chow hypersurface, 171
$Ch_{gr}, Ch_{in}, Ch_{fl}, Ch_{pr}$, versions of
    Cayley–Chow hypersurface, 284
$Ch_{sch}^{-1}(\ )$, Cayley–Chow inverse, 177
Chow
    equations, ideal of, 173
    hull of Mumford divisor, 175
    hull, of a cycle, 174
    variety, 5
CM, Cohen–Macaulay, xv, 372
coeff($\ $), set of coefficients, 418
coeff$_E(\ )$, coefficient of E in, 418
Coefficient
    of a prime divisor, 418
    vector of marking, 310
    floating, 317
    generic, 216
    major, 216
    standard, 216
Cohen–Macaulay, CM, xv, 372
Completion, $\widehat{\ }$, xvii, 397
Component-wise dominant, 155
Cone
    affine, 81
    deformation to, 39, 82
Continuous choice, 221
Contraction, xiv
    crepant, 318
    simultaneous, 323
    simultaneous, crepant, 323
$CoSp(\ )$, $\mathbb{Q}$-vector space spanned by
    coefficients, 434
$C^{mp^s} \mathcal{MSch}(\ )$, functor of canonically
    polarized, marked schemes, 333
Crepant, contraction, 318

Curve
  stable, 10
  stable extension, 11
Curvilinear scheme, 402
Cycle
  degree of, 171
  effective, 170
  fundemantal ~, [ ], 170
  geometrically reduced, 171
  on a scheme, 170
  width of, 174
Cyclic cover, 426

$\Delta^{>c}, \Delta^{=c}, \Delta^{<c}$, 418
Δ_c or Δ_c^{div}, divisorial fiber, 108
$\mathbb{D}$, unit disc, 32
Decomposition
  locally closed, 415
  partial, 415
Deformation, 237
  $\mathbb{W}$-~, 247
  $\lesssim$-~, 247
  hypersurface singularities, 76
  KSB-~, 247
  locally trivial, 237
  nonalgebraic, 31
  of quotients, 73
Demi-normal, 430
  open condition, 390
Depth, 371
  along a subscheme, 371
  and flatness, 374
  and push forward, 372
  of a sheaf, 371
  of slc scheme, 425
  semicontinuity, 380
Descent, 338
  and functorial polarization, 339
  for flat, projective morphism, 340
  for rigid, projective morphism, 341
Dévissage, 381
Diff( ), different, 423
Different, 423
  properties of ~, 424
Differentiation, 238
  cohomological, 238
Discrepancy
  log ~, 420
  of a divisor, 420
Discrete choice, 220
Divergence, ∇, 242

Divisor
  big, xvi
  canonical, 17
  Cartier, 146
  Cartier, relative, 150
  Cartier, valuative criterion, 156
  generically $\mathbb{Q}$- or $\mathbb{R}$-Cartier, relative, 141
  generically Cartier, 146
  generically Cartier, relative, 152
  Mumford, xv, 146
  Mumford, relative, 175
  Mumford, universal family, 179
  on a scheme, xv
  over a scheme, xv
  $\mathbb{Q}$-~, 147
  $\mathbb{R}$-~, 147
  reduced, xv
  Weil, 147
Divisorial
  fiber, Δ_c or Δ_c^{div}, 108
  log terminal, 421
  pull-back, 141
  restriction, D_t or D_t^{div}, xvii, 106
  sheaf, 129
  sheaf, flat family, 129
  sheaf, generically flat, 130
  sheaf, mostly flat, 130
  sheaf, valuative criterion, 155
  subscheme, 146
  subscheme, family of ~, 149
  support and Fitting ideal, 275
  support, DSupp( ), 274
  support, Bertini theorem, 278
  support, final definition, 276
dlt, divisorial log terminal, 421
  is CM, 425
DSupp( ), divisorial support, 274
Du Bois singularity, 97
  cohomology and base change, 98
Du Val singularity, 68
Dual graph, 67
Dualizing
  sheaf, 17, 61
  sheaf, base change, 102
  sheaf, construction, 101
  sheaf, other deformations, 65
  sheaf, relative, 419

Elementary étale, 66
emb( ), embedded subsheaf, 370
Embedded point, 370

Embedding, locally closed, 414
$\mathcal{E}^s\mathcal{M}\mathcal{S}ch(\ )$, $E^s MSch(\ )$ functor and moduli of
   embedded, marked schemes, 331
Enough one-parameter families, 416
Envelope, affine or linear, 435
Etale, elementary, 66
Ex( ), exceptional set, xv

F^{**}, reflexive hull, 348
$F^{[**]}$, hull or S_2-hull, 348
F^H, relative hull, 350
Family
   algebraic, 120
   canonically embedded, 333
   canonically polarized, 332
   Cartier, normal base, 141
   Cayley–Chow, 120
   divisorial sheaves, flat, 129
   divisorial sheaves, mostly flat, 130
   divisorial subschemes, 149
   generically Cartier divisors, 152
   Hilbert–Grothendieck, 121
   locally stable, 59, 118
   marked pairs, 311
   mostly flat of line bundles, 193
   non-projective, 22
   of polarized schemes, 327
   one-parameter, 59
   pairs, 59, 138
   polarized with K-flat divisors, 330
   stable, 89
   stable over smooth base, 167
   stable, extension of, 169
   universal, 121
   varieties, 58
   well-defined, 140
   well-defined, reduced base, 141
Fano pair, xvi
Fiber, divisorial, Δ_c or Δ_c^{div}, 108
Fiber-wise ample, 193
Field of moduli, 52
   hyperelliptic curve, 53
Fine moduli space, 7
   universal family, 341
Fitting ideal, 273
Flat
   family, divisorial sheaves, 129
   generically, 130
   mostly, 130
FlatCM( ), flat and CM locus, 276

Flatness
   associated points, 398
   Bertini theorem, 400
   curvilinear fibers, 403
   Hironaka's theorem, 409
   is open, 374
   nodal fibers, 406
   relative codimension ≥ 3, 410
   relative codimension 0, 199, 401
   relative codimension 1, 405
   relative codimension 2, 408
   residue field extension, 400
   with reduced fibers, 198
Flattening decomposition, 127
Floating coefficient, 317
Formally K-flat, 281
Framing, projective, 329
Free group action, 335
Frobenius power, $I^{[q]}$, 173
Full subscheme, 294
Functorial polarization, 339
Fundamental cycle [ ], 170

General type, xiv, 23
Generically
   Cartier divisor, 146
   flat, 130
   flat and pure, 276
   $\mathbb{Q}$- or $\mathbb{R}$-Cartier divisor, relative, 141
Genus, sectional, 123
Geometric quotient
   by free group action, 337
   by group action, 336
   existence, 336
Geometrically injective, 414
Grothendieck–Lefschetz theorem, 116

Henselisation, 66
   strict, 67
Hilbert
   function, of ω_X, 132
   function, of divisorial sheaf, 132
   functor, 6
   ~-Grothendieck family, 121
   scheme, 6
   ~-to-Chow map, 124
Hilb( ), Hilbert scheme, 122
Hilb^{str}, strongly embedded part of Hilb, 329
$Hom_X(\ ,\ )$, $\mathcal{H}om_X(\ ,\ )$, $\mathbf{Hom}_S(\ ,\ )$, xvii, 360
Homeomorphism, universal, 412

$\mathcal{H}ull(\ )$, Hull( ), functor and moduli space of
  universal hulls, 363
Hull, 130
  $S_2 \sim$, 348
  algebraic spaces, 368
  Bertini theorem, 351, 377
  of a sheaf, 348
  pull-back, 130
  pure, 348
  reflexive, 348
  relative, 350
  universal, 353
  universal, characterization, 353
  universal, fine moduli space, 363
Hurwitz formula, 425
$\mathcal{H}usk(\ )$, Husk( ), functor and moduli of husks,
  357
Husk, 355
  algebraic spaces, 368
  Bertini theorem, 355
  quotient, 355
  quotient, relative, 356
  relative, 356
  tight, 355
Hypersurface
  Cayley–Chow, 171, 284
  Cayley–Chow, flag, Grassmann, incidence,
    product versions, 284, 285
  K-flatness, 295

$\mathcal{I}(\ ,\ )$, intersection form, 321
$I(\ ,\ )$ intersection numbers, 187
$I^{[q]}$, Frobenius power, 173
Index
  Cartier $\sim$, xv
  of ω_X, 247
  of a variety, xv
Intersection number, xv
Inversion of adjunction, 424
$\mathbf{Isom}_S(\ ,\ )$, $\mathbf{Isom}_S(\ ,\ )$, xvii, 337
  finiteness of, 90, 338
Isotrivial family, 52

K_X, canonical class, 17, 419
$K_{X/C}$, relative canonical class, 61
$\kappa(X)$, Kodaira dimension, xiv, 23
$\mathcal{K}Div(\ )$, KDiv( ), functor and moduli of
  K-flat divisors, 261
K-flat, 259
  additive, 262
  base change, 281

Bertini theorem, 263
  equals stable C-flat, 283
  family, polarized, 330
  flat implies $\sim$, 262
  formal nature of, 283
  formally $\sim$, 281
  functor of $\sim$ pull-backs, 293
  hypersurface singularities, 295
  implies C-flat, 282
  linear equivalence, 262
  locally $\sim$, 281
  multiplicative, 262
  over reduced base, 262
  push forward of, 262
  reasons for definition, 260
  seminormal curves, 301
Kodaira dimension, xiv, 23
  jump of, 29–31
Kodaira lemma, 440
KSB, Kollár–Shepherd-Barron, 229
  good moduli theory, 230–232
  stable, 229, 230
  stable, major coefficients, 232
KSB-deformation, 247
KSBA, Kollár–Shepherd-Barron–Alexeev
  good moduli theory, 308, 313, 315, 316
  stable strong form, 316
  stable, general coefficients, 315
  stable, rational coefficients, 313

$L^{[m]}$, reflexive power of divisorial sheaf, 129
lc, log canonical, 421
  center, 422
LEnv( ), linear envelope, 435
$\mathrm{LEnv}_{\mathbb{Z}}(\ )$, integral points of LEnv, 314
Lexicographic order, $\leq$, 187
Lie derivative, 239
Linear
  equivalence, $\sim$, xv
  $\mathbb{Q}$-$\sim$ equivalence, $\sim_{\mathbb{Q}}$, xv
  $\mathbb{R}$-$\sim$ equivalence, $\sim_{\mathbb{R}}$, 434
  system, 4
Link, 158
Local
  morphism, 155
  numerical criterion of $\sim$ stability, 184
  Picard group, 116
  stability, Bertini, 64
  stability, reduced base, 118
  stability, representable, reduced base, 143

Locally
  C-flat, 288
  closed decomposition, 415
  closed embedding, 414
  closed partial decomposition, 415
  K-flat, 281
  stable, 28
  stable morphism, 59, 118
  stable pair, 58
  stable, equivalent conditions for, 142
  stable, KSB version, 135
  stable, reduced base, 118
Locus
  flat or flat-CM, 276
  non-Cartier, 152
Log big, 429
Log canonical, 27
  lc, 421
  center, 422
  modification, 428
Log center, 422
  mld bounds, 422
Log discrepancy, 420
Log resolution, xvi

Major coefficient, 216
Map, rational, xiv
  birational, xiv
Marked
  family of ~ pairs, 311
  pair, with divisors, 310
  reasons for, 309
Marking, of pair or family, 310, 311
Matsusaka inequality, 440
$\mathcal{MDiv}(\ )$, MDiv( ), functor and moduli of
    Mumford divisors, 175
Minimal log discrepancy, mld, 421
mld, minimal log discrepancy, 421
Model, canonical, xvi, 23
Modification
  canonical, xvi, 428
  finite, 410
  log canonical, 428
  semi-log-canonical, slc, 211
  simultaneous, canonical, 188
Moduli
  boundary, 217
  embedded pairs, 221
  embedded varieties, 221
  enough one-parameter families, 416
  field of, 52

interior, 217
  representable, 13
  separated, 15
Moduli space
  categorical, 8
  categorical quotient, 334, 335
  coarse, 8
  fine, 7
  fine for universal hull, 363
  genus 2 curves, 45
  husks, 357
  hypersurfaces, 38, 40
  irreducible components proper, 169
  KSBA, exists, 308
  non-separated, 18, 21, 41, 48
  projectivity of, 225
  quotient by group action, 333
  quotient husks, 357
  reduced version, 138
  stable varieties, 120
Moduli theory
  Alexeev, 235
  good, 226
  KSB, 230
  KSB, standard coeffs, 231
  KSBA, rational coeffs, 313
  KSBA, real coeffs, 315
  KSBA, strong form, 316
  V^+, 234
Monomorphism, 414
$\mathrm{Mor}_S(\ ,\ )$, $\mathbf{Mor}_S(\ ,\ )$, xvii, 337
Morphism, xiv
  dominant, 155
  Hilbert-to-Chow, 124
  locally stable, 59
  locally stable, reduced base, 118
  pure dimensional, 104
  scheme, **Mor**, 337
  small, xv
  stable, 13
Morse lemma, 390
Mostly flat, 130
  divisorial sheaf, 130
  family of line bundles, 193
  S_2 sheaf, 130
Mumford divisor, xv, 146
  along subscheme, 146
  Cayley–Chow correspondence, 176
  Chow hull of, 175
  flat, 227
  functor and moduli, 175

relative, 175
relative class group, 265
universal family, 179

Nagata openness criterion, 375
Nakai–Moishezon criterion, 441
Nef, xv
Negativity lemma, 444
Node, 430
    deformation of, 407
Noether normalization
    étale version, 395
    fails for affine morphism, 394
    local version, 396
Norm, 117
Normal, 411
    crossing, simple, xvi
    pair, 411
Numerical
    criterion for relative line bundles, 184
    criterion of local stability, 184
    criterion of stability, 181, 183
    criterion of simult. canonical model, 186
    equivalence, $\equiv$, xv
    pull-back, 442, 443
Numerically
    log canonical, 163
    polarized, 328
    $\mathbb{Q}$-Cartier, 163
    $\mathbb{R}$-Cartier, 163
    $\mathbb{R}$-Cartier, lc modification of, 428
    relatively trivial, 163
    semi-log-canonical, 163
    slc is slc, 164

$\omega_{X/S}$, relative dualizing sheaf, 419
ω_X, canonical or dualizing sheaf, 17
$\omega_X^{[m]}$, reflexive power of canonical sheaf, 18
Obstruction theory, 218

$\mathbb{P}_k^n$ or $\mathbb{P}_x^n$, projective n-space, xiv
Pair, xvi, 418
    Calabi–Yau, xvi
    family of, 59, 138
    family, marked, 311
    Fano, xvi
    locally stable, 58
    marked with divisors, 310
    normal, 411
    rigid and universal family, 341
    seminormal, 411

stable, 58
weakly normal, 411
well-defined family, 140
Partial decomposition, 415
PGL, group scheme, 333
Pic( ), **Pic**( ), Picard group and scheme, xvii
Pic^{loc}( ), **Pic**^{loc}( ), local Picard group and
    scheme, 116, 164
Picard group, xvii
    for smooth morphisms, 159
    local, 116, 164
$\mathcal{P}^s\mathcal{MSch}$( ), functor of polarized, marked
    schemes, 331
    stack version, 332
Poincaré residue map, 423
Pointed scheme, 330
Polarization
    family of schemes, 327
    functorial, 339
    numerical, 328
    scheme, 327
    strong, 327
Potentially, slc or ..., 99
Pre-polarization, 327
Preserve residue fields, 414
Projection
    approximation of, 279
    various versions, 280
Projective
    framing, 329
    moduli space, 225
Proper
    group action, 335
    valuative-~, 15
$\mathcal{P}^s\mathcal{Sch}$( ), functor of polarized schemes, 327
    étale sheafification of, 328
    stack version, 330
Pull-back
    C-flat, 294
    Cartier, 154
    divisorial, 141
    generically Cartier, 140
    hull-~, 130
    K-flat, 293
    locally stable, representable, 160
    numerical, 442, 443
    $\mathbb{Q}$- or $\mathbb{R}$-Cartier divisors, 141
    stable, representable, 160
    Weil-divisor, 139
Pure
    quotient, 147, 371

relatively, 365
scheme, 371
sheaf, 370
vertically, 352
Purely log terminal, plt, 421

$q^{[*]}(D)$, pull-back (generically Cartier), 140
$q^*_{\mathrm{Wdiv}}(D)$, Weil-divisor pull-back, 139
$\mathbb{Q}$-Cartier divisor, xv
    valuative criterion, 156
$\mathbb{Q}$-divisor, xv, 147
QHusk( ), QHusk( ), functor and moduli of
    quotient husks, 357
Quasi-étale, 62
Quot( ), Quot( ), functor and scheme, 359
Quot-scheme, 359
Quotient
    categorical, by group action, 334
    geometric, by free group action, 337
    geometric, by group action, 336
    geometric, existence, 336
    husk, 355
    singularity, 27

$R($, $)$, canonical ring, 23, 26
$\mathcal{R}$, restriction or Poincaré residue map, 423
Rational
    double point, 68
    map, xiv
    singularity, xv
$\mathbb{R}$-Cartier divisor, 434
    class, 440
    valuative criterion, 156
$\mathbb{R}$-divisor, xv, 147, 434
    ample, 439
    convex approximation, 436, 437
    depth of pluricanonical sheaf, 438
$\mathbb{R}$-line bundle, 440
Reduced
    normal form, 230
    divisor, xv
Regular sequence, 371
Relatively
    ample, xv
    generically Cartier, 141
    isomorphic, 364
    pure, 365
Representable
    C-flat pull-back, 294
    Cartier pull-back, 154
    flat, divisorial pull-back, 154

flatness, 127
functor, 126
hull of divisorial sheaves, 131
invertible hull of sheaves, 131
K-flatness, 261, 293
local stability, 119, 160
local stability, reduced base, 143
moduli theory, 13
pull-backs, 126
stability, 119, 160
stability, over reduced base, 143
Residue map, 423
Resolution, xv
    dual graph of $\sim$, 67
    log $\sim$, xvi
    of Du Val singularities, 68
Restriction
    divisorial, D_t or D_t^{div}, xvii, 106
    map, 107, 353
Riemann-Roch, asymptotic, 439
Rigid, scheme or pair, 337
    and universal family, 341
Ring, canonical, 23, 26

S_2, Serre's condition, 371
    divisorial sheaf, 129
    family of varieties, 118
    for families, 107
    for restriction, 106
    Hilbert-to-Chow map, 124
    mostly flat, 130
Semi-log-canonical, slc, 28, 431
    and depth, 425
    modification, 211
Seminormal, 411
    K-flatness for $\sim$ curves, 301
    pair, 411
Separatedness
    for husks, 358
    for stable maps, 90, 338
    moduli spaces, 15
    valuative criterion, 15
Serre's condition S_m, 371
    along a subset, 105
Seshadri criterion, 441
Sheaf
    canonical, 17
    divisorial, 129
    dualizing, 17
Simple normal crossing, snc, xvi

Singularity
  canonical, 24, 421
  cyclic quotient $\mathbb{A}^2/\frac{1}{n}(1,q)$, 27, 248
  dlt, 421
  Du Bois, 97
  Du Val, 68
  klt, 421
  lc, log canonical, 27, 421
  log terminal, 421
  plt, 421
  potentially slc or ..., 99
  quotient, 27
  slc, semi-log-canonical, 28
  terminal, 421
slc, semi-log-canonical, 28, 431
  characterization using normalization, 431
  depth, 425
  potentially ~, 99
S_m, Serre's condition, 371
  along a subset, 105
  Bertini theorem, 377
  is open, 374
Small morphism, xv
snc, simple normal crossing, xvi
$S\mathcal{P}(\ )$, SP( ), functor and moduli of stable,
    marked pairs, 137, 223, 308
$S\mathcal{P}^{\text{rigid}}(\ )$, SP$^{\text{rigid}}(\ )$, functor and moduli of
    rigid pairs, 341
SSupp, scheme-theoretic support, 274
Stability
  automatic in codimension ≥ 3, 62, 115, 184
  local, representable, 119
  numerical criterion, 181, 183
  representable, 119
  representable over reduced base, 143
Stabilization functor, 34, 35
  over nodal curves, 36
Stable
  Alexeev ~, 234
  Alexeev–Filipazzi–Inchiostro, 320
  curve, 10
  equivalent conditions for, 142
  extension for curves, 11
  extension, weak, 96
  family over smooth base, 167
  family, 1 parameter, 89
  family, extension of, 169
  KSB, 135, 229, 230
  KSB, major coefficients, 232
  KSBA, general coefficients, 315
  KSBA, rational coefficients, 313

  KSBA, strong form, 316
  locally ~, 28
  morphism, 13
  one parameter family, 59
  pair, 58
  V^+-~, 233
  variety, 28
Stably C-flat, 281
  equals K-flat, 283
  independence of embedding, 293
Standard coefficient, 216
Stratum, of an snc pair, 422
Strongly ample, 326
Subscheme, divisorial, 146
Support
  divisorial, 274
  divisorial, final definition, 276
  scheme-theoretic, 274
$S\mathcal{V}(\ )$, SV( ), functor and moduli of stable
    varieties, 120

Tight husk, 355
tors( ), torsion subsheaf, 370
Trace map, 101
Transform, birational, xv
Tree, 67
Twig, 67

Univ( ), universal family, 121
Universal
  family, 121
  family for rigid pairs, 341
  family of flat Mumford divisors, 227
  homeomorphism, 412
  hull, 353
  hull, characterization, 353
Universally flat, 353

V^+, strict Viehweg stability, 233
V-deformation, 247
Valuative criterion
  for $\mathbb{Q}$-Cartier divisor, 156
  for flat, divisorial sheaf, 155
  for relative Cartier divisor, 156
  locally closed embedding, 415
  morphism, 413
  section, 413
Valuative-proper, 15
  stable map, 90
Vanishing theorem
  Ambro-Fujino version, 429
  Fujita version, 440

Variety, xiv
  general type, xiv, 23
  stable, 28
Vertically pure, vpure( ), 352
Volume, 16, 385
  and push-forward, 385
  birational models, 387, 388
  finite maps, 388
  perturbations, 389
vpure( ), vertically pure, 352

W-deformation, 247

Weakly normal, 411
  pair, 411
Weil divisor, 147
  pull-back, 139
  relative, 148
Well-defined
  family, 140
  family, reduced base, 141
Width of a cycle, 174

$\mathbb{Z}$-divisor, xv

CPSIA information can be obtained
at www.ICGtesting.com
Printed in the USA
LVHW041753060523
746307LV00001B/1